Nanoparticles and Catalysis

Edited by
Didier Astruc

Further Reading

Schmid, G. (Ed.)

Nanoparticles

From Theory to Application

2004
ISBN: 978-3-527-30507-0

Rao, C. N. R., Müller, A.,
Cheetham, A. K. (Eds.)

Nanomaterials Chemistry

Recent Developments and New
Directions

2007
ISBN: 978-3-527-31664-9

Rao, C. N. R., Müller, A.,
Cheetham, A. K. (Eds.)

**The Chemistry of
Nanomaterials**

Synthesis, Properties and Applications

2 Volumes
2004
ISBN: 978-3-527-30686-2

Cornils, B., Herrmann, W. A., Muhler, M.,
Wong, C.-H. (Eds.)

Catalysis from A to Z

A Concise Encyclopedia

Third, Completely Revised and Enlarged Edition
3 Volumes
2007
ISBN: 978-3-527-31438-6

R. A. Sheldon, I. Arends, U. Hanefeld

Green Chemistry and Catalysis

2007
ISBN 978-3-527-30715-9

Nanoparticles and Catalysis

Edited by
Didier Astruc

WILEY-VCH Verlag GmbH & Co. KGaA

The Editor

Prof. Dr. Didier Astruc
Université de Bordeaux I
Institut des Sciences
Moléculaires, UMR CNRS
N° 5255
351 Cours de la Liberation
33405 Talence Cedex
France

Library of Congress Card No.:
applied for

British Library Cataloguing-in-Publication Data
A catalogue record for this book is available from the British Library.

Bibliographic information published by the Deutsche Nationalbibliothek
Die Deutsche Nationalbibliothek lists this publication in the Deutsche Nationalbibliografie; detailed bibliographic data are available in the Internet at <http://dnb.d-nb.de>.

© 2008 WILEY-VCH Verlag GmbH & Co. KGaA, Weinheim

Typesetting   SNP Best-set Typesetter Ltd., Hong Kong
Printing   Strauss GmbH, Mörlenbach
Binding   Litges & Dopf GmbH, Heppenheim
Cover Design   Adam-Design, Weinheim

Printed in the Federal Republic of Germany
Printed on acid-free paper

ISBN: 978-3-527-31572-7

Contents

Nanoparticles and Catalysis. Edited by Didier Astruc
Copyright © 2008 WILEY-VCH Verlag GmbH & Co. KGaA, Weinheim
ISBN: 978-3-527-31572-7

Preface

Catalysis is a central concept in chemistry, playing, for instance, a key role in biological and industrial processes. For a century, catalysis has been divided into
homogeneous and heterogeneous reactions, and scientific communities have crystallized around each aspect. The advent of nanosciences has now clearly promoted
the bottom-up strategy over the top-down one, making this traditional frontier
obsolete. Thus, the molecular approach is presently most useful for the definition
of selective and efficient heterogeneous catalysts that can be removed from reaction media and re-used. Nanocrystals of size one to only a few nanometers present
the best catalytic efficiency, yet their support is most important for the synergistic
activation of substrates, as best illustrated by Haruta's famous catalysis of CO
oxidation by O_2 at 200 K by oxide-supported gold nanoparticles (NP).

Therefore, this book is timely in gathering together the best experts in catalysis,
coming originally from both homogeneous and heterogeneous catalysis communities, who have imagined a large number of synthetic approaches to catalytically
active transition-metal NPs and their derivatives. All of them now focus on aspects
that promote selectivity and efficiency for a broad variety of molecular-activation
processes with goals ranging from organic synthesis to hydrocarbon reforming
and environmental problems.

Thus the first part of the book (Chapters 1 to 9) deals with NP catalysis, empha-
sizing the key role of NP supports; the second part (Chapters 10 to 12) concerns
specific metals (namely Pd, Ru, Ir and Au), and the last part (Chapters 13 to 18)
focuses on specific substrates of particular interest for organic chemistry, hydrocarbon reforming and environmental aspects. Among the metals, Pd and Au
are the most effective catalysts. Palladium is the most efficient catalyst for
carbon–carbon bond formation thus, besides Chapter 10 that is devoted to PdNPs,
their catalytic properties also spread over the first part of the book. Gold is the
most efficient NP catalyst for a variety of aerobic (thus low-cost) oxidation reactions, and AuNP catalysis is covered in four chapters at the end of the second part
and beginning of the third part. Each chapter is introduced in more detail in
Chapter 1.

All the chapters have been reviewed by two to three independent referees. For
their invaluable refereeing assistance and expertise (some have served twice), we
are grateful to:

Nanoparticles and Catalysis. Edited by Didier Astruc
Copyright © 2008 WILEY-VCH Verlag GmbH & Co. KGaA, Weinheim
ISBN: 978-3-527-31572-7

Markus Antonietti (Postdam, Germany), Jean-Claude Bertolini, (Villeurbanne, France), Harry Bitter (Utrecht, The Netherland), Geoffrey Bond (Salford, U.K.), Pierre Braunstein (Strasbourg, France), Mingshu Chen (College Station, USA), Bert Chandler (Trinity Univ., USA), Runo Chaudret (Toulouse, France), Mingshu Chen (College Statin, Texas, US), Carmen Claver (Taragone, Spain), Avelino Corma (Valencia, Spain), Gabriele Centi (Messina, Italy), Richard M. Crooks (Austin, Texas, USA), Bernard Coq (Montpellier, France), De Chen (Trondheim, Norway), D. Samuel Deutsch (Columbia, SC, USA), Daniel Duprez (Poitiers, France), W. Nicholas Delgass (Purdue Univ., USA), D. Wayne Goodman, (College Station, USA), Pascal Granger (Lille, France), Masatake Haruta (Toyo, Japan), Claude R. Henry (Marseille, France), Xander Nijhuis (Utrecht, The Netherlands), Carlos Moreno-Castilla (Granada, Spain), Ulrich Heiz (TU Munich, Germany) José M. Parera (Santa Fe, Argentina), Miquel Pericas (Tarragona, Spain), Laura Prati (Milano, Italy), Manfred T. Reetz (Mühlheim, Germany), Alain Roucoux (Rennes, France), Catherine Santini (Lyon, France), Guenter Schmid (Essen, Germany), Ulrich Schubert (Vienna, Austria), Sandor Szabo (Budapest, Hongrie), Serge Thorimbert (Paris, France), Dmitry Yu. Murzin (Turku, Finland), Marco Zecca (Padova, Italy).

We strongly believe that this book will greatly catalyze research in this key field in the forthcoming years for the benefit of our Society. Have a nice read!

Bordeaux, July 2007   *Didier Astruc*

List of Contributors

Didier Astruc
Institut des Sciences
Moléculaires, UMR CNRS
N° 5255
Université Bordeaux I
351 Cours de la Liberation
33405 Talence Cedex
France

Jean-Marie Basset
LCOMS – C2P2 – UMR 5265
(CNRS/CPE Lyon/UCBL)
Bât 308F – CPE Lyon
43, bd du 11 Novembre 1918
69616 Villeurbanne Cedex
France

Philippe Bazin
Laboratoire Catalyse &
Spectrochimie
CNRS-ENSICAEN
Université de Caen
6, boulevard Maréchal Juin
14050 Caen
France

Helmut Bönnemann
Max-Planck-Institut für
Kohlenforschung
Kaiser-Wilhelm-Platz 1
45470 Mülheim an der Ruhr
Germany

Lyudmila M. Bronstein
Department of Chemistry
Indiana University
Department of Chemistry
800 E. Kirkwood Av.
Bloomington, IN 47405
USA

Jean-Pierre Candy
LCOMS – C2P2 – UMR 5265 (CNRS/
CPE Lyon/UCBL)
Bât 308F – CPE Lyon
43, bd du 11 Novembre 1918
69616 Villeurbanne Cedex
France

Bert D. Chandler
Department of Chemistry
Trinity University
1 Trinity Place
San Antonio, TX 78212
USA

Avelino Corma
Instituto de Tecnología Química
UPV-CSIC
Universidad Politécnica de Valencia
Avda. de los Naranjos
46022 Valencia
Spain

Nanoparticles and Catalysis. Edited by Didier Astruc
Copyright © 2008 WILEY-VCH Verlag GmbH & Co. KGaA, Weinheim
ISBN: 978-3-527-31572-7

Marco Daturi
Laboratoire Catalyse &
Spectrochimie
CNRS-ENSICAEN
Université de Caen
6, boulevard Maréchal Juin
14050 Caen
France

Cristina Della Pina
Università degli Studi di Milano
Dipartimento di Chimica
Inorganica, Metallorganica ed
Analitica
Via Venezian, 21
20133 Milano
Italy

Jairton Dupont
Laboratory of Molecular Catalysis
Institute of Chemistry
UFRGS
Av. Bento Gonçalves
9500 Porto Alegre
91501-970 RS
Brazil

Laurent Djakovitch
IRCELYON, Institut de
Recherches sur la
Catalyse et l'Environnement de
Lyon
UMR 5256 CNRS, Université
Lyon 1
2, avenue Albert Einstein
69626 Villeurbanne
France

Florence Epron
Laboratoire de Catalyse en Chimie
Organique
UMR6503 CNRS
Université de Poitiers
Faculté des Sciences
Fondamentales et Appliquées
Bâtiment de chimie
40, avenue du recteur Pineau
86022 Poitiers Cedex
France

Ovidiu Ersen
Institut de Physique et Chimie des
Matériaux de Strasbourg (IPCMS)
Université Louis Pasteur
23, rue du Loess
67037 Strasbourg Cedex 08
France

Catherine Especel
Laboratoire de Catalyse en Chimie
Organique
UMR6503 CNRS
Université de Poitiers
Faculté des Sciences
Fondamentales et Appliquées
Bâtiment de chimie
40, avenue du recteur Pineau
86022 Poitiers Cedex
France

Ermelinda Falletta
Università degli Studi di Milano
Dipartimento di Chimica Inorganica,
Metallorganica ed Analitica
Via Venezian, 21
20133 Milano
Italy

Hermenegildo Garcia
Instituto de Tecnología Química
CSIC-UPV
Universidad Politécnica de
Valencia
Av. de los Naranjos
46022 Valencia
Spain

François Garin
Université Louis Pasteur
LMSPC, UMR 7515 CNRS ECPM
25, rue Becquerel
67087 Strasbourg
France

John D. Gilbertson
Department of Chemistry
Trinity University
1 Trinity Place
San Antonio
TX 78212
USA

Gregory Godard
LCOMS – C2P2 – UMR 5265
(CNRS/CPE Lyon/UCBL)
Bât 308F – CPE Lyon
43, bd du 11 Novembre 1918
69616 Villeurbanne Cedex
France

Masatake Haruta
Department of Applied Chemistry
Graduate School of Urban
Environmental Sciences
Tokyo Metropolitan University
1-1 Minami-Osawa
Hachioji
192-0397 Tokyo
Japan

Jun Kawahara
Department of Applied Chemistry
Graduate School of Urban
Environmental Sciences
Tokyo Metropolitan University
1-1 Minami-Osawa
Hachioji
192-0397 Tokyo
Japan

Klaus Köhler
Department Chemie
Anorganische Chemie
Technische Universität München
Lichtenbergstraße 4
85747 Garching
Germany

Gwendoline Lafaye
Laboratoire de Catalyse en Chimie
Organique
UMR6503 CNRS
Université de Poitiers
Faculté de Sciences
Fondamentales et Appliquées
Bâtiment de chimie
40, avenue du recteur Pineau
86022 Poitiers Cedex
France

Marc-Jacques Ledoux
Laboratoire des Matériaux, Surfaces et
Procédés pour la Catalyse (LMSPC)
UMR 7515 du CNRS
ECPM-Université Louis Pasteur
25, rue Becquerel
67087 Strasbourg Cedex 02
France

Pierre Légaré
Université Louis Pasteur
LMSPC, UMR 7515 CNRS ECPM
25, rue Becquerel
67087 Strasbourg
France

Catherine Louis
Laboratoire de Réactivité de
Surface
UMR 7609 CNRS
Université Pierre et Marie Curie
Paris 6
4, place Jussieu
75252 Paris Cedex 05
France

Patrice Marécot
Laboratoire de Catalyse en
Chimie Organique
UMR6503 CNRS
Université de Poitiers
Faculté de Sciences
Fondamentales et Appliquées
Bâtiment de chimie
40, avenue du recteur Pineau
86022 Poitiers Cedex
France

Olivier Marie
Laboratoire Catalyse &
Spectrochimie
CNRS-ENSICAEN
Université de Caen
6, boulevard Maréchal Juin
14050 Caen
France

Valentina G. Matveeva
Tver Technical University
Department of Biochemistry
Str. A. Nikitin, 22
170026, Tver
Russia

Elies Molins
Institut de Ciència de Materials de
Barcelona (CSIC)
Department of Crystallography and
Solid State Chemistry
Campus de la UAB
08193 Bellaterra
Spain

Kyatanahalli S. Nagabhushana
Forschungszentrum Karlsruhe GmbH
ITC – CPV
Hermann-von-Helmholtz-Platz 1
76344 Eggenstein-Leopoldshafen
Germany

Audrey Nowicki
Ecole Nationale Supérieure de Chimie
de Rennes
UMR 6226 "Sciences Chimiques de
Rennes"
Equipe Chimie Organique et
Supramoleculaire
Avenue Gal Leclerc
35700 Rennes
France

Katrin Pelzer
Fritz-Haber-Institut der MPG
Department for Inorganic Chemistry
Faradayweg 4–6
14195 Berlin
Germany

Cuong Pham-Huu
Laboratoire des Matériaux, Surfaces et
Procédés pour la Catalyse (LMSPC)
UMR 7515 du CNRS
ECPM-Université Louis Pasteur
25, rue Becquerel
67087 Strasbourg Cedex 02
France

Karine Philippot
Equipe "Nanostructures et
Chimie Organométallique"
Laboratoire de Chimie de
Coordination du CNRS, UPR
8241
205, route de Narbonne
31077 Toulouse Cedex 04
France

Manfred T. Reetz
Max-Planck-Institut für
Kohlenforschung
Kaiser-Wilhelm-Platz 1
45470 Mülheim/Ruhr
Germany

Ryan M. Richards
Jacobs University Bremen GmbH
Campus Ring 1
28759 Bremen
Germany

Michele Rossi
Università degli Studi di Milano
Dipartimento di Chimica
Inorganica, Metallorganica e
Analitica
Via Venezian, 21
20133 Milano
Italy

Alain Roucoux
Ecole Nationale Supérieure de
Chimie de Rennes
UMR 6226 "Sciences Chimiques
de Rennes"
Equipe Chimie Organique et
Supramoleculaire
Avenue Gal Leclerc
35700 Rennes
France

Dagoberto de Oliveira Silva
Laboratory of Molecular Catalysis
Institute of Chemistry
UFRGS
Av. Bento Gonçalves
9500 Porto Alegre
91501-970 RS
Brazil

Esther M. Sulman
Tver Technical University
Department of Biochemistry
Str. A. Nikitin, 22
170026, Tver
Russia

Frédéric Thibault-Starzyk
Laboratoire Catalyse et Spectrochimie
CNRS-ENSICAEN-Université de Caen
6 boulevard Maréchal Juin
14050 Caen Cedex
France

Adelina Vallribera
Universidad Autónoma de Barcelona
Facultat de Ciències, Departament de
Química Edificic, Campus UAB
08193 Bellaterra
Spain

Johannes G. de Vries
DSM Research, DSM Pharma
Chemicals
Advanced Synthesis, Catalysis &
Development
P. O. Box 18
6160 MD, Geleen
The Netherlands

1
Transition-metal Nanoparticles in Catalysis: From Historical Background to the State-of-the Art

Didier Astruc

1.1
Introduction

The nanosciences have recently evolved as a major research direction of our modern Society resulting from an ongoing effort to miniaturize at the nanoscale processes that currently use microsystems. Towards this end, it is well admitted that the bottom-up approach should now replace the classic top-down one, a strategic move that is common to several areas of nanosciences including opto-electronics, sensing, medicine and catalysis. The latter discipline certainly is the key one for the development of starting chemicals, fine chemicals and drugs from raw materials. During the twentieth century, chemists have made considerable achievements in heterogeneous catalysis [1], whereas homogeneous catalysis [2] progressed after the second world war (hydroformylation) and especially since the early 1970s (hydrogenation). Heterogeneous catalysis, that benefits from easy removal of catalyst materials and possible use of high temperatures, suffered for a long time from lack of selectivity and understanding of the mechanistic aspects that are indispensable for parameter improvements. Homogeneous catalysis is very efficient and selective, and is used in a few industrial processes, but it suffers from the impossibility of removal of the catalyst from the reaction media and its limited thermal stability. *Green catalysis* aspects now obviously require that environmentally friendly (for instance phosphine-free) catalysts be designed for easy removal from the reaction media and recycling many times with very high efficiency. These demanding conditions bring a new research impetus for catalyst development at the interface between homogeneous and heterogeneous catalysis, gathering the sophisticated fulfilment of all the constraints that were far from being fully taken into account by the pioneers and even the specialists in each catalytic discipline in the former decades. Yet the considerable knowledge gained from the past research in homogeneous, heterogeneous, supported and biphasic catalysis, including also studies in non-classical conditions (solvent-free, aqueous, use of ionic liquids, fluorine chemistry, microemulsions, micelles, reverse micelles, vesicles, surfactants, aerogels,

Nanoparticles and Catalysis. Edited by Didier Astruc
Copyright © 2008 WILEY-VCH Verlag GmbH & Co. KGaA, Weinheim
ISBN: 978-3-527-31572-7

polymers or dendrimers), should now help establish the desired optimized catalytic systems.

In this context, the use of transition-metal nanoparticles (NPs) in catalysis [3] is crucial as they mimic metal surface activation and catalysis at the nanoscale and thereby bring selectivity and efficiency to heterogeneous catalysis. Transition-metal NPs are clusters containing from a few tens to several thousand metal atoms, stabilized by ligands, surfactants, polymers or dendrimers protecting their surfaces. Their sizes vary between the order of one nanometer to several tens or hundreds of nanometers, but the most active in catalysis are only one or a few nanometers in diameter, i.e. they contain a few tens to a few hundred atoms only [4]. This approach is also relevant to homogeneous catalysis, because there is a full continuum between small metal clusters and large metal clusters, the latter being also called colloids, sols or NPs. NPs are also well soluble in classic solvents (unlike metal chips in heterogeneous catalysis) and can often be handled and even characterized as molecular compounds by spectroscopic techniques that are well known to molecular chemists, such as ^{1}H and multinuclear NMR, infrared and UV–vis spectroscopy and cyclic voltammetry. Molecular mechanisms involving the NP surfaces in catalytic reactions are much more difficult to elucidate, however, than those of monometallic catalysts, and the size and shape of the NP catalysts are key aspects of the catalytic steps. NPs themselves can also be used as catalysts in homogeneous systems or alternatively they can be heterogenized by fixation onto a heterogeneous support such as silica, alumina, other oxides or carbon, for instance carbon nanotubes. Thus, the field of NP catalysis involves both the homogeneous and heterogeneous catalysis communities, and these catalysts are sometimes therefore called "semi-heterogeneous" [3, 5]. This field has attracted a considerable amount of attention recently, as demonstrated by the burgeoning number of publications in all kinds of catalytic reactions, because NP catalysts are selective, efficient, and recyclable and thus meet the modern requirements for *green catalysts*. Applications are already numerous, and the use of these catalysts in industry will obviously considerably expand in the coming years. Table 1.1 shows the impressive number of catalytic reactions that have been achieved using transition-metal NPs under rather mild conditions.

The stabilization of NPs during their synthesis can be electrostatic, steric, electrosteric (combination of steric and electrostatic, see Fig. 1.1) or by ligands [4, 5, 8, 11]. The NP synthesis can also occasionally be carried out from metals by atomic metal vaporization or from metal(0) complexes [11]. In view of the catalyst recycling, NP catalysts are often immobilized or grafted onto inorganic or organic polymer supports [4, 5, 8, 11]. The mechanism of transition-metal NP self-assembly has recently been subjected to detailed studies by Finke's group with a proposal of a four-step nucleation mechanism including two autocatalytic steps [11k]. Such mechanistic studies are of fundamental interest for NP catalysis overall.

There are many reviews on the multiple NP synthetic modes [4–11], and here we will not systematically detail this aspect *per se*. We concentrate our attention on catalysis, from the pioneering studies to the present state of the art.

Table 1.1 Reactions catalyzed by transition-metal nanoparticles.

Reaction	References (see also general Refs. 3–11)
Hydrogenation	
Simple olefins and dienes	2, 6b, 10d–i, 12, 14d,f,j,k,l, 16, 17a,b,d, 23, 26, 31l,o–q, 32, 33, 35a,i, 36a, 40a, 41, 43d, 44b,h, 46a–d, 58, 61
Alkynes	14c, 37b, 39, 40b, 62, 63
CO_2	10, 82
Arene ring	12a, 29, 33k,m, 34, 37b, 65–70
Arene rings of dibenzo-18-crown-6-ether	64
Acrolein	12b, 78
Methylacrylate	16
Allylic alcohols	14d, 17b
N-isopropylacrylamide	17b
Ethylpyruvate	30a, 63
Citral	32c,d,e
Styrene	18i
Trans-stilbene	32h,f
Opening of epoxides	14a
Dehydrolinalol	14b
Citronellal	59b
9-Decen-1-ol	35n
Various olefins including functional ones	9f,g; 14e, 22b,d, 32h, 33k,m
Polar olefins	22b
Nitroaromatics	9a, 59c, 60, 61
Ketones, benzonitrile	14h, 33l, 35d,i
Cinnamaldehyde	47, 59a
Asymmetric hydrogenation	29
Heck C—C coupling (ArX + olefin → arylolefin)	7, 8b, 14f, 17f, 18i, 22, 24a, 31a–n, 32n, 33e–g,i,j, 35g,n, 40b, 43a–f, 44a–k, 53, 71, 72, 73
Suzuki C—C coupling (ArX + Ar'B(OH)$_2$→ Ar–Ar')	13a,b, 14i, 17i–k, 21, 22, 23, 24a, 25, 26, 27, 31m,n, 33e,h,i 34b, 35b,o, 43c, 44c–e,l, 54, 72f, 74a, 76
Sonogashira C—C coupling (ArX + alkyne → arylalkyne)	28c, 43g,h,i
Stille C—C coupling (ArX + Bu$_3$SnR → Ar-R)	33h
Negishi C—C coupling (ArCl+RZnX → Ar-R)	44m
Kumada C—C coupling (ArCl+RMgX(Ar-R)	44j,k
Dehydrohalogenation of aryl halides	44i,n
Amination of aryl halides and sulfonates (ArX + RNHR'→ Ar-N(R)R')	44k–o, 74b, 77
Hydrosilylation	9h, 49a
Coupling of silanes	28a,b
Hydroxycarbonylation of olefins	75

Table 1.1 *Continued*

Reaction	References (see also general Refs. 3–11)
[3 + 2] Cycloaddition	55
McMurry coupling	56, 57
Oxidation	
CO	8a, 9d,e, 11j, 17, 34, 36a,b,e, 37c,d, 45, 51b, 80
Dihydrogen	46e
Aromatic amines	33g, 46f
Alkyl amines	9d
1-Phenylethanol	38
CH_3OH and alcohol electro-oxidation	47
Cyclooctane	49b
Cyclohexane	46g
Ethene and propene epoxidation	50
Glucose	52, 46h
Diol, Glycerol, ethylene glycol	46c,e, 81
Oxalate	46h
Amination	24b, 43j
Carbonylation	
Aryl halides	24b
Methanol	53a
Allylic alkylation	30f, 31, 35c, 57
Mannich	57
Pauson-Khand	11h, 35c
Hydroconversion of hydrocarbons	40a
Combustion: alkanes, arenes, alcohols	32e, 35j, 79
Methanol reforming	36c, 39h

1.2
Historical Background

Soluble AuNPs appeared about two thousand years ago and were used as pigments for esthetic and curative purposes. On the materials side, their use to make ruby glass and for coloring ceramics was known in these ancient times, as exemplified by the famous Lycurgus cup (dated 4th century AD, British Museum) [8a]. Modern syntheses of NPs are often inspired by the 150-year old method of Faraday who demonstrated the formation of red solutions of AuNPs by reduction of tetrachloroaurate $[AuCl_4]^-$ using phosphorus as the reducing agent [8a,b]. This strategy has been popularized again by Schiffrin's group in 1993 [8a,c], using $NaBH_4$ reduction of a metal precursor such as $HAuCl_4$ in a biphasic organic solvent–water system in the presence of the phase-transfer reagent $[N(C_8H_{17})_4]Br$ followed by the addition of a thiol that stabilizes the NPs as a thiolate ligand [8a,c]. Likewise,

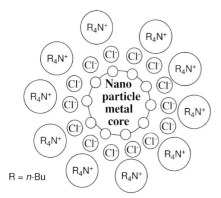

R = *n*-Bu

Fig. 1.1 "Electrosteric" (i.e. *electrostatic* with the halide anions located between the positively charged NP surface and the tetra-*N*-butyl ammonium cations and *steric* with the tetra-*N*-butyl ammonium cation) stabilization of metal NPs obtained by reduction of a metal chloride salt in the presence of a tetra-*N*-butyl ammonium cation (Bönnemann-type synthesis of Eq. (1.1)). The presence of chloride or other anions (rather than ammonium cations) near the NP surface was demonstrated. Finke showed that the order of stabilization of IrNPs by anions followed the trend: polyoxometallate > citrate > polyacrylate ~ chloride. Thus, the stabilization of metal NPs by anions can also have an important steric component [25f,g].

Caruso reported the synthesis and stabilization of PdNPs using Na_2PdCl_4 and (4-dimethylamino)pyridine [8d]. Since the 1980s, metal salts, a molecular stabilizer and a reductant [4–7, 11] have been used by Bönnemann, as represented in Eq. (1.1) [6].

$$\text{THF}$$
$$MX_n(NR_4)_m + (n - m)\,Red \rightarrow M_{NP} + (n - m)\left(Red^+X^-\right) + m\left(NR_4^+X^-\right) \qquad (1.1)$$

$$M = \text{Group 8–10 metal}, X = \text{Cl or Br}, R = C_{4-12}\text{ alkyl},$$
$$Red = M'H\,(M' = H, Li, LiBEt_3, NaBEt_3, KBEt_3).$$

In Chapter 2, Helmut Bönnemann, Kyatanahalli S. Nagabhushana and R. M. Richards review these pioneering studies as well as recent advances in catalysis with both homogeneous and heterogeneous reactions. Typical successful catalytic reactions include the hydrogenation of C=C, C=O and C=N bonds, the reduction of N—O bonds and the formation of C—C bonds (typically Suzuki). The use by Bönnemann's group of metal NPs as electrocatalysts (i.e. the catalysis of anodic oxidation or cathodic reduction reactions, essentially methanol and CO oxidation) is also addressed, and this concept is widely applied to the design of fuel cells whereby the combination of two or several metals provides a positive synergy for catalytic activation.

Another early popular NP synthetic method used the thermal decomposition of metal(0) precursors in the form of, for instance, metal carbonyls (Fe, Co, Ni, Ru, Rh, Ir) in the presence of stabilizing polymers [8e–g]. This method was not only useful for catalytic purposes, but in 1996 Hess and Parker [8h] and Thomas [8i] also reported in on the ferromagnetic properties (ferrofluids) and used the thermal decomposition of dicobaltoctacarbonyl as a versatile method. Other zero-valent metal complexes such as Pd(dba)$_2$ and M$_2$(dba)$_3$ (M = Pd, Pt) were reported in 1970 by Takahashi *et al.*, then, in 1991, the Gallezot group produced efficient NP catalysts upon reaction with either H$_2$ or CO. The Bradley–Chaudret group reported the hydrogenation of zerovalent complexes of olefinic ligands as early as 1992 [8j]. The metal-vapor technique to produce metal NPs, conceptually (but not practically) an ideal one, was first published in 1927 by Roginski and Schalnikoff [8l] and was made popular in modern times by work from the groups of M. L. H. Green, Timms and Ozin [8m]. Physical synthetic means [6–8, 10, 11] such as electrochemistry, developed by Reetz [7], became numerous in the 1980s for the synthesis of transition-metal NPs that were subsequently used in catalysis.

Considering now catalysis with NPs that are directly connected to the NP surfaces, Oswald's papers in 1907 that focused on the dramatic increase of NP surface when a given cube was divided into small cubes were especially seminal. A useful pioneering review on early transition-metal NP synthesis and catalysis has been published by John S. Bradley in the book on colloids and clusters edited by G. Schmid in 1994 [5d].

It has been well known since the nineteenth century that photography involves AgNPs whereas the decomposition of hydrogen peroxide was carried out by Bredig in 1899 using PtNPs [5d]. Catalytic studies using transition-metal NP catalysts became popular in the second half of the twentieth century with a small but important group of reactions, namely hydrogenation, hydrosilylation and hydration of unsaturated organic substrates and redox reactions including water photosplitting and photocatalytic hydrogenation. Thus, pioneering catalytic applications of NPs were reported in 1940 by Nord on nitrobenzene reduction [9a], in 1970 by Parravano on hydrogen atom transfer between benzene and cyclohexane and oxygen atom transfer between CO and CO$_2$ using AuNPs [9b]. Then, Haruta's seminal and famous studies on oxide supported AuNP-catalyzed CO oxidation by O$_2$ at low temperatures were a real breakthrough, resulting from his understanding that it was small, oxide-supported AuNPs that were active and that the nm-size was crucial (see Chapter 15 and [9c–e]). In the 1970s, Bond and Sermon [9f] and Hirai *et al.* [9g] disclosed AuNP-catalyzed olefin hydrogenation. Hirai's contribution in the 1970s and 1980s was especially impressive with the use of RhNPs that were generated using aqueous methanol or NaOH in methanol as a reducing agent for RhCl$_3 \cdot$ 3H$_2$O, producing PVA-stabilized RhNPs that were more active than previously reported RhNPs/PVA due to NP size reduction to one or only a few nm [9g]. The Maire group showed the efficiency of micelles for the NP catalytic hydrogenation of unsaturated substrates in the early 1980s [9h]. The preparation of NP in constrained environments including microemulsions, micelles, inverse micelles and vesicles was pioneered by the seminal work from the group of Fendler in the early 1980s [9i]. On these lines, Hirai collaborated with Toshima in the 1980s in work involving the use of various surfactants to stabilize catalytically active PtNPs produced by reduction of H$_2$PtCl$_6$, either photochemically or by reaction with dihydrogen. Toshima actively pursued his research on metal NPs in catalysis in the early 1990s with polymer-stabilized well-mixed bimetallic NPs generated either by direct co-reduction of two salts of two different metals (such as Pt and Au), by sequential reduction (the first reduced NPs serving as seeds for the surface condensation of the second metal) or by galvanometric reduction of a metal salt by initially produced NPs of another more easily reduced metal [5f,g]. In 1986, another well-known work appeared by Lewis who demonstrated the colloidal mechanism of olefin hydrosilylation catalysis by silanes using organometallic complexes of Co, Ni, Pd or Pt including Speier catalyst (alcoholic H_2PtCl_6) [9j], whereas these catalysts were formerly believed to follow the classic monometallic organometallic mechanism (i.e. oxidative addition of the Si–H bond of the silane to the transition-metal center, followed by alkene insertion and reductive elimination). That decade saw the beginning of extended NP catalytic studies, especially in the fields of redox catalysis, photocatalysis (photo-water splitting and photo-hydrogenation of alkenes, alkynes and CO_2) [10a–g], hydrogenation of unsaturated substrates and oxidation [10h,i]. NPs synthesized as indicated above using the Bönnemann-type synthesis from metal salts were further used in various hydrogenation reactions and C—C coupling reactions such as the Heck reactions between butyl acrylate and iodobenzene or aryl bromides and styrene [4, 5, 7, 11].

The first years of this twentyfirst century have seen an exponential number of publications in the NP field with goals of both (i) *improving* catalyst activities and selectivities and (ii) *understanding* the catalytic mechanisms [11]. The modes of preparation of catalytically active NPs have been diverse and currently include impregnation [12a], co-precipitation [12a,b], deposition-precipitation [12c], sol–gel [12a,d], gas-phase organometallic deposition [12f], sonochemical [12g], microemulsion [12h], laser ablation [12i], electrochemical [12j], and cross-linking [12k]. We will classify and discuss the categories of NP catalysts by the type of support, and then the various reactions will be collected by references in Table 1.1. The field of metal NP catalysis is now spreading in several directions around the interface between homogeneous catalysis and heterogeneous catalysis with mutual benefits.

1.3
Polymers as NP Stabilizers

Polymers provide metal NP stabilization not only because of the steric bulk of their framework, but also by weak binding to the NP surface by the heteroatom, playing the role of ligands. Poly(N-vinyl-2-pyrrolidone) (PVP) is the most commonly used polymer for NP stabilization and catalysis, because it fulfils both steric and ligand requirements [5f]. For instance Pt-, Pd- and RhNPs stabilized by PVP, are synthesized by refluxing ethanolic reduction of the corresponding metal halide and immobilized in an ionic liquid, 1-n-butyl-3-methylimidazolium hexafluorophosphate ([BMI][PF$_6$]), and are very efficient olefin and benzene hydrogenation catalysts at 40 °C that can be recycled without loss of activity (see Chart 1.1 for the two major polymer formulas used for NP catalysis) [12k].

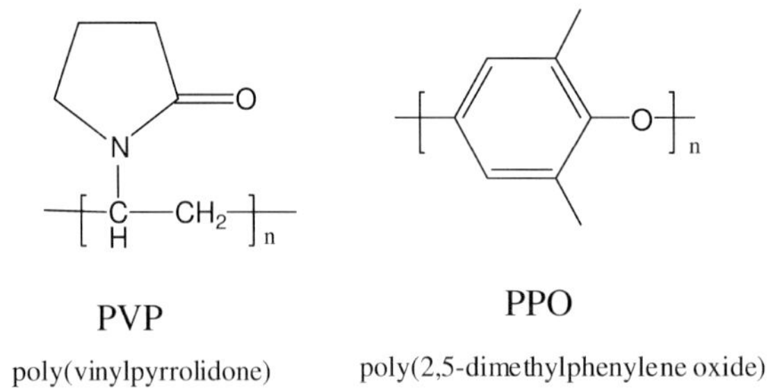

PVP

poly(vinylpyrrolidone)

PPO

poly(2,5-dimethylphenylene oxide)

Chart 1.1 Two major polymer families used as metal NP supports for catalysis

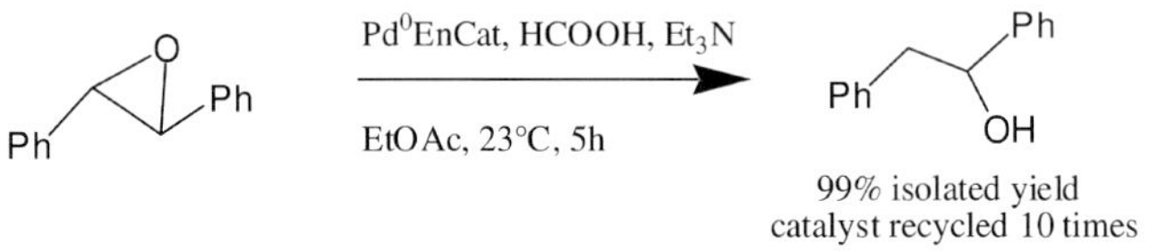

Pd⁰EnCat, HCOOH, Et₃N

EtOAc, 23°C, 5h

99% isolated yield
catalyst recycled 10 times

Scheme 1.1 Ring-opening hydrogenolysis of epoxides catalyzed by PdNPs (2 nm) microencapsulated in polyurea. Recycling experiments can be carried out at least ten times with 97–99% yield. (Ref. [14a], Yu group, *Org. Lett.* **2003**, 4665).

With standard PVP-stabilized NP catalysts, parameters such as size and stability during the catalytic process have been examined. For instance, decreasing the PdNP size down to 3 nm in the Suzuki reaction improved the catalytic activity, suggesting that the low-coordination number vertex and edge atoms on the particle surface are active catalytic sites [13]. Many other polymers have been used recently for efficient catalysis: polyurea (Scheme 1.1) [14a], polyacrylonitrile and /or polyacrylic acid (Fig. 1.2) [14b], multilayer polyelectrolyte films (Fig. 1.3) [14c], polysilane shell-cross-linked micelles (Fig. 1.4) [14d], polysiloxane (Fig. 1.5) [14e], oligosaccharides [14f], copolymers synthesized by aqueous reversible addition-fragmentation chain-transfer polymerization[14g], π-conjugated conducting polypyrrole [14h], poly(4-vinylpyridine) [14h], poly(*N,N*-dialkylcarbodiimide[14i], polyethylene glycol [14j], chitosan [14k] and hyperbranched aromatic polyamides (aramids) [14l]. Classic surfactants such as sodium dodecylsulfate (SDS) are also used as NP stabilizers for catalysis [14m]. Water-soluble polymers have been used with success for selective hydrogenation of cyclic vs. non-cyclic olefins [5d].

A very important concept pioneered in the 1970s is that of catalysis using two different metals such as Au and Pd in the same NP [15]. This idea has been beautifully developed by Toshima's group who used PVP to stabilize core–shell bimetallic Au-PdNPs, i.e. for instance NPs in which the core is Au whereas Pd atoms are located on the shell (Fig. 1.6) [16]. Subsequent to co-reduction, this structure is controlled by the order of reduction potentials of both ions and the coordination abilities of both atoms to PVP. The location of Au in the core and Pd on the shell was demonstrated by EXAFS, and it was shown that such heterobimetallic

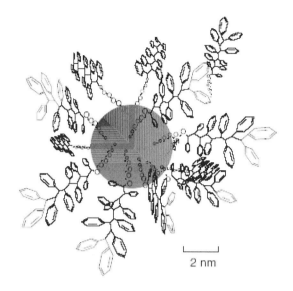

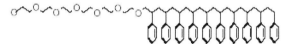

Fig. 1.2 PdNP adsorbed on polyacrylic acid particles as hydrogenation catalyst: stabilizing effect of a PdNP due to adsorbed block copolymer (Reprinted with permission from Ref. [14c]; Gröschel group, *Catal. Lett.* **2004**, *95*, 67).

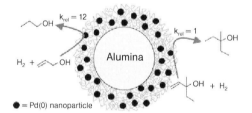

Fig. 1.3 Principle of the formation of PdNPs in multilayer polyelectrolyte films for selective hydrogenation (the layer-by-layer deposition is both convenient and versatile) (Reprinted with permission from Ref. [14d]; Bruenning group, *J. Am. Chem. Soc.* **2004**, *126*, 2658).

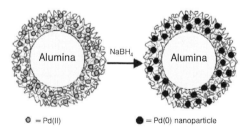

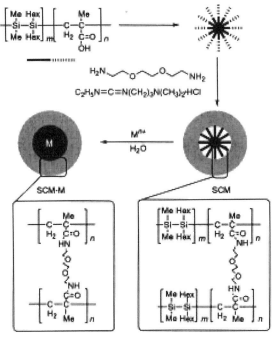

Fig. 1.4 Schematic illustration of the synthesis of metal NPs derived from polysilane shell cross-linked micelle templates (Reprinted with permission from Ref. [14f]; Sakurai group, *Chem. Lett.* **2003**, 32, 980).

Au-cored PdNPs are more active in catalysis than simple PVP-stabilized PdNPs. Thus, the Au core enhances the catalytic properties of PdNPs at the PdNP surface [5f,g]. Conversely, design strategies can lead to the opposite core–shell structure (Pd core, Au shell), and specific catalytic properties were obtained for methylacrylate hydrogenation [16].

Cyclohexene hydrogenation was catalyzed with PdNPs stabilized by highly branched amphiphilic polyglycerol (75% esterified with palmitoyl chloride) and this system was submitted to a continuously operating membrane reactor for recovery and recycling of the PdNP catalyst [16c].

In Chapter 3, Lyudmila M. Bronstein, Valentina G. Matveeva and Esther M. Sulman review metal NP catalysis using polymers, in particular, work in Bronstein's group concerning the hydrogenation of chain acetylene alcohols and direct oxidation of L-sorbose. These authors stress the importance of and interest in block copolymers such as polystyrene-block-poly-4-vinylpyridine, PS-b-4VP, and even better poly(ethylene oxide)-block-poly-2-vinylpyridine, PEO-b-P4VP (the latter being used in water). The catalytic efficiency is optimal for the smallest NPs and decrease as the NP size decreases.

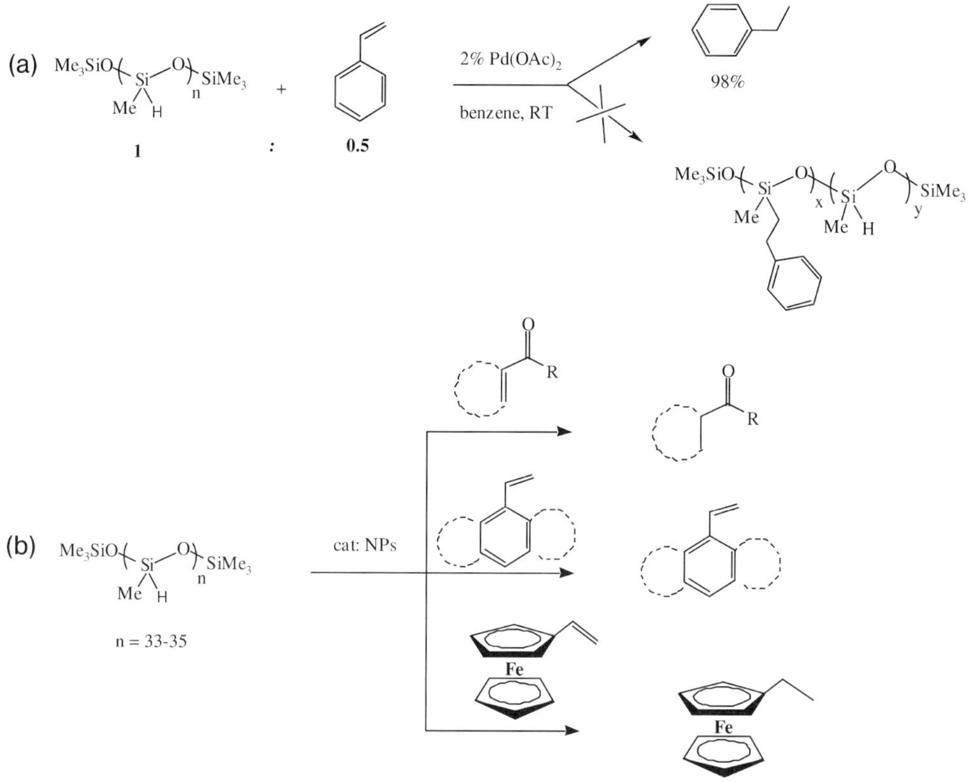

Fig. 1.5 Polysiloxane-PdNPs, generated by reduction of
Pd(OAc)$_2$ with polymethylhydrosiloxane, as recyclable
chemoselective hydrogenation catalysts: selective reduction
of styrene (a) and polysiloxane-PdNP catalyzed reduction of
alkenes (b). (Ref. [14e]; Caudhan group, *J. Am. Chem. Soc.*
2004, *126*, 8493).

1.4
Dendrimers as NP Stabilizers

Dendrimers are hyperbranched macromolecules that are constructed around a
core and are well defined by regular branching generation after generation. Dendrimers, as polymers, are macromolecules; but unlike polymers, they are perfectly
defined on the molecular level with a polydispersity of 1.0 [17]. Having shapes of
molecular trees or cauliflowers, they become globular beyond low generations, and
thus behave as molecular boxes [17c] that can entrap and stabilize metal NPs,
especially if they contain heteroatoms in their interiors [17d,18]. The dendritic
branches and termini can serve as gates to control access of small substrates inside
the dendrimer to the encapsulated NP. Finally, the dendrimer terminal groups

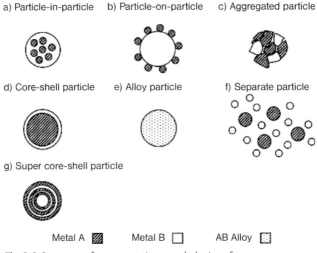

a) Particle-in-particle   b) Particle-on-particle   c) Aggregated particle

d) Core-shell particle   e) Alloy particle   f) Separate particle

g) Super core-shell particle

Metal A   Metal B   AB Alloy

Fig. 1.6 Summary of representative morphologies of
bimetallic nanoparticles (Reprinted with permission from ref.
16a, Kunitake and Toshima groups, *J. Am. Chem. Soc.* **2003**,
125, 11034).

provide the desired solubility in organic, aqueous or fluorous media. The formation of NPs stabilized by dendrimers was proposed in 1998 by three research
groups Crooks [11b,18a], Tomalia [19a,b] and Esumi [19c–e]. The first two groups
introduced the metal NPs inside the dendrimers whereas the last one stabilized
them at the dendrimer periphery. The former strategy has proved very successful
because of the molecular definition of dendrimers that encapsulate NPs and their
ability to serve as a box and generation-dependent filter of substrates. Crooks [18a]
showed that complexation of the inner nitrogen atoms of tertiary amines by metal
cations (Cu^{2+}, Au^{3+}, Pt^{2+}, Pd^{2+}, Fe^{3+} Ru^{3+}) could be followed by reduction by $NaBH_4$
to metal(0) leading to the agglomeration of metal atoms to NPs inside the PAMAM
dendrimers. Crooks also pioneered the field of catalysis using these dendrimer-
encapsulated nanoparticles (DENs), and some examples of the work of his research
group follow [18a]. When the terminal amino groups were protonated at pH 2
prior to complexation by metal ions, the latter proceeded selectively onto the inner
nitrogen atoms, resulting in water solubility and subsequent catalytic activity in
water. For instance, selective hydrogenation of allylic alcohol and *N*-isopropyl
acrylamide was catalyzed in water by such PAMAM dendrimer-PdNPs (Fig. 1.7
and Chart 1.2). The addition of decanoic acid solubilizes the dendrimer-NP catalyst
in toluene by terminal amino group–carboxylic acid reaction, and the catalyst
hydrogenates the substrates more rapidly than in water. Alternatively, a perfluorinated polyether "ponytail" can be covalently grafted in order to solubilize the
PAMAM-dendrimer PdNP catalyst in supercritical CO_2, and this catalyst was
shown to perform classic Pd-catalyzed Heck coupling between aryl halides and
methacrylate, yielding predominantly (97%) *trans*-cinnamaldehyde. Oxidation

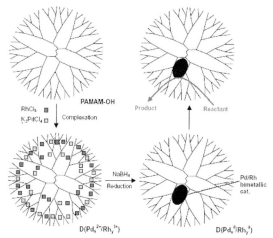

Fig. 1.7 Strategy pioneered by Crooks for the catalysis by NPs encapsulated in PAMAM or PPI dendrimers: complexation of the inner nitrogen atoms of tertiary amines by a metal cation, then reduction to metal(0) by NaBH$_4$ leading to the formation of NPs inside the dendrimer, followed by the catalyzed reaction. The use of PIP dendrimers requires control of the pH before metal ion complexation in order to selectively protonate the terminal amino group (pK_a = 9.5), not the inner ones (pK_a = 5.5) whereas in the PAMAM series, OH-terminated dendrimers are used (Ref. [18f]; Crooks, *Acc. Chem. Res.* **2001**, *34*, 181). Schematic diagram for the preparation of dendrimer-encapsulated bimetallic NPs (Reprinted with permission from Ref. [23]; Rhee group, *J. Mol. Catal. A: Chem.* **2003**, *206*, 291).

[18a] and reduction [18b] catalysis could be achieved using such dendrimer-encapsulated NPs. *Crooks' work on homogeneous catalysis using DENs is also detailed in the first part of Chapter 4 by Bert Chandler and John D. Gilbertson that is also devoted to heterogeneous catalysis using DEN precursors [20] (vide infra).*

El Sayed investigated in details the effect of the PAMAM dendritic generation on the catalytic activity in the Suzuki C—C coupling reaction between phenyl iodide and phenylboronic acid at 80 °C. Generations 3 and 4 were found to be good stabilizers (contrary to generation 2), because the dendrimers stabilize the metal NPs by preventing their agglomeration but they do not fully passivate the metal surface. The PAMAM dendrimer-stabilized PdNPs (1.3 ± 0.1 nm) were compared to PVP-stabilized PdNPs (2.1 ± 0.1 nm) for this Suzuki reaction carried out in 3 : 1 MeCN : H$_2$O at 100 °C, and the mechanism was found to be similar with phenylboronic acid adsorption onto the NPs, but the 2nd cycle/1st cycle ratio was higher for the dendrimer-PdNP catalyst [11a, 21]. Using a different mode of synthesis of these 4th-generation PAMAM dendrimer-stabilized PdNPs (3.2 ± 1 nm) as catalysts, Christensen found Suzuki coupling to occur with iodobenzene in EtOH at 78 °C, whereas bromobenzene requires a temperature of 153 °C in DMF. The amount of catalyst was only 0.055%, which is significantly less than with traditional catalysts. It was suggested that, since the G4-dendrimer diameter is only 4.5 nm, the PdNPs are stabilized by the dendrimer rather than encapsulated [22a]. In studies with 3rd- to 5th-generation poly(propylene imine) (PPI) functionalized

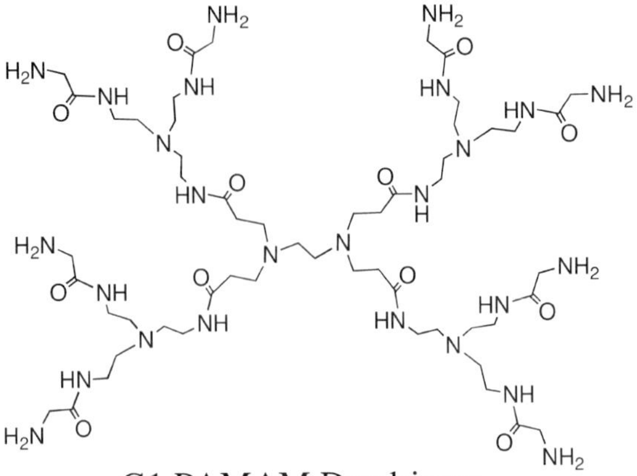

G1 PAMAM Dendrimer

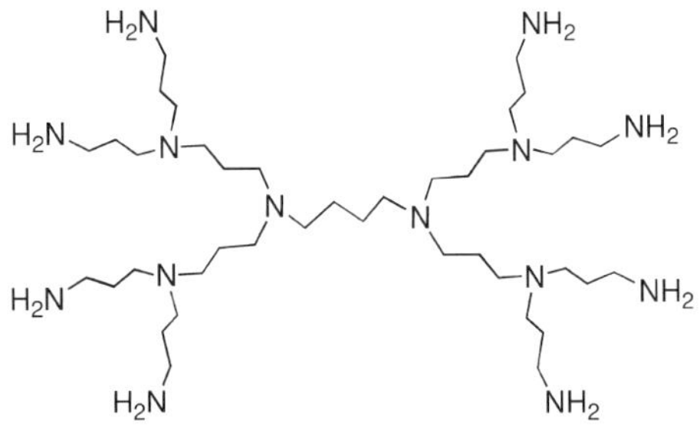

G1 PPI Dendrimer

Chart 1.2 The two families of commercial dendrimers considered as metal NP support for catalysis: PAMAM and PPI dendrimers (only the first generation G1 is represented). PPI dendrimers are smaller then PAMAM (2.8 nm vs. 4.5 nm for G4, respectively), but more stable (470 °C vs. 100 °C respectively) [11b].

by reaction with triethoxybenzoic acid chloride, the dendrimer-stabilized PdNP catalyst led to substrate specificity for the hydrogenation of polar olefins, due to the strong interaction between polar substrates and the inner tertiary amino groups. For instance, in competitive hydrogenations of 3-cyclohexene-1-methanol and cyclohexene, G5-PdNPs gave only reduction of the former while the traditional Pd/C catalyst gave incomplete hydrogenation of both compounds under the same conditions (Fig. 1.8) [22b,c]. Catalytic activity has been found for the Heck reaction of iodobenzene with ethylacrylate in refluxing toluene (75% yield) and Suzuki reactions of iodo- and bromobenzene with PhB(OH)$_2$ in refluxing ethanol (42–47% yield) have been observed for PdNP-cored dendrimers of the third generation. However, no activity was obtained for hydrogenation reactions [22d].

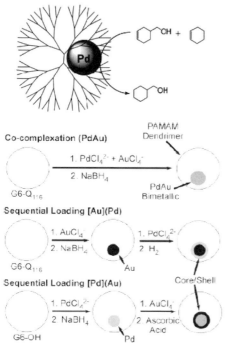

Fig. 1.8 Modes of synthesis of PAMAM-dendrimer-encapsulated heterobimetallic Pd-AuNPs (Reprinted with permission from Ref. [18k]; Crooks group, *J. Am. Chem. Soc.* **2004**, *126*, 15583).

This encapsulation strategy has recently been extended to bimetallic NP catalysts [22, 23], evidence that these NPs are bimetallic being provided by single-particle X-ray dispersive spectroscopy (EDS) [17a,b]. It was shown that the G4-PAMAM Pd/M NPs (M = Pt or Au) catalyze allylic alcohol hydrogenation more efficiently than the analogous monometallic Pt or Pd catalyst or their mixture, due to favorable synergistic effects [18]. When the inner PdNP was located at the dendritic core of G3-NPs, Heck and Suzuki coupling could be obtained in 38% to 90% yield in refluxing toluene or ethanol for a day [20] (Fig. 1.9). Suzuki catalysis in PPI dendrimer-supported PdNPs was also carried out for comparison with PAMAM dendrimers [18c]. The form of the NPs within the dendrimer is not clear despite TEM studies. Are the NPs really completely inside the dendrimer? Is the dendritic core encapsulated in the NP? Are there several close NPs in the dendrimer cavities or are they connected? Likewise, more work is called for in order to understand the very nature of the catalytically active Pd species.

The other dendrimer stabilizing strategy carried out by Esumi involves coordination of the NPs by surface amino groups of PAMAM and PPI dendrimers [19c–e], and these catalysts were used for various catalytic reactions including the reduction of 4-nitrophenol. In this case, one may visualize PdNPs surrounded by a number

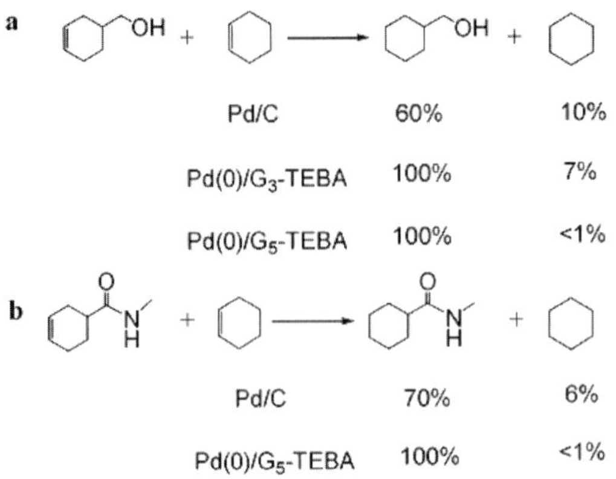

Fig. 1.9 Competitive hydrogenation of (a) 3-cyclohexene-1-methanol and cyclohexene (b) N-methyl-3-cyclohexene-1-carboxamide and cyclohexene using various Pd catalysts (Reprinted with permission from Ref. [22b], Kaneda group, *Nano Lett.* **2002**, *2*, 999).

of dendrimers that can also bridge PdNPs. In both situations, the dendrimers clearly stabilize small nanoparticles by a combination of polyligand and steric effects. Whether these dendrimer-stabilized PdNPs are the active species in Pd catalysis or reservoirs of much smaller, very active, Pd fragments, is unclear.

In Chapter 4, Bert Chandler and John D. Gilbertson also review, in addition to homogeneous DEN catalysis, their use of dendrimer-protected NPs heterogeneous systems. Indeed, Chandler's group has grafted DENs (including bimetallic ones) onto a variety of solid supports and used the resulting solids as heterogeneous catalysts, for instance for CO oxidation. The dendrimer protection of these NPs is removed at high temperature (150°C to 300°C), the PAMAM dendrimers being thermally unstable.

1.5
Ligand Stabilization of NPs

The introduction of ligands as NP stabilizers is of special interest, because it focuses on the precise molecular definition of the catalytic materials. In principle, this strategy potentially allows a good control of the molecular modulation in order to optimize the parameters that govern the efficiency in catalytic reactions, including enantioselective ones.

Gladysz showed that a thermomorphic fluorous palladacycle acts as a PdNP catalyst precursor for the Heck reaction at 80–140 °C in DMF with very high turnover numbers [24a]. Molecular palladium complexes such as palladacycles and other palladium salts have also been used as PdNP precursors upon treatment with CO in DMF or toluene at room temperature, and these PdNPs catalyzed nucleophilic substitution/carbonylation/amination affording iso-indolinones at room temperature [24b]. PdNPs capped with special ligands such as polyoxometallates [25] and cyclodextrins [26] were shown to be active for the catalysis of the hydrogenation of unsaturated substrates and of the Suzuki, Heck and Stille reactions. For instance, iodo- and bromoarenes and iodoferrocene are coupled to phenyl boronic acid by refluxing in MeCN : H_2O, 1:1 (v/v) in the presence of K_2CO_3 or $Ba(OH)_2$ and 1% perthiolated β-cyclodextrin-PdNPs [26a]. These 3-nm PdNPs are also active for the hydrogenation of water-soluble alkenes [26b,c]. In fact, the simplest dodecathiolate-PdNPs catalyze the Suzuki reaction of halogenoarenes, including chloroarenes, with phenylboronic acid, even at ambient temperature, and recycling several times has been achieved [27]. Another very simple mode of stabilization involved addition of silanes R_3SiH such as *tert*-butyldimethylsilane to PdX_2 (X = Cl⁻, OAc⁻) in *N, N*-dimethylacetamide. The black NP solution formed in this way catalyzed silane alcoholysis of sugars [28a] and selective cross-coupling of the silane with phenyl and vinyl thioethers giving the corresponding thiosilanes and silthianes (Eq. (1.2)) [28b]:

$$\text{RSR}' + \text{HSi}(t-\text{Bu})\text{Me}_2 \quad \xrightarrow[\text{DMA}]{\text{PdNPs}} \quad \text{HSi}(t-\text{Bu})\text{Me}_2 + \text{R}'\text{H} \quad \text{at } 25°\text{C} \quad (1.2)$$

R and R′ = alkyl, aryl; DMA = *N,N*′-dimethylacetamide

In view of atom economy, the synthesis of core–shell NPs having a cheap metal core such as Ni and a noble metal shell such as Pd has been achieved by thermal decomposition (235 °C) of Pd and Ni precursors ([Ni(acac)$_2$] + [Pd(acac)$_2$] + trioctylphosphine), the Ni complex decomposing before the Pd one. The Ni-cored PdNPs showed a much better activity than PdNPs without Ni, and having the same amount of Pd atoms, for the Sonogashira coupling of *p*-bromoacetophenone with phenylacetylene in toluene at 80 °C, although *p*-chloroacetophenone was unreactive [28c]. Enantioselective reactions have been carried out with metal NPs [29, 30]. The first example of an asymmetric reaction catalyzed by metal NPs was reported by the group of Lemaire, Besson and Gallezot in 1994 with the RhNP catalyzed hydrogenation of 2-methylanisole *o*-cresol trimethylsilyl ether induced by a chiral amine, *R*-dioctylcyclohexyl-1-ethylamine as a RhNP ligand [29]. The hydrogenation of ethyl pyruvate was found by Bönnemann [30a] to be efficiently catalyzed by cinchonidine-Pt- or PdNPs (75–80% ee, Scheme 1.2), and the ee was later improved (up to 95–98%) [30b–d]. Fujihara reported 2,2′-bis-(diphenylphosphino)-1,1′-binaphtyl [BINAP] – stabilized PdNPs with a diameter of 2.0 ± 0.5 nm and narrow size distribution. It was found that these BINAP-PdNPs catalyzed asymmetric hydrosilylation of styrene under mild conditions (ee: 95% at 0 °C), in contrast to inactive monometallic BINAP-Pd complexes [30e]. Recently, enantioselective allylic alkylation was reported by the Gomez and Chaudret groups to be catalyzed by PdNPs stabilized by a chiral xylofuranide diphosphite with 97% ee [30f]. The authors of these few NP reports support catalysis by the NP themselves. The very nature of the catalytically active species (and also whether catalysis actually occurs on the NP surface) remains unclear, however, and the catalytically active species may equally well be much smaller Pd fragments, leaching from the

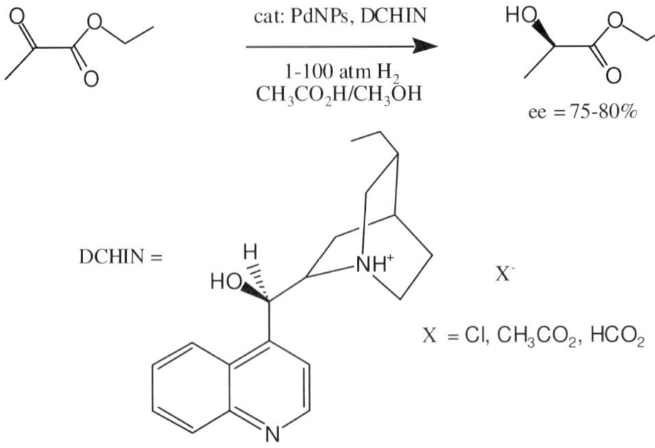

Scheme 1.2 Enantioselective hydrogenation of ethyl pyruvate catalyzed by cinchonidine-Pt- or PdNPs (Ref. [30a], Bönnemann group, *Chem. Eur. J.* **1997**, 2, 1200).

NP, to which the asymmetric ligand is bound. Anyway, the selectivity obtained is remarkable.

1.6
"Ligand-free" Heck Reactions using Low Pd-Loading

The original reports by Mizoroki [31a], then by Heck [31b], on the Pd-catalyzed coupling reaction of aryl iodides with olefin used a Pd salt (PdCl$_2$ resp. Pd(OAc)$_2$), a base (NaOAc resp. NBu$_3$) and a solvent (methanol resp. *N*-methylpyrolidone), but no phosphine or other ligand. Beletskaya reported a similar phosphine-free reaction of iodo- and bromoarenes in water, and the Pd-loading was as low as 0.0005 mol% (the term "homeopathic dose was used) with 3-iodobenzoic acid [31c]. The Reetz [31d,n] and de Vries [31e,n] groups reported extremely efficient Heck catalysis of coupling between aryl bromides and styrene in organic solvents with such very low Pd-loading. Reetz found that PdNPs are formed when PdCl$_2$, Pd(OAc)$_2$ or Pd(NO$_3$)$_2$ is warmed in THF in the presence of a tetrabutylammonium carboxylate that functions as a reducing and stabilizing agent [31f]. Polar solvents such as propylene carbonate also generated such PdNPs upon heating Pd(OAc)$_2$. PdNPs generated in this way from Pd(OAc)$_2$ [31d–f] or palladacycles [31g–i] are active catalysts in the Heck reaction, as demonstrated by following the reactions using TEM. [31j] Very interestingly, it was found that the Pd catalyst "improves" upon lowering the Pd-loading, which was accounted for by an equilibrium between PdNPs serving as a catalyst reservoir and small (monomeric or dimeric) catalytically active Pd species [31d,h,k]. When the catalyst concentration is too high, inactive Pd black forms (Scheme 1.3).

This points to the fact that the rate of the catalytic reaction must be extremely high, since most of the Pd is in the form of PdNPs. This type of Heck reaction

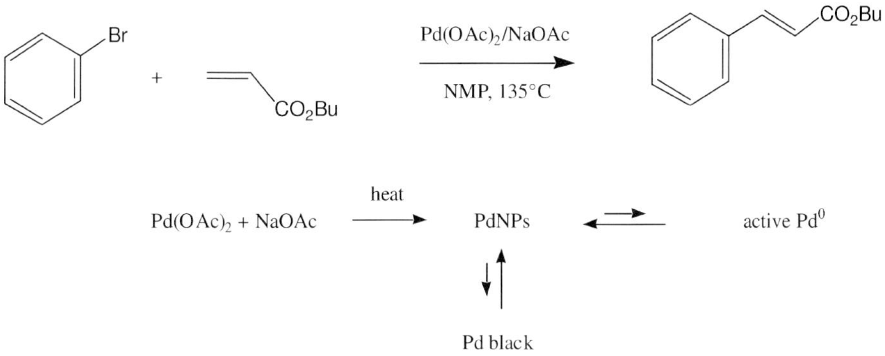

Scheme 1.3 Heck-type reactions catalyzed by extremely low loading of Pd salt proceeding via PdNPs (Refs. [31d,f,j,n] Reetz group).

seems quite general with aryl bromides and has been scaled up by DSM to the kg scale in order to prepare a drug intermediate [31e]. Likewise, a range of enantio-pure substituted *N*-acetyl-phenylalanines was obtained from methyl *N*-acetamido-acrylate and various bromoarenes at very low Pd-loading in the absence of other ligands, followed by Rh-catalyzed hydrogenation [31l]. De Vries reported similar behavior for the Suzuki reaction of aryl bromides with TOFs up to 30 000 [31m]. The precise nature of the active species in these Pd-catalyzed C–C coupling reactions is not known, and it may well be an anionic mono- or dimeric Pd^0 species to which an anionic ligand (Cl^- or OAc^-) is bound [32–34]. This means that the anionic Amatore–Jutand mechanism that has now met general acceptance for homogeneous Heck catalysis is also operating with many heterogeneous systems (Scheme 1.4). This type of catalysis is very important in term of "*Green Chemistry*" as waste is largely minimized here in the absence of added ligand and such low Pd-loading [31n]. It also suggests that supported PdNP catalysts (heterogeneous catalysts, *vide infra*) could well behave in a related way. The Jones group has recently published an important comprehensive review on the nature of active species in Pd-catalyzed Heck and Suzuki–Miyaura reactions [31o]. This concept could be extended to other types of catalysis, and indeed other examples of PdNP pre-catalysts are known for C-C bond formation for which Pd is the best metal catalyst [31p] (Scheme 1.5). Other metal NP catalysts are also known, for instance Pt-catalyzed hydrosilylation [5c] and Ru-catalyzed hydrogenation using transition-metal complexes. RuNPs have also been shown to catalyze the Heck reaction [31q].

In Chapter 10, Laurent Djakovitch, Klaus Köhler and Johannes G. de Vries review the role of PdNPs as pre-catalysts for C—C coupling reactions. This chapter represents a timely general review from these experts embracing both homogeneous and heterogeneous systems. Emphasis is placed on homeopathic PdNP catalysis that was rationalized and used by de Vries earlier and on the homogeneous type of catalysis with heterogeneously supported PdNP catalysts with leaching of catalytically active Pd atoms returning after their catalytic acts onto the PdNP reservoir.

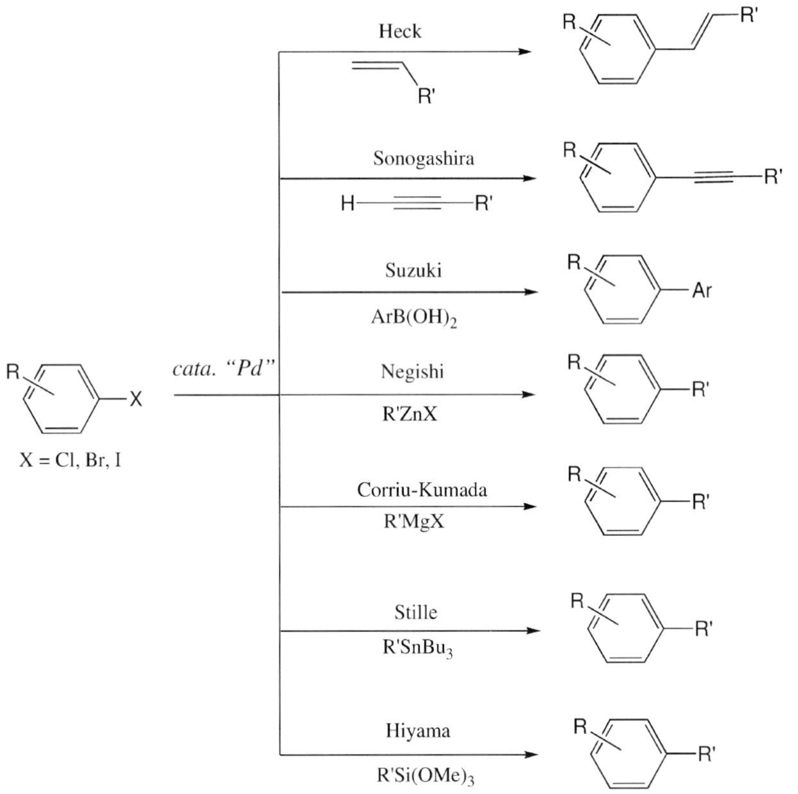

Heck

Sonogashira

Suzuki

ArB(OH)$_2$

Negishi

R'ZnX

Corriu-Kumada

R'MgX

Stille

R'SnBu$_3$

Hiyama

R'Si(OMe)$_3$

cata. "Pd"

X = Cl, Br, I

H————R'

Scheme 1.4 Pd-catalyzed C—C bond formation with aryl halides and triflates.

1.7
The Roles of Micelles, Microemulsions, Surfactants and Aerogels

"Fluorous" strategies [32a,b] have been used on various occasions for the NP catalysis mentioned above as for instance by the Crooks [17] and Gladysz [24a] groups. Fluoro surfactants can also serve as micellar stabilizers for PdNPs in water-in-scCO$_2$ microemulsions that were used as hydrogenation catalysts for simple olefins [32b–d] and citral [32e]. In such systems, hydrogen can work both as a reductant for Pd salts and for the unsaturated substrate. Ultrafine PdNPs in reverse micelles using KBH$_4$ as a reducing agent of Pd^{II} precursors led to catalytic hydrogenation of allylic alcohol and styrene in iso-octane, although the *bis*(2-ethylhexyl) sulfosuccinate surfactant inhibited the hydrogenation activity (Fig. 1.10) [32f]. The oxidation of N, N, N', N'-tetramethyl-*p*-phenylenediamine by [Co(NH$_3$)$_5$Cl]$^{2+}$ was catalyzed by PdNPs in an aqueous/AOT/*n*-heptane microemulsion [32g]. Functional olefins such as 4-methoxy-cinnamic acid were selectively hydrogenated as well as nitrobenzene (to aniline) in supercritical CO$_2$ using PdNPs in a water-in-CO$_2$ microemulsion [32h,i]. Oxidation of cyclooctane by

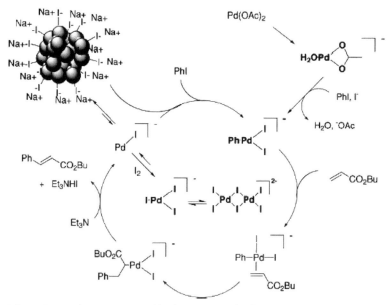

Scheme 1.5 Mechanism proposed by de Vries [31n] for the "homeopathic" Heck reaction between phenyl iodide and butyl acrylate catalyzed by Pd(OAc)$_2$ [31c–f, l–o]. Reprinted with permission from Ref. [31n].

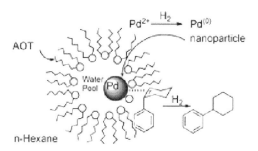

Fig. 1.10 Hydrogenation of 10-(3-propenyl) anthracene catalyzed by PdNPs in water-in-oil microemulsion (much faster than Pd/C catalysis). (Reprinted with permission from Ref. [32f]; Wai group, *Chem. Commun,* **2003**, 1041).

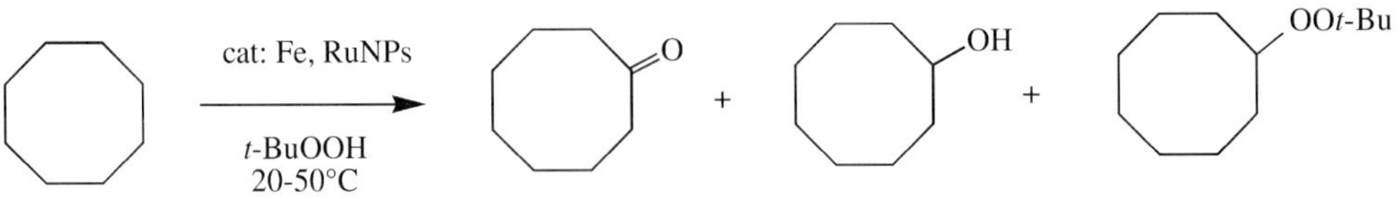

Scheme 1.6 Cyclo-octane oxidation catalyzed by FeNPs in reverse microemulsions or RuNPs in biphasic water–organic media (Ref. [49b], Roucoux-Patin group).

tert-butylhydroperoxide (*t*BHP) was catalyzed by FeNPs in a reverse microemulsion or with RuNPs in a biphasic water/cyclooctane solvent with recycling without loss of activity (Scheme 1.6), and microemulsions were found to be efficient in ligand-free Heck catalysis [32j].

Aerogels, less dense solid materials (with densities around $0.2\,g\,cm^{-3}$ and high surface area around $600\,m^2\,g^{-1}$), are excellent supports for NPs in catalysis, because they are porous and thus leave the active NP surface exposed. Aerogels are prepared by supercritical drying or liophilization of gels, and the NP-ated aerogels are prepared before, during or after the sol–gel process.

Aerogel-supported NPs in catalysis are reviewed by Adelina Vallriebera and Elies Mollins in Chapter 5. In particular, the catalysis that is reviewed in Chapter 5 with NP-ated arerogels includes both classic liquid-phase reactions (olefin oxidation, epoxidation and dihydroxylation, various C–C bond formation reactions including Heck coupling, hydrocarbonylation, Claisen–Schmidt condensation, Michael addition) and high-temperature heterogeneously catalyzed processes (Fischer–Tropsch, steam reforming, CO, methanol and butane oxidation, dechlorination of chlorinated volatile organic compounds, oxygen and cyclohexene reduction by H_2 and isomerizaton of 1-butene).

1.8
Ionic Liquid Media for Catalysis by NPs

Ionic liquids (ILs) were introduced in catalysis by Yves Chauvin in the 1990s [33a] and have received considerable attention in this field [33b,c]. Yves Chauvin introduced the imidazolium salts that are the most frequently used ILs in catalysis. ILs are valuable media for catalysis with PdNPs because the substituted imidazolium cation is bulky, favoring the electrosteric stabilization of NPs, as do the t-Bu_4N^+ salts in Fig. 1.1. The size of the cation (that can eventually be tuned by the choice of the *N*-alkyl substituents) also has an important influence on the stabilization, size and solubility of the NPs, these factors playing a role in catalysis. ILs are also non-innocent, however, as they readily produce Pd-*N*-heterocyclic carbene complexes upon deprotonation of the imidazolium salt at sufficiently high temperature. Thus, these carbene ligands can be bound to the NP surface or give mononuclear mono- or bis-carbene complexes subsequent to leaching of Pd atoms from the PdNP surface (*vide infra*) [33].

IrNPs in 1-*n*-butyl-3-methylimidazolium hexafluorophosphate ([BMIM][PF$_6$]) were used directly for the hydrogenation of olefins, and good results were obtained [33b]. Phenanthroline-stabilized PdNPs prepared in [BMIM][PF$_6$] according to the method reported by Schmid [33c] but without using acetic acid as the solvent, (Pd(OAc)$_2$ + Phen. H$_2$O + 1 atm H$_2$ in the IL at room temperature), efficiently catalyzed the hydrogenation of olefins and the selective hydrogenation of cyclohexadiene to cyclohexene under mild conditions (1 atm H$_2$, 40 °C). Under these conditions, formation of Pd-carbene complexes from the IL does not occur, thus the IL simply plays the role of a PdNP-stabilizing solvent like other salts known to provide electrostatic NP stabilization. This catalyst could be re-used several times [33d], and it was much more active than Phen-protected PdNPs supported on TiO$_2$ for the hydrogenation of 1-hexene [6b]. It was found that PdNPs, formed by reaction of Pd(OAc)$_2$ with tetrabutylammonium acetate dissolved in tetrabutylammonium bromide, efficiently catalyzed the stereospecific reaction of cinnamates with aryl halides to give β-aryl-substituted cinnamic esters.

The role of the IL is crucial in both the PdNP formation and stereospecifity of C–C coupling that could not be obtained in previous studies of PdNP-catalyzed Heck reactions [33e–g]. Salts of *N*-butyronitrile pyridinium cation react with PdCl$_2$ to give dinitrile complexes that turn black upon addition of phenyltributylstannane, and the PdNPs formed catalyze Stille and Suzuki C–C coupling reactions. It is believed that the nitrile groups coordinate to the PdNP surface, which results in PdNP stabilization [33h].

Palladium acetate led to the formation of PdNPs in the presence of the IL 1,3-dibutylimidazolium salts. It was suggested that formation of *N*-heterocyclic Pd-carbene complexes is at the origin of the formation of PdNPs that catalyze Suzuki coupling [33i]. Such carbene complexes were shown to form and also catalyze the Heck reaction, formation of PdNPs under these conditions being highly suspected to be responsible for catalysis [33e,j]. Indeed, heating these *N*-heterocyclic Pd-carbene complexes led to PdNP formation subsequent to ligand loss (Scheme 1.7). The selectivity of the reactions in such IL media also depends on the solubility, and the solubility difference can be used for the extraction of the product [33k–m].

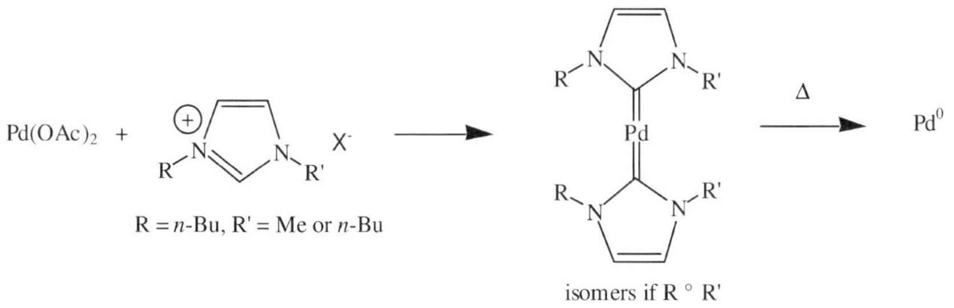

Pd(OAc)$_2$ + [imidazolium salt, R–N⁺=N–R', X⁻] → [Pd-carbene complex] $\xrightarrow{\Delta}$ Pd^0

R = *n*-Bu, R' = Me or *n*-Bu

isomers if R ° R'

Scheme 1.7 Formation of Pd-carbene complexes by reaction between palladium acetate and imidazolium salts followed by decomplexation at high temperature and formation of catalytically active PdNPs for Heck-type reactions [33e,i,j].

In conclusion, IL are favorable media for the electrostatic stabilization of pre-formed NPs at room temperature and subsequent catalysis, whereas they give for instance Pd-carbene complexes upon deprotonation of the imidazolium cation, yielding, at high temperature, PdNP catalysts whose mechanism of action is discussed in the ligand-free catalysis section.

The field of ionic-liquid-stabilized NPs is reviewed by Jairton Dupont and D. Dagoberto de Oliveira Silva in Chapter 6, including the supramolecular aspects of ILs (multiple H-bonding) and the most frequent examples of reactions that are catalyzed by IL-stabilized NPs (hydrogenation of olefins, diolefins, arenes and ketones, hydroformylation, Heck reaction and Negishi coupling).

1.9
Oxide Supports for NP Catalysts

There is a fast growing and now large body of literature reports concerning NPs supported on metal oxides with a variety of supports, i.e. oxides of Si [34, 35], Al [36], Ti [37], ZrO_2 [43], Ca [38], Mg [39, 43], and Zn [39h]. These oxides are in the form of SiO_2 aerogels or sol–gels such as Gomasil G-200, high-surface silica (see for instance Scheme 1.8), M41S silicates and alumimosilicates, MCM-41 mesoporous silicates such as HMS and SBA-15 silica, silica spheres [35b], microemulsions (SiO_2), hydroxyapatite (Ca^{2+}) [38], hydrotalcite (Mg^{2+}, Al^{3+}) [39], zeolites (SiO_2, Al_2O_3) [40], molecular sieves (Fig. 1.11) [41] and alumina membranes (Fig. 1.12) [36a]. Thus, despite the large variety of supports, the majority of them involve a form of silica. The catalytic reactions examined with these supported NPs are hydrogenation reactions of unsaturated substrates including selective ones, Heck and other C—C coupling reactions and oxidation of CO and alcohols using molecular oxygen. The heterogenization of polymer- or dendrimer-stabilized NPs on a solid support such as silica permits one to benefit from the classic advantages of heterogeneous catalysis, i.e. stability to high temperatures and easy removal from the reaction medium and from the bottom-up approach of NP synthesis. A few remarkable recent examples are discussed below, and Table 1.1 gathers references for each NP-catalyzed reaction.

PtNPs and bimetallic dendrimer-stabilized Pd-AuNPs were adsorbed onto a high-surface silica support and thermally activated to remove the dendrimers

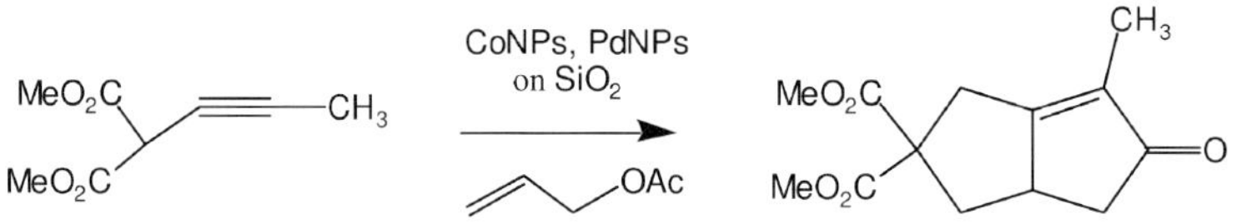

Scheme 1.8 Sequential allylic alkylation and Pauson–Khand reactions, both catalyzed by Co and PdNPs supported on silica for one-pot syntheses of bicyclic enones (Ref. 35c, Chung group, *Org. Lett.* **2002**, *4*, 4361).

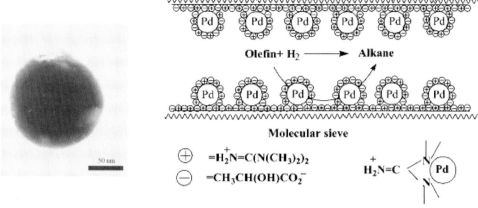

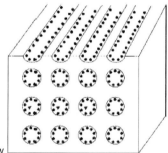

Fig. 1.11 Transmission electron micrograph (left) of a molecular sieve with supported PdNPs (right). The catalyst, active in hydrogenation of olefins, contains 20 wt.% IL whose average layer thickness is 0.4 nm. (Reprinted with permission from Ref. [41b], Han group, *Angew.Chem. Int. Ed.* **2004**, *43*, 1397).

Gas flow

Fig. 1.12 Sketch of a piece of alumina membrane with metal NP elucidating its application for gas phase catalysis of 1,3-butadiene hydrogenation (PdNPs) and CO oxidation (RuNPs) (Reprinted with permission from Ref. [36a], Schmid and Chaudret groups, *Z. Anorg. Allg. Chem.* **2004**, *630*, 1913).

(Fig. 1.13). The Chandler group showed that these NPs were smaller than 3 nm and highly active for CO oxidation catalysis near room temperature, and hydrogenation of toluene was also efficiently carried out [34]. The fabrication of uniform hollow spheres with nanometer to micrometer dimensions having tailored properties has recently been intensively studied using various procedures [42a]. Monodisperse Pd nanospheres of 300-nm size catalyzed Suzuki coupling of iodothiophene with phenylboronic acid using 3 mol% Pd catalyst in ethanol under reflux whereas 15 mol% Pd catalyst was used to couple bromobenzene under these conditions (Fig. 1.13) [35b]. Hydroxyapatite, $[Ca_{10}(PO_4)_6(OH)_2]$, with $[PdCl_2(CH_3CN)_2]$ gives a

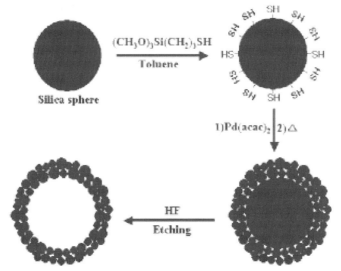

Fig. 1.13 Principle of the formation of hollow PdNP spheres used for the catalysis of Suzuki reactions in refluxing ethanol with K_3CO_4 as a base. With 2-iodothiophene and phenylboronic acid (catalyst: 3% Pd spheres), at least seven cycles could be achieved with 95–97% yields (Reprinted with permission from Ref. [35b], Hyeon group, *J. Am. Chem. Soc.* **2002**, *124*, 7642).

monomeric $PdCl_2$ species chemisorbed on the hydroxyapatite surface that is readily transformed into supported PdNPs with narrow size distribution in the presence of alcohol. These PdNPs catalyze the oxidation of 1-phenylethanol under an atmospheric O_2 pressure in solvent-free conditions with a very high turnover up to 236 000 and a remarkable turnover frequency of $9800 h^{-1}$. The work up is easy and the catalyst is recyclable without requiring additives to complete the catalytic cycles [38]. Hydrotalcite anionic clays are layered double hydroxides of formula $M^{2+}{}_{1-x}M^{3+}{}_x(OH)_2(An^-)_{x/n} \cdot yH_2O$ with $A^{n-} = CO_3^{2-}$, Cl^- or NO_3^- that have high anion exchange capacities. These materials, after calcination at temperatures over 723 K, serve as supports for noble-metal catalysts. For instance, C–C coupling [39d] and selective semi-hydrogenation of alkynes [39e] have been obtained. Immobilization of a PdNP catalyst on a solid surface, such as molecular sieves, was achieved by the IL 1,1,3,3-tetramethylguadinidium lactate, and this system was used for solvent-free hydrogenation of alkenes with high activity and stability (for instance with a cyclohexene/Pd mol ratio of 12 000, 100% conversion was obtained in 10 h at 20 °C with a TOF of 20 min^{-1}) [41b]. The PdNPs of size 1–2 nm before and after catalysis, are stabilized by guanidinium ions [33d]. Ag-PdNPs prepared directly in ultrathin TiO_2 gel films by a stepwise ion-exchange/reduction method showed activity in methyl acrylate hydrogenation 267 times higher than commercial Pd black and 1.6 times higher than PdNPs that did not contain Ag. This outstanding activity was explicable by the large fraction of the surface-exposed Pd atoms [37c]. Polyelectrolyte multilayers serve as supports for PdNP-catalyzed selective hydrogenation of allylic alcohols, isomerization being suppressed. Thus, polyacrylic acid and polyethyleneimine-Pd(II) complex were alternately adsorbed on 150-mm diameter alumina particles, and subsequent reduction of Pd(II) to PdNPs was carried out using NaBH$_4$ [14d].

The mechanisms of oxide-supported PdNP catalysis are far from being understood. The oxide support has of course a strong influence on the activity. For instance, in the Heck reaction, the activity is dominated by the support according to:

$$C(84\%) \sim \text{H-Mordenite} (83\%) > \text{ZrO}_2 (49\%) \sim \text{TiO}_2 (45\%) >$$
$$\text{MgO}(37\%) \text{ ZnO} (37\%) > \text{SiO}_2 (7\%)$$

The good activity of zeolites in this reaction is seemingly due to better stabilization of active species in the cavities and to better dispersion of PdNPs on the oxide support. There are many heterogeneous catalysis studies discussing the influence of parameters (solvent, catalyst, base, temperature, recycling activity, NP size). Indeed, Djakovitch and Köhler have proposed that, since the results are often similar for homogeneous and heterogeneous systems in terms of selectivity, the heterogeneous mechanism proceeds by leaching of molecular Pd species into the solution, a phenomenon favored in DMF (see Chapter 10). This "leaching mechanism", first proposed by Julia *et al.* in 1973 [42b], then shown by Shmidt and Mametova [42c] and subjected to detailed experimental evidence by Arai [42d], is now firmly established [43a]. On the other hand, dehalogenation is favored on heterogeneous supports compared to homogeneous catalysis. Thus, it has been proposed that supported PdNPs are responsible for dehalogenation [43]. Along this line, the useful 3-phase test (also called Collman test, first proposed by Rebeck) has been used by Davies to discriminate homogeneous and heterogeneous Pd-catalyzed carbon–carbon coupling reactions [43k].

The heterogeneous Pd-catalyzed Heck reaction has been extended to important α-arylated carbonyl derivatives, and the model arylation of diethylmalonate has been examined. The NaY-PdNP catalyst disclosed a limited activity, giving yields comparable to those obtained with the homogeneous [Pd(OAc)$_2$/4 PPh$_3$] system, but the Pd concentration used is only 2% and the recyclability is good. The amination of halogenoarenes has also been investigated with MgO-PdNPs and ZrO$_2$-PdNPs, the amphoteric supports giving the best yields, which would indicate that they favor the rate-limiting C—N coupling in the reductive-elimination step. For this latter reaction, the zeolite supports give a better *para* selectivity, supposedly because of the optimal control of the "shape selectivity" [43, 44].

This leaching concept established for PdNPs applies to Ni/C for which catalysis of a variety of C—C and C—N coupling reactions (Kumada, Suzuki, Negishi, Heck, amination of bromoarenes) was also found by Lipshutz's group with leaching Ni species leading to a mixture of homogeneous and heterogeneous catalysis. However, the amount of leaching metal is much more important in the PdNP case than with NiNP for which the amount of leaching remained essentially constant and extremely low along the reactions [44k–o].

The Heck and Suzuki reactions have been catalyzed by other NPs such as RuNPs, including alumina-supported RuNPs [43c]. Hydroxyapatite-supported RuNPs were recently found to be efficient and recyclable catalysts for *cis*-dihydroxlation and oxidative cleavage of alkenes [43d]. Supported RuNPs in the pores of mesoporous Al-MCM-41 materials were prepared by H_2 reduction of adsorbed $[Ru(NH_3)_6]^{2+}$, and the activity in benzene hydrogenation was studied in the absence and presence of 330 ppm H_2S in H_2. By comparison, zeolite-supported catalysts were found to be more efficient when H_2S was present [43d].

Oxide supports in transition-metal NP catalysis are discussed in this book in detail in Chapters 4, 5, 9, 10, 15, 16 and 17 dealing with heterogeneous catalysis. In particular, Florence Epron, Catherine Especel, Gwendoline Lafaye and Patrice Marécot review multimetallic NPs prepared by redox processes applied in catalysis in Chapter 9. This chapter is authored by the successors of Jacques Barbier who headed the Poitiers laboratory and pioneered oxide-supported multimetallic heterogeneous catalysis. It focuses on the advantages of silica- or alumina-supported multimetallic NP catalysts of organic transformations (particularly improvement in the selectivity of hydrogenation and hydrogenolysis and competition between C=C and C=O bond hydrogenation), naphtha reforming, fuel cells design and environmental catalysis. The strength of this chapter is also the systematic analysis of the redox approach to bi- and tri-metallic NPs coupled with compared catalytic efficiency. Practical aspects (including NP synthesis, stability in water, deposition reactions) and characterization methods and techniques are emphasized therein.

1.10
Carbon Supports for NP Catalysts

Charcoal is a classic commercial support for catalysts such as Pd/C. Bönnemann applied his general synthetic method using the reduction in THF of quaternary ammonium salts of metal cations to simple impregnation upon stirring the NP suspension on charcoal, and used these charcoal-supported metal NPs for a variety of catalyzed reactions [7, 29a, 34, 35a, 41a,c]. This procedure has also been used by Reetz for his electrochemically prepared metal NPs including catalytically active bimetallic NPs [8, 35a]. Activated carbons that are suitable as support materials in catalytic processes need to be prepared and modified in order to obtain adequate surface area, porosity and pore size distribution. Purification by acid treatment and elution processes is required in order to remove ash, extractable sp^3 material and contaminants. These supports need pre-treatment and conditioning for the preparation of suitable surface chemistry to optimize precious metal/support interactions during impregnation and dispersibility in the reaction media. *Pd/C catalysis (as well as overall PdNP catalysis) is extensively reviewed in Chapter 10 by Laurent Djakovitch, Klaus Köhler and Johannes G. de Vries.* The mechanism is quasi-homogeneous, and small Pd species in solution act as the catalytically active ones.

The Pd is leaching and is redeposited at the end of the reaction, which provides an excellent recovery of the precious metal from the reaction mixture. The precipitation of the catalyst at the end of the reaction significantly changes its state and decreases its activity, however, making its re-use unattractive. Köhler *et al.* also showed that optimization of the Pd/C catalyst (temperature, solvent, base and Pd loading) allows turnover frequencies (TOFs) of up to 9000 to be reached and Pd concentration down to 0.005 mol% to be developed for the Heck reaction of unactivated bromobenzene at 140 °C [44]. The turnover numbers (TONs) are surpassed, however, by those of the best homogeneous catalysts.

Heterogeneous Ni/C catalysis is also essentially known for hydrogenation of unsaturated compounds. Recent reports concern hydrodehalogenation of aryl halides (including aryl chlorides), Kumada, Suzuki and Negishi-type C—C coupling and aromatic amination. Whereas the mechanism for heterogeneous hydrogenation follows true surface chemistry with dihydrogen chemisorption onto the metal surface, C—C and C—N coupling reactions involve nickel bleed from the carbon support (whose % can be very high), a one-time event at the very beginning of the reactions. There exists an equilibrium for this homogeneous Ni species between Ni located inside or outside the C pores, strongly favoring the former. Thus, unlike in the Pd/C case, only traces of metal are detectable in solution. The Ni is completely recovered on charcoal, making this heterogeneous catalyst important for applications. These reactions appear to be due to a combination of heterogeneous and homogeneous catalysis [44].

Recently, new supports such as high-area carbon have been used to prepare bimetallic Pt–Ru NPs that catalyze methanol electro-oxidation with enhanced activities compared to commercial catalysts [46i–l]. Pd, Rh and RuNPs, deposited onto functionalized carbon nanotubes through hydrogen reduction of metal-β-diketone precursors are effective catalysts for hydrogenation of olefins in sc.CO$_2$. High electroactivity was also found in oxygen reduction for potential fuel cell application. Such PdNPs on carbon nanotubes in supercritical CO$_2$ are efficient, for instance, in *trans*-stilbene hydrogenation [32g], and comparison between carbon nanotubes and activated carbon as support was carried out for Heck and Suzuki reactions, aerobic alcohol oxidation and selective hydrogenation [44].

Cuong Pham-Huu and Marc-Jacques Ledoux review carbon and silicon nanotubes containing catalysts in Chapter 7. The very high external surface of these carbon supports reduces the mass transfer limitations, and the specific interactions between the deposited phase and the support surface can explain the catalytic behavior. This chapter presents general background on the development of the catalytic synthesis of one-dimensional (1-D) carbon and carbide materials that are remarkably illustrated by outstanding images. It also compares carbon and silicon nanotubes containing catalysts to other supported catalysts. In particular, hydrogenation and dehydrogenation reactions of various substrates and desulfuration reactions are surveyed. Metal NPs and related materials supported on carbon nanotubes are also reviewed in a recent article by Compton's group with emphasis on the methods of functionalizing CNTs with metal NPs (including electrochemical, chemical and physical methods) and applications of CNT-supported NPs in catalysis, hydrogen sensing, sensing and energy storage [44r].

1.11
NPs of Noble Metals (Ru, Rh, Pd, Pt and Their Oxides) in Catalysis

Pioneers such as Nord in the 1940s and Turkevitch in the 1950s already used Pd as the metal of choice for the synthesis of NPs. Then, the reduction of halides of the transition metals and rare earth using MBR$_3$H in THF by Bönnemann in the 1990s covered all the periodic table. The active NP catalysts that have been most studied, however, are those of the noble metals Ru, Rh, Pd, Pt and Au. Thus, the second part of this book includes a number of chapters that are devoted to one or two of these specific noble metals (a specific place is reserved thereafter for gold, *vide infra*). The most studied metal by far is Pd, because not only is it an excellent hydrogenation catalyst, but mainly it also catalyzes the formation of a large variety of C—C bonds. *Laurent Dajakovitch, Klaus Köhler and Johannes G. de Vries detail PdNP catalysis in Chapter 10. Alain Roucoux, Audrey Nowicki and Karine Philippot in Chapter 11 deal with RhNPs and RuNPs in catalysis with emphasis on soluble NPs in various liquid media (hydrogenation of alkenes, alkynes and aromatics and C—C bond formations such as hydroformylation, methanol carbonylation, coupling reactions and Pauson–Khand reactions, oxidation, dehydrocoupling and hydrosilylation).*

The seminal work by Reetz in the 1990s used the Bönnemann-type reduction of metal salts although cathodic reduction was the synthetic method of choice, and the catalytic activity of these and many other transition metals was demonstrated and used. *Manfred T. Reetz details in Chapter 8 these studies devoted to size-selective synthesis of nanostructured metal- and metal-oxide colloids and their use as catalysts. It is shown in this Chapter 8 how Pd^{II} is reduced to Pd^0 by the acetate anion (refer to Heck catalysis with PdNPs formed from Pd(OAc)$_2$). Of particular interest is the clear relationship between the oxidation potential of the carboxylate anion RCO$_2^-$ and the size of the PdNPs formed. The more negative this potential, i.e. the larger the driving force for Pd^{II} reduction, the smaller the PdNPs formed. Since it is known from Marcus theory that a larger driving force results in a faster reaction, this finding is in accord with the long-known observation that the NPs are larger if they are grown slowly using a weaker reductant whereas they are smaller using a stronger reductant. This is important, because small NPs are sought given their higher catalytic efficiency than that of larger ones. Serendipitous discovery of the aerobic oxidation of CoNPs to Co oxide NPs led the author to investigate and generalize to many metal NP oxides the formation of the PtO$_2$ NPs (Adams catalyst) from PtCl$_4$ and a base in water in the presence of a stabilizer. This aspect concerning many mono- and di-metal NP oxides and their catalytic functions (chemical catalysis, electrocatalysis and corrosion) is the subject of the second part of Manfred Reetz' Chapter 8.*

1.12
Gold Nanoparticle-based Catalysts

AuNPs occupy a special place given their great success and present developments. Thus, four chapters of the book are devoted to the various aspects of AuNP catalysis. In the AuNP-catalyzed CO oxidation to CO$_2$ by O$_2$ that can occur down to 200 K

(!) [9c,d], the oxide support (Fe_2O_3, TiO_2 or Co_3O_4) is indispensable, a seminal discovery by Haruta in 1987 that was due to his recognition of the crucial requirement for the small size of the AuNPs (<5 nm) [9d,e, 45]. This surprising discovery, given the long believed chemical inertia of gold, has completely changed the way chemists look at this magic metal. The finding of low-temperature catalysis of CO oxidation is also of great interest, because the Pt/Pd catalysts that are used in cars for CO oxidation work only at temperatures above 200 K. Thus, CO pollution essentially occurs during the first five minutes after starting the engines. The low-temperature supported-AuNP catalyzed CO oxidation obviously could solve this problem. In addition, the selective oxidation of CO by O_2 in the presence of H_2 allows the purification of H_2 from residual CO. There are many other challenges in catalytic oxidation chemistry that can be addressed using this type of catalyst (*vide infra*). It is considered that, among the various ways to prepare supported AuNP catalysts, Haruta's deposition-precipitation procedure is the most suitable [9d]:

$HAuCl_{4aq}$ + NaOH → $[Au(OH)_4]^-$ Na^+_{aq} (pH 6–10) → $Au(OH)_3$/Support → wash, dry then calcinate at 563–673 K → AuNP/Support (optimal AuNP size: 3 nm, stable hemispherical NPs, the size being controlled by the calcination temperature whose optimum is 570 K, optimal support: TiO_2 for which the addition of Mg citrate is necessary during or after co-precipitation for a good dispersion of tiny AuNPs). The mechanism of CO oxidation is not clear, however [45a].

It has been shown that oxidation of CO by O_2 to CO_2 can also be catalyzed by the cluster Au_6^- in the gas phase (Fig. 1.14) [45c], and calculations with Au_{10} predict CO oxidation by O_2 below room temperature [45d]. Recent studies by Haruta's

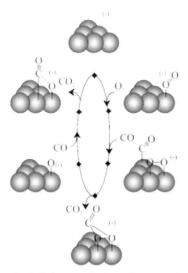

Fig. 1.14 Proposed schematic mechanism of the Au_6^--catalyzed formation of CO_2 from CO and O_2 in the gas phase (Reprinted with permission from Ref. [44c], Wallace and Whetten, *J. Am. Chem. Soc.* **2002**, *124*, 1999).

group indicate that the TOF increases together with a decrease in AuNP diameter. The enhancing effect of moisture on the metal oxide support has been demonstrated. Kinetics showed that the rate of CO oxidation is independent of the CO concentration and only slightly dependent on the O_2 concentration (of the order of 0 to 0.25) in the low concentration range down to 0.1 vol%. Although the mechanism was thus eventually depicted with AuNPs alone [45c,d] (Fig. 1.14), activation by the oxide support is also needed and must be involved in the mechanism. Haruta suggested that CO probably adsorbs on the edge and step sites of AuNP surfaces and O_2 adsorbs on the support surfaces [50b]. Both CO and O_2 must be adsorbed on the catalyst surface near saturation and the rate-determining step must be the reaction between the two adsorbed species. [9d,e, 45b]. The support may also have a crucial role in the cleavage of the superoxide species or negatively polarized oxygen atom of the O_2 molecule before, during or after CO binding [45]. Haruta recently depicted his mechanistic proposal as follows [11j]:

$$O_2 + Au_{NP}/TiO_2 \quad\quad \rightarrow \quad Au_{NP}/TiO_2 \ldots O_2$$
$$Au/TiO_2 \ldots O_2 + 2Au_{NP}CO \quad \rightarrow \quad OCAu_{NP} = O + CO_2 + TiO_2$$
$$OCAu_{NP} = O \quad\quad \rightarrow \quad Au_{NP} + CO_2$$

Intriguingly, the Corma group reported that Au^{3+} on CeO_2 catalyzed the homocoupling of phenylboronic acid, and the mechanism was proposed to involve Au^{3+} and Au^+ interconversion in the catalytic cycle [449]. Spectroscopic studies, by the same research group, of Au supported on nanocrystalline CeO_2 showed that CO was bonded to Au^{3+}, Au^+ and Au^0 species, whereas the active form of O_2 was bonded to CeO_2 as superoxide η^1-O_2^- confirming the nucleophilic attack of the electron-deficient carbon atom of CO by superoxide on the way to CO_2 [449]. Another open question concerns the oxide support: is it best to use a redox active, thus reducible, oxide such as TiO_2 or a redox inactive one such as silica or alumina?

Applications are expected in both the catalytic removal of CO produced at ambient temperature by engines and the removal of CO traces from dihydrogen streams feeding the fuel cells. There is a tremendous recent increase in the number of reports focusing on this area of supported AuNP-catalyzed CO oxidation, TiO_2 becoming the dominant support [45].

The field of CO oxidation, including Masatake Haruta' systems, recent advances and key mechanistic issues are reviewed by Catherine Louis in Chapter 15. This chapter details the basic features of CO oxidation catalyzed by AuNPs, the preparation of the oxide supports, conditions for efficient catalysis, electronic and morphological properties of the catalytically active supported AuNPs. It also offers deep and rational mechanistic discussions including the influence of the nature of the support.

Besides CO oxidation, other supported-AuNP catalyzed reactions have recently been disclosed, confirming that supported AuNPs are now a very popular means to catalytically activate dihydrogen and dioxygen:

- *hydrogenation* of 1,4-butadiene to butenes [46a], acrylaldehyde to allylic alcohol [46b], citral to geraniol + nerol [46c] and benzalacetone to phenyl-3-butene-2-ol [46d].

- *oxidation (using O_2 or air)* of alcohols to aldehydes [46c], *o*-hydroxybenzyl alcohol to salicylic aldehyde [46d], ethane-1,2-diol to glycolate [46c], other diols to hydroxymonocarboxylates [46e], β-amino alcohols to β-amino-acids [46e], aliphatic aldehydes to carboxylic acids [46e], D-glucose to gluconic acid [46e] or oxalate [46m], dihydrogen to hydrogen peroxide [45f], aromatic amines (with CO) to carbamate [46g], propene to propene oxide [45h] and cyclohexane to cyclohexol and cyclohexanone [46n-o]. Alcohols, in particular methanol are also oxidized electrochemically, supported AuNPs being more active electrocatalysts than AuNPs alone [47].

Other applications of supported-AuNP catalyzed reactions are numerous. They include:

- Oxidative decomposition by Fe_2O_3-supported AuNPs of bad-smelling alkylamines responsible for the unpleasant atmosphere in toilets [9d].
- Oxidative decomposition of dioxin coming from incinerator outlet gases by La_2O_3-supported AuNPs integrated with Pd/SnO_2 and Ir/La_2O_3 [9d].
- Direct epoxidation of propylene to propylene oxide by TiO_2 (MCM48)-supported AuNPs [9d], sensors able to simultaneously detect H_2 and CO at low level using Co_2O_3-supported AuNPs [9e].
- CO safety masks for efficient removal of CO from contaminated atmospheres [9e, 45f].
- Various liquid-phase synthetic processes.

Chapter 12 by Avelino Corma and Hermenegildo Garcia offers a remarkably broad and insightful overview and review of the general properties of AuNPs as catalysts for a variety of oxidation reactions including the role of templates and supports, with emphasis on ceria for which the Corma group has introduced seminal concepts. Sustainability, green chemistry aspects, comparison between PdNPs and AuNPs as aerobic oxidation catalysts, reaction mechanisms and prospects are also discussed in this chapter.

Chapter 13 by Cristina Della Pina, Ermelinda Falletta and Michele Rossi reviews AuNP catalyzed oxidations in organic chemistry including kinetic and mechanistic studies of the AuNP catalyzed oxidation of glucose, alcohols and aldehydes with emphasis on the original work emanating from Rossi's laboratory including seminal studies that have indeed largely contributed to these findings. Perspectives for industrial applications are also underlined in the conclusion.

Chapter 14 by Jun Kawahara and Masatake Haruta reviews AuNP-catalyzed propene epoxidation by dioxygen and dihydrogen. Propylene oxide, which is processed to polyurethane polyols and propylene glycols, is one of the most important chemical feedstocks whose efficient, free of side product and selective synthesis is highly challenging, the present world annual production being several million tons per year. This chapter includes catalyst preparation and catalytic tests from Masatake Haruta's laboratory with the AuNP-TiO₂ and AuNP-Ti-SiO₂ systems, and detailed mechanistic discussions are presented in the conclusion.

1.13
Environmental Problems: NOx Pollution and How to Remove NOx Using NP Catalysis

Nitrogen oxides, in particular NO and NO_2, are produced by fossil fuel burning, simple strong heating of air and, in particular, car exhaust systems. Thus, the two sources of NOx pollution are factories and cars. Therefore, depollution from nitrogen oxides is a must for our modern Society that still needs considerable improvement, as can be frequently considered by exaggerated air pollution in specific regions with a high density of factories and/or urban population and car traffic. The "DeNOx" process, i.e. the removal of nitrogen oxides, is possible using catalysis by transition-metal NPs, and the nanosize of metal NPs is actually a crucial parameter in this catalysis. Today industrial catalysts in use for "selective catalytic reduction" (SCR) are based on TiO_2-supported V_2O_5-WO_3 and/or V_2O_5-MoO_3, but deNOx of car exhaust is still unresolved and under investigation. The three-way catalyst (TWC) based on combinations of Pt, Pd and Rh is efficient to remove CO and NOx from gasoline engines at 400 °C to 800 °C, but the period of time needed to reach these temperature involves pollution, and TWCs are ineffective for diesel and lean-burn gasoline engines.

Thus, Chapter 16 by Frédéric Thibault-Starzyk, Marco Daturi, Philippe Bazin and Olivier Marie is very important with regard to this problem, as it reviews NO heterogeneous catalysis viewed from the angle of NPs. In particular, this chapter presents the up-to-date research in progress concerning the size, stability, sintering and morphology of the NPs, metal NPs in zeolites, the role of TEM studies and of the ceria and zirconia supports, new catalytic NPs and materials such as Ga-Al-Zn complex oxides and carbon nanotubes supports for phosphotungstic acid.

For various other environmental problems, see also Chapters 9 and 18.

1.14
Hydrocarbon Reforming: Activation of Hydrocarbons by NP Catalysts

Hydrocarbon reforming is an essential industrial process that was carried out 50 years ago using simple oxide materials. For instance, the first performed reactions using Pt-supported alumina in a flow reactor were carried out by Gault's group in Strasbourg, and mechanistic studies of hexane isomerization using such catalysts were pioneered at that time. Sophisticated noble metal NPs are now used, and up-to-date research connects selectivity to mean NP size and morphological aspects. Model studies using nanocrystals as well as supported NPs are helpful in order to understand and optimize these selectivities, for instance in alkane isomerization reactions. *Chapter 17 by François Garrin and Pierre Legaré, the successors of Gault in the Strasbourg laboratory, reviews hydrocarbon catalytic reactivity of supported metal NPs with emphasis on the progress concerning the understanding of the catalytic behavior under a reductive atmosphere and the influence of NP size, electronic structure and morphology.*

It is also interesting to compare this "heterogeneous" NP approach to that involving organometallic surfaces catalysis developed by Basset's group. In this approach, the catalyst is molecular and anchored to silica via Si–O–metal bridges. For instance, [silica(SiO)$_2$Ta^{III}-H] catalyzes propane metathesis to ethane and *n*-butane with turnover numbers between 25 and 60 at only 150 °C. This surface catalyst is generated from organometallic surface-bound precursors such as [silica(SiO)xTa^V(=CHt-Bu)(CH$_2$$t$-Bu)$_{3-x}$], $x = 1$ or 2 that are the silica-surface analogs of Schrock's seminal Ta alkylidene complexes. The mechanism is suggested to proceed by a series of σ-bond metathesis of C—H bonds and α and β-eliminations whereas direct σ-bond metathesis of C—C bonds in not favored. The α-elimination from d^2 metal-methyl or metal-alkyl species generates, respectively, HTa=CH$_2$ or HTa=CHR, and it is proposed that the mechanism then follows an alkene metathesis pathway with olefins generated by β-elimination (including metallacyclobutane intermediates as in the Chauvin mechanism).

Thus, only σ-bond metathesis of two kinds of σ bonds is observed in these systems: H—H, and C—H involving Ta···C···H interaction but not Ta···H···C, because a negatively polarized pentacoordinated carbon intermediate is not favorable (Scheme 1.9) [82].

As in Pd catalysis, molecular and NP catalysts are complementary rather than competitive, because molecular catalysts are either excellent models allowing us to improve our understanding or their ligands can still so far improve on the performances of NP catalysis in terms of selectivities. This comparison thus represents a challenge for the heterogeneous community.

| H-H bond activation giving a Ta-hydride species | C-H bond activation of alkanes giving a Ta-alkyl and H$_2$ | reverse C-H bond activation **not observed** due to a negatively polarized penta-coordinated β carbon |

Scheme 1.9 Hydrogenolysis of Ta-alkyl and methanolysis of Ta-H by σ-bond metathesis for alkane disproportionation disclosed by Basset's group [50] with the silica-supported catalyst [(SiO)xTa^V(=CH-t-Bu)(CH$_2$$t$Bu)$_{3-x}$].

1.15
Surface Organometallic Chemistry on Metal NPs

In pursuing the introduction of the contributions by Basset's group, it should be emphasized that organometallic chemistry and catalysis on oxides (*vide supra*) has its counterpart here with the related concept involving organometallic chemistry and *catalysis using organometallic complexes deposited on metal NPs.* Indeed, in the last (but not least!) chapter of this book, *Chapter 18, Katrin Pelzer, Gregory Godard, Jean-Pierre Candy and Jean-Marie Basset* offer an overview of this area with three types of materials devoted to both catalysis (hydrogenation, hydrogenolysis, dehydrogenation, isomerization) and environmental chemistry (dearsenification and demercurisation of crude-oil, extraction of heavy metals from contaminated water):

1. A material in which a surface organometallic fragment is linked to the metallic nanoparticle via "covalent" bonds.
2. A material in which a "naked" adatom is located on the metal NP surface.
3. A material in which the "adatoms" are incorporated in the NP yielding a bulk alloy.

Rh, Pt, Ru, and Ni NPs with Sn, Ge, As, Hg organometallic complexes are discussed with these processes.

1.16
Application and Perspectives in Organic Chemistry

The finding of catalytic formation of C–C and C–N bonds by Pd catalysts has been a considerable progress for organic chemistry in the last five years of the 20th century [34]. It became possible to easily functionalize olefins, alkynes and aromatics. Yet, the problem of catalyst recovery and pollution by phosphine was unsolved. These aspects are crucial for the pharmaceutical industry, the inclusion of metal and phosphine contaminants in drugs being unacceptable. NP-supported catalysis with Pd, Ni and eventually other metals now provides a very satisfying solution to catalyst removal from the reaction mixture by simple filtration, because the carbon and NaY zeolite support (the best ones) are reservoirs of metal species that retain these metal species subsequent to reaction in solution. The amount of metal left in solution is of the order of ppm. Moreover, such type of catalysis is phosphine free. Applications have started at DSM with the synthesis of kg-scale pharmaceutical intermediates, and it is likely that this type of procedure will now spread at a high speed in the pharmaceutical and other industries. The Heck reaction, for instance, is a key reaction for the production of fine chemicals on a multi-ton scale per year [34a] such as the herbicide Prosulfuron™ [34b], the anti-inflammatory Naproxen™ [34c] or the anti-asthma Singulair™ [34d]. On the other hand, in terms of "*Green Chemistry*", it is likely that procedures involving chemicals such as ionic liquids, micelles, surfactants and other additives in homogeneous solution will less retain the attention of those interested in atom economy problems and applicable organic synthesis.

So far, high-performance molecular catalysts, reported by the groups of Buchwald, Hartwig and Fu, that are able to activate and functionalize chloroaromatics are complementary to PdNPs that cannot carry out this task, seemingly because the stereoelectronic requirement provided by the ligand (with electron donor groups to increase electron density needed in the oxidative addition step and the large bulk of these ligands required to facilitate the reductive elimination step) is essentially missing in NPs. Likewise, olefin and alkyne metathesis that has been a success story with molecular catalysts cannot be achieved with NPs, probably because alkylidenes bridge metal surface atoms instead of presenting monohapto doubly-bonded alkylidene ligands on single metal centers, which suppresses their metathesis reactivity in the presence of unsaturated hydrocarbons (see Garin's mechanism in *Chapter 17* and Basset's concept for alkane activation, however).

Given the extraordinary present progression of AuNP-catalyzed reactions, it is obvious that many applications in oxidation catalysis will very soon penetrate the everyday world of organic chemists (as illustrated in this book). Thus, altogether, NP catalysis embraces an increasingly large part of organic reactions and is respectful of the "*Green Chemistry*" aspects.

1.17
Conclusion

The state of the art of metal NP design and catalysis is now well advanced thanks to the efforts of the pioneers and the advent of modern characterization techniques. A large variety of NP preparation modes and materials serve as supports or grafting cores, and several media compete for efficient catalytic processes. Monodisperse, small (1–10 nm) supported metal NPs, including bi-and trimetallic NP catalysts, most often more efficient than those containing only one type of metal, are available for many current reactions with great catalytic efficiency due to the enhanced available NP surface. Compared surface (Pd(1,1,1) single crystal) and PdNP studies of catalysis of alkene hydrogenation have even shown that only PdNPs catalyze this reaction [32i]. This progress has considerably improved the selectivity of NP catalyzed reactions, especially the hydrogenation of unsaturated substrates that proceeds truly heterogeneously. High enantioselectivity has been obtained, although the demonstration that asymmetric induction occurs at the metal NP surface rather than on the more reactive leaching liganded monometallic fragment is at least not obvious. For example, in the Pd/C and Ni/C catalyzed C—C and C—N bond-forming reactions, the mechanism involves such a leaching pathway with recovery of the metal on the support at the end of the reaction. More mechanistic studies are called for in order to understand the nature of the very active metal species in solution. On the side of efficiency, the activation of aryl chlorides for C—C coupling reactions still remains a key challenge to match organometallic catalysts that are presently more efficient than metal NP catalysts. A major recent finding was the removal of these catalysts by filtration, although recycling and efficient re-use many times of supported NP catalysts remains a challenge. Along this line, magnetic separation of magnetic NP catalysts appears as a challenging solution [48].

Altogether, the field of metal NP catalysis is a fascinating one, as exemplified by the new gold rush and its very attractive perspectives in oxidation chemistry and by the ligand-free Pd and Ni catalysis using "homeopathic" amounts of catalyst. The metal NP field is presently burgeoning, and it is anticipated that these key challenges will be met in the close future, and that this area of nanoscience will be much more applied in tomorrow's laboratory and industry.

References

1 *Catalysis by Metals and Alloys*, V. Ponec, G. C. Bond (Eds.), Elsevier, Amsterdam, **1995**, Vol. 95; G. C. Bond, *Heterogeneous Catalysis: Principles and Applications*, Oxford Science Publications, Clarendon Press, Oxford, **1987**; *Handbook of Heterogeneous Catalysis*, G. Ertl, H. Knözinger, J. Vertkamp (Eds.), Wiley-VCH, Weinheim, **1997**; R. J. Farauto, C. H. Bartholomew, *Fundamentals of Industrial Catalytic Processes*, Blackie Academic and Professional, London, **1997**; J. M. Thomas, W. J. Thomas, *Principle and Practice of Heterogeneous Catalysis*, VCH, Weinheim, **1997**; C. R. Henry, *Appl. Surf. Sci.* **2000**, *164*, 252; T. P. St Clair, D. W. Goodman, *Top. Catal.* **2000**, *13*, 5; M. Bowker, R. A. Bennet, A. Dickinson, D. James, R. Smith, P. Stone, *Stud. Surf. Sci. Catal.* **2001**, *133*, 3; M. Kralik, A. Biffis, *J. Mol. Catal. A: Chem.* **2001**, *177*, 113; J. Thomas, R. Raja, *Chem. Rec.* **2001**, *1*, 448; C. Mohr, P. Claus, *Sci. Prog.* **2001**, *84*, 311; J. M. Thomas, B. F. G. Johnson, R. Raja, G. Sankar, P. A. Migley, *Acc. Chem. Res.* **2003**, *36*, 20; O. Alexeev, B. C. Gates, *Ind. Eng. Chem. Res.* **2003**, *42*, 1571.

2 *Applied Homogeneous Catalysis with Organometallic Compounds*, B. Cornils, W. A. Herrmann (Eds.), Wiley-VCH, Weinheim, Vol. 1 and 2, **1996**; W. A. Herrmann, B. Cornils, *Angew.Chem. Int. Ed. Engl.* **1997**, 36, 1049; *Metal-Catalyzed Cross-Coupling Reactions* F. Diederich, P. J. Stang (Eds.), Wiley-VCH, Weinheim,

1998; *Metal-Catalyzed Cross-Coupling Reaction*, Vol. 1 and 2, A. de Mejere, F. Diederich (Eds.), Wiley-VCH, Weinheim, **2004**; J. A. Gladysz, *Pure Appl. Chem.* **2001**, 73, 1319; J. Gladysz, Introduction to the special issue of *Chem. Rev.* **2002**, 102(10) 3215–3892, J. Gladysz (Ed.) dedicated to recoverable catalysts; D. Astruc, *Organometallic Chemistry and Catalysis*, Springer, Heidelberg, **2007** (Parts 4 and 5).

3 For recent reviews on transition-metal NP-catalyzed reactions, see: D. Astruc, F. Lu, J. Ruiz Aranzaes, *Angew. Chem. Int. Ed.* **2005**, *44*, 7399; J. G. De Vries, *Dalton Trans.* **2006**, 421; D. Astruc, *Inorg. Chem.* **2007**, *46*, 1884.

4 *Nanoparticles and Nanostructured Films. Preparation, Characterizations and Applications*, J. H. Fendler (Eds.), Wiley-VCH, Weinheim, Germany, **1998**; G. Schmid in *Nanoscale Materials in Chemistry*, K. J. Klabunde (Eds.), Wiley-Interscience, New York, **2001**, pp. 15–59; R. G. Finke in *Metal Nanoparticles. Synthesis, Characterizations and Applications*, D. L. Feldheim, C. A. Foss, Jr., (Eds.), Marcel Dekker, New York, **2002**, Ch. 2, pp. 17–54; A. Wieckowski, E. R. Savinova, C. G. Vayenas, *Catalysis and Electrocatalysis at Nanoparticle Surfaces*, Marcel Dekker, New York, **2003**, p. 970; *Nanoparticles*, G. Schmid (Eds.), Wiley-VCH, Weinheim, **2004**.

5 Early reviews on NP catalysis: (a) J. B. Michel, J. T. Scharz, in *Catalyst Preparation*

Science, IV, B. Delmon, P. Grange, P. A. Jacobs, G. Poncelet (Eds.), Elsevier, Amsterdam, **1987**, pp. 669–687; (b) G. Schmid, *Chem. Rev.* **1992**, *92*, 1709; (c) L. N. Lewis, *Chem. Rev.* **1993**, *93*, 2693–2730; (d) J. S. Bradley in *Clusters and Colloids*, Ed. G. Schmid (Ed.), VCH, Weinheim, **1994**, Ch. 6, pp. 459–544; (e) *Catalysis by Di-and PolynuclearMetal-Cluster Complexes*, L. N. Lewis, R. D. Adams, F. A. Cotton (Eds.), Wiley-VCH, New York, **1998**, p. 373; (f) N. Toshima in *Fine Particles Sciences and Technology – From Micro- to New Particles*, E. Pellizzetti (Ed.), Kluwer: Dordrecht, **1996**, pp. 371–383; (g) N. Toshima, T. Yonezawa, *New J. Chem.* **1998**, 1179–1201; Y. Shiraishi, N. Toshima, *J. Mol. Catal. A: Chem.* **1999**, *141*, 187; Y. Shiraishi, N. Toshima, *Colloid Surf. A* **2000**, *169*, 59.

6 (a) H. Bönnemann, W. Brijoux, E. Dinjus, T. Fretzen, B. Joussen, J. Korall, *Angew. Chem. Int. Ed. Engl.* **1990**, *29*, 273; (b) H. Bönnemann, W. Brijoux, R. Brinkmann, E. Dinjus, T. Fretzen, B. Joussen, J. Korall, *Angew. Chem. Int. Ed. Engl.* **1992**, *31*, 323; H. Bönnemann, W. Brijoux in *Active Metals: Preparation, Characterization, Applications*, A. Fürstner (Ed.), VCH-Weinheim, **1996**, pp. 339–379.

7 (a) M. T. Reetz, W. Helbig, *J. Am. Chem. Soc.* **1994**, *116*, 7401; (b) M. T. Reetz, S. A. Quaiser, *Angew. Chem. Int. Ed. Engl.* **1995**, *34*, 2240; (c) M. T. Reetz, W. Helbig, S. A. Quaiser, in *Active Metals: Preparation, Characterization, Applications*, A. Fürstner (Ed.), VCH-Weinheim, **1996**, pp. 279–297; (d) M. T. Reetz, R. Breinbauer, K. Wanninger, *Tetrahedron Lett.* **1996**, *37*, 4499; (e) M. T. Reetz, G. Lohmer, *Chem. Commun.* **1996**, 1921; (f) for recent reports by Reetz's group on PdNP catalysis, see Refs. [31d,f,j,n, 41c].

8 (a) M.-C. Daniel, D. Astruc, *Chem. Rev.* **2004**, *104*, 293; (b) M. Faraday, *Philos. Trans.* **1857**, *151*, 183. (c) M. Brust, M. Walker, D. Bethell, D. J. Schiffrin, R. J. Whyman, *J. Chem. Soc. Chem. Commun.* **1994**, 801; (d) D. I. Gittins, F. Caruso, *Angew. Chem.* **2001**, *113*, 3089; *Angew. Chem. Int. Ed.* **2001**, *40*, 3001; (e) T. W. Smith, *US Patent 4252671, 4252672, 4252673, 4252674 and 4252678*, **1981**; (f) C. H. Griffiths, H. P. O'Horo, T. W. Smith, *J. Appl. Phys.* **1979**, *50*, 7108; (g) M. Kilner, G. J. Russell, S. Hoon, B. K. Tanner, *J. Mag. Mater.* **1983**, *39*, 107; (h) P. H. Hess, , P. H. Parker, *J. Appl. Polym. Sci.* **1966**, *10*, 1915; (i) J. R. Thomas, *J. Appl. Phys.* **1966**, *37*, 2914; (j) J. S. Bradley, J. M. Millar, E. W. Hill, S. Behal, B. Chaudret, A. Duteil, *Faraday Discuss.* **1991**, *92*, 255; J. S. Bradley, E. W. Hill, S. Behal, C. Klein, B. Chaudret, A. Duteil, *Chem. Mater.* **1992**, *4*, 1234; (k) S. Roginsky, A. Schalnikoff, *Kolloid Z.* **1927**, *43*, 67; (l) J. R. Blackborrow, D. Young, *Metal Vapor Synthesis*, Springer Verlag, New York, **1979**.

9 (a) L. D. Rapino, F. F. Nord, *J. Am. Chem. Soc.* **1941**, *63*, 2745 and 3268; K. E. Kavanagh, F. F. Nord, *J. Am. Chem. Soc.* **1943**, *65*, 2121; (b) D. Y. Cha, G. Parravano, *J. Catal.* **1970**, *18*, 320; G. Parravano, *J. Catal.* **1970**, *18*, 320; (c) M. Haruta, T. Kobayashi, H. Sano, N. Yamada, *Chem. Lett.* **1987**, 405; M. Haruta, N. Yamada, T. Kobayashi, S. Lijima, *J. Catal.* **1989**, *115*, 301; observation of CO oxidation at 200 K: M. Haruta, S. Tsuboda, T. Kobayashi, H. Kagehiama, M. J. Genet, B. Demon, *J. Catal.* **1993**, *144*, 175; (d) M. Haruta, CATTECH, **2002**, *6*, 102; (e) for other excellent recent reviews on AuNP catalyzed CO oxidation by O$_2$, see ref. 11j, 50b and: T. V. Choudary, D. W. Goodman, *Top. Catal.* **2002**, *21*, 25; G. Schmid, B. Corain, *Eur. J. Inorg. Chem.* **2003**, 3081; see also Ref. [8a] (f) G. C. Bond, P. A. Sermon, *Gold Bull.* **1973**, *6*, 102; (g) H. Hirai, Y. Nakao, N. J. Toshima, *Macromol. Sci. Chem.* **1978**, *A12*, 1117 and **1979**, *A13*, 727; (h) M. Boutonnet, J. Kizling, P. Stenius, G. Maire, *Colloids Surf.* **1982**, *5*, 209; M. Boutonnet, J. Kizling, R. Touroude, G. Maire, P. Stenius, *Appl. Catal.* **1986**, *20*, 163; (i) K. Kurihara, J. H. Fendler, *J. Am. Chem. Soc.* **1983**, *105*, 6152; J. H. Fenler, P. Tundo, *Acc. Chem. Res.* **1984**, *17*, 3; J. H. Fendler, K. Kurihara, in *Metal Containing Polymeric Systems*, J. E. Sheats, C. E. Carraher, Jr., C. U. Pittman, Jr. (Eds.), Plenum Press, New York, **1985**, pp. 341–353; (j) L. N. Lewis, N. Lewis, *J. Am. Chem. Soc.* **1986**, *108*, 7228.

10 (a) A. Moradpour, E. Amouyal, P. Keller, H. Kagan, *Nouv. J. Chem.* **1978**, *2*, 547; (b) A. Henglein, *J. Phys. Chem.* **1979**, *83*, 2858; (c) A. Henglein, J. Lillie, *J. Am. Chem. Soc.* **1981**, *103*, 1059; (d) K. Kurihara, J. H. Fendler, I. J. Ravet, *Mol. Catal.* **1986**, *34*, 325; (e) M. Boutonnet, J. Kizling, R. Touroude, G. Maire, P. Stenius, *Appl. Catal.* **1986**, *20*, 163; (f) Y. Degani, I. J. Willner, *Chem. Soc., Perkin Trans. II* **1986**, 37; (g) I. Willner, R. Maidan, D. Mandler, H. Dürr, G. Dörr, K. Zengerle, *J. Am. Chem. Soc.* **1987**, *109*, 6080; (h) J. S. Bradley, E. W. Hill, M. E. Leonowitz, H. Witzke, *J. Mol. Catal.* **1987**, *41*, 59; (i) C. Larpent, H. Patin, *J. Mol. Catal.* **1988**, *44*, 191.

11 Recent reviews on NP catalysis: Ref. [3] and (a) M. A. El-Sayed, *Acc. Chem. Res.* **2001**, *34*, 257; (b) R. M. Crooks, M. Zhao, L. Sun, V. Chechik, L. K. Yeung, *Acc. Chem. Res.* **2001**, *34*, 181; (c) T. Yonezawa, N. Toshima, *Polymer-Stabilized Metal Nanoparticles: Preparation, Characterization and Applications*, in *Advanced Functional Molecules and Polymers*, H. S. Nalwa (Ed.), OPA N.V., **2001**, Vol. 2, Ch. 3, pp. 65–86; A. B. R. Mayer, *Polym. Adv. Technol.* **2001**, *12*, 96; (d) H. Bönnemann, R. Richards, *Synth. Methods Organomet. Inorg. Chem.* **2002**, *10*, 209; (e) A. Roucoux, J. Schulz, H. Patin, *Chem. Rev.* **2002**, *33*, 27–36. (f) I. I. Moiseev, M. N. Vargaftik, *Russ. J. Chem.* **2002**, *72*, 512; (g) A. T. Bell, *Science* **2003**, *299*, 1688; (h) M. Moreno-Manas, R. Pleixats, *Acc. Chem. Res.* **2003**, *36*, 638–643; (i) B. F. G. Johnson, *Top. Catal.* **2003**, *24*, 147 (j) AuNP-catalyzed CO oxidation: M. Haruta, *J. New Mater. Electrochem. Syst.* **2004**, *7*, 163; (k) J. A. Widegren, R. G. Finke, *J. Mol. Catal. A Chem.* **2003**, *198*, 317; J. A. Widegren, R. G. Finke, *Inorg. Chem.* **2002**, *41*, 1558; (b) C. Besson, E. E. Finney, R. R. Finke, *J. Am. Chem. Soc.* **2005**, *127*, 8179; C. Besson, E. E. Finney, R. R. Finke, *Chem. Mater.* **2005**, *17*, 4925.

12 (a) X.-D. Mu, D. G. Evans, Y. Kou, *Catal. Lett.* **2004**, *97*, 151; (b) P. Claus, A. Brückner, C. Möhr, H. Hofmeister, *J. Am. Chem. Soc.* **2000**, *122*, 11430; (c) A.-I. Kozlov, A. P. Kozlova, K. Asakura, Y.

Matsui, T. Kogure, T. Shido, Y. Iwazawa, *J. Catal.* **2000**, *196*, 56; (d) A. Martino, S. A. Yamanaka, J. S. Kawola, D. A. Ly, *Chem. Mater.* **1997**, *9*, 423; (e) T. Li, J. Moon, A. A. Morrone, J. J. Mecholsky, D. R. Talham, J.-H. Adair, *Langmuir* **1999**, *15*, 4328; (f) U.-A. Paulus, U. Endruschat, G.-J. Feldmeyer, T.-J. Schmidt, H. Bönnemann, J.-J. Behm, *J. Catal.* **2000**, *195*, 383; (g) Y. Mizukoshi, R. Oshima, Y. Mizukoshi, Y. Nagata, *Languir.* **1999**, *8*, 2733. (h) S. Papp, I. Dekany, *Colloid Polym. Sci.* **2001**, *279*, 449; (i) C. B. Hwang, Y.-S. Fu, Y.-L. Lu, S.-W. Jang, P.-T. Chou, C.-R. Wang, S.-J. Yu, *J. Catal.* **2000**, *195*, 336; (j) K.-T. Wu, Y.-D. Yao, C.-R. C. Wang, P. F. Chen, E.-T. Yeh, *J. Appl. Phys.* **1999**, *85*, 5959; (k) R. P. Andres, J.-D. Bielefeld, J.-I. Henderson, D.-B. Janes, V.-R. Kolagunta, C.-P. Kubink, W. Mahoney, R.-G. Osifchin, R. Reifenberger, *Science* **1996**, *273*, 1690.

13 (a) R. Narayanan, M. A. El-Sayed, *J. Am. Chem. Soc.* **2003**, *125*, 8340; (b) Y. Li, E. Boone, M. A. El-Sayed, *Langmuir* **2002**, *18*, 4921.

14 (a) S. V. Ley, C. Mitchell, D. Pears, C. Ramarao, J.-Q. Yu, W.-Z. Zhou, *Org. Lett.* **2003**, *5*, 4665; (b) M. M. Demir, M. A. Gulgun, Y. Z. Menceloglu, B. Erman, S. S. Abramchuk, E. E. Makhaeva, A. R. Khokhlov, V. G. Matveeva, M. G. Sullman, *Macromolecules* **2004**, *37*, 1787. (c) L. R. Gröschel, A. Haidar, K.-H. Beyer, R. Reichert, R. Schomäcker, *Catal. Lett.* **2004**, *95*, 67; (d) S. Kidambi, J.-H. Dai, J. Lin, M. L. Bruening, *J. Am. Chem. Soc.* **2004**, *126*, 2658; (e) B. P. S. Chauhan, J. S. Rathore, T. Bandoo, *J. Am. Chem. Soc.* **2004**, *126*, 8493; (f) T. Sanji, Y. Ogawa, Y. Nakatsuka, M. Tanaka, H. Sakurai, *Chem. Lett.* **2003**, *32*, 980; (g) A. B. Lowe, B. S. Sumerlin, M. S. Donovan, C. L. McCormick, *J. Am. Chem. Soc.* **2002**, *124*, 11562; (h) A. Drelinkiewicza, A. Waksmundzka, W. Makowski, J. W. Sobczak, A. Krol, A. Zieba, *Catal. Lett.* **2004**, *94*, 143; (i) Y.-B. Liu, C. Khemtong, J. Hu, *Chem. Commun.* **2004**, 398; (j) U. R. Pillai, E. Sahle-Demessie, *J. Mol. Catal.* **2004**, *222*, 153; (k) M. Adlim, M. Abu Bakar, K. Y. Liew, J. Ismail, *J. Mol. Catal. A.* **2004**, *212*, 141; (l) D. Tabuani, O. Monticelli, A. Chincarini, C. Bianchini, F. Vizza, S. Moneti, S. Russo,

Macromolecules **2003**, *36*, 4294; (m) C. C. Yang, C. C. Wan, Y. Y. Wang, *J. Colloid Interface Sci.* **2004**, *279*, 433.

15 J. H. Sinfelt, *Acc. Chem. Res.* **1977**, *10*, 15; J. H. Sinfelt, *Bimetallic Catalyst: Discoveries, Concepts and Applications*, Wiley, New York, **1983**; J. H. Sinfelt, *Int. Rev. Phys. Chem.* **1988**, *7*, 281.

16 (a) J.-H. He, I. Ichinose, T. Kunitake, A. Nakao, Y. Shiraishi, N. Toshima, *J. Am. Chem. Soc.* **2003**, *125*, 11034; (b) Y. Shiraishi, D. Ikenaga, N. Toshima, *Aust. J. Chem.* **2003**, *56*, 1025; (c) R. Sablong, U. Schlotterbeck, D. Vogt, S. Mecking, *Adv. Syn. Catal.* **2003**, *345*, 333.

17 (a) *Dendrimers and Nanosciences*, D. Astruc (Ed.), C. R. Chimie, Elsevier, Paris, **2003**, 6 (special issue: vol. 8–10); (b) G. R. Newkome, C. N. Moorefield, F. Vögtle, *Dendrimers and Dendrons: Concepts, Synthesis, Applications*, Wiley-VCH, Weinheim, **2001**; (c) J. F. G. A. Jansen, E. M. M. de Brabander-van den Berg, E. W. Meijer, *Science*, **1994**, 266, 1226; (d) D. A. Tomalia, A. M. Naylor, W. A. Goddard III, *Angew. Chem. Int. Ed. Engl.* **1990**, *29*, 138.

18 (a) M. Zhao, L. Sun, R. M. Crooks, *J. Am. Chem. Soc.* **1998**, *120*, 4877; M. Zhao, R. Crooks, *Angew. Chem.* **1999**, *111*, 375; *Angew. Chem. Int. Ed.* **1999**, *38*, 364; V. Chechik, M. Zhao, R. M. Crooks, *J. Am. Chem. Soc.* **1999**, *121*, 4910; M. Zhao, R. M. Crooks, *Adv. Mater.* **1999**, *11*, 217; V. Chechik, M. Zhao, R. M. Crooks, *J. Am. Chem. Soc.* **2000**, *122*, 1243; R. M. Crooks, M. Zhao, L. Sun, V. Chechik, L. K. Yeung, *Acc. Chem. Res.* **2001**, *34*, 181; L. K. Yeung, R. M. Crooks, *Nano Lett.* **2001**, *1*, 14; L. K. Yeung, C. T. Lee, K. P. Jonston, R. M. Crooks, *Chem. Commun.* **2001**, 2290; R. W. Scott, A. F. Datye, R. M. Crooks, *J. Am. Chem. Soc.* **2003**, *125*, 3708; Y. Niu, R. M. Crooks, in *Dendrimers and Nanosciences*, D. Astruc (Ed.), *C. R. Chimie*, Elsevier, Paris, **2003**, 6, 1049; R. W. Scott, O. M. Wilson, S.-K. Oh, E. A. Kenik, R. M. Crooks, *J. Am. Chem. Soc.* **2004**, *126*, 15583; Y.-G. Kim, S.-K. Ho, R. M. Crooks, *Chem. Mater.* **2004**, *16*, 167; R. W. J. Scott, C. Sivadinarayana, O. M. Wilson, Z. Yan, D. W. Goodman, R. M. Crooks, *J. Am.*

Chem. Soc. **2005**, *127*, 1380; J.-H. Liu, A.-Q. Wang, Y.-S. Shi, H.-P. Lin, C.-Y. Mou, *J. Phys. Chem. B* **2005**, *109*, 40; O. M. Wilson, M. R. Knecht, J. C. Garcia-Martinez, R. M. Crooks, *J. Am. Chem. Soc.* **2006**, *128*, 4510: (b) E. Kunio, M. Keiko, Y. Tomokazu, *J. Colloid Interfac. Sci.* **2002**, *254*, 402; (c) J. Lemo, K. Heuze, D. Astruc, *Inorg. Chim. Acta* **2006**, *359*, 4909.

19 (a) L. Balogh, D. A. Tomalia, *J. Am. Chem. Soc.* **1998**, *120*, 7355; (b) L. Balogh, D. R. Swanson, D. A. Tomalia, G. L. Hagnauer, A. T. McManus, *Nano Lett.* **2001**, *1*, 18; (c) K. Esumi, A. Susuki, N. Aihara, K. Usui, K. Torigoe, *Langmuir* **1998**, *14*, 3157; (d) K. Esumi, K. Satoh, A. Suzuki, K. Torigoe, *Shikizai Kyokaishi* **2000**, *73*, 434; (e) K. Esumi, R. Isono, T. Yoshimura, *Langmuir* **2004**, *20*, 237.

20 (a) H.-F. Lang, R. A. May, B. L. Iversen, B. D. Chandler, *J. Am. Chem. Soc.* **2003**, *125*, 14832; (b) H.-F. Lang, S. Maldonado, K. J. Stevenson, B. D. Chandler, *J. Am. Chem. Soc.* **2004**, *126*, 12949; L. W. Beakley, S. E. Yost, R. Cheng, B. D. Chandler, *Appl. Catal. A: General* **2005**, *292(1–2)*, 124; B. D. Chandler, J. D. Gilbertson, *Top. Organomet. Chem.*, **2006**, *20*, 97; *Dendrimer Catalysis*, L. Gade (Ed.), Springer-Verlag, Berlin, **2006**.

21 (a) Y. Li, M. A. El Sayed, *J. Phys. Chem. B* **2001**, *105*, 8938; (b) R. Narayanan, M. A. El-Sayed, *J. Phys. Chem. B* **2004**, *108*, 8572; (c) R. Narayanan, M. A. El-Sayed, *Langmuir* **2005**, *21*, 2027; R. Narayanan, M. A. El-Sayed, *J. Catal.* **2005**, *234*, 348.

22 (a) M. Pittelkow, K. Moth-Poulsen, U. Boas, J. B. Christensen, *Langmuir* **2003**, *19*, 7682; (b) M. Ooe, M. Murata, T. Mizugaki, K. Ebitani, K. Kaneda, *Nano Lett.* **2002**, *2*, 999; (c) K. Mori, T. Hara, T. Mizugaki, K. Ebitani, K. Kaneda, *J. Am. Chem. Soc.* **2004**, *126*, 10657; (d) K. R. Gopidas, J. K. Whitesell, M.-A. Fox, *Nano Lett.* **2003**, *3*, 1757.

23 Y.-M. Chung, H.-K. Rhee, *J. Mol. Catal. A.: Chem.* **2003**, *206*, 291.

24 (a) C. Rocaboy, J. A. Gladysz, *Org. Lett.* **2002**, *4*, 1993; (b) R. Grigg, L.-X. Zhang, S. Collard, A. Keep, *Tetrahedron Lett.* **2003**, *44*, 6979.

25 Polyoxometallate-stabilized metal NPs and their catalytic function: (a) J. D. Aiken III,

R. G. Finke, *J. Am. Chem. Soc.* **1999**, *121*, 8803; (b) S. Ozkar, R. G. Finke, *J. Am. Chem. Soc.* **2002**, *124*, 5796; (c) A. M. Fargo, J. F. Odzak, F. S. Lai, B. C. Gates, *Nature* **2002**, *415*, 623; (d) V. Kogan, Z. Aizenshtat, R. Popovitz-Biro, R. Neumann, *Org. Lett.* **2002**, *4*, 3529; (e) M. Ohde, H. Yukiohde, C. M. Wai, *Chem. Commun.* **2002**, 2388; (f) anion-mediated adsorption of tetra-*n*-butyl ammonium cation on surfaces and metal NPs: Z. Deng, D. E. Irish, *J. Phys. Chem.* **1994**, *98*, 11169; S. Özkar, R. G. Finke, *J. Am. Chem. Soc.* **2002**, *124*, 5796; (g) ealiest report of ammonium-stabilized metal NPs: J. Kiwi, M. Grätzel, *J. Am. Chem. Soc.* **1979**, *101*, 7214.

26 (a) L. Strimbu, J. Liu, A. E. Kaifer, *Langmuir* **2003**, *19*, 483; (b) J. Alvarez, J. Liu, E. Roman, A. E. Kaifer, *Chem. Commun.* **2000**, 1151; (c) J. Liu, J. Alvarez, W. Ong, E. Roman, A. E. Kaifer, *Langmuir* **2001**, *17*, 6762.

27 F. Lu, J. Ruiz, D. Astruc, *Tetrahedron Lett.* **2004**, *45*, 9443.

28 (a) M.-K. Chung, G. Orlova, J. D. Goddart, M. Schlaf, R. Harris, T. J. Beveridge, G. White, F. R. Hallett, *J. Am. Chem. Soc.* **2002**, *124*, 10508; (b) M.-K. Chung, M. Schlaf, *J. Am. Chem. Soc.* **2004**, *126*, 7386; (c) S. U. Son, Y. Jang, J. Park, H. B. Na, H. M. Park, H. J. Yun, J. Lee, T. Hyeon, *J. Am. Chem. Soc.* **2004**, *126*, 5026 and references cited therein.

29 K. Nasar, F. Fache, M. Lemaire, M. Draye, J. C. Béziat, M. Besson, P. Galez, *J. Mol. Catal.* **1994**, *87*, 107.

30 (a) H. Bönnemann, G. A. Braun, *Chem. Eur. J.* **1997**, *3*, 1200; (b) J. U. Köhler, J. S. Bradley, **1998**, *14*, 2730; (c) X. Zuo, H. Liu, D. Guo, X. Yang, *Tetrahedron* **1999**, *45*, 203; (d) M. Studer, H.-U. Blaser, C. Exner, *Adv. Synth. Catal.* **2003**, *345*, 45; (e) M. Tamura, H. Fujihara, *J. Am. Chem. Soc.* **2003**, *125*, 15742; (f) S. Jansat, M. Gomez, K. Philippot, G. Muller, E. Guiu, C. Claver, S. Castillon, B. Chaudret, *J. Am. Chem. Soc.* **2004**, *126*, 1592.

31 (a) T. Mizoroki, K. Mori, A. Ozaki, *Bull. Chem. Soc. Jpn.* **1971**, *44*, 581; (b) R. F. Heck, J. P. Nolley, Jr., *J. Org. Chem.* **1972**, *37*, 2320; (c) I. P. Beletskaya, A. V. Cheprakov, *Chem. Rev.* **2000**, *100*, 3009;

(d) M. T. Reetz, E. Westermann, R. Lomer, G. Lohmer, *Tetrahedron Lett.* **1998**, *39*, 8449; (e) A. H. M. de Vries, J. M. C. A. Mulders, J. H. M. Mommers, H. J. W. Henderckx, J. G. de Vries, *Org. Lett.* **2003**, *5*, 3285; (f) M. T. Reetz, M. Maase, *Adv. Mater.* **1999**, *11*, 773; (g) C. Rocaboy, J. A. Gladysz, *New J. Chem.* **2003**, *27*, 39; (h) M. Nowotny, U. Hanefeld, H. van Koningsveld, T. Maschmeyer, *Chem. Commun.* **2000**, 1877; (i) I. P. Beletskaya, A. N. Kashin, N. B. Karlstedt, A. V. Mitin, A. V. Chepakov, G. M. Kazankov, *J. Organomet. Chem.* **2001**, *622*, 89; (j) M. T. Reetz, E. Westermann, *Angew. Chem.* **2000**, *112*, 170; *Angew. Chem. Int. Ed.* **2000**, *39*, 165 (see also E. Westermann, Dissertation, Ruhr-Universität Bochum, 1999); (k) , T. Rosner, J. Le Bars, A. Pfaltz, D. G. Blackmond, *J. Am. Chem. Soc.* **2001**, *123*, 1848; (l) C. E. Williams, J. M. C. A. Mulders, J. G. de Vries, A. H. M. de Vries, *J. Organomet. Chem.* **2003**, *687*, 494; (m) J. G. de Vries, A. H. M. de Vries, *Eur. J. Org. Chem.* **2003**, 799; (n) focus articles on ligand-free Heck reactions using extremely low Pd loading: A. H. M. de Vries, F. J. Parlevliet, L. Schmeder-van de Vondervoort, J. H. M. Mommers, H. J. W. Henderickx, M. A. N. Walet, J. de Vries, *Adv. Synth. Catal.* **2002**, *344*, 996; M. T. Reetz, J. G. de Vries, *Chem. Commun.* **2004**, 1559; (o) N. T. S. Phan, M. Van der Sluys, C. J. Jones, *Adv. Syn. Catal.* **2006**, *348*, 609; (p) T. Sugihara, T. Satoh, M. Miura, *Tetrahedron Lett.* **2005**, *46*, 8269; E. Alacid, C. Nájera, *Adv. Synth. Catal.* **2006**, *348*, 945; R. Tatumi, T. Akita, H. Fujihara, *Chem. Commun.* **2006**, 3349; I. Ryjomska, A. T. Trzeciak, J. J. Ziolkowski, *J. Mol. Catal. A: Chem.* **2006**, *257*, 3 (q) K. Pelzer, O. Vidoni, K. Philippot, B. Chaudret, V. Colliere, *Adv. Funct. Mater.* **2003**, *13*, 118; P. J. Dyson, D. J. Ellis, G. Laurenczy, *Adv. Synth. Catal.* **2003**, *345*, 211; J. A. Widegren, R. G. Finke, *J. Mol. Catal. A: Chem.* **2003**, *198*, 317; Y. Na, S. Park, S. B. Han, H. Han, S. Ko, S. Chang, *J. Am. Chem. Soc.* **2004**, *126*, 250.

32 (a) I. T. Horvath, J. Rabai, *Science* **1994**, *266*, 72; I. T. Horvath, *Acc. Chem. Res.* **1998**, *31*, 641; L. P. Barthel-Rosa, J. A. Gladysz, *Coord. Chem. Rev.* **1999**, *578*, 190; (b) H. Ohde, C. M. Wai, H. Kim, M. Ohde,

J. Am. Chem. Soc. **2002**, *124*, 4540; (c) P. Meric, K. M. K. Yu, S. C. Tsang, *Catal. Lett.* **2004**, *95*, 39; (d) P. Meric, K. M. K. Yu, S. C. Tsang, *Langmuir* **2004**, *20*, 8537; (e) K. M. K. Yu, C. M. Y. Yeung, D. Thompsett, S. C. Tsang, *J. Phys. Chem. B* **2003**, *107*, 4515; (f) B.-H. Yoon, H. Kim, C. M. Wai, *Chem. Commun.* **2003**, 1040; (g) H. Ohde, C. M. Wai, H. Kim, J. Kim, M. Ohde, *J. Am. Chem. Soc.* **2002**, *124*, 4540; (h) X. R. Ye, Y-H. Lin, C. M. Wai, *Chem. Commun.* **2003**, 642; (i) A. M. Doyle, S. K. Shaikhutdinov, S. D. David Jackson, H.-J. Freund, *Angew. Chem. Int. Ed.* **2003**, *42*, 5240; (j) J.-Z. Jiang, C. Cai, *J. Coll. Interfac. Sci.* **2006**, *299*, 938 (see also Ref. [43] for Scheme 8).

33 (a) Y. Chauvin, L. O. Mussmann, H. Olivier, *Angew. Chem. Int. Ed. Engl.* **1995**, *34*, 2698; (b) J. Dupont, R. F. de Souza, P. A. Z. Suarez, *Chem. Rev.* **2002**, *102*, 3667; (c) J. Dupont, G. S. Fonseca, A. P. Umpierre, P. F. P. Fitchtner, S. R. Teioxera, *J. Am. Chem. Soc.* **2002**, *124*, 4228; G. Schmid, M. Harm, *J. Am. Chem. Soc.* **1993**, *115*, 2047; (d) J. Huang, T. Jiang, H. Gao, Y. Chang, G. Zhao, W. Wu, *Chem. Commun.* **2003**, 1654; (e) V. Calo, A. Nacci, A. Monopoli, S. Laera, N. Cioffi, *J. Org. Chem.* **2003**, *68*, 2929; (f) V. Calo, A. Nacci, A. Monopoli, A. Detomaso, P. Illiade, *Organometallics* **2003**, *22*, 4193; (g) M. Spiro, D. M. De Jesus, *Langmuir* **2000**, *16*, 2664 and 4896; G. Battistuzzi, S. Cacchi, G. Fabrizi, *Synlett* **2002**, 439. (h) D. Zhao, Z. Fei, T. Geldbach, R. Scopelliti, P. J. Dyson, *J. Am. Chem. Soc.* **2004**, *126*, 15876; (i) R. R. Deshmuhk, R. Rajagopal, K. V. Srinivasan, *Chem. Commun.* **2001**, 1544; (j) L. Xu, W. Chen, J. Xiao, *Organometallics* **2000**, *19*, 1123; (k) C. W. Scheeren, G. Machado, J. Dupont, P. F. P. Fichtner, S. R. Texeira, *Inorg. Chem.* **2003**, *42*, 4738; (l) E. T. Silveira, A. P. Umpierre, L. M. Rossi, G. Machado, J. Morais, G. V. Soares, I. J. R. Baumvol, S. R. Teixeira, P. F. P. Fichtner, J. Dupont, *Chem. Eur. J.* **2004**, *10*, 3734; (m) G. S. Fonseca, J. D. Scholten, J. Dupont, *Synlett* **2004**, *9*, 1525; (n) C. S. Consorti, F. R. Flores, J. Dupont, *J. Am. Chem. Soc.* **2005**, *127*, 12054.

34 (a) C. E. Tucker, J. G. de Vries, *Top. Catal.* **2002**, *19*, 111; (b) M. Beller, A. Zapf, in *Handbook of Organopalladium Chemistry for Organic Synthesis*, Vol. 1, E.-i. Negishi (Ed.), Wiley, Hoboken, **2002**, p. 1209; (c) P. Baumeister, W. Meyer, K. Oertle, G. Seifert, H. Seifert, H. Steiner, *Chimia*, **1997**, *51*, 144; (d) J. McChesney, *Spec. Chem.* **1999**, *6*, 98.

35 (a) J. P. M. Niederer, A. B. J. Arnold, W. F. Hölderich, B. Tesche, M. Reetz, H. Bönnemann, *Top. Catal.* **2002**, *18*, 265; (b) S.-W. Kim, M. Kim, W. Y. Lee, T. Hyeon, *J. Am. Chem. Soc.* **2002**, *124*, 7642; (c) K. H. Park, S. U. Son, Y. K. Chung, *Org. Lett.* **2002**, *4*, 4361; (d) R. Abu-Reziq, D. Avnir, J. Blum, *J. Mol. Catal. A.* **2002**, *187*, 277; (e) A. Horvath, A. Beck, Z. Koppany, A. Sarkany, L. Guczi, *J. Mol. Catal. A.* **2002**, *182*, 295; (f) A.-M. Huang, Z.-F. Liu, C. Lai, J.-D. Hua, *J. Appl. Polym. Sci.* **2002**, *85*, 989; (g) A. Molnar, A. Papp, K. Miklos, P. Forgo, *Chem. Commun.* **2003**, 2626; (h) T. Sanji, Y. Ogawa, Y. Nakatsuka, M. Tanaka, H. Sakurai, *Chem. Lett.* **2003**, *32*, 980; (i) O. Dominguez-Quintero, S. Martinez, Y. Henriquez, L. D'Ornelas, H. Krentzien, J. Osuna, *J. Mol. Catal. A.* **2003**, *197*, 185; (j) I. Yuranov, P. Moeckli, E. Suvorova, P. Buffat, L. Kiwi-Minsker, A. Renken, *J. Mol. Catal. A.* **2003**, *192*, 239; (k) B. Corain, P. Guerriero, G. Schiavon, M. Zapparoli, M. Kralik, *J. Mol. Catal. A.* **2004**, *211*, 237; M. Kralik, V. Ktratky, P. Centomo, P. Guerriero, S. Lora, B. Corain, *J. Mol. Catal. A: Chem.* **2003**, *195*, 219; (l) P. Canton et al. *Catal. Lett.* **2003**, *88*, 141; (m) V. Johank et al. *Surf. Sci.* **2004**, *561*, L218; (n) T. G. Galow, U. Dreshler, J. A. Hanson, V. M. Rotello, *Chem. Commun.* **2002**, 1076; (o) R. B. Bedford, U. G. Singh, R. I. Walton, R. T. Williams, S. A. Davis, *Chem. Mater.* **2005**, 17, ASAP.

36 (a) H.-P. Kormann, G. Schmid, K. Pelzer, K. Philippot, B. Chaudret, *Z. Anorg. Allg. Chem.* **2004**, *630*, 1913; (b) M. Yashima, L. K. L. Falk, A. E. C. Palmqvist, K. Holmberg, *J. Colloid Interfac. Sci.* **2003**, *268*, 348; (c) S. Schauermann, J. Hoffmann, V. Johanek, J. Hartmann, J. Libuda, H.-J. Freund, *Angew. Chem.* **2002**, *114*, 2643; *Angew. Chem. Int. Ed.* **2002**, *41*, 2532; (d) M. Heemeier, A. F. Carlsson, M. Naschitzki, M. Schmal, M. Bäumer, H.-J.

Freund, *Angew. Chem.* **2002**, *114*, 4242; *Angew. Chem. Int. Ed.* **2002**, *41*, 4073; (e) V. Johanek, M. Laurin, J. Hoffmann, S. Schauermann, A. W. Grant, B. Kasemo, J. Libuda, H.-J. Freund, *Surf. Sci.* **2004**, *561*, L218.

37 (a) K. Ebitani, Y. Fujie, K. Kaneda, *Langmuir* **1999**, 1907; K. Kaneda, M. Higushi, T. Himanaka, *J. Mol. Cat. A: Chem.* **2001**, *63*, L33; (b) K.-M. Choi, T. Akita, T. Mizugaki, K. Ebitani, K. Kaneda, *New J. Chem.* **2003**, *27*, 324; (c) L. Guczi, A. Beck, A. Horvath, Zs. Koppany, G. Stefler, K. Frey, I. Sajo, O. Geszti, J. Lynch, *J. Mol. Catal. A.* **2004**, *204*, 545.

38 K. Mori, T. Hara, T. Mizugaki, K. Ebitani, K. Kaneda, *J. Am. Chem. Soc.* **2004**, *126*, 10657.

39 (a) T. Nishimura, N. Kakiuchi, M. Inoue, S. Uemura, *Chem. Commun.* **2000**, 1245; (b) N. Kakiuchi, Y. Maeda, T. Nishimura, S. Uemura, *J. Org. Chem.* **2001**, *66*, 6220; (c) S. Schauermann, J. Hoffmann, V. Johanek, J. Hartmann, J. Libuda, H.-J. Freund, *Angew. Chem. Int. Ed.* **2002**, *41*, 2532; (d) M. Heemeier, A. F. Carlsson, M. Naschitzki, M. Schmal, M. Bäumer, H.-J. Freund, *Angew. Chem. Int. Ed.* **2002**, *41*, 4073; (e) B. M. Choudary, S. Mahdi, N. S. Chowdari, M. L. Kantam, B. Streedhar, *J. Am. Chem. Soc.* **2002**, *124*, 14127; (f) A. Mastalir, Z. Kiraly, *J. Catal.* **2003**, *220*, 372; (g) A. Mastalir, Z. Kiraly, *J. Catal.* **2003**, *220*, 372; (h) S. Bertarione, D. Scarano, A. Zecchina, V. Johanek, J. Hoffmann, S. Schauermann, J. Libuda, G. Rupprechter, H.-J. Freund, *J. Catal.* **2004**, *223*, 64; (i) P. Pfeifer, K. Schubert, M. A. Liauw, G. Emig, *Appl. Catal. A* **2004**, *270*, 165.

40 G. Riahi, D. Guillemot, M. Polisset-Tfoin, A. A. Khodadadi, J. Fraissard, *Catal. Today*, **2002**, *72*, 115; M. Mtelkar, C. V. Rode, R. V. Chaudhari, S. S. Joshi, A. M. Nalawade, *Appl. Catal. A: General* **2004**, *273*, 11.

41 (a) N. Toshima, Y. Shiiraishi, T. Teranishi, M. Mitake, T. Tominaga, H. Watanabe, W. Brijoux, H. Bönnemann, G. Schmid, *Appl. Organomet. Chem.* **2001**, *15*, 178; (b) J. Huang, T. Jiang, H.-X. Gao, B.-X. Han, Z.-M. Liu, W.-Z. Wu,

Y.-H. Chang, G.-Y. Zhao, *Angew. Chem. Int. Ed.* **2004**, *43*, 1397 (c) M. T. Reetz, H. Schulenburg, M. Lopez, B. Spliethoff, B. Tesche, *Chimia* **2004**, *58*, 896.

42 (a) F. Caruso, *Adv. Mater.* **2001**, *13*, 11; F. Caruso, R. A. Caruso, H. Möhhwald, *Science* **1998**, *282*, 1111; Y. Lin, Y. Lu, B. Gates, Y. Xia, *Chem. Mater.* **2001**, *13*, 1146; M. Julia, M. Duteil, C. Grard, E. Kunz, *Bull. Soc. Chim. Fr.* **1993**, 2791; (b) M. Julia, M. Duteil, *Bull. Soc. Chim. Fr.* **1993**, 2791; (c) A. F. Shmidt, L. V. Mametova, *Kinet. Katal.* **1996**, *37*, 406; (d) F. Zhao, M. Arai, *React. Kinet. Catal. Lett.* **2004**, *81*, 281; F. Zhao, B. M. Bhanage, M. Shirai, M. Arai, *Chem. Eur. J.* **2000**, *6*, 843; see also Arai's Refs. [73 c,g, 86a–c].

43 (a) M. Wagner, K. Köhler, L. Djakovitch, M. Mühler, *Top. Catal.* **1999**, *13*, 319; (b) K. Köhler, L. Djakovitch, M. Wagner, *Catal. Today*, **2001**, *66*, 105; (c) L. Djakovitch, K. Köhler, *J. Am. Chem. Soc.* **2001**, *123*, 5990; (d) A. Nejjar, C. Pinel, L. Djakovitch, *Adv. Synth. Catal.* **2003**, *345*, 612; (e) L. Djakovitch, M. Wagner, C. G. Hartung, M. Beller, K. Köhler, *J. Mol. Catal. A: Chem.*, **2004**, *219*, 121; (f) S. Pröckl, W. Kleist, M. A. Gruber, K. Köhler, *Angew. Chem. Int. Ed. Engl.* **2004**, *43*, 1881; (g) L. Djakovitch, P. Rollet, *Adv. Syn. Catal.* **2004**, *346*, 1782; (h) L. Djakovitch, P. Rollet, *Tetrahedron Lett.* **2004**, *45*, 1367; (i) S. Chouzier, M. Gruber, L. Djakovitch, *J. Mol. Catal. A: Chem.* **2004**, *212*, 43; (j) J. Penzien, C. Haessner, A. Jentys, K. Köhler, T. E. Muller, J. A. Lercher, *J. Catal.* **2004**, *221*, 302; (k) J. P. Collman, K. M. Kosydar, M. Bressan, W. Lamanna, T. Garrett, *J. Am. Chem. Soc.* **1984**, *106*, 2569; J. Rebek, *Tetrahedron*, **1979**, *35*, 723; I. W. Davies, L. Matty, D. L. Hugues, P. J. Reider, *J. Am. Chem. Soc.* **2001**, *123*, 10139 (for a recent detailed analysis of this test in the context of Pd catalysis, see Jones' review [31o]).

44 (a) P. W. Albers, J. G. E. Krauter, D. K. Ross, R. G. Heiodenreich, K. Köhler, S. F. Parker, *Langmuir*, **2004**, *20*, 8254; (b) P. Rylander. *Catalytic Hydrogenation in Organic Synthesis*, Academic Press, New York, **1979**; (c) M. L. Toebes, J. A. van Dillen, K. P. de Jong, *J. Mol. Catal. A: Chem.* **2001**, *173*, 75; (d) C. R. Le Blond,

A. T. Andrews, Y. Sun, J. R. Sowa, Jr., *Org. Lett* **2001**, *3*, 1555; (e) E. B. Mobufu, J. H. Clark, D. J. Macquarrie, *Green Chem.* **2001**, *3*, 23; (f) R. G. Heidenreich, J. G. E. Krauter, J. Pietsch, K. Köhler, *J. Mol. Catal. A: Chem.* **2002**, *182-183*, 499; K. Köhler, R. G. Heidenreich, J. G. E. Krauter, J. Pietsch, *Chem. Eur. J.* **2002**, *8*, 622; (g) F. Zhao, K. Kurakami, M. Shirai, M. Arai, *J. Catal.* **2000**, *194*, 479; (h) B. Coq, F. Figueras, P. Geneste, C. Moreau, P. Moreau, M. Warawdekar, *J. Mol. Catal.* **1993**, *78*, 211; A. Cwik, Z. Hell, F. Figueras, *Org. Biol. Chem.* **2006**, *3*, 4307; A. Cwik, Z. Hell, F. Figueras, *Tetrahedron Lett.* **2006**, *47*, 3023; (i) V. A. Yakovlev, V. V. Terskikh, V. I. Simagina, V. A. Likholobov, *J. Mol. Catal. A* **2000**, *153*, 231; (j) P. Styring, C. Grindon, C. M. Fisher, *Catal. Lett.* **2001**, *77*, 219; (k) B. H. Lipshutz, S. Tasler, W. Chrisman, B. Spliethoff, B. Tesche, *J. Org. Chem.* **2003**, *68*, 1177; (l) B. H. Lipshutz, J. A. Sclafani, P. A. Blomgren, *Tetrahedron*, **2000**, *56*, 2139; (m) B. H. Lipshutz, P. A. Blomgren, *J. Am. Chem. Soc.* **1999**, *121*, 5819; (n) B. H. Lipshutz, *Adv. Synth. Catal.* **2001**, *343*, 313; (o) B. H. Lipshutz, H. Ueda, *Angew. Chem.* **2000**, *112*, 4666; *Angew. Chem. Int. Ed.* **2000**, *39*, 4492; (p) F. Zaera, *Acc. Chem. Res.* **2002**, *35*, 129. K. Anderson, S. C. Fernandez, C. Hardacre, P. C. Marr, *Inorg. Chem. Commun.* **2004**, *7*, 73; (q) S. Carrettin, J. Guzman, A. Corma, *Angew. Chem. Int. Ed.* **2005**, *44*, 2242; J. Gurman, S. Carretin, A. Corma, *J. Am. Chem. Soc.* **2005**, *127*, 3286; A. Corma, H. Garcia, A. Leyva, *J. Catal.* **2006**, *240*, 87; A. Corma, H. Garcia, A. Leyva, *J. Mol. Catal. A: Chemical* **2005**, *230*, 97; (r) G. G. Wildgoose, C. E. Banks, R. G. Compton, *Small* **2006**, *2*, 182.

45 45. (a) For AuNP-catalyzed CO oxidation by O₂, see Chapter 15, Haruta's reviews in Refs. [9e, 11j, 50b] and G. J. Hutchings, *Catal. Today* **2005**, *100*, 55.

46 For oxide-AuNP-catalyzed oxidation reactions, see Ref. [3] and Chapters 12–15. For specific reactions, see: (a) S. Okumura, S. Nakamara, T. Tsubota, M. Nakamura, M. Azuma, *Catal. Lett.* **1998**, *51*, 53; M. Okumura, T. Akita, M.

Haruta, *Catal. Today* **2002**, *74*, 265; (b) C. Mohr, H. Hofmeister, P. Claus, *J. Catal.* **2003**, *213*, 86; (c) S. Biella, M. Rossi, *Chem. Commun.* **2003**, 378; S. Biela, G. L. Castiglioni, C. Fumagalli, L. Prati, M. Rossi, *Catal. Today* **2002**, *72*, 43; (d) C. Milone, R. Ingoglia, G. Neri, A. Pistone, S. Galvagno, *Appl. Catal. A* **2001**, *211*, 251; C. Milone, M. L. Tropeano, G. Gulino, G. Neri, R. Ingoglia, S. Galvano, *Chem. Commun.* **2002**, 868; (e) P. Landon, P. J. Collier, A. J. Papworth, C. J. Kiely, G. J. Hutchings, *Chem. Commun.* **2002**, 2058; (f) F. Shi, Y. Deng, *J. Catal.* **2002**, *211*, 548; S.-W. Kim, S. U. Son, S. S. Lee, T. Hyeon, Y. K. Chung, *Chem. Commun.* **2001**, 2212; (g) G. Lue, D. Ji, G. Qian, Y. Qi, X. Wang, J. Suo, *Appl. Catal., A: General*, **2005**, *280*, 175; (h) K. Zhu, J. Hu, R. Richards, *Catal. Lett.* **2005**, *100*, 195.

47 M. M. Maye, J. Luo, Y. Lin, M. H. Engelhard, H. Mark, H. Hepel, C.-J. Zhong, *Langmuir* **2003**, *19*, 125 and references cited therein; C. Roth, I. Hussain, M. Bayati, R. J. Nichols, D. J. Schiffrin, *Chem. Commun.* **2004**, 1532; H. Tang, J. H. Chen, M. Y. Wang, L. H. Nie, Y. F. Kuang, S. Z. Yao, *Appl. Catal. A* **2004**, *275*, 43; for a recent review, see Ref. [8a]; Z. Liu, X. Y. Ling, X. Su, J. Y. Lee, *J. Phys. Chem. B* **2004**, *108*, 8234; J. Solla-Gullon, F. J. Vidal-Iglesias, V. Montiel, A. Aldaz, *Electrochem. Acta* **2004**, *49*, 5079; X. Zhang, F. Zhang, K.-Y. Chan, *J. Mater. Sci.* **2004**, *39*, 5845; Z. He, J. Chen, D. Liu, H. Zhou, Y. Kuang, *Diamond Relat. Mater.* **2004**, *13*, 1764; E. V. Spinace, A. O. Neto, M. Linardi, *J. Power Sources* **2004**, *129*, 121; C. Bock, C. Paquet, M. Couillard, G. A. Botton, B. R. MacDougall, *J. Am. Chem. Soc.* **2004**, *126*, 8028; K. Zhu, J. Hu, R. Richards, *Catal. Lett.* **2005**, *100*, 195; G. M. Veith, A. R. Lupini, S. J. Pennycook, G. W. Ownby, N. J. Dudney, *J. Catal.* **2005**, *231*, 151; S.-H. Baeck, T. F. Jaramillo, A. Kleiman, E. W. McFarland, *Measur. Sci. Technol.* **2005**, *16*, 54.

48 I. Shinkai, A. O. King, R. D. Larsen, *Pure Appl. Chem.* **1994**, *66*, 1551; A. P. Philipse, M. P. B. van Bruggen, C. Pathmamanoharan, *Langmuir* **1994**, *10*, 92; W. Teunissen, M. F. de Groot, J. Geus, O. Stephan, M. Tence, C. Colliex, *J. Catal.*

2001, *204*, 169; P. Tarjaj, C. J. Serna, *J. Am. Chem. Soc.* **2003**, *125*, 15754; A.-H. Lu, W. Schmidt, N. Matoussevitch, H. Bönnemann, B. Spliethoff, B. Tesche, E. Bill, W. Kiefer, u. F. Schüth, *Angew. Chem. Int. Ed.* **2004**, *43*, 4303; S. C. Tsang, V. Caps, I. Paraksevas, D. Chadwick, D. Thompsett, *Angew. Chem. Int. Ed.* **2004**, *43*, 5645; P. Wang Xiao, B. Shen, N. He, *Colloid Surf. A: Physicochem. Eng. Asp.* **2006**, *276*, 116.

49 (a) G. Schmid, H. West, H. Mehles, A. Lehnert, *Inorg. Chem.* **1997**, *36*, 891; A. M. Caporusso, L. A. Aronica, E. Schiavi, G. Martra, G. Vitulli, P. Salvadori, *J. Organomet. Chem.* **2005**, *690*, 1063; (b) C. Larpent, F. Brisse-Le Menn, H. Patin, *New. J. Chem.* **1991**, *15*, 361; C. Larpent, F. Brisse-Le Menn, H. Patin, *J. Mol. Catal.* **1991**, *65*, L35; C. Larpent, E. Bernard, F. Brisse-Le Menn, H. Patin, *J. Mol. Catal. A: Chem.* **1997**, *116*, 277; F. Launay, H. Patin, *New J. Chem.* **1997**, *21*, 247; F. Launay, A. Roucoux, H. Patin, *Tetrahedron Lett.* **1998**, *39*, 1353; see also ref. 11e.

50 D. Soulivong, C. Copéret, J. Thivolle-Cazat, J.-M. Basset, B. Maunders, R. B. A. Pardy, G. J. Sunley, *Angew. Chem. Int. Ed.* **2004**, *43*, 5366; C. Copéret, F. Lefebvre, J.-M. Basset, in *Handbook of Metathesis*, R. H. Grubbs (Ed.), Wiley-VCH, Weinheim, **2003**, Vol. 1, Ch. 1.12, pp.190–2004.

51 M. Haruta, *Stud. Surf. Sci. Catal.* **2003**, *14* (*Sci. Technol. Catal.* **2002**, *31*); H. Tang, J. H. Chen, M. Y. Wang, L. H. Nie, Y. F. Kuang, S. Z. Yao, *Appl. Catal. A* **2004**, *275*, 43; A. K. Sinha, S. Seelan, M. Okumura, T. Akita, S. Tsubota, M. Haruta, *J. Phys. Chem. B* **2005**, *109*, 3956.

52 H. Bönnemann, W. Brijoux, A. Schulze Tilling, K. Siepen, *Top. Catal.* **1997**, *4*, 217; H. Bönnemann, W. Brijoux, R. Brinkmann, A. Schulze Tilling, K. Siepen, H. Bönnemann, W. Brijoux, A. Schulze Tilling, T. Shilling, B. Tesche, K. Seevogel, R. Franke, J. Hormes, G. Köhl, J. Pollmann, J. Rothe, W. Vogel, *Inorg. Chim. Acta* **1998**, *270*, 95.

53 (a) Q. Wang, H. Liu, M. Han, X. Li, D. Jiang, *J. Mol. Catal. A: Chem.* **1997**, *118*, 145; (b) M. Beller, H. Fischer, K. Köhlein, C.-P. Reisinger, W. A. Herrmann, *J. Organomet. Chem.* **1996**, *520*, 257.

54 M. T. Reetz, R. Breinbauer, K. Wanninger, *Tetrahedron Lett.* **1996**, *37*, 4499.

55 M. T. Reetz, R. Breinbauer, P. Wedemann, P. Binger, *Tetrahedron* **1998**, *54*, 1233.

56 M. T. Reetz, S. A. Quaiser, C. Merk, *Chem. Ber.* **1996**, *129*, 741.

57 K. Manabe, Y. Mori, T. Wakabayashi, S. Nagayama, S. Kobayashi, *J. Am. Chem. Soc.* **2000**, *122*, 7202.

58 A. B. R. Mayer, J. E. Mark, R. E. Morris, *Polym. J.* **1997**, *275*, 333; A. Borsla, A. M. Wilhelm, H. Delmas, *Catal. Today* **2001**, *66*, 389.

59 (a) W. Liu, H. Liu, X. An, X. Ma, Z. Liu, L. Quiang, *J. Mol. Catal. A: Chem.* **1999**, *147*, 73 and refs cited therein. (b) W. Yu, H. Liu, M. Liu, Z. Liu, *React. Funct. Polym.* **2000**, *44*, 21. (c) X. Yang, Z. Deng, H. Liu, *J. Mol. Cat. A Chem.* **1999**, *144*, 123.

60 N. Pradhan, A. Pal, T. Pal, *Langmuir* **2001**, *17*, 1800.

61 H. Hirai, N. Yakura, Y. Seta, S. Hodoshima, *React. Funct. Polym.* **1998**, *37*, 121.

62 G. Schmid, V. Maihack, F. Lantermann, S. Peschel, *J. Chem. Soc., Dalton Trans.* **1996**, 589.

63 H. Bönnemann, G. A. Braun, *Angew. Chem. Int. Ed.* **1996**, *35*, 1992.

64 P. Drogna-Landré, M. Lemaire, D. Richard, P. Gallezot, *J. Mol. Catal.* **1993**, *78*, 257; P. Drogna-Landré, D. Richard, M. Draye, P. Gallezot, M. Lemaire, *J. Catal.* **1994**, *147*, 214.

65 W. D. Harman, *Chem. Rev.* **1997**, *97*, 1953.

66 Y. Lin, R. G. Finke, *Inorg. Chem.* **1994**, *33*, 4891.

67 (a) H. Yang, H. Gao, R. Angelici, *Organometallics* **2000**, *19*, 622; (b) J.-L. Pellegatta, C. Blancy, V. Collière, R. Choukroun, B. Chaudret, P. Cheng, K. Philippot, *J. Mol. Catal. A: Chemical* **2002**, *178*, 55.

68 J. Blum, I. Amer, A. Zoran, Y. Sasson, *Tetrahedron Lett.* **1983**, *24*, 4139; J. Foise, R. Kershaw, K. Dwight, A. Wold, *Mater. Res. Bull.* **1985**, *20*, 147.

69 R. W. Albach, M. Jautelat, in *Two-Phase Hydrogenation Method and Colloidal*

Catalysts for the Preparation of Cyclohexanes from Benzenes, **1999**, Ger. Offen. (A. G. Bayer), Germany.

70 J. Schulz, A. Roucoux, H. Patin, *Chem. Commun.* **1999**, 535; J. Schulz, A. Roucoux, H. Patin, *Chem. Eur. J.* **2000**, *6*, 618; J. Schulz, S. Levigne, A. Roucoux, H. Patin, *Adv. Synth. Catal.* **2002**, *344*, 266; A. Roucoux, J. Schulz, H. Patin, *Adv. Synth. Catal.* **2003**, *345*, 222; V. Mévellec, A. Roucoux, E. Ramirez, K. Philippot, B. Chaudret, *Adv. Synth. Catal.* **2004**, *346*, 72.

71 T. Jeffery, *Tetrahedron* **1996**, *52*, 10113.

72 S. Klingelhöfer, W. Heitz, A. Greiner, S. Oestreich, S. Förster, M. Antonietti, *J. Am. Chem. Soc.* **1997**, *119*, 10116; J. Walter, J. Heiermann, G. Dyker, S. Hara, H. Shioyama, *J. Catal.* **2000**, *189*, 449; B. M. Choudary, N. S. Chowdari, K. Jyoti, N. S. Kumar, M. L. Kantam, *Chem. Commun.* **2002**, 586; J. Le Bars, U. Specht, J. S. Bradley, D. G. Blackmond, *Langmuir* **1999**, *15*, 7621; M. Moreno-Manas, R. Pleixats, S. Villaroya, *Organometallics* **2001**, *20*, 4524.

73 For recent reviews on Heck reaction, see Ref. [31c,n] and: S. Bräse, A. de Meijere, in *Metal-Catalyzed Cross-Coupling Reactions*, F. Diederich, P. J. Stang (Eds.), Wiley-VCH, Weinheim, **1998**, pp. 99–166; I. P. Beletskaya, A. V. Cheprakov, *Chem. Rev.* **2000**, 100, 3009; N. J. Whitcombe, K. K. Hii, S. E. Gibson, *Tetrahedron* **2001**, *57*, 7449; B. M. Bhanage, M. Arai, *Catal. Rev.* **2001**, *43*, 315; A. F. Littke, G. C. Fu, *Angew. Chem. Int. Ed.* **2002**, *41*, 4176; A. Hillier, P. Nolan, *Plat. Met. Rev.* **2002**, 46, 50; A. F. Little, G. C. Fu, *Angew. Chem.* **2002**, *114*, 4350; *Angew. Chem. Int. Ed.* **2002**, *41*, 4176; M. Larhed, A. Hallberg, in *Handbook of Organopalladium Chemistry for Organic Synthesis*, Vol. *1*, E.-I. Negishi (Ed.), Wiley, Hoboken, **2002**, p. 1133; W. Hieringer, in *Applied Homogeneous Catalysis with Organometallic Compounds*, Vol. *2*, 2nd edn., B. Cornils, W. A. Herrmann (Eds.), Wiley-VCH, Weinheim, **2002**, p. 721; B. M. Bhanage, S.-i. Fujita, M. Arai, *J. Organomet. Chem.*

2003, *687*, 211; M. Shibasaki, E. M. Vogl, T. Oshima, in *Comprehensive Asymmetric Catalysis*, Vol. sup. **2004**, E. N. Jacobsen, A. Pfaltz, H. Yamamoto, (Eds.), Springer Verlag, Heidelberg, **2004**, p. 73; J. Tsuji, *Palladium Reagents and Catalysis*, Wiley, West Sussex, **2004**; *Handbook of Organopalladium Chemistry for Organic Synthesis*, Negishi, E. (Ed.), Wiley, Hoboken, **2002**; K. Ferré-Filmon, L. Delaude, A. Demonceau, A. F. Noels, *Coord. Chem. Rev.* **2004**, *248*, 2323.

74 (a) For a recent review on the Suzuki reaction, see: A. Suzuki in *Modern Arene Chemistry*, D. Astruc (Eds.), Wiley-VCH, Weinheim, **2003**, pp. 53–106; (b) for a recent review on amination of aryl halides and sulfonates, see: J. F. Hartwig in *Modern Arene Chemistry*, D. Astruc (Eds.), Wiley-VCH, Weinheim, **2003**, pp. 107–168.

75 F. Bertoux, E. Monflier, Y. Castanet, A. Mortreux, *J. Mol. Cat. A* **1999**, *143*, 23.

76 M. T. Thathagar, J. Beckers, G. Rothenberg, *J. Am. Chem.* **2002**, *124*, 11858; A. F. Thathagar, P. J. Kooyman, R. Boerleider, E. Jansen, C. J. Elsevier, G. Rothenberg, *Adv. Synth. Catal.* **2005**, *347*, 1965; M. B. Thathagar, J. E. ten Elshof, G. Rothenberg, *Angew. Chem. Int. Ed.* **2006**, *45*, 2886; L. D. Pachón, C. J. Elsevier, G. Rothenberg, *Adv. Synth. Catal.* **2006**, *348*, 1705.

77 L. Djakovitch, M. Wagner, K. Kohler, *J. Organomet. Chem.* **1999**, *592*, 225; B. H. Lipshutz, H. Ueda, *Angew. Chem. Int. Ed.* **2000**, *39*, 4492.

78 C. Mohr, H. Hofmeister, J. Radnik, P. Claus, *J. Am. Chem. Soc.* **2003**, *125*, 1905; K. Chen, Z. Zhang, Z. Cui, D. Yang, *Gaofenzi Xuebao* **2000**, 180, *Chem. Abstr.* **133**, 159490.

79 S. Scire, S. Minico, C. Crisafulli, C. Satriano, A. Pistone, *Appl. Catal. B* **2003**, *40*, 43.

80 C. T. Campbell, *Science* **2004**, *306*, 234.

81 (a) H. Berndt, A. Martin, I. Pitsch, U. Prusse, K.-D. Vorlop, *Catal. Today* **2004**, *91*, 191; S. Carrettin, P. McMorn, P. Johnston, K. Griffin, C. J. Kiely, G. A. Attard, G. J. Hutchings, *Top. Catal.* **2004**, *27*, 131; (b) S. Biella, F. Porta, L. Prati,

M. Rossi, *Catal. Letters* **2003**, *90*, 23; (c)
A. Venugopal, M. S. Scurrel, *Appl. Catal.,
A: General* **2004**, *258*, 241.

82 (a) R. L. Augustine, S. T. O'Leary, *J. Mol.
Catal. A: Chem.* **1995**, *277*, 95; (b) J. Le

Bars, U. Specht, J. S. Bradley, D. G.
Blackmond, *Langmuir* **1999**, *15*, 7621;
(c) B. M. Choudary, S. Madhi, N. S.
Chowdari, M. L. Kantam, B. Sreedhar,
J. Am. Chem. Soc. **2002**, *124*, 14127.

2
Colloidal Nanoparticles Stabilized by Surfactants or Organo-Aluminum Derivatives: Preparation and Use as Catalyst Precursors

Helmut Bönnemann, Kyatanahalli S. Nagabhushana, and Ryan M. Richards

2.1
Background

During the last few decades a considerable body of knowledge has been gained regarding the synthesis, characterization, and potential application of nanosized metal particles [1–26]. Highly dispersed mono- and bi-metallic colloids have been used as precursors for a new type of catalyst that is applicable both in the homogeneous [27] and heterogeneous phases [28]. Besides the obvious applications in chemical catalysis, recent studies have examined the great potential of nanostructured metal colloids as fuel cell catalyst precursors [29, 30]. To date, two major industrial applications of nanocatalysis have emerged from the benches of the research laboratories: Headwaters NanoKinetix and Degussa have developed and patented [31] a direct synthesis method for the production of H_2O_2 from hydrogen and oxygen using size-defined nanoclusters of palladium (6 nm) anchored to a catalyst support. This simplified process offers multiple advantages over the current multi-step cost-intensive anthraquinone process: It is environmentally friendly, eliminating the use of a toxic substance while producing only water as a by-product, and has the potential to reduce production costs from 25–35 cents per pound to less than 15, while reducing the capital cost required for large-scale H_2O_2 production plants by up to 50%.

Even more spectacular results in terms of the increasing importance of nanocatalysis for bulk industrial processes have recently been reported by Kuipers and de Jong [32, 33]. By dispersing metallic cobalt nanoparticles of specific sizes on inert carbon nanofibers the authors were able to prepare a new nano-type Fischer–Tropsch catalyst. A combination of X-ray absorption spectroscopy, electron microscopy, and other methods has revealed that zerovalent cobalt particles are the true "active centers" which convert CO and H_2 into hydrocarbons and water. Further, a profound size effect on activity, selectivity, and durability was observed. Via careful pressure–size correlations, Kuipers and de Jong have found that or cobalt particles of 6 or 8 nm are the optimum size for Fischer–Tropsch catalysis. The Fischer–Tropsch process (invented in 1925 at the Kaiser-Wilhelm-Institute for

Nanoparticles and Catalysis. Edited by Didier Astruc
Copyright © 2008 WILEY-VCH Verlag GmbH & Co. KGaA, Weinheim
ISBN: 978-3-527-31572-7

Coal Research,[1] Mülheim, Germany) is currently regarded as one of the key steps in the huge gas-to-liquid (GTL) or biomass-to-liquid (BTL) units [34, 35] which are currently debated in industry to substitute ca. 10% of the global supply of diesel fuel by sulfur- and nitrogen-free "Synfuel" within a decade. "Nano for Energy" is the relevant key phrase here.

2.2
General Introduction

The advantage of using surfactants as colloidal stabilizers for metal nanoparticles resides in the fact that surface-active chemicals have sufficient strength to limit the particle growth at certain stages (size control) but can easily be detached from the metallic surface to provide catalytically active sites, either in the liquid phase ("quasi-homogeneous") or – after deposition on supports – as heterogeneous catalysts. Surfactants as colloidal stabilizers not only help to control the growth to give particles with "monodisperse" size distribution (i.e. size deviations <5%) but also efficiently prevent agglomeration of the metallic cores acting as steric "spacers", as shown in Fig. 2.1.

Metal organo- and hydro-sols can be obtained with exceptionally high metal concentrations (<1 mol metal l^{-1}), and with excellent preparative reproducibility in multi-gram amounts. Further, surfactant-stabilized metal alloy nanoparticles are easily accessible via the co-reduction of different metal salts. As shown in Fig. 2.1, even the inner structure of the colloidal alloy particles can be tailored by adjusting the reduction conditions and/or preparative pathways to prepare homogeneous [8, 36, 37], gradient [38–40] or egg-shell type [41] nano-alloys. "Reductive stabilization"

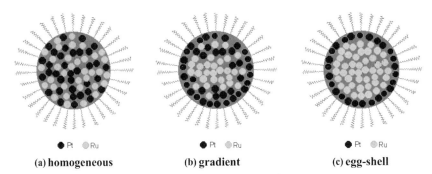

● Pt　◉Ru　　　　● Pt　◉Ru　　　　● Pt　◉Ru
(a) **homogeneous**　　　(b) **gradient**　　　(c) **egg-shell**

Fig. 2.1 Homogeneous, segregated and core–shell structured bimetallic nanoparticles that can be generated by colloidal synthesis during the reduction.

1) In 1949 the name changed to Max-Planck-Institut für Kohlenforschung.

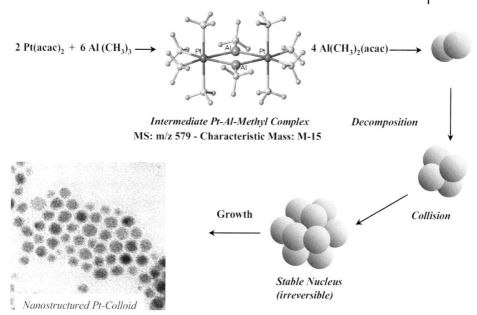

$$2\ Pt(acac)_2\ +\ 6\ Al\,(CH_3)_3 \longrightarrow$$

$$\longrightarrow\ 4\ Al(CH_3)_2(acac)\longrightarrow$$

Intermediate Pt-Al-Methyl Complex
MS: m/z 579 - Characteristic Mass: M-15

Decomposition

Collision

Growth

Stable Nucleus
(irreversible)

Nanostructured Pt-Colloid

Fig. 2.2 Nanoscopic Pt colloids in the "embryonic state" during "reductive stabilization" [43, 44]. Applied Methods: NMR, MS, XANES, EXAFS, ASAXS.

using an excess of triorganoaluminum as the reductive agent for transition metal salts and simultaneously as the protective shell is a different approach for multi-gram preparations of nanometallic organosols. This method has opened straight-forward access to a plethora of catalytically active nanoparticles with tunable size, morphology, and structure [42–44]. The key feature is the formation of a reactive metal–organic colloidal protecting shell around the particles, as shown in Fig. 2.2.

Reactive alkylaluminum groups in the protective shell allow substitution of the alkyl groups by a large variety of organic molecules. This "modification" of the aluminumorganic protecting shell can be used to convert the dispersion properties of the original organosols to give hydrosols with a vast spectrum of solubilities of the colloidal metals in alcohol and/or water [45], Fig. 2.3. With respect to applications in heterogeneous catalysis it should be mentioned that inorganic surfaces bearing OH groups can also react with the active Al–C bonds via protonolysis. This opens new pathways to anchor the nanoparticles firmly on the carrier surface and suppress unwanted agglomeration.

Decomposition of preformed transition metal complexes - where the metals are already in a low valent state - in the presence of surfactants or other types of sta-bilizers is another route to colloidal metal nanoparticles that was pioneered by Bradley and Chaudret [6, 7, 46–55]. Organometallic complexes and several organic salt derivatives of transition metals easily decompose when energy is applied, e.g.

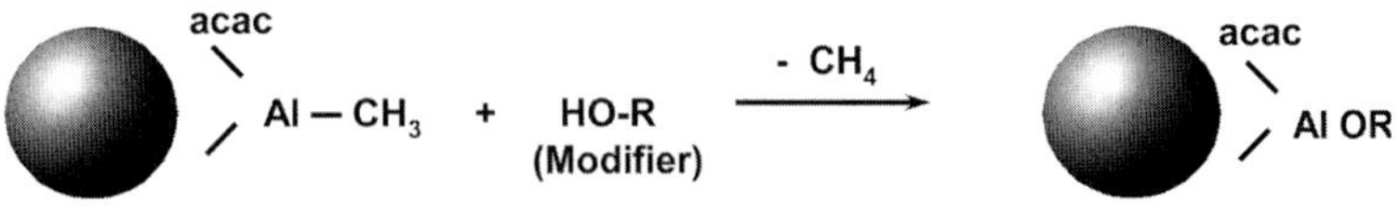

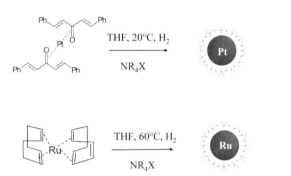

Organosol + Modifier = Modified Sol

Fig. 2.3 Hydrosols via "modification" of the Al–organic protective shell. As modifiers can be used: e.g. alcohols, carbonic acids, silanols, sugars, polyalcohols, polyvinylpyrrolidone, surfactants, silica, alumina. Advantages: Tailoring of the dispersion in lipophilic and/or hydrophilic solvents (e.g. water); Anchoring onto surfaces via d-bonds.

Fig. 2.4 Nanoparticle synthesis through hydrogenolysis of zerovalent metal complexes.

in the form of heat, light or ultrasound (Fig. 2.4). As shown in Fig. 2.4, hydrogenolysis of zerovalent metal complexes in the presence of stabilizers yields very clean surfaces which are especially suitable for mechanistic studies e.g. in fuel cell catalysis. In addition, this method is well suited for the subsequent decoration or coating of preformed nanoparticulate metal cores to form an egg-shell type structure (Fig. 2.1(c)). This pathway is however restricted to the availability of appropriate "starting complexes".

Microwave heating has been shown to provide a more homogeneous particle nucleation and shorter aggregation times. This modification was introduced to generate polymer-stabilized Pt-colloids (spheres, 2–4 nm, having narrow size distribution) [56–58]. Suslick and Gedanken [59–62] have successfully developed the sonochemical decomposition of metal salts and organometallic complexes. The limitation of photo-, γ-radiolysis, and laser irradiation for introducing energy, however, lies in the restriction to low metal concentrations in solution [63–65] which makes this variation unsuitable for scale-up for the manufacture of nanostructured metal colloids for practical applications.

Controlled thermolysis of metal carbonyls in the presence of aluminium alkyls is a special case of nanoparticle formation via the decomposition of zerovalent metal complexes. This led to a production process for monodisperse, magnetic Co-, Fe/Co, and Fe-nanocolloids having particle sizes adjustable between 2 and

$$Co_2(CO)_8 + Al(octyl)_3$$

Molar ratio Co : Al = 10 : 1

1. **16 h, 110 °C, Toluene**

$\longrightarrow$

2. **16 h Synthetic Air**

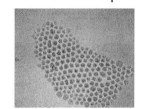

Fig. 2.5 Size-selective formation of airstable Co particles via thermolysis of Co_2CO_8 controlled by aluminum alkyl.

15 nm and a narrow size distribution with a standard variance of 1.6 nm [64–75] (Fig. 2.5).

The particle size can easily be precisely adjusted by the chain length of the alkyl radical and the applied concentration of the trialkyl-aluminum compound. If mixtures of low-valency carbonyl compounds of different metals are used, polymetallic alloy particles form. For example: With a molar Co:Al ratio of about 10:1, the thermolysis of Co carbonyl yields a Co particle size of 10 nm in the presence of $Al(C_8H_{17})_3$, 6 nm in the presence of $Al(C_2H_5)_3$ and 3.5 nm in the presence of $Al(CH_3)_3$. If, in the case of $Al(C_8H_{17})_3$, the molar Co:Al ratio in the batch is changed from 12:1 to 0.5:1, the particle size decreases from 10 to 5.4 nm.

The isolated, monodisperse, magnetic nanocolloids resulting from this process do not have long-term stability in air but can easily be protected from total oxidation by a post treatment called "smooth oxidation". The magnetic particles are treated in organic solution by exposure to a stream of synthetic air. Characterization by XANES, UPS, MIES, and FTIR [75] revealed that CO molecules adjacent to the nano-cobalt surface are oxidized during the "smooth oxidation" process to form a dense coating of Co-oxide-carbonate around the metallic core of the particles which are then stable in air and can be handled safely under ambient conditions. It was also shown that e.g. cobalt nanoparticles can be protected against oxidation with a highly stable carbon coating [74]. The use of these highly magnetic nanomaterials in catalysis has recently been demonstrated by Schüth et al., in the "nanoengineering" of a magnetically separable hydrogenation catalyst [76].

A special case of controlled decomposition of preformed organometallic complexes is the first wet chemical preparation of a Pt_{13}-cluster from $[(COD)Pt(CH_3)_2]$ in the presence of triorganoaluminum [77] (Fig. 2.6). The starting material was recently made readily accessible via a one-pot-synthesis [78] (Eq. (2.1)

$$Pt(acac)_2 + Al(CH_3)_3 + COD \rightarrow (CH_3)_2PtCOD \quad yield \sim 90\% \qquad (2.1)$$

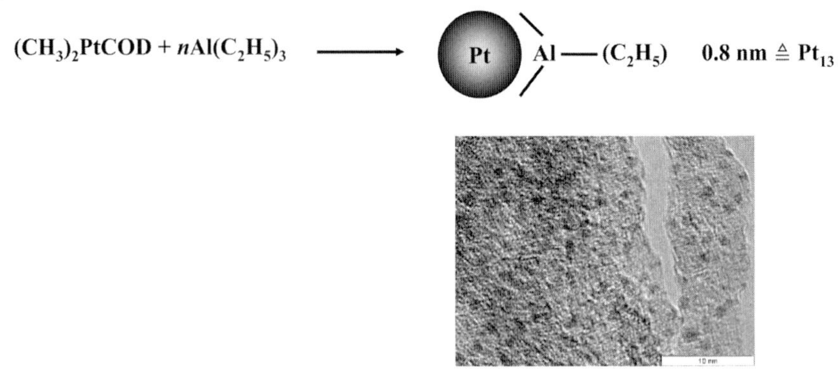

$$(CH_3)_2PtCOD + nAl(C_2H_5)_3 \longrightarrow \quad Pt \quad Al-(C_2H_5) \quad 0.8 \text{ nm} \triangleq Pt_{13}$$

Fig. 2.6 Formation of Pt_{13}-clusters stabilized by trialkylaluminum.

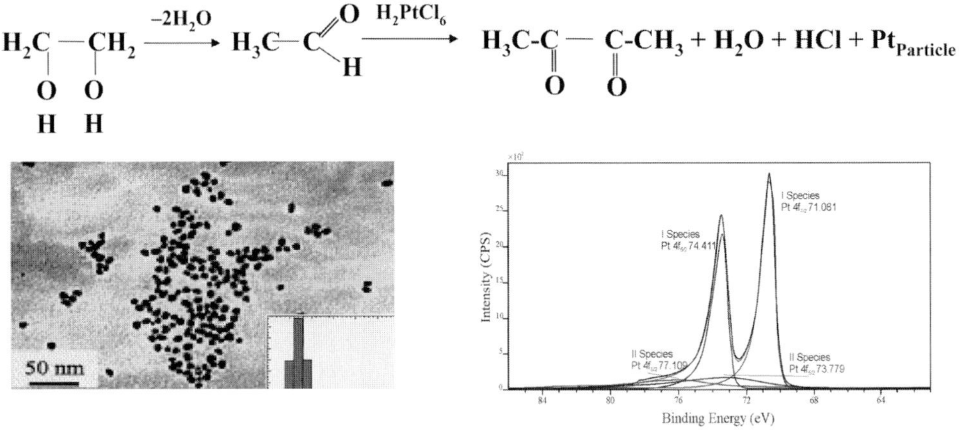

$$H_2C-CH_2 \xrightarrow{-2H_2O} H_3C-C \overset{O}{\underset{H}{\diagup}} \xrightarrow{H_2PtCl_6} H_3C-C-C-CH_3 + H_2O + HCl + Pt_{Particle}$$

Fig. 2.7 Schematic representation of the Polyol process exemplified with Pt. Left: TEM image of Pt particles prepared via Polyol method (without surfactant). Right: XPS analysis of platinum nonoparticles. The TEM shows a narrow particle size distribution (ca. 3 nm). Experimental XPS curves fit sufficiently with the Pt(0) standard. TEM image reproduced from F. Bonet, V. Delmas, S. Grugeon, R. Herrera Urbina, P.-Y. Siebert, and K. Tekaia-Elhsissen, *Nanostructured Materials*, *Vol. 11, No. 8* (1999), p. 1280, Fig. 2a, © 2000 with permission from Elsevier Science.

The one-shell structure of the Pt_{13}-cluster (0.75 ± 0.10 nm) was confirmed by X-ray absorption near-edge spectroscopy (XANES).

In order to control the particle morphology and size, surfactants may also be applied to the Polyol process [79, 80].

In a comparative study [81], we have been able to show that – according to XANES and XPS studies – Pt is predominantly present in the zerovalent state (Fig. 2.7). However, there is a strong indication that a small layer at the particle surface is slightly oxidized. For most applications in catalysis this is only a minor drawback because re-reduction of the oxidized surface occurs easily in the temperature range

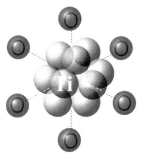

Fig. 2.8 Ti$_{13}$ cluster with 6 THF Oxygen atoms in an octahedral configuration.

applied during most catalytic reactions and/or under the reductive conditions the catalyst may be exposed to e.g. during hydrogenation.

Small metal nanoclusters from the early transition metal series can best be stabilized in the zerovalent state by THF. The structure of the regular Ti$_{13}$ cluster of the formula [Ti$_{13}$· 6 THF] shown in Fig. 2.8 was thoroughly characterized by a combination of NMR, TEM, XPS, XANES, and EXAFS [82–84].

2.3
Synthetic Methodologies

2.3.1
Stabilization via Surfactants

2.3.1.1 **Organosols**
The role of protecting agents in stabilizing nanostructured metals in colloidal form against agglomeration and oxidation is pivotal. Owing to their large surface to mass ratio, nanoparticles possess excess surface free energy that is comparable to their lattice energy, leading to structural instabilities. Protective agents are therefore essential to control the size and even crystal growth and shape during particle synthesis (see e.g. Refs. [85–87]). Surfactants are well suited as protecting agents because they bind strongly enough to the metallic surface to prevent agglomeration of the nanoparticles but are easily detached from the metal surface by catalytic substrates.

The Tetraalkylammoniumhydroborate Method Early on, the relatively bulky tetra-octylammonium halides were recognized as powerful "electrostatic" stabilizers for nanoparticles (mono- and even pluri-metallic) of nearly all transition metals of the Groups 6–11 of the periodic table [4, 88, 89]. Further, via the combination of [NR$_4$]$^+$ with [Bet$_3$H]$^-$ in the same molecule a species is formed that is capable of both reducing the metal salt and immediately providing the stabilizing agent. Thus, the surface-active [NR$_4$]$^+$ species are formed immediately at the reduction centre in high local concentrations (see Eq. (2.2)). Based on these findings we have developed a general route for the multigram synthesis of 1–10 nm transition metal organosols isolable in dry form and redispersible in high concentration in organic phases [13, 90] (Fig. 2.9).

$$MX_v + NR_4(Bet_3H) \rightarrow M_{colloid} + vNR_4X + vBet_3 + v/2\, H_2 \uparrow$$

$$M = \text{metals of the Groups } 6-11;\ X = Cl, Br;\ v = 1, 2, 3;$$
$$\text{and } R = \text{alkyl}, C_6 - C_{20} \tag{2.2}$$

These materials can be converted to "pure" metal colloids containing 70–85 wt.% of metal via workup with ethanol or ether and subsequent re-precipitation by a solvent of different polarity (compare Table IX in Ref. [8]). Later, a careful examination using UPS and MIES methods [91] has revealed that the NR_4X is bound to the metal surface through the negatively charged chloride while the long alkyl chains shield the metallic core in an umbrella-like mode (Fig. 2.10).

The pre-preparation of $NR_4(Bet_3H)$ can be avoided by coupling the NR_4X agent to the metal salt prior to the reduction step. Again, high local concentration of the

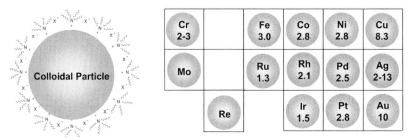

Cr 2-3			Fe 3.0	Co 2.8	Ni 2.8	Cu 8.3
Mo			Ru 1.3	Rh 2.1	Pd 2.5	Ag 2-13
	Re			Ir 1.5	Pt 2.8	Au 10

Diameters of colloidal particles in nm

Fig. 2.9 Metal organosols accessible via the $NR_4(Bet_3H)$ route (Eq. (2.2)). As synthesized, the $[NR_4]^+$-stabilized "raw" organosols typically contain 6–12 wt.% metal.

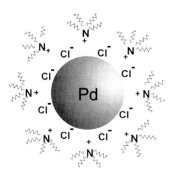

Fig. 2.10 Structural representation of NR_4X-stabilized metal organosols. Reproduced with permission from Elsevier Science [91].

protecting agent is provided right at the reduction center which allows application to the whole range of conventional reducing agents (Eq. (2.3)).

$$(NR_4)_w \, MX_v Y_w + v \, Red \rightarrow M_{colloid} + vRedX + wNR_4Y$$

$$M = Metals \ 4\text{--}11; \ Red = H_2, \ HCOOH, \ K, \ Zn, \ LiH, \ LiBEt_3H,$$
$$NaBEt_3H, \ KBEt_3H;$$

$$X, Y = Cl, \ Br; \ v, \ w = 1\text{--}3 \ \text{and} \ R = alkyl \ (C_6 - C_{12}) \tag{2.3}$$

THF as Colloidal Stabilizer for Early Transition Metal Colloids Solvents such as THF or propylenecarbonate can act as powerful colloidal stabilizers. Small sized clusters of zerovalent early transition metals Ti, Zr, V, Nb, and Mn have been stabilized by THF after [BEt$_3$H$^-$]-reduction of the pre-formed THF-adducts (Eq. (2.4)) [82–84, 92].

$$x[TiBr_4 \cdot 2THF] + x \cdot 4K[BEt_3H] \xrightarrow{\text{THF, 6h, 20°C}} [Ti \cdot 0.5THF]_x +$$
$$x \cdot 4BEt_3 + x \cdot 4KBr\downarrow + x \cdot 2H_2 \tag{2.4}$$

Table 2.1 summarizes the results.

By analogy, [Mn · 0.3 THF] particles (1–2.5 nm) were prepared [93] and their physical properties reported [94]. The THF in Eq. (2.4) has been successfully replaced by tetrahydrothiophene for Mn-, Pd-, and Pt- in the organosols but attempts to stabilize Ti and V using this approach led to decomposition [11]. XAS

Table 2.1 THF-stabilized organosols of early transition metals.

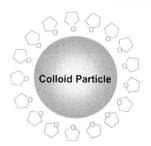

	Ti	V	Cr	Mn
	Zr	Nb	Mo	
	Hf	Ta	W	

Product	Starting material	Reducing agent	T [°C]	t [h]	Metal content [%]	Size [nm]
[Ti · 0.5 THF]x	TiBr$_4$· 2 THF	K[Bet$_3$H]	rt	6	43.5	–
[Zr · 0.4 THF]x	ZrBr$_4$· 2 THF	K[Bet$_3$H]	rt	6	42	–
[V · 0.3 THF]x	VBr$_3$· 3 THF	K[Bet$_3$H]	rt	2	51	–
[Nb · 0.3 THF]x	NbCl$_4$· 2 THF	K[Bet$_3$H]	rt	4	48	–
[Mn · 0.3 THF]x	MnBr$_2$· 2 THF	K[Bet$_3$H]	50	3	70	1–2.5

studies on Mn nanoparticle syntheses confirmed that the nanoparticles are in their metallic state with no counter ions from the starting material [92]. THF-stabilized $Mn_{(0)}$ particles which exhibit super-paramagnetism below 20 K were identified as the first example of an antiferromagnetic metal colloid [94]. Along similar lines, the super-paramagnetic properties of nanostructured Ni particles have also been investigated [95].

2.3.1.2 Hydrosols

Since the preferred solvent in catalyst technology is water, for economical and ecological reasons, we have developed a process which permits pre-stabilization of mono- and bi-metallic nanoparticles in the form of hydrosols using highly hydrophilic tensides as the colloidal stabilizers [11]. In order to achieve a high local concentration of the protective agents the metal salts are pre-treated with a given surfactant prior to the reduction step. The metal salts are subsequently reduced using conventional agents such as hydrogen, formic acid, ethylene glycol, etc., yielding colloidal, nanoscopic hydrosols which can be isolated in the form of highly water soluble powders. Aqueous $metal_{(0)}$ solutions of at least 100 mg metal l^{-1} which exhibit excellent long term stability are readily accessible. The core particle size of these hydrophilic versions of surfactant-stabilized mono- and pluri-metallic colloids can be tailored between 1 and 10 nm and statistical TEM examinations have shown that the resulting materials are generally monodisperse. The use of colloidally stabilized mono- and pluri-metallic nanoparticles as separately isolable precursors of heterogeneous catalysts is an advantageous alternative to the traditional *in situ* formulation of active metal components on support surfaces [14, 96]. Using long-chain alkylsulfobetaines as the stabilizer, a number of highly water-soluble nanometal colloids are accessible in excellent yields (see Fig. 2.11). The zerovalent nature of the metal core has been confirmed by a combination of spectroscopic methods [97].

A broad variety of hydrophilic surfactants may also be used as alternatives to amphiphilic sulfobetaines. Isolable nanometal colloids soluble in water with at least 100 mg of metal l^{-1} have been obtained with a wide range of cationic, anionic, and nonionic surfactants. Even, environmentally benign sugar soaps have been successfully applied [see Table 2 in Ref. [11]].

$$PtCl_2 \ + \ C_{12}H_{25}N(CH_3)_2(C_3H_6\text{-}SO_3) \ + \ Li_2CO_3 \ + \ H_2 \ \longrightarrow$$

Fig. 2.11 Preparation of sulfobetain-stabilized hydrosols.

2.3.2
"Reductive Stabilization" with Aluminum Alkyls

Organoaluminum compounds have been used for the "reductive stabilization" of mono- and bi-metallic nanoparticles [42–44]. From Fig. 2.2 it is clear that by using this route colloids of zerovalent elements of Groups 6–11 of the periodic table – in mono- or pluri-metallic form – are exclusively obtained as organosols stabilized by an organometallic protective shell (which by definition is highly sensitive to air and/or moisture). However, under truly anaerobic conditions these materials can be isolated in the dry form as powders with long-term stability and having excellent solubility in dry organic solvents.

2.3.2.1 Organosols
Depending on the amount of trialkylaluminum used during the synthetic procedure, very small particle sizes are accessible via "reductive stabilization". Figure 2.12 shows an example where $Al(CH_3)_3$ was applied to $Pt(acac)_2$ as the "stabilizing reductant" in the molar ratio of 4:1. The size of the resulting $Pt_{(0)}$ particles was 1.2 nm on average and the HRTEM micrographs clearly show crystalline structures [98].

Table 2.2 summarizes some typical examples of mono- and bi-metallic organosols prepared via the "reductive stabilization" pathway.

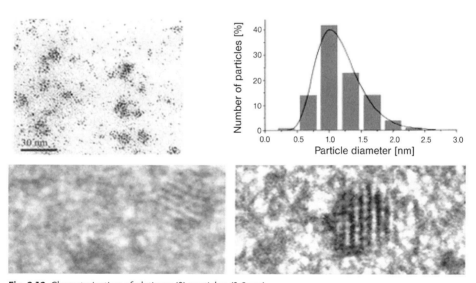

Fig. 2.12 Characterization of platinum(0) particles (1.2 nm). Top: TEM analysis of aluminum-organic-stabilized Pt nanoparticles: TEM image and corresponding particle-size distribution. Bottom: HRTEM images of larger Pt particles showing lattice fringes.

Table 2.2 Mono- and bimetallic nanocolloids prepared via "reductive stabilization".

Metal salt		Reducing agent		Solvent toluene	Conditions		Product m [g]	Metal content	Particle size
Formula	g/mmol	Formula	g/mmol	ml	T [°C]	t [h]		wt.%	F [nm]
Ni(acac)$_2$	0.275/1	Al(i-but)$_3$	0.594/3	100	20	10	0.85	Ni: 13.8	2–4
Fe(acac)$_2$	2.54/10	Al(me)$_3$	2.1/30	100	20	3	2.4	n.d.	
RhCl$_3$	0.77/3./1	Al(oct)$_3$	4.1/11.1	150	40	18	4.5	Rh: 8.5	2–3
								Al: 6.7	
Ag-neodecanoate	9.3/21.5	Al(oc)t$_3$	8.0/21.8	1000	20	36	17.1	Ag: 11.8	8–12
								Al: 2.7	
Pt(acac)$_2$	1.15/3	Al(me)$_3$	0.86/7.6	150	20	24	1.45	Pt: 35.8	2.5
								Al: 15.4	
PtCl$_2$	0.27/1	Al(me)$_3$	0.34/3	125	40	16	0.47	Pt: 41.1	2.0
								Al: 15.2	
Pd(acac)$_2$	0.54/1.8	Al(et)$_3$	0.46/4	500	20	2	0.85	Pd: 22	3.2
Pt(acac)$_2$	0.09/0.24							Pt: 5.5	
								Al: 12.7	
Pt(acac)$_2$	7.86/20	Al(me)$_3$	8.64/120	400	60	21	17.1	Pt: 20.6	1.3
Ru(acac)$_3$	7.96/20							Ru: 10.5	
								Al: 19.6	
Pt(acac)$_2$	1.15/2.9	Al(me)$_3$	0.86/12	100	60	2	1.1	Pt: 27.1	n.d.
	0.19/1							Sn: 5.2	
SnCl$_2$								Al: 14.4	

2.3.2.2 Hydrosols

As outlined above, the technically advantageous universal use of nanostructured metal particles requires decomposition-free redispersibility of the metal colloids in sufficiently high concentrations in a wide range of hydrophilic solvents, particularly water. Although we found ourselves unable to reveal the "exact structure" of the colloidal organoaluminum protective shell formed during the process of "reductive stabilization" depicted in Fig. 2.2, it is clear from quantitative protonolysis experiments that unreacted organoaluminum groups such as Al–CH$_3$, Al–C$_2$H$_5$, or Al–C$_8$H$_{17}$ are still present in the stabilizer. Table 2.3 summarizes a number of mono- and bimetallic organosols obtained via "reductive stabilization" which have been transformed via subsequent protonolysis by the "modifiers" listed in Table 2.4 to achieve hydrosols (cf. Fig. 2.3). Via this approach the "solubility-spectrum" of zerovalent nanometals can be tailored properly in a wide range of hydrophilic media including water (cf. Table 2.5).

Further, a careful TEM study has confirmed that the transformation of the Al–R into Al–OR groups in the protecting shell from the starting organosols to the resulting hydrosols leaves the size of the metallic core virtually untouched. This is exemplified in Fig. 2.13 with Pt/Ru, a relevant precursor to fuel cell catalysts.

The proton-active modifiers which were used for controlled protonolysis of the active Al–C bonds include: long chain alcohols, carboxylic acids, silanols, proton-active surfactants, and sugars (see Table 2.4).

Table 2.3 Starting materials: organometallic-prestabilized nanometal colloids.

No.	Metal salt Formula	g/mmol	Reductant Formula	g/mmol	Solvent Formula	ml	Conditions T [°C]	t [h]	Product* m [g]	Metal content, % by weight	Particle size [nm]	Id. #
1	Cr(acac)$_3$	2.5/7.2	AlMe$_3$	3.5/50	toluene	100	20	3	2.9			MK1
2	Fe(acac)$_2$	2.54/10	AlMe$_3$	2.1/30	toluene	100	20	3	2.4			MK2
3	Co(acac)$_2$	2.57/10	AlMe$_3$	3.5/50	toluene	100	60	3	4.3			MK3
4	Ni(acac)$_2$	2.57/10	AlMe$_3$	2.1/30	toluene	100	20	3	2.6			MK4
5	Ru(acac)$_3$	1.99/5	AlMe$_3$	1.05/15	toluene	100	60	24	2.0	Ru: 16.7 Al: 11.4		MK5
6	Ru(acac)$_3$	0.4/1	AlEt$_3$	0.51/45	toluene	125	20	16	0.8	Ru: 12.6 Al: 15.2		MK6
7	RuCl$_3$	0.21/1	AlEt$_3$	0.51/45	toluene	125	20	16	0.6	Ru: 16.8 Al:20.2		MK7
8	Rh(acac)$_3$	0.4/1	AlMe$_3$	0.63/9	toluene	100	60	22	0.5			MK8
9	Rh(acac)$_3$	0.2/0.5	AlEt$_3$	0.26/2.3	toluene	65	20	16	0.4	Rh: 12.9 Al: 15.2		MK9
10	RhCl$_3$	0.11/0.5	AlMe$_3$	0.16/2.3	toluene	65	40	19	0.2	Rh: 25 Al: 30.4		MK10
11	RhCl$_3$	0.21/1	AlEt$_3$	0.51/4.5	toluene	125	20	16	0.62	Rh: 16.6 Al: 19.6		MK11
12	RhCl$_3$	0.77/3.1	AlOct$_3$	4.1/11.1	THF	150	40	18	4.5	Rh: 8.5 Al: 6.7	2–3	MK12
13	Pd(acac)$_2$	0.3/1	AlMe$_3$	0.14/2	THF	300	20	5	0.39	Pd: 27 Al: 14		MK13
14	Pd(acac)$_2$	0.29/0.94	AlEt$_3$	0.21/1.9	toluene	250	20	18	0.4	Pd: 22 Al: 13		MK14
15	PdCl$_2$	0.18/1	AlEt$_3$	0.26/2.25	toluene	250	20	4	0.42	Pd: 23.2 Al: 21.3		MK15
16	Ag neodecanoate	9.3/21.5	AlOct$_3$	8.0/21.8	toluene	1000	20	36	17.1	Ag: 11.8 Al: 2.7	8–12	MK16
17	ReCl$_5$	0.36/1	LiBut	0.32/5	THF	100	60	36	0.5			MK17
18	ReCl$_5$	0.364/1	NaAlEt$_4$	0.83/5	toluene	150	60	90	0.6			MK18

Table 2.3 *Continued*

No.	Metal salt Formula	g/mmol	Reductant Formula	g/mmol	Solvent Formula	ml	Conditions T [°C]	t [h]	Product* m [g]	Metal content, % by weight	Particle size [nm]	Id. #
19	Ir(acac)$_3$	0.25/0.5	AlMe$_3$	0.16/2.25	toluene	65	60	16	0.35	Ir: 27.5 Al: 17.4		MK19
20	Ir(acac)$_3$	0.49/1	AlEt$_3$	0.51/4.5	toluene	125	80	16	0.9	Ir: 21.4 Al: 13.5		MK20
21	IrCl$_3$	0.3/1	AlEt$_3$	0.51/4.5	toluene	125	80	16	0.7	Ir: 27.5 Al: 17.4		MK21
22	Pt(acac)$_2$	7.88/20	AlMe$_3$	4.32/60	toluene	200	40	24	8.3	Pt: 42.3 Al: 17.5		MK22
23	Pt(acac)$_2$	3.9/10	AlEt$_3$	3.4/30	toluene	1000	20	16	6.4	Pt: 32.7 Al: 10.6	1.0	MK23
24	Pt(acac)$_2$	0.39/1	AlBut$_3$	0.59/3	toluene	125	20	16	0.86	Pt: 24.5 Al: 12.9		MK24
25	Pt(acac)$_2$	0.38/1	HAlEt$_2$	0.26/3	toluene	100	20	23	0.3			MK25
26	Pt(acac)$_2$	0.38/1	NaAlEt$_4$	0.50/3	toluene	100	60	12	0.8			MK26
27	Pt(acac)$_2$	0.38/1	MgEt$_2$	1.2/14.6	toluene	100	20	21	1.2	Pt: 14.9 Mg: 20.8		MK27
28	Pt(acac)$_2$	0.38/1	ZnEt$_2$	0.37/3	toluene	100	20	27	0.5			MK28
29	PtCl$_2$	0.27/1	AlMe$_3$	0.21/3	toluene	100	20	22	0.4			MK29
30	PtCl$_2$	0.27/1	AlMe$_3$	0.34/3	toluene	125	40	16	0.47	Pt: 41.1 Al: 15.2	2.0	MK30
31	PtCl$_2$	0.27/1	AlEt$_3$	0.34/3	toluene	125	20	16	0.52	Pt: 43 Al: 13.6	2.0	MK31
32	PtCl$_2$	0.27/1	AlBut$_3$	0.59/3	toluene	125	20	16	0.74	Pt: 26.4 Al: 10.9		MK32
33	PtCl$_2$	1.0/3.75	AlOct$_3$	2.7/7.5	THF	300	20	16	3.5	Pt: 20.9 Al: 5.8		MK33
34	Fe(acac)$_2$ Co(acac)$_2$	2.54/10 1.29/5	AlMe$_3$	5.4/75	toluene	200	20	3	4.9			MK34

Table 2.4 Modifiers.

No.	Substance class	Name	Trade name
1	alcohol	1-decanol	
2	carboxylic acid	2-hydroxypropionic acid	DL-lactic acid
3	carboxylic acid	cis-9-octadecenoic acid	oleic acid
4	silanol	triphenylsilanol	
5	sugar	D-(+)-glucose	grape sugar
6	polyalcohol	polyethylene glycol 200	PEG 200
7	vinyl pyrrolidone polymerizate	polyvinyl pyrrolidone K30	PVP, Polyvidon, Povidon
8	surfactant, cationic	di(hydrotallow)dimethylammonium chloride	Arquad 2HT-75
9	surfactant, cationic	3-chloro-2-hydroxypropyldimethyl-dodecylammonium chloride	Quab 342
10	surfactant, amphiphilic betaine	lauryldimethylcarboxymethylammonium betaine	Rewoteric AM DML
11	surfactant, anionic	Na cocoamidoethyl-N-hydroxyethylglucinate	Dehyton G
12	surfactant, non-ionic	decaethylene glycol hexadecyl ether	Brij 56
13	surfactant, non-ionic	polyethylene glycol dodecyl ether	Brij 35
14	surfactant, non-ionic	polyoxyethylene sorbitane monolaurate	Tween 20
15	surfactant, non-ionic	polyoxyethylene sorbitane monopalmitate	Tween 40
16	active charcoal		
17	silica		silica gel 60
18	alumina		

Even active OH groups present on inorganic surfaces have been shown to cleave the Al–C bonds in the colloidal protecting shell. This finding allows anchoring of nanosized metal particles of well defined sizes on inorganic catalyst supports.

2.3.2.3 Formation of Nanoparticle Networks

Bifunctional spacer molecules of different sizes have been used to build nanoparticle networks formed via self-assembly of arrays of metal colloid particles prepared via "reductive stabilization" [98–100]. The particles are interlinked through rigid spacer molecules with proton-active functional groups to bind at the active aluminum–carbon sites in the metal–organic protecting shells (Fig. 2.14).

The resulting networks have been characterized by various methods such as TEM, X-ray absorption spectroscopy (XAS), anomalous small angle X-ray scattering (ASAXS), metastable impact electron spectroscopy (MIES), and ultraviolet photoelectron spectroscopy (UPS). These investigations have been summarized in a recent microreview [98].

Table 2.5 Modification of organometallic-prestabilized metal colloids.

No.	Metal colloid Metal	Id. #	mmol	m [g]	Solvent Name	ml	Modifier Table 2, No.	m [g]	Temp. T [°C]	Time t [h]	Product [a] m [g]	Metal content %	A	B	C	D	E	F	G
1	Cr	MK 1	1	0.52	THF	200	13	2.0	60	16	3.2		–	+	+	+	–	–	–
2	Fe	MK 2	0.5	0.12	THF	100	1	2.0	60	16	2.0		+	+	+	+	–	+	–
3	Co	MK 3	1	0.43	THF	100	13	2.0	60	16	2.1		–	+	+	+	–	–	–
4	Ni	MK 4	1	0.39	THF	100	13	2.0	60	16	1.1		–	+	+	+	–	–	–
5	Rh	MK 8	0.5	0.25	THF	100	13	1.0	20	16	1.3		–	–	+	+	+	–	–
6	Pd	MK 13	1	0.39	THF	300	13	1.0	20	16	1.4		–	+	+	+	+	–	–
7	Pt	MK 22	0.25	0.1	THF	25	1	2.5	60	16	2.6		+	+	+	+	+	+	–
8	Pt	MK 22	0.5	0.21	THF	100	2	1.5	60	16	1.2		–	–	–	+	–	–	–
9	Pt	MK 22	0.5	0.21	THF	100	3	1.5	60	16	1.4		+	+	+	+	–	+	+
10	Pt	MK 22	0.5	0.21	THF	100	5	1.5	60	16	1.0		–	–	–	–	–	–	+
11	Pt	MK 22	0.5	0.21	THF	100	6	0.8	60	16	0.9		–	–	–	–	–	–	+
12	Pt	MK 22	0.5	0.21	THF	100	7	1.5	60	16	1.2		–	–	–	–	–	–	+
13	Pt	MK 22	0.2	0.08	THF	25	8	2.0	60	16	2.0		–	+	+	+	–	–	–
14	Pt	MK 22	0.5	0.21	THF	100	9	1.5	60	16	1.2		–	+	+	+	–	–	+
15	Pt	MK 22	0.2	0.08	THF	25	10	2.0	60	16	2.1		–	–	+	+	–	–	+
16	Pt	MK 22	0.2	0.08	THF	25	11	2.0	60	16	2.05		–	–	–	+	–	–	–
17	Pt	MK 22	0.25	0.105	THF	25	12	2.5	20	16	2.8	Pt: 9.3 Al: 5.6	–	–	–	+	+	–	+
18	Pt	MK 22	0.5	0.21	THF	100	13	0.4	60	16	0.5		–	+	+	+	–	–	+
19	Pt	MK 22	0.5	0.21	THF	100	14	0.8	60	16	0.81	Pt: 8.5 Al: 2.4	–	–	+	+	–	–	+
20	Pt	MK 22	0.2	0.08	THF	25	15	2.0	60	16	2.03		–	+	+	+	–	–	+
21	Pt	MK 23	0.33	0.2	THF	100	13	0.53	60	16	0.51		–	+	+	+	–	–	+
22	Pt	MK 25	0.33	0.1	THF	100	13	1.0	20	16	1.7		–	+	+	+	–	–	+
23	Pt	MK 26	0.5	0.35	THF	100	13	2.0	60	16	1.0		–	+	+	+	–	–	+
24	Pt	MK 27	0.5	0.56	THF	100	13	2.0	60	16	2.6	Pt: 4.6 Mg: 5.6	–	+	+	+	–	–	+
25	Pt	MK 29	0.9	0.15	THF	100	1	1.2	60	16	1.5		+	+	+	+	–	+	–
26	Pt	MK 30	1.0	0.47	toluene	100	4	1.0	60	3	1.3	Pt: 11.0 Al: 3.9	–	–	+	–	–	–	–
27	Fe₂Co	MK 34	0.5	0.136	THF	100	1	1.5	60	16	1.6		+	+	+	–	–	+	–
28	FeAu	MK 41	0.5/0.17	0.26	THF	100	13	0.8	60	16	2.17		–	+	+	+	+	–	–
29	PtRu	MK 36	1.0/1.0	0.94	THF	100	13	2.0	60	16	3.2	Pt: 6.3 Ru: 3.0 Al: 5.1	–	+	+	+	+	–	+[b]
30	Pt₃Sn	MK 39	0.5/0.17	0.36	THF	200	13	1.0	60	16	1.4	Pt: 6.8 Sn: 1.2 Al: 3.2	–	+	+	+	+	–	+

a May contain residual solvent;
b ethanol–water mixture (25% by volume of ethanol).
A=hydrocarbons, B=aromatics, C=ethers, D=alcohols, E=ketones, F=pump oils (Shell Vitrea Oil 100, Shell), G=water and aqueous solutions, +=solubility>100 mmol l⁻¹, –=insoluble.

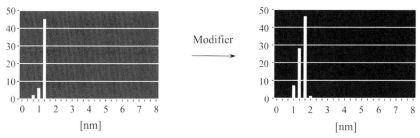

Fig. 2.13 Size conservation of colloidal Pt/Ru particles under the hydrophilic modification of the $(CH_3)_3$-Al-acac protecting shell using polyethylene glycol-dodecylether(*Brij 35*(r)). A good solubility can be observed: >100 mg/AT metal in aromatics, hydrocarbons, ethers, alcohols, ketones, and water.

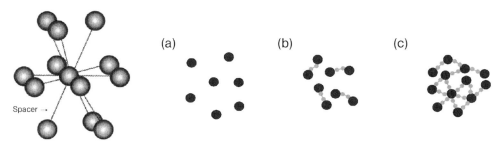

Fig. 2.14 Formation of particle networks via protolytic cross-linking with bifunctional spacers.

2.3.3
Shape-Selective Particle Synthesis

Shape-selective syntheses of nanosized metal particles are significant both from the fundamental point of view and for specific applications in nanocatalysis. Since the early work of Henglein and El-Sayed [101] until the present [85–87, 102] much work has been devoted to controlling the shape of the resulting nickel or platinum nanoparticles via slow diffusional growth in the presence of effective "capping agents" such as polyvinylpyrrolidone (PVP) in aqueous solution. Initially, small nuclei of the respective metals are formed during the (rapid) reduction step. Once this "seed" is formed, "controlled particle shaping" can be brought about in the presence of effective "capping agents" because the subsequent process of growth and ripening can be slowed down significantly, depending on the seed to PVP concentration ratio. In our experience [87], the most important factor in achieving shape and size control is to find the optimum concentration ratio of metal "seed" to "capping agent". "Seeding" is an important factor: without any Pt(0) nuclei added as "seed" with a Pt:PVP ratio of 10:1 we have obtained relatively large Pt

Pt-Precursors + PVP

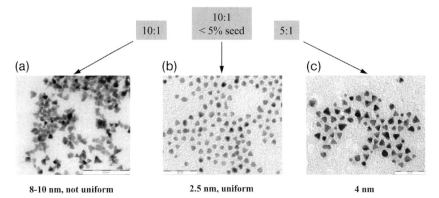

Fig. 2.15 Shape selective synthesis of Pt nanoparticles using PVP as the capping agent.

particles with very little shape control (Fig. 2.15(a)). After the addition, prior to reduction, of <5 % of the Pt in the form of small sized "nuclei" (= the seed) at the same Pt:PVP ratio, monodispersed Pt particles (2.5 nm) were obtained possessing exclusively a tetrahedral shape (Fig. 2.15(b)). When the amount of capping agent was substantially increased (Pt:PVP=5:1) tetrahedral Pt particles were produced with a size of 4 nm (Fig. 2.15(c)).

The chemical nature of the reducing agent also plays a (minor) role in shape control: Hydrogen – as a weak reducing agent – causes slow reduction kinetics, which favors a slow diffusional growth of the nanoparticles leading to a good monodispersity. Moreover, keeping the initial metal concentration low was shown to be favorable for growth of very monodisperse Pt tetrahedra via a uniform diffusional particle ripening process.

2.4
Applications in Catalysis

Lipophilic or hydrophilic nanostructured metal colloids dissolved in the form of organo- or hydro-sols can serve as homogeneous – or "quasi-homogeneous"[2] [103] catalysts either in organic, aqueous, or biphasic phases. Pre-prepared nanocolloids can readily be deposited on a diverse range of supports to give high-performance heterogeneous catalysts. To differentiate from classic heterogeneous catalysis we have coined the nanochemical bottom-up approach the "precursor method" [96]. In a recent state-of-the-art article, Astruc [27] thoroughly reviewed and evaluated the latest developments at the frontiers of "nanoparticles as recyclable catalysts". In this context, Astruc used the term "semi-heterogeneous" to describe the specific

2) Also termed soluble heterogeneous catalysts.

function of nanoparticulate catalyst systems [27, p. 7853]. The focus of this overview, however, is somewhat narrower covering the catalytic use of surfactant-, solvent- and organo-aluminum- stabilized nanoparticles in specific fields such as fine chemicals production, hydrogen storage, and electrocatalysis.

2.4.1
Quasi-homogeneous Catalysts

Early in the development of the field there were hopes that unsupported nanoparticles stabilized with surfactants and/or organo-aluminum-shells in liquid phases should work as "soluble surfaces" exhibiting very high catalytic turnover frequencies associated with reasonable lifetimes and, potentially, be recyclable [8–10, 104, 105].

In fact, Finke has reported remarkable catalytic lifetimes for the polyoxoanion- and tetrabutylammonium-stabilized transition metal nanoclusters in the homogeneous phase [26, 106–113].

Pd nanoparticles, formed via salt reduction with tetrabutyl ammonium carboxylates were reported by Reetz as highly efficient Heck reaction catalysts [114, 115].

Propylene carbonate-stabilized palladium nanoparticles were also shown to be active catalysts for the Heck reaction [116, 117]. The formation of olefins from aldehydes and ketones via McMurry-type coupling reactions was reported using Bu_4NBr-stabilized Ti colloids (3 nm) [22]. The THF-protected Ti_{13}-nanocluster (Fig. 2.8) was shown to hydrogenate Ti and Zr sponges in the "quasi-homogeneous" phase [82, 83]. Further, the THF-protected Ti_{13}-nanocluster has been found, by several groups, to be one of the best available catalysts for reversible hydrogen storage in alanates [118, 119]. The enantioselective hydrogenation of ethylpyruvate in HOAc/MeOH solution was also performed using cinchonidine-stabilized Pt colloids [120, 121]. The wide range of "homogeneous catalysis" applications for transition-metal nanoparticles stabilized by ligands, polymers or micelles has recently been reviewed by Astruc [27].

2.4.2
Heterogeneous Catalysts

Mono- and pluri-metallic nanometal colloids of defined size, intermetallic structure, and even shape can readily be deposited on supports to give heterogeneous "*egg-shell* nanometal-catalysts" which contain the active metal particles as a thin layer (<250 nm) on the surface of the respective carriers [12]. This new type of catalyst is manufactured simply by dipping the supports into dispersions of the above mentioned organo- or hydrosols (cf. Sections 2.3.1.1, 2.3.1.2, 2.3.2.1, and 2.3.2.2) in organic or aqueous media at ambient temperature to adsorb the pre-prepared particles. This has been exemplified with charcoal, graphene, various oxidic supports, and even low-surface materials such as quartz, sapphire, and highly oriented pyrolitic graphite (HOPG). This method of nanocatalyst preparation was first demonstrated using organosols in organic media [8, 10, 104, 122]. However, since due to both economical and ecological reasons the preferred solvent in catalyst technology is water, we have expanded our method using highly concentrated metal hydrosols in aqueous dispersion [11]. Heterogeneous *"egg-shell* nanocatalysts" are generated by a three step synthetic protocol:

1. Synthesis of nanometallic colloids having tailor-made size, mono- or pluri-metallic composition, intermetallic structure, and – eventually – specific shape ("precursors")
2. Deposition of the colloidal particles on a given support – usually via "dip-coating"
3. Final activation via reductive annealing at 300 °C to produce the active catalyst (so-called "conditioning").

As a result, high-performance heterogeneous nanocatalysts are obtained which have successfully been applied to hydrogenate C—C double bonds (stereo- and regio-selective), organic carbonyl groups, unsaturated C—N bonds, partial oxidation, and to reduce N—O bonds for the manufacture of fine chemicals (cf. Section 2.4.3.1) [123, 124].

Further, this technology was demonstrated to be especially advantageous for the production of high-performance fuel cell catalysts (cf. Section 2.4.3.2). An obvious advantage resides in the fact that both the size and the composition of the colloidal metal organo- or hydro-sols ("precursors") may be tailored independent of the support. The metal particle surface may subsequently be modified by "surface doping" (cf. Section 2.4.3.2). The industrial feasibility of preparing *"egg-shell* nano-metal catalysts" in this way has been demonstrated by Degussa [125].

2.4.2.1 The "Precursor Concept" [126]

The three-step "precursor concept" for manufacture of heterogeneous *"egg-shell* nanocatalysts" was developed in the 1990s [8–11]. The catalyst precursor is manufactured by dipping the supports into an organic or aqueous media containing the dispersed precursor at ambient temperature to adsorb the pre-prepared particles. A standard procedure for the manufacture of an *"egg-shell* Pt 3-nm nanocatalyst" via adsorption of a pre-prepared Pt hydrosol on e.g. Degussa active carbon 196 [127, 128] is:

1. Precursor synthesis:
   3-(N,N-Dimethyldodecylammonio)-propanesulfonate (SB12)-protected Pt colloid in water: A mixture of 1.4 g (5.3 mmol) of $PtCl_2$, 7.2 g (21.2 mmol) of 3-(N,N-dimethyldodecylammonio)-propanesulfonate (SB12), and 0.4 g (5.3 mmol) of Li_2CO_3 was stirred in 100 ml of H_2O; and H_2 was introduced under ambient pressure for 3 h at 20 °C. After approximately 30 min, a clear black solution was formed and all volatile compounds were evaporated in vacuum (0.1 Pa, 40 °C). The resulting black powder was fully redispersible in water. Yield was 8.4 g with 10.7% Pt content and a mean particle size of 3 ± 0.2 nm according to TEM.

Nano Particle   Protecting Shell

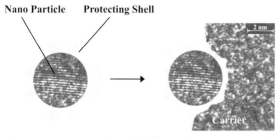

Fig. 2.16 Deposition of the colloidal "catalyst precursor" on the carrier.

2. Deposition on an active carbon support (cf. Fig. 2.16):
   The Pt hydrosol precursor (7.95 g) containing 0.850 g $Pt_{(0)}$
   was dissolved in 100 ml of cold UHQ water producing a
   clear, dark colloidal solution. 16.15 g of "active carbon carrier
   196 (Degussa)" was oxidized with NaOCl and suspended in
   100 ml UHQ water. Using a magnetic stirrer at a low
   rotational speed, the precursor solution was added dropwise
   over 1.5 h to the suspension of the carrier. The mixture was
   kept stirring for another 2 days and subsequently filtered.
   The filtrate was completely colorless indicating that the Pt
   precursor was quantitatively adsorbed on the carrier. The
   raw catalyst material was dried in vacuo for 3 h at 100 °C to
   yield 25.2 g (containing the colloidal stabilizer SB12 and
   residual moisture).
3. "Conditioning" to generate the active catalyst:
   A combination of AFM, STM and XPS has revealed [14, 96]
   the interaction of platinum hydrosols with oxide (sapphire,
   quartz) and graphite single crystal substrates. The metal
   core is immediately adsorbed e.g. on the HOPG surface
   when dipped into aqueous Pt colloid solutions at 20 °C. As
   shown in Fig. 2.17, the protecting shell (composed of either
   residual surfactants or al-organic residues) forms a carpet-
   like coat which cannot
   be removed from the particle surface, even by intense
   washing with solvents. The organic protecting shell
   decomposes on annealing at 280 °C and above in UHV. The
   thermal degradation was monitored by XPS up to 800 °C and
   by STM.

It was shown in this study [14, 96] that the Pt particles remain virtually unchanged
until ca. 800 °C. It should be mentioned here that the al-organic protecting shells
formed through "reductive stabilization" (cf. Section 2.3.2.1) are transformed into
thin layers of Al_2O_3 covering the nanoparticle surface during the "conditioning"
process. This effectively prevents particle agglomeration and sintering processes
(cf. Ref. [10] for details).

Supporting

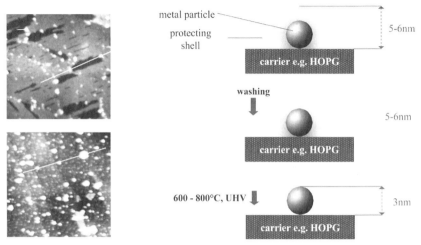

metal particle

protecting shell

carrier e.g. HOPG

5-6nm

washing

carrier e.g. HOPG

5-6nm

600 - 800°C, UHV

carrier e.g. HOPG

3nm

Fig. 2.17 STM investigation of the "precursor" approach to catalyst preparation.

In order to remove excess protecting shell the supported colloidal pre-catalysts – as obtained in step 2 – are activated via heat treatment at 300 °C in an *intermittent* stream of oxidizing and reducing gases. We have coined the term "conditioning" (i.e. reactive annealing) for this process [126].

Practically, 25 g of the SB12-stabilized Pt-pre-catalyst obtained in step 2 was treated in a tube furnace at 300 °C, first for 30 min under a flow of pre-heated argon and secondly for 30 min under a flow of pre-heated argon containing 3.5 vol.% oxygen. After an intermission of 10 min under flowing argon, pre-heated hydrogen was applied for another 30 min in order to re-reduce the particle surface to the zerovalent state.

In the given example, 17 g of the desired "egg-shell Pt nanocatalyst" was isolated after the "conditioning" procedure. Elemental analysis showed a Pt-loading of 5 wt.% on carbon while TEM confirmed the average particle size to be unchanged compared to the precursor (3 nm) and the zerovalent state of the Pt surface was confirmed by XPS.

The example shows that the three-step preparation procedure described above produces true "nanocatalysts" having naked metal particles of defined size deposited on the support surface. Generally, carbon-supported colloidal pre-catalysts are conditioned at 300 °C. However, individual heating and gas flow conditions may be optimized for every catalyst system on the basis of TGA-MS analysis data. For example, the optimum temperatures for "conditioning" supported nanometallic pre-catalysts having tetraoctylammonium or aluminum-organic protective shells are 280 °C and 250 °C respectively [96, 126].

Inspection of Fig. 2.18 shows the dramatic increase in catalyst activity effected by "conditioning" monitored by H_2-sorption. Using the "precursor concept" homo-

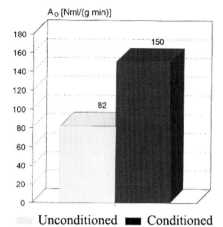

A_O [Nml/(g min)]

Unconditioned ▬ Conditioned

Fig. 2.18 Effect of "conditioning" on catalytic activity as determined by H_2 sorption. 20% Pt on vulcan relative activity determined by H_2-Sorption.

geneous and gradient structure alloys, layered bimetallics and decorated particles are all readily accessible on a wide range of supports. Since precise tailoring of size, structure, and composition is already done during the precursor synthesis, laborious manipulations and post-treatments to achieve specific particle sizes, structures, and compositions often applied in conventional catalyst preparation have become redundant.

2.4.3
Applications

To date, nanocatalysis has not found its way into bulk chemical synthesis applications. It can, however, be expected that true control of catalyst structure at the nanoscale, associated with a very uniform dispersion of the active nanoparticles in the homogeneous phase or with strong anchoring of the particles on heterogeneous supports, will prove successful sooner or later also at the ton-scale. Using TEM and other analytic tools we have recently found evidence for the formation of colloidal rhodium during catalytic hydroformylation – which is still the largest industrial application of homogeneous catalysis [129]. Recent findings using size-selective nano-type Fischer–Tropsch catalysts [32, 33] to convert CO and H_2 into hydrocarbons have yielded increased expectations for novel, advantageous heterogeneous catalyst systems based on nanochemical approaches.

2.4.3.1 **Fine Chemicals**
Ti_{13}-nanoclusters (Fig. 2.8) have been used as powerful activators for heterogeneous noble metal hydrogenation catalysis [9, 130]. In the so-called "butyronitrile hydrogenation standard test": the activity of surfactant-stabilized colloidal rhodium

(5 wt.% on charcoal) was found to surpass that of conventional salt impregnation catalysts of the same metal loading. The addition of 0.2% of colloidal Ti(0) to the supported noble metal resulted in a significant enhancement in activity. Additionally, the hydrogenation of acrylic acid can be promoted significantly by the addition of neodymium ions to the palladium particles [131]. Finally, the selective transformation of 3,4-dichloronitrobenzene to the corresponding aniline has been selected to test pre-prepared Pt hydrosols as heterogeneous catalyst precursors [132].

In batch and continuous tests the performance of the colloidal catalyst systems has been compared to conventional Pt/C-systems. The potential of the colloidal heterogeneous catalyst lies in the possibility of "fine tuning" the properties for specific applications by the addition of special dopants or "poisons" to the precursor. The influence of metal ions on the hydrogenation of *o*-chloronitrobenzene over platinum colloids, and the effect of metal complexes on the catalytic performance of metal clusters have also been demonstrated [133–135].

Since industrial processes often rely on alloy-like bimetallic catalysts [136–138] nanostructured bimetallic colloid catalysts have opened the possibility of differential studies on the mutual influence of the two different metals on the catalytic properties. The controlled co-reduction of two different metal ions has made bimetallic colloids readily accessible on the multigram scale. Bimetallic particles having a gradient metal distribution or a layered structure are most interesting for catalytic applications. In the catalytic hydrogenation of crotonic acid to butanoic acid a clear synergistic effect of Pt and Rh was observed [8] when bimetallic colloidal precursors ($Pt_{20}Rh_{80}$) were applied that have a gradient core–shell structure with an increase in the Rh concentration from the core towards the surface of the particle [38]. Toshima has discussed a similar effect in the partial hydrogenation of 1,3-cyclooctadiene with $Pt_{80}Pd_{20}$- and $Pd_{80}Au_{20}$- colloidal catalysts [139]. Synergistic effects in bimetallic precursors (e.g. Pt/Ru or Pt/Cu have been found to be most important in the broad field of electrocatalysis [cf. Section 2.4.3.2]).

A surfactant was found to control the selectivity in the cis-selective partial hydrogenation of 3-hexyn-1-ol giving leaf alcohol, which is a valuable fragrance (Eq. (2.5)) [140].

$$(2.5)$$

The performance in this reaction (Eq. (2.5)) of heterogeneous Pd colloid catalysts on $CaCO_3$ modified by a number of surfactants was compared with conventional Pd/C and Lindlar catalysts. The selectivity was found to depend on the support and various promoters. The highest activity and the best selectivity (98.1%) towards the desired *cis*-3-hexen-1-ol was found when employing a lead-acetate-promoted palladium colloid on $CaCO_3$ modified by the zwitterionic surfactant SB-12 (*N,N*-dimethyl-dodecylammoniopropanesulfonate). According to chemisorption results, small residues of the surfactant are still present on the surface of the immobilized particles. This colloid catalyst, being twice as active as a conventional Lindlar catalyst, surpassed its selectivity by 0.5%. Hydrophilic protecting shells strongly improve the contact of heterogenized metal colloid surfaces with substrates in aqueous media. A hydrophilic ruthenium colloid precursor supported on lanthanum oxide suspended in an aqueous solution of sodium hydroxide converts benzene into cyclohexene with 59% selectivity at 50% benzene conversion [141].

Superior catalytic oxidation catalysts were obtained when surfactant-stabilized Pd–Pt-precursors were supported on charcoal and promoted by bismuth. By comparison with industrial heterogeneous Pd/Pt catalysts, charcoal-supported NOct$_4$Cl Pd$_{88}$/Pt$_{12}$ alloy particles (1.5 to 3 nm), show an excellent activity combined with high selectivity in the glucose oxidation to gluconic acid by molecular oxygen (see Eq. (2.6)) [142].

$$
\begin{array}{c}
\text{CHO} \\
\text{—OH} \\
\text{HO—} \\
\text{—OH} \\
\text{—OH} \\
\text{CH}_2\text{OH} \\
\text{D-(+)-glucose}
\end{array}
\quad
\begin{array}{c}
+ \, 1/2 \, O_2 \quad \text{Pd-Pt-Bi/C} \\
+ \, \text{NaOH} \quad \xrightarrow{\qquad\qquad} \\
H_2O \\
\text{pH 9.5}
\end{array}
\quad
\begin{array}{c}
\text{COONa} \\
\text{—OH} \\
\text{HO—} \\
\text{—OH} \\
\text{—OH} \\
\text{CH}_2\text{OH} \\
\text{D-gluconic acid Na-salt}
\end{array}
\qquad (2.6)
$$

Contrary to a common prejudice, the colloidal Pd/C oxidation catalysts demonstrated substantially enhanced durability as compared to conventionally manufactured Pd/C catalysts under identical conditions [104]. Obviously, the lipophilic NOct$_4$Cl surfactant layer prevents the colloid particles from coagulating and being poisoned in the alkaline aqueous reaction medium.

Chiral molecules on the surface of the metal colloid can induce enantioselective control. Following this concept a new type of enantioselective platinum sol catalyst stabilized by the alkaloid dihydrocinchonidine was designed [120, 121]. Chirally modified Pt catalyst precursors have been prepared in different particle sizes by the reduction of platinum tetrachloride with formic acid in the presence of different amounts of the chiral alkaloid. Optical yields up to 80% *ee* were obtained in the hydrogenation of ethyl pyruvate. This type of catalyst was demonstrated to be structure insensitive since turnover frequencies (ca. 1 s^{-1}) and enantiomeric excess are independent of the particle size.

2.4.3.2 Electrocatalysis

During the last few decades Pt-based nanocatalysts have been introduced to Fuel Cell Technology to improve the exploitation of the expensive noble metal component via a substantial enhancement of the active catalyst surface using particles in the 2–5 nm range. [143, 144]. In essence, a fuel cell is an electrochemical reactor for the direct conversion of hydrogen or methanol into electricity [144]. Among the wide-ranging applications of fuel cells are low-emission transport systems, stationary power stations, and combined heat and power plants. While initial studies were carried out in the early 1900s and major innovations have been achieved in the last few decades, further developments are still needed, notably in the sectors of engineering, membranes, and, not least, optimized catalysts, before fuel cells will become an everyday commodity and the first electric cars powered by fuel cells can roll off the production lines. This is not expected to be seen before the year 2020. However, in the automotive sector, very promising innovations have recently been presented jointly by Volkswagen AG and the Swiss Paul-Scherrer-Institute (PSI) where a high-temperature phosphoric acid fuel cell (PAFC) operating at 160–220 °C is used with orthophosphoric acid as the electrolyte and a special high-temperature membrane [145]. In PAFCs the anode catalyst is Pt and e.g. Pt/Cr/Co has been proposed for the cathode [146].

For application in PAFC we have prepared a trimetallic colloidal precursor (size 3.8 nm) of composition $Pt_{50}Co_{30}Cr_{20}$ via the "reductive stabilization pathway" starting from the corresponding metal salts [147]. Conventional proton exchange membrane fuel cells (PEMFC) work at lower temperatures (80 °C). Pure Pt nanocatalysts can only be used here if ultra-pure hydrogen is available as the feed. Since industrial-grade H_2 (for thermodynamic reasons) is always contaminated with (even small) amounts of carbon monoxide CO-tolerant electrocatalysts had to be developed which rely on Pt-alloy electrocatalysts. These catalyst systems can also be applied for the conversion of reformer gas or methanol into electricity in so-called direct methanol fuel cells (DMFC). The final source of feed to be used for fuel cell based energy production is still under debate. It is, however, clear, that coal (through the water gas shift reaction), solar cells (for making H_2 via electrolysis), and the conversion of for example biomasses [148] will all contribute to a future "hydrogen economy". Initially, the focus of our investigations was nanometallic CO-tolerant bi- and tri-metallic fuel cell catalyst systems for the anode of PEM and DMFC [8, 14, 15, 29, 37, 126, 130, 149–161]. Recently, we have turned our attention to providing effective nanocatalysts for the oxygen reduction at the cathode [162]. Since an average automobile fuel cell would need to generate about 70 kW, requiring about 19 g of Pt, the question has been raised whether there are enough platinum reserves in the world to meet the future needs of fuel cell technologies [163]. To provide fuel cell technology with platinum-free nanocatalysts, we have synthesized a carbon-supported Ru catalyst where the 2.7 nm Ru particles are surface-doped with an *egg-shell* layer of Se. Modification of Ru nanoparticles with Se enhances their electrocatalytic activity in the O_2-reduction because Se inhibits surface oxidation. The Se-modified Ru/C samples with Se : Ru ratios from 0 to 1 were prepared by reacting carbon-supported Ru nanoparticles with SeO_2, followed by reductive annealing, and were characterized using high-resolution transmission electron microscopy, energy-dispersive X-ray, X-ray diffraction analysis, X-ray photoelectron spectroscopy, and extended X-ray absorption fine structure. The results suggest that Se interacts strongly with Ru, resulting in a chemical bond between Ru and Se and formation of Ru selenide clusters whose core, at low Se content, can be described as $Ru_2Se_2O_{0.5}$. At $Se:Ru=1$, high-resolution electron microscopy shows evidence of the formation of core–shell particles comprising a hexagonally packed Ru core and a Ru selenide shell with lamellar morphology [41, 164]. Size- and shape-selective electrocatalysts are amongst the focus of our current research. One of the best examples for particle size effects in oxygen reduction reactions catalyzed by Pt nanoparticles (1–30 nm) was recently published by Arenz [165 and references therein]. Applying cubo-octahedral Pt particles (average size: 4 nm), obtained via shape-selective synthesis as described above [87, 166], we have been able to find a real case of shape dependence in fuel cell catalysis [167] (cf. Fig. 2.19).

Size- and shape-controlled cubo-octahedral platinum$(_0)$ nanoparticles of 4 nm average size stabilized by sodium polyacrylate showing {111} and {100} surfaces were used to prepare Vulcan supported electrocatalysts. Cyclic voltammetric CO oxidation studies carried out by the thin film rotating disc method show two different sites of CO oxidation (Fig. 2.19). This can be assigned to differences in the activity of the crystal surfaces and is in agreement with single crystal studies. TEM results after cyclic voltammetric characterization show a complete absence of agglomerations.

Theoretical calculations have proven to be very useful in designing efficient FC catalysts based on high-throughput experiments generating trimetallic nanosystems having PtRu alloyed with Co, Ni, or W [168]. For example, the activity trend observed with real Pt, PtRu, PtRuNi, and PtRuCo catalysts generated by sputtering techniques corresponds exactly to the trends predicted by theory (see Fig. 2.20(a) and (b)).

In the following sections three typical examples of our nanochemical methodology for the preparation of advantageous fuel cell catalysts are discussed in the form of "case studies".

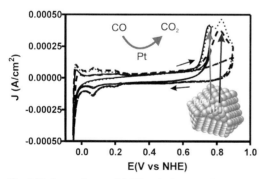

Fig. 2.19 Stepped potential CO-stripping experiment, 100 mV s^{-1} in 1 M H_2SO_4, argon-purged solution. The sweep window was increased from 0.75 V vs. NHE (continuous line) to 0.9 V after the first two cycles (dashed line). The dotted line is the complete CV going directly to 0.9 V without the pre-step at 0.75 V.

(a)

(b)

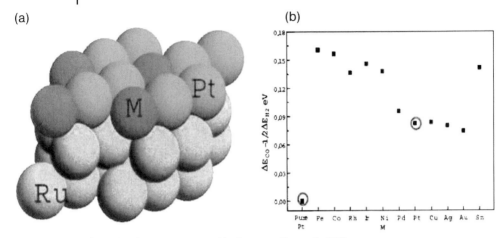

Fig. 2.20 Schematic structure (a) of ternary alloy on Ru(001) surface, MPt₂/Ru, with M selected from the set {Fe, Co, Rh, Ir, Ni, Pd, Pt, Cu, Ag, Au, Sn}.The two bottom layers represent Ru atoms while Pt atoms form the top surface atoms. The predicted surface activities of various ternary alloys are shown in the plot (b). [Reproduced with permission from the American Chemical Society (168)].

Fig. 2.21 Preparation of 30 wt.% Pt₅₀Ru₅₀/Vulcan XC 72 (Cat. 1) via the ammoniohydroborate co-reduction pathway.

Case Study 1: Pt/Ru@Carbon Three types of 30 wt.% Pt₅₀Ru₅₀/Vulcan XC 72 electrocatalysts for Direct Methanol Oxidation Fuel Cells (DMFC) were prepared using NR₄Bet₃H (Cat. 1), LiBet₃H (Cat 2.), and Al(Me)₃ (Cat 3.) for the co-reduction of Pt- and Ru-salts [169]:

Experimental: The three catalyst precursors were synthesized using Pt- and Ru-chlorides as the starting material for the co-reduction with N(oct)₄Bet₃H (Cat.1), LiBet₃H (Cat.2), respectively and the acetylacetonate derivatives for Al(me)₃ core-duction (Cat. 3). The as-prepared colloidal catalyst precursors were supported on Vulcan XC72, as exemplified for (Cat 1) in Fig. 2.21, and activated via "conditioning" at 300 °C (cf. Section 2.4.2.1).

All three catalysts were characterized *ex situ* by scanning electron microscopy (SEM) coupled with energy dispersive X-ray analysis (EDX) and X-ray diffraction

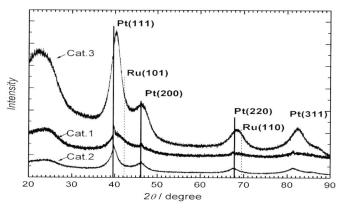

Fig. 2.22 Comparison of the XRD powder pattern for Cat.1, Cat.2, and Cat.3 obtained by the respective co-reduction routes.

(XRD) and, further, by *in situ* CO stripping and cyclic voltammetry. Subsequently, the activity towards methanol electro-oxidation was tested in steady-state experiments at 22 °C and 60 °C. Thus, a structure–activity profile was developed based on the results.

Results: EDX indicated that all three catalysts had identical metal loadings and confirmed a uniform distribution of the bimetallic particles on the Vulcan XC72 surface. Particle size and Pt and Ru content were determined via XRD, as shown in Fig. 2.22. The first peak at $2\theta \approx 25°$ originates from Vulcan XC-72 carbon support while the other peaks are reflections of the face centered cubic (f.c.c.) crystal lattice of Pt (vertical lines are pure Pt references).

For Cat. 1 and 2 all peaks appear at similar 2θ values, indicative of pure platinum (for example, the f.c.c. (220) structure. The diffraction peak of Cat.1, Pt f.c.c (220) was observed at $2\theta \approx 67.8°$ and for Cat. 2 it appears at a 2θ value of 67.6° while the Pt f.c.c. (220) reference is at $2\theta \approx 67.5°$. In the case of Cat. 3, the reflections are shifted towards higher angles compared to that of pure Pt but appear below the Ru (101) phase. The Pt f.c.c. (111) diffraction peak of Cat. 3 appears at $2\theta \approx 40.4°$ and is between the diffraction peaks for Pt f.c.c. (111) at $2\theta \approx 39.8°$ and the diffraction peak for Ru h.c.p. (101) at $2\theta \approx 42.2°$. The f.c.c. (220) diffraction peak of the sample appears at $2\theta \approx 68.8°$ which is between the diffraction peaks for Pt f.c.c. (220) at $2\theta \approx 67.5°$ and the diffraction peak for Ru h.c.p. (110) at $2\theta \approx 69.4°$. The lattice parameter values are given in Table 2.6.

For Cat. 1 and Cat. 2, the lattice parameter values are between the lattice parameter of pure Pt and that of the PtRu alloy with reflections corresponding to 90.3 at.% Pt, indicating a low extent of alloying. In the case of Cat. 3, the lattice parameter corresponds to a Pt content of approximately 48.3 at.% Pt, which indicates a true alloy composition. In all samples, no peak reflections of the h.c.p. crystal structure of Ru were seen. The average particle sizes were estimated by using the Scherrer equation (Table 2.6). The average particle sizes were in the

Table 2.6 Characterization of the prepared PtRu catalysts: Pt content (total, EDX; surface, XRD), mean particle sizes (*d*, XRD), lattice parameters (*a*, XRD).

Catalyst	Reducing agent	Pt (at%)		d (nm) XRD	a (nm) XRD
		EDX	XRD		
Cat. 1	NR₄Bet₃H	52.8	90.3	3.6	0.391
Cat. 2	LiBet₃H	50.3	90.3	3.9	0.391
Cat. 3	Alme₃	54.0	~48.3	2.9	0.3859

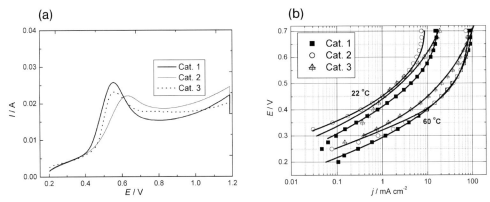

Fig. 2.23 (a) Comparison of the CO stripping of the three colloidal catalysts. (b) Steady-state polarization curves along with simulated curves for methanol oxidation of the three different 30 wt.% PtRu/Vulcan XC-72 catalysts (symbols) at two different temperatures (22 and 60 °C). Conditions: fixed delay of 5 min, methanol concentration 1 M in the working electrode compartment, flow rate 10 L h⁻¹.

range 2.9–3.9 nm leading to approximately the same surface areas for all catalysts. The electrochemical performances initially indicated that all three catalysts were very active for methanol oxidation. The CO stripping experiment and the steady-state curves, along with simulated curves for methanol oxidation at room temperature and 60 °C for the different catalysts are shown in Fig. 2.23.

The CO stripping peak (Fig. 2.23(a)) potential is directly dependent on the PtRu surface composition. In addition, concerning the CO stripping peak potential, it was found that for the same overall PtRu composition, the alloyed PtRu catalyst has a more positive peak potential than a non-alloyed Pt + Ru catalyst obtained by physical mixing of Pt and Ru colloidal catalysts. In short, first, the most negative peak potential towards CO oxidation appears at a surface with a Ru-content of approximately 50 at.% and second, the alloyed catalyst is less active than the non-alloyed catalyst. Based on the CO stripping analysis, one can conclude that Cat. 2 has a higher Ru surface composition than the other two catalysts.

Experimental steady-state curves (symbols), along with simulated curves (dark lines) for methanol oxidation at room temperature and 60 °C for the different catalysts are shown in Fig. 2.23(b). The values of potential (E in Volts) versus the current density (mA cm^{-2}) shows the most active catalyst at both room temperature and 60 °C is Cat. 1 (i.e. more current density at lower voltages), while the least active is Cat. 3. The activity of Cat. 2 at room temperature is less than that of Cat. 1 but at 60 °C shows a remarkably similar activity to that of Cat. 1. This indicates that under the set electrochemical conditions, there is a change in the composition of this catalyst at elevated temperatures.

Conclusions: All three 30 wt.% Pt$_{50}$Ru$_{50}$/Vulcan XC 72 electrocatalysts (Cat.1–3) generated by the respective reduction agents via the colloidal precursor route showed good electrocatalytic activity in methanol oxidation. Small differences in the surface structures, which are a direct consequence of the synthetic methodology used, is likely the reason for the differences in their electrochemical properties. Cat. 2 which exhibited segregated individual metal atoms to a greater extent demonstrated the worst activity while Cat. 1 with intermetallic Pt–Ru exhibited pronounced activity. These comparisons strongly suggest that a real Pt/Ru "alloy" composition in the nanoparticles is not a necessary prerequisite to accomplishing good DMFC activity. It is enough to generate a "bimetallic" Pt–Ru particle structure [161].

Case Study 2: Carbon (Vulcan XC72) Supported Bimetallic Electrocatalysts. Preparation by the borate method of Pt–Cu colloidal catalysts having structural variations and comparative study of their electrochemical performance.

In order to study the effect of structural variations on the electrochemical performance of bimetallic nano-electrocatalysts we have carried out a comparative examination of two types of Pt/Cu@Vulcan XC 72 systems in the hydrogen oxidation reaction (HOR) [170].

Experimental: Pt/Cu salt co-reduction or, alternatively, consecutive reduction of Cu(acac)$_2$ and PtCl$_2$ allows generation of either alloyed Pt/Cu and/or "onion-type" Pt-on-Cu@Vulcan XC72 (20 wt.%, Pt:Cu = 50:50 a/o) electrocatalysts. XRD, XPS, TEM, SEM, and electrochemical measurements were used for thorough catalyst characterization. Alloyed Pt/Cu precursors (average particle size 3.6 nm) were obtained via co-reduction of copper acetylacetonate and platinum chloride with alkalitriethylborohydride giving bimetallic nanoparticles well dispersed on Vulcan XC 72 and ready for "conditioning" (for experimental details see Ref. [170]. For generating onion-type Pt@Cu core–shell structured electrocatalysts a stepwise reduction process in the presence of Vulcan XC72 was performed. The support was first well dispersed in dry THF, then copper acetylacetate was added to the carbon suspension, followed by LiBet$_3$H dissolved in THF. The Vulcan XC-72-supported copper nanoparticles on carbon were isolated, then re-dispersed in fresh THF, followed by introduction of $PtCl_2$ to the sonicated suspension. Finally, $LiBet_3H$ was added to produce $Pt(0)$. Lithium chloride was washed away by applying excess THF repeatedly. The raw catalyst was activated by "conditioning" at 300 °C using argon, oxygen and hydrogen for 30 min each.

Results: TEM characterization indicated a uniform distribution of metal particles on Vulcan XC 72 (Fig. 2.24). The carbon-supported "homogeneously alloyed" Cu/Pt nanoparticles (Fig. 2.24(a)) exhibit a mean particle size of 3.6 nm (occasionally agglomerated); the "onion-type" Pt@Cu type catalyst (Fig. 2.24(b)), in contrast, shows a broad particle size distribution (up to 12 nm).

Figure 2.25 provides a comparative XRD study of "homogeneously alloyed" Cu/Pt/C system (a) and the "onion-type" Pt@Cu electrocatalyst (b).

(a)   (b)

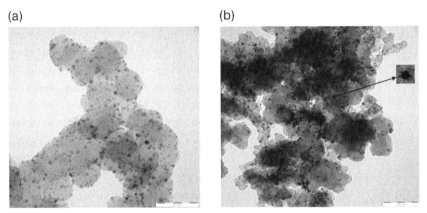

Fig. 2.24 TEM of "homogeneously alloyed" Cu/Pt nanoparticles on carbon (a) and the "onion-type" Pt@Cu electrocatalyst (b). The inset on micrograph (b) shows small Pt nanoparticles (1–2 nm) decorating larger copper nanoparticles (6–8 nm).

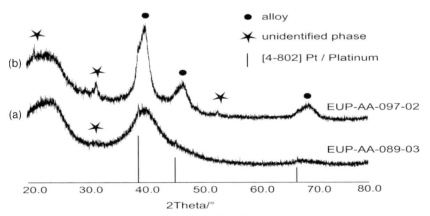

Fig. 2.25 XRD patterns of (a) "homogeneously alloyed" Cu/Pt nanoparticles on carbon and (b) an onion-type Pt@Cu electrocatalyst (20 wt.% metal loading in either case).

The characteristic diffractions peaks of pure Pt (111) at $2\theta = 39.76°$, Pt (200) 46.24° and Pt (220) at 67.46° are slightly shifted to higher 2θ values, indicating the successful reduction of Pt to its metallic state. The modification of the crystal lattice indicates alloying with Cu. The broad peaks that appear at $2\theta = 25°$ are attributed to the presence of Vulcan XC 72 carbon as the supporting material and the scattering contribution of the amorphous glass capillary. The XRD patterns show that the "onion-type" Pt@Cu catalyst contains larger metal particles than the "homogeneously alloyed" Pt–Cu catalyst. The full-width-half-maximum of the reflections in the "onion-type" catalyst is much smaller whereas for the "homogeneously alloyed" type catalyst the reflections are very broad with overlapping of signals. Moreover, a clearly pronounced shoulder on the low angle side of the main peak reveals that the "onion-type" catalyst material contains both reduced Pt^0 and alloyed Pt/Cu nanoparticles. From the powder pattern of the "homogeneously alloyed" catalyst sample, a slightly different result is obtained. Very sharp but weak reflections, exactly at the reflection positions for pure platinum, are positioned on top of the broad reflections of the alloy. This indicates the presence of a small number of larger platinum crystals amongst much smaller Pt/Cu nanoparticles. Determination of the particle sizes was not possible due to extensive overlapping of the reflections belonging to two different phases. It is well understood that Cu and Pt are completely miscible, and at low temperatures three ordered phases (Cu_3Pt, CuPt and $CuPt_3$) are found. Since a temperature of 300 °C is applied during the "conditioning" process, the presence of non-alloyed intermetallic Pt–Cu phases is highly improbable. During the sequential reduction of Cu- and Pt-salts generating the "onion-type" structure, alloy formation can occur via the diffusion of Pt atoms from the surface into the Cu core resulting in the formation of a disordered Pt–Cu "surface alloy" during the "conditioning" procedure.

In the case of Pt, XPS analysis confirmed the metallic state while both metallic and oxidized forms of Cu are found to be present in both catalyst systems. The HOR reactions (linear sweep voltammetry) are presented in Fig. 2.26. These curves confirm the powerful catalytic action of both types of Pt–Cu catalyst.

Conclusions: The presence of Cu in the vicinity of Pt affects the electronic structure of Pt. It can be noticed that the relative activities (cf. Fig. 2.26) of the Cu-containing catalysts are more than 50% higher than the industrial benchmark (28.6 wt.% Pt/C (HP). The "onion-type" 20 wt.% Pt@Cu/Vulcan XC 72 catalyst (50:50 a/o) showed the best performance as was confirmed by the analysis of the Koutecky–Levich plots (cf. Ref. [170]). The intercepts of the Cu-containing catalysts are lower than the intercepts of the industrial benchmark Pt/C catalyst although the Pt loading in the benchmark catalyst is nearly twice that in the discussed Cu-containing systems. Also, the slopes for the Cu-containing catalysts are substantially higher than those of the industrial benchmark. This again confirms the superiority of the Cu-containing catalysts to Pt/C systems applied in HOR. Both types of Cu-containing catalysts exhibit improved CO tolerance compared to "pure" Pt@Vulcan catalysts. The CO stripping peak of the "homogeneously alloyed" Pt/Cu catalyst is lower by 0.2 V than the CO stripping peak found with a pure Pt standard catalyst. Further, the CO oxidation ("ignition") potential of the

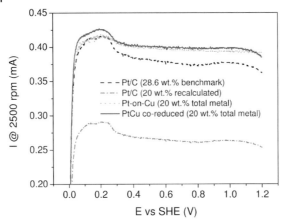

Fig. 2.26 Comparison of anodic polarization curves for three electrocatalysts. The dashed line represents the current industrial benchmark catalyst (28.6 wt.% Pt/Vulcan XC 72); the dash-dot line represents the identical catalyst normalized to 20 wt.% Pt; the dotted line represents the "onion-type" 20 wt.% Pt@Cu/Vulcan XC 72 catalyst (50:50 a/o); the full line represents the "homogeneously alloyed" type 20 wt.% Pt/Cu/Vulcan XC 72 catalyst.

"homogeneously alloyed" Pt/Cu catalyst is lower by 0.1 V than that found with a pure Pt standard catalyst. The "onion-type" Pt@Cu catalyst shows, however, less CO tolerance than the comparative "homogeneously alloyed" Pt/Cu system. Thus, it can be concluded that with "homogeneously alloyed" Pt/Cu catalysts, CO oxidation is enhanced by the affinity of CO to Cu. Further, the affinity of Cu for adjacent OH may enable a Langmuir–Hinshelwood type mechanism to occur during CO oxidation. In addition, gradual leaching of Cu may create a "sub-nanoporous" roughness at the Pt nanoparticle surface, which in turn can drastically increase the catalytically active surface area. The sum of all these sub-effects which are introduced by Cu into nanosized Pt electrocatalysts obviously helps to optimize the HOR performance.

Case Study 3: Surface-doped Pt/Ru/Co@Carbon   Based on the above-mentioned DFT calculations performed by Nørskov [168] we have prepared trimetallic electro-catalysts having PtRu/C "surface-doped" with Co(0) in order to produce highly active but at the same time CO tolerant electrocatalysts. For example, Pt/Ru/Fe/C, Pt/Ru/Ni/C, and Pt/Ru/Co/C systems were manufactured with the metal ratios being 45:45:10 a/o and a total metal loading of 20 wt.% on Vulcan XC 72. The resulting catalysts were compared with the industrial $Pt_{50}Ru_{50}$ standard catalyst under identical conditions. Full characterization was done via a combination of TEM, XRD, XPS, and XAS measurements, further BET, and electrochemical tests [171].

Experimental: Initially a PtRu/C catalyst was generated using $RuCl_3$ and $N(oct)_4Bet_3H$ in THF solution to form the PtRu-precursor which was supported on the Vulcan XC 72 carrier. After the "conditioning" step the PtRu/C material was redispersed in THF. Then, the respective amount of zerovalent Fe-, Ni-, or

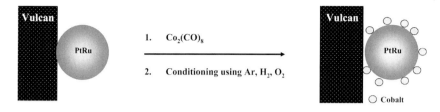

Limiting Currents

Pt:Ru (50:50 a/o)	100%
Pt:Ru:Fe (45:45:10 a/o)	108.4%
Pt:Ru:Ni (45:45:10 a/o)	118.3%
Pt:Ru:Co (45:45:10 a/o)	136.4%

Fig. 2.27 Preparation scheme for surface-doped trimetallic electrocatalysts (e.g. Co@Pt/Ru/Vulcan XC 72); comparison of electrocatalyst performances ("limiting currents").

Co-carbonyl compounds, e.g. $Co_2(CO)_8$, dissolved in THF was added and the whole mixture was stirred vigorously for 6 h at room temperature. The solvent was evaporated and the material "conditioned" at 300 °C in the following order: 10 min under flowing argon, 30 min under hydrogen, 5 min intermittent sweep with argon, 10 min under oxygen, 30 min under hydrogen, and finally 10 min under argon (cf. Fig. 2.27).

Although a detailed evaluation of the benefits of "surface doping" for the optimization of anode and cathode electrocatalysts to efficiently convert CO-containing H_2 (reformate-gas) or methanol into electricity is reserved for an *in extenso* paper which is currently in preparation, it can already be stated that the method of "surface-doping" using zerovalent transition metals or e.g. selenium [41, 164] is a versatile tool that affords the ability to "tailor" catalyst surfaces. Thus, applying "additives" will be the subject of future efforts in "tailoring" catalyst properties specific to FC applications (high temperatures, tolerant to poisons in the feed, e.g. CO, CO_2, sulfur etc.) and may provide an opportunity to bridge the gap towards real world conditions with real feedstocks with acceptable long term stabilities. For successful application in vehicles, FC power sources have to compete with the 3000 working hours currently achieved with conventional combustion motors. Attaining this goal will require a significant breakthrough in the field of FC catalyst development and will most likely involve the nanoscale "tailoring" of the catalyst system.

2.5
Miscellaneous Applications and Future Outlook

Nanoscale syntheses will be the key components in the ultimate economic viability of photovoltaic electricity production via dye sensitized solar cells (DSSC) [172].

This technology is primarily based on nano-TiO_2, nano-SnO_2 and nano-Pt. Recent progress in the manufacture of cost-effective, long-term stable DSSC has been achieved using screen-printing techniques [173]. Optimized production conditions of screen-printed cells, however, require the accessibility of 1–2 nm Pt nanoparticles where the surface is truly in the zerovalent state. Using our preparation pathways we have been able – in close cooperation with the Fraunhofer Institute ISE (Freiburg, Germany) in the frame of the "Dye Solar Cell Network" – to develop SnO_2 doped with a nano-Pt layer which could be transformed into a paste to give an effective catalytic layer for iodide/triiodide reduction [81]. Using nanosized $Pt_{(0)}$ precursors as the catalysts has advantages over conventional platinum catalyst layers. According to our comparative study [81], the best electrochemical performance for I^-/I_3^- reduction was obtained using catalytic layers which were fabricated using nanoclusters gained from either the "polyol process" ($0.4 \Omega cm^2$) or via "hydrogen reduction" ($1.1 \Omega cm^2$) [174, 175]. Currently, we have focused our attention on the manufacture of a new type of "nano solar cells" based on low cost electrode layers based entirely on nano-metal oxides. Through this approach, the costly Ru-based dye and any precious metals will be completely avoided in the future.

Conventionally prepared cobalt and other elemental magnetic nanoparticles (Fe, Ni, etc.) are too costly due to their relatively low stability toward oxidation under ambient conditions [176]. Noble metal-alloyed particles such as Co_3Pt or Fe_xPt_{1-x}, metal-core cobalt or Fe/Co alloy particles prepared via controlled decomposition in the presence of aluminum alkyls and stabilized against air and moisture via subsequent "smooth oxidation" have been successfully applied and harnessed in a variety of applications [75, 177]. For example, the improved magnetorheological properties of these types of magnetic colloids are to be exploited in high-performance bearings and seals. Moreover, manipulating these nanoparticles – eventually covered with biocompatible coats such as L-cysteine or dextrane – by applying an external field opens up a broad field for application in biomedicine. Preliminary tests have confirmed that metallic-core magnetic particles after "smooth oxidation" and coating the surface with e.g L-cysteine or dextrane are even less toxic than conventional magnetite systems [178]. This conjures up hopes of applying metallic-core magnetic nanomaterials coupled with drugs, antibodies [179] or with oligonucleotides for cell-specific cancer-diagnosis and also therapy, namely via hyperthermia [180] or as specific "nanosensor" systems [181]. With the help of such nanometallic tools, diseases or predispositions for diseases could potentially be discovered earlier than at present.

All in all, we can say that the scientific, technical, and economical opportunities have just started to emerge from a fruitful collaboration of the classical disciplines such as physics, chemistry, engineering, biology, and medicine at the nanoscale.

Acknowledgments

The authors want to express their sincere thanks to all co-workers and scientific collaborators mentioned in our group's references below for their valuable contributions and discussions over the years. Namely, we want to thank Dr. R. Mynott, Dr. B. Tesche, and Dr. C. Weidenthaler and their group members at the Max-Planck-Institut für Kohlenforschung, Mülheim a. d. Ruhr, Germany, for the measurement and interpretation of NMR, Electron Microscopy, XRD, and XPS data, respectively. We are deeply indebted to Prof. J. Hormes, P. D. Dr. H. Modrow and their co-workers at the Institute of Physics, University of Bonn, Germany and the CAMD at the University of Louisiana, Baton Rouge, USA for numerous XAS investigations. The contribution of significant electrochemical results by Prof. R. J. Behm and his team at the Institute of Surface Chemistry, University of Ulm, Germany, Dr. Stanko Hocevar's group at the National Institute of Chemistry, Ljubljana, Slovenia, Dr. Stylianos Neophytides at the ICEHT-FORTH, Patras, Greece, Drs. A. L. Antozzi and G. N. Martelli at DeNora Tecnologie Elettrochimiche, S.r.l., Milano, Italy, and Prof. Dr. K. Sundmacher with Dr. T. Vidakovic at the Max-Planck-Institut für Dynamik Komplexer Technischer Systeme, Magdeburg, Germany, is also highly acknowledged. Last, but not least we want to thank Dipl.-Ing. R. Brinkmann for the illustration of this chapter (Figures 2.1–2.19, Figures 2.21 and 2.27). This work would not have been possible without financial support of DFG through the Priority programs SPP 1060 (1998–2004), SPP 1072 (1998–2004), and SPP 1104 (2000–2006), the BMBF Networks "Dye Solar Cells" (2004–2006) and "O_2 RedNet" (2004–2006) and the European Union 5th FWP acronym APOLLON (2002–2004) under contract Nr. ENK-CT-2001-00572. All these contributions are highly acknowledged.

References

1 G. Schmid (Ed.), *Clusters and Colloids*, Wiley-VCH, Weinheim, **1994**.

2 G. Schmid (Ed.), *Nanoparticles: From Theory to Application*, Wiley-VCH, Weinheim, **2004**.

3 G. Schmid (Ed.), *Nanotechnology*, Springer, Berlin, **2006**.

4 H. Bönnemann, W. Brijoux, R. Brinkmann, E. Dinjus, R. Fretzen, T. Joussen, R. Korall, *J. Mol. Catal.*, **1992**, *74*, 323.

5 M.-P. Pileni, *J. Exp. Nanosci.*, **2006**, *1(1)*, 13.

6 B. Chaudret, *C. R. Phys.*, **2005**, *6*, 117.

7 B. Chaudret, Surface and Interfacial Organometallic Chemistry and Catalysis *Top. Organomet. Chem.*, Springer Verlag, **2005**, *16*, 233.

8 H. Bönnemann, W. Brijoux, R. Brinkmann, R. Fretzen, T. Joussen, R. Köppler, P. Neiteler, J. Richter, *J. Mol. Catal.*, **1994**, *86*, 129.

9 H. Bönnemann, G. Braun, W. Brijoux, R. Brinkmann, A. Schulze Tilling, K. Seevogel, K. Siepen, *J. Organomet. Chem.*, **1996**, *520*, 143.

10 H. Bönnemann, W. Brijoux, in *Active Metals*, A. Fürstner (Ed.), Wiley-VCH, Weinheim, **1996**, p.339.

11 H. Bönnemann, W. Brijoux, in *Advanced Catalysts and Nanostructured Materials*, W. Möser (Ed.), Academic Press, San Diego, **1996**, Ch. 7, pp. 174, 189.

12 H. Bönnemann, W. Brijoux, in *Metal Clusters in Chemistry*, Vol. 2, P. Braunstein, L. A. Oro, P. R. Raithby (Eds.), Wiley-VCH, Weinheim, **1999**, 913.

13 U. Kreibig, H. Bönnemann, J. Hormes, in *Handbook of Surfaces and Interfaces of Materials*, H. S. Nalwa (Ed.), Academic Press, San Diego, **2001**, Vol. 3, p. 1.

14 R. Richards, R. Mörtel, H. Bönnemann, *Fuel Cell Bull.*, **2001**, *37*, 7.

15 H. Bönnemann, K. S. Nagabhushana, in *Dekker Encyclopedia of Nanoscience and Nanotechnology*, J. A. Schwarz, C. I. Contescu, K. Putyera (Eds.), Vol. 1, Marcel Dekker, New York, **2004**, Vol. 1, p. 739.

16 H. Bönnemann, K. S. Nagabhushana, in *Encyclopedia of Nanoscience and Nanotechnology*, H. S. Nalwa (Ed.), American Scientific Publishers, Stevenson Ranch, CA, **2004**, Vol. 1, p. 777 and references therein

17 M. T. Reetz, W. Helbig, S. A. Quaiser, in *Active Metals*, A. Fürstner (Ed.), VCH, Weinheim, **1996**, p. 279.

18 M. T. Reetz, W. Helbig, *J. Am. Chem. Soc.* **1994**, *116*, 7401.

19 M. T. Reetz, W. Helbig, S. Quaiser (to Studiengesellschaft Kohle), U.S. Pat. 5 620 564 (Apr. 15, **1997**) and U.S. Pat. 5 925 463 (Jul. 20, **1999**).

20 M. A. Winter, PhD Thesis, Verlag Mainz, Aachen, ISBN 3-89653-355, **1998**.

21 J. A. Becker, R. Sscäfer, W. Festag, W. Ruland, J. H. Wendorf, Pebler, S. A. Quaiser, W. Helbig, M. T. Reetz, *J. Chem. Phys.* **1995**, *103*, 2520.

22 M. T. Reetz, S. A. Quaiser, C. Merk, *Chem. Ber.*, **1996**, *129*, 741.

23 M. T. Reetz, W. Helbig, S. A. Quaiser, *Chem. Mater.*, **1995**, *7*, 2227.

24 M. T. Reetz, S. A. Quaiser, *Angew. Chem.* **1995**, *107*, 2461; *Angew. Chem. Int. Ed. Engl.* **1995**, *34*, 2240.

25 U. Kolb, S. A. Quaiser, M. Winter, M. T. Reetz, *Chem. Mater.* **1996**, *8*, 1889.

26 B. J. Hornstein, R. G. Finke, *Chem. Mater.*, **2003**, *15*, 899.

27 D. Astruc, F. Lu, J. R. Aranzaes, *Angew. Chem., Int. Ed.*, **2005**, *44*, 7852.

28 K. S. Nagabhushana, H. Bönnemann, in *Nanotechnology in Catalysis*, Vol. 1, B. Zhou, S. Hermans, G. A. Somorjai (Eds.), Kluwer Academic/Plenum Press, New York, **2004**, p. 51.

29 H. Bönnemann, R. Richards, in *CAtalysis and ElectrocAtalysis At Nanoparticle Surfaces*, A. Weikowski, E. R. Savinova, C. G. Vayenas (Eds.), Marcel Dekker, New York, **2003**, p. 343.

30 H. Bönnemann, K. S. Nagabhushana, *J. New Mater. Electrochem. Sys.*, **2004**, *7*, 93.

31 R. Micheal, for Headwaters Nanokinetix, Inc., US Pat. application No. 10/401 351, March 28, **2003**.

32 M. Jacoby, *Chem. Eng. News*, June 5, **2006**.

33 G. L. Bezemer, J. H. Bitter, H. P. C. E. Kuipers, H. Oosterbeek, J. E. Holewijn, X. Xu, F. Kapteijn, A. J. Van Dillen, K. P. De Jong, *J. Am. Chem. Soc.*, **2006**, *128*, 3956.

34 H. Schulz, *Appl. Cat. A: General*, **1999**, *186*, 3.

35 M. J. A. Tijmensen, *Biomass Bioenergy*, **2002**, *23*, 129.

36 L. E. Aleandri, H. Bönnemann, D. J. Jones, J. Richter, J. Roziere, *J. Mater. Chem.*, **1995**, *5*, 749.

37 W. Vogel, P. Britz, H. Bönnemann, J. Rothe, J. Hormes, *J. Phys. Chem. B*, **1997**, *101*, 11029.

38 H. Bönnemann, W. Brijoux, J. Richter, R. Becker, J. Hormes, J. Rothe, *Z. Naturforsch b*, **1995**, *50*, 333.

39 J. Rothe, G. Köhl, J. Hormes, H. Bönnemann, W. Brijoux, K. Siepen, *J. Phys. IV France, Colloq. C2*, **1997**, *7*, 959.

40 K. Siepen, H. Bönnemann, W. Brijoux, J. Rothe, J. Hormes, *Appl. Organomet. Chem.*, **2000**, *14*, 549.

41 V. I. Zaikovskii, K. S. Nagabhushana, V. V. Kriventsov, S. V. Cherepanova, R. I. Kvon, H. Bönnemann, D. I. Kochubey, E. R. Savinova, *J. Phys. Chem. B*, **2006**, *110*, 6881.

42 H. Bönnemann, W. Brijoux, R. Brinkmann, U. Endruschat, W. Hofstadt, K. Angermund, *Rev. Roum. Chim.*, **1999**, *44*, 1003.

43 K. Angermund, M. Bühl, U. Endruschat, F. T. Mauschick, R. Mörtel, R. Mynott, B. Tesche, N. Waldöfner, H. Bönnemann, G. Köhl, H. Modrow, J. Hormes, E. Dinjus, F. Gassner, H-G. Haubold, T. Vad, *Angew. Chem. Int. Ed.*, **2002**, *41(21)*, 4041.

44 K. Angermund, M. Bühl, U. Endruschat, F. T. Mauschick, R. Mörtel, R. Mynott, B. Tesche, N. Waldöfner, H. Bönnemann, G. Köhl, H. Modrow, J. Hormes, E. Dinjus, F. Gassener, H-G. Haubold,

T. Vad, M. Kaupp, *J. Phys. Chem. B*, **2003**, *107*(*30*), 7507.

45 H. Bönnemann, W. Brijoux, R. Brinkmann, WO 99/59713, (Nov. 25, **1999**), (to Studiengesellschaft Kohle).

46 C. Amiens, D. De Caro, B. Chaudret, J. S. Bradley, *J. Am. Chem. Soc.* **1993**, *115*(*24*), 11638.

47 D. De Caro, H. Wally, C. Amiens, B. Chaudret, *J. Chem. Soc., Chem. Commun.*, **1994**, 1891.

48 A. Rodriguez, C. Amiens, B. Chaudret, M. J. Casanove, P. Lecante, J. S. Bradley, *Chem. Mater.*, **1996**, *8*, 1978.

49 M. Bardaji, O. Vidoni, A. Rodriguez, C. Amiens, B. Chaudret, M. J. Casanove, P. Lecante, *New. J. Chem.*, **1997**, *21*, 1243.

50 J. S. Bradley, E. W. Hill, S. Behal, C. Klein, B. Chaudret, A. Duteil, *Chem. Mater.*, **1992**, *4*, 1234.

51 A. Duteil, R. Queau, B. Chaudret, R. Mazel, C. Roucau, J. S. Bradley, *Chem. Mater.*, **1993**, *5*, 341.

52 D. De Caro, V. Agelou, A. Duteil, B. Chaudret, R. Mazel, Ch. Roucau, J. S. Bradley, *New. J. Chem.* **1995**, *19*, 1265.

53 C. Pan, F. Dassenoy, M. Casanove, K. Philippot, C. Amiens, P. Lecante, A. Mosset, B. Chaudret, *J. Phys. Chem. B*, **1999**, *103*, 10098.

54 M. Veelst, T. Ely, C. Amiens, E. Snoeck, P. Lecante, A. Mosset, M. Respaud, J. Broto, B. Chaudret, *Chem. Mater.*, **1999**, *11*, 2702.

55 T. O. Ely, C. Pan, C. Amiens, B. Chaudret, F. Dassenoy, P. Lecante, M. J. Casanove, A. Mosset, M. Respaud, J. M. Broto, *J. Phys. Chem. B.*, **2000**, *104*, 695.

56 D. Boxall, G. Deluga, E. Kenik, W. King, C. Lukehart, *Chem. Mater.* **2001**, *13*(*3*), 891.

57 S. Komarneni, D. Li, B. Newalkar, H. Katsuki, A. S. Bhalla, *Langmuir* **2002**, *18*(*15*), 5959.

58 K. E. Gonsalves, H. Li, R. Perez, P. Santiago, M. J. Yacaman, *Co-ord. Chem. Rev.*, **2000**, *206–207*, 607.

59 K. S. Suslick, J. Prince, *Annu. Rev. Mater. Sci.*, **1999**, *29*, 295.

60 A. Dhas, A. Gedanken, *J. Mater. Chem.*, **1998**, *8*, 445.

61 Y. Koltypin, A. Fernandez, C. Rojas, J. Campora, P. Palma, R. Prozorov, A. Gedanken, *Chem. Mater.*, **1999**, *11*, 1331.

62 R. A. Salkar, P. Jeevanandam, S. T. Aruna, Y. Koltypin, A. Gedanken, *J. Mater. Chem.*, **1999**, *9*, 1333.

63 A. Henglein, D. Meisel, *Langmuir*, **1998**, *14*, 7392.

64 J. Belloni, M. Mostafavi, H. Remita, J. L. Marignier, M. O. Delcourt, *New J. Chem.*, **1998**, *22*, 1239.

65 A. Henglein, *J. Phys. Chem. B.*, **2000**, *104*, 2201.

66 H. Bönnemann, W. Brijoux, R. Brinkmann, N. Matoussevitch, N. Waldöfner, PCT EP 2003/003814 (**2003**).

67 H. Bönnemann, W. Brijoux, R. Brinkmann, N. Matoussevitch, N. Waldofner, *Magnetohydrodynamics*, **2003**, *39*, 29.

68 H. Bönnemann, W. Brijoux, R. Brinkmann, N. Matoussevitch, N. Waldöfner, N. Palina, H. Modrow, *Inorg. Chim. Acta*, **2003**, *350*, 617.

69 H. Bönnemann, W. Brijoux, R. Brinkmann, M. Feyer, W. Hofstadt, G. Khelashvili, N. Matoussevitch, K. Nagabhushana, *Strem Chemiker*, **2004**, *21*, 1–18. (visit www.strem.com)

70 A. Heinmann, A. Wiedermann, M. Kammel, H. Bönnemann, N. Matoussevitch, *Appl. Organomet. Chem.*, **2004**, *18*, 561.

71 E. Romanus, N. Matoussevitch, S. Prass, J. Heinrich, R. Müller, D. V. Berkov, H. Bönnemann, P. Webber, *Appl. Organomet. Chem.*, **2004**, *18*, 548.

72 S. Rudenkiy, M. Frerichs, F. Volgts, W. Maus-Friedrichs, V. Kempter, R. Brinkmann, N. Matoussevitch, W. Brijoux, H. Bönnemann, N. Palina, H. Modrow, *Appl. Organomet. Chem.*, **2004**, *18*, 553.

73 H. Bönnemann, R. A. Brand, W. Brijoux, H.-W. Hofstadt, M. Frerichs, V. Kempter, W. Maus-Friedrichs, N. Matoussevitch, K. S. Nagabhushana, F. Voigts, V. Caps, *Appl. Organomet. Chem.*, **2005**, *19*, 790.

74 A.-H. Lu, W.-C. Li, N. Matoussevitch, B. Spliethoff, H. Bönnemann, F. Schüth, *Chem. Commun.*, **2005**, 98.

75 S. Behrens, H. Bönnemann, N. Matoussevitch, E. Dinjus, H. Modrow,

N. Palina, M. Frerichs, V. Kempter, W. Maus-Friedrichs, A. Heinemann, M. Kammel, A. Wiedenmann, *Z. Phys. Chem.*, **2005**, *220*, 3.

76 A. Lu, W. Schmidt, N. Matoussevitch, B. Spliethoff, B. Tesche, E. Bill, W. Kiefer, F. Schüth, *Angew. Chem. Int. Ed.*, **2004**, *43*, 4303.

77 F. Wen, H. Bönnemann, R. J. Mynott, B. Spliethoff, C. Weidenthaler, N. Palina, S. Zinoveva, H. Modrow, *Appl. Organomet. Chem.*, **2005**, *19(7)*, 827.

78 F. Wen, H. Bönnemann, *Appl. Organomet. Chem.*, **2005**, *19(7)*, 94.

79 C. Ducamp-Sanguesa, R. Herrera-Urbina, M. Figlarz, *J. Solid State Chem.*, **1992**, *100(2)*, 272.

80 P. Y. Silvert, R. Herrera-Urbina, N. Duvauchelle, V. Vijaykrishnan, K. Tekaia-Elhsissen, *J. Mater. Chem.* **1996**, *6(4)*, 573.

81 G. Khelashvili, S. Behrens, C. Weidenthaler, C. Vetter, A. Hinsch, R. Kern, K. Skupien, E. Dinjus, H. Bönnemann, *Thin Solid Films*, **2006**, *511–512*, 342.

82 H. Bönnemann, B. Korall, *Angew. Chem. Int. Ed. Engl.*, **1992**, *31*, 1490.

83 H. Bönnemann, W. Brijoux, *Nanostruc. Mater.*, **1995**, *5*, 135.

84 R. Franke, J. Rothe, J. Pollmann, J. Hormes, H. Bönnemann, W. Brijoux, T. Hindenburg, *J. Am Chem. Soc.*, **1996**, *118*, 12090.

85 J. S. Bradley, B. Tesche, W. Busse, M. Maase, M. T. Reetz, *J. Am. Chem. Soc.*, **2000**, *122*, 4631.

86 Y.-T. Yu, B.-Q. Xu, *Appl. Organomet. Chem.*, **2006**, *20*, 638.

87 S. Kinge, H. Bönnemann, *Appl. Organomet. Chem.*, **2006**, *20*, 784.

88 H. Bönnemann, W. Brijoux, R. Brinkmann, E. Dinjus, T. Joussen, B. Korall, *Angew. Chem. Int. Ed.*, **1991**, *30*, 1312.

89 H. Bönnemann, R. Brinkmann, R. Köppler, P. Neiteler, J. Richter, *Adv. Mater.*, **1992**, *4*, 804.

90 H. S. Nalwa (Ed.), *Encyclopedia of Nanoscience and Nanotechnology, Vol. 1– 10*, American Scientific Publishers, Stevenson Ranch, CA, **2004**.

91 J. Hormes, H. Modrow, R. Brinkmann, N. Waldöfner, H. Bönnemann, L. Beuermann, S. Krischok, W. M. Friedrichs, V. Kempter, *Surf. Sci.*, **2002**, *497*, 321.

92 R. Franke, J. Rothe, R. Becker, J. Pollmann, J. Hormes, H. Bönnemann, W. Brijoux, R. Köppler, *Adv. Mater.*, **1998**, *10*, 126.

93 R. Franke, J. Rothe, R. Becker, J. Pollmann, J. Hormes, H. Bönnemann, W. Brijoux, R. Köppler, *Adv. Mater.*, **1998**, *10*, 126.

94 J. Sinzig, L. J. De Jongh, H. Bönnemann, W. Brijoux, R. Köppler, *Appl. Organomet. Chem.*, **1998**, *12*, 387.

95 Y. Volokitin, J. Sinzig, G. Schmid, H. Bönnemann, L. J. De Jongh, *Z. Phys. D, At., Mol. Clusters*, **1997**, *40*, 136.

96 H. Bönnemann, R. Richards, in *Synthetic Methods of Organometallic and Inorganic Chemistry*, W. A. Herrmann, G. Brauer, (Eds.), Thieme Verlag, Stuttgart, **2002**, Vol. 10, Ch. 20, p. 209.

97 J. Rothe, J. Pollmann, R. Franke, J. Hormes, H. Bönnemann, W. Brijoux, K. Siepien, J. Richter, *J. Anal. Chem.*, **1996**, *355*, 372.

98 F. Wen, N. Waldöfner, W. Schmidt, K. Angermund, H. Bönnemann, S. Modrow, S. Zinoveva, H. Modrow, J. Hormes, L. Beuermann, S. Rudenkiy, W. M.-Friedrichs, V. Kempter, T. Vad, H-G. Haubold, *Eur. J. Inorg. Chem.*, **2005**, *2005(18)*, 3625.

99 H. Bönnemann, N. Waldöfner, H. G. Haubold, T. Vad, *Chem. Mater.*, **2002**, *14*, 1115.

100 T. Vad, H.-G. Haubold, N. Waldöfner, H. Bönnemann, *J. Appl. Crystallogr.*, **2002**, *35*, 459.

101 T. S. Ahamadi, Z. I. WanG. T. C. Green, A. Henglein, M. A. El-Sayed, *Science*, **1996**, *272*, 1924.

102 T. Herricks, J. Y. Chen, Y. N. Xia, *Nano Lett.*, **2004**, *4*, 2367.

103 H. A. Wicrenga, L. Soethout, I. W. Gerritsen, B. E. C. Van Do Leemput, H. Van KEMPEN, G. Schmid, *Adv. Mater.*, **1990**, *2*, 482.

104 H. Bönnemann, R. Brinkmann, P. Neiteler, *Appl. Organomet. Chem.*, **1994**, *8*, 361.

105 M. Beller, H. Fischer, K. Kühlein, C.-P. Reisinger, W. A. Herrmann, *J. Organomet. Chem.*, **1996**, *520*, 257.

106 Y. Lin, R. G. Finke, *J. Am. Chem. Soc.*, **1994**, *116*, 8335.

107 J. D. Aiken III, Y. Lin, R. G. Finke, *J. Mol. Catal. A.*, **1996**, *114*, 29.

108 M. A. Watzky, R. G. Finke, *J. Am. Chem. Soc.*, **1997**, *119*, 10382.

109 M. A. Watzky, R. G. Finke, *Chem. Mater.*, **1997**, *9*, 3083.

110 T. Nagata, M. Pohl, H. Weiner, R. G. Finke, *Inorg. Chem.*, **1997**, *36*, 1366.

111 J. D. Aiken III, R. G. Finke, *J. Am. Chem. Soc.*, **1998**, *120*, 9545.

112 J. D. Aiken III, R .G. Finke, *Chem. Mater.*, **1999**, *11*, 1035.

113 J. D. Aiken III, R. G. Finke, *J. Mol. Catal. A*, **1999**, *145*, 1.

114 M. T. Reetz, M. Maase, *Adv. Mater.*, **1999**, *11*, 773.

115 M. T. Reetz, E. Westermann, *Angew. Chem. Int. Ed.*, **2000**, *39*, 165.

116 M. T. Reetz, E. Westermann, R. Lohmer, G. Lohmer, *Tetrahedron Lett.*, **1998**, *39*, 8449.

117 A. H. M. De VRIES, J. M. C. A. Mulders, J. H. W. Mommers, H. J. W. Henderckx, J. H. De Vries, *Org. Lett.*, **2003**, *5*, 3285.

118 M. Fichtner, O. Fuhr, O. Kircher, J. Rothe, *Nanotechnology*, **2003**, *14*, 778.

119 B. Bogdanovic, M. Felderhoff, S. Kaskel, A. Pommerin, K. Schlichte, F. Schüth, *Adv. Mater.*, **2003**, *15*, 1012.

120 H. Bönnemann, G. A. Braun, *Angew. Chem. Int. Ed. Engl.* **1996**, *35*, 1992.

121 H. Bönnemann, G. A. Braun, *Chem. Eur. J.* **1997**, *3*, 1200.

122 H. Bönnemann, in Preparation of Catalysts VI, G. Poncelet, J. Martens, B. Delmon, P. A. Jacobs, P. Grange (Eds.), *Stud. Surf. Sci. Catal.*, Amsterdam, **1995**, 91, S.185.

123 H. Bönnemann, W. Brijoux, T. Joussen (to Studiengesellschaft Kohle mbH), U.S. Pat. 5 580 492 (Aug. 26, **1993**).

124 H. Bönnemann, W. Brijoux, R. Brinkmann, J. Richter (to Studiengesellschaft Kohle mbH), U.S. Pat. Appl. 849 482 (Aug. 29, **1997**), U. S. Patent No. 6 090 746 (Jul. 18, **2000**).

125 A. Freund, P. Panster (to Degussa AG), European Patent EP 715 889 A2 (**1995**).

126 H. Bönnemann, U. Endruschat, J. Hormes, G. Köhl, S. Kruse, H. Modrow, R. Mörtel, K. S. Nagabhushana, *Fuel Cells*, **2004**, *4(4)*, 297.

127 G. Witek, M. Noeske, G. Mestl, S. Shaikhutdinov, R. J. Behm, *Catal. Lett.*, **1996**, *37*, 35.

128 S. K. Shaikhutdinov, F. A. Möller, G. Mestl, R. J. Behm, *J. Catal.*, **1996**, *163*, 492.

129 F. Wen, H. Bönnemann, J. Jiang, D. Lu, Y. Wang, Z. Jin, *Appl. Organomet. Chem.*, **2005**, *19*, 81.

130 H. Bönnemann, R. M. Richards, *Eur. J. Inorg. Chem.*, **2001**, *10*, 2455.

131 T. Teranishi, K. Nakata, M. Miyake, N. Toshima, *Chem. Lett.*, **1996**, 277.

132 H. Bönnemann, W. Wittholt, J. D. Jentsch, A. S. Schulze Tilling, *New J. Chem.*, **1998**, *22*, 713.

133 X. Yang, H. Liu, *Appl. Catal. A: General*, **1997**, *164*, 197.

134 W. Yu, H. Liu, X. An, X. Ma, Z. Liu, L. Qiang, *J. Mol. Catal. A.*, **1999**, *147*, 73.

135 H. Feng, H. Liu, *J. Mol. Cat. A: Chemical*, **1997**, *126*, L5.

136 J. H. Sinfelt, *Acc. Chem. Res.*, **1987**, *20*, 134.

137 J. H. Sinfelt, G. H. Via, F. W. Lytle, *J. Chem. Phys.*, **1980**, *72*, 4832.

138 J. H. Sinfelt, G. H. Via, F. W. Lytle, *J. Chem. Phys.*, **1981**, *75*, 5527.

139 N. Toshima, T. Yonezawa, *New J. Chem.*, **1998**, 1179.

140 H. Bönnemann, W. Brijoux, A. S. Schulze Tilling, K. Siepen, *Top. Catal.*, **1997**, *4*, 217.

141 H. Bönnemann, P. Britz, H. Ehwald, *Chem. Technik*, **1997**, *49*, 189.

142 H. Bönnemann, W. Brijoux, R. Brinkmann, A. S. Schulze Tilling, T. Schilling, B. Tesche, K. Seevogel, R. Franke, J. Hormes, G. Köhl, J. Pollmann, J. Rothe, W. Vogel, *Inorg. Chim. Acta*, **1998**, *270*, 95.

143 K. Kodesch, G. Simader, *Fuel Cells and Their Applications*, Wiley-VCH Weinheim, **1996**.

144 W. Vielstich, A. Lamm, H. A. Gasteiger (Eds.), *Handbook of Fuel Cells*, John Wiley and Sons, West Sussex, England, **2003**, Vol. 1–4.

145 Visit www.volkswagen-media-services. com and www.innovations-report.de/ html/berichte/energie_elektrotechnik/ bericht-73229.html; Last visited, 10th November **2006**.

146 G. J. K. Acres, J. C. Frose, G. A. Hards, R. J. Potter, T. R. Ralph, D. Thompsett, G. T. Burstein, G. J. Hutchings, *Catal. Today*, **1997**, *38*, 393.

147 W. Wittholt, PhD Thesis, *Mehrmetallische Pt-Kolloidkatalysatoren und ihre Anwendung in der Halonitroaromatenreduktion bzw. in Phosphorsäure-Brennstoffzellen*, RWTH Aachen, **1996**.

148 N. Boukis, U. Galla, H. Bönnemann, E. Dinjus, H_2 tec, **2005**, 14.

149 T. J. Schmidt, M. Noeske, H. A. Gasteiger, R. J. Behm, P. Britz, W. Brijoux, H. Bönnemann, *Langmuir*, **1997**, *13*, 2591.

150 H. Bönnemann, P. Britz, W. Vogel, *Langmuir*, **1998**, *14*, 6654.

151 T. J. Schmidt, M. Noeske, H. A. Gasteiger, R. J. Behm, P. Britz, H. Bönnemann, *J. Electrochem. Soc.*, **1998**, *145*, 925.

152 H. Bönnemann, in *New Materials for Electrochemical Systems III*, Extended Abstract., Third International Symposium, Montreal, Quebec, Canada, **1999**, p. 31.

153 H. Bönnemann, R. Brinkmann, P. Britz, U. Edruschat, R. Mörtel, U. A. Paulus, G. J. Feldmeyer, T. J. Schmidt, H. A. Gasteiger, R. J. Behm, *J. New Mater. Electrochem. Sys.*, **2000**, *3*, 199.

154 U. A. Paulus, U. Endruschat, G. J. Feldmeyer, T. J. Schmidt, H. Bönnemann, R. J. Behm, *J. Catal.*, **2000**, *195*, 383.

155 J. Kaiser, Z. Jusys, R. J. Behm, R. Mörtel, H. Bönnemann, in *International Conference with Exhibition PEFC, Ist European PEFC Forum, Proceedings*, F. N. Buchi, G. G. Scherer, A. Wokaun (Eds.), Kinzel, Göttingen, **2001**, p. 59.

156 T. J. Schmidt, Z. Jusys, H. A. Gasteiger, R. J. Behm, U. Endruschat, H. Bönnemann, *J. Electroanal. Chem.*, **2001**, *501*, 132.

157 H. Bönnemann, R. Brinkmann, S. Kinge, T. O. Ely, M. Armand, *Fuel Cells*, **2004**, *4*, 289.

158 H. Bönnemann, K. S. Nagabhushana, *Chem. Ind.*, (Belgrade, Yogosl.) **2004**, *58*, 271.

159 B. Brendenbach, S. Bucher, J. Hormes, H. Bönnemann, K. S. Nagabhushana, R. Brinkmann, H. Modrow, *Phys. Scr.*, **2005**, *115*, 773.

160 H. Modrow, G. Köhl, J. Hormes, H. Bönnemann, U. Endruschat, R. Mörtel, *Phys. Scr.*, **2005**, *115*, 671.

161 J. Kaiser, L. Colmenares, Z. Jusys, R. Mörtel, H. Bönnemann, G. Köhl, H. Modrow, J. Hormes, R. J. Behm, *Fuel Cells*, **2006**, *6(3–4)*, 190.

162 R. J. Behm, K. S. Nagabhushana, H. Bönnemann, L. Colmenares, Z. Jusys, *Journal of Applied Electro Chemsitry*, **2007**, accepted.

163 R. G. Cawthorn, *South African J. Sci.*, **1999**, *95*, 481.

164 L. Colemares, Z. Jusys, S. Kinge, H. Bönnemann, R. J. Behm, *J. New Mater. Electrochem. Syst.*, **2006**, *9*, 107.

165 M. Arenz, K. J. J. Mayrhofer, V. Stamenkovic, B. B. Blizanac, T. Tomoyuki, P. N. Ross, N. M. Markovic, *J. Am. Chem. Soc.*, **2005**, *127*, 6819.

166 J. W. Yoo, S. M. Lee, H. T. Kim, M. A. El-Sayed, *Bull. Korean Chem. Soc.* **2004**, *25*, 395.

167 S. Kinge, C. Urgeghe, A. De Battisti, H. Bönnemann, *Fuel Cells*, **2007**, in preparation.

168 P. Strasser, Q. Fan, M. Devenney, W. H. Weinberg, P. Liu, J. K. Norskov, *J. Phys. Chem. B.*, **2003**, *107*, 11013.

169 T. Vidakovic, M. Christov, K. Sundmacher, K. S. Nagabhushana, W. Fei, S. Kinge, H. Bönnemann, *Electrochim. Acta*, **2006**, *52*, 2277.

170 K. S. Nagabhushana, C. Weidenthaler, S. Hocevar, D. Stremcnik, M. Gaberseck, A. L. Antozzi, G. N. Martelli, *J. New Mater. Electrochem. Syst.*, **2006**, *9*, 73.

171 H. Bönnemann, K. S. Nagabhushana, C. Weidenthaler, B. Spliethoff, J. Norskov, H. Modrow, N. Palina, et al., manuscript in preparation.

172 M. Grätzel, *Nature*, **2001**, *414*, 338.

173 M. Späth, P. M. Sommeling, J. A. M. Van Roosmalen, H. J. P. Smit, N. P. G. Van DER BURG, D. R Mahieu, N. J. Bakker, J. M. Kroon, *Prog. Photovolt.*, **2003**, *11*, 207.

174 H. Bönnemann, G. Khelashvili, S. Behrens, A. Hinsch, K. Skupien, E. Dinjus, *J. Clust. Sci.*, **2007**, 141.

175 G. Khelashvili, A. Hinsch, S. Behrens, W. Habicht, D. Schild, A. Eichhöfer, R. Sastrawan, K. Skupien, E. Dinjus, H. Bönnemann, *Thin Solid Films*, **2007**, 4074.

176 D. V. Talapin, E. V. Shevchenko, H. Weller, in *Nanoparticles*, G. Schmid (Ed.), VCH-Wiley, Weinheim, **2004**, p. 207.

177 N. Matoussevitch, A. Gorschinski, W. Habicht, J. Bolle, E. Dinjus, H. Bönnemann, S. Behrens, *J. Phys.: Condeus. Matter*, **2006**, *18*, 2543.

178 O. Mikhalyk, Institute of Experimental Oncology, Uni-Klinikum, LMU München, personal communication.

179 H. Gu, K. Xu, B. Xu, *Chem. Commun.*, **2006**, 941.

180 A. Jordan, R. Scholz, P. Wust, H. Fähling, R. J. Felix, *Magn. Mater.* **1999**, *201*, 413.

181 L. Josephson, J. M. Perez, R. Weissleder, *Angew. Chem. Int. Ed.*, **2001**, *40*, 3204.

3
Nanoparticulate Catalysts Based on Nanostructured Polymers

Lyudmila M. Bronstein, Valentina G. Matveeva, and Esther M. Sulman

3.1
Introduction

The catalytic nanoparticles possess unique catalytic properties due to their large surface area and considerable number of surface atoms leading to an increased amount of active sites [1–3]. The catalytic properties of nanoparticles depend on the nanoparticle size, nanoparticle size distribution, and nanoparticle environment [4]. Moreover, the surface of nanoparticles plays an important role in catalysis, being responsible for their selectivity and activity. As was demonstrated in the last decade, the formation of nanoparticles in a nanostructured polymeric environment allows enhanced control over nanoparticle characteristics, yet the stabilizing polymer (its functionality) is of great importance, determining the state of the nanoparticle surface [5–8].

Recently, catalytic nanoparticles formed in various types of nanostructured polymers have been extensively studied in major catalytic reactions [5–7]. Although traditional polymer-based catalysts may be faulted by such shortcomings as uncontrolled swelling or poor accessibility of nanoparticles in the polymer bulk, the nanostructured polymers are mainly free from these limitations due to the small size of the nanoparticle-containing zones. Moreover, some swelling can even be advantageous, providing better access to the nanoparticle surface. The majority of nanostructured polymers for catalytic applications are amphiphilic block copolymers and dendrimers while some other polymers have also been explored. The ability of amphiphilic block copolymers to form micelles in dilute solutions in selective solvents [9, 10] has been used to stabilize catalytic nanoparticles in the functionalized micelle cores [11, 12]. Pd nanoparticles formed in PS-*b*-P4VP micelles were used as a catalyst for the cross-coupling reaction between aryl halides and alkenes (Heck reaction) [12]. These hybrid systems exhibited nearly the same efficiency as low-molecular-mass palladium complexes (conventional catalysts of the Heck reaction), being at the same time much more stable. The PS-*b*-P4VP micelles containing Pd and Pd/Au nanoparticles were studied in the hydrogenation of cyclohexene, 1,3-cyclooctadiene, and 1,3-cyclohexadiene [11]. A strong influence of the synthetic pathway to form nanoparticles and the type of

Nanoparticles and Catalysis. Edited by Didier Astruc
Copyright © 2008 WILEY-VCH Verlag GmbH & Co. KGaA, Weinheim
ISBN: 978-3-527-31572-7

reducing agent on the catalytic activity of the colloids was found. The lowest activity was observed for the $N_2H_4 \cdot H_2O$ reduction which is related to a morphology where only a small number of noble-metal colloids is embedded in the micelle core. The highest activity was obtained for the super-hydride reduction where the data suggest the existence of many metal clusters per micelle.

Unlike regular block copolymer micelles which are well permeable for reagents, triblock nanospheres with hydroxylated polyisoprene coronas, cross-linked poly(2-cinnamoyloxyethyl methacrylate) shells, and poly(acrylic acid) cores, filled with Pd nanoparticles, showed slower hydrogenation of alkenes than Pd blacks due to the need for the reactant(s) to diffuse into and the products to diffuse out of the encapsulating nanospheres [13]. On the other hand, microspheres formed by diblock poly(*t*-butyl acrylate)-*block*-poly(2-cinnamoyloxyethyl methacrylate) and filled with Pd nanoparticles demonstrated good permeability and higher catalytic activity in the hydrogenation of methyl methacrylate than the commercial Pd black catalyst [14].

Au nanoclusters formed in the PS-*b*-P2VP micelles displayed catalytic activity in the electro-oxidation of CO [15, 16]. The smallest particles studied (1.5 nm) were the most active for electro-oxidation of CO and had the largest fraction of oxygen associated with gold at the surface, as measured by Au^{3+}/Au^0 X-ray photoemission intensities.

When nanoparticles are formed in the block copolymer corona [17, 18], allowing better access to the nanoparticle surface, the catalytic systems are highly active in cyclohexene hydrogenation, but stabilization of nanoparticles is poor, leading to their aggregation.

The homogeneous catalysts based on block copolymers with nanoparticles (or strictly speaking microheterogeneous catalysts as the catalytic reactions occur on the surface of the nanoparticles) were used to form heterogeneous catalysts. Amphiphilic polystyrene–poly(ethylene glycol) resin-supported Pd nanoparticles exhibited high catalytic performance in the hydrogenation of olefins and the hydrodechlorination of chloroarenes under aqueous conditions [19]. Dechlorination was found to take place smoothly using aqueous ammonium formate with high recyclability, which would provide a safe and clean detoxification protocol for aqueous chloroarene pollutants.

Heterogenization of catalytic nanoparticles stabilized by block copolymers can be carried out by their incorporation into porous membranes based on poly(acrylic acid) crosslinked with a difunctional epoxide [20–22]. Membranes with defined porosities and amounts of palladium were studied in the selective hydrogenation of propyne to propene as a model reaction. The porosity of the polymer membrane, the content of catalyst, and the residence time of the reaction mixture were found to influence the conversion and selectivity. The main advantage of these membranes compared to other heterogeneous catalysts is simple adjustment to reaction conditions and facilitated mass transfer.

Another most popular nanostructured system for the stabilization of catalytic nanoparticles is dendrimers. These catalysts are prepared via sorption of metal ions into dendrimers (normally using commercially available poly(amido-amines),

PAMAM) followed by complexation with amino groups. Subsequent chemical reduction results in the formation of dendrimer-encapsulated metal nanoparticles comprising exactly the same number of atoms as pre-loaded into the dendrimer. If no exchange of metal ions between dendrimer macromolecules occurs, the nanoparticle size distribution is narrow and fine control of the particle growth is possible [23–25]. Such nanocomposites are well soluble and retain their stability for months. Both mono- and bimetallic nanoparticles can be formed [25–27]. Dendrimer-encapsulated nanoparticles have been extensively explored in the Heck and Suzuki reactions [28–30], electrochemical reduction of oxygen [24], hydrogenation of allyl alcohol [31–33], hydrogenation of cyclohexene [34], the Stille reaction [35], and the hydrogenation of *p*-nitrophenol (Au–Si binary particles) [36]. Polyaryl ether aminediacetic acid dendrons have been synthesized and used as stabilizer for the preparation of Pt nanoparticles (2–5 nm) exhibiting catalytic activity in the hydrogenation of phenyl aldehydes to phenyl alcohols [37]. In a number of papers it was demonstrated that the catalytic activity can be controlled by using dendrimers of different generations which show different sizes and morphologies [31–34, 37]. Dendrimer encapsulated nanoparticles can be used for heterogenization of nanoparticles on an inorganic support (titania) followed by dendrimer removal [26, 38]. Such a supported PdAu catalyst demonstrated enhanced CO oxidation activity compared to monometallic nanoparticles [26].

One more nanostructured polymeric system for the stabilization of catalytic nanoparticles is formed by layer-by-layer deposition of oppositely charged polyelectrolytes [39]. Such a multilayer system can be formed not only on flat surfaces but also on colloids [40, 41]. Silver and platinum nanoparticles synthesized in multilayered polyelectrolyte films [42] possessed electrocatalytic properties in methylene bromide reduction; however, these catalysts are less efficient than a pure silver electrode because polymer films are insulators. These data are consistent with the results obtained by Zhao and Crooks [43] who used dendrimer-encapsulated platinum nanoparticles deposited on an electrode as a catalyst for an electrocatalytic reaction. Clearly, in spite of some catalytic activity these polymeric systems with encapsulated metal nanoparticles show no promise as electrode coatings in electrocatalytic reactions, because the insulating properties of the polymer reduce the catalyst efficiency. Conversely, silver nanocomposite multilayer films based on the sequential electrostatic deposition of a positively charged third-generation PAMAM, negatively charged poly(styrenesulfonate) (PSS) and poly(acrylic acid) (PAA) showed potential in the catalytic reduction of 4-nitrophenol with sodium borohydride [22], where the insulating nature of the polymers is of no importance. On the other hand, unlike block copolymer micelles or dendrimers, nanoparticles buried within multilayer films are expected to have lower catalytic activity due to a permeability issue.

We studied the behavior of catalytic nanoparticles formed in a nanostructured polymeric environment in the hydrogenation of long chain acetylene alcohols and the direct oxidation of a monosaccharide (L-sorbose). These reactions were chosen because of their industrial relevance and also because of the special importance of high selectivity, which can be achieved using a polymeric environment in mild conditions (below 100 °C). As nanostructured polymeric systems we used block copolymer micelles, nanoporous hypercrosslinked polystyrene, crosslinked functionalized polymer colloids and microgels leading to inorganic mesoporous solids.

In this chapter we will discuss the influence of such key parameters as nanoparticle morphology, polymeric environment and inorganic support on the catalytic behavior of nanoparticulate catalysts. The metal content in the nanoparticulate catalysts discussed here was assessed using X-ray fluorescence spectroscopy [44], while the particle analysis was carried out using mainly transmission electron microscopy (TEM), X-ray powder diffraction (XRD) and X-ray photoelectron spectroscopy (XPS) [44]. It is noteworthy that XPS, although a valuable tool for determining the oxidation state of a metal, should be used with caution for nanoparticles in a nonconductive polymeric environment because, in some cases, high binding energy values can be due to charging of the nanoparticles. As a result, a higher oxidation state is demonstrated than that actually present in a real sample [45]. Normally, comparison with the XRD data allows one to clarify this issue.

3.2
Catalytic Nanoparticles in a Nanostructured Environment

3.2.1
Hydrogenation with Catalytic Nanoparticles

Catalytic nanoparticles formed in nanostructured polymers can be successfully used in hydrogenation [11] and, in particular, in selective hydrogenation of long chain acetylene alcohols: 3,7-dimethyl-6-octen-1-yn-3-ol (dehydrolinalool, DHL, C10), 2-methyl-3-butyn-2-ol, (dimethylethynylcarbinol, DMEC, C5); and 3,7,11,15-tetramethyl-1-hexadecyn-3-ol, (dehydroisophytol, DHIP, C20) to the corresponding alcohols with a double bond: 3,7-dimethyl-1,6-octadien-3-ol (linalool, LN), 2-methyl-3-buten-2-ol (dimethylvinylcarbinol, DMVC), and 3,7,11,15-tetramethyl-1-hexadecen-3-ol (isophytol, IP). Hydrogenation of these acetylene compounds is of particular interest because the reaction products are intermediates in the synthesis of vitamins A, E and K and fragrant substances (Scheme 3.1). Because the natural resources cannot supply the growing needs, the necessity for synthetic hydrogenation products has been increasing.

3.2.1.1 **Block Copolymer-based Catalysts**

3.2.1.1.1 **PS-*b*-P4VP-based Catalysts** In the mid-1990s we developed nanoparticulate catalysts based on block copolymer micelles derived from the polystyrene-*block*-poly-4-vinylpyridine (PS-*b*-P4VP) [11, 46]. In selective solvents (toluene and THF) these block copolymers form micelles with the P4VP cores, the latter serve as nanoreactors for metal nanoparticle formation. The Pd nanoparticles of 2.6 nm

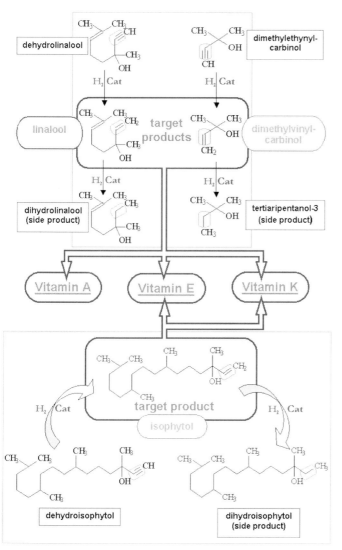

Scheme 3.1 Catalytic hydrogenation of dehydrolinalool, dimethylethynylcarbinol and dehydroisophytol.

in diameter (Fig. 3.1) were prepared by incorporation of $Pd(CH_3COO)_2$ followed by reduction with $NaBH_4$ [11]. According to XPS, the Pd binding energy $E_b(3d_{5/2})$ $= 335.15$ eV for this sample exactly matches that of $Pd_{(0)}$. We studied the DHL hydrogenation with Pd nanoparticles stabilized in the PS-*b*-P4VP micelles [47]. We discovered that the most crucial feature of nanoparticulate catalysts formed in block copolymer micelles is a polymeric environment in the micelle cores [47]. The P4VP micelle cores not only control the nanoparticle size and size distribution

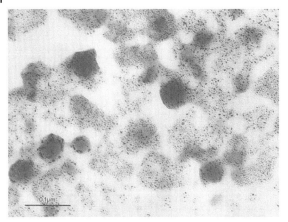

Fig. 3.1 Pd nanoparticles formed in PS-*b*-P4VP micelles. Reprinted with permission from Ref. [11]. Copyright (1997) American Chemical Society.

[7, 11, 46, 48, 49] but also provide a modification of the nanoparticle surface due to electron donation to Pd from pyridine units, resulting in decreased strength of LN adsorption [50].

It is well established that low molecular weight modifiers such as quinoline, pyridine, etc. [51] increase the selectivity of the hydrogenation of acetylene alcohols, but often the modifiers leach and selectivity deteriorates. In the case of pyridine units of the P4VP block, the modification is fairly permanent. The stability of modification, which governs the stability of catalytic properties and high selectivity, is one of the important advantages of catalytic nanoparticles stabilized in the polymeric media [47].

Along with monometallic Pd nanoparticles, we studied the synthesis and properties of PdAu, PdPt, and PdZn bimetallic nanoparticles prepared at a molar ratio of 4/1 in PS-*b*-P4VP micelles and their catalytic behavior in DHL hydrogenation [44]. Because the amount of Pd in all the bimetallic nanoparticles is much higher than that of the second metal, Pd mainly determines the catalytic properties.

The choice of second metal or metal-modifier for Pd was based on its electronic properties. Au and Pt have higher electronegativity than Pd, so they can be acceptors of electrons from Pd. In contrast, Zn, with very low electronegativity, is a donor to Pd, so its influence should be opposite. The morphology of the bimetallic nanoparticles was governed by the method of their preparation: two metal compounds were incorporated in block copolymer micelles and their reduction was carried out simultaneously. Thus, the structure of the bimetallic nanoparticles depended directly on the ability of the metal to be reduced in the particular conditions. According to the standard potentials [44], $AuCl_4^-$ can be reduced more easily than Pd^{2+} leading to a core(Au)–shell(Pd) structure. In contrast, the values of the standard potentials for Pd and Pt are close so a core–shell structure is hardly possible. The negative value for Zn^{2+} (from Zn acetate) shows zinc ions can be barely reduced with common reducing agents (super-hydride) [44]. These facts indicate that a core–shell structure is possible only for the PdAu pair, though the polymeric environment could influence the ability of ions to be reduced.

To clarify the structure of bimetallic nanoparticles we used FTIR. An FTIR study of carbon monoxide adsorbed on nanoparticles can give useful information about the structure and active sites of metallic systems [52]. Presumably, hydrogen is activated on one kind of reactive center, which we call "metallic" (Z), and then diffuses along the surface to substrate molecules (S) adsorbed on other centers called "organometallic" (Z_0). Organometallic centers might be formed due to coordination of the metal with polymer ligands (in the P4VP micelle core) [44]. For traditional heterogeneous catalysts, organometallic centers are formed due to coordination of a metal with a substrate, solvent, or support or can be deliberately synthesized [53]. These two kinds of active centers should exist on each single nanoparticle. Their presence can be confirmed by FTIR data on CO adsorption: the existence of terminal and bridging CO.

In the case of bimetallic particles, adsorbed carbon monoxide is also regarded as a useful infrared probe of surface composition [54]. Based on that, we studied CO adsorption on monometallic Pd and bimetallic PdAu, PdPt and PdZn nanoparticles formed in the PS-*b*-P4VP micellar solutions (Fig. 3.2). The samples, except PdAu, showed both terminal and bridging CO adsorption [55, 56], revealing the

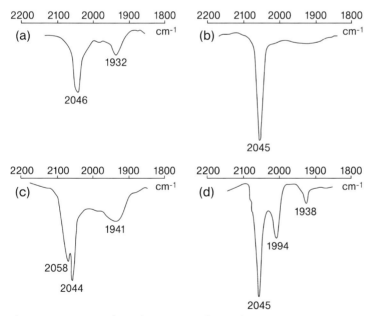

Fig. 3.2 FTIR spectra of CO adsorption on Pd (a) and bimetallic PdAu (b), PdPt (c) PdZn (d) nanoparticles formed in the PS-*b*-P4VP micelles. Reprinted from Ref. [51], Copyright (2000), with permission from Elsevier.

presence of two types of active centers. Moreover, PdPt and PdZn nanoparticles display two bands in the region of terminal absorption, while Pd and PdAu samples exhibit only one band in this region. The latter should indicate linear CO molecules adsorbed on the Pd surface. The lack of the bridging adsorption and the only peak being the terminal one reveal the presence of only Pd atoms on the PdAu nanoparticle surface and the existence of only one type of active center. Based on these findings we believe PdAu to have a core–shell structure, while PdPt and PdZn nanoparticles form cluster-in-cluster structures.

The XRD data for bimetallic nanoparticles show that in all cases small particles with diameter of 1.5–2.0 nm are formed with diffraction pattern typical for Pd$_{(0)}$. No crystalline structure for Au, Pt or Zn was found. The TEM image of PdPt nanoparticles formed in the PS-*b*-P4VP micelles is presented in Fig. 3.3.

FTIR spectroscopy on CO adsorption and XPS show that the second metal (Au, Pt, or Zn) acts as a modifier for Pd, changing both its electronic structure and surface geometry.

Table 3.1 contains the data on catalytic properties and kinetic parameters for these reactions [44]. For all the catalysts, the optimal conditions were found, providing high selectivity of hydrogenation (up to 99.8%) (# 1–4, Table 3.1), which is mainly determined by the modifying influence of the pyridine groups in the P4VP cores [44]. The selectivity is also dependent on the reaction conditions. We found that the highest selectivity is achieved at low agitation rates (with diffusion limitation) due to slow diffusion of the hydrogen to the reaction sites. Table 3.1 includes data on selectivity for reactions occurring both with and without diffusion limitations (low and high agitation rates).

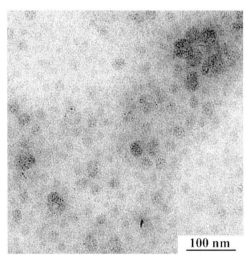

100 nm

Fig. 3.3 TEM image of PdPt colloids formed in PS-*b*-P4VP micelles. Reprinted from Ref. [51], Copyright (2000), with permission from Elsevier.

Table 3.1 Catalytic properties and kinetic parameters of selective hydrogenation of acetylene alcohols with micellar catalysts based on PS-b-P4VP[a].

#	Catalyst	wt.% Pd	Substrate	S (A)[b]	Reaction at 960 shakings min⁻¹						
					TOF, s⁻¹	S (A)[c]	Kinetic model	k (mol/mol)s⁻¹	Q	E_a	k_o
1	PS-b-P4VP-Pd	0.036	DHL	99.8 (100)	18.5	99.0 (100)	$W = k$	2.08×10^{-2}		23	66
2	PS-b-P4VP-PdAu				36.9		$W = \dfrac{kx_1}{(x_1 + Qx_2)^2}$	$0.14^{[d]}$	0.14	55	107
3	PS-b-P4VP-PdZn				34.4	98.5 (100)	$W = \dfrac{kx_1}{x_1 + Qx_2}$	3.8×10^{-2}	1.3×10^{-2}	26	356
4	PS-b-P4VP-PdPt				49.2			5.8×10^{-2}	5.2×10^{-2}	26	659
5	PS-b-P4VP-Pd/Al₂O₃	0.035		99.5 (100)	34.0	98.0 (100)	$W = \dfrac{kx_1}{x_1 + Qx_2}$	1.35×10^{-2}	2.6×10^{-2}	27	790
6	PS-b-P4VP-Pd/Al₂O₃	0.035	DHIP	99.5 (100)	9.8	98.0 (100)	No data				
7	PS-b-P4VP-PdAu/Al₂O₃		DMEC		37.0		No data				

a Reaction conditions: 90°C, 480 (with diffusion limitations) or 960 (without diffusion limitations) shakings min⁻¹, toluene (30 mL), C_o 0.44 mol L⁻¹, C_c 2.3×10⁻⁵ mol Pd L⁻¹.

b S(A)* is the selectivity (conversion) in the diffusion limitation regime.

c S(A)** is the selectivity (conversion) in the regime without diffusion limitations (kinetic measurements).

d For PS-b-P4VP-PdAu, k is in (mol/mol)² s⁻¹.

In Table 3.1 C_o is the initial acetylene alcohol concentration, C_c is the catalyst concentration, S is the selectivity (%), A is the acetylene alcohol conversion (%), turnover frequency (TOF) is the mole of substrate converted over a mole (Pd) of the catalyst per second, X_i is the relative concentration $X_i = C_i / C_o$ (where C_i is the current concentration of the substrate at $i=1$ and product at $i=2$). Strictly speaking TOF should be calculated per Pd atoms participating in the catalytic reaction (available surface atoms), but for the sake of comparison with literature data, in this chapter we will use the TOF definition given above. To find the kinetic relationships, we have studied the reaction kinetics at different substrate-to-catalyst ratio $SCR = C_0 / C_c$. Kinetic curves for DHL hydrogenation with Pd and bimetallic catalysts are presented in Fig. 3.4.

For the computation we have used the integral method using cubic spline and the combined gradient method of Levenberg–Marquardt [57, 58]. The kinetic models chosen describe well the hydrogenation kinetics. In the formulas presented in Table 3.1 k is the kinetic parameter of the reaction and Q takes into account the coordination (adsorption) of the product (LN) and substrate (DHL) with the catalyst (the ratio of the adsorption–desoprtion equilibrium constants for LN and DHL). Parameters of the Arrhenius equation, apparent activation energy E_a, kJ mol^{-1}, and frequency factor k_o, have been determined from the data on activities at different temperatures. The frequency factor is derived from the ordinate intercept of the Arrhenius dependence and provides a measure of the number of collisions or active centers on the surface of catalytic nanoparticles.

The models obtained are a formal description of the DHL hydrogenation kinetics with mono- and bimetallic colloidal catalysts (#1–4, Table 3.1). One can see

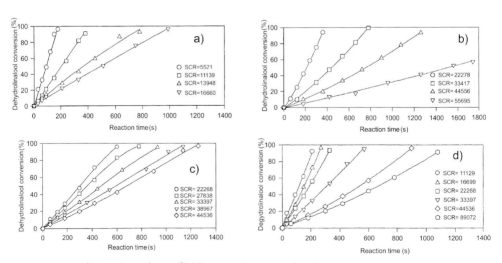

Fig. 3.4 Dependence of DHL conversion on reaction time with Pd (a) and bimetallic PdAu (b), PdZn (c) PdPt (d) catalysts for different SCR at 90 °C and 960 shakings min^{-1}. Reprinted from Ref. [51], Copyright (2000), with permission from Elsevier.

that, for the Pd nanoparticulate catalyst, the rate of hydrogenation does not depend on the current substrate concentration, while for bimetallic catalysts a complex dependence of the hydrogenation rate on substrate concentration is demonstrated. For bimetallic catalysts, in the equation of the model the denominator containing the coordination (adsorption) parameter is introduced, which shows a more pronounced impact of the coordination (adsorption) processes on DHL hydrogenation compared to the monometallic one. On the other hand, because the coordination (adsorption) parameter Q is very small for PdPt and PdZn nanoparticulate catalysts, the second component in the denominator can be omitted so the equation becomes $W=k$, i.e., identical to that found for Pd nanoparticles. Among all the catalysts, PdAu stands apart: its model equation contains a squared term in the denominator [44], revealing the high impact of the substrate/product coordination (adsorption).

Another interesting feature of all bimetallic catalysts is a significant increase in frequency factor (k_o) compared to monometallic Pd nanoparticles. For PdPt and PdZn nanoparticulate catalysts, this can be explained by an increase in defects in bimetallic particles [59] since these particles have "cluster-in-cluster" morphology. However, it is noteworthy that for the PdAu catalyst, k_0 increases even more, by several orders of magnitude. At the same time, the activation energy E_a increases approximately by a factor of 2 compared to that of other catalysts. An increase in the activation energy for PdAu is in agreement with two facts: Au, being a strong acceptor in relation to Pd and being located in the core of the particle (which is confirmed by CO adsorption), pulls electron density from Pd. The low electron density of the nanoparticles provides favorable conditions for hydrogenation of double bonds, as described in Refs. [59, 60], while for triple-bond hydrogenation, high electron density on the catalyst surface is normally preferable [61]. Then, low electron density increases the energy barrier of the reaction. On the other hand, at a nanoparticle size of 1–2 nm for bimetallic core–shell particles, the majority of the Pd atoms should be located on the nanoparticle surface [59, 60], while for a Pd nanoparticle, a fraction of Pd atoms is at the center of the particle. Thus, an increase in the frequency factor for PdAu colloids (compared to Pd nanoparticles) can be explained both by defects on the nanoparticle surface generated by the formation of core–shell structures and by the fact that the majority of the Pd atoms are on the surface and available for the substrate.

The TOF values (# 1–4, Table 3.1) for mono- and bi-metallic colloidal catalysts show that, for bimetallics, the catalytic activity is higher than that for the Pd ones, which can be explained by the modifying influence of gold, platinum, and zinc. It is known that the addition of a second metal can change the electronic properties of catalyst (ligand effect) [62–66]. As discussed in Ref.[67], this change can influence the energy of metal–hydrogen and metal–substrate bonds and the amount of hydrogen adsorbed. The addition of a modifying metal to a basic component can also change the surface geometry (ensemble effect) [64, 68–73]. Thus, a combination of all these factors determines the change in catalytic activity of bimetallic catalysts, which is reflected in the TOF values and in the change in hydrogenation mechanisms.

Homogeneous catalysts based on PS-*b*-P4VP (# 1–4, Table 3.1) reveal very high selectivity (99.8% at 100% conversion) at high reaction rates but the separation of the catalyst from the reaction solution is a very tedious task so technologically these catalysts do not hold great promise.

Heterogenization (deposition) of the Pd-nanoparticle-containing PS-*b*-P4VP micelles on alumina (# 5–7, Table 3.1) allowed us to solve two problems: to facilitate separation of the product from the reaction solution by mere filtration of the catalyst and to increase the stability of the catalytic system without selectivity loss. It is worth noting that for Pd-based catalysts (compare # 1 and 5, Table 3.1) heterogenization leads to increased catalytic activity. We believe this is caused by the increase in the amount of active centers when micelles are adsorbed on the alumina surface so the catalytic centers are more available for substrate. This is confirmed by the increase in the frequency factor while the activation energy remains nearly unchanged (Table 3.1). Thus heterogenization increases the activity of the nanoparticles formed in block copolymer micelles, making them good candidates for commercial applications. An increase in the length of the hydrocarbon tail (C5, C10, C20) in acetylene alcohols (see Scheme 3.1) leads to a decrease in the hydrogenation rate while selectivity is retained (Table 3.1, #5–7). We think this is caused by slower diffusion of larger reagents towards catalytic sites, while interaction with the active sites is not altered.

3.2.1.1.2 PEO-*b*-P4VP-based Catalysts

The main disadvantage of the catalysts based on PS-*b*-P4VP is the solvent, toluene, for solubilization of the PS corona. This solvent is hazardous and may pollute the final product due to its comparatively low vapor pressure and difficult removal. To overcome this problem, we used poly(ethylene oxide)-*block*-poly-2-vinylpyridine (PEO-*b*-P2VP) block copolymer to form catalytic nanoparticles as PEO-*b*-P2VP forms micelles in water and its mixtures with alcohols [74]. An obvious advantage of an aqueous medium is the environmental factor [75, 76]. Similar to PS-*b*-P4VP, in PEO-*b*-P2VP, Pd nanoparticles grow surrounded by pyridine species, as the core-forming block is P2VP. At the same time, there is an important difference between these two block copolymer systems. As the P2VP block is pH sensitive [77], in a polar medium (water or its mixtures) pH may strongly influence the micellar characteristics which should correlate with catalytic properties giving an important tool for tuning the catalytic activity and selectivity. The amphiphilic nature of DHL complies with the chosen system although water is a poorer solvent for long chain acetylene alcohols than toluene.

Here Pd nanoparticles have been prepared by incorporation of Na_2PdCl_4 in the PEO-*b*-P2VP micellar solution followed by reduction with $NaBH_4$ [74]. Figure 3.5 shows a PEO-*b*-P2VP micelle filled with Pd nanoparticles. The latter have a mean diameter of 2.5 nm so they are slightly larger than the particles formed in the PS-*b*-P4VP micelles. This may be due to partial swelling of the micelle core in water.

For these Pd nanoparticles, we studied the relationships between the catalyst structure and catalytic properties at different pH and solvent compositions in the selective hydrogenation of DHL [74]. As a second solvent (additional to water), isopropanol (IPrOH) was chosen, because it is a good solvent for DHL [78] and,

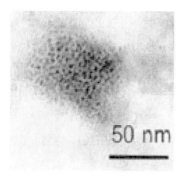

Fig. 3.5 Pd nanoparticles formed in a PEO-*b*-P2VP micelle. Reprinted from Ref. [74], Copyright (2004), with permission from Elsevier.

Table 3.2 Catalytic properties and kinetic parameters of selective hydrogenation of acetylene alcohols with micellar catalysts based on PEO-*b*-P2VP[a].

#	Catalyst	Substrate	H₂O content; pH	TOF, s⁻¹	S (A)	Mean micelle diameter[b], nm
1	PEO-*b*-P2VP-Pd	DHL	N1: 5% H_2O, pH 9.4	5.7	99.4 (100)	256
			N2: 50% H_2O, pH 13.4	9.2	95.9 (100)	34
			N3: 30% H_2O pH 13.4	9.6	98.0 (100)	70
			N4: 25% H_2O pH 13.4	6.8	98.7 (100)	28
			N6: 30% H_2O pH 13.9	7.3	96.8 (100)	86
			N7: 30% H_2O pH 14.3	9.4	98.0 (100)	—
2	PEO-*b*-P2VP-Pd/Al₂O₃		30% H_2O, pH 13.4	10.3	98.5 (100)	—
3	PEO-*b*-P2VP-Pd	DHIP		1.6	99% (100%)	—
		DMEC		10.6	91% (100%)	—
4	PEO-*b*-P2VP-Pd/Al₂O₃			12.0	91% (100%)	—

a   Pd content is 0.06 wt.%; Reaction conditions: 70 °C, 960 shakings min⁻¹, solvent (30 mL): IPrOH + H_2O, C_o 0.4 mol L⁻¹, C_c 1.7×10^{-5} mol Pd L⁻¹. TOF is calculated at 100% substrate conversion.
b   From AFM measurements.

in combination with water, allows one to keep the micelles intact. As previously reported [51], an alkali medium (KOH addition) favors the acetylene alcohol hydrogenation, so the reaction was studied in the pH range 9.4–14.3.

Effects of pH and Solvent Composition   Table 3.2 contains data on catalytic activity and selectivity in DHL hydrogenation at different pH and IPrOH content in the solvent. Runs N2–N4 present the data obtained at a fixed pH value (13.4). The selectivity values increase with increasing IPrOH content. For run N2, the catalyst displays practically the same selectivity but the activity increases with increasing IPrOH content, while subsequent alcohol addition leads both to the selectivity and activity increasing (runs N2 and N3).

To clarify the pH influence, we carried out the hydrogenation experiments at the optimal IPrOH content (70 vol.%) varying the KOH solution loading (pH value). One can see (runs N3 and N 6–7, #1, Table 3.2) that the catalytic activity passes through a minimum plateau (at pH 13.9 . . . 14.1), while the selectivity varies with no direct compliance with the activity change.

Since the micellar characteristics of the PEO-*b*-P2VP-Pd block copolymer aggregates in a given solvent and at a given pH may influence the catalytic properties, this was another important variable for the present system. PEO-*b*-P2VP forms micelles in water whose characteristics are mainly preserved after metallation (incorporation of a metal salt and metal nanoparticle formation) [49], but the behavior of this system in the presence of isopropanol is more complicated. To clarify the effect of solvent composition and pH on micellar characteristics, TEM and atomic force microscopy (AFM) studies of the key reaction mixtures were performed. Table 3.2 summarizes the catalytic data and mean diameters of the micelles and micellar aggregates obtained from AFM.

One can assume that the size and density of micellar aggregates should influence the catalytic activity and selectivity. The larger the size and the higher the density of the micellar structures, the slower the rate of the substrate and IPrOH penetration to the active sites (internal diffusion limitations), hence, the slower the reaction should proceed. (Density of the micelles and micellar aggregates can be derived from the electron density [darkness in the TEM image] of these structures). At the same time, when micelle cores are denser, the local concentration of the pyridine ligands on the nanoparticle surface is higher, so the particle modification is stronger. As discussed above [44, 47], modification of the metal nanoparticle surface with pyridine units causes the selectivity gain, thus, increase in the micelle density may increase the selectivity of DHL hydrogenation. The data presented in Table 3.2 and obtained from AFM and TEM measurements confirmed these hypotheses.

To illustrate this phenomenon, AFM images and corresponding histograms of the micellar structures observed for two samples, N1 and N2, are presented in Figs. 3.6 and 3.7. The AFM image sample N1 (Fig. 3.6) confirms the formation of large micellar aggregates with mean diameter 256 nm. This sample has a comparatively low activity due to the large aggregates but high selectivity due to the high density of the P2VP cores. The N2 sample shows nearly spherical micelles measuring only about 34 nm (Fig. 3.7). The density of the majority of these micelles estimated by TEM (not shown here) is rather low. These two events result in high activity but rather low selectivity.

Thus, micelle characteristics depending on pH and "IPrOH:water" ratio play a crucial part in the catalytic properties as they may control the substrate and product transport towards active centers and their modification.

Kinetics of DHL Hydrogenation   The kinetics of selective DHL hydrogenation with Pd nanoparticles formed in PEO-*b*-P2VP was studied for the reactions carried out at a pH of 14.3 and 70 vol.% IPrOH solvent, as these conditions provide comparatively high selectivity (98.0% at 100% DHL conversion) and TOF values (9.4 mol

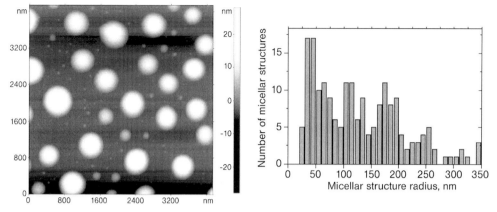

Fig. 3.6 AFM image and size histogram for the sample N1 from Table 3.2. Reprinted from Ref. [74], Copyright (2004), with permission from Elsevier.

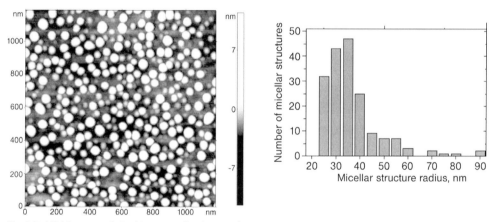

Fig. 3.7 AFM image and size histogram for the sample N2 from Table 3.2. Reprinted from Ref. [74], Copyright (2004), with permission from Elsevier.

DHL mol Pd^{-1} s^{-1}) (see Table 3.2, # 1, N7). The apparent activation energy was determined from the initial reaction rate dependence [79] on the temperature varied in the range 50–70 °C. Its value, deduced from the slope of the Arrhenius dependence, was 31 kJ mol^{-1}. Other kinetic experiments were carried out at 70 °C varying the SCR values (C_o and C_c were varied in the ranges 0.21–0.58 mol L^{-1} and 1.6×10^{-5}–1.8×10^{-5} mol of Pd/L, respectively).

DHL conversion dependences on the reaction time obtained at different SCRs are presented in Fig. 3.8a. These data allowed us to obtain kinetic expressions describing the hydrogenation kinetics in the range of the studied SCRs and the reaction conditions. Fig. 3.8 b shows the kinetic data in dependence on DHL conversion, $(1 - XDHL) \times 100\%$, on relative time . One can see the existence of an initial phase characterized by a low reaction rate (up to approx. 20% DHL conversion) unlike PS-*b*-P4VP where no initial phase was observed. Remarkably, during this phase the selectivity remained 100% followed by a fall to $98.0 \pm 0.5\%$; after that the selectivity remained constant.

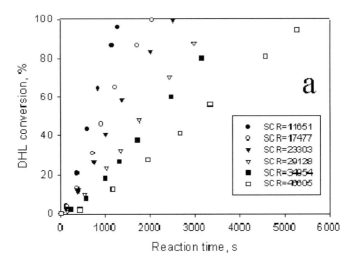

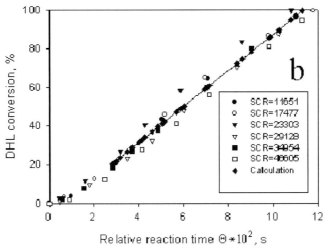

Fig. 3.8 Dependences of DHL conversion on reaction time (a) and relative time (b), at different SCR values. The solid line indicates the calculated linear dependence; the initial phase (up to 20% DHL conversion) was not taken into consideration. Reaction conditions: 960 shakings min^{-1}, 70 °C, pH14.3, 70 vol% IPrOH. Reprinted from Ref. [74], Copyright (2004), with permission from Elsevier.

This fact is in agreement with conclusions that during the initial phase of the alkyne hydrogenation over Pd, both reactive and less reactive surface species are formed and that less reactive ones are responsible for high selectivity[50, 80]. Thus, we can assume that during the initial phase (similar to an induction period), the reaction proceeds via less reactive species yielding 100% selectivity, while highly reactive species responsible for the selectivity drop are being formed. After the initial phase, one can see a linear dependence indicating zero order with respect to DHL, which is often related to the strong adsorption of the substrate on the catalyst surface [81]. The TOF value was confirmed to be constant and equal to $9.4 \pm 0.3\,\mathrm{s}^{-1}$ for all the studied SCRs.

The best kinetic model which describes well the hydrogenation kinetics of the computation curve from the experimental data is presented below:

$$W = \frac{k}{X_1 + QX_2}$$

where X_i is the relative concentration $X_i = C_i / C_o$ (C_i is the current concentration of the substrate at $i=1$ and product at $i=2$). Here, k and Q are equal to 7.3 (mol/ mol) s^{-1} and 0.61, respectively. Based on the experimental data obtained at different temperatures, the parameters of the Arrhenius equation: $E_a = 33\,\mathrm{kJ\,mol^{-1}}$ and $k_o = 102$ were calculated. One can see a good agreement between the experimental and calculated values of E_a. The higher activation energy for this system (PEO-*b*-P2VP-Pd) compared to PS-*b*-P4VP-Pd at nearly the same frequency factor (k_o) results in lower catalytic activity: TOF values are 18.5 and 9.6 for PS-*b*-P4VP-Pd and PEO-*b*-P2VP-Pd, respectively. The difference in the catalytic activity of PS-*b*-P4VP-Pd and PEO-*b*-P2VP-Pd can be explained by the hydrophobic–hydrophilic interactions between substrates and catalysts in a particular solvent. In the PS-*b*-P4VP-Pd system the hydroxy group (adjacent to a triple bond) of the acetylene alcohol is primarily directed toward the polar P4VP core containing catalytic nanoparticles due to the unfavorable hydrophobic environment of the PS corona. In the PEO-*b*-P2VP-Pd system, the corona, solvent and core are hydrophilic so there is no primary orientation of the substrate towards the active sites. Another interesting difference between the two block copolymer systems is the much higher coordination (adsorption) parameter ($Q=0.61$) for the PEO-*b*-P2VP-Pd system, revealing that adsorption of DHL and LH are nearly equal, while for PS-*b*-P4VP-Pd, $Q=10^{-3}$, i.e., the LN coordination (adsorption) is negligible compared to that of DHL. It is noteworthy that in this case, the increased adsorption of LN does not lead to loss in selectivity (the non-selective hydrogenation might be facilitated).

Similar to the PS-*b*-P4VP-Pd system, we studied the heterogeneous catalysts obtained by deposition of PEO-*b*-P2VP-Pd micellar solution on alumina. The data on activity and selectivity of the PEO-*b*-P2VP-Pd/Al2O3 are presented in Table 3.2. One can see that again heterogenization results in an increase in the catalytic activity (although less pronounced than for PS-*b*-P4VP-Pd/Al2O3) without selectivity loss. Kinetic studies allowed calculation of a kinetic model for PEO-*b*-P2VP-Pd/Al2O3 which is identical to that of the homogeneous catalyst. This reveals that the mechanism of the selective hydrogenation of the triple bond is similar for both heterogeneous and homogeneous catalytic systems based on PEO-*b*-P2VP. Using the experimental data on temperature dependence of catalytic properties and computation, the activation energy and a frequency factor were found to be 33 kJ mol^{-1} and 103, respectively. Thus increase in catalytic activity upon heterogenization of PEO-*b*-P2VP-Pd can be attributed to the increase in the frequency factor by an order of magnitude while the activation energy remained unchanged.

It should be noted that similar to the catalysts based on PS-*b*-P4VP, for the PEO-*b*-P2VP based systems, increase in substrate length leads to a decrease in the reaction rate (Table 3.2) indicating diffusion limitations for the reagents within block copolymer micelles. This diffusion should be directly dependent on the size of the reagent.

Thus, heterogenization of the block copolymer-based nanoparticulate catalysts makes them more technologically favorable and better suited for industrial applications while creating the opportunity for a repetitive use of these catalysts.

3.2.1.2 Nanoparticulate Catalysts Formed via Microgel Templating

Other homogeneous nanostructured polymeric matrices are cationic and anionic microgels based on sulfonated polystyrene and poly(ethylmethacryltetramethyl ammonium chloride) which can sequester metal ions from reaction solutions and serve as nanoreactors for nanoparticle formation [82]. To realize the advantages of heterogeneous systems, we used these microgels containing Pd nanoparticles along with polystyrene-*block*-poly(ethylene oxide) block copolymers as templates for casting mesoporous alumina with nanoparticles. Metal-particle-containing aluminas templated both over cationic and anionic microgels consist of an interpenetrating pore system and alumina nanowires (2–3 nm in diameter and about 40 nm in length); the latter are intimately embedded in the porous network. It also includes Pd nanoparticles (Fig. 3.9) [83]. It is noteworthy that both block copolymer and microgels are removed during the calcination procedure as a mesoporous solid is formed. Here A-Pd-NCM-[Al] stands for anionic (A) microgel with Pd (Pd) nanoparticles used for templating of alumina ([Al]). C-Pd-NCM-[Al] stands for cationic (C) microgel with Pd (Pd) nanoparticles used for templating of alumina ([Al]). From the TEM data, in A-Pd-NCM-[Al] the Pd nanoparticles are in the size range 6–19 nm, while for C-Pd-NCM-[Al], the particle diameter varies from 6 to 27 nm. At the same time, the micrographs show smaller nanoparticles (of about 2 nm in diameter) included in the alumina nanofibers. According to the ^{27}Al MAS NMR spectra, alumina samples, independentl of microgel type, contain mainly six-coordinated Al species which are normally responsible for the low acidity of a mesoporous solid. The latter factor is favorable for the hydrogenation reaction [83].

The catalytic properties of Pd-containing aluminas were studied in the selective hydrogenation of all three long chain acetylene alcohols: DMEC, DHL and DHIP (Scheme 3.1, Table 3.3). Because the structure of aluminum sites in both A-Pd-

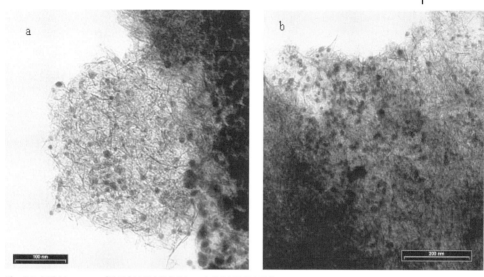

Fig. 3.9 TEM images of A-Pd-NCM-[Al] (a) and C-Pd-NCM-[Al] (b). Reprinted with permission from Ref. [83]. Copyright (2003) American Chemical Society.

Table 3.3 Catalytic activity and selectivity of microgel-templated mesoporous alumina in the hydrogenation of acetylene alcohols[a].

#	Catalyst	wt.% Pd	Substrate	TOF, s^{-1}	S (A)
1	A-Pd-NCM-[Al]	1.4	DHL	2.6	90 (99%)
2	C-Pd-NCM-[Al]	0.5		3.7	90 (99%)
3	A-Pd-NCM-[Al]	1.4	DMEC	2.1	86 (98)
4	C-Pd-NCM-[Al]	0.5		5.0	90 (99)
5	A-Pd-NCM-[Al]	1.4	DHIP	2.6	90 (98)
6	C-Pd-NCM-[Al]	0.5		3.5	98 (99)
7	Pd/Al$_2$O$_3$ commercial non-modified	0.5		2.5	96 (99)
8	Pd/Al$_2$O$_3$ modified with Zn acetate, pyridine and alkali	0.5		1.7	97–98 (100)

a   Reaction conditions: 70 °C, 960 shakings min^{-1} (regime without diffusion limitations), IPrOH (30 mL), C_o 0.43 mol L^{-1}, C_c 1.4×10^{-5} mol Pd L^{-1}.

NCM-[Al] and C-Pd-NCM-[Al] catalysts is the same (according to ^{27}Al NMR data), the support influence should also be similar. The higher activity of C-Pd-NCM-[Al] can be explained by a lower fraction of the large particles and a lower Pd content compared to A-Pd-NCM-[Al]. The latter factor allows one to avoid shielding the particle surface by other particles located in its vicinity. When a shorter substrate is used (C_5), the same trend is observed (Table 3.3): higher activity with C-Pd-NCM-[Al] catalyst than with A-Pd-NCM-[Al]. At the same time, for C_{20} alcohol hydrogenation, the difference in activity between the above catalysts is rather negligible, but the difference in selectivity is pronounced (Table 3.3). The higher selectivity of C-Pd-NCM-[Al] can be explained by the much greater pore size, which prevents diffusion limitations for C_{20} acetylene alcohol molecules and facilitates desorption and removal of isophytol (hydrogenation product) from the Pd particle surface. This prevents further hydrogenation of isophytol to side products, resulting in the decreased selectivity. Thus, the C-Pd-NCM-[Al] catalyst allows one to achieve high selectivity for DHIP hydrogenation at a reasonable catalyst activity. These values are especially remarkable when compared with a commercial Pd/Al$_2$O$_3$ catalyst (Table 3.3). When this catalyst was modified with Zn acetate, pyridine and alkali, the selectivity increased to 97–98%, but the TOF was only $1.7\,s^{-1}$ [51]. DHIP hydrogenation with Lindlar catalyst (Pd/CaCO$_3$ modified with quinoline or Cu^{2+}, Cd^{2+}, Mn^{2+}, Fe^{3+}, and Co^{2+} ions) also results in a selectivity increase to 97.2–98.0%, but the activity is nearly halved [84] compared to the non-modified Pd/Al$_2$O$_3$ catalyst (see Table 3.3). Thus the advantages of the C-Pd-NCM-[Al] catalyst in DHIP hydrogenation are high selectivity and much higher activity than for the commercial catalysts, although without using the additional modifiers which might pollute the final product and be hazardous for the environment. We believe that C-Pd-NCM-[Al] can also be an excellent catalyst for other reactions involving substrates of a similar size or structure.

On the other hand, in the hydrogenation of DMEC and DHL these catalysts are much less active and selective than PS-*b*-P4VP-Pd or PEO-*b*-P2VP-Pd. The lower activity may be due to the formation of comparatively large particles (compare ~2 nm nanoparticles in block copolymer-based catalysts and the nanoparticles formed in microgel-templated catalysts). The lower selectivity can be caused by the absence of modifying groups on the nanoparticle surface.

3.2.1.3 Nanoparticulate Catalysts Formed in Highly-functionalized Polymer Colloids

Recently, we developed a new type of self-supported system based on a novel family of polysilsesquioxane colloids [85, 86]. They were synthesized by the hydrolytic condensation of a functionalized precursor: *N*-(6-aminohexyl)aminopropyltrimethoxysilane (AHAPS), H$_2$N(CH$_2$)$_6$NHCH$_2$CH$_2$CH$_2$Si(OCH$_3$)$_3$, to form highly crosslinked poly(aminohexyl)(aminopropyl)silsesquioxane (PAHAPS) [85, 86]. The PAHAPS structure is highly functionalized and its functionality can be varied by protonation of the amino groups. These features make PAHAPS a robust polymer material obtained in a one-pot reaction procedure. We reported a synthesis of PAHAPS in various reaction conditions, allowing tailoring of the PAHAPS micro-, and macro-structure, and the correlation between PAHAPS and metal nanoparticle characteristics and reaction variables. As amino groups easily form complexes with the majority of transition metal compounds, PAHAPS colloids can template a variety of particles of interest. This system may thus have wide application in chemistry and materials science.

For the catalytic study, the hydrolytic condensation of AHAPS was carried out in two ways: with excess water (water serves both as a solvent and a hydrolyzing agent) and with water sufficient only to provide hydrolysis of methoxy groups, while THF is used as a solvent (it is a good solvent for the AHAPS tails). Hydrolysis in water leads to lamellar ordering within polymer colloids while in THF the polymer colloids are disordered. Incorporation of Pd compounds is driven by complexation with Pd compounds. Reduction of metallated PAHAPS was carried out using two reducing agents: $NaBH_4$ providing fast nucleation and $N_2H_4 \cdot xH_2O$ providing slow nucleation. In the majority of cases, when metal particle formation is not restricted by the walls of the small cavity ("cage" effect) [87–89], and metal ions are able to diffuse through the polymer stabilizer, slow nucleation results in large particles, while fast nucleation produces small particles [46]. Metallation of the PAHAPS colloids is envisioned in Scheme 3.2. Pd species were sequestered by PAHAPS colloids from aqueous solutions of K_2PdCl_4 followed by reduction. The degree of metallation is very high: by elemental analysis data Pd content reached 17–21 wt.%, depending on the reaction conditions. It is noteworthy that here the particle size only depends slightly on the strength of the reducing agent. For both reducing agents, the majority of the particles measure about 1.3–1.5 nm. In addition, the sample reduced with $N_2H_4 \cdot xH_2O$ contains also a fraction of particles measuring about 3.5 nm and located at the edges of the polymer colloids (Fig. 3.10) [86]. These facts reveal that particle growth is mainly controlled by the cavity size and high functionality of PAHAPS, but at the edges of the polymer colloids the particle growth is less restricted.

Table 3.4 presents the catalytic properties of PAHAPS colloids containing Pd nanoparticles in selective hydrogenations of DHL and DMEC. Polymer notations in Table 3.4 are consistent with those presented in Ref.[86]. One can see that the higher activity was observed for the PAHAPS catalysts where nanoparticles were formed by $N_2H_4 \cdot xH_2O$ reduction revealing that the particles formed at the edges

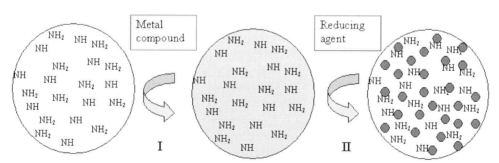

Scheme 3.2 Schematic image of PAHAPS colloid metallation. Step I is incorporation of metal compounds inside the PAHAPS colloids due to interaction with amino groups. Step II is metal particle formation due to reduction of metal species. Reprinted with permission from Ref. [86]. Copyright (2003) American Chemical Society.

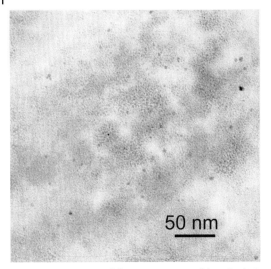

Fig. 3.10 TEM image of the cross-section of the PAHAPS-H4 sample obtained after incorporation of K_2PdCl_4 and metal particle formation using $N_2H_4 \cdot xH_2O$.

Table 3.4 Catalytic properties and kinetic parameters in the selective hydrogenation of DHL and DMEC with PAHAPS-Pd[a].

#	Catalyst	Wt.% Pd	Substrate	TOF, s^{-1}	S (A)
1	PAHAPS-H4-Pd1-NaBH$_4$[b]	20.60	DHL	0.17	96.7 (67)
			DMEC	0.3	98.0 (74)
2	PAHAPS-H4-Pd1-N$_2$H$_4$[b]	20.60	DHL	2.44	93.7 (96)
			DMEC	1.8	98.4 (92)
7	PAHAPS-F1-Pd1-N$_2$H$_4$[c]	17.55	DHL	2.94	91.9 (92)
			DMEC	2.01	96.0 (94)

a Reaction conditions: 70 °C, 960 shakings min^{-1} (regime without diffusion limitations), isopropanol (30 mL), C_o 0.43 mol L^{-1}, C_c 2×10^{-5} mol Pd L^{-1};
b PAHAPS was prepared in water;
c PAHAPS was prepared in THF.

are more available for a substrate. There was no dependence of the catalyst activity on the internal polymer ordering, demonstrating that highly crosslinked PAHAPS colloids are impenetrable for a substrate. This is also confirmed by the absence of the dependence of the catalytic activity on the substrate length (DHL and DMEC), indicating the absence of substrate diffusion through the catalyst. We think that, in general, the lower activity of these systems compared to catalysts based on block copolymers is due to the large amount of "wasted" Pd nanoparticles located within the PAHAPS colloids. It is noteworthy that the PAHAPS-Pd catalysts also display lower selectivity than the block copolymer-based catalysts due to inefficient modification of the nanoparticle surface with primary and secondary amines compared to the pyridine units of the P4(2)VP blocks.

3.2.2
Oxidation with Catalytic Nanoparticles

An example of a catalytic reaction of fundamental, as well as technological, interest is the oxidation of L-sorbose to 2-keto-L-gulonic acid (2-KGA). Several strategies based on chemical, electrochemical, biotechnological and catalytic methods are currently available for the oxidation of L-sorbose, which is an intermediate in Vitamin C production. While the catalytic route (this reaction is catalyzed with Pt catalysts) seemed to be most promising, it had a number of complications. Due to acetonation, carbohydrate functional groups are initially protected from oxidation, in which case the reaction selectivity increases. The initial functional groups must, however, be recovered after oxidation, resulting in remarkable loss of end-product [90, 91]. For the commercial Pt/C (5% Pt) catalyst, this provides only 28–37% 2-KGA yield at 100% L-sorbose conversion. Direct catalytic oxidation of L-sorbose to 2-KGA (Scheme 3.3) [92–95] does not require acetonation. This reaction has been performed in the presence of O_2 over Pt or Pd catalysts deposited

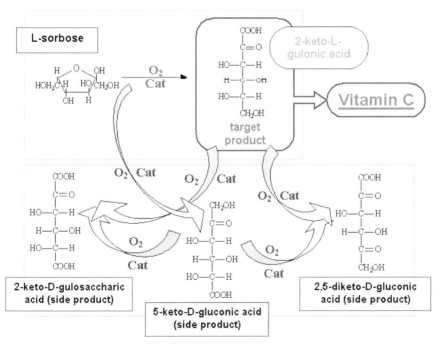

Scheme 3.3 Catalytic oxidation of L-sorbose.

Table 3.5 Catalytic properties of L-sorbose oxidation with Pt-containing nanostructured catalysts [a].

#	Catalyst	Wt. % Pt	TOF 103, s^{-1}	S (A)
1	Pt/γ-Al$_2$O$_3$ (H$_2$)[b]	3.00	0.17	76.7(63)
2	PS-*b*-P4VP-Pt/Al$_2$O$_3$ (NaBH$_4$)	0.35	0.14	45(23)
3	PEO-*b*-P2VP-Pt/Al$_2$O$_3$ (NaBH$_4$)	1.50	5.0	60 (84)
4	HPS-Pt2-H$_2$	3.02	1.03	88.0(89)
5	HPS-Pt1	7.51	0.54	95.7(100)
6	HPS-Pt2	3.02	1.34	95.8(100)
7	HPS-Pt3	1.87	1.11	94.9(100)
8	MN-270-Pt-THF	1.72	2.5	96.8(100)

a   Reaction conditions: water (25 mL) as solvent; atmospheric pressure of oxygen with flow rate of 14×10^6 m^3 s^{-1}, stirring of 1000 rpm (no diffusion limitation); the reaction time was 1.2×10^4 s. Alkaline (NaHCO3) quantity is equal to initial L-sorbose concentration C_o.

b   At gradual alkaline loading; parentheses indicates a reducing agent used.

on activated carbon or alumina. In this case, the reaction must be conducted in neutral or low-alkali media. Selectivity may be increased through the addition of modifying (promoting) agents such as phosphine and aminophosphine complexes, as well as aromatic and cycloaliphatic amines [92–95], at optimal loading levels. Despite the use of modifiers, the resultant selectivity decreases from 95% at 30% conversion to 40% at 100% conversion. Thus, the two principal disadvantages of this procedure are (i) low selectivity at high conversion and (ii) contamination of the end-product with reaction modifiers.

To address these issues, we studied the catalytic behavior of Pt nanoparticles formed in several nanostructured polymeric systems in the selective oxidation of L-sorbose to 2-keto-L-gulonic acid (Table 3.5) [89]. Commercial Pt/γ-Al$_2$O$_3$ (3% Pt, Degussa AG) was used for comparison. The reaction has been conducted in alkali media (NaHCO$_3$), yet gradual alkaline loading (NaHCO$_3$) provides the highest selectivity, while one-shot results in the highest TOF.

From Table 3.5 one can see that the platinum catalyst based on PS-*b*-P4VP shows low activity and conversion because access to catalytic nanoparticles located in the micelle cores is impeded by the PS corona which is insoluble in water. When a block copolymer contains a water soluble corona (PEO), activity significantly increases due to accessibility of the catalytic particles, however selectivity and L-sorbose conversion are still low. Although quaternary ammonium compounds are known to improve activity and selectivity in L-sorbose oxidation [92–96], apparently pyridine units adsorbed on the Pt nanoparticle surface promote too high activity which normally results in poorly controlled reaction. This leads to decreased selectivity and formation of side products such as 5-keto-D-gluconic acid, 2-keto-L-gulosaccharic acid, and 2,5-diketo-D-gluconic acid (Scheme 3.3). Thus for L-sorbose oxidation, we were looking for nanoparticulate catalysts with a less-functionalized nanoparticle surface. We think that nanoparticles formed in cationic and anionic microgels and later templated in mesoporous alumina might be prospective catalysts for L-sorbose oxidation. However, these catalysts have several disadvantages.

First, the particles formed are too large so efficiency of noble metal usage is too low. Secondly, the catalyst synthesis is too complex, involving several stages. In our earlier work we studied the formation of Co nanoparticles in nanoporous hypercrosslinked polystyrene (HPS) which showed great promise in controlling the nanoparticle size [97].

HPS is the first representative of a new class of cross-linked polymers characterized by unique topology and unusual properties[98, 99]. The internal morphology of HPS is shown schematically in Fig. 3.11. Due to its high crosslink density, which can exceed 100%, HPS consists of nanosized rigid cavities of about 2 nm in size (micropores). It is easily produced by chemically incorporating methylene groups between adjacent phenyl rings in dissolved polystyrene homopolymer or gelled poly(styrene-*r*-divinylbenzene) copolymer in the presence of ethylene dichloride. Good solvent and conformational network rigidity are generally necessary to prepare hypercrosslinked polymers possessing extremely high inner surface areas (typically in the range of $1000 \, m^2 \, g^{-1}$). HPS is an interesting object for the creation of novel catalytic systems. This material can act as a nanostructured matrix, which governs the particle growth. As was demonstrated, HPS perfectly controls formation of Co nanoparticles [97]. We surmised this material can be an excellent carrier of catalytically active particles, especially because HPS can swell in almost all solvents, including precipitants for HPS. This can provide access to catalytic sites in virtually all reaction media. Another advantage of HPS is low cost.

To form Pt nanoparticles, HPS was impregnated in an inert atmosphere with a solution of H_2PtCl_6 in THF followed by reduction with H_2. It is noteworthy that incorporation of the THF solution of platinic acid in microporous HPS results in partial reduction of the Pt(IV) species and formation of Pt(II) complexes where the ligands are the THF oxidation products [89]. The accumulation of Pt(II) complexes, as well as subsequent Pt nanoparticle formation, is effectively restricted by

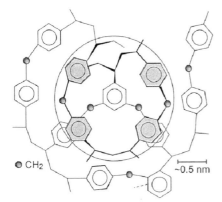

Fig. 3.11 Schematic illustration of the internal network of HPS. The speckled phenyl rings reside in a different plane relative to the unspeckled ones in the cross-linked material, and the circle identifies a postulated cavity in which metal nanoparticles could grow. Reprinted with permission from Ref. [97]. Copyright (1999) American Chemical Society.

the size of the HPS nanocavities. After H_2 reduction in dry, solid polymer, the nanoparticles possess a narrow size distribution and a mean diameter of 1.3–1.4 nm (Fig. 3.12).

As can be seen from Table 3.5 (#4), Pt nanoparticles formed in HPS are more selective in L-sorbose oxidation than the catalysts based on block copolymers but still the selectivity is not sufficient. We believe that low selectivity may be attributed to saturation of the nanoparticle surface with H_2 during reduction, which alters the sorption–desorption equilibria of the L-sorbose, oxygen and 2-keto-L-gulonic acid [89]. As was demonstrated in Ref. [96], L-sorbose oxidation can be catalyzed not only with $Pt_{(0)}$ but also in the presence of Pt ions. This prompted us to investigate the catalytic properties of HPS-Pt without Pt ion reduction.

The selectivity of 95.8% at 100% L-sorbose conversion is achieved with the HPS-Pt catalyst. As can be seen from Table 3.5 (#5), decrease in Pt content from 7.51 to 3.02 wt.% results in an increase in catalytic activity due to a decrease in shielding of the neighboring particles. On further decrease in the Pt content (to 1.87 wt.%) the activity decreases slightly due to decrease in the amount of the active centers. The kinetic curves for HPS-Pt-2 (Fig. 3.13) reveal the existence of an induction period of about 100 min, during which time L-sorbose conversion is negligible. The presence of the induction period should be attributed to formation of active catalytic species *in situ* while the initial Pt species should be practically inactive.

A sample of the HPS-Pt-2 catalyst collected after the induction period was isolated and subjected to examination by XRD, TEM and XPS. The XRD pattern acquired from this material indicates the presence of Pt crystallites possessing a mean diameter of ca. 4 nm, along with some unresolved structure that might be

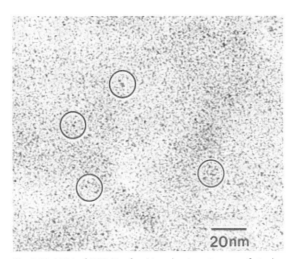

Fig. 3.12 TEM of HPS-Pt after H_2 reduction. Groups of single Pt nanoparticles are highlighted by circles. Reprinted with permission from Ref. [89]. Copyright (2001) American Chemical Society.

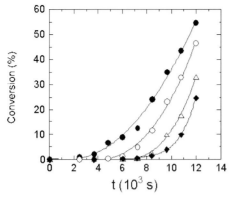

Fig. 3.13 Dependence of L-sorbose conversion on reaction
time (*t*) in the presence of (a) HPS-Pt-2 at different L-sorbose
concentrations (C_0, in M): 0.05 (·), 0.11 (○), 0.16 (▲), 0.22
(△), and 0.32 (○).The solid lines are guides for the eye.
Reprinted with permission from Ref. [89]. Copyright (2001)
American Chemical Society.

assigned to amorphous Pt nanoclusters [89]. An electron micrograph of this
sample (Fig. 3.14) shows the coexistence of small nanoparticles with a mean
diameter approaching the resolution of the microscope (0.5 nm) and larger particles with a mean diameter of up to 6.0 nm. The overall mean diameter of the
nanoparticles in this specimen is measured to be 1.8 ± 1.1 nm.

According to the XPS data, the Pt $4f_{7/2}$ binding energy changes from 73.6 to
72.4 eV after the induction period in the HPS-Pt2 catalyst. Deconvolution of this
spectrum reveals four different species with different binding energies. While one
of them possesses a Pt $4f_{7/2}$ binding energy of 71.3 eV and can be ascribed to Pt(0),
additional species with higher binding energies also appear to exist. We think that
the presence of "oxidized" Pt species may reflect the strong chemical interaction
between the surface atoms of the Pt nanoparticles and 2-KGA. Such interaction
can strongly change the binding energy of Pt. The catalytic species are definitely
formed *in situ* during the induction period. Therefore, HPS appears to serve two
crucial roles, first as a support for the Pt catalyst during L-sorbose oxidation, and
secondly as the nanostructured matrix responsible for controlling nanoparticle
growth [89].

In addition to selectivity and activity, another important feature of any promising
catalyst is the stability of its catalytic properties. Unfortunately, catalysts often
lose their activity after only two or three repeated uses, typically due to either loss
of the catalytic metal [96] or contamination of the catalytic surface by reaction
products [100]. As is seen from the data presented in our work [89], neither the
activity nor the selectivity of the HPS-Pt-2 catalyst deteriorates, even after 15
catalytic cycles. This result demonstrates that nanoparticles can exhibit remarkably
stable catalytic properties. Such stability is attributed to what we refer to as "double

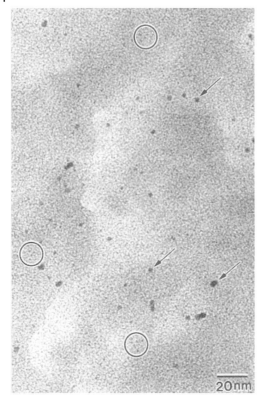

Fig. 3.14 TEM image of HPS-Pt-2 after the induction period during the direct oxidation of L-sorbose in an aqueous medium. Groups of single Pt nanoparticles are highlighted by circles, whereas substantially enlarged nanoparticles are identified by arrows. Reprinted with permission from Ref. [89]. Copyright (2001) American Chemical Society.

protection," in which (i) the location of the nanoparticles within HPS micropores of comparable size prevents leaching of catalytic metal; and (ii) chemical interaction of the nanoparticle surface with 2-KGA acid prevents surface contamination.

The only drawback of the studied system was the fully microporous nature of HPS (pores with diameters of about 2 nm), limiting the transport of substrate within the metallated HPS and hence the catalytic activity.

This motivated us to study metallation and catalysis with a commercially available HPS developed by the Purolite Co., as this material combines both micropores and macropores [101]. We believed that this should allow better mass transfer within HPS and higher catalytic activity. On the other hand, the presence of macropores might weaken the nanoparticle size control. We studied the structure and catalytic properties of the nanocomposite based on Purolite HPS and containing

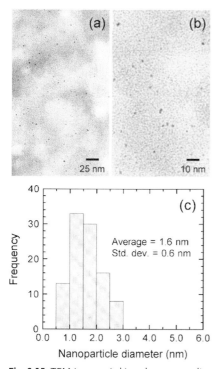

Fig. 3.15 TEM images (a,b) and corresponding particle
size histogram (c) of MN-270-Pt containing 1.72 wt.% Pt.
Reprinted with permission from Ref. [101]. Copyright (2004)
American Chemical Society.

Pt compound clusters (denoted as "MN-270-Pt"). No hydrogen reduction was
applied to form Pt nanoparticles prior to oxidation.

The TEM image of the "MN-270-Pt" nanocomposite after incorporation of platinic acid is presented in Fig. 3.15. One can see that the material contains nanoparticles with a diameter exceeding 0.7 nm (the estimated resolution of the microscope).
The mean particle diameter calculated from a particle size histogram is 1.6 nm
and standard deviation is 0.6 nm. Evidently, since HPS is a very hydrophobic
matrix, Pt species do not dissipate within a polymer matrix but form well defined
clusters. Light colored areas in the TEM image indicate macropores. The XRD
profile of this sample confirms the absence of the Pt-containing crystalline phase,
but suggests the presence of Pt amorphous scatterers.

At the highest selectivity of the "MN-270-Pt" catalyst measured in this work
(96.8% at 100% conversion), the value of TOF is found to be 2.5×10^{-3} mol mol
$Pt^{-1} s^{-1}$, which demonstrates an increase in the catalyst activity compared to HPS-
Pt-2 [89] and an even higher increase over other oxidation catalysts. We also attribute this increase in catalytic activity to the presence of macropores along with
micropores.

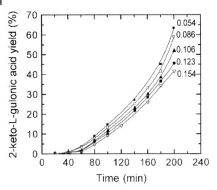

Fig. 3.16 Dependence of 2-keto-L-gulonic acid yield on reaction time in the presence of "MN-270-Pt" at different L-sorbose molar concentrations (displayed in the figure). Pt content is 1.72 wt.%. Reprinted with permission from Ref. [101]. Copyright (2004) American Chemical Society.

The kinetic curves (Fig. 3.16) also reveal the existence of an induction period of about 60 min, during which time the 2-keto-L-gulonic acid yield is negligible. Comparison of this value with the data on HPS-Pt-2 (fully microporous HPS) shows that the induction period is significantly diminished (from 100 min to 60 min) for "MN-270-Pt". The shortening of the induction period should be attributed to the presence of macropores and facilitation of L-sorbose diffusion[101].

Comparison of the TEM images and particle size histograms shows that particle sizes and particle size distributions are similar in the "MN-270-Pt" and "MN-270-Pt-induction" samples, suggesting that mixed Pt(0/II/IV) clusters in "MN-270-Pt" serve as nuclei for formation of Pt(0) nanoparticles; yet the nanoparticle size is controlled by the pore size. The higher value of a mean particle size obtained from the XRD data indicates that small Pt nanoparticles are amorphous. Clearly, reduction of Pt species in "MN-270-Pt" by L-sorbose during the induction period results in stabilization of the nanoparticle surface while macropores provide easy access of the substrate to the growing nanoparticles. This facilitates particle nucleation, while oxidation products may be adsorbed on the nanoparticle surface. Both these events contribute to the control over the particle size [101]. By contrast, in the fully microporous HPS, access of L-sorbose to the Pt species is impeded by a small pore size, thus nucleation occurs slowly and unevenly, while 2-keto-L-gulonic acid (a nanoparticle stabilizer) might be in short supply. This yields larger and broader distributed nanoparticles after the induction period [89].

We studied the effect of temperature on the kinetics of L-sorbose oxidation over the range 60–80 °C. Increasing temperature yields an increase in the rate of L-sorbose oxidation both for HPS-Pt-2 and "MN-270-Pt". The temperature dependence of L-sorbose oxidation can be further scrutinized if it follows Arrhenius behavior. To establish the applicability of the Arrhenius equation for "MN-270-Pt", we have measured the reaction time yielding 10% L-sorbose conversion ($\tau_{0.1}$). The computation yields an apparent activation energy E_a value of $35 \pm 3\,kJ\,mol^{-1}$ for "MN-270-Pt" (Table 3.5). This value is comparable with the $39 \pm 3\,kJ\,mol^{-1}$ found for HPS-Pt-2. Since the activation energy for both catalysts is nearly the same, while the catalytic activities are very different, we calculated an apparent frequency factor. Comparison of the apparent frequency factors shows that k_o of 3.2×10^6 for "MN-270-Pt" is two orders of magnitude higher than k_o of HPS-Pt-2 (2.5×10^4). We think that this can be due to the presence of macropores facilitating mass transfer and better accessibility of Pt nanoparticle active sites for a substrate.

The enhanced stability of Pt nanoparticles within MN-270 was amazing [101]. On the contrary, one might expect easier aggregation of the nanoparticles in the presence of macropores. This stability reveals that the clusters of the Pt species formed in MN-270 after incorporation of platinic acid are preferentially situated in the micropores, not at the macropore interfaces. We believe that such preferential positioning of the clusters could be caused by their hydrophobic shell due to organic ligands in the $Pt_{(II)}$ complex [89]. In addition, Pt nanoparticle nucleation and growth also occur in the micropores, restricting the nanoparticle size. Moreover, while one might expect that the some aggregation would increase upon repeated exposure to the reaction medium (during repeated use), the TEM and XRD results reported in Ref. [101] indicate that consecutive exposure of the Pt nanoparticles to the reaction medium may actually reduce the mean nanoparticle size.

Unlike hydrogenation of acetylene alcohols, for L-sorbose oxidation the kinetic studies have not yet been extended to computation of the kinetic models because of the high complexity of the mechanism of L-sorbose oxidation.

3.3
Conclusions and Outlook

We have studied the catalytic behavior of nanoparticulate catalysts based on nanostructured polymers in the hydrogenation of acetylene alcohols (including long chain compounds) and in the oxidation of L-sorbose. Both reactions are of fundamental and technological significance so the development of new types of selective and efficient catalysts is an important mission. We demonstrated that catalytic nanoparticles formed in block copolymer micelles, such as PS-*b*-P4VP and PEO-*b*-P2VP, are very selective and efficient catalysts for the hydrogenation of acetylene alcohols. These enhanced properties are due to the high surface area of catalytic nanoparticles, the easy accessibility of the catalytic centers in the block copolymer micelles, and proper modification of the nanoparticle surface with pyridine units of the micelle cores. However, the activity of the catalysts is dependent on the length of the reagent: the larger the hydrophobic tail of the acetylene alcohol, the slower the reaction. We believe this is caused by diffusion of reacting molecules within block copolymer micelles. Heterogenization of the homogeneous (micellar) catalysts on alumina leads to increased activity due to an increase in the amount of available active centers and makes the block copolymer-based catalysts more technologically favorable. At the same time, when a heterogeneous catalyst is formed by placing nanoparticles in highly crosslinked, although highly functionalized, polymer colloids, inaccessibility of catalytic nanoparticles located away from the catalyst surface makes these catalysts ineffective in the hydrogenation of even small substrates (DMEC).

Although we greatly emphasize the tremendous importance of the proper polymeric environment for efficient functioning of the catalytic nanoparticles in these systems, the structure and properties of the nanoparticles themselves should not be overlooked. We found that bimetallic particles, where Pd is modified with Pt, Au or Zn, are much more efficient catalysts than monometallic Pd nanoparticles formed in the same type of block copolymer micelles. Although the structure of bimetallic particles is different depending on the type of metal-modifier, the general feature of bimetallic particles is the following: modification increases the amount of active centers, increasing the catalytic activity. In a more traditional vein, the increase in the particle size leads to a decrease in catalytic activity, as was observed for the Pd nanoparticulate catalyst formed via templating of Pd-nanoparticle-containing microgels in mesoporous alumina.

While nanoparticles formed in the above block copolymers are excellent catalysts for the hydrogenation of acetylene alcohols, another reaction of interest, oxidation of L-sorbose with the PEO-*b*-P2VP-based catalyst resulted in high activity, but very low selectivity. We attributed this phenomenon to a "wrong" kind of particle modification with pyridine units of the P2VP block. Conversely, the particles formed in a non-functionalized HPS matrix during the induction period demonstrated both high activity and high selectivity at 100% conversion. We think that, for this reaction, formation of catalytic centers *in situ* during oxidation of L-sorbose creates the most favorable environment of catalytic nanoparticle surface. Moreover, when HPS contains not only micropores (~2 nm), but also large macropores (50–80 nm), this facilitates transport of reagents and reaction products and creates excellent conditions for enhanced activity, selectivity and stability.

We believe catalytic nanoparticles formed in micro/macroporous HPS can be promising catalysts for a number of important catalytic reactions, including cross-coupling, asymmetric hydrogenation of alkyl pyruvates, selective hydrogenation of dienes, allyl alcohols, allene ketones, polar olefins, etc. Fundamentally, nanoparticulate catalysts based on nanostructured polymers can be used in any catalytic reactions occurring at ambient temperatures. Modification of catalytic systems by using metal-modifiers and/or by design of polymer functional groups may allow fine tuning of the catalytic properties. By variation of ligands (solvent, support, polymer, modifier, reactant) forming catalytic active centers on the nanoparticle surface one can adapt to environmental catalysis (removal of pollutants), such as degradation of phenols or hydrogenation of nitrates. In our future work we plan to focus on these reactions.

References

1 J. H. Fendler, *Nanoparticles and Nanostructured Films: Preparation, Characterization and Applications*, Wiley-VCH, New York, **1998**.

2 A. Wieckowski, E. R. Savinova, C. G. Vayenas, *Catalysis and Electrocatalysis at Nanoparticle Surfaces*, Marcel Dekker, New York, **2003**, pp. 970.

3 G. Schmid, *Nanoparticles: From Theory to Application*, Wiley-VCH, Weinheim, **2004**, pp. 434.

4 G. A. Somorjai, A. M. Contreras, M. Montano, R. M. Rioux, *Proc. Nat. Acad. Sci.* **2006**, *103*, 10577.

5 D. Astruc, F. Lu, J. R. Aranzaes, *Angew. Chem. Int. Ed.* **2005**, *44*, 7852.

6 C. Mueller, M. G. Nijkamp, D. Vogt, *Eur. J. Inorg. Chem.* **2005**, *20*, 4011.

7 L. M. Bronstein, in *Encyclopedia of Nanoscience and Nanotechnology*, Vol. 7, H. S. Nalwa (Ed.), APS, Stevenson Ranch, CA, **2004**, 193.

8 L. M. Bronstein, in *Dekker Encyclopedia of Nanoscience and Nanotechnology*, Vol., J. A. Schwarz, C. I. Contescu, K. Putyera (Eds.), Marcel Dekker, New York, **2004**, 2903.

9 F. S. Bates, G. H. Fredrickson, *Phys. Today* **1999**, *52*, 32.

10 S. Foerster, M. Antonietti, *Adv. Mater.* **1998**, *10*, 195.

11 M. V. Seregina, L. M. Bronstein, O. A. Platonova, D. M. Chernyshov, P. M. Valetsky, J. Hartmann, E. Wenz, M. Antonietti, *Chem. Mater.* **1997**, *9*, 923.

12 S. Klingelhoefer, W. Heitz, A. Greiner, S. Oestreich, S. Förster, M. Antonietti, *J. Am. Chem. Soc.* **1997**, *119*, 10116.

13 R. S. Underhill, G. Liu, *Chem. Mater.* **2000**, *12*, 2082.

14 Z. Lu, G. Liu, H. Phillips, J. M. Hill, J. Chang, R. A. Kydd, *Nano Lett.* **2001**, *1*, 683.

15 T. F. Jaramillo, S.-H. Baeck, B. R. Cuenya, E. W. McFarland, *J. Am. Chem. Soc.* **2003**, *125*, 7148.

16 B. Roldan Cuenya, S.-H. Baeck, T. F. Jaramillo, E. W. McFarland, *J. Am. Chem. Soc.* **2003**, *125*, 12928.

17 A. B. R. Mayer, J. E. Mark, R. E. Morris, *Polym. J* **1998**, *30*, 197.

18 A. B. R. Mayer, J. E. Mark, *Colloid Polym. Sci.* **1997**, *275*, 333.

19 R. Nakao, H. Rhee, Y. Uozumi, *Org. Lett.* **2005**, *7*, 163.

20 L. Groeschel, R. Haidar, A. Beyer, H. Coelfen, B. Frank, R. Schomaecker, *Ind. Eng. Chem. Res.* **2005**, *44*, 9064.

21 L. Groeschel, R. Haidar, A. Beyer, K.-H. Reichert, R. Schomaecker, *Catal. Lett.* **2004**, *95*, 67.

22 Z. Liu, X. Wang, H. Wu, C. Li, *J. Colloid Interface Sci.* **2005**, *287*, 604.

23 F. Gröhn, B. J. Bauer, Y. A. Akpalu, C. L. Jackson, E. J. Amis, *Macromolecules* **2000**, *33*, 6042.

24 M. Zhao, R. M. Crooks, *Angew. Chem. Int. Ed.* **1999**, *38*, 364.

25 R. M. Crooks, B. I. Lemon, L. Sun, L. K. Yeung, M. Zhao, *Top. Curr. Chem.* **2001**, *212*, 81.

26 R. W. J. Scott, C. Sivadinarayana, O. M. Wilson, Z. Yan, D. W. Goodman, R. M. Crooks, *J. Am. Chem. Soc.* **2005**, *127*, 1380.

27 R. W. J. Scott, O. M. Wilson, S.-K. Oh, E. A. Kenik, R. M. Crooks, *J. Am. Chem. Soc.* **2004**, *126*, 15583.

28 Y. Li, M. A. El-Sayed, *J. Phys. Chem. B* **2001**, *105*, 8938.

29 R. Narayanan, M. A. El-Sayed, *J. Phys. Chem.* **2004**, *108*, 8572.

30 J. Lemo, K. Heuze, D. Astruc, *Org. Lett.* **2005**, *7*, 2253.

31 Y. Niu, L. K. Yeung, R. M. Crooks, *J. Am. Chem. Soc.* **2001**, *123*, 6840.

32 S.-K. Oh, Y. Niu, R. M. Crooks, *Langmuir* **2005**, *21*, 10209.

33 O. M. Wilson, M. R. Knecht, J. C. Garcia-Martinez, R. M. Crooks, *J. Am. Chem. Soc.* **2006**, *128*, 4510.

34 K. Esumi, K. Sato, A. Suzuki, K. Torigoe, *Shikizai Kyokaishi* **2000**, *73*, 434.

35 J. C. Garcia-Martinez, R. L. Lezutekong, R. M. Crooks, *J. Am. Chem. Soc.* **2005**, *127*, 5097.

36 T. Endo, T. Yoshimura, K. Esumi, *J. Colloid Interface Sci.* **2005**, *286*, 602.

37 Y. Du, W. Zhang, X. Wang, P. Yang, *Catal. Lett.* **2006**, *107*, 177.

38 R. W. J. Scott, O. M. Wilson, R. M. Crooks, *Chem. Mater.* **2004**, *16*, 5682.

39 Y. Lvov, G. Decher, H. Möhwald,
Langmuir **1993**, *9*, 481.

40 G. B. Sukhorukov, E. Donath,
H. Lichtenfeld, E. Knippel, M. Knippel,
A. Budde, H. Mohwald, *Colloids Surf.,
A* **1998**, *137*, 253.

41 R. A. Caruso, A. Susha, F. Caruso,
Chem. Mater. **2001**, *13*, 400.

42 J. Dai, M. L. Bruening, *Nano Lett.* **2002**,
2, 497.

43 M. Zhao, R. M. Crooks, *Adv. Mater.*
1999, *11*, 217.

44 L. M. Bronstein, D. M. Chernyshov,
I. O. Volkov, M. G. Ezernitskaya, P. M.
Valetsky, V. G. Matveeva, E. M.
Sulman, *J. Catal.* **2000**, *196*, 302.

45 C. Wagner, W. Riggs, L. Davis, G.
Mullenberg, *Handbook of X-ray
Photoelectron Spectroscopy*, Perkin-Elmer
Corporation, Minnesota, **1978**.

46 M. Antonietti, E. Wenz, L. Bronstein,
M. Seregina, *Adv. Mater.* **1995**, *7*, 1000.

47 E. Sulman, Y. Bodrova, V. Matveeva,
N. Semagina, L. Cerveny, V. Kurtc,
L. Bronstein, O. Platonova, P. Valetsky,
Appl. Catal. A **1999**, *176*, 75.

48 O. A. Platonova, L. M. Bronstein, S. P.
Solodovnikov, I. M. Yanovskaya, E. S.
Obolonkova, P. M. Valetsky, E. Wenz,
M. Antonietti, *Colloid Polym. Sci.* **1997**,
275, 426.

49 L. M. Bronstein, S. N. Sidorov, P. M.
Valetsky, J. Hartmann, H. Coelfen,
M. Antonietti, *Langmuir* **1999**, *15*, 6256.

50 A. Molnar, A. Sarkany, M. Varga,
J. Mol. Catal. A: Chem **2001**, *173*, 185.

51 E. M. Sulman, *Russ. Chem. Rev.* **1994**,
63, 923.

52 G. Schmid, *Clusters and Colloids: From
Theory to Applications*, VCH, Weinheim,
Germany, **1994**, pp. 555.

53 M. Agnelli, H. M. Swaan,
C. Marquezalvarez, G. A. Martin,
C. Mirodatos, *J. Catal.* **1998**, *175*, 117.

54 J. S. Bradley, J. M. Millar, E. W. Hill,
S. Behal, B. Chaudret, A. Duteil,
Faraday Discuss. **1991**, *92*, 255.

55 J. S. Bradley, E. W. Hill, S. Behal,
C. Klein, B. Chaudret, A. Duteil, *Chem.
Mater.* **1992**, *4*, 1234.

56 A. Duteil, R. Queau, B. Chaudret,
R. Mazel, C. Roucau, J. S. Bradly, *Chem.
Mater.* **1993**, *5*, 345.

57 Y. Bard, *SIAM J. Numer. Anal.* **1970**, *7*,
157.

58 D. W. Marquardt, *J. Soc. Ind. Appl. Math.*
1963, *11*, 431.

59 M. Harada, K. Asakura, Y. Ueki,
N. Toshima, *J. Phys. Chem. B* **1992**, *96*,
9730.

60 N. Toshima, M. Harada, Y. Yamazaki, K.
Asakura, *J. Phys. Chem. B* **1992**, *92*, 9927.

61 J. Rajaram, A. P. S. Narula, H. P. S.
Chawla, D. Sukh, *Tetrahedron* **1983**, *39*,
2315.

62 S. Recchia, C. Dossi, N. Poli, A. Fusi,
L. Sordelli, R. Rzaro, *J. Catal.* **1999**, *11*.

63 R. D. Adams, T. S. Barnard,
Organometallics **1998**, *17*, 2885.

64 J. P. Candy, C. C. Santini, J. M. Basset,
Top. Catal. **1997**, *4*, 211.

65 J. C. Bertolini, *Surf. Rev. Lett.* **1996**, *3*,
1857.

66 A. Lopezganova, N. Martin, M. Viniegra,
J. Mol. Catal. A: Chem **1995**, *96*, 155.

67 K. R. Krishnamurthy, *Stud .Surf. Sci.
Catal.* **1998**, *113*, 139.

68 D. A. G. Aranda, F. B. Noronha, A. P.
Ordine, M. Schmal, *Phys. Status Solidii A:
Appl. Res.* **1999**, *173*, 109.

69 S. Naito, T. Hasebe, T. Miyao, *Chem. Lett.*
1998, *11*, 1119.

70 A. Vieira, M. A. Tovar, C. Pfaff,
B. Mendez, C. M. Lopez, F. J. Machado,
J. Goldwasser, M. M. R. Agudelo, *J.
Catal.* **1998**, *177*, 60.

71 M. C. J. Brandford, M. A. Vannice, *Catal.
Lett.* **1997**, *48*, 31.

72 F. M. T. Mendes, M. Schmal, *J. Appl.
Catal. A* **1997**, *163*, 153.

73 D. A. G. Aranda, M. Schmal, *J. Catal.*
1997, *171*, 398.

74 N. V. Semagina, A. V. Bykov, E. M.
Sulman, V. G. Matveeva, S. N. Sidorov,
L. V. Dubrovina, P. M. Valetsky, O. I.
Kiselyova, A. R. Khokhlov, B. Stein, L. M.
Bronstein, *J. Mol. Catal. A* **2004**, *208*, 273.

75 R. E. Fredricks, *J. Appl. Polym. Sci.* **1995**,
57, 509.

76 H. C. B. Chen, J. Czupski, *J. Adh. Seal.
Council* **1996**, *1*, 205.

77 T. J. Martin, K. Prochazka, P. Munk, S. E.
Webber, *Macromolecules* **1996**, *29*, 6071.

78 A. M. Pak, O. I. Kartonozhkina, Y. L.
Sheluduakov, *Book of Abstracts, 9th
International Symposium on Relations*

between Homogeneous and Heterogeneous Catalysis (Southampton, UK) **1998**, p.123.

79 L. Fabre, P. Gallezot, A. Perrard, *Catal. Commun.* **2001**, *2*, 249.

80 P. Maetz, R. Touroude, *Appl. Catal. A* **1997**, *149*, 189.

81 P. A. Rautanen, J. R. Aittamaa, A. O. I. Krause, *Ind. Eng. Chem. Res.* **2000**, *39*, 4032.

82 M. Antonietti, F. Gröhn, J. Hartmann, L. Bronstein, *Angew. Chem. Int. Ed.* **1997**, 36, **2080**.

83 L. M. Bronstein, D. M. Chernyshov, R. Karlinsey, J. W. Zwanziger, V. G. Matveeva, E. M. Sulman, G. N. Demidenko, H.-P. Hentze, M. Antonietti, *Chem. Mater.* **2003**, *15*, 2623.

84 J. Rajaram, A. P. S. Narula, H. P. S. Chawla, S. Dev, *Tetrahedron* **1983**, *39*, 2315.

85 L. M. Bronstein, C. Linton, R. Karlinsey, B. Stein, D. I. Svergun, J. W. Zwanziger, R. J. Spontak, *Nano Lett.* **2002**, *2*, 873.

86 L. M. Bronstein, C. Linton, R. Karlinsey, E. Ashcraft, B. Stein, D. I. Svergun, M. Kozin, I. A. Khotina, R. J. Spontak, U. Werner-Zwanziger, J. W. Zwanziger, *Langmuir* **2003**, *19*, 7071.

87 D. I. Svergun, M. B. Kozin, P. V. Konarev, E. V. Shtykova, V. V. Volkov, D. M. Chernyshov, P. M. Valetsky, L. M. Bronstein, *Chem. Mater.* **2000**, *12*, 3552.

88 L. M. Bronstein, D. M. Chernyshov, P. M. Valetsky, E. A. Wilder, R. J. Spontak, *Langmuir* **2000**, *16*, 8221.

89 S. Sidorov, I. Volkov, V. Davankov, M. Tsyurupa, P. Valetsky, L. Bronstein, R. Karlinsey, J. Zwanziger, V. Matveeva, E. Sulman, N. Lakina, E. Wilder, R. Spontak, *J. Am. Chem. Soc.* **2001**, *123*, 10502.

90 H. A. Luazidi, M. Z. Benabdallah, J. Berlan, C. Kot, P. L. Fabre, M. Mestre, J. F. Fauvarque, *Can. J. Chem. Eng.* **1996**, *74*, 105.

91 P. Cognet, J. Berlan, G. Lacoste, P. L. Fabre, J. M. Jud, *J. Appl. Electrochem.* **1995**, *25*, 1105.

92 T. Mallat, C. Bronnimann, A. Baiker, *Appl. Catal., A* **1997**, *149*, 103.

93 T. Mallat, C. Broennimann, A. Baiker, *J. Mol. Catal. A: Chem* **1997**, *117*, 425.

94 C. Broennimann, Z. Bodnar, R. Aeschimann, T. Mallat, A. Baiker, *J. Catal.* **1996**, *161*, 720.

95 C. Broennimann, T. Mallat, A. Baiker, *J. Chem. Soc., Chem. Commun.* **1995**, *13*, 1377.

96 I. Bacos, T. Mallat, A. Baiker, *Catal. Lett.* **1997**, *43*, 201.

97 S. N. Sidorov, L. M. Bronstein, V. A. Davankov, M. P. Tsyurupa, S. P. Solodovnikov, P. M. Valetsky, E. A. Wilder, R. J. Spontak, *Chem. Mater.* **1999**, *11*, 3210.

98 V. A. Davankov, M. P. Tsyurupa, *React. Polym.* **1990**, *13*, 27.

99 M. P. Tsyurupa, V. A. Davankov, *J. Polym. Sci.: Polym. Chem. Ed.* **1980**, *18*, 1399.

100 Y. Schuurman, B. F. M. Kuster, K. van der Wiele, G. B. Marin, *J. Appl. Catal. A* **1992**, *89*, 47.

101 L. Bronstein, G. Goerigk, M. Kostylev, M. Pink, I. A. Khotina, P. M. Valetsky, V. G. Matveeva, E. M. Sulman, M. G. Sulman, A. V. Bykov, N. V. Lakina, R. J. Spontak, *J. Phys. Chem. B* **2004**, *108*, 18234.

4

PAMAM Dendrimer Templated Nanoparticle Catalysts

Bert D. Chandler and John D. Gilbertson

4.1
Introduction and Background

4.1.1
Traditional Routes to Heterogeneous Catalysts

An important class of industrial catalysts consists of an active component dispersed in the form of very small particles over high surface area solids [1]. As the field of industrial heterogeneous catalysis has developed, catalyst formulations have evolved such that state-of-the-art catalysts often contain two or more metals and/or main-group elements. The additives may promote a desired reaction, prevent undesirable side reactions, or enhance catalyst longevity [1–3]. Bimetallic nanoparticle catalysts, in particular, are widely employed in industry and will similarly be vital to the success and economic viability of hydrogen energy technologies [4], particularly fuel cells and hydrogen production from photochemical and biorenewable resources [5, 6].

In spite of the commercial importance of bimetallic catalysts, catalyst formulations, particularly the active sites or phases, are not necessarily well characterized or well understood. This is particularly true for bimetallic catalysts, largely due to the difficulties associated with preparing, characterizing, and proving the presence and activity of well-defined supported bimetallic nanoparticles. Traditional preparative routes involve impregnating metal salts (typically inexpensive chloride complexes) onto high surface area supports, followed by various thermal activation steps [7]. Most important industrial supports can also be employed as separation media; consequently, chromatographic separation of salt precursors as they pass through the support pore structure is unavoidable. Further, nanoparticle preparation via traditional routes depends on the surface and even gas phase mobility of the species present during the thermal treatments [2, 7]. These processes vary widely for different metals and are poorly understood, at best [2, 7]. Metal salt precursors also introduce ions (e.g., K^+, Cl^-) which may inadvertently affect catalyst properties or have deleterious effects on catalyst performance [8–10].

Nanoparticles and Catalysis. Edited by Didier Astruc
Copyright © 2008 WILEY-VCH Verlag GmbH & Co. KGaA, Weinheim
ISBN: 978-3-527-31572-7

4.1.2
Molecular Cluster and Colloid Routes

Ligand-stabilized inorganic and organometallic molecular clusters[1] offer potential advantages over traditional methods for preparing supported bimetallic catalysts [2]. Using well-defined molecular precursors allows the deposition of particles that are initially consistent in composition and limits the introduction of ions that may adversely affect catalyst performance. The potential for low thermolysis temperatures (e.g. for CO ligands) may also prevent particle agglomeration. Gates' group and others have a number of excellent reviews detailing the cluster method, particularly as it pertains to preparing extremely well-defined supported structures consisting of a few atoms [2, 11].

A central problem with the molecular cluster route, however, is that under reaction conditions clusters are likely to lose their molecular integrity [11]. Although clusters have played an important role in understanding metallic bonding and selectivity in catalysis [12], in general, they have not proven to be particularly good models for nanoparticle catalysts and are more appropriately considered as potential precursors to nanoparticle catalysts [11]. As precursors, clusters provide strict initial compositional control and the possibility of preparing nanoparticles with compositions that are not formed via traditional methods [2].

Even as catalyst precursors, clusters have drawbacks due to inflexibility in metal stoichiometries. Metal ratios are set by cluster stoichiometry and cannot be systematically varied without an extensive cluster library. Cluster preparation is often time consuming and expensive, so such libraries are seldom available. Further, using clusters (nominally 4–12 atoms) to prepare nanoparticles (nominally 10–1000 atoms) relies on the same atom/particle mobility that plagues traditional routes, although perhaps to a lesser degree. Essentially, one must hope that cluster precursors stay intact long enough during activation to impart their precise stoichiometry to growing nanoparticles. In terms of understanding the properties of nanoparticles, this situation is less than desirable. Even worse, ligands that stabilize many noble metals often contain elements that can be severe poisons for catalysts (e.g. phosphines, arsines, S donors) [13].

Colloidal nanoparticles can be employed as heterogeneous catalyst precursors in the same way as molecular clusters. The recent explosion of interest in nanoparticles for nanotechnology applications has fueled the development of several new nanoparticle preparative techniques [14–16]. In some cases, careful syntheses allow control over nanoparticle size, composition, and even morphology [14–16]. Several synthetic routes to colloidal nanoparticles in the 1–5 nm range are now

---

1) The term "cluster" is used to describe different entities in various communities, in particular, the terms "cluster" and "nanoparticle" are often used interchangeably. We use the word "cluster" exclusively to describe discrete molecular compounds that have known coordination spheres, atomically precise compositions, and metal–metal bonds in a three-dimensional arrangement. We distinguish this from "nanoparticles", which are either supported or colloidal dispersions of metal atoms that have ranges in both particle sizes and distributions.

becoming available [14–16]. For large atoms like platinum, this size range is particularly important for catalysis as it corresponds to dispersions of 95–20%, respectively. This not only maximizes the fraction of metal on particle surfaces, but also coincides with a transition from large particles with primarily metallic character to nanoparticles composed of tens of atoms that may have characteristics that are more molecular.

In many respects, colloidal nanoparticles offer opportunities to combine the best features of the traditional and cluster-based catalyst preparation routes to prepare uniform bimetallic catalysts with controlled particle properties [17]. In general, colloidal metal ratios are reasonably variable and controllable. Further, the application of solution and surface characterization techniques may ultimately help correlate solution synthetic schemes to catalytic activity. This synergism between solution phase nanoparticles, stabilized by interactions with various complexing agents, and supported nanoparticles, which are stabilized by strong metal-support interactions has important potential for both fields. New synthetic methodologies offer new routes to heterogeneous catalyst preparation, while heterogeneous catalysis and surface characterization may offer important new insights into the surface chemistry and properties of new nanoparticle systems. The remainder of this chapter discusses advances in using PAMAM dendrimer templated nanoparticles at the interface of these important research areas.

4.1.3
PAMAM Dendrimers

Dendrimers are a specific class of hyperbranched polymers that emanate from a single core and ramify outward with each subsequent branching unit [15, 16, 18, 19]. They have well-defined and nearly monodisperse architectures that arise from the regular nature of their repeat units. In the commonly employed "divergent" synthesis, dendrimers can be prepared through sequential, alternating reactions of two smaller units, one of which has a focal point or bifurcation. Several classes of dendrimers are known, including polypropyleneimine (PPI), polyamidoamine (PAMAM), and Fréchet-type polyether dendrimers [15, 16]. Starburst® PAMAM dendrimers (see Fig. 4.1), a specific class of commercially available dendrimers that have repeating amine/amide branching units, have drawn considerable interest in recent years due to their potential applications in medicine, nanotechnology, and catalysis [19–23]. These dendrimers are readily functionalized to terminate in diverse moieties such as primary amines, carboxylates, hydroxy groups, or hydrophobic alkyl chains. Because dendrimer size and end groups can be varied, they are typically named by their generation (e.g., G1, G2, etc.) and exterior functionality (e.g., $-NH_2$, $-OH$).

Higher generation dendrimers (G4 and larger) adopt roughly spherical or globular structures with exterior branches becoming increasingly intertwined. The interweaving of dendrimer branches is distinct from the crosslinked branches that are found in many polymeric colloid stabilizers. The dendritic macromolecular architecture gives rise to a relatively open interior pocket for high generation

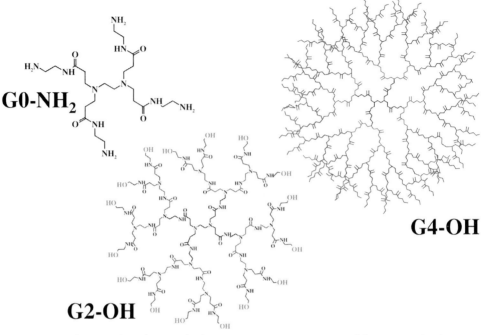

Fig. 4.1 Polyamidoamine dendrimers. Heteroatoms were removed from G4-OH for clarity.

dendrimers, while maintaining a closed, but porous exterior. The presence of these open spaces within the dendrimer interior and the synthetic control over their composition, architecture, and interior/exterior functionalities creates an environment that facilitates trapping guest species [24, 25].

4.2
Synthesis and Characterization of Dendrimer Encapsulated Nanoparticles

4.2.1
Synthetic Schemes and Nomenclature

PAMAM dendrimers can be used to template and stabilize reduced (zero valent) metal nanoparticles in solution [22, 26]. A number of synthetic strategies to prepare dendrimer encapsulated nanoparticles (DENs) are available. The primary routes are described here; readers are directed to recent reviews for more thorough discussions of dendrimer mediated nanoparticle syntheses [14, 25].

The most common and straightforward preparative route for DENs is analogous to the "ship in a bottle" synthesis (see Fig. 4.2). In the first step, metal precursors (e.g. $Cu(OH_2)_6^{2+}$, $PdCl_4^{2-}$, $PtCl_4^{2-}$) are intercalated through the porous exterior of a PAMAM dendrimer and complexed to the interior amine and/or amide groups. Electrons are added via a reducing agent, typically BH_4^-, but other reagents may

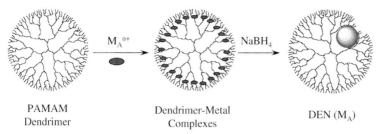

Fig. 4.2 Schematic representation of direct complexation synthesis for monometallic DENs.

be employed [14, 25]. The reduced metal atoms, which are effectively trapped within the interior cavity, coalesce into a nanoparticle. Hence, the dendrimer plays multiple roles, serving both as a template for the number (and type) of metal atoms, and as a colloid stabilizer, preventing agglomeration after the particles form. Purification is straightforward and reaction byproducts are readily removed with dialysis.

DENs are typically named by the dendrimer from which they are prepared along with the metal:dendrimer stoichiometry used in the nanoparticle synthesis, e.g. G5-OH(Pt$_{50}$). With careful synthetic techniques, nanoparticles with very narrow particle size distributions (1.3 ± 0.3 nm) can be selectively prepared inside dendrimer interior cavities [14]. Provided that metal reoxidation is prevented, DENs are stable for long periods of time and do not agglomerate, since the nanoparticles are trapped within the dendrimer framework [14, 25]. Further, this general synthetic methodology is quite flexible, having been used to prepare monometallic Pt, Au, Cu, Pd, and Ru nanoparticles [14] as well as a number of bimetallic systems (see below).

Some metals do not readily form complexes with PAMAM dendrimers, especially in aqueous solution. Employing alkyl chain functionalized dendrimers in organic solution is one means of circumventing these difficulties [27]. DENs of these metals can also be prepared by reacting the salt of a more noble metal (e.g. AuCl$_4^-$) with a nanoparticle of a less noble metal (e.g. Cu) [14, 25]. These "galvanic displacement" syntheses, which are shown schematically in Fig. 4.3, are synthetically more challenging than directly loading metal complexes into dendrimer interiors. Cu^0 nanoparticles are air-sensitive and will oxidize to Cu^{2+} in several hours [28], so syntheses must be done under inert atmosphere with degassed solvents. Cu^{2+} also readily forms complexes with PAMAM dendrimers, making Cu removal an important consideration. For DENs in aqueous solution, this can be accomplished by dialysis at pH 4.30 [29].

4.2.2
Bimetallic Nanoparticles

Bimetallic DENs can be prepared using methodologies comparable to those for monometallic DENs. Important additional considerations are the interactions between metal salt precursors and the potential to control particle morphology

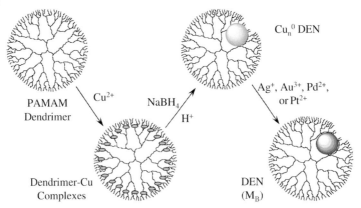

Fig. 4.3 Schematic representation of galvanic displacement synthesis for DENs.

Table 4.1 Bimetallic DEN systems: synthetic methods and catalytic reactions.

DEN	Synthesis[a]	Dendrimer(s)	Catalysis	Ref.
PdPt	Co-complex	G4-OH	allyl alcohol hydrogenation	30
PdPt	Co-complex	G4-OH	1,3-COD hydrogenation	31
PdRh	Co-complex	G4-OH	1,3-COD hydrogenation	33
PdAu	Co-complex	G6-Q116	allyl alcohol hydrogenation	32
PdAu	Co-complex	G4-NH$_2$	CO oxidation[b]	79
PtAu	Galvanic	G5-OH	CO oxidation[b]	37
PtCu	Co-complex	G5-OH	CO oxidation[b] toluene hydrogenation[b]	36
PdAg	Co-complex	G4-NH$_2$		34
AuAg	Seq. Red.	G3-NH$_2$, G3.5-NH$_2$, G5-NH$_2$, G5.5-NH$_2$	*p*-nitrophenol reduction	41
AuAg	Seq. Red.	G6-OH, G8-OH		40
[Au](Pd)[c]	Seq. Red.	G6-Q116	allyl alcohol hydrogenation	32
[Pd](Au)[c]	Seq. Red.	G6-OH		32
[Au](Ag)[c]	Seq. Red.	G6-OH		40
[AuAg](Au)[c]	Seq. Red.	G6-OH		40

a   Co-complex = co-complexation; Galvanic = galvanic displacement; Seq. Red. = sequential reduction.
b   Heterogeneous catalysis.
c   Core–shell nanoparticles where [ ] = core metal(s) and ( ) = shell metal

(well mixed or core–shell nanoparticles). Co-complexation has been used to prepare a variety of bimetallic DENs, including PdPt [30, 31], PdAu [32], PdRh [33], and PdAg [34] (see Table 4.1). One of the first studies was Scott, Datye, and Crooks' work on the PdPt system [30]. Using K$_2$PdCl$_4$ and K$_2$PtCl$_4$ as metal precursors, the researchers held the metal:dendrimer loading constant while varying the metal:metal stoichiometry. The relative complexation rates for different precursors are an important consideration for this synthetic scheme. $PdCl_4^{2-}$ complexes quickly, while $PtCl_4^{2-}$ requires several days to react [35], so, in practice, these syntheses likely involve sequential complexation of the ions with the more reactive complex binding first. This method was employed for PtCu DENs, allowing $PtCl_4^{2-}$ to bind G5-OH for 2 days before adding $Cu(NO_3)_2$, which binds in minutes. Elemental analysis of the resulting nanoparticles showed them to be consistently enriched in Cu, suggesting that either Pt(II) complexation was incomplete or that Cu(II) may displace some of the Pt(II)-dendrimer complexes [36].

Galvanic displacement is synthetically useful when the metal precursors have the potential to react with one another. This is the case for Pt–Au syntheses using $AuCl_4^-$ and $PtCl_4^{2-}$ salts, because Pt(II) complexes are unstable with respect to oxidation by Au(III) species [37]. To minimize the potential for this reaction, solutions of each precursor can be combined immediately prior to being added to freshly prepared Cu DENs. The metal ratios reported through Cu displacement are consistent with ratios set in the initial syntheses, but this synthetic scheme has the general drawback of preparing particles with wider size distributions than other methods. The Cu displacement synthesis is likely to be most useful for rapidly screening metal ratios as it is a much faster route to Pt-based systems, leading to nanoparticles in several hours rather than several days.

Unless specific synthetic steps are undertaken to prepare well-defined morphologies (see below) bimetallic DENs prepared by co-complexation and galvanic displacement are generally described as "well-mixed" bimetallic nanoparticles or "alloy-type" particles. The critical aspect of these descriptions is the random, intimate mixing of the two metals within individual nanoparticles. We prefer the "well-mixed" term as the word alloy also refers to thermodynamically stable bulk solid solutions. In many cases, including nanoparticles prepared with dendrimers [37], bimetallic nanoparticles can be prepared throughout bulk immiscibility gaps [38, 39]. The "well-mixed" particle terminology avoids potential misconceptions associated with differing interpretations of "alloy".

4.2.3
Core–Shell Nanoparticles

An additional method for preparing DENs can be described as the "sequential reduction" method. This method, shown in Fig. 4.4, involves the initial complexation and reduction of a "seed" metal (M_A), followed by the complexation and subsequent reduction of the second metal (M_B) to produce the $M_A M_B$ system. The synthetic utility of this method is that it provides the means to access both well mixed and core–shell type DENs. When the reducing agent used for M_B is a mild reductant, such as H_2 or ascorbic acid, core–shell nanoparticles with an M_A core and M_B shell can be selectively prepared (see Table 4.1). In the nomenclature for these core–shell nanoparticles, [M_A] denotes the core metal(s) and (M_B) indicates the exterior metal shell. Several bimetallic DENs have been prepared via this route, including AuAg [40, 41], [Au](Pd) [32], [Pd](Au) [32], [Au](Ag) [40] and [AuAg](Au) [40].

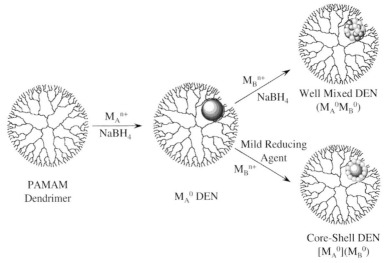

Fig. 4.4 Schematic representation of sequential reduction syntheses for bimetallic DENs.

4.2.4
Characterization

Particle sizes and distributions, an important property of DENs, are generally determined by high resolution transmission electron microscopy (HRTEM). Figure 4.5 shows a representative HRTEM micrograph and particle size distributions for several Ru DENs and Table 4.2 compares the observed diameters with calculated values. The particle size and distributions shown in Fig. 4.5 are typical for DENs, although the metal:dendrimer ratio and dendrimer generation used in the synthesis ultimately control these distributions. In some cases, using "magic numbers" (numbers of metal atoms for closed shell sphere packing around a central atom in a cubooctahedral arrangement) for the metal:dendrimer ratio may help to reduce the distribution of particle sizes [14]. An additional method of determining particle sizes, employed by Kim, Garcia-Martinez, and Crooks, is also worth noting. In this study, the investigators used differential pulse voltammetry to estimate the size of dendrimer templated Pd and Au nanoparticles [42]. Their study estimated particle sizes to be very close to the ideal sizes calculated from the metal:dendrimer stoichiometry and the metallic radius of each metal. Further, the study concluded that TEM measurements overestimated the size of the smallest Pd nanoparticles due to inadequate point-to-point resolution [42].

UV–visible spectroscopy has also proven to be a valuable tool for characterizing DENs. Rayleigh scattering gives rise to the monotonic increase in absorption as wavelength decreases, showing the initial preparation of nanoparticles [43]. Au and Ag nanoparticles have intense surface plasmon bands that are valuable additional spectroscopic tools [43–45]. These bands, shift with particle size and

Table 4.2 Bimetallic DEN systems: synthetic methods and catalytic reactions. Reprinted with permission from *J. Phys. Chem. B* **2006**, *110*, 7725–7731. Copyright 2006 American Chemical Society.

DEN	Measured mean particle diameter[a]	Calculated theoretical diameter[b]
G4OH (Ru$_{20}$)	0.9	0.81
G4OH (Ru$_{40}$)	1.2	1.02
G5OH (Ru$_{60}$)	1.4	1.17
G5OH (Ru$_{100}$)	1.4	1.39

a   Obtained from the particle size distributions of Fig. 4.5.
b   Calculated from Ru cell parameters assuming a spherical particle.

composition, and are therefore useful handles for the physical characterization of bimetallic nanoparticle composition.

X-ray photoelectron spectroscopy (XPS) provides further evidence for the complete reduction of the metal ions to nanoparticles. A representative XPS study from Rhee's group explored the degree of metal reduction in bimetallic nanoparticles using the G4-OH(Pd^{2+}/Pt^{2+}) system [31]. In this study, the Pd/Pt ratio was varied while maintaining a constant metal:dendrimer ratio of 55:1. The peaks corresponding to Pd($3d_{5/2}$) and Pd($3d_{3/2}$) at 337.6 and 342.7 eV, respectively, were assigned to Pd^{2+}. After reduction, these peaks shift to 334.9 and 340.5 eV, respectively, and are consistent with Pd^0 [46]. Comparable shifts were observed for Pt. The Pt^{2+} peaks at 72.5 eV (Pt($4f_{7/2}$)) and 75.7 eV (Pt($4f_{5/2}$)) shift to 71.3 and 74.4 eV, respectively, upon reduction and are consistent with Pt^0. Peaks for unreduced Pd and Pt were not reported, suggesting that complete reduction of both metals occurred.

Particle composition is more difficult to evaluate. Bulk elemental analysis (atomic absorption spectroscopy (AA) or inductively coupled plasma mass spectrometry (ICP-MS) are most common for metals) is useful in confirming the overall bimetallic composition of the sample, but provides no information regarding individual particles. Microscopy techniques, particularly energy dispersive spectroscopy (EDS), has supported the assertion that bimetallic DENs are bimetallic nanoparticles, rather than a physical mixture of monometallics [14]. Single particle EDS studies of G4-OH(Pd$_{30}$Pt$_{10}$) DENs [30] and activated dendrimer templated PtAu nanoparticles [37] show both systems to be composed of two metals intimately mixed within individual nanoparticles. EDS spectra collected over large areas, which sample tens or hundreds of particles, generally agree well with the bulk composition measurements [37] and with stoichiometries set in nanoparticle synthesis [30, 32, 37].

Just as DENs particle sizes have some distribution (albeit relatively narrow), there is surely some distribution in particle compositions for bimetallic DENs.

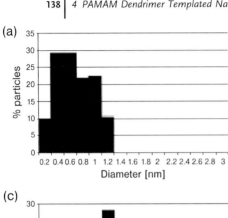

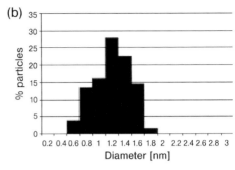

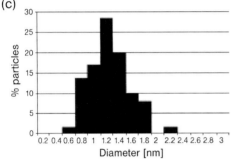

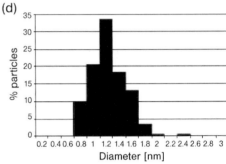

Fig. 4.5 Metal particle size distributions obtained by HRTEM measurements of (A) $Ru_{20}G4OH$, (B) $Ru_{40}G4OH$, (C) $Ru_{60}G5OH$, and (D) $Ru_{100}G5OH$ solution deposited on a carbon-coated copper grid. The representative HRTEM image is of $Ru_{60}G5OH$. Reprinted with permission from *J. Phys. Chem. B* **2006**, *110*, 7725–7731. Copyright 2006 American Chemical Society.

This is a fundamentally important aspect of DENs, particularly with regard to their catalytic properties; however, there are presently no reliable characterization methods for evaluating particle composition distributions. One method that has been applied to PdAu [32] and PtPd [30] DENs, as well as dendrimer templated PtAu [37] is to collect single particle EDS spectra from several (15–20) nanoparticles. These experiments indicate that individual particle composition distributions may vary widely, but the difficulty obtaining data from the smallest particles may skew the results somewhat.

4.3
Homogeneous Catalysis

4.3.1
Monometallic Catalysts

The area of homogeneous catalysis by monometallic DENs, which was reviewed by Scott, Wilson, and Crooks [14] in 2005, has largely focused on a broad range of "proof of concept" experiments. As such, catalysis has largely been used for test reactions to show changes and modifications to the nanoparticles or dendrimers; we are aware of no in-depth mechanistic studies using DENs to date. Nonetheless, the reported studies highlight the flexibility of the dendrimer motif and begin to probe some important structural considerations for these systems.

The ability to functionalize DENs allows them to be used as homogeneous hydrogenation catalysts in a variety of reaction media inluding water, organic solvents, supercritical CO_2, and biphasic fluorous solvents [14]. Pd DENs have been the most widely studied homogeneous catalysts, especially for carbon–carbon coupling catalysts. In Heck coupling reactions, Pd DENs show higher activities and selectivites relative to other colloidal Pd catalysts [47]. Pd catalyzed Stille [48] and Suzuki [49, 50] couplings have also been explored. Pd DENs also compared favorably with polymer stabilized nanoparticles for stability in Suzuki couplings, although the DENs exhibited lower TOFs [50].

As homogeneous hydrogenation catalysts, the PAMAM dendrimer surface functions as a size- and shape-selective membrane for reactant molecules [14, 51–53]. Polypropyleneimine DENs also impart substantial selectivity towards polar substrates in competitive hydrogenation reactions [53]. In a clever set of experiments, the size selectivity of Pd DENs for alkene hydrogenation, coupled with molecular "rulers" (alkenes tethered to large cyclodextrin "stoppers") has been used to estimate the distance of the Pd nanoparticle from the surface dendrimer surface [14]. This distance (0.7 ± 0.2 nm, on average) indicates that the nanoparticles are substantially displaced from the G4-OH dendrimer center (G4-OH radius $\approx$ 2.2 nm) [14, 54]. Pd DENs have also been used to investigate particle size effects for allyl alcohol hydrogenation [55].

4.3.2
Bimetallic Catalysts

Catalysis is a potentially sensitive probe for nanoparticle properties and surface chemistry, since catalytic reactions are ultimately carried out on the particle surface. In the case of bimetallic DENs, catalytic test reactions have provided clear evidence for the modification of one metal by another. DENs also provide the opportunity to undertake rational control experiments, not previously possible, to evaluate changes in catalytic activity as a function of particle composition.

As Table 4.1 shows, several bimetallic DENs have been employed as homogeneous catalysts, predominately for hydrogenations. The most detailed studies have been with allyl alcohol hydrogenation and the partial hydrogenation of 1,3-cyclooctadiene (1,3-COD) test reactions. In these studies, turnover frequencies (TOFs), which can be normalized per mole of nanoparticles, can be compared as a function of the metallic atomic ratio in the DENs. Comparison with physical mixtures of monometallic DENs with the same net atomic ratio allows investigators to directly compare both the magnitude and direction of changes to rationally prepared control materials.

Figure 4.6 shows a plot of TOFs for the partial hydrogenation of 1,3-COD by PdRh DENs compared to TOFs for physical mixtures of Pd and Rh monometallic DENs as a function of mol% Rh [33]. As the mol% of Rh in the bimetallic DENs was increased, an increase in the TOF was observed that was *greater* than that of the physical mixtures. Importantly, the average particle size and distribution did not change as the mol% Rh increased, which was used to rule out the possibility that the TOF enhancement was a consequence of a systematic decrease in particle size. This allows the conclusion that the bimetallic DENs truly are intimately mixed bimetallic nanoparticles and that a "synergistic" effect is responsible for the catalytic rate enhancement.

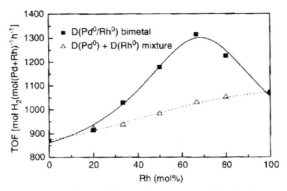

Fig. 4.6 Dependence of the catalytic activity of the dendrimer-encapsulated PdRh bimetallic nanoparticles on their composition in the partial hydrogenation of 1,3-cyclooctadiene. Reprinted with permission from *J. Mol. Catal., A* **2003**, *206*, 291–298. Copyright 2003 Elsevier.

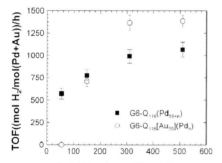

Fig. 4.7 Turnover frequencies (TOFs) for the hydrogenation of allyl alcohol using G6-Q116(Pd$_{55+n}$) and G6-Q116[Au$_{55}$](Pd$_n$), which was prepared using the sequential-loading method, for n = 0, 95, 255, 455. Conditions: 22 °C, substrate:metal = 3300:1, [Pd + Au] = 150 µM. The ■ represent TOF data for Pd-only DENs, while the ○ represent data for the bimetallic DENs. Reprinted with permission from *J. Am. Chem. Soc.*, **2004**, *126*, 15583–15591. Copyright 2004 American Chemical Society.

Most of the homogeneous catalysis studies have reported some degree of catalytic rate enhancement when metals are intimately mixed in bimetallic nanoparticles. This synergistic effect was observed in the hydrogenation of allyl alcohol by PdAu [32] and PdPt [30] DENs, as well as the reduction of *p*-nitrophenol by AuAg [41] DENs. One particularly noteworthy study of this synergistic effect compared allyl alcohol hydrogenation by G6-Q116(Pd$_{55+n}$) to G6-Q116[Au$_{55}$](Pd$_n$) for values of *n* from 0 to 455 (see Fig. 4.7). This study examines the effect of particle size and morphology on catalytic activity, and highlights the type of structural study for which DENs are uniquely suited.

The first interesting observation from this study is that TOF, normalized for the total number of metal atoms, actually increased with *n*. As particle size increases, the fraction of surface atoms decreases, so faster reaction rates are actually catalyzed by fewer surface atoms. It is tempting to conclude that the smallest nanoparticles are inherently less active for alkene hydrogenation in solution. However, the presence of the dendrimer makes it difficult to draw this conclusion. The dendrimer clearly plays an important role in mass transfer to the nanoparticle catalyst (this can be used advantageously to selectively hydrogenate substrates with different steric properties) [14]. Additionally, as the extraction experiments show, bonding between the metal surface and dendrimer interior amine and amide groups is important for nanoparticle stabilization. Smaller nanoparticles, which have a higher fraction of surface atoms but fewer total surface sites, could simply be more fully passivated by the dendrimer. In other words, as particle size increases, the number of free surface atoms may increase, making for a more active catalyst. The ultimate origin of the rate enhancement observed for these DENs is unclear, and some combination of structural and electronic effects is likely. Experimental procedures that could distinguish the relative impact of the structural and electronic influences are unknown to us, and the development of an appropriate methodology would contribute substantially to this field.

Figure 4.7 also shows that the Au core nanoparticles are more active catalysts than pure Pd particles containing the same total number of atoms. Since pure Au nanoparticles are inactive for alkene hydrogenation, it is difficult to attribute this enhancement to anything other than a synergistic modification of Pd by Au. The nature of the synergistic rate enhancements, which have been well documented in the homogeneous and heterogeneous catalysis literature, are of fundamental interest and importance [3]. Particle size affects both surface geometries (e.g. curvature, size of extended planar surface) and electronic structure [56]. A dopant metal can potentially affect both of these parameters by donating or withdrawing electron density or epitaxially templating altered surface arrangements. Homogeneous catalysis studies have highlighted the magnitude and direction of these effects (at least for aqueous alkene hydrogenation), but assessing the relative importance of structural and electronic effects in the presence of the PAMAM dendrimer is difficult in these systems.

4.4
Supported Dendrimer Templated Nanoparticles

In the preceding section, we have labeled DENs exclusively as homogeneous catalysts; however there is considerable debate as to whether they are appropriately considered homogeneous catalysts or "soluble heterogeneous catalysts". In our opinion, the ability to rapidly and easily separate a heterogeneous catalyst from a solution is the key component to any heterogeneous system. Indeed, this ease of separation is a primary reason why the chemical industry utilizes heterogeneous catalysts far more regularly than homogeneous systems. Therefore, soluble nanoparticles, regardless of their origin, ought to be considered homogeneous catalysts when they are in the same phase as the substrate(s).

Additionally, the "ligand sphere" or colloid stabilizer (i.e. the dendrimer for DENs) plays an important role that is not present in the traditional supported nanoparticle catalysts that are discussed below. Dendrimers clearly act as membranes that influence substrate access to the reactive nanoparticle. Additionally, it is not yet possible to determine the strength and number of interactions between nanoparticles and dendrimers. The strength of these interactions almost certainly changes with different metals; changes in the number of interactions are less clear as different metals may interact with either amine or amide moieties. Conventionally prepared heterogeneous catalysts have no similar ligand sphere, save perhaps the oxide support on one side of the particle and the reactive medium (substrates, solvent, etc.) on the particle surface. Based on both the solubility arguments and the importance of this "ligand sphere," soluble nanoparticles in general, and DENs in particular, appear to be more similar to ligand stabilized clusters than they are to heterogeneous nanoparticle catalysts.

4.4.1
Immobilized Intact DENs

The preceding discussions have hopefully convinced readers that the PAMAM dendrimer template offers a variety of new opportunities for studying catalysis. These properties also present challenges for evaluating nanoparticle properties, particularly reaction kinetics. Nanoparticle surface geometric and electronic properties are extremely difficult to probe in solution, especially when the dendrimer inhibits access by various probe molecules. Further, the number of "bonds" between nanoparticle surfaces and dendrimer amine and amide groups is essentially unknown. In cases where the dendrimer may preferentially bind one metal over another, stoichiometries and activities are difficult to evaluate, thus making it extremely difficult to interpret catalysis results in terms of particle composition.

Evaluating dendrimer templated nanoparticles in the absence of the dendrimer provides opportunities for insights into these new materials. In order to pursue these investigations, it is first necessary to immobilize DENs onto an appropriate substrate and to gently remove the dendrimer shell (see Fig. 4.8). Opportunities for controlling nanoparticle size and composition make DENs potentially important precursors for heterogeneous catalysts and electrocatalysts, and DEN deposition and thermolysis are similarly critically important steps in pursuing these applications.

4.4.1.1 Electrocatalysis by Immobilized Intact DENs
DENs can be immobilized on electroactive substrates to prepare "heterogenized" homogeneous catalysts. Electrochemical grafting of hydroxyl-terminated Pt DENs yields covalently linked DENs that are active solid–liquid heterogeneous

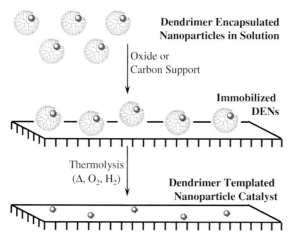

Fig. 4.8 Schematic preparation for dendrimer templated heterogeneous catalysts.

electrocatalysts [57]. These heterogenized systems are quite robust, withstanding numerous electrochemical cycles and sonication in 0.5 M H_2SO_4. Thiophene terminated PAMAM dendrimers can also be electrocopolymerized with 3-methylthiophene [58]. Coordination of $PtCl_4^{2-}$ and reduction yields DENs within this matrix. Similarly, amine terminated Pt and Pd DENs have been immobilized on planar Au surfaces through self-assembled monolayer (SAM) chemistry [59, 60]. Amine terminated dendrimers can also be deposited onto Au films and then used to template Pt nanoparticles which are active electrocatalysts [61].

4.4.1.2 Construction of DENs on Oxide Surfaces

Several studies have examined constructing PAMAM [62, 63] and Frechet-type polyether dendrimers [64] on silica surfaces. These can subsequently be used to bind Pd(II) and catalyze intramolecular carbonylations en route to the preparation of fused heterocycles containing O, N, and S heteroatoms [65] as well as the selective hydrogenation of dienes to monoolefins [66]. Additionally, dendrimers constructed on mesoporous SBA silicas have been used to template Pd nanoparticles and catalyze allyl alcohol hydrogenation (see Fig. 4.9) [67]. As with studies of DENs in solution [52], this study showed that higher generation dendrimers slow the overall activity of the Pd nanoparticles. At the same time, higher generation dendrimers improved selectivity towards 1-propanol by preventing isomerization of the hydroxy group.

Another means of templating nanoparticles in oxide pores is to add dendrimers into sol–gel syntheses. PAMAM and PPI dendrimers can be added to sol–gel silica

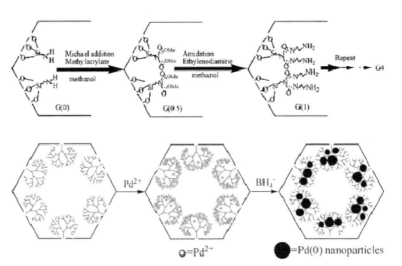

Fig. 4.9 Schematic preparation of DENs constructed in the pores of an SBA silica. Reprinted with permission from *J. Am. Chem. Soc.*, **2006**, *128*, 716–717. Copyright 2006 American Chemical Society.

[68–70] and zinc arsenate [71] syntheses to template mesopores within the oxide material. There are relatively few reports of catalytic applications of these materials, although titanosilicates have been explored as oxidation catalysts [72, 73]. In one report, the dendrimer bound Cu^{2+} ions were added to sol–gel silica and calcined to yield supported copper oxide nanoparticles. After reduction with hydrogen, these particles were active for the N_2O + CO test reaction [74]. Sol–gel templating has also been applied to preformed DENs, although nanoparticle occlusion within the sol–gel matrix may be a problem for some systems [75].

4.4.1.3 DEN Deposition onto Oxide Supports

Wetness impregnation methods can be used to deposit DENs onto a variety of porous oxide supports, although this often requires concentrating DENs solutions to the point where dendrimer agglomeration may become problematic. Depending on the desired substrate (inorganic oxides [36, 76, 77], electroactive carbons [57], planar Au [60]) and the dendrimer terminal group, a variety of chemical deposition options are also available. DENs prepared using hydroxy-terminated PAMAM dendrimers can be deposited by a variety of "slow adsorption" techniques, in which DENs are stirred with an oxide support at appropriate pH for approximately 24 h [36, 76, 77]. The adsorption process appears to be controlled by the oxide nanoparticle interactions. Alternately, sol–gel chemistry has also been used to immobilize hydroxy-terminated DENs [75, 78, 79].

In the absence of a solvent, supported, intact DENs are completely inactive as catalysts and do not bind CO. Presumably, upon drying, the organic dendrimer collapses onto the nanoparticle, poisoning the metal surface [80, 81]. The activity of immobilized DEN electrocatalysts in water and a study by Williams and coworkers have provided the strongest evidence in support of dendrimer poisoning of supported dried DENs. Using a solid-solution *in situ* ATR-IR spectroscopy technique, Williams and coworkers have shown that alumina-supported intact Pt DENs are capable of binding CO in the presence of water. Control experiments showed that the supported DENs do not migrate into the solution and stay adsorbed on the oxide. The supported DENs were active for CO oxidation in water and showed activity comparable to a Pt/Alumina catalyst prepared via wetness impregnation of H_2PtCl_6. This study also showed that the presence of the Pt nanoparticle has a substantial effect on the stretching frequencies associated with the amide bonds that make up the dendrimer backbone [81].

4.4.2
High Temperature Dendrimer Removal

Identifying appropriately mild activation conditions for supported DENs is a prerequisite for utilizing them as precursors to heterogeneous nanoparticle catalysts, because PAMAM dendrimers are not thermally stable [16]. Ideal activation conditions should be forcing enough to remove or passivate the organic material, yet mild enough not to induce particle agglomeration. Surface particle agglomeration or sintering processes are extremely temperature dependent [82, 83], so minimizing activation temperatures is critical for ultimately correlating supported catalyst properties with synthetic methodologies and particle properties.

Initial work with Pd DENs immobilized on mica showed the importance of carefully determining decomposition protocols as even a short (10 min) treatment at high temperature (630 °C) caused substantial particle growth [84]. Thermogravimetric analysis (TGA) studies indicate that, under the conditions of the TGA experiment, high temperatures (500 °C) are required to completely remove organic matter from Pd and Au DENs immobilized in porous sol–gel TiO₂ [78]. This treatment resulted in substantial particle agglomeration, although pore templating by the dendrimer mitigated the particle growth.

Similar to the TGA experiments, *in situ* infrared spectroscopy has been used to follow the amide bond stretching frequencies while heating under various atmospheres [37, 77, 80, 85–87]. These experiments (see Fig. 4.10) suggest that dendrimer removal requires relatively forcing conditions to maximize CO adsorption and catalytic activity on supported Pt catalysts prepared from DENs. A variety of activation conditions have been chosen based on these experiments, generally involving some combination of oxygen and/or hydrogen treatments.

The specific activation conditions required for an individual catalyst likely depend on the metal and support, but 300 °C appears to be somewhat of a watershed temperature. Activation at temperatures above 300 °C generally coincides with loss of metal surface area due to sintering [77, 80, 85]. The metal loading, dendrimer loading, and metal:dendrimer ratios also impact activation conditions,

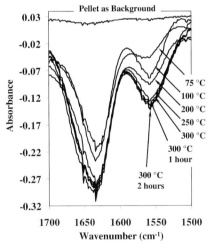

Fig. 4.10 Typical *in situ* dendrimer decomposition experiment in 20% O₂. The supported DENs were pressed into a self-supporting wafer, loaded into the IR cell, and the temperature was increased at approximately 5 °C min⁻¹. The first five spectra shown are at intervals of approximately 30 °C; the bottom three spectra were collected after soaking at 300 °C for 2 h.

suggesting that it may be necessary to optimize activation conditions for individual catalysts. Using temperatures at or near 300 °C, supported Pt [77, 80, 85, 86], Au [88], Pt–Au [37, 76], Pt–Cu [36], and Ru [87, 89] nanoparticles have been prepared. In most cases, treatments at 300 °C resulted in little to no particle agglomeration and the resulting supported nanoparticles have remained in the 1–3 nm range (see Fig. 4.11).

The different conclusions based on TGA and IR experiments are substantial; however, they are understandable based on the conditions of each experiment. Although TGA provides explicit information regarding when removal of organic species is complete, it is most effective for monitoring rapid changes in mass as a function of temperature. Even at a relatively slow temperature ramp (e.g.

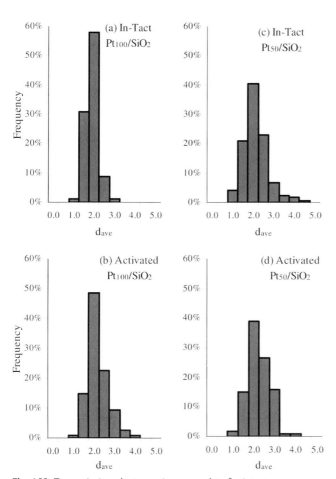

Fig. 4.11 Transmission electron microscopy data for intact (a) and (c) and activated (b) and (d) Pt/SiO₂ catalysts. Reprinted with permission from *J. Am. Chem. Soc.* **2003**, *125*, 14832–14836. Copyright 2003 American Chemical Society.

$1\,°C\,min^{-1}$), standard TGA experiments are ineffective in evaluating decomposition processes that take place over the course of tens of minutes or hours.

In situ infrared spectroscopy has been used in much the same fashion as TGA, but temperature profiles have been combined with monitoring changes at constant temperature [77, 80, 85–87]. Infrared spectroscopy does not yield the same direct information about the complete removal of organic residues that TGA provides. On the other hand, CO adsorption experiments performed along with dendrimer decomposition experiments provide direct information regarding metal availability. Further, IR experiments provide important information regarding dendrimer decomposition products and residues that can act as poisons for supported metal nanoparticle catalysts.

4.4.2.1 Models for Dendrimer Removal

Both TGA and IR experiments show that the PAMAM dendrimer backbone begins decomposing at temperatures as low as $75\,°C$, although more forcing conditions are required to fully activate the catalysts. The low onset temperature of dendrimer decomposition is not surprising given that PAMAM dendrimers can undergo retro-Michael addition reactions at temperatures above $100\,°C$ [14]. To avoid this, PAMAM dendrimer synthesis and modification are typically carried out at or near ambient temperature using reaction times as long as several days [16]. Further, Bard and coworkers have shown that PAMAM dendrimers are susceptible to oxidation by molecular oxygen in aqueous solution at near ambient temperatures, even in the absence of a nanoparticle catalyst [90].

Pt and Pd are good oxidation catalysts and have been clearly implicated in catalyzing dendrimer oxidation during decomposition: the dendrimer amide bonds degrade upon heating in He, even in the absence of a metal nanoparticle [78, 85, 91]. Since the dendrimer architecture is relatively unstable at elevated temperatures, it is perhaps surprising that such forcing conditions were initially suggested for complete activation of DENs. The necessity of forcing activation conditions has been attributed to the production of various carbonaceous species during activation. Under oxidizing atmospheres, the appearance of stretching frequencies consistent with the formation of surface carboxylates has been reported [86]. Similarly, several persistent surface bands consistent with coke-type species were identified when activating supported Ru DENs under H_2 [87].

A working model for dendrimer thermolysis during calcination involves the PAMAM dendrimer backbone initially reacting with oxygen (which may or may not be activated by a nanoparticle) in a relatively facile process to generate carboxylates and other surface species [86]. Removal of carbonaceous species closely associated with the nanoparticle is required for complete activation of the catalyst. For Pt DENs, the surface carboxylates may be strongly adsorbed to the nanoparticle surface and extended O_2 treatments are required for deep oxidation of the hydrocarbon to reach reasonably volatile species [86]. Once formed, however, it appears that they can be removed more readily with a hydrogen treatment than with further oxidation [80]. A high-resolution TEM study of supported Ru DENs indicated that substantial sintering occurs during activation, but that relatively small particle sizes (2.5 nm) and distributions were still accessible from DENs [89].

4.4.2.2 Low Temperature Dendrimer Removal

Given that it may be difficult to remove surface carboxylates from supported DENs, the question arises as to whether it is inherently necessary to remove all organic material to prepare clean, active supported nanoparticles. In a practical sense, except when dealing with freshly calcined supports, carbon species from atmospheric sources are always present on oxide surfaces. Indeed, C—H stretching vibrations are readily observable in infrared spectra of supports taken directly from a manufacturer's container. The critical species are those directly adsorbed to the metal surface and in close proximity to the nanoparticles.

Using silica supported Pt-DENs, we showed that activation temperatures could be reduced to as low as 150°C by employing CO oxidation reaction conditions [86]. In this activation protocol, CO essentially acts as a protecting group: strong CO adsorption prevents Pt from participating in dendrimer oxidation while at the same time preventing fouling of the nanoparticle surface by dendrimer fragments (see Fig. 4.12). Supported Pt catalysts pretreated in 1% CO and 25% O_2 at 150°C for 16 h had essentially the same infrared spectra of adsorbed CO and CO oxidation activity as catalysts oxidized at 300°C for 16 h (see Figs. 4.13 and 4.14). The third data set in Fig. 4.13 is for a catalyst activated in CO/O_2 at 200°C. CO does not bind Pt strongly enough at this temperature to protect the surface from the dendrimer decomposition products and thus does not impart the same protecting effect.

This low temperature CO/O_2 treatment was also used for Au/TiO_2 DENs; for some pretreatment temperatures, the Au/TiO_2 catalysts were more than an order of magnitude more active for CO oxidation than Pt/SiO_2 [88]. The Au catalysts also showed substantial room temperature activity that was extremely reproducible. However, the catalysts were not stable over time, presumably due to dendrimer decomposition products migrating onto the Au nanoparticles and poisoning the catalysis.

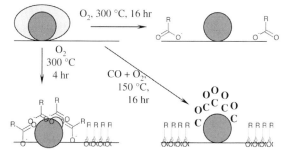

Fig. 4.12 Schematic description of various pretreatment protocols. Reprinted with permission from *Langmuir* **2005**, *21*, 10776–10782. Copyright 2005 American Chemical Society.

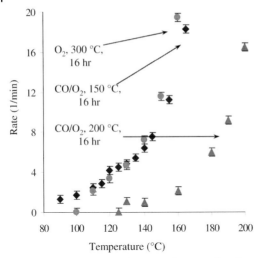

Fig. 4.13 CO oxidation catalysis as a function of oxidation treatments. A rate vs. temperature graph following oxidation in 20% O_2 and in 1% CO + 20% O_2. Reprinted with permission from *Langmuir* **2005**, *21*, 10776–10782. Copyright 2005 American Chemical Society.

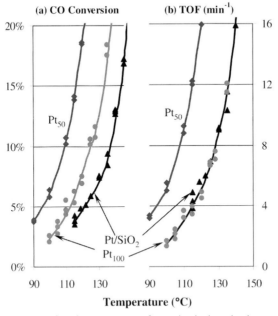

Fig. 4.14 Infrared spectroscopy of CO adsorbed on dendrimer templated and traditionally prepared catalysts. Reprinted with permission from *Langmuir* **2005**, *21*, 10776–10782. Copyright 2005 American Chemical Society.

4.4.3
Monometallic Dendrimer Templated Nanoparticle Catalysts

Dendrimer templated iron oxide nanoparticles have been used as precursors for growing carbon nanotubes [92, 93], but silica-supported Pt catalysts have been examined in the most depth, largely to identify appropriate dendrimer thermolysis protocols. The first study with these materials involved catalysts prepared from G5-OH(Pt_{50}) and G5-OH(Pt_{100}), calcined under O_2 at 300 °C for 4 h and reduced at 300 °C for 2 h [80]. The resulting catalysts are active for both oxidation and hydrogenation reactions. The turnover frequencies (TOFs) for catalytic CO oxidation (see Fig. 4.14) and toluene hydrogenation (at 60 °C) by the Pt_{100}/SiO_2 catalyst were indistinguishable from a traditionally prepared catalyst, as was the infrared spectrum of CO adsorbed on Pt_{100}/SiO_2. Beyond an important proof of concept, this study showed that dendrimer-derived catalysts can serve as good models for traditionally prepared catalysts.

Figure 4.15 shows spectra of CO adsorbed on Pt_{20}/SiO_2 catalysts compared to a catalyst prepared from a simple Pt precursor. The Pt(acetylacetonate)$_2$ catalyst is consistent with standard Pt catalysts, and the full coverage CO stretching frequency (2085 cm^{-1}) is consistent with CO linearly adsorbed on Pt [1, 94]. Dendrimer templated catalysts tend to have somewhat lower primary stretching

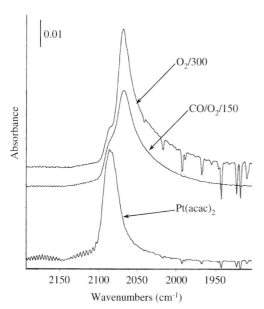

Fig. 4.15 CO oxidation catalysis for Pt_{50} and Pt_{100} DENs after oxidation at 300 °C (20% O_2/He, 4 h) and reduction at 300 °C (20% H_2/He, 2 h). Reprinted with permission from *J. Am. Chem. Soc.* **2003**, *125*, 14832–14836. Copyright 2003 American Chemical Society.

frequencies: fully activated Pt_{20}/SiO_2 catalysts, for example, have a primary stretching frequency near $2068\,cm^{-1}$ with a distinguishable shoulder at $2085\,cm^{-1}$. These bands remain consistent after a number of activation protocols, including calcination at 300 °C for up to 24 h [86]. A number of studies have loosely correlated lower CO stretching frequencies with adsorption onto low coordination edge and corner sites that dominate small nanoparticles [94]. Similarly, high coverage bands near $2085\,cm^{-1}$ have been assigned to terrace or face sites on extended surfaces [94]. The relative intensity of these adsorption bands has also been correlated to the presence of small (<2 nm) supported Pt particles; however, the complexity of factors affecting band intensity (intensity borrowing, dipole coupling, dephasing) makes quantitative interpretation of these bands unreliable.

The consistency of the high ($2068\,cm^{-1}$) and low ($2058\,cm^{-1}$) coverage values for the Pt_{20}/SiO_2 catalyst suggests this is a diagnostic band for fully activated dendrimer templated catalysts. The same $2068\,cm^{-1}$ high coverage band is also found for a Pt_{50}/SiO_2 catalyst prepared with sol–gel chemistry, along with the 10 wavenumber red shift during thermal desorption [75]. Dendrimer-derived catalysts that are not fully activated tend to have slightly blue shifted high coverage CO stretching frequencies. All of these catalysts have relatively similar activities for CO oxidation, indicating that this reaction is relatively fast, even on partially poisoned Pt surfaces. This is consistent with suggestions that the rate limiting step for CO oxidation by smaller nanoparticles involves desorption of a key surface intermediate [95]. In general, toluene hydrogenation appears to be a more sensitive probe reaction for dendrimer-derived catalysts as activity for this reaction is closely tied to the cleanliness of Pt nanoparticle surfaces [75, 80].

4.4.4
Supported Bimetallic Dendrimer Templated Nanoparticles

4.4.4.1 Infrared Spectroscopy of Supported Bimetallic Dendrimer Templated Nanoparticles

Infrared spectroscopy of adsorbed CO is a useful characterization tool for dendrimer templated supported nanoparticles, because it directly probes particle surface features. Because adsorbed CO stretching frequencies are sensitive to surface geometric and electronic effects, it is potentially possible to evaluate the relative effects of each on nanoparticle properties. Infrared spectroscopy of adsorbed CO has been used to investigate several dendrimer templated PtAu and PtCu catalysts. PtAu nanoparticles from $G5\text{-}OH(Pt_{16}Au_{16})$ prepared via Cu displacement have been prepared on a variety of oxide supports (silica, alumina, titania) [37, 76]. For all the supports, bands assigned for atop Au–CO and Pt–CO were observed (see Fig. 4.16). Heating the samples under He flow caused substantial changes in the Pt–CO bands as Au–CO desorbed, providing conclusive evidence for the intimate mixing of the two metals in individual particles.

A complementary study evaluated composition effects on dendrimer templated PtCu nanoparticles [37]. Although Cu–CO bands were not observed, a similar red shift in the Pt–CO stretching frequency to the PtAu system was observed,

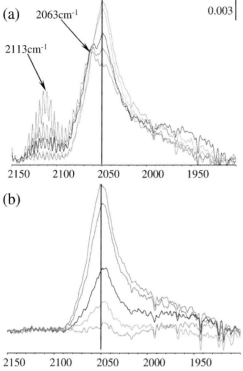

Fig. 4.16 Infrared spectroscopy during CO desorption from Pt$_{16}$Au$_{16}$ (a) 30 [blue], 70, 90, and 120 [red] °C and (b) 120 [blue], 150, 170, 180, and 190 [red] °C. Reprinted with permission from *J. Am. Chem. Soc.*, **2004**, *126*, 12949–12956. Copyright 2004 American Chemical Society.

indicating the presence of well-mixed bimetallic nanoparticles throughout the composition range. Infrared spectroscopy of CO adsorbed on both the PtAu and PtCu catalysts showed that the shifts in the CO stretching frequency upon Cu or Au incorporation were small relative to the magnitude of dipole coupling effects [36, 37, 76]. These results indicate that electronic effects (electron donation from one metal to another) are likely to be minimal for these systems [36].

4.4.4.2 Catalysis by Supported Bimetallic Dendrimer Templated Nanoparticles

Heterogeneous catalysis also directly probes the surface properties of supported nanoparticles, and has been employed for dendrimer templated PtAu [37], PdAu [79], and PtCu [36] nanoparticles. Figure 4.17 shows CO oxidation for Pt$_{16}$Au$_{16}$/SiO$_2$ catalysts compared with monometallic catalysts. Similar to the homogeneous catalysis studies, all three metal systems show synergism in catalytic activity CO oxidation catalysis, with the bimetallic catalysts being more active than any of the corresponding monometallic catalysts.

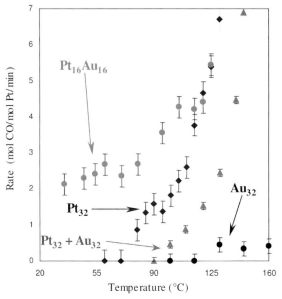

Fig. 4.17 CO oxidation catalysis by silica supported Pt_{32}, Au_{32}, $Pt_{16}Au_{16}Pt$, and $Pt_{32}+Au_{32}$ NPs. Rate is reported as moles CO converted per total moles Pt per minute; for Au_{32}, the rate is in moles CO converted per total mole Au per minute. Reprinted with permission from *J. Am. Chem. Soc.*, **2004**, *126*, 12949–2956. Copyright 2004 American Chemical Society.

The dendrimer templating route also provides opportunities for important new control experiments in evaluating supported catalysts. Using DENs, it is possible to compare catalysis over well-mixed bimetallic nanoparticles with "co-metallic" catalysts composed of monometallic nanoparticles codeposited on the same support. For the PtAu system, these experiments assisted in characterizing the bimetallic materials and shed light onto the nature of the catalytic enhancement. Monometallic Au particles sinter badly on silica (see Table 4.3) and do not bind CO. However, when Pt is mixed into the nanoparticle, Au sintering is greatly diminished and surface Au atoms are observed in the IR. The infrared spectra, coupled with TEM data and CO oxidation catalytic activity indicated that bimetallic Pt–Au particles might exchange surface and subsurface atoms to maximize particle–substrate interactions. Under activation conditions, Pt appears to concentrate at the particle–support interface, anchoring the nanoparticles. In the presence of CO, some Pt appears to migrate to the surface of small nanoparticles to take advantage of the strong Pt–CO interactions (see Fig. 4.18). This type of exchange has also been suggested by computational studies for the PtAu system [96].

The ambient temperature catalytic activity of the PtAu catalyst was subsequently attributed to surface Au atoms and higher temperature activity attributed to surface Pt. Further studies with these catalysts supported on other oxides (alumina, titania)

Table 4.3 Catalyst preparation and characterization data for Pt–Au catalysts. Reprinted with permission from *J. Am. Chem. Soc.*, **2004**, *126*, 12949–12956. Copyright 2004 American Chemical Society.

	Pt$_{32}$	Au$_{32}$	Pt$_{16}$Au$_{16}$	Pt$_{32}$ + Au$_{32}$
Nominal Cu:dendrimer[a]	32	48	40	32, 48
Pt loading (wt.%)[b]	0.25	—	0.14	0.14
Au loading (wt.%)[b]	—	0.29	0.15	0.15
Pt:Au ratio[c]	—	—	15:17	15:17
d_{ave} (nm)[d]	3.3	14.5	2.6	4.2
Standard deviation[e]	0.8	14.8	1.1	4.2
d_{MP} (nm)[f]	3.5	12	2.0	3.0

a   Cu:dendrimer stoichiometry.
b   Supported catalyst % Pt and Au, determined by atomic absorption spectroscopy. Cu content was also measured for all catalysts and was always below detection limits (0.01%).
c   Determined from Pt and Au loadings.
d   Arithmatic mean diameter of all imaged particles.
e   Standard deviation in d_{ave}.
f   Most probable diameter: The observed particle diameter that occurs with the highest frequency, i.e. the highest bar on the particle size distribution. This quantity offers an additional means of comparing different types of distributions (e.g. Gaussian vs.

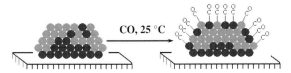

Fig. 4.18 Restructuring of Pt-Au nanoparticles in the presence of CO. Reprinted with permission from *J. Am. Chem. Soc.*, **2004**, *126*, 12949–2956. Copyright 2004 American Chemical Society.

support this conclusion and indicate that catalysis occurs on the nanoparticle. Arrhenius plots prepared from DENs also showed enhanced CO oxidation activity upon intimate mixing of the two metals [79].

We have also studied PtCu nanoparticles, examining the catalytic properties as a function of particle composition. For this system, rate enhancement does not show the same maximum in activity as a function of metal content as do the homogeneous hydrogenation studies described above [36]. CO oxidation catalysis data, shown in Table 4.4, indicate a small enhancement in activity and small drop in apparent activation energy upon incorporating Cu into Pt nanoparticles. These changes are remarkably consistent throughout the composition range, indicating that the changes are primarily due to small modifications in Pt activity (Cu is inactive for CO oxidation under these conditions). This activity change could be due to breaking up larger Pt islands through simple dilution, or by providing Cu sites for either CO or O$_2$ binding/activation.

Table 4.4 Catalytic activity of dendrimer templated heterogeneous PtCu catalysts.

Catalyst	% Pt	Pt:Cu[a]	CO oxidation		Toluene hydrogenation
			rate @ 60 °C[b]	E_{app}[c] (kJ mol^{-1})	rate @ 60 °C[d]
Pt$_{45}$	0.19	—	180	67	1300
Pt$_{30}$Cu$_{15}$	0.17	1.7	350	46	460
Pt$_{23}$Cu$_{23}$	0.12	0.85	360	49	230
Pt$_{15}$Cu$_{30}$	0.057	0.43	370	47	<5

a determined via AA spectroscopy. All corrected rates are per total mol of Pt in the catalyst, i.e. no assumptions regarding, or corrections for, the fraction of exposed metal were used.
b mol CO/mol Pt/min × 10^3.
c Apparent activation energies from Arrhenius plots (kJ mol^{-1}).
d mol CO/mol Pt/min × 10^3.

In contrast to homogeneous hydrogenations (the PdAu system is most similar), incorporation of Cu into Pt nanoparticles had a substantial poisoning effect on toluene hydrogenation catalysis Pt [36]. As Table 4.4 shows, catalytic rates, which were normalized per total mole of Pt, drop by more than two orders of magnitude as the Cu content of the particles increases. This also contrasts with heterogeneous CO oxidation studies for the same catalysts, which showed enhanced activity upon Cu incorporation. The disparate catalysis data were interpreted in terms of a surface enrichment in Cu under the hydrogenation reaction conditions, driven by the lower heat of sublimation of Cu [36, 97, 98].

Correlating the hydrogenation catalysis data with the CO oxidation data and infrared studies is more difficult, but two models explain the shifts in activity. The infrared data clearly indicate that Pt is on the surface in the presence of CO, and that the surface is active for CO oxidation. As was the conclusion for the PtAu system, one explanation is that the nanoparticle surface is substantially different under reducing atmospheres (enriched in Cu) than it is under CO; i.e. CO induces surface restructuring for the heterogeneous systems. It is also possible that the effect of diluting Pt with Cu shuts off toluene hydrogenation activity, which would be expected if toluene hydrogenation requires a minimum Pt ensemble. At this point, it is not possible to discern between these two possibilities; nor is it possible to evaluate the relative importance of these effects on the homogenous systems.

4.5
Summary, Outlook, and Links Between Homogeneous and Heterogeneous Catalysis

In a sense, PAMAM dendrimers can be considered "nanoreactors" for preparing metal nanoparticles. Several synthetic methodologies are now available for preparing a wide variety of mono and bimetallic nanoparticles of the order of 1–3 nm. Beyond well-mixed bimetallic nanoparticles, it is also possible to selectively prepare core–shell particle morphologies, with either mono- or bi-metallic cores. Electron microscopy techniques, as well as UV–visible and X-ray photoelectron spectroscopies have been valuable tools in characterizing these particles.

The chemical and catalytic properties of DENs are of particular interest. Due to their readily functionalized exteriors, DENs can be used as catalysts in a variety of solvents or reaction media. They have been employed for several types of reactions, particularly hydrogenations and carbon–carbon coupling chemistry. Homogeneous catalysis studies generally show that bimetallic nanoparticles are more active hydrogenation catalysts than their monometallic counterparts. Synergism is also observed in heterogeneous CO oxidation catalysis. Synergism is not general, however, as heterogeneous toluene hydrogenation catalysis is severely poisoned when Cu is incorporated into Pt nanoparticles.

The source of changes in catalytic activity is of fundamental importance, and dendrimer templated nanoparticles offer new means of evaluating the relative influence of structural and electronic effects. Solution-phase particle size studies on alkene hydrogenation indicate anti-pathetic (activity increasing with particle size) dependence. Such homogenous studies are uncommon, but have long standing precedent in the supported catalyst literature [99]. Heterogeneous alkene hydrogenation can be either structure sensitive (surface reaction rate depends on particle size and/or surface geometry) or structure insensitive (surface reaction rates similar over wide particle size ranges and surface geometries), depending on the alkene [100]. The presence of the dendrimer makes it difficult to unambiguously attribute the nature of the particle size effect on catalysis, but future combined homogeneous and heterogeneous catalytic studies may offer opportunities to sort this out.

Additionally, DENs offer new opportunities to investigate, understand, and evaluate the relative influence of structural and electronic effects in bimetallic catalysts. This fundamental understanding is critical if the goal of controllably tuning bimetallic catalyst properties is to be realized. Studies of CO adsorbed on heterogenized nanoparticles provide preliminary evidence that, at least for the systems studied thus far, electron donation from one metal to another may be small. This, in turn, indicates that rate enhancements may be due to changes in surface or particle geometries when a second metal is incorporated. There are still relatively few studies, however, and these conclusions can only be considered preliminary at this stage.

Another interesting and potentially important property of these particles is their potential to exchange surface and subsurface atoms. CO appears to be able to draw Pt atoms to the surface of heterogenized PtAu and PtCu nanoparticles. It is unclear how general this property is, or if it is only applicable to certain metal systems, but it is clearly important to understanding nanoparticle dynamics. Over time, and with deeper understanding of these exchange dynamics, it may be possible to develop synthetic methodologies to prepare new supported nanoparticle catalysts with core–shell morphologies and improved catalytic properties.

Acknowledgements

The authors gratefully acknowledge the Robert A. Welch Foundation (Grant number W-1552) and the U.S. National Science Foundation (Grant number CHE-0449549) for financial support of our work. We also thank Professors Keith Stevenson and Dick Crooks, as well as their research groups, for a variety of experimental assistance and valuable discussions.

References

1 V. Ponec, G. C. Bond (Eds.), *Catalysis by Metals and Alloys*, Elsevier, Amsterdam, **1995**, Vol. 95.

2 O. Alexeev, B. C. Gates, *Ind. Eng. Chem. Res.* **2003**, *42*, 1571.

3 J. H. Sinfelt, *Bimetallic Catalysts: Discoveries, Concepts, and Applications*, John Wiley & Sons, New York, **1983**.

4 Y. Xu, A. V. Ruban, M. Mavrikakis, *J. Am. Chem. Soc.* **2004**, *126*, 4717.

5 C. S. Song, *Catal. Today* **2002**, *77*, 17.

6 J. M. Thomas, R. Raja, B. F. G. Johnson, S. Hermans, M. D. Jones, T. Khimyak, *Ind. Eng. Chem. Res.* **2003**, *42*, 1563.

7 V. Ponec, G. C. Bond, *Stud. Surf. Sci. Catal.* **1995**, *95*.

8 A. Wolf, F. Schuth, *Appl. Catal. A: General* **2002**, *226*, 1.

9 H.-S. Oh, J. H. Yang, C. K. Costello, Y. M. Wang, S. R. Bare, K. H. H, M. C. Kung, *J. Catal.* **2002**, *210*, 375.

10 S. D. Lin, M. Bollinger, M. A. Vannice, *Catal. Lett.* **1993**, *17*, 245.

11 A. Brenner, in *Metal Clusters*, Moskovits, M. (Ed.) John Wiley & Sons, New York, **1986**, pp. 249–282.

12 D. M. P. Mingos, D. J. Wales, *Introduction to Cluster Chemistry*; Prentice-Hall, London, **1990**.

13 B. D. Chandler, L. I. Rubinstein, L. H. Pignolet, *J. Mol. Catal. A: Chemical* **1998**, *133*, 267.

14 R. W. J. Scott, O. M. Wilson, R. M. Crooks, *J. Phys. Chem. B* **2005**, *109*, 692.

15 M. Fisher, F. Vogtle, *Angew. Chem. Int. Ed.* **1999**, *38*, 884.

16 J. M. J. Fréchet, D. A. Tomalia (Eds.), *Dendrimers and other Dendritic Polymers*, John Wiley & Sons, West Sussex, UK, **2001**.

17 D. Astruc, F. Lu, J. Ruiz Aranzaes, *Angew. Chem. Int. Ed.* **2005**, *44*, 7852.

18 G. R. Newkome, H. He, C. N. Moorefield, *Chem. Rev.* **1999**, *99*, 1689.

19 A. W. Bosman, H. M. Janssen, E. W. Meijer, *Chem. Rev.* **1999**, *99*, 1665–1966.

20 R. Kreiter, A. W. Kleij, R. J. M. Gebbink, G. van Koten, *Top. Curr. Chem.* **2001**, *217*, 163.

21 F. Zeng, S. C. Zimmerman, *Chem. Rev.* **1997**, *97*, 1681.

22 R. M. Crooks, B. I. Lemon, L. Sun, L. K. Yeung, M. Zhao, *Top. Curr. Chem.* **2001**, *212*, 82.

23 L. J. Twyman, A. S. H. King, I. K. Martin, *Chem. Soc. Rev.* **2002**, *31*, 69–82.

24 A. I. Cooper, J. D. Londono, G. Wignall, J. B. McClain, E. T. Samulski, J. S. Lin, A. Dobrynin, M. Rubinstein, A. L. C. Burke, J. M. J. Frechet, J. M. DeSimone, *Nature* **1997**, *389*, 368.

25 R. M. Crooks, M. Zhao, L. Sun, V. Chechik, L. K. Yeung, *Acc. Chem. Res.* **2001**, *34*, 181.

26 M. F. Ottaviani, F. Montalti, N. J. Turro, D. A. Tomalia, *J. Phys. Chem. B* **1997**, *101*, 158.

27 M. R. Knecht, J. C. Garcia-Martinez, R. M. Crooks, *Langmuir* **2005**, *21*, 11981.

28 M. Q. Zhao, R. M. Crooks, *Chem. Mater.* **1999**, *11*, 3379.

29 H. Lang, B. D. Chandler, unpublished results.

30 R. W. J. Scott, A. K. Datye, R. M. Crooks, *J. Am. Chem. Soc.* **2003**, *125*, 3708.

31 Y.-M. Chung, H.-K. Rhee, *Catal. Lett.* **2003**, *85*, 159.

32 R. W. J. Scott, O. M. Wilson, S.-K. Oh, E. A. Kenik, R. M. Crooks, *J. Am. Chem. Soc.* **2004**, *126*, 15583.

33 Y.-M. Chung, H.-K. Rhee, *J. Mol. Catal. A: Chemical* **2003**, *206*, 291.

34 Y. M. Chung, H. K. Rhee, *J. Colloid Interface Sci.* **2004**, *271*, 131.

35 P. J. Pellechia, J. Gao, Y. Gu, H. J. Ploehn, C. J. Murphy, *Inorg. Chem.* **2003**, *43*, 1421.

36 N. N. Hoover, B. J. Auten, B. D. Chandler, *J. Phys. Chem. B* **2006**, *110*, 8606.

37 H. Lang, S. Maldonado, K. J. Stevenson, B. D. Chandler, *J. Am. Chem. Soc.* **2004**, *126*, 12949.

38 C. W. Hills, N. H. Mack, R. G. Nuzzo, *J. Phys. Chem. B.* **2003**, *107*, 2626.

39 J. Luo, M. M. Maye, V. Petkov, N. N. Kariuki, L. Wang, P. Njoki, D. Mott, Y. Lin, C.-J. Zhong, *Chem. Mater.* **2005**, *17*, 3086.

40 O. M. Wilson, R. W. J. Scott, J. C. Garcia-Martinez, R. M. Crooks, *J. Am. Chem. Soc.* **2005**, *127*, 1015.

41 T. Endo, T. Yoshimura, K. Esumi, *J. Colloid Interface Sci.* **2005**, *286*, 602.

42 Y.-G. Kim, C. Garcia-Martinez Joaquin, M. Crooks Richard, *Langmuir* **2005**, *21*, 5485.

43 U. Kreibig, M. Vollmer, Optical Properties of Metal Clusters, **1995**, Vol. 25, Springer, Berlin.

44 J. A. Creighton, D. G. Eadon, *J. Chem. Soc. Faraday Trans.* **1991**, *87*, 3881.

45 P. Mulvaney, *Langmuir* **1996**, *12*, 788–800.

46 C. D. Wagner, W. M. Riggs, *Handbook of X-Ray Photoelectron Spectroscopy*, Perkin-Elmer Co., Minnesota, **1979**.

47 L. K. Yeung, R. M. Crooks, *Nano Lett.* **2001**, *1*, 14–17.

48 J. C. Garcia-Martinez, R. Lezutekong, R. M. Crooks, *J. Am. Chem. Soc.* **2005**, *127*, 5097.

49 M. Pittelkow, K. Moth-Poulsen, U. Boas, J. B. Christensen, *Langmuir* **2003**, *19*, 7682.

50 Y. Li, M. A. El-Sayed, *J. Phys. Chem. B* **2001**, *105*, 8938.

51 M. Zhao, R. M. Crooks, *Angew. Chem. Int. Ed.* **1999**, *38*, 364.

52 Y. H. Niu, L. K. Yeung, R. M. Crooks, *J. Am. Chem. Soc* **2001**, *123*, 6840.

53 M. Ooe, M. Murata, T. Mizugaki, K. Ebitani, K. Kaneda, *Nano Lett.* **2002**, *2*, 999.

54 F. Gröhn, B. J. Bauer, Y. A. Akpalu, C. L. Jackson, E. J. Amis, *Macromolecules* **2000**, *33*, 6042.

55 O. M. Wilson, M. R. Knecht, C. Garcia-Martinez Joaquin, M. Crooks Richard, *J. Am. Chem. Soc.* **2006**, *128*, 4510.

56 R. Schloegl, S. B. Abd Hamid, *Angew. Chem. Int. Ed.* **2004**, *43*, 1628.

57 H. Ye, R. M. Crooks, *J. Am. Chem. Soc.* **2005**, *127*, 4930–4934.

58 J. Alvarez, L. Sun, R. M. Crooks, *Chem. Mater.* **2002**, *14*, 3995.

59 S.-K. Oh, Y.-G. Kim, H. Ye, R. M. Crooks, *Langmuir* **2003**, *19*, 10420.

60 H. Ye, R. W. J. Scott, R. M. Crooks, *Langmuir* **2004**, *20*, 2915.

61 S. Rahghu, S. Berchmans, K. L. N. Phani, V. Yegnaraman, *Pramana* **2005**, *65*, 821.

62 S. C. Bourque, F. Maltais, W. J. Xiao, O. Tardif, H. Alper, *J. Am. Chem. Soc.* **1999**, *121*, 3035.

63 S. C. Bourque, H. Alper, L. E. Manzer, P. Arya, *J. Am. Chem. Soc.* **2000**, *122*, 956.

64 Z.-X. Guo, J. Yu, *J. Mater. Chem.* **2002**, *12*, 468.

65 S.-M. Lu, H. Alper, *J. Am. Chem. Soc.* **2005**, *127*, 14776.

66 P. P. Zweni, H. Alper, *Adv. Synth. Catal.* **2006**, *348*, 725.

67 Y. Jiang, Q. Gao, *J. Am. Chem. Soc.* **2006**, *128*, 716.

68 G. Larsen, E. Lotero, M. Marquez, *Chem. Mater.* **2000**, *12*, 1513.

69 G. Larsen, E. Lotero, *J. Phys. Chem. B* **2000**, *104*, 4840.

70 M. C. Rogers, B. Adisa, D. A. Bruce, *Catal. Lett.* **2004**, *98*, 29.

71 G. Larsen, R. Spretz, E. Lotero, *Chem. Mater.* **2001**, *13*, 4077.

72 M. C. Rogers, B. Adisa, D. A. Bruce, *Catal. Lett.* **2004**, *98*, 29.

73 C. G. Howard, D. A. Bruce, *Catal. Lett.*, in preparation.

74 G. Larsen, S. Noriega, *Appl. Catal. A: General* **2004**, *278*, 73.

75 L. Beakley, S. Yost, R. Cheng, B. D. Chandler, *Appl. Catal. A: General* **2005**, *292*, 124.

76 B. Auten, H. Lang, B. D. Chandler, **2007**, submitted.

77 H. Lang, R. A. May, B. L. Iversen, B. D. Chandler, in *Catalysis of Organic Reactions*; Sowa, J. (Ed.), Taylor & Francis Group/CRC Press: Boca Raton, FL, **2005**, pp. 243.

78 R. W. J. Scott, O. M. Wilson, R. M. Crooks, *Chem. Mater.* **2004**, *16*, 5682.

79 R. W. J. Scott, C. Sivadinarayana, O. M. Wilson, Z. Yan, D. W. Goodman, R. M. Crooks, *J. Am. Chem. Soc.* **2005**, *127*, 1380.

80 H. Lang, R. A. May, B. L. Iversen, B. D. Chandler, *J. Am. Chem. Soc.* **2003**, *125*, 14832.

81 D. X. Liu, J. X. Gao, C. J. Murphy, C. T. Williams, *J. Phys. Chem. B.* **2004**, *108*, 12911.

82 P. Forzatti, L. Lietti, *Catal. Today* **1999**, *52*, 165.

83 C. H. Bartholomew, *Appl. Catal. A: General* **2001**, *212*, 17–60.

84 L. Sun, R. M. Crooks, *Langmuir* **2002**, *18*, 8231–8236.

85 S. D. Deutsch, G. Lafaye, D. Liu, B. D. Chandler, C. T. Williams, M. D. Amiridis, *Catal. Lett.* **2004**, *97*, 139.

86 A. Singh, B. D. Chandler, *Langmuir* **2005**, *21*, 10776.

87 G. Lafaye, C. T. Williams, M. D. Amiridis, *Catal. Lett.* **2004**, *96*, 43.

88 B. Auten, C. J. Crump, A. R. Singh, B. D. Chandler, Catalysis of Organic Reactions, Steve R., Schmidt (Eds.) **2006**, 315–323.

89 G. Lafaye, A. Siani, P. Marecot, M. D. Amiridis, C. T. Williams, *J. Phys. Chem. B* **2006**, *110*, 7725.

90 W. I. Lee, Y. Bae, A. J. Bard, *J. Am. Chem. Soc.* **2004**, *126*, 8358.

91 O. Ozturk, T. J. Black, K. Perrine, K. Pizzolato, C. T. Williams, F. W. Parsons, J. S. Ratliff, J. Gao, C. J. Murphy, H. Xie, H. J. Ploehn, D. A. Chen, *Langmuir* **2005**, *21*, 3998.

92 P. B. Amama, M. R. Maschemann, T. S. Fisher, T. D. Sands, *J. Phys. Chem. B* **2006**, *110*, 10636.

93 H. C. Choi, W. Kim, D. Wang, H. Dai, *J. Phys. Chem. B* **2002**, *106*, 12361.

94 P. Hollins, *Surf. Sci. Rep.* **1992**, *16*, 51.

95 A. Bourane, S. Derrouiche, D. Bianchi, *J. Catal.* **2004**, *228*, 288.

96 L. D. Kieken, M. Neurock, D. Mei, *J. Phys. Chem. B* **2005**, *109*, 2234.

97 V. Y. Borovkov, D. R. Luebke, V. Kovalchuk, J. L. d'Itri, *J. Phys. Chem. B* **2003**, *107*, 5568.

98 D. Chakraborty, P. P. Kulkarni, V. I. Kovalchuk, J. L. d'Itri, *Catal. Today* **2004**, *88*, 169.

99 M. Che, C. O. Bennett, *Adv. Catal.* **1989**, *36*, 55.

100 G. A. Somorjai, A. L. Marsh, *Philos. Trans. R. Soc. London, Ser. A: Math., Phys., Eng. Sci.* **2005**, *363*, 879.

5
Aerogel Supported Nanoparticles in Catalysis

Adelina Vallribera and Elies Molins

Dedicated to the memory of Prof. Marcial Moreno-Mañas

5.1
Introduction

An important aspect of catalysis is the possibility to design more efficient and environmentally friendly processes in what is called green chemistry. For this purpose catalysts should be recyclable without leaching or contaminating the resulting products. Heterogeneous catalysis dominates the industrial scenario due to the easy recovery of the catalysts by simple filtration or decantation. However, homogeneous catalysts possess all the theoretical advantages when compared with heterogeneous catalysts, i.e. high activity per mol of metal, high selectivity, mild reaction conditions, and absence of diffusion problems, easy steric and electronic modulation, and good mechanistic knowledge. Therefore, obtaining catalysts possessing simultaneously the advantages of the homogeneous and heterogeneous variants is a topic of high interest to which industrial and academic groups are dedicating efforts worldwide. Figure 5.1 summarizes the diverse strategies explored at present.

Within the currently ubiquous nanoscience and nanotechnology efforts, the use of nanoparticles and nanostructured materials in catalysis appears as one of the most successful approaches [1]. It has been said that the most important drawback in nanoparticle applications is their tendency to aggregate and their dispersion in porous media has been shown to be a good way to prevent this. In this chapter a particularly interesting porous supporting material for nanoparticles, the aerogels, will be reviewed and their use in catalytical processes analyzed.

The precursory gel of an aerogel, prior to the drying step, is usually obtained by a series of hydrolysis and condensation reactions of the initial reactants to form a sol. This complex process is affected by many parameters such as the nature and concentration of the alkoxide and the solvent, the water concentration, the temperature and the presence of acid or base catalysts. All will greatly influence the characteristics of the material obtained. For instance, the alkoxide concentration will define the

Nanoparticles and Catalysis. Edited by Didier Astruc
Copyright © 2008 WILEY-VCH Verlag GmbH & Co. KGaA, Weinheim
ISBN: 978-3-527-31572-7

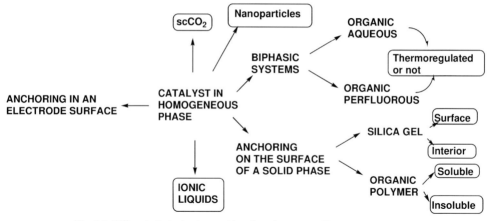

Fig. 5.1 Different strategies to combine the advantages of both heterogeneous and homogeneous catalysis.

density and porosity of the aerogel, the nature of the solvent changes the pore size distribution (methanol gives rise to smaller pores and, consequently, more transparent aerogels), and the amount of water will influence the reaction time. The "two-step" aerogels, as compared to the "one-step" ones, are obtained from a prepolymerized alkoxide obtained in acid conditions, followed by a base catalyzed condensation reaction during which gelation occurs. This alternative synthesis method gives rise to materials with smaller and narrower pore size distribution. The gel consists of a three-dimensional network of concatenated nanoparticles and a filling solvent. It can be removed from its containing mold and can stand on its own, although to avoid damage it is necessary to cover it with solvent. At this time, it is very important to allow the gels to age to complete reaction and also to wash them to eliminate the remaining water which will disturb the drying process.

Although regarded as less dense solid materials, aerogels usually present moderate densities ranging from 0.1 to $0.3\,\mathrm{g\,cm^{-3}}$. The most interesting characteristic in the current context is their high surface area (about $600\,\mathrm{m^2\,g^{-1}}$) related to a fractal type network of pores with a wide distribution of diameters, often centered in the mesoporous range. It is also noticeable that their amorphous structure is constituted of branch-linked colloidal particles. Aerogels are prepared by supercritical drying of the precursory gels, although alternative methods have been claimed, such as lyophilization. High surface area, i.e. larger than $500\,\mathrm{m^2\,g^{-1}}$, xerogels, can behave similarly to aerogels, at least for catalysis, so they can also be suitable as catalyst substrates. Although the high pressure process makes the material more expensive than desired, it allows the complete disappearance of the liquid surface tension embedding gels achieving its exchange by a gas without stressing the fragile structure network of the material. Aerogel materials have been excellently reviewed by Husing and Schubert [2], Gesser and Goswami [3], and Pierre and Pajonk [4].

Nanocomposite aerogels mainly consist as homogeneous dispersions of nanoparticles in an aerogel matrix. The preparation of nanocomposite aerogels can be

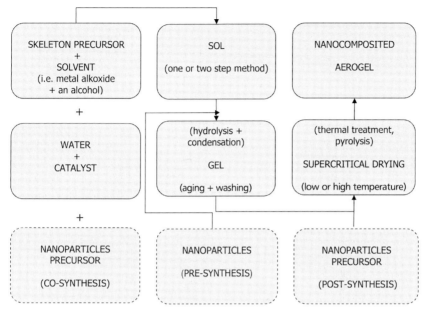

Fig. 5.2 Schematics of the preparation of nanocomposite aerogels.

accomplished by several routes, but, as a general classification, we can distinguish between methods involving the synthesis of the nanoparticles *before, during* or *after* the sol–gel process (Fig. 5.2). If the nanoparticles were previously prepared, the difficulty is to avoid their aggregation and, also, to achieve their homogeneous dispersion [5]. On the other hand, the particles can be prepared by any existing method, separately characterized, and probably, they will not greatly interfere with the sol–gel process. The particles can also be surface treated in order to avoid premature aggregation and/or sedimentation, and they are usually added to the sol just before gelation. At this stage, soft sonication can help to keep the medium homogeneous.

Another possibility is to introduce a metal precursor into the sol during the gelling stage, which would be followed by a reduction step. This method has been used successfully to prepare some magnetic [6, 7] and catalytic aerogels [8]. Although the homogeneity obtained by this method is usually good, the nanoparticle load cannot be high and, in the case of iron, mixtures of phases are obtained. On the other hand, the particle growth is limited by the pore size, which may be desirable. Either salts or organometallic complexes can be used as precursors of the nanoparticles. Once dried, the resulting aerogels can be thermally treated either to decompose the metal precursors and/or to allow increasing particle size, mainly if the drying is done at low temperature, i.e. exchanging the solvent by liquid carbon dioxide and drying at the supercritical conditions of CO_2.

Finally, nanoparticles can also be hosted after the gelling step by impregnation with suitable metal precursors [9]. Successive soaking of the gels in concentrated

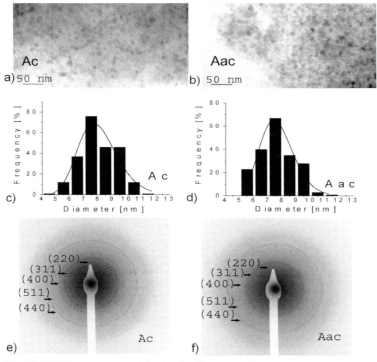

Fig. 5.3 TEM images, nanoparticle size distributions and electron diffraction patterns of γ-Fe$_2$O$_3$ silica aerogel nanocomposites prepared by the impregnation method using as metal precursors anhydrous ferrous acetate (Ac) and anhydrous ferrous acetylacetonate (Aac).

solutions of metal precursors permits saturation or even precipitation in the gel pores. A further thermal treatment, either in the supercritical drying stage or after it, will form the nanoparticles. This method gives rise to quality materials retaining high surface areas (Fig. 5.3). On the other hand, the nanoparticle content is limited by the solubility of the metal precursor.

The use of different metal alkoxides to form the initial sol results in mixed aerogels (binary solid solutions of silica, titania, alumina, zirconia, vanadia, rutenia, etc.) that cannot be considered as nanocomposited. This field has also been largely explored for catalytic applications [10].

Supercritical drying can be performed at high temperature, the lower limit being the critical point of the filling solvent (i.e. methanol P_c=80.9 bar, T_c=512.6 K, ethanol P_c=61.4 bar, T_c=513.9 K, acetone P_c=47.0 bar, T_c=508.1 K) or at low temperature, after exchanging the filling solvent with liquid carbon dioxide ($P_c=73.8$ bar, $T_c=304.1$ K). In any case, the reactor is first pressurized up to values higher than the critical pressure, and then heated until reaching supercritical conditions, i.e. pressure and temperature greater than P_c, T_c. At this point, there is no meniscus interface between liquid and gas and the supercritical fluid fills the reactor and the material homogeneously. When the reactor is slowly vented and the pressure is lower than P_c, the fluid in the reactor becomes gas, so the filling liquid has been removed from the material without much stress. The aerogel can be recovered after a final cooling step. Thermal treatments can be further applied to decompose precursors, to grow larger nanoparticles or to increase their crystallinity, also to produce oxidation or reduction depending on the atmosphere, although care has to be taken to avoid too large temperatures due to the progressive collapse of the porous network. Basic characterization of the material includes the determination of density and porosity (large runtimes for BET measurements are necessary due to the slow diffusion), X-ray diffraction (useful for crystalline nanoparticles) and transmission electron microscopy.

Nanocomposited carbon aerogels represent a different approach because of the organic nature of the precursory gel and the pyrolysis step [11]. In this case, high metal content can be achieved (up to 46% in the case of Pd-doped carbon aerogel) (Fig. 5.4).

It is impossible to list all the investigated systems so we have selected a series of important transformations. We aim to give a general overview of the different possibilities of aerogels in catalysis. Most of the reviewed systems correspond to active nanoparticles hosted in aerogels, but other related active aerogels have been included for comparison and to give a more complete view. A first classification between processes carried out in gaseous or liquid media has been used. In general, a quantitative comparison of catalysts from different groups will not be easy, since reactant concentrations, temperature, analytical methods etc. often vary.

We recommend readers to look at previous reviews involving the catalytic applications of aerogels: Pajonk [10], Baiker [12] and Moreno-Castilla [13].

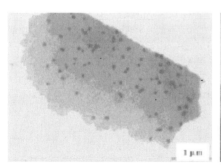 

Fig. 5.4 TEM images of (a) Cu- and (b) Pd-nanocomposited carbon aerogels.

5.2
Aerogel Nanocomposites as Catalysts of Important Reactions in Gaseous Media

5.2.1
Fischer–Tropsch Synthesis

The Fischer–Tropsch synthesis has recently attracted renewed interest; the process of the production of petroleum-like products from CO and H_2 (this is termed gas-to-liquid) is of high importance since one estimate indicates that gas reserves, expressed as oil equivalents, are four times as great as oil reserves [14]. Classically, two different Fischer–Tropsch processes were used based on iron catalysts: the Arge and Syntol processes. More recently some cobalt catalysts supported on mesoporous silica have been studied with success by the group of Ohtsuka [15]. With respect to the use of aerogels, in 1982 Teichner et al. [16] reported that iron oxide (Fe_2O_3 particles having a mean particle size of $110 nm$) dispersed on silica aerogels ($690 m^2 g^{-1}$ BET surface area) and submitted to an oxidative pretreatment (O_2 at $500 °C$) exhibited a catalytic activity about 300 times higher than the activity of the conventional fused iron reduced catalyst in the synthesis of hydrocarbons from the catalytic hydrogenation of carbon monoxide. Recently, the group of Eyring [17] reported the use of cobalt catalysts on silica aerogel for Fischer–Tropsch activity in a laboratory scale packed-bed reactor (pressure maintained at $100 psi$). The cobalt catalyst supported on sol–gel derived silica aerogel was prepared from $Co(NO_3)_2$ and tetramethoxysilane. The gel was dried under supercritical conditions, calcined and reduced in a stream of H_2 to obtain metallic cobalt. The CO conversion was found to be 5.3, 19.8, and 22.3 for the 2, 6, 10% loadings of cobalt, respectively. Between 6 and 10 wt.% of cobalt spherical nanoparticles were formed (27.1 and 36.9 nm, respectively), whereas above 10% larger nanoneedles [17] appeared.

5.2.2
Steam Reforming

An application of the steam reforming reaction is the production of methane. Methane is obtained fairly readily from any hydrocarbon feedstock that can be vaporized, for example naphtha. Naphtha fractions with a final boiling point of less than $220 °C$ are generally considered suitable for catalytic steam reforming. In 1959 ICI started up the first large-scale pressure steam reformer using naphtha as a feedstock.

Considering nonane as an example the reaction is

$$C_9H_{20} + 4H_2O \rightarrow 7CH_4 + 2CO_2$$

Nanobinary (Mg–Al) and ternary metal (Ni–Mg–Al) oxy/hydroxides were synthesized by aerogel protocols from magnesium methoxide, aluminum isopropoxide and nickel acetylacetonate [19]. After supercritical drying the material obtained had a high surface area >500 m^2 g^{-1} and small particle size <20 nm. The catalyst was reduced at 550 °C at atmospheric pressure of hydrogen and exhibited activity in the pre-reforming of naphtha.

5.2.3
Oxidation Processes

5.2.3.1   Oxidation of Carbon Monoxide

The ability of nanosized gold supported on titania (Au/TiO$_2$) as catalyst for ambient temperature oxidation of carbon monoxide has been extensively described in the literature. The typical Au/TiO$_2$ catalyst is derived from nanocrystalline titania, such as Degusa P25, onto which a preformed gold colloid is deposited or a gold salt is absorbed and reduced. Haruta et al. [20] reported high catalytic activity at 8 wt.% Au/TiO$_2$ by using a high-surface titania precursor (100 m^2 g^{-1} titanium hydroxide) and calcining at 300 °C to form well dispersed 3 nm gold particles.

In 2002, the group of Rolinson [21] reported the use of gold-titania composite aerogels for room temperature oxidation of carbon monoxide. Alkanethiolate monolayer-protected gold clusters with a well controlled size distribution were added to the Ti(IV) precursor during the sol–gel chemistry. After calcination the Au particles aggregate to average diameters of 5–10 nm.

Another approach [22] consisted in the preparation of titania-coated silica aerogels from aged silica wet-gel, soaking the gel in a toluene solution of titaniumtetraisopropoxide (impregnation method). Then Au nanoparticles were incorporated to the wet-gel by adding thiol-passitivated Au nanoparticles. The composites were then dried under supercritical CO$_2$ conditions, followed by a calcination process [23]. Small Au nanoparticles of 2.2 nm were obtained, maintaining the size of the original particles (Fig. 5.5). The CO oxidation activity was better than with a pure titania aerogel support.

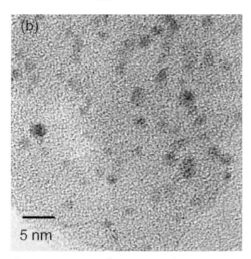

Fig. 5.5 TEM image of Au nanoparticles in a titania-coated silica composite aerogel.

5.2.3.2 **Methanol Oxidation**

Partial oxidation of methanol has many commercial applications for the production of formaldehyde, methyl formate and dimethyl ether (dehydration):

$$2CH_3OH \rightarrow CH_3OCH_3 + H_2O$$

$$CH_3OH + \frac{1}{2}O_2 \rightarrow HCOH + H_2O$$

$$2CH_3OH + O_2 \rightarrow HCOOCH_3 + H_2O$$

Some producers of formaldehyde use ferric oxide–molybdenum oxide catalyst in a ratio of about 1:4 and reaction takes place at 300–400 °C. With this precedent in mind, partial oxidation of methanol over Fe_2O_3-supported (SiO_2 or MoO_3) aerogels at temperatures between 225 and 300 °C was investigated by the group of Wang [24]. Fe_2O_3–SiO_2 was prepared from ferric acetylacetonate and tetramethyl ortosilicate and Fe_2O_3–MoO_3 from ferric acetylacetonate and molybdenum acetylacetonate. After supercritical drying and calcination the particles obtained had a mean particle size of 50–200 nm. The reaction evaluations were done using a 1:6 methanol to oxygen ratio in supercritical carbon dioxide (pressure 90 bar) obtaining mixtures of dimethyl ether, formaldehyde, methyl formate and carbon dioxide. The wide product selectivity pattern was associated with different processes and the authors provide an insight into the mechanism. Other studies with silica supported-iron oxide aerogel (Fe_2O_3–SiO_2), prepared from ferric acetylacetonate and tetramethyl ortosilicate (1–5 nm particles) were carried out by the same authors [25a,b]. Here the aerogels were not calcinated, and were evaluated using an ambient fixed-bed flow. A feed mixture was prepared by flowing nitrogen gas through liquid methanol and then injecting oxygen. The reaction temperature was incrementally raised from 225 to 300 °C. Low temperature and reaction time favored methanol oxidation to formaldehyde.

Recently, by combination of ferric acetylacetonate and gold acetate in a sol-to-aerogel process, preparation of gold/iron oxide aerogel was achieved. The oxidation activity enhanced with decreasing catalyst pre-treatment temperature and with increasing gold loading up to 5 wt.%. The oxidized gold exhibited higher activity than the metallic gold towards the total oxidation to CO_2:

$$CH_3OH + \frac{3}{2}O_2 \rightarrow CO_2 + 2H_2O$$

5.2.3.3 **n-Butane Oxidation**

Tsang et al. [26] described the use of Pt nanoparticles (average size 5 nm) encapsulated in porous silica aerogel coating (5% of metal), and then deposited on conventional γ-alumina as a composite catalyst for butane oxidation (Fig. 5.6). Preparation of the material was carried out from a microemulsion obtained by using ammonium tetrachloroplatinate(II), cetyltrimethylamonium bromide in a toluene/water mixture, and then the reduction of platinum anion was accom-

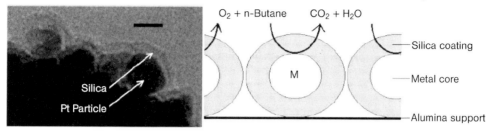

Fig. 5.6 TEM micrograph and a simplified model showing the alumina-supported silica aerogel Pt nanocomposite.

plished with hydrazine hydrate. The reduction process did not interfere with the stability of the microemulsion. TEOS and NaOH were then added to form the silica gel coating. Finally, aluminum oxide was used to immobilize a substantial amount of Pt-silica coated nanoparticles through attachment with external OH groups. The temperature range for oxidation of *n*-butane to CO_2 and H_2O was about 200–250 °C.

5.2.3.4 Volatile Organic Compounds Oxidation

Volatile organic compounds (VOC) are emitted to the atmosphere from a wide range of industrial processes. The commercial catalysts for VOC combustion contain noble metals, transition metal oxides, or their combinations, stabilized in highly porous alumina or silica. It has been demonstrated that chromia displays the highest activity in catalytic combustion amongst transition metal oxides. Chromia aerogel was prepared by Landau et al. [27] from aqueous chromium nitrate in the presence of urea followed by supercritical drying of the gel. Promotion of high surface area aerogel with Pt, Au, Mn and Ce increases its activity in ethylacetate combustion used as a model compound.

On the other hand, chlorinated compounds are hazardous to human health and the environment. For completely chlorinated volatile organic compounds (CVOC), their decomposition through catalytic pathways towards complete oxidation to produce CO_2, H_2O and HCl is desired. Several comparative studies have confirmed that supported chromium oxide is the most effective catalyst. Landau et al. [28] described the high efficiency of a Pt-promoted CrOOH aerogel with a surface area of $500 \, m^2 \, g^{-1}$ in the full combustion of 1,2-dichloroethane and chlorobenzene. Pt was deposited by impregnation of an aqueous solution of H_2PtCl_6 on the Cr aerogel. The synthesis is similar to that mentioned before in the sol–gel processing of an aqueous solution of urea and $Cr(NO_3)_3 \cdot 9H_2O$.

Another approach, due to the group of Moreno-Castilla [29] is based in the use of platinum containing monolithic carbon aerogels. Carbon materials are widely used as VOC adsorbents; thereby carbon aerogels produced from the polymerization of resorcinol–formaldehyde mixtures by sol–gel chemistry can also be used in these applications. Platinum-containing carbon aerogels were prepared either by impregnating the carbon aerogel with $[Pt(NH_3)_4]Cl_2$ or by dissolving this platinum precursor in the organic mixture. Metallic Pt was found after He or H_2 treatments in the impregnated series, whereas it was present due to reduction of the precursor during carbonization of the aerogel. These platinum-containing aerogels were applied as catalysts in the combustion of toluene.

5.2.4
Dehydrochlorination of Chlorinated Volatile Organic Compounds

Pd–Ag/SiO$_2$ aerogel [30] was used as catalyst for selective dehydrochlorination of 1,2-dichloroethane to ethylene (200–350 °C). This bimetallic sample was prepared by applying a methodology previously described by Schubert et al. consisting of a cogelation of tetraethoxysilane with organically substituted alkoxides capable of forming chelates with palladium and silver ions. First the palladium and silver complexes were prepared from a palladiumacetylacetonate and 3-(2-aminoethyl)a minopropyltrimethoxysilane mixture and a silver acetate and 3-(aminopropyl)triet hoxysilane mixture. TEOS was then added. After gelation and aging, the gels were dried under supercritical conditions, followed by calcination and reduction at atmospheric pressure in flowing H_2 at 350 °C. TEM micrographs showed that the bimetallic aerogel consists of small and large particles of 2 to 3 nm and 7 to 10 nm. The reaction was conducted in a stainless steel tubular reactor at a pressure of 0.3 MPa. The reaction mixture consisted of $CH_2Cl–CH_2Cl$, H_2 and He.

More recently Klabunde [31a,b] studied the role of nanocrystalline MgO in the reaction of 1-chlorobutane and 1-bromobutane to isomers of butene. Nanocrystalline MgO was prepared by a modified aerogel procedure, yielding a white powder of 400–500 m^2 g^{-1} and 4 nm average crystallite size. The procedure consisted in the use of Mg turnings that were allowed to react with methanol. The methanolic solution of $Mg(OCH_3)_2$ was added to toluene and water with the consequent formation of a gel. After heating the gel at 265 °C in an autoclave a $Mg(OH)_2$ aerogel was obtained. Heating at 500 °C overnight produced the MgO aerogel. Recently, the same researchers [32] have used the commercially available NanoActive Magnesium Oxide PlusTM (AP-MgO), a material that is guaranteed to have a surface area >600 m^2 g^{-1} with crystallite size of 4 nm.

5.2.5
Oxygen Reduction Reaction

The reaction between oxygen and hydrogen normally produces a great deal of heat:

$$2H_2(g) + O_2(g) \rightarrow 2H_2O(l)$$

However, if the reaction is carried out in a fuel cell, electricity is produced. A simple fuel cell is one in which hydrogen gas (fuel) is passed over one electrode (anode), oxygen is passed over the other (cathode) and the electrolyte is aqueous potassium hydroxide. The cell Pt/H2(g)/OH-(aq)/O2(g)/Pt which produces about 1.2 V at 25 °C has been used to produce electric power on some space missions.

Carbon aerogels have been investigated as a promising alternative to conventional carbon blacks as supports for fuel cell electrocatalysis because of their high conductivity, high mesoporosity and high surface area (for facile fuel and oxidant transport). High dispersion of platinum in the catalytic layer of proton-exchange membrane fuel cells (PEMFC) is necessary in order to reduce the cost-to-efficiency ratio. Several methods [33] have been described to obtain platinum-doped carbon aerogels. In the first the electrocatalytical material was synthesized from the sol–gel reaction precursors 2,4-dihydroxybenzoic acid and formaldehyde. The platinum precursor was $[Pt(NH_3)_4]Cl_2$, ion exchanged onto the gel surface. The gel was supercritically dried and pyrolized. In the second, a carbon aerogel was first prepared from resorcinol and formaldehyde, supercritically dried and pyrolized. The ground carbon aerogel was then impregnated with a platinum precursor H_2PtCl_6 chemically reduced with $NaBH_4$. More recently, the group of Smirnova [34] has described the introduction of the metallic precursor via organometallic precursors using supercritical carbon dioxide, followed by a secondary pyrolysis to form Pt nanoparticles on the surface of the aerogel carbon support.

Another approach consisted of the preparation of thiophene-modified resorcinol–formaldehyde aerogels prepared using a two stage sequence (Scheme 5.1)

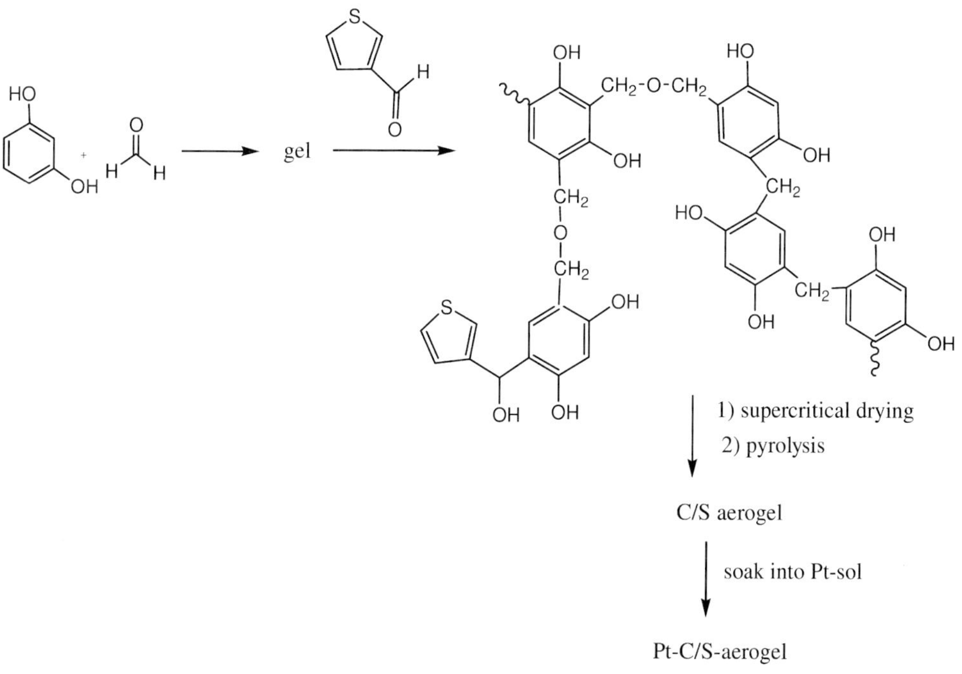

Scheme 5.1 Schematics of the preparation of Pt supported on carbon aerogels.

[35]. Resorcinol–formaldehyde gels were generated in the first step using conventional sol–gel chemistry, followed by a second reaction step in which the gel was reacted with 3-thiophene carboxaldehyde. After pyrolysis the noble metal (Pt) is immobilized on the carbon aerogel, mimicking the thiophene-mediated binding in Vulcan carbon (a commercial porous carbon support). A Pt sol was previously prepared by reducing H_2PtCl_6, the resulting Pt colloids exhibited a size distribution of 1–5 nm. Platinum nanoparticles spontaneously absorb onto C/S aerogels with a high degree of dispersion. In general, this Pt-modified aerogel exhibited an oxygen reduction activity similar to Pt-modified Vulcan carbon.

The group of Ye [36a,b] reported the synthesis of electrocatalytic materials by mixing an inorganic salt or organometallic compound of platinum or non-noble metals ($Fe(NO_3)_3$, $CoCl_2$ or $Co(acac)_2$) with polyacrylonitrile in DMF/water. After gelation the solvent was removed by CO_2 supercritical extraction. The aerogel was then pyrolysed at 900 °C. The electrode structure was fabricated as follows: the aerogel powder was ultrasonically dispersed in a mixture of alcohol/water. A small amount of this suspension was spread several times onto a glassy carbon electrode. Stability of the coating was improved by a heat treatment at 75 °C. These metal-based materials show good electrocatalytic activity for oxygen reduction in acidic solutions.

5.2.6
Cyclohexene Hydrogenation

As we have mentioned, supercritical carbon dioxide can be used to prepare aerogel nanocomposites. It has been demonstrated that SCFs allow the impregnation of catalytic metals into pre-formed supports. In 2004 the group of Howdle [37] described the preparation of palladium-aerogel nanocomposite by placing the complex $Pd(hfpd)_2$ and a previously prepared silica aerogel in an autoclave filled with CO_2 (27.5 MPa and 40 °C). Loading of the metal complex into the matrix of aerogel was obtained. Then, full reduction of the metal complex was carried out with H_2 at 6.9 MPa and 40 °C. Atomic absorption analysis indicated a loading of Pd of 8% and the particles formed were in two discrete size ranges, <6 nm and 15–20 nm. The hydrogenation of cyclohexene was studied under continuous flow conditions employing $scCO_2$ as reaction medium. The application of SCFs in catalytic hydrogenation is well documented, and the advantages are well known. On the one hand catalytic hydrogenation is normally carried out in an organic solvent and the reaction rate depends on the rate of transportation of hydrogen to the metal surface, hydrogen being completely miscible with SCF this limitation is reduced. On the other hand SCF systems have negligible surface tension which makes them excellent solvents for catalysts matrices. The reaction was performed by adding H_2 (80 °C and a pressure of 10 MPa). The aerogel catalyst exhibited excellent activity and was found to be re-usable, after routine pre-treatment with hydrogen.

5.2.7
Isomerization of 1-Butene

Studies on the behavior of chromium, molybdenum and tungsten oxide-doped monolithic carbon aerogels in the isomerization of 1-butene have been carried out by the group of Moreno-Castilla [38]. Samples were prepared by polymerization of a resorcinol–formaldehyde mixture which contained metallic salts of the elements and then the polymer was carbonized at both 500 and 1000 °C. The catalytic performance of the samples was studied at temperatures ranging from 50 to 425 °C. The tungsten oxide containing aerogel was much more active than those with chromium and molybdenum oxide, probably due to the high acidity of this oxide. Isobutene and *trans*-2-butene were the main reaction products, and the selectivity to *cis*-2-butene was very low.

5.3
Aerogel Nanocomposites as Catalysts of Important Reactions in Liquid Media

5.3.1
Oxidation of Olefins

Oxidation of olefins affords interesting oxygenated compounds such as 1,2-diols and epoxides. By performing the reaction enantioselectively on prochiral substrates, optically active products can be obtained.

5.3.1.1   Dihydroxylation of Olefins

Osmium-catalyzed olefin dihydroxylation, represented by the well-known Upjohn [39] procedure and its asymmetric version developed by Sharpless et al. [40] is one of the most useful transformations for the functionalization of alkenes. The reaction proceeds in the presence of a catalytic amount of OsO_4 using a co-oxidant such as *N*-methylmorpholine *N*-oxide (NMO) in *tert*BuOH and water (Upjohn conditions). The high cost, toxicity, and possible contamination of osmium catalysts in the products restrict the use in industry. Heterogenization via microencapsulation (OsO_4 was immobilized into polystyrene, see entry 1 of Table 5.1) [41], covalent anchoring (addition of OsO_4 to a tetrasubstituted olefin that is covalently linked to a silica support, followed by reoxidation, (see entries 2 and 3 of Table 5.1) [42], and use of ionic liquids (OsO_4–DMAP complex is soluble in [bmim]PF$_6$, see entries 4–6 of Table 5.1) [43] has been used in order to address this problems.

   Recently, a recoverable and reusable catalyst AP-Mg-OsO_4 was developed by the group of Choudary [44]. It was found that commercially available aerogel prepared AP-MgO has defects sites on the surface, for example Mg^{2+} sites which are Lewis acids. This situation presents an opportunity to prepare new materials. So, the modified aerogel was prepared via a counterionic stabilization of OsO_4 with the

Table 5.1 Different approaches to heterogenization of the dihydroxylation of olefins.

Entry	Olefin	Conditions	Product	Yield (%consecutive cycles)	Ref.
1		MC OsO$_4$ (5 mol%)-NMO H$_2$O/-acetone-CH$_3$CNrt		84, 84, 83, 84, 83	41
2	1-hexene	Os-SiO$_2$-60 (0.25 mol%)-NMO H$_2$O/tBuOH-CH$_2$Cl$_2$rt	CH$_3$-(CH$_2$)$_3$-CH—CH$_2$ OH OH	99, 98	42
3		Os-SiO$_2$-60 (0.25 mol%)-NMO H$_2$O/tBuOH-CH$_2$Cl$_2$rt		99	42
4	Ph	OsO$_4$ (2 mol%)-DMAPrt. [bmim]PF$_6$/H$_2$O/tBuOH		95, 93, 96, 95, 93, 93	43
5		OsO$_4$ (2 mol%)-DMAPrt. [bmim]PF$_6$/H$_2$O/tBuOH		93, 89, 91, 87, 85, 82	43
6	Ph, Ph	OsO$_4$ (2 mol%)-DMAPrt. [bmim]PF$_6$/H$_2$O/tBuOH		93, 87, 89, 80, 93	43
7	Ph, Ph	AP-Mg-OsO$_4$-NMO (2 mol%) H$_2$O-CH$_3$CN-acetone		94, 90, 89, 87 ,85	44
8	Ph	PEM-MC OsO$_4$ (5 mol%) (DHQD)$_2$PHAL Triton X-405 K$_3$Fe(CN)$_6$, K$_2$CO$_3$, H$_2$O		76 (74 ee)	48

Mg^{2+} cations present on the edge of commercially available nanocrystalline MgO through treatment with K$_2$OsO$_4$ (Scheme 5.2). The heterogeneous dihydroxylation was performed successfully, even in relatively large substrates such as stilbene (entry 7 of Table 5.1). In most of these cases the catalysts could be recovered and reused successfully several times.

The homogeneous Sharpless asymmetric dihydroxylation (AD) discovered in 1988 is based on biscinchona alkaloids such as 1,4-bis(9-O-dihydriquinidinyl)pht halazine ((DHQD)$_2$-PHAL). The generality of this asymmetric reaction has prompted several research groups to investigate the preparation of supported versions of cinchona alkaloid chiral ligands. Alkaloids such as (DHQD)$_2$-PHAL and its derivatives have been incorporated into various insoluble and soluble organic polymers and onto inorganic supports such as silica gel. The results of the groups of Lohray [45], Bolm [46] and Crudden [47] are remarkable. They demonstrated

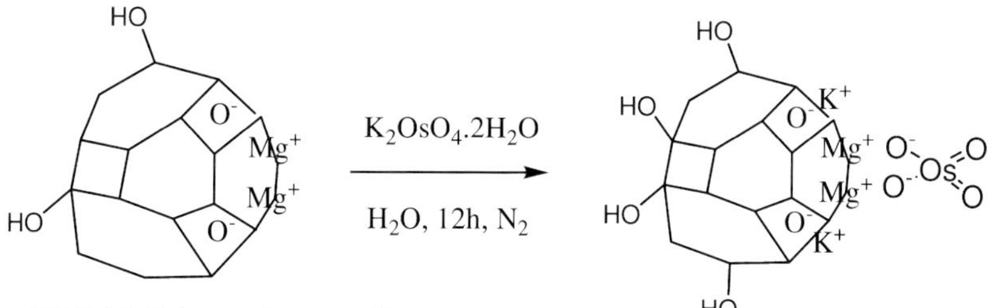

NAP-MgO (aerogel prepared)

Scheme 5.2 Preparation of AP-Mg-OsO$_4$ [44].

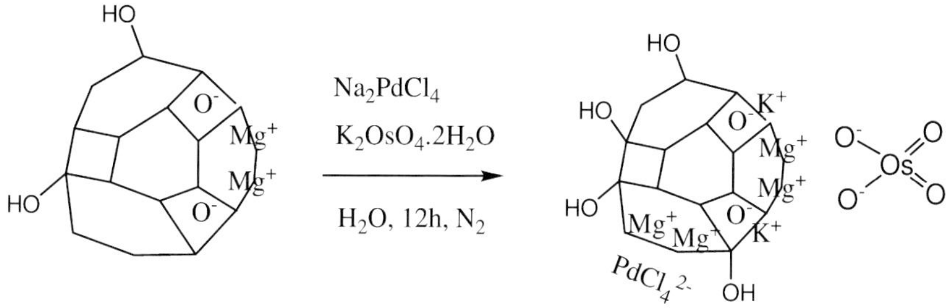

Scheme 5.3 Preparation of bifunctional catalyst NAP-Mg-PdOs [50].

that supporting alkaloids on structured silicates (chloropropyl functionalized silica, mesoporous molecular sieves or amorphous silica gel) gave enantioselectivities nearly identical to those observed in the homogeneous case. Another interesting approach is the heterogenization of OsO$_4$. Kobayashi [48] has demonstrated that using PEM-MC OsO$_4$ (5 mol%, PEM-MC: phenoxyethoxymethylpolystyrene microencapsulated) and (DHQD)$_2$-PHAL high ees are obtained, even in water (entry 8, Table 5.1). Interesting results were obtained by Choudary by an ion exchange technique using layered double hydroxides and quaternary ammonium salts covalently bound to resin and silica as ion exchangers. The study was done with various co-oxidants [49].

Based on achiral dihydroxylation methodology through the use of commercially available AP-MgO as support to immobilize osmium tetroxide by counterionic stabilization, Choudary used new bifunctional catalysts such as NAP-Mg-PdOs to perform tandem Heck-AD of olefins. The catalyst was prepared from NAP-MG (SA 600 m^2 g^{-1}) via counterionic stabilization using Na$_2$PdCl$_4$ and K$_2$OsO$_4$ as metal sources (Scheme 5.3) [50]. First the prochiral olefins were obtained from the corresponding aryl halides by a Mizoroki–Heck reaction in the presence of 3 mol% of catalyst, then the AD was carried out (Scheme 5.4, Table 5.2). The catalyst can be used for relatively larger substrates such as stilbene and methyl cinnamate. After completion of the reaction the catalyst was recovered by simple filtration. The recovered catalyst was reused and consistent enantioselectivity was obtained even after five cycles (85, 85, 83, 82, 80 %ee) in the AD of *trans*-stilbene.

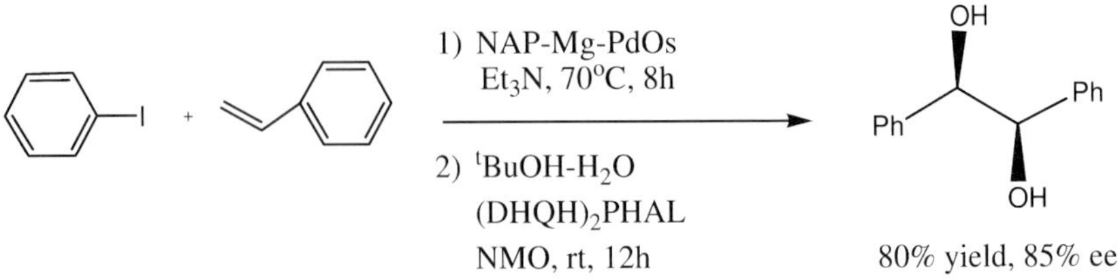

Scheme 5.4 Example of a tandem Heck asymmetric dihydroxylation reaction.

Table 5.2 Tandem Heck asymmetric dihydroxylation reaction [50].

Entry	Aryl halide	Olefin	Product	Yield % (ee %)
1	Ph-I	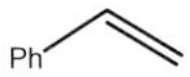	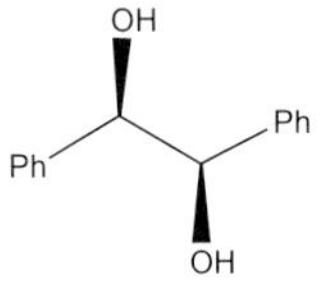	80 (85)
2	Ph-I	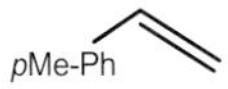	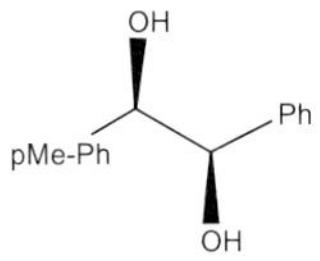	75 (65)
3	pMe-Ph-I	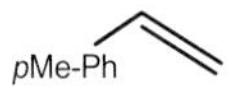	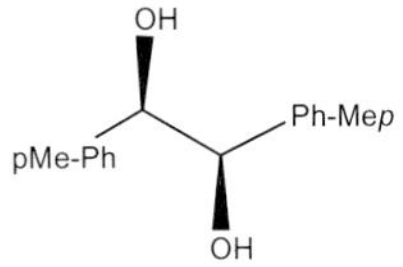	70 (82)
4	Ph-I	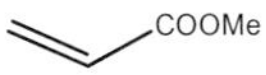	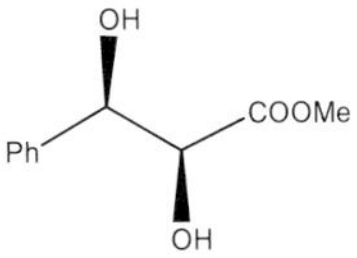	85 (73)
5	pOMe-Ph-I	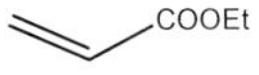	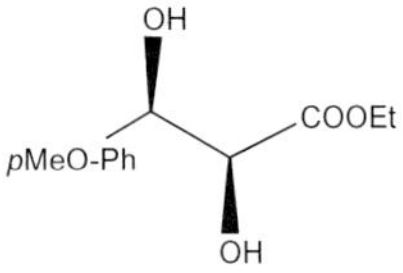	82 (78)

5.3.1.2 Epoxidation of Olefins

Epoxides are versatile intermediates in organic synthesis. The most general reagents for conversion of alkenes to epoxides in homogeneous conditions are peroxycarboxylic acids. The oxidation is believed to be a concerted process. Moreover, a process to avoid acidic conditions involves reaction with hydrogen peroxide. There is a clear demand for solid materials that catalyze epoxidations, therefore heterogeneous epoxidation remains a very active field of research [51].

In this chapter only titanium-catalyzed reactions with organic peroxides are compared.

The classical Ti-SiO$_2$ catalyst was initially prepared from TiCl$_4$ and pyrogenic SiO$_2$ in 1969 [52]. Almost 30 years later, Maier et al. [53] synthesized calcinated xerogels via a sol–gel process with TEOS and various Ti-cyclopentadienyl complexes (entry 1, Table 5.3). In 1995, Baiker [54] demonstrated that sol–gel prepared titania–silica mixed aerogels showed better catalytic behavior in epoxidation of different bulky olefins than Ti zeolites [55] and silica supported titania described at that time [56]. The most common oxidant was cumene peroxide. The drying method, the titanium content and the calcination temperature were the most important parameters. Aerogels dried by semicontinuous extraction with supercritical CO$_2$ at low temperature were found to be more efficient (entry 2, Table 5.3). In 2001 Baiker described the preparation of a series of titania–silica mixed

Table 5.3 Different titanium-based materials tested as catalysts in epoxidation reactions of olefins.

Entry	Catalyst	Titanium source	Ti content	Substrates	Yield (%)	Ref.
1	Ti-SiO$_2$ xerogel	Cp$_2$TiCl$_2$ (CpTiCl$_2$)$_2$O	0.08–03 mol% Ti	1,3-cyclooctadiene cyclooctene cyclohexene 1-octene	17–42 (GLC)	53
2	TiO$_2$-SiO$_2$ aerogel	Ti(O-i-Pr)$_4$ + acac (1 : 1)	2–20 wt.% TiO$_2$	cyclohexene 1-hexene cyclododedecene norbornene α-terpineol	95 (conversion)	54
3	TiO$_2$-SiO$_2$ aerogel	Ti (acac)$_2$(O-i-Pr)$_2$	1–20 wt.% TiO$_2$	2-cyclohexen-1-ol	95 (conversion)	57
4	TiO$_2$-SiO$_2$ aerogel-sil	Ti (acac)$_2$(O-i-Pr)$_2$	1–20 wt.% TiO$_2$	2-cyclohexen-1-ol	95	57
5	TiO$_2$-4SiO$_2$ aerogel	Ti[OSi(OtBu)$_3$]$_4$	25 wt.% TiO$_2$	cyclohexene	49	58
6	Ti-grafted aerosil	Ti[OSi(OtBu)$_3$]$_4$	1.01 wt.% Ti	cyclohexene	95	60
7	Ti-grafted MCM-41	Ti(C$_5$H$_5$)$_2$Cl$_2$	1.79 wt.% Ti	cyclohexene	not given	59
8	Ti-grafted TUD-1	Ti(OnBu)$_4$	1.87 wt.% Ti	cyclohexene	not given	59
9	Ti on SBA-15	(iPrO)Ti[OSi(OtBu)$_3$]$_3$	0.25–1.77 wt.% Ti	cyclohexene	not given	60
10	Ti-MCM-41 silyl	Ti(OEt)$_4$	2 wt.% TiO$_2$	cyclohexene	87	61
11	Ti-MCM-41 dry	Ti(OEt)$_4$	2 wt.% TiO$_2$	cyclohexene	91	61
12	Ti-ITQ-2	TiCp$_2$Cl$_2$	0.1–1 wt.% TiO$_2$	cyclohexene	68–72	62
13	Ti-SiO$_2$	several	0.8–2 mol% Ti	methyl oleate	38 reusable	63

aerogels with 0–100 wt.% TiO_2 content. The study of these aerogels indicated the presence of various active Ti sites: from tetrahedral Ti surrounded by only SiO ligands to octahedral Ti surrounded by TiO ligands in the titania domains. The abundance of isolated tetrahedral Ti sites is directly related to the higest epoxidation activity and selectivity. The productivity of the catalysts went to a maximum at 10 wt.% of TiO_2 (entry 3, Table 5.3) [57]. Moreover, silylation of the aerogels was carried out with *N*-methyl-*N*-(trimethylsilyl)trifluoroacetamide and this modification enhanced the epoxidation rate and improved the selectivity, confirming that hydrophobization is an efficient strategy for improving the performance of titania–silica mixed oxides (entry 4, Table 5.3).

In 2000, Tilley et al. converted the tris(*tert*-butoxy)siloxy complex $Ti[OSi(OtBu)_3]_4$, through a pyrolytic process, to $TiO_2.4SiO_2$ materials consisting of roughly spherical particles with an average diameter of ca. 25 nm (Fig. 5.7). Supercritical drying in CO_2 and calcination afforded an aerogel with a surface area of $677 \, m^2 g^{-1}$, that exhibited catalytic activity for the epoxidation of cyclohexene with cumene or *tert*-butyl-hydroperoxide (Scheme 5.5, entry 5, Table 5.3) [58].

Futhermore, a series of catalysts have been prepared based on the mesoporous siliceous sieve MCM-4159,61 or related mesoporous based structures such as TUD,59 SBA-1560 and delaminated zeolite ITQ-2.62 A selection of the results is presented in Table 5.3.

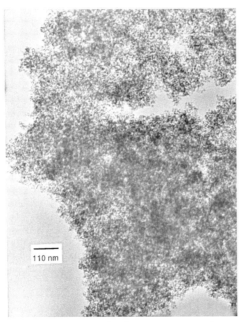

Fig. 5.7 TEM micrograph of $TiO_2 \cdot 4SiO_2$ material.

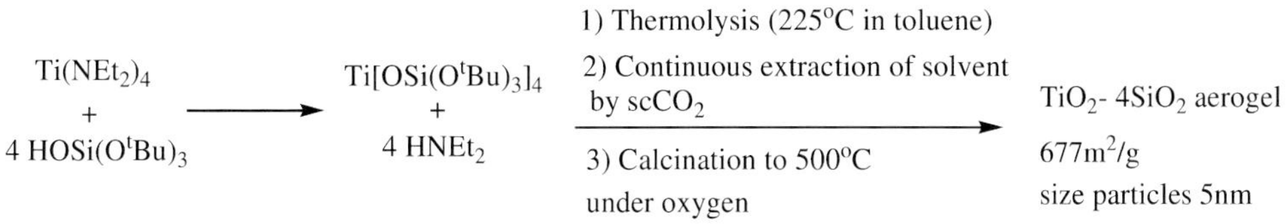

Ti(NEt$_2$)$_4$ $\xrightarrow{\hspace{2cm}}$ Ti[OSi(O^{t}Bu)$_3$]$_4$ 1) Thermolysis (225°C in toluene)

+ + 2) Continuous extraction of solvent by scCO$_2$

4 HOSi(O^{t}Bu)$_3$ 4 HNEt$_2$

3) Calcination to 500°C under oxygen $\xrightarrow{\hspace{2cm}}$ TiO$_2$- 4SiO$_2$ aerogel

677m^2/g

size particles 5nm

Scheme 5.5 Preparation of TiO$_2$.4SiO$_2$ aerogel.

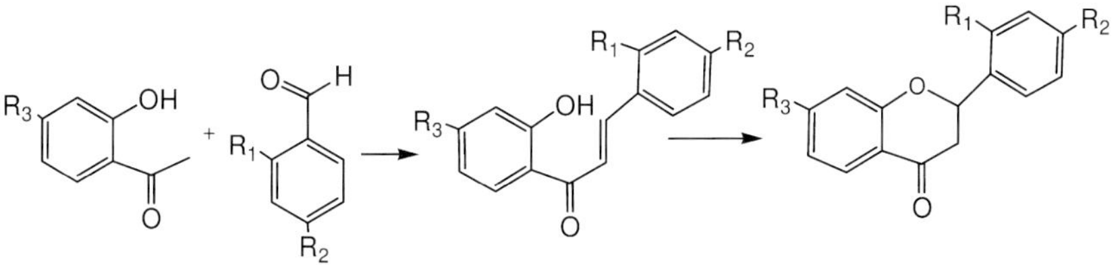

Scheme 5.6 Synthesis of flavonone type natural compounds through the use of a Claisen–Schmidt condensation.

5.3.2
Claisen–Schmidt Condensation

The Claisen–Schmidt condensation is a valuable C–C bond forming reaction yield-ing the preparation of α,β-unsaturated ketones. Among them the chalcones are interesting intermediates in the synthesis of the flavonoid family. Traditionally the Claisen–Schmidt condensation is carried out using 10–60% of alkaline hydroxides or sodium ethoxide. From the point of view of using a solid-base catalyst MgO has been extensively used because it exhibits strong Brönsted basicity in comparison with other bulk metal oxides. Amiridis et al. [64] described the use of pure com-mercially available MgO calcined at 475 °C to remove any absorbed impurity as catalyst for the synthesis of flavanone (Scheme 5.6). Recently, Choudary [65] reported that nanocrystalline MgO (aerogel prepared MgO, NanoActiveTM MgO Plus, 509 m^2 g^{-1}) was found to be more active than conventionally prepared MgO (NanoActiveTM MgO, 250 m^2 g^{-1}) or commercial MgO (30 m^2 g^{-1}). All the samples were activated at 250 °C before use.

5.3.3
Mizoroki–Heck C–C Coupling Reactions

This universally known [66] reaction was discovered independently by Heck and Mizoroki about 30 years ago. Basically it consists of the arylation or vinylation of alkenes and is generally catalyzed in solution by palladium species (Scheme 5.7). One of the major problems of the early homogeneous systems was the precipita-tion of palladium black. Addition of phosphanes improves the stability however oxidation of this ligand is a drawback for easy purification of the products. Con-sequently, development of heterogeneous catalysis [67] through supported palla-dium or stabilized colloidal palladium catalysts is an area of great interest.

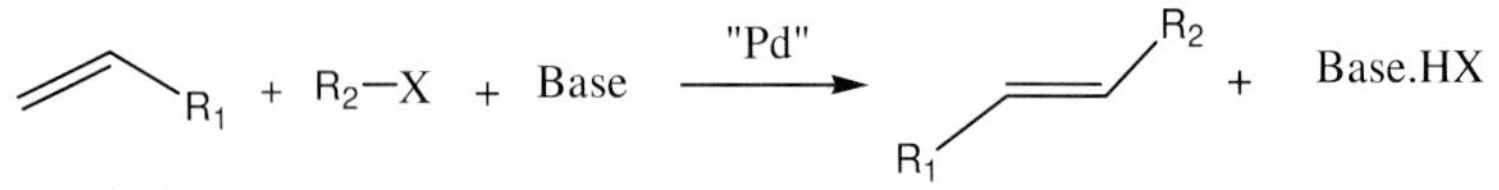

$$\diagup\!\!\!\diagdown_{R_1} + R_2\text{—}X + \text{Base} \xrightarrow{\text{"Pd"}} \diagup\!\!\!\diagdown^{R_2}_{R_1} + \text{Base.HX}$$

R_1: aryl, vinyl

X: I, Br, Cl, OTf, OTs, N_2^+, COCl...

R_2: Ph, OR, COOR, $CONR_2$, CN...

Base: NR'_3, MOAc (M = alkaline metal), K_3PO_4...

Scheme 5.7 Mizoroki–Heck reaction.

Table 5.4 Supported Pd catalysts on carbon tested in Mizoroki–Heck reactions.

Entry	Catalyst % Pd	Alkene	R_2-X	Temperature, °C (time, h)	Yield %	Ref.
1	Pd/C, 5 wt.%	methylacrylate	iodobenzene	150 (4)	70	68
2	Pd/C, 5 wt.%	styrene	chlorobenzene	120 (4)	62	68
3	Pd/CGr 33 wt.%	ethylacrylate	iodobenzene	100 (6)	87	72
4	Pd(3W)/C 5 wt.%	styrene	bromobenzene	140 (2)	82	73
5	Pd/C 5 wt.% [OMIm]BF$_4$	iodobenzene	butylacrylate	225 W microwave irradiation (0.5 min)	86,80,72,67,71	76
6	C/Pd 20% Pd	iodobenzene	styrene	100	89	78
7	Pd-carbon aerogel 35 wt.%	iodobenzene	ethylacrylate	80 (24)	100/100/100/100/100	79
8	Pd-carbon aerogel 35 wt.%	iodobenzene	styrene	80 (72)	70	79
9	Pd-carbon aerogel 35 wt.%	iodobenzene	3-buten-2-one	80 (30)	90	79

Among supported palladium catalysts palladium on carbon (Pd/C) is one of the most studied (Table 5.4). One of the initial and more interesting results is the reaction of styrene and chlorobenzene reported by Julia et al. in 1973 (entry 2, Table 5.4) [68]. The groups of Hallberg [69] and Augustine and O'Leary [70] studied the effect of different solvents. Palladium on carbon was also employed by Beller [71] for the first reaction involving aryl diazonium salts. In this work the authors pointed out that aryl diazonium salts require a higher catalyst amount than halides. In consequence they tried to reuse the catalyst but even in the first recycle a low yield was obtained. Ronchi [72] reported the use of Pd-graphite (Pd/CGr, entry 3,

Table 5.4), prepared from reduction of Pd(II) salts with potassium graphite. The results suggested that this catalyst was not very active. However, some years later Jikei and Kakimoto [73] prepared a more active Pd/CGr based on a smaller crystallite size. In 2002, Köhler et al. [74] studied a variety of Pd/C catalysts with different properties (Pd dispersion, oxidation state, water content, conditions of catalysts preparation etc.) in the Heck reaction of aryl bromides with olefins (entry 4, Table 5.4). The authors pointed out the hypothesis that the leached Pd from the support is the active species and the solid Pd/C catalyst acts as a reservoir that delivers catalytically active Pd species into solution. All catalysts were obtained by wet impregnation (5% Pd loading). The Heck reaction can also be conducted in ionic liquids through promotion by microwave irradiation. Moreover the reaction of iodobenzene with methylacrylate in NMP was reported to be accelerated by ultrasound [75]. The ionic liquid containing the catalyst system was used five consecutive times with only a slight loss of activity (entry 5, Table 5.4) [76]. Perosa [77] reported the addition of a phase transfer catalyst to an ionic liquid as a method to accelerate the C–C coupling reaction. As far as we know, only by using ionic liquids has Pd on carbon been recovered and reused with success.

Another approach is to support palladium nanoparticles in porous carbon supports. Activated carbons are typically microporous materials with broad pore size distribution, which may limit their catalytic applications involving large organic molecules. Mesoporous carbon materials may overcome this obstacle. Lu et al. reported the preparation of palladium-containing mesoporous carbon beginning with TEOS sucrose and palladium nitrite. The silica/palladium/nitrite/sucrose nanocomposites were carbonized at 900 °C to form palladium/carbon/silica composites, and then the silica templates were removed by HF solution. The study with the prepared catalysts indicated that the smaller palladium particle favors a faster reaction (entry 6, Table 5.4) [78]. Moreover, we [79] have contributed with the synthesis of organic and carbon aerogels doped with metallic Pd through sol–gel processes. Organic aerogels were prepared by sol–gel copolymerization of formaldehyde with the potassium salt of 2,4-dihydroxybenzoic acid, followed by ion-exchange with Pd(OAc)$_2$ and supercritical drying (Fig. 5.8). Organic aerogels were further carbonized to obtain carbon aerogels. As a result of the substrates used for copolymerization each repeating unit of the organic polymer contains a binding site for the metal ions, securing a uniform dispersion of the dopant metal. Carbon aerogels could be reutilized several times in the case of ethylacrylate (entry 7, Table 5.4). Polymerization of the styrene and 3-buten-2-one inside the active sites of the aerogel could be an explanation for the decreased activity in these cases (entries 8 and 9, Table 5.4).

Palladium can also be supported on different metal oxides. In 1995, Augustine and coworkers used Pd/SiO$_2$ as catalyst for the reaction of aryl chlorides and alkylenolethers [80]. The group of Arai studied the reaction of iodobenzene and methylacrylate in the presence of several bases. They presented evidence that leached soluble Pd species were the true catalysts [81].

Khöler et al. concluded that the activity of the palladium supported catalyst depends on the nature of the oxide support and the palladium dispersion, and in

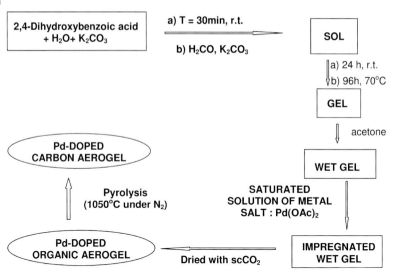

Fig. 5.8 Synthesis of metal-doped organic and carbon aerogels (post-synthesis, impregnation).

inert SiO_2 the activity was directly connected with surface area [82]. The group of Shimizu prepared the silica supported mercaptopropylsiloxane Pd(II) complex (Pd-SH-FSM) as an ordered mesoporous material through the use of $NaHSi_2O_5 \cdot 3H_2O$ and $C_{16}H_{33}NMe_3Cl$ as a template (entry 1, Table 5.5). They first compared the activity with Pd-FSM not having the sulfur coordinating ligand but impregnated with $Pd(OAc)_2$ (entry 3, Table 5.5). Then with Pd-SH-SiO_2 prepared by sol–gel from TEOS and 3-mercaptoethyltriethoxysilane followed by impregnation with $Pd(OAc)_2$ [83]. The Pd-SH-FSM and Pd-SH-SiO_2 showed excellent recyclability for the Heck reaction. The proportion of palladium leaching to the solution was higher in the case of Pd-FSM. It was concluded that sulfur ligands in the size-restricted mesopores are effective in preventing the aggregation of coordinated palladium complexes and this results in durability. Deactivation of the Pd complex via aggregation has been described to be significant for conventional Pd catalysts.

Park et al. described the use of tetra(ethylene glycol) to stabilize Pd(0) nanoparticles derived from $Pd(PPh_3)_4$ as metal source in the presence of MEOS and after gelation in water (entry 4, Table 5.5). The nanoparticles were fairly regular in size, ranging from 2 to 5 nm. This material was effective for activation of aryl iodides and bromides [84]. Recently, the group of Lu used ordered organic functionalized mesoporous silica containing covalent bonded diphenylphosphinoethylligands that were synthesized using a surfactant-templating approach. These silicas were bonded with palladium ions (using $Pd(OAc)_2$) resulting in the formation of catalytically active organometallic complexes that showed good activity and could be reused [85]. Butyl acrylate was reacted with iodobenzene in DMF in the presence of the supported palladium aerogel (entry 5, Table 5.5). Following each reaction the catalyst was removed by filtration, washed with the reaction solvent and dried

Table 5.5 Mizoroki–Heck reactions based on the use of Pd supported on metal oxide catalysts.

Entry	Catalyst % Pd	R$_2$-X Alkene	Alkene	Temperature,°C (time, h)	Yield %	Ref.
1	Pd-SH-FSM 3.4 wt.%	4-bromoacetophenone	ethylacrylate	130	92/95/97/ 97/99/97	83
2	Pd-SH-SiO$_2$ 3.6 wt.%	4-bromoacetophenone	ethylacrylate	130	84/93/88/99/88/95	83
3	Pd-FSM 1.2 wt.%	4-bromoacetophenone	ethylacrylate	130	96	83
4	Pd/TEG/SiO$_2$ 0.59 wt.%	iodobenzene	methylacrylate	120 (17)	80	84
5	Pd/DPPE/SiO$_2$	3-iodotoluene	styrene	100 (24)	87/82/85/83	85
6	Pd/SiO$_2$ aerogel 0.4 wt.%	iodobenzene	butylacrylate	80 (2)	100	88
7	Pd-SiO$_2$ 3.7 wt.% aerogel	iodobenzene	ethylacrylate	80 (30)	70/60	79
10	Pd(a) aerogel 11.4 wt.%	iodobenzene	ethylacrylate	(24/35/45/40)	97/90/85/98	79
11	Pd(a)R aerogel 16.3 wt.%	iodobenzene	ethylacrylate	(8/11/13/20)	99/93/90/83	79
12	Pd(a)OR aerogel 15.8 wt.%	iodobenzene	ethylacrylate	(7/8/9/12)	94/93/90/93	79
13	Pd(a) aerogel 11.4 wt.%	iodobenzene	styrene	(72)	82	79
14	Pd(a)R aerogel 16.3 wt.%	iodobenzene	styrene	(10/25/23/48)	74/65/56/79	79
15	Pd(a)OR aerogel 15.8 wt.%	iodobenzene	styrene	(48)	5	79
16	Pd(a) aerogel 11.4 wt.%	iodobenzene	3-buten-2-one	(72)	98	79
17	Pd(a)R aerogel 16.3 wt.%	iodobenzene	3-buten-2-one	(48)	99	79
18	Pd(a)OR aerogel 15.8 wt.%	iodobenzene	3-buten-2-one	(48)	99	79

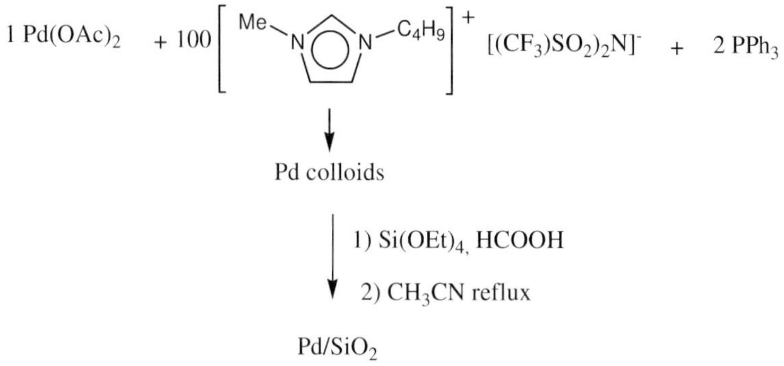

$$1 \ Pd(OAc)_2 \quad + \ 100 \left[Me\text{-}N\overset{\frown}{\underset{\smile}{N}}N\text{-}C_4H_9 \right]^+ \quad [(CF_3)SO_2)_2N]^- \quad + \quad 2 \ PPh_3$$

↓

Pd colloids

1) Si(OEt)$_4$, HCOOH

2) CH$_3$CN reflux

Pd/SiO$_2$

Scheme 5.8 Preparation of Pd/SiO$_2$ aerogels using an ionic liquid.

at 20 °C under vacuum. On recycling a reduction in the rate of reaction was observed.

It is reported in the literature that ionic liquids can be used in the manufacture of aerogels [86] and can stabilize colloids [87]. Based on this knowledge Marr first prepared Pd colloids from Pd(OAc)$_2$, triphenylphosphine and 1-butyl-3-methyl imidazolium bistrifluorosulphonylimide. The metal colloid suspension was then submitted to incubation with Si(OEt)$_4$ and formic acid. After aging of the formed gel, the ionic liquid was extracted by refluxing in acetonitrile. This process yields a grey green aerogel that was used as catalyst (entry 6, Table 5.5 and Scheme 5.8). The authors [88] described the formation of narrow nanoparticles of Pd(0), 2 nm in diameter, and a low loading of Pd(0) of 0.37 wt.%. The process of reduction of Pd(II) to Pd(0) was not explained in the article. Others have studied the role of the anion present in the ionic liquid and the triphenylphosphine in the transformation of Pd(II) to the metal during the formation of the colloid [87]. In general, addition of PPh$_3$ to Heck reactions avoids palladium black formation, and the systems show better recyclability.

We have prepared silica aerogel nanocomposites by impregnation with Pd(acac)$_2$ of the silica wet gels followed by EtOH supercritical drying (Fig. 5.9) [79]. The Pd-SiO$_2$ aerogel exhibited low activity in the Mizoroki–Heck reaction and analysis of the crude reaction mixture showed a leaching of 11% of Pd. Moreover, different nanocomposite silica aerogels were synthesized using Pd(OAc)$_2$ as metal source (Table 5.6, Fig. 5.10). They were synthesized by tethering the metal to the silica matrix with the use of the complexing silane AEAPTS [*N*-(aminoethyl)aminopropyl] trimethoxysilane] (Fig. 5.11) [89]. In the first step the metal complex [Pd(AEAPTS)$_2$]$^{2+}$ 2AcO$^-$ was formed by reaction of palladium acetate with 2 molar equivalents of AEAPTS. In the second step TEOS was added as the silica source. Sol–gel processing of the mixtures resulted in different gels that were dried with scCO$_2$ to afford aerogels Pd(a). Some of the aerogels were submitted to calcination and then reduced in a stream of hydrogen (Pd(a)OR). Other were directly reduced without prior calcination (Pd(a)R). The materials were tested as catalysts. In general the best results were obtained for the Pd(a)R aerogel. The analysis of the crude

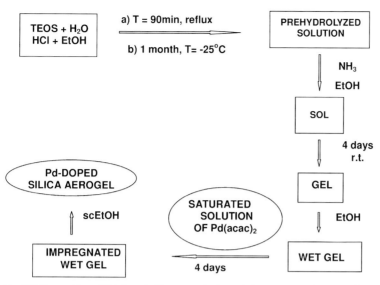

Fig. 5.9 Preparation of Pd-doped silica aerogels through impregnation (post-synthesis).

Table 5.6 Synthesis features and characteristics of Pd-silica aerogels.

Material	Ref.	Metal source	Surface area $(m^2 g^{-1})$	Pore volume $(cm^3 g^{-1})$	Density $(g\,cm^3)$	Metal (%)	Mean particle size (nm)	XRD phase
Pd/SiO$_2$ aerogel	87	Pd(OAc)$_2$	—	—	—	0.37	2	—
Pd-SiO$_2$	77	Pd(acac)$_2$	825	1.6	0.05	3.7	39	fcc-Pd
Pd(a)	78	Pd(OAc)$_2$	516	1.7	0.4	11.4	1.3	amorphous
Pd(a)R	78	Pd(OAc)$_2$	690	2.0	0.4	16.3	1.4	fcc-Pd
Pd(a)OR	78	Pd(OAc)$_2$	1020	2.4	0.3	15.8	8.3	fcc-Pd

reaction mixture of the first reaction indicates a leaching of 3%. The advantages of using a functionalized organo alkoxysilane to tether the metal to the gel matrix are control of the homogeneity of the metal distribution and reduction in leaching of the metal

5.3.4
Hydrocarbonylation of Aryl Halides

Palladium-catalyzed hydrocarbonylation of aryl and vinyl halides or triflates is a well known reaction used as a key step in the synthesis of complex biologically active molecules. The aryl and vinyl halide or triflate is converted into the

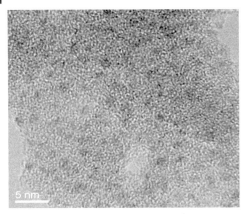

Fig. 5.10 HRTEM image of Pd(a) (Table 5.6).

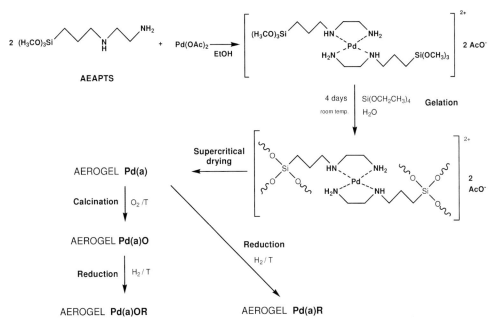

Fig. 5.11 Preparation of Pd-silica aerogels using an organofunctional alkoxysilane with a group capable of coordinating the metal.

corresponding carboxylic acid. Carbon monoxide is readily available; however, it is a highly toxic gas. Therefore the development of techniques where carbon monoxide is generated *in situ* is a target of interest. The group of Cacchi has developed a methodology based on the generation of carbon monoxide from lithium formate and acetic anhydride [90].

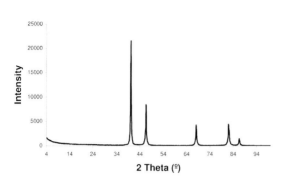
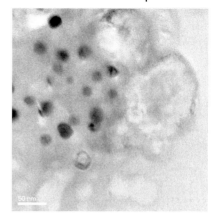

Fig. 5.12 XRD patterns and TEM image of Pd-carbon aerogel.

$$Ar\text{-}I \quad + \quad HCOOLi \quad + \quad Ac_2O \quad \xrightarrow{\text{Pd cat}} \quad Ar\text{-}COOH$$

Pd cat.: Pd-carbon aerogel, Ar = Iodotoluene, yields: 88, 93, 93, 95, 95, 95, 91, 93, 96, 95, 96, 95

Pd cat.: Pd-carbon aerogel, Ar = p-iodobenzoate, yields: 90, 93, 93, 93, 94, 91, 94, 97, 97, 93

Scheme 5.9 Hydrocarbonylation of aryl halides. Results using a Pd-carbon aerogel.

Different palladium catalysts such as $Pd_2(dba)_3$, $PdCl_2(dpp)$ and $Pd(OAc)_2$ were active in homogeneous conditions. Interestingly, Pd/C was also effective. With the aim of recovering and reusing the catalyst a Pd-carbon aerogel (35 Pd wt.%) prepared by some of us [79] was used (Fig. 5.12). The TEM images of the aerogel showed nanoparticles with diameters between 15 and 23 nm and the X-ray diffraction pattern characteristic of fcc-Pd. The BET surface area was $422\,m^2\,g^{-1}$. The catalyst system could be reused several times without any appreciable loss of activity. In Scheme 5.9 the excellent yields of consecutive cycles using the same batch of aerogel for two different reactions are shown [91].

5.3.5
Michael Additions

The Michael addition of active methylene (and methine) compounds to activated π-systems is one of the more useful C—C bond-forming reactions (Scheme 5.10). Classical basic activation of the nucleophile can generate by-products from competing side reactions. Therefore, much effort has been dedicated to the development of catalysts for Michael reactions, mainly transition metals and lanthanides [92]. The current challenge is the development of heterogeneous catalytic systems.

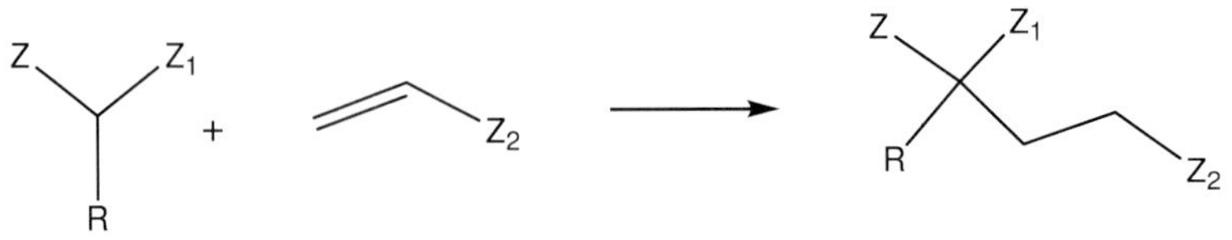

Z, Z_1,Z_2: electron releasing group

Scheme 5.10 Michael reaction.

The available information about supported metals as catalysts for Michael reactions will be commented upon, with emphasis on selected silica supports [93].

In 1997, the group of Kotzuki described that Yb(OTf)$_3$ dispersed on chromatography silica gel by mixing catalyzed Michael additions at high pressure. Some examples were given of reactions of a series of β-dicarbonyl compounds and unsaturated ketones and ethylacrylate [94]. Some years later, in 2003, Kaneda et al. described the use of Sc(III) enclosed in the interlayer space of montmorillonites (mont), prepared by treatment of commercially available Na$^+$-mont through cation exchange using Sc(OTf)$_3$. This catalyst showed excellent catalytic activity for the Michael reaction of 1,3-dicarbonyls with enones and ethylacrylate [95a,b]. Moreover, for the reaction of ethyl 2-oxocyclopentanecarboxylate and 3-buten-2-one the catalyst was reused during 3 cycles without loss of activity. Lately, they have reported the use of Cu^{2+}-mont. Other examples based on the use of clays were reported by Shimizu et al., Fe^{3+} was immobilized by ion exchange using as inorganic supports Na-fluorotetrasilicic mica and Na$^+$-mont. These systems could be used as recoverable catalysts of β-dicarbonyl compound with active Michael acceptors such as enones [96].

In the group of of Moreno-Mañas, we have been interested in different aspects of Michael reactions: diastereoselective Ni-catalyzed additions [97a,b], the use of highly fluorinated Ni-based catalysts under biphasic organic-perfluorinated conditions [97c], the use of triphenylphosphane as an organocatalyst [97d,e] and mechanistic insights into Michael additions [97f]. Recently, as a new approach for heterogeneous catalysts we prepared nanoparticles of ferrihydrite anchored in a silica aerogel [9]. The material was synthesized from sol–gel processing of Fe(NO$_3$)$_2$ and TEOS. The gels were then dried at supercritical conditions of ethanol. The features and characteristics of the Fe-aerogel can be found in entry 1 of Table 5.7. This Fe-aerogel exhibited good catalytic activity in the reaction of acetylacetone to diethylazodicarboxylate. The material could be reused 5 consecutive times obtaining excellent yields of pure product after recrystallization (70, 74, 76, 71, 80%). Analysis of the reaction crude (first cycle) indicated a leaching of 3% of iron (Scheme 5.9).

Although the use of lanthanides as Lewis acid catalysis has been extensively described [98], entrapment of lanthanide species in aerogels has been scarcely studied. We have prepared [99] Eu-doped materials using the copolymerization of a resorcinol exchange derivative containing an ion exchange moiety and formaldehyde. This methodology ensures a homogeneous dispersion of the doping metal.

Table 5.7 Synthesis features and characteristics of metal doped gels and aerogels used as catalysts in Michael reactions.

Entry	Aerogel	Metal source	Metal wt.%	Bulk density (g cm^{-3})	BET surface area (m^{2}g^{-1})	Mean particle size (nm)	XRD
1	Fe-silica aerogel	Fe(NO$_3$)$_2$·9H$_2$O	13	0.65	603	3	six-line ferrihydrite
2	Eu-gel	Eu(OTf)$_3$	12.03	0.044	—	particles are differentiated	amorphous
3	Eu-organic	Eu(OTf)$_3$	13.70	0.33	308	particles are clearly seen	amorphous
4	Eu-carbon	Eu(OTf)$_3$	27.3	0.53	497	particles are clearly seen	Eu$_2$O$_2$S, EuOF and Eu$_3$O$_2$F$_5$

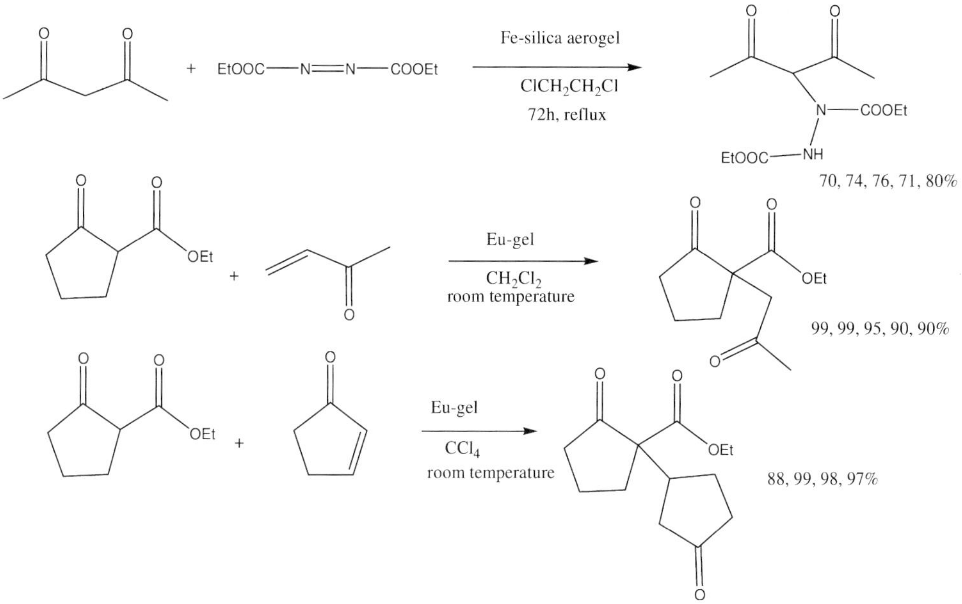

Scheme 5.11 Conjugate additions of β-dicarbonyl compounds and activated π-systems.

Thus, europium loaded gels were prepared. Some of them were dried by solvent evacuation at CO$_2$ supercritical conditions to obtain the Eu-organic aerogel. After pyrolysis at 1050 °C a Eu-carbon aerogel was obtained (entries 2–4 of Table 5.7). The materials were tested as catalysts in Michael additions (Scheme 5.11). We performed the reaction of ethyl 2-oxocyclopentanecarboxylate and 3-buten-2-one five consecutive times using the same batch of Eu-gel (99, 99, 95, 90, 90%). The reaction time increased for the fifth cycle from 5 to 6h at room temperature. We also used cyclopentenone as an electrophile. The reaction could be carried out four consecutive times with excellent yields although the time of the reaction increased for each consecutive run. Analysis of reaction crude (first cycle) indicated a leaching of 0.08% of the europium. Eu-organic and Eu-carbon aerogels exhibited low catalytic activity probably due to the presence of mixtures of different europium oxides that are not active as Lewis acids. In the case of Eu-carbon aerogels several species have been identified in an XRD study of the sample, corresponding to a mixture of europium oxide-sulfide and europium oxide-fluorides.

5.3.6
Synthesis of Carbon Nanotubes

Since their discovery by Iijima in 1991 [100] carbon nanotubes have been one of the more studied materials. Single-walled nanotubes (SWNT) have shown many unique electrical and mechanical properties. Chemical vapor deposition (CVD) is currently the best method for large-scale production of nanotubes. Initially, preparation by CVD of carbon dioxide, benzene, ethylene or methane had the disadvantages of low yield and amorphous carbon impurities. In 1995 Song [101] demonstrated that silica aerogels could be used as a matrix for deposition of carbon from decomposition of acetylene at 670 °C. Carbon nanostructures were formed. Single-walled carbon nanotubes of high quality were synthesized using methane [102] or by disproportionation of carbon monoxide [103] on an Al_2O_3 aerogel-supported Fe/Mo catalyst. The aerogel was prepared from $Al(OsecBu)_3$ and HNO_3 in EtOH. The sources of metal were the following commercial salts: $Fe_2(SO_4) \cdot 4H_2O$ and $MoO_2(acac)_2$. After gelation the gels were dried under supercritical conditions of carbon dioxide ($600 \, m^2 \, g^{-1}$). The powder was calcined at 500 °C. Single-walled carbon nanotubes were then prepared in a simple CVD set-up heating the catalyst at 850–1000 °C under Ar. Then the Ar was switched to H_2 and finally to CH_4 [102]. The same Fe/Mo aluminium aerogel was used to grow SWNT nanotubes switching from Ar to CO [103]. The obtained nanotubes were free of amorphous-carbon coating. Others have used Ni-doped aluminum aerogels catalysis [104] to induce methane decomposition. In this case the aerogel was prepared from a mixture of $Al(NO_3)_2$ and NH_3, the metal source was $Ni(NO_3)_2$ and drying was carried out in supercritical conditions of ethanol.

As we have seen, there have been discovered a large number of interesting applications of nanocomposited aerogels in the catalysis domain. Although the applications of aerogels to catalysis has been a research subject for a long time, the number of publications continues increasing each year. Due to the suitability of the substrate aerogel to host nanoparticles whilst keeping them accessible to the reaction media and maximizing the reactive surface, we expect new developments in gaseous as well as liquid media. The search for more selective and efficient processes producing less undesired products and the exploration of new energy power sources due to the exhausting of fossil reserves are vitalizing the catalysis concept. Chimio-, regio- and enantio-selective synthesis, steam reforming and electrocatalysis are hot topics that, if efficient and scalable processes are discovered, can have tremendous impact in the near future.

Acknowledgments

The authors have been funded by the Comisión Interministerial de Ciencia y Tecnología (MAT2003-1052, CTQ2005-04968-C02-01, MAT2006-13572-C02-01) and Generalitat de Catalunya (2005SGR452, 2005SGR00305).

References

1 D. Astruc, F. Lu, J. Ruiz, *Angew. Chem. Int. Ed.* **2005**, *44*, 7852–7872.

2 N. Husing, U. Schubert, *Angew. Chem. Int. Ed.* **1998**, *37*, 22–45.

3 H. D. Gesser, P. C. Goswami, *Chem. Rev.* **1989**, *89*, 765–788.

4 A. C. Pierre, G. M. Pajonk, *Chem. Rev.* **2002**, *102*, 4243–4265.

5 M. Gich, L. Casas, A. Roig, E. Molins, J. Sort, S. Suriñach, M. D. Baró, J. S. Muñoz, L. Morellón, M. R. Ibarra, J. Nogués, *Appl. Phys. Lett.* **2003**, *82*, 4307–4309.

6 L. Casas, A. Roig, E. Rodríguez, E. Molins, J. Tejada, J. Sort, *J. Non-Cryst. Solids* **2001**, *285*, 37–43.

7 L. Casas, A. Roig, E. Molins, J. M. Greneche, J. Asenjo, J. Tejada, *Appl. Phys. A-Mater. Sci. Process.* **2002**, *74*, 591–597.

8 S. Martínez, M. Meseguer, L. Casas, E. Rodríguez, E. Molins, M. Moreno-Mañas, A. Roig, R.M. Sebastián, A. Vallribera, *Tetrahedron* **2003**, *59*, 1553–1556.

9 M. Popovici, M. Gich, A. Roig, L. Casas, E. Molins, C. Savii, D. Becherescu, J. Sort, S. Suriñach, J. S. Muñoz, M. D. Baró, J. Nogués, *Langmuir* **2004**, *20*, 1425–1429 and references therein.

10 G. M. Pajonk, *Catal. Today*, **1997**, *35*, 319–337.

11 L. C. Cotet, M. Gich, A. Roig, I. C. Popescu, V. Cosoveanu, E. Molins, V. Danciu, *J. Non-Cryst. Solids* **2006**, *352*, 2772–2777.

12 M. Schneider, A. Baiker, *Catal. Today*, **1997**, *35*, 339–365.

13 C. Moreno-Castilla, F. J. Maldonado-Hódar, *Carbon*, **2005**, *43*, 455–465.

14 M. M. Green, H. A. Wittcoff, *Organic Chemistry Principles and Industrial Practice*, Wiley-VCH, Weinheim, 2003.

15 Y. Ohtsuka, T. Arai, S. Takasaki, N. Tsubouchi, *Energy Fuels* **2003**, *17*, 804–809.

16 F. Blanchard, J. P. Reymond, B. Pommier, S. J. Teichner, *J. Mol. Catal.*, **1982**, *17*, 171–181.

17 B. C. Dunn, P. Cole, D. Covington, M. C. Webster, R. J. Puigmire, R. D. Ernst, E. M. Eyring, N. Shah, G. P. Huffman, *Appl. Catal. A-Gen*, **2005**, *278*, 233–238.

18 P. Dutta, B. C. Dunn, E. M. Eyring, N. Shah, G. P. Huffman, A. Manivannan, M. S. Seehra, *Chem. Mater.* **2005**, *17*, 5183–5186.

19 B. M. Choudary, V. S. Jaya, R. Reddy, M. L. Kantam, M. M. Rao, S. S. Madhavendra, *Chem. Mater.* **2005**, *17*, 2740–2743.

20 M. Haruta, A. Ueda, S. Tsubota, R. M. Torres-Sánchez, *Catal. Today*, **1996**, *29*, 443–447.

21 J. M. Pietron, R. M. Stroud, D. R. Rolinson, *Nano Lett.* **2002**, *5*, 545–549.

22 Y. Tai, J. Murakami, K. Tajiri, F. Ohashi, M. Daté, S. Tsubota, *Appl. Catal. A-Gen* **2004**, *268*, 183–187.

23 Y. Tai, Y. Ochi, F. Ohashi, K. Tajiri, J, Murakami, M. Daté, S. Tsubota, *Eur. Phys. J. D* **2005**, *34*, 125–128.

24 C. T. Wang, R. J. Willey, *J. Catal.* **2001**, *202*, 211–219.

25 (a) C.-T. Wang, S.-H. Ro, *Appl. Catal. A-Gen*, **2005**, *285*, 196–204; b) C.-T. Wang, S.-H. Ro, *J. Non-Cryst. Solids*, **2006**, *352*, 35–43.

26 K. M. K. Yu, C. M. Y. Yeung, D. Thompsett, S. C. Tsang *J. Phys. Chem. B* **2003**, *107*, 4515–4525.

27 H. Rotter, M. V. Landau, M. Carrera, D. Goldfarb, M. Herskowitz *Appl. Catal. B-Environ.* **2004**, *47*, 111–126.

28 H. Rotter, M. V. Landau, M. Herskowitz *Environ. Sci. Technol.* **2005**, *39*, 6845–6850.

29 F. J. Maldonado-Hóldar, C. Moreno-Castilla, A. F. Pérez-Cadenas *Appl. Catal. B-Enviromental* **2004**, *54*, 217–224.

30 B. Heinrichs, P. Delhez, J.-P. Schoebrechts, J.-P. Pirard *J. Catal.,* **1997**, *172*, 322–335.

31 (a) V. D. Fenelonov, M. S. Mel'gunov, I. V. Mishakov, R. M. Richards, V. D. Chesnokov, A. M. Volodin, K. J. Klabunde, *J. Phys. Chem. B.* **2001**, *105*, 3937–1941; (b) I. V. Mishakov, A. F. Bedilo, R. M. Richards, V. D. Chesnokov, A. M. Volodin, V. I Zaikovskii, R. A. Buyanov, K. J. Klabunde *J. Catal.* **2002**, *206*, 40–48.

32 P. P. Gupta, K. L. Hohn, L. E. Erikson, K. J. Klabunde, A. F. Bedilo *AIChE J,* **2004**, *50*, 3195–3205.

33 J. Marie, S. Berthon-Fabry, P. Achard, M. Chatenet, A. Pradourat, E. Chaine, *J. Non-Cryst. Solids*, **2004**, *350*, 88–96.

34 A. Smirnova, X. Dong, H. Hara, A. Vasiliev, N. Sammes *Int. J. Hydrogen Energ.,* **2005**, *30*, 149–158.

35 W. S. Baker, J. W. Long, R. M. Stroud, D. R. Rolison, *J. Non-Cryst. Solids*, **2004**, *350*, 80–87.

36 (a) S. Ye, A.K. Vijh, *Electrochem. Commun.,* **2003**, *5*, 272–275; (b) S. Ye, A. K. Vijh, *J. Solid State Electrochem.,* **2005**, *9*, 146–153.

37 K. S. Morley, P. Licence, P. C. Marr, J. R. Hyde, P. D. Brown, R. Mokaya, Y. Xia, S. M. Howdle, *J. Mater. Chem.* **2004**, *14*, 1212–1217.

38 C. Moreno-Castilla, F. J. Maldonado-Hódar, J. Rivera-Utrilla, E. Rodríguez-Castellón, *Appl. Catal. A-Gen* **1999**, *183*, 345–356

39 V. Van Rheenen, R. C. Kelly, D. Y. Cha, *Tetrahedron Lett.* **1976**, *23*, 1973–76.

40 E. N. Jacobsen, I. Markó, W. S. Mungall, G. Schröder, K. B. Sharpless, *J. Am. Chem. Soc.* **1988**, *110*, 1968–1970.

41 S. Nagayama, M. Endo, S. Kobayashi, *J. Org. Chem.* **1998**, *63*, 6094–6095.

42 A. Severeyns, E. D. De Vos, L. Fiermans, F. Verpoort, P. J. Grobet, P. A. Jacobs, *Angew. Chem. Int. Ed.* **2001**, *40*, 586–589.

43 Q. Yao, *Org. Lett.* **2002**, *4*, 2197–2199.

44 B. M. Choudary, K. Jyothi, M. L. Kantam, B. Sreedhar, *Adv. Synth. Catal.* **2004**, *346*, 45–48.

45 B. B. Lohray, E. Nandanan, V. Bhushan, *Tetrahedron: Asymm.* **1996**, *7*, 2805–2808.

46 C. Bolm, A. Maischak, A. Gerlach, *Chem. Commun.* **1997**, 2353–2354.

47 I. Motorina, C. M. Crudenn, *Org. Lett.* **2001**, *3*, 2325–2328.

48 (a) S. Kobayashi, M. Endo, S. Nagayama, *J. Am. Chem. Soc.* **1999**, *121*, 11229–11230; (b) S. Kobayashi, T. Ishida, R. Akiyama, *Org. Lett.* **2001**, *3*, 2649–2652; (c) T. Ishida, R. Akiyama, S. Kobayashi, *Adv. Synth. Catal.* **2003**, *345*, 576–579.

49 B. M. Choudary, N. S. Chowdari, K. Jyothi, M. L. Kantam, *J. Am. Chem. Soc.* **2002**, *124*, 5341–5349.

50 B. M. Choudary, K. Jyothi, M. Roy, M. L. Kantam, B. Sreedhar, *Adv. Synth. Catal.* **2004**, *346*, 1471–1480.

51 D. E. de Vos, B. F. Sels, P. A. Jacobs, *Adv. Synth. Catal*, **2003**, *345*, 457–473.

52 F. Wattimena, H. P. Wulff, UK *Patent* 1 249 079, **1969**.

53 S. Thorimbert, S. Klein, W. F. Maier, *Tetrahedron*, **1995**, *51*, 3787–3792.

54 R. Hutter, T. Mallat, A. Baiker, *J. Catal.* **1995**, *153*, 177–189.

55 M. A. Camblor, A. Corma, A. Martínez, J. Pérez-Pariente, *J. Chem. Soc., Chem. Commun.*, **1992**, 589–590.

56 A. Corma, M. T. Navarro, J. Pérez-Martínez, *J. Chem. Soc., Chem. Commun.* **1994**, *2*, 147–148.

57 C. Beck, T. Mallat, T. Bürgi, A. Baiker, *J. Catal.,* **2001**, *204*, 428–439.

58 M. P. Coles, C. G. Lugmair, K. W. Terry, T. D. Tilley, *Chem. Mater.* **2000**, *12*, 122–131.

59 Z. Shan, E. Gianotti, J. C. Jansen, J. A. Peters, L. Marchese, T. Maschmeyer, *Chem. Eur. J.* **2001**, *7*, 1437–1443.

60 A. Corma, M. Domine, J. A. Gaona, J. L. Jorda, M. T. Navarro, F. Rey, J. Pérez-Pariente, J. Tsuji, B. McCulloch, L. T. Nemeth, *Chem. Comun.* **1998**, 2211–2212.

61 J. Jarupatrakorn, T. Don Tilley, *J. Am. Chem. Soc.* **2002**, *124*, 8380–8388.

62 A. Corma, U. Diaz, V. Fornes, J. L. Jorda, M. Domine, F. Rey, *Chem. Commun.* **1999**, 779–780.

63 L. A. Rios, P. Weckes, H. Schuster, W. F. Hoelderich, *J. Catal.* **2005**, *232*, 19–26.

64 (a) M. T. Drextler, M. D. Amiridis, *Catal. Lett.* **2002**, *17*, 175–181; (b) Z. Liu, J. A. Cortés-Concepción, M. Mustian, M. D. Amiridis, *Appl. Catal. A-Gen* **2006**, *302*, 232–236.

65 (a) B. M. Choudary, K. V. S. Ranganath, J. Yadav, L. Kantam, *Tetrahedron Lett.* **2005**, *46*, 1369–1371; (b) B. M. Choudary, M. L. Kantam, K. V. S. Ranganath, K. Mahendar, B. Sreedhar, *J. Am. Chem. Soc.* **2004**, *126*, 3396–3397.

66 For Heck reaction reviews and important articles see: (a) T. Mizoroki, K. Mori, A. Ozaki, *Bull. Chem. Soc. Jpn.* **1971**, *44*, 581–581; b) R. F. Heck, J. P. Nolley, *J. Org. Chem.* **1972**, *37*, 2320–2322; (c) A. de Meijere, F. E. Meyer, *Angew. Chem. Int. Ed. Engl.* **1994**, *33*, 2379–2411; (d) W. Cabri, I. Candiani *Acc. Chem. Res.* **1995**, *28*, 2–7; (e) E. Negishi, C. Copéret, S. Ma, S. Liou, F. Liu, *Chem. Rev.* **1996**, *96*, 365–393; (f) G. T. Crisp, *Chem. Soc. Rev.* **1998**, *27*, 427–436; (g) I. P. Beletskaya, A. V. Cheprakov, *Chem. Rev.* **2000**, *100*, 3009–3066; (h) N. J. Whitcombe, K. Kuok Hii, S. E. Gibson, *Tetrahedron* **2001**, *57*, 7449–7473; (i) M. T. Reetz, J.G. de Vries, *Chem. Commun.* **2004**, 1559–1562; (j) N. T. S. Phan, M. Van der Sluys, C. W. Jones, *Adv. Synth. Catal* **2006**, *348*, 609–679.

67 A. Biffis, M. Zecca, M. Basato, *J. Mol. Catal. A: Chem* **2001**, *173*, 249–274.

68 (a) M. Julia, M. Duteil, *Bull. Soc. Chim. Fr.*, **1973**, 2790; (b) M. Julia, M. Duteil, *Bull. Soc. Chim. Fr.*, **1973**, 2791–2794.

69 A. Hallberg, L. Westfelt, *J. Chem. Soc., Perkin Trans. 1*, **1984**, 933–935; (b) C.-M. Andersson, A. Hallberg, *J. Org. Chem.* **1988**, *53*, 235–239.

70 R. L. Augustine, S. T. O'Leary, *J. Mol. Catal.* **1992**, *72*, 229–242.

71 M. Beller, K. Külhein, *Synlett*, **1995**, 441–442.

72 D. Savoia, C. Trombini, A. Umani-Ronchi, G. Verardo, *J. Chem. Soc., Chem. Commun.* **1981**, 540–541.

73 M. Jikei, Y. Ishida, Y. Seo, M.-A. Kakimoto, Y. Imai *Macromolecules*, **1995**, *28*, 7924–7929.

74 (a) K. Köhler, R. G. Heidenreich, J. G. E. Krauter, J. Piertsch, *Chem. Eur. J.* **2002**, *8*, 622–631; (b) R. G. Heidenreich, J. G. E. Krauter, J. Pietsch, K. Köhler, *J. Mol. Catal A: Chem*, **2002**, *182–183*, 499–509; (c) L. Djakovtch, M. Wagner, C. G. Hartung, M. Beller, K. Köhler, *J. Mol. Catal. A: Chem.* **2004**, *219*, 121–130.

75 G. V. Ambulgekar, B. M. Bhanage, S. D. Samant, *Tetrahedron Lett.* **2005**, *46*, 2483–2485.

76 X. Xie, J. Lu, B. Chen, J. Han, X. She, X. Pan, *Tetrahedron Lett.* **2004**, *45*, 809–811.

77 A. Perosa, P. Tundo, M. Selva, S. Zinovyev, A. Testa, *Org. Biomol. Chem.* **2004**, *2*, 2249–2252.

78 Q. Hu, J. Pang, N. Jiang, J. E. Hampsey, Y. Lu, *Micropor. Mesopor. Mater.* **2005**, *81*, 149–154.

79 S. Martínez, A. Vallribera, C. L. Cotet, M. Popovici, L. Martin, A. Roig, M. Moreno-Mañas, E. Molins, *New J. Chem.* **2005**, *29*, 1342–1345.

80 R. L. Augustine, S. T. O'Leary, *J. Mol. Catal. A: Chem.* **1995**, *95*, 277–285.

81 (a) F. Zhao, K. Murakami, M. Shirai, M. Arai, *J. Catal.* **2000**, *194*, 479–483; (b) F. Zhao, B. M. Bhanage, M. Shirai, M. Arai, *Chem. Eur. J.* **2000**, *6*, 843–848.

82 M. Wagner, K. Khöler, L. Djakovitch, S. Weinkauf, V. Hagen, M. Muhler, *Top. Catal.*, **2000**, *13*, 319–326.

83 K. Shimizu, S. Koizumi, T. Hatamachi, H. Yoshida, S. Komai, T. Komada, Y. Kitayama, *J. Catal.* **2004**, *228*, 141–151.

84 N. Kim, M. S. Kwon, C. M. Park, J. Park, *Tetrahedron Lett.* **2004**, *45*, 7057–7059.

85 Q. Hu, J. E. Hampsey, N. Jiang, C. Li, Y. Lu *Chem. Mater.* **2005**, *17*, 1561–1569.

86 S. D. Dai, Y. H. Ju, H. J. Gao, J. S. Lin, S. J. Pennycook, C. E. Barnes, *Chem. Commun.* **2000**, 243–244.

87 N. A. Hamill, C. Hardacre, S. E. J. McMath, *Green Chem.*, **2002**, *4*, 139–142.

88 K. Anderson, S. Cortiñas, C. Hardacre, P. C. Marr, *Inorg. Chem. Commun.* **2004**, 73–76.

89 S. Martínez, M. Moreno-Mañas, A. Vallribera, U. Schubert, A. Roig, E. Molins, *New J. Chem.* **2006**, *30*, 1–7.

90 S. Cacchi, G. Fabrizi, A. Goggiamani, *Org. Lett.* **2003**, *5*, 4269–4272.

91 S. Cacchi, C. L. Cotet, G. Fabrizi, G. Forte, A. Goggiamani, L. Martín, S. Martinez, E. Molins, M. Moreno-Mañas, F. Petrucci, A. Roig, A., Vallribera *Tetrahedron* **2007**, *63*, 2519–2523.

92 For reviews on transition-metal and lanthanide species catalyzed Michael reactions see: (a) J. Christoffers, *Eur. J. Org. Chem.* **1998**, 1259–1266; (b) J. Comelles, M. Moreno-Mañas, A. Vallribera, *Arkivoc*, **2005**, *9*, 207–238.

93 For a review on the use of iron compounds as catalysts in organic reactions including Michael additions see: C. Bolm, J. Legros, J. Le Paid, L. Zani, *Chem. Rev.* **2004**, *104*, 6217–6254.

94 H. Kotsuki, K. Arimura, *Tetrahedron Lett.* **1997**, *43*, 7583–7586.

95 (a) T. Kawabata, T. Mizugaki, K. Ebitani, K. Kaned, *J. Am. Chem. Soc.* **2003**, *125*, 10486–10487; (b) T. Kawabata, M. Kato, T. Mizugaki, K. Ebitani, K. Kaneda, *Chem. Eur. J.* **2005**, *11*, 288–297.

96 (a) K. Shimizu, M. Miyagi, T. Kan-no, T. Komada, Y. Kitayama, *Tetrahedron Lett.* **2003**, *44*, 7421–7424; (b) K. Shimizu, M. Miyagi, T. Kan-ho, T. Hatamachi, T. Kodama, Y. Kitayama, *J. Catal.* **2005**, *229*, 470–479.

97 (a) J. Clariana, N. Gálvez, C. Marchi, M. Moreno-Mañas, A. Vallribera, E. Molins, *Tetrahedron*, **1999**, *55*, 7331–7344; (b) C. Marchi, E. Trepat, M. Moreno-Mañas, A. Vallribera, E. Molins, *Tetrahedron*, **2002**, *58*, 5699–5708; (c) M. Meseguer, M. Moreno-Mañas, A. Vallribera, *Tetrahedron Lett.* **2000**, *41*, 4093–4095; (d) M. Lumbierres, C. Marchi, M. Moreno-Mañas, R. M. Sebastián, A. Vallribera, E. Lago, E. Molins, *Eur. J. Org. Chem.*, **2001**, 2321–2328; (e) C. Gimbert, M. Llumbierres, C. Marchi, M. Moreno-Mañas, R. M. Sebastián, A. Vallribera, *Tetrahedron* **2005**, *61*, 8598–8605; (f) J. Comelles, M. Moreno-Mañas, E. Pérez, A. Roglans, R. M. Sebastián, A. Vallribera, *J. Org. Chem.* **2004**, *69*, 6834–6842.

98 M. Shibasaki, K.-I. Yamada, N. Yoshikawa, in *Lewis Acids in Organic Synthesis*, H. Yamamoto (Ed.) Wiley-VCH, Weinheim, **2000**, Vol. 2, pp. 911–942.

99 S. Martínez, L. Martín, E. Molins, M. Moreno-Mañas, A. Roig, A. Vallribera, *Monatsh. Chem.* **2006**, *137*, 627–637.

100 S. Ijima, *Nature*, **1991**, *354*, 56–58.

101 X.-Y. Song, W. Cao, M. R. Ayers, A. J. Hunt, *J. Mater. Res.* **1995**, *10*, 251–254.

102 M. Su, B. Zheng, J. Liu, *Chem. Phy. Lett.* **2000**, *322*, 321–326.

103 B. Zheng, Y. Li, J. Liu, *Appl. Phys.* **2002**, *74*, 345–348.

104 L. Piao, Y. Li, J. Chen, L. Chang, J. Y. S. Lin, *Catal. Today*, **2002**, 145–155.

6
Transition-metal Nanoparticle Catalysis in Imidazolium Ionic Liquids

Jairton Dupont and Dagoberto de Oliveira Silva

6.1
Introduction

Modern transition-metal NPs differ from classical colloids in several important respects [1] such as size and stability in solution. In most cases transition-metal NPs are smaller (1–10 nm in diameter) and display a narrow size distribution compared with classical colloids, which are typically >10 nm in diameter. In this range of sizes, metal NPs have unique properties that are based on their inherent large surface-to-volume ratio and quantum size effect. Indeed, the density of states in the valence and conductive bands of these "strange morsels of matter" – with reduced size down to a few hundred atoms – decreases and the electronic and magnetic properties change dramatically [2, 3]. Moreover, transition-metal NPs are isolable, re-dissolvable and possess well defined compositions, unlike classical colloids. The synthesis and catalytic properties of transition-metal NPs are reproducible and these particles are soluble in organic solvents, unlike classical colloid chemistry that is typically aqueous. However, transition-metal NPs are only kinetically stable because the formation of bulk metal is thermodynamically favored. Therefore, NPs that are freely dissolved in solution must be stabilized in order to prevent their agglomeration i.e. diffusing together and coalescing, eventually leading to the formation of the bulk metal.

The stabilization of NPs in solution can be achieved by electrostatic and/or steric protection by, for example, the use of water-soluble polymers, quaternary ammonium salts, surfactants, or polyoxoanions [4–7]. In both models the slowing of the NP agglomeration rate is essentially due to the kinetic stabilization provided by a layer of surface-adsorbed anions or polymers.

The catalytic activity and selectivity of soluble metal-particle catalysts depends not only on the relative abundance of different types of active sites, but also on the concentration and type of stabilizers present in the medium [6]. For example, good capping ligands – that stabilize robust nanocrystals with very narrow size distributions – are almost inactive catalysts for the hydrogenation of olefins [8]. Moreover, NPs containing stabilizers that bind less strongly to the metal surface than other anions generate higher catalytic activity in the hydrogenation of arenes [9].

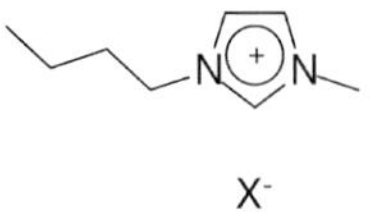

X= BF_4, BMI.BF_4
X= PF_6, BMI.PF_6
X= $N(CF_2CF_3)_2$, BMI.N(Tf)$_2$
X= CF_3SO_3, BMI.OTf

X^-

Fig. 6.1 Examples of imidazolium based ILs [19, 21, 22].

The choice of stabilizer allows one to tune the solubility of the NPs; for example, the transfer of NPs from water to organic solvents (and vice versa) has been demonstrated using N- or P-containing stabilizers [10]. Moreover, it is possible to fine tune the surface properties by the use of ligands for asymmetric catalysis for instance [11, 12]. At this point it is important to clarify that the definition of homogeneous and heterogeneous catalysts refers herein to the active sites and not the classical solubility criterion. Thus, heterogeneous catalysts are those that have multiple types of active sites and homogeneous catalysts those that have typically a single type of active site. Although these soluble (i.e., dispersible) NPs are sometimes referred to as "soluble heterogeneous catalysts" we prefer to designate soluble transition-metal NPs as oligo-site catalysts since, in many cases, their catalytic activity and selectivity are quite different to those observed with classical heterogeneous catalysts.

Quaternary ammonium salts are one of the most popular and investigated classes of stabilizing agents for soluble transition-metal NPs catalysts [13–16]. It is assumed that the stabilization in these cases is essentially due to the positive charge on the metal surface which is ultimately induced by the adsorption of the anions to the coordinatively unsaturated, electron-deficient, and initially neutral metal surface [17]. In the 1990s a new class of quaternary ammonium salts based on the 1,3-dialkylimidazolium cation associated with weakly coordinating anions such as tetrafluoroborate and hexafluorophosphate (Fig. 6.1) was introduced as liquid immobilizing agents for organometallic catalysis [18–20].

More recently imidazolium ILs have been used as a template for the synthesis of a plethora of stable transition-metal NPs with small size and narrow size distributions [23, 24]. These particles immobilized in ILs constitute highly active multiphase catalytic systems for various reactions. The main goal of this chapter is to disclose the mechanistic and structural aspects of the formation and stabilization of transition-metal NPs in imidazolium ILs, especially those that have been used in catalytic reactions.

6.2
Ionic Liquids: Structural Aspects

1,3-Dialkylimidazolium salts possess very interesting properties, such as a very low vapor pressure, are non-flammable, have high thermal and electrochemical stability, etc. Numerous reviews and books have been dedicated to their synthesis, physico-chemical properties, structural aspects, applications in organic synthesis and organometallic catalysis, enzymatic catalysis, separation processes, nano-materials, electrochemistry and new materials [25, 26].

Although imidazolium ILs are, by definition, quaternary ammonium salts, they differ from the classical salts in at least one very important aspect: they possess pre-organized structures, mainly through hydrogen bonds [27–29], that induce structural directionality (see below) in contrast to classical salts in which the aggre-gates display charge-ordering structures. Surprisingly, in most of these cases imidazolium ILs can still be regarded as simple solvents – although different and sometimes contradictory solvent properties have been derived from several studies – with polarities comparable to DMF, acetonitrile and short chain alcohols and a coordination ability similar to that of dichloromethane [30–39]. We prefer, however, to consider ILs as liquid supports (nanostructured materials) in which the intro-duction of other molecules occurs with the formation of inclusion-type com-pounds [40–42]. This model is based on the fact that, in particular, imidazolium-based ILs display a pronounced self-organization in the solid, liquid and even in the gas phase [43–51]. This approach has been historically used in classical molten salt chemistry even if significant randomness in organization is necessary to describe the structure of a liquid [52–54]. However, in most cases there is only 10–15% volume expansion on going from the crystalline to the liquid state and the ion–ion or atom–atom distances are similar in both the solid and liquid states. For example, various X-ray and neutron scattering studies of solid and liquid NaCl suggest that the structural organization observed in the crystal exists in the liquid phase [53, 54]. Furthermore, while long-range order is lost on going from the crystal to the liquid, similarities remain as a consequence of the Coulombic forces between the cations and anions of the ILs [52]. It is also clear that the long-range Coulombic interactions in ionic organic liquids can lead to longer spatial correlations than those in comparable classical van der Waals organic liquids [55]. In a recent overview of the X-ray studies reported over recent years on the structure of 1,3-dialkylimidazoilum salts a typical trend was revealed: they form in the solid state an extended network of cations and anions connected by hydrogen bonds. The monomeric unit is always constituted of one imidazolium cation surrounded by at least three anions and in turn each anion is surrounded by at least three imid-azolium cations [44]. The three-dimensional arrangement of the imidazolium ILs is generally formed through chains of the imidazolium rings (Fig. 6.2).

This molecular arrangement generates channels in which the anions are gener-ally accommodated as chains [57]. This structural pattern depends on the anion geometry, and the internal arrangements along the imidazolium columns vary with the type of the *N*-alkyl substituents. These IL structures can adapt or are adaptable to many species, as they provide hydrophobic or hydrophilic regions and a high directional polarizability which can be oriented parallel or perpendicular to the included species.

ILs form extended hydrogen bond systems in the liquid state and are conse-quently highly structured i.e. they are, by definition, "supramolecular" fluids. This

(a)   (b)

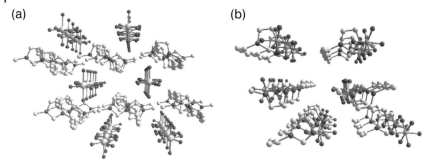

Fig. 6.2 A three-dimensional simplified schematic view
of the arrangements of 1-*sec*-butyl-3-methylimidazolium cation
associated with SbF_6^- (a) and PF_6^- (b) showing the channels
formed [56].

$$[Ir(cod)Cl]_2 \xrightarrow[H_2]{IL} [Ir(0)]_n \quad (a)$$

$$[Ru(cod)(cot)] \xrightarrow[H_2]{IL} [Ru(0)]_n \quad (b)$$

Scheme 6.1 Preparation of transition-metal NPs in ILs:
(a) reduction of transition metal compounds [23] or (b)
decomposition of organometallic compounds in a formal zero
oxidation state [80].

is one of the special qualities of ILs that differentiates them from the classical ion
aggregates of which ion pairs and ion triplets are widely recognized examples. This
structural organization of ILs can be used as "entropic drivers" for spontaneous,
well-defined, and extended ordering of nanoscale structures [58]. Indeed, the
unique combination of adaptability towards other molecules and phases plus the
strong H-bond-driven structure makes ILs potential key tools in the preparation
of a new generation of chemical nanostructures [58–60].

6.3
Formation of Nanoparticles in Imidazolium Ionic Liquids

Transition-metal NPs are easily prepared by the simple reduction of metal compounds or the decomposition of organometallic compounds in the zero oxidation
state dissolved in the imidazolium ILs (Scheme 6.1 and Table 6.1). Functionalized
imidazolium or pyridinium ILs such as those containing thiol, alcohol or cyano
groups have also been used for the formation and stabilization of Ni, Ag and Au
NPs [61–75]. Bimetallic nanorods, hyperbranched nanorods, and NPs with different CoPt compositions have also been easily prepared in $BMI.NTf_2$ [76].

Table 6.1 Examples of metal NPs prepared in ILs by reduction or controlled decomposition of transition-metal compounds with hydrogen.

Entry	Precursor	IL	T (°C)	P (bar)	Time (h)	Size TEM (nm)	Ref.
1	$[IrCl(cod)]_2$	$BMI.BF_4$	75	5	0.17	2.9 ± 0.4	23, 84
		$BMI.PF_6$	75	5	0.17	2.0 ± 0.6	
		BMI.OTF	75	5	0.17	2.6 ± 0.6	
2	$[Ir(cod)(MeCN)_2]^+$	$BMI.NTf_2$	22	40	22	2.1 ± 0.6	85
		$BMI.BF_4$	75	4	18	2.1 ± 0.5	
3	$[Ru(cod)(cot)]$	$BMI.PF_6$	75	4	18	2.6 ± 0.4	80
		BMI.OTF	75	4	18	1.9 ± 0.6	
		$BMI.BF_4$	75	4	1.7	2.5 ± 0.4	
4	RuO_2	$BMI.PF_6$	75	4	0.7	2.0 ± 0.2	86
		BMI.OTF	75	4	4.5	2.5 ± 0.4	
		$BMI.BF_4$	75	4	1.5	3.4 ± 0.8	
5	$Pt_2(dba)_3$	$BMI.PF_6$	75	4	1.5	2.4 ± 0.7	29
		BMI.OTF	75	4	1.5	2.8 ± 0.5	
6	$RhCl_3$	$BMI.PF_6$	75	4	1	2.3 ± 0.6	87
7	$Pd(acac)_2$	$BMI.BF_4$	75	4	0.1	4.9 ± 0.8	88

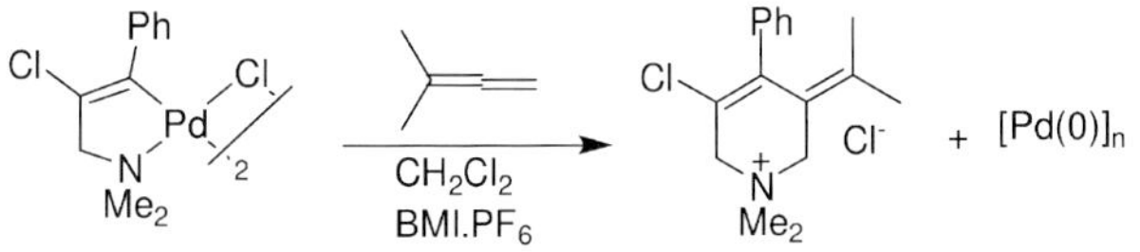

Scheme 6.2 Pd(0) NPs prepared in dichloromethane and transferred to the IL [79].

Alternatively, transition-metal NPs in ILs may be obtained by the simple transfer of the freshly prepared MNP in water [77, 78] or "classical" organic solvents to the ILs, as for example from the reaction of palladacycles with allenes (Scheme 6.2) [79].

In both cases (Scheme 6.1) the formation of the NPs follows the auto-catalytic mechanism developed by Finke and coworkers [81]. For example, for Ir or Pt(0) NPs, the catalytic hydrogenation (H_2 uptake) of 1-decene and cyclohexene, respectively, is used as a reporter reaction via the pseudo-elementary step concept, Scheme 6.3 (where A is the precatalyst $[Ir(cod)Cl]_2$ or $Pt_2(dba)_3$ and B is the catalytically active Ir(0) or Pt(0) nanoclusters, Scheme 6.3).

If the alkene hydrogenation (step (c), Scheme 6.3) is a fast reaction, on the timescale of steps (a) and (b), it can serve as a reporter reaction for the Ir(0) or Pt(0) formation, i.e., the kinetics of the overall reaction are represented only by the steps (a) and (b) in Scheme 6.3. Moreover, the sum of all three steps leads to a "kinetically equivalent" elementary step which relates the overall H_2 consumption (cyclohexene) with the formation of Ir(0) or Pt(0) NPs, the so-called pseudo-elementary step (d) (Scheme 6.3).

(a) A $\xrightarrow{k_1}$ B

(b) A + B $\xrightarrow{k_2}$ 2B

(c) α [B + *alkene* + H$_2$ $\xrightarrow{\text{fast}}$ B + *alkane*]

(d) A + α *alkene* + α H$_2$ $\xrightarrow{k_{obs}}$ B + α *alkane*

Scheme 6.3 Pseudo-elementary step concept [82].

This mechanism is equivalent to the two-step mechanism for transition-metal nanocluster self-assembly, from metal salts under reductive conditions, proposed by Finke [81, 83] with one main conceptual difference, in the case of Pt, since Pt$_2$(dba)$_3$ is a Pt(0) nanocluster precursor in the zero oxidation state, the first step (a) in Scheme 6.3, must be considered to be the organometallic complex decomposition, and not a metal nucleation step as occurs, for example, when using a Pt precursor in an oxidation state different to zero.

The hydrogenation kinetics of cyclohexene catalyzed by Pt$_2$(dba)$_3$ dispersed in BMI.PF$_6$, BMI.BF$_4$ and BMI.OTf are shown in Fig. 6.3. The kinetics curves were treated using the pseudo-elementary step and fitted (Eq. (6.1) by the following integrated rate equation for metal-salt decomposition (A $\rightarrow$ B, k_1) and autocatalytic nanocluster surface growth (A + B $\rightarrow$ 2B, k_2). For a more detailed description of the use of the pseudo-elementary step for the treatment of hydrogenation kinetic data and derivation of the kinetic equations see elsewhere [81–83].

$$[\text{cyclohexene}]_t = \frac{(k_1/k_2) + [\text{cyclohexene}]_0}{1 + \dfrac{k_1}{k_2[\text{cyclohexene}]_0} * \exp^{(k_1 + k_2[\text{cyclohexene}]_0)t}} \tag{6.1}$$

As expected, the kinetic curve of Fig. 6.3 shows no induction period, indicating that the first step, the Pt$_2$(dba)$_3$ decomposition, is fast and the catalyst nanocluster B is readily available from the onset of the hydrogenation reaction. The rate constants values obtained from the fit of Fig. 6.3 are summarized in Table 6.2.

The kinetic curves shown in Fig. 6.3 can also be fit by a simple exponential equation. This is possible because the surface growth step (k_2) is quite slow when compared with the decomposition step (k_1) – one must take into consideration the species concentration when comparing first and second order rate constants – and therefore, k_2 has a very small influence on the global rate for the NPs formation.

Indeed, *in situ* analysis of the metal NPs in the ILs by transmission electron microscopy revealed an average NP size, after reaction completion, of 3.4 nm in BMI.BF$_4$, 2.3 nm in BMI.PF$_6$ and 2.9 nm in BMI.OTf corroborating the idea that agglomeration is not dominant in the formation of these nanoclusters and the

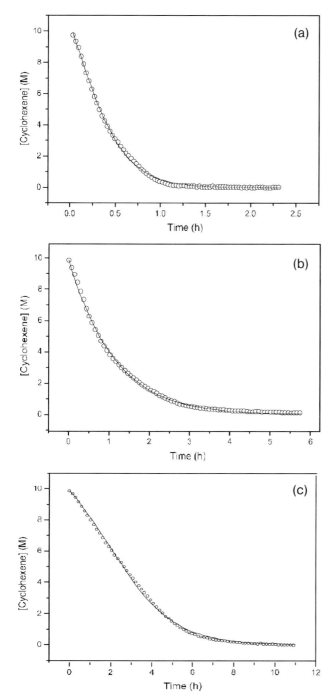

Fig. 6.3 Cyclohexene hydrogenation curve with the catalyst precursor $Pt_2(dba)_3$ dispersed in $BMI.BF_4$ (a), $BMI.PF_6$ (b) and $BMI.OTf$ (c) under 6 atm constant hydrogen pressure at 75 °C. Cyclohexene molar ratio/$Pt_2(dba)_3$ = 250 [89].

Table 6.2 Rate constants, k_1 and k_2, obtained for the hydrogenation of cyclohexene[a] by the catalyst precursor $[IrCl(cod)]_2$ and $Pt_2(dba)_3$ dispersed in ILs.

IL	k_1 (h^{-1})		k_2 (M^{-1} h^{-1})[b]	
	Ir	Pt	Ir	Pt
BMI.OTf	0.35 ± 0.03	0.53 ± 0.05	$1.0 \times 10^{-3} \pm 0.1 \times 10^{-3}$	11.5 ± 0.3
BMI.PF$_6$	2.04 ± 0.05	0.86 ± 0.02	$3.2 \times 10^{-3} \pm 0.2 \times 10^{-3}$	2.5 ± 0.2
BMI.BF$_4$	0.14 ± 0.02	1.34 ± 0.07	$1.2 \times 10^{-4} \pm 0.2 \times 10^{-4}$	70.5 ± 0.5

a 75 °C and under 6 atm of hydrogen (constant pressure) for $Pt_2(dba)_3$ and 4 atm for $[Ir(cod)Cl]_2$.
b Rate constants obtained considering a substrate/catalyst ratio = 250.

small sized Pt(0) NPs are the true catalyst responsible for the catalytic hydrogenation reaction.

Moreover, attempts to fit the curve with a mechanism including a third step, the bimolecular aggregation (B + B → C, k_3) [90, 91], or with the recently discovered double autocatalytic mechanism [92, 93], which includes a fourth step for the formation of bulk metal (B + C → 1.5C, k_4), did not converge the fit, which again strongly suggests that agglomeration is not significant in these cases.

6.4
Stabilization of the Metal Nanoparticles in Imidazolium Ionic Liquids

The intrinsic high charge of imidazolium salts, which creates an electrostatic colloid-type protection (DLVO-type stabilization) [94] for the transition-metal NPs similar to those proposed for quaternary ammonium salts [95–97] may be, as a first approximation [98], adequate for the description of the stabilizing effect. However, as discussed in the introduction, imidazolium salts differ in various aspects from classical quaternary ammonium salts. ILs possess pre-organized structures mainly through hydrogen bonds [40, 43] that induce structural directionality, in opposition to classical quaternary ammonium salts in which the aggregates display mainly charge-ordering structures [99]. It is noticeable that the anions and cations even in the gas phase form aggregates of the type $[(BMI)_x(X)_{x-n})]^{n+}[(BMI)_{x-n}(X)_x)]^{n-}$, where BMI is the 1-n-butyl-3-methylimidazolium cation and X is the anion, [43, 48–50]. Therefore, it is quite probable that the anion coordination to the NP surface in imidazolium ILs occurs via anionic aggregates of the type $[(BMI)_{x-n}(X)_x)]^{n-}$ rather than isolated X anions. Therefore the DLVO model cannot be used in these systems, as recently pointed out by Finke [98], since the DLVO model was not designed to account for counterions with multiple charges nor was it designed to account for sterically stabilized systems [100, 101]. Moreover, it has been recently demonstrated that Ir(0) nanoclusters may react with imidazolium-based ILs to form surface-attached carbenes [85]. The structure of the metal NPs, either dispersed in the ILs or after isolation, has been studied using various physico-chemical techniques such as XPS, EXAFS, SAXS, MET and XRD. Moreover, H/D and D/H labeling experiments during the formation of Ir(0) NPs have been recently performed using either D_2 or deuterated imidazolium ILs and indicate the intervention of *N*-carbenes at the metal surface, at least as transient species.

6.4.1
Analysis of the Isolated Nanoparticles (XRD and XPS)

The XRD of metal (Ir, Rh, Ru, Pd, Pt, Ni and Co) NPs isolated from the ILs confirm the crystalline nature of the material and the mean diameter can be estimated from the XRD diffraction pattern by means of the Debye–Scherrer equation calculated from the full width at half maximum (FWHM) of the crystalline planes obtained with Rietveld's refinements. The simulations of Bragg reflections and Rietveld's refinement were performed with a pseudo-Voigt function using the FULLPROF code. It is worth pointing out that the use of full width at half maximum (FWHM) of a peak to estimate the size of crystalline grain by means of the Scherrer equation has serious limitations since it does not take into account the existence of a distribution of sizes and the presence of defects in the crystalline lattice. Therefore, the calculation of the diameter of grain from FWHM of the peak can overestimate the real value since the larger grains give a strong contribution to the intensity, while the smaller grains just enlarge the base of the peak. Moreover, the presence of defects in a significant amount causes an additional enlargement of the diffraction line. Considering this enlargement, the size obtained can be smaller or larger than the real size of the grains. These problems can be minimized by the use of Rietveld's refinement method. Indeed, these discrepancies are confirmed by the values found for the average diameter of the NPs without the structural refinement, which significantly differs from those found by means of Rietveld's refinement. The values obtained with Rietveld's refinement are invariably much closer to those determined by other techniques such as TEM and SAXS [89]. In cases in which exposure of the isolated NPs to air may be a problem, such as in the case of Ru, Co and Ni NPs that form an oxide layer, the analysis can be performed *in situ* [86].

XPS analysis of the isolated NPs prepared in ILs shows in all cases, besides the M—M (M = Ir, Rh, Ru, Pt and Pd) and a small M—O contribution, an M—F contribution for the NPs prepared in ILs containing PF_6^- and BF_4^- anions. The contribution of the M—F component is more pronounced in the case of the PF_6 anion in which a F 1s and P contribution are also present in the XPS spectra [80, 88, 89, 102]. The XPS analysis of the samples after an Ar^+ sputtering shows that the F 1s signal is eliminated as well as the M—F component in the Ir 4f region, which displays mainly the M—M component, showing that only the external surface metal atoms were bound to F (Fig. 6.4) [84].

It is worth mentioning that the purity, in particular the water and halide contents [104], of the imidazolium ILs play an important role in the MNPs' chemistry, since

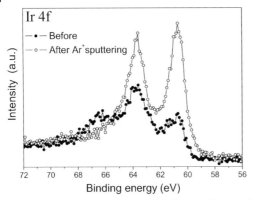

Fig. 6.4 X-ray photoelectron spectra of the Ir 4f region for Ir$_{(0)}$ nanoparticles prepared in BMI.PF$_6$ before and after Ar$^+$ sputtering. The Ar$^+$ sputtering eliminates the outermost layers of the particles and therefore results in the Ir 4f region having mainly the Ir—Ir bond component [103].

these impurities may influence the stability and catalytic properties of the material [105]. However, the XPS results clearly indicate that even relatively low coordinating anions, such as PF$_6^-$ and BF$_4^-$, coordinate with the metal surface and no other IL contaminant is present on the metal surface [84, 88, 89]. Moreover, these experiments indicate that the MNPs' surfaces are more susceptible to oxidation in air than the bulk metal and that this oxide layer may also be a significant source of stability of the metal NPs [106].

6.4.2
Analysis of the MNPs Dispersed in Ionic Liquids

SAXS analysis of Pt$_{(0)}$ and Ir$_{(0)}$ NPs dispersed in ILs indicated the formation of a layer surrounding the metal particles. It is important to note that calculations using the Guinier approximation law in the small angle region gave inconsistent results, indicating that the assumption that the NPs are simply diluted in the ILs cannot be applied in these cases. In opposition, the calculations using Porod's law gave consistent results. The Porod's region gives information about the interfaces present in the sample, where the electronic contrast is only at the interface between the two media (Fig. 6.5).

In this model we assume the presence of two phases: a crystalline phase (nanocrystal or NPs) and a semi-ordered phase (IL) [8, 28, 107, 108]. This model can be used since ILs are not considered as statistical aggregates of anions and cations but instead as a three-dimensional network of anions and cations i.e. polymeric supramolecular structures constituted from aggregates of the type $[(BMI)_x(X)_{x-n})]^{n+}[(BMI)_{x-n}(X)_x)]^{n-}$. In this context, the two-phase model was adopted to represent the nanocrystals dispersed in the IL [109]. Applying the interface distribution function to the experimental data, the extended molecular lengths of pure

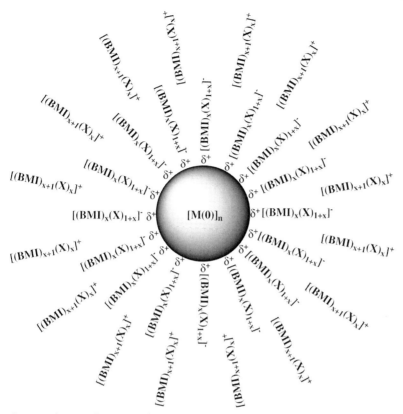

Fig. 6.5 Schematic illustration of the proposed two-phase model: $M_{(0)}$ stands for the metal NP and $[(BMI)_x(X)_{1+x}]^-$ and $[(BMI)_{x+1}(X)_x]^+$ for the anionic and cationic imidazolium aggregates, respectively [84].

ILs (semi-ordered phase) were estimated to be 2.8, 4.0, and 4.0 nm for BMI.BF$_4$, BMI.PF$_6$ and BMI.OTF, respectively. Note that the extended molecular length of 2.8 nm obtained for BMI.BF$_4$ is around twice that of the calculated value for a monomeric BMI.BF$_4$ unit [110, 111]. Moreover, the calculated molecular lengths (AM1) for $[(BMI)_2(BF_4)_3]^-$ and $[(BMI)_4(PF_6)_5]^-$ supramolecular clusters are 2.4 nm and 4.1 nm, respectively.

Inasmuch as most of the 1,3-dialkylimidazolium ILs have extremely low vapor pressure and relatively high viscosity at room temperature, *in situ* TEM observations in dispersed ILs can be carried out. Indeed, this method was applied for size and shape *in situ* analysis of various transition-metal NPs dispersed in the ILs (see Table 6.1). The size and shape determined by this *in situ* technique are the same as those obtained with NPs mixed with an epoxy resin distributed between two silicon wafer pieces and dried at 50 °C. In these cases the samples were pre-thinned mechanically to a thickness of about 20 nm and then ion milled to an electron transparency using 3 kV Ar$^+$ ion beams.

More interestingly, detailed analysis of the Pt(0) NPs dispersed in the ILs, shows a rather strong fluctuation of the contrast density. These fluctuations of contrast density are characteristic of amorphous substances. The sample regions containing particles embedded in the IL should not present a high contrast density fluctuation if the surrounding IL were to maintain its liquid features as in the pure liquid observation field. The perimeter and the core of the particle images also show strong contrast density fluctuations under larger under-focus conditions, and with increasing under-focus, it becomes difficult to observe, offering strong evidence for the interaction features of the IL with the Pt(0) NPs (Fig. 6.6).

It was also possible, by HRTEM measurements, to observe defects in the crystalline lattice of the Pt(0) NPs prepared in BMI.BF$_4$ (Fig. 6.7). Note that twin defects are typical for small particles.

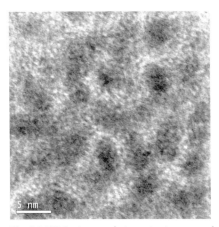

Fig. 6.6 TEM micrograph (negative image, under-focus) of the Pt$_{(0)}$ in BMI.PF$_6$ IL showing the contrast density fluctuation around the metal NPs [89, 112].

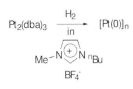

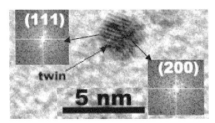

Fig. 6.7 Experimental HRTEM images of single twinned platinum NPs in BMI·BF$_4$, the planar defects and interplanar distances are indicated by arrows [89].

6.4.3
H/D and D/H Labeling Experiments

Using D_2 and ^2H NMR spectroscopy in the formation of Ir(0) NPs from the reduction of $[Ir(cod)(MeCN)_2]^+$ in BMI.NTf$_2$/acetone and proton sponge, Finke and coworkers have detected surface-ligand-coordinated *N*-heterocyclic carbenes.

The Ir(0) NP catalyzed H/D exchange, in which deuterium incorporation in the 2-H and 4-H positions of the imidazolium occurs only after an induction period (Scheme 6.4). The kinetics are well-fit by the analytical equations corresponding to the auto-catalytic mechanism which is a diagnostic of nanocluster formation [85].

Moreover, we have also observed that the D/H exchange only occurs after the formation of Ir(0) NPs and the D/H exchange occurs preferentially at the less acidic C-4 and C-5 imidazolium positions in the reaction of [Ir(cod)Cl]$_2$ in deuterated BMI.NTf$_2$ with molecular hydrogen (Scheme 6.4 and Figure 6.8) [113].

The D/H exchange only occurs after the complete consumption of the alkene and no D-incorporated alkane was detected, indicating that the coordinated NHC carbene is easily displaced by the alkene and that these carbenes are less strongly bound to the metal surface than is observed in mononuclear metal compounds [114].

6.4.4
MNPs in Ionic Liquid/Additive Ligands or Polymeric Stabilizers

MNPs in simple imidazolium ILs tend to agglomerate/aggregate after some reactions such as hydrogenation of aromatic compounds or ketones. However, more stable catalytic systems in ILs can be obtained by the addition of ligand or polymeric stabilizers such as poly[(*N*-vinyl-2-pyrrolidone)-co-(1-vinyl-3-alkylimidazo-lium halide)] copolymers [115], poly(*N*-vinyl-2-pyrrolidone) (PVP) [116], carbon, montmorillonite (MMT) or mesoporous SBA-15 [117, 118]. The NP/IL/extra-stabilizer combination usually exhibits an excellent synergistic effect that enhances the activity and durability of the catalyst for the hydrogenation of olefins. Another

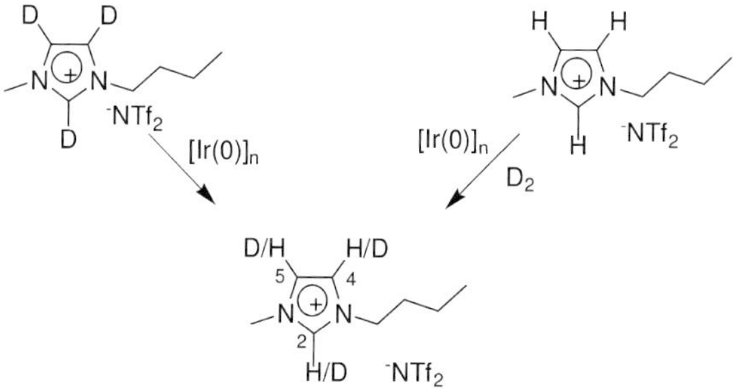

Scheme 6.4 H/D and D/H labeling experiments in the hydrogenation of alkenes by the Ir(I) NP precursor in IL.

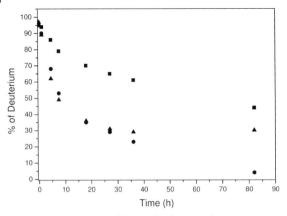

Fig. 6.8 D/H exchange of the imidazolium nucleus (■ = C-2, ▲, ● = C-4 and C-5) during the cyclohexene (13 mmol) hydrogenation (4 atm, constant pressure) by [Ir(cod)Cl]$_2$ (0.026 mmol) NP precursor in BMI.NTf$_2$-d_3 (0.5 mL) [113].

highly interesting approach consists of the use of ligands on the metal surface, akin to classical mononuclear organometallic chemistry, as for example, phenanthroline ligand-protected palladium NPs [119]. Moreover, elegant investigations by Dyson and coworkers on functionalized ILs [26, 120] such as imidazolium or pyridinium salts with the nitrile functional group attached to the alkyl side demonstrated that the ILs could act as both solvent and ligand for metal-catalyzed reactions and recycling and reuse is considerably superior in the nitrile-functionalized IL [64, 121].

Moreover, the NPs prepared in ILs may be deposited on other supports. For example, Pt(0) and Rh(0) NPs prepared in 1,1,3,3-tetramethylguanidinium trifluoroacetate or 1,1,3,3-tetramethylguanidinium lactate ILs, via reduction of their halide salts with glycol and with the aid of microwave heating, may be decorated onto carbon nanotubes in the presence of ILs [122].

6.5
Catalytic Properties of Transition-metal Nanoparticles in Ionic Liquids

The transition-metal NPs dispersed in imidazolium ILs are active catalysts for the hydrogenation of alkenes, arenes and ketones (Table 6.3). Moreover, Pd(0) NPs are active catalyst precursors for C–C coupling reactions, serving as reservoirs of mononuclear catalytically active species. In most cases, the catalytic reactions are typically multiphase systems in which the NPs dispersed in the ILs form the denser phase and the substrate and product remain in the upper-phase. In these cases the ionic catalytic solution is easily recovered by simple decantation and can be reused several times without any significant loss in catalytic activity.

Table 6.3 Examples of catalytic reactions promoted by transition-metal NPs in ILs.

Entry	Reaction	IL	Precursor	Ref.
1	hydrogenation of olefins and diolefins	BMI.BF$_4$, BMI.F$_6$ BMI.PF$_6$ [BMI.BF$_4$], BMI.PF$_6$, BMI.OTF BMI.PF$_6$	Pd Pt$_2$(dba)$_3$ RuO$_2$ [IrCl(cod)]$_2$	88 112 86 23, 123
2	arene hydrogenation	BMI.PF$_6$	[IrCl(cod)]$_2$, RhCl$_3$ [Ru(cod)(cot)]	124 80
3	hydrogenation of ketones	BMI.PF$_6$	[IrCl(cod)]$_2$	125
3	olefin hydroformylation	BMI.BF$_4$	RhCl$_3$	126
4	Negishi cross-coupling	BDMI.BF$_4$	Pd(dba)$_2$	127
6	Heck reaction	BMI.PF$_6$	palladacycle	79

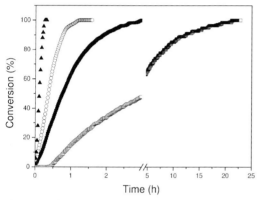

Fig. 6.9 Hydrogenation of olefins (▲ = 1-hexene, ○ = cyclohexene, ● = 1-methylcyclohexene and △ = 2,3-dimethyl-2-butene) by Ir(0) NPs dispersed in BMI.PF$_6$ at 5 atm of hydrogen (constant pressure) and 75 °C [103].

6.5.1
Hydrogenation of Alkenes and Ketones

The hydrogenation of simple olefins by Ir(0) NPs dispersed in ILs depends on steric hindrance at the C=C double bond, the reactivity follows the order terminal > disubstituted > tri-substituted > tetrasubstituted (Fig. 6.9). Such an order of reactivity for olefins is the same as that of classical iridium complexes in homogeneous conditions [128].

The olefin hydrogenation by the Ir(0) NPs in ILs follows the classical monomolecular surface reaction mechanism $v = k_c K[S]/1 + K[S]$. The reaction rate is a mass controlled process under hydrogen pressure <4 atm. The catalytic kinetic constant

(k_c) and the adsorption constant (K) under hydrogen pressures ≥4 atm are independent of the hydrogen concentration, indicating zero order dependence on hydrogen pressure and the reaction depends only on the olefin concentration in the IL [123].

The hydrogenation rate by transition-metal NPs is apparently more sensitive to the nature of the IL anion than to the size of the metal NPs. For example, the hydrogenation of 1-hexene by Pt(0) NPs in BMI.BF$_4$, in which larger NPs are formed, is much faster than that performed in BMI.PF$_6$ where smaller particles are formed. This is at first glance contradictory since it is expected that smaller NPs should, in principle, display higher catalytic activity than larger ones. However, in this case the NP surface is more exposed since the BF$_4^-$ anion is much less coordinating than the PF$_6^-$ anion. Indeed, the NPs prepared in BMI.BF$_4$ were isolated and re-dispersed in BMI.PF$_6$ and those isolated from BMI.PF$_6$ were re-dispersed in BMI.BF$_4$ (Table 6.4). These NP dispersions were used in the hydrogenation of 1-hexene and the reactions performed in the IL containing the more coordinating anion (PF$_6^-$) are much slower than those performed in BMI.BF$_4$. This result provides another indication that the NP is surrounded by the anionic species of the IL.

Interestingly, in multiphase catalytic processes the primary products can be extracted during the reaction thus modulating the product selectivity (using different substrates and reaction products' solubility with the catalyst containing phase, such as dienes/monoenes and arenes/cycloalkenes). Indeed, this approach can constitute a suitable method for avoiding consecutive reactions of primary products and has been used in the partial hydrogenation of dienes and arenes by transition-metal NPs dispersed in ILs.

In as much as 1,3-butadiene is at least four times more soluble in BMI.BF$_4$ than butenes, the selective partial hydrogenation could be performed by Pd(0) NPs embedded in the IL. Selectivities up to 72% in 1-butene were achieved at 99% 1,3-butadiene conversion, 40 °C and 4 atm of constant pressure of hydrogen

Table 6.4 1-Hexene hydrogenation by Pt(0) NPs isolated from BMI.BF$_4$ (3.4 nm) and re-dispersed in BMI.PF$_6$ and those isolated from BMI.PF$_6$ (2.3 nm) and re-dispersed in BMI.BF$_4$ at 75 °C under 6 atm of hydrogen (constant pressure) and 1-hexene/Pt = 250.

Entry	Pt($_0$) medium preparation	Pt($_0$) size (nm)	Reaction medium	TOF (h^{-1})[a]
1	BMI.BF$_4$	3.4 ± 0.3	BMI.PF$_6$	404 (1300)
2	BMI.PF$_6$	2.3 ± 0.3	BMI.BF$_4$	1000 (2380)

a Turnover frequency = mol(1-hexene)/mol(Pt$_2$(dba)$_3$). h and in parentheses the corrected turnover frequency considering only the exposed atoms on the NP surface (3.4 nm, 32% and 2.3 nm, 42%) using the magic number approach [5].

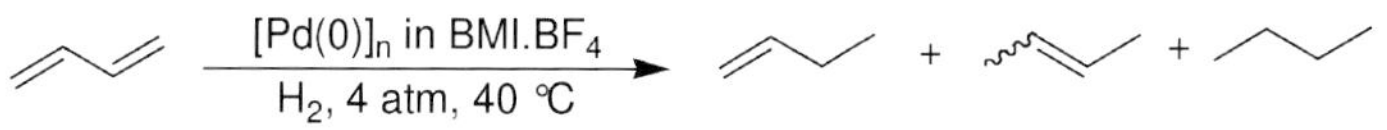

Scheme 6.5 Products resulting from the hydrogenation of 1,3-butadiene by Pd(0) NPs [88].

(Scheme 6.5). The amounts of butane (fully hydrogenated 1,3-butadiene) and *cis*-2-butene products are marginal and the butenes do not undergo an isomerization process, indicating that the soluble Pd(0) NPs possess pronounced surface-like rather than single-site-like catalytic properties [88].

Pd(0) NPs dispersed in simple ILs are not stable and tend to agglomerate after the hydrogenation of alkenes or dienes [88]. However, phenanthroline ligand-protected Pd(0) NPs in $BMI.PF_6$ IL are very active and selective for the hydrogenation of olefins, and the NPs/IL system can be reused many times without reducing the activity [119]. The immobilization of Pd(0) NPs onto the surface of a molecular sieve by 1,1,3,3-tetramethylguanidinium lactate IL also results in a more stable catalytic system. The NP/IL/molecular sieve combination exhibits an excellent synergistic effect that enhances the activity and durability of the catalyst for the hydrogenation of olefins [129]. Similarly, Pt(0), Pd(0) and Rh(0) NPs stabilized by poly(*N*-vinyl-2-pyrrolidone) (PVP) have also been prepared by simple ethanolic reduction of the corresponding metal halide salts immobilized in $BMI.PF_6$. These materials are very effective olefin hydrogenation catalysts which can be recycled by a simple decantation procedure and reused without loss of their catalytic activity [116].

6.5.2
Hydrogenation of Ketones

The Ir(0) NPs prepared in ILs are also effective catalysts for the hydrogenation of ketones and a TOF (mol of acetone/mol of Ir) up to $116 h^{-1}$ has been observed in the reduction of acetone at 75 °C under 5 atm of hydrogen after 2 h [125]. The hydrogenation of ketones containing aryl groups such as benzylmethylketone by the Ir(0) NPs is highly selective towards the reduction of the aromatic ring (Scheme 6.6). In this case a selectivity of 92% in the saturated ketone was attained at 97% benzylmethylketone conversion, indicating the preferential coordination of the aromatic ring to the metal surface rather than the ketone moiety.

6.5.3
Hydrogenations of Arenes

The hydrogenation of aromatic compounds by transition metal NPs apparently follows the classical Horiuti–Polanyi mechanism [130]. A series of the initial reaction rate constants was obtained from various competitive toluene/benzene and toluene/monoalkylbenzene hydrogenation experiments catalyzed by transition-metal NPs prepared in the presence of imidazolium ILs [Ir(0), Rh(0) and Ru(0)]. The reaction constants for the alkyl substituents can be expressed by steric factors

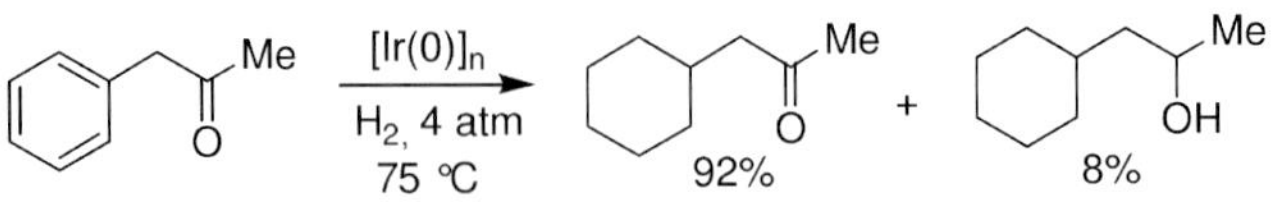

Scheme 6.6 Selective hydrogenation of benzylmethylketone (97% conversion) by Ir$_{(0)}$ NPs prepared in BMI.PF$_6$ [125].

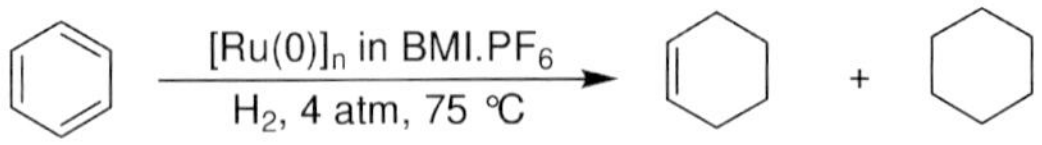

Scheme 6.7 Partial hydrogenation of benzene to cyclohexene by Ru$_{(0)}$ NPs in BMI.PF6 at 75 °C and under 4 atm of hydrogen (constant pressure) [80].

and are independent of any other non-steric factors. It was suggested that bulky alkylbenzene substituents, for transition metal NPs hydrogenation reactions, lower the overall hydrogenation rate, implying a more disturbed transition state compared to the initial state of the hydrogenation (in terms of the Horiuti–Polanyi mechanism) [131]. The hydrogenation of arenes containing functional groups, such as anisole, by the Ir(0) NPs occurs with concomitant hydrogenolysis of the C—O bond, suggesting that these NPs behave as multi-site catalysts rather than single-site catalysts [87].

Ru(0) NPs dispersed in simple ILs are efficient multiphase catalysts for the partial hydrogenation of benzene under mild reaction conditions (4 atm, 75 °C). The ternary diagram (benzene/cyclohexene/BMI.PF$_6$) indicated a maximum of 1% cyclohexene concentration in BMI.PF$_6$, which is attained at 4% benzene concentration in the ionic phase. This solubility difference in the IL was used for the extraction of cyclohexene during benzene hydrogenation by Ru catalysts suspended in BMI.PF$_6$. Selectivity up to 39% in cyclohexene can be attained at very low benzene conversion (Scheme 6.7). Although the maximum yield of 2% in cyclohexene is too low for technical applications, it represents a rare example of partial hydrogenation of benzene by soluble transition metal NPs [80].

It is interesting to note that most transition-metal NPs in simple imidazolium ILs tend to agglomerate after the hydrogenation of arenes [124]. However, more efficient catalytic systems have been obtained by immobilizing the NPs in polymeric ILs such as poly[(N-vinyl-2-pyrrolidone)-co-(1-vinyl-3-alkylimidazolium halide)] copolymers. Rhodium NPs stabilized by the ionic copolymer showed high lifetime and activity in arene hydrogenation under forcing conditions (75 °C and a hydrogen pressure of 40 bar) with a total turnover number of 20 000 and a TOF of 250 h^{-1} [115]. Another alternative involves immobilization of Ru NPs onto inorganic supports such as montmorillonite (MMT) [117] or mesoporous SBA-15 [118] using the 1,1,3,3-tetramethylguanidinium trifluoroacetate ([TMG][TFA]) IL. Using this approach Ru(0) NPs have been prepared on an MMT support such that they are very stable catalysts for the hydrogenation of benzene. These catalysts were reused for four runs for the hydrogenation of benzene (110 °C, 80 atm of H$_2$ for 2.5 h) giving TOF of 4000 h^{-1} in each run [117].

6.5.4
Miscellaneous Reactions

Bimetallic Pt-Ru/C NPs of around 2.5–3.5 nm have been prepared by reduction of H_2PtCl_6, $RuCl_3$ and Vulcan XC-72 carbon black dispersed in $BMI.BF_4$ and they possess high electrocatalytic activity for methanol oxidation [132].

Well-defined and *in situ* prepared Pd(0) NPs in ILs have been reported to promote C—C coupling reactions such as Heck and Suzuki processes [133–135]. In most cases, however, the Pd(0) NPs are reservoirs of homogeneous Pd(0) catalytically active species [79].

The same behavior was observed in the hydroformylation of olefins catalyzed by Rh(0) NPs prepared in $BMI.PF_6$. A strong influence of the NP size on the hydroformylation reaction was observed. Aldehydes were generated obtained preferentially when 5.0 nm Rh(0) NPs were used in the hydroformylation of 1-alkenes. Although small NPs also generate catalytically active species the chemoselectivity decreases (around 11–17% of olefin isomerization) compared to those performed with 5.0 nm NPs. In contrast, the large sized NPs (15 nm) produce only small amounts of aldehydes, similar to what is observed with a classical heterogeneous Rh/C catalyst precursor. TEM, XRD, IR and NMR experiments indicated that these Rh(0) NPs degraded under the reaction conditions into soluble monuclear Rh-carbonyl catalytically active species [126].

6.6
Conclusions and Perspectives

There is no doubt that 1,3-dialkylimidazolium ILs are best described as hydrogen-bonded polymeric supramolecules of the type $\{[(DAI)_x(X)_{x-n})]^{n+} [(DAI)_{x-n}(X)_x)]^{n-}\}n$ where DAI is the 1,3-dialkylimidazolium cation and X the anion. The formation and stabilization of NPs in these fluids occurs with the re-organization of the hydrogen bond network and the generation of nanostructures with polar and non-polar regions where the NPs are included. The IL forms a protective layer surrounding the transition-metal NPs surface with an extended molecular length of around 2.8–4.0 nm depending on the type of anion, suggesting the presence of semi-organized anionic species composed of supramolecular aggregates of the type $[(BMI)_{x-n}(X)_x)]^{n-}$. This protective layer is probably composed of imidazolium aggregate anions located immediately adjacent to the NP surface – providing the Columbic repulsion – and counter-cations that provide the charge balance i.e. quite close to DLVO type stabilization [94]. However, the DLVO model cannot completely explain the stabilization properties of imidazolium ILs towards the metal NPs since it treats counterions as mono-ionic point charges and was not designed to account for sterically stabilized systems [100, 101]. Therefore, together with the electrostactic stabilization provided by the intrinsic high charge of the IL, a steric type stabilization can also be envisaged due to the presence of anionic and cationic supramolecular aggregates of the type $[(BMI)_x(X)_{x-n})]^{n+}[(BMI)_{x-n}(X)_x)]^{n-}$ where BMI is the 1-n-butyl-3-methylimidazolium cation and X is the anion. These supramolecular aggregates may even be present in cases in which the stabilization may also involve surface attached carbenes as transient species [85].

Moreover, transition-metal NPs dispersed in these fluids are stable effective catalysts for reactions in multiphase conditions. The catalytic properties (activity and selectivity) of these soluble metal NPs indicate that they possess a pronounced surface-like (multi-site) rather than single-site-like catalytic properties. The catalytic properties of these dispersed metal NPs are influenced by the nature of the IL anions. In other cases the metal NPs are not stable and tend to aggregate or they serve as a reservoir of mononuclear catalytically active species. However, the IL provides a favorable environment for the formation of metal NPs with, in most cases, a small size and size distribution under very mild conditions. These NPs can be easily transferred to other organic and inorganic supports to generate more stable and active catalysts. It is also important to point out that only a few simple ILs have been used so far for the formation of NPs for catalysis using only classical synthetic methods i.e. chemical reduction of metal salts or controlled decomposition of organometallic compounds. Thus, the use of other functionalized ILs may provide stronger stabilizing fluids for the metal NPs without losing their catalytic properties. Furthermore, the use of other transition-metal NP synthetic methods such as electrochemical reduction in ILs may provide other nanomaterials with different catalytic properties.

Acknowledgment

Thanks are due to the following Brazilian agencies: CNPq, CAPES and CT-PETRO, and PETROBRAS for partial financial support.

References

1 J. D. Aiken, Y. Lin, R. G. Finke, *J. Mol. Catal. A: Chem.* **1996**, *114*, 29.

2 O. A. Belyakova, Y. L. Slovokhotov, *Russ. Chem. Bul.* **2003**, *52*, 2299–2327.

3 J. Jortner, U. Even, A. Goldberg, I. Schek, T. Raz, R. D. Levine, *Surf. Rev. Lett.* **1996**, *3*, 263–280.

4 G. Schmid, *Clusters and Colloids: From Theory to Applications*, VCH, New York, **1994**.

5 R. G. Finke, D. V. Feldheim, C. A. Foss Jr. (Eds.), *Metal Nanoparticles*, Marcel Dekker, New York, **2002**, pp. 17–54.

6 J. D. Aiken, R. G. Finke, *J. Mol. Catal. A: Chem.* **1999**, *145*, 1.

7 R. G. Finke, S. Ozkar, *Coord. Chem. Rev.* **2004**, *248*, 135.

8 C. A. Stowell, B. A. Korgel, *Nano Lett.* **2005**, *5*, 1203–1207.

9 A. Roucoux, J. Schulz, H. Patin, *Adv. Synth. Catal.* **2003**, *345*, 222.

10 D. I. Gittins, F. Caruso, *Angew. Chem. Int. Ed.* **2001**, *40*, 3001–3004.

11 S. Jansat, M. Gomez, K. Philippot, G. Muller, E. Guiu, C. Claver, S. Castillon, B. Chaudret, *J. Am. Chem. Soc.* **2004**, *126*, 1592.

12 B. Chaudret, *Compt. Rend. Phys.* **2005**, *6*, 117–131.

13 A. Roucoux, J. Schulz, H. Patin, *Chem. Rev.* **2002**, *102*, 3757.

14 D. Astruc, F. Lu, J. R. Aranzaes, *Angew. Chem. Int. Ed.* **2005**, *44*, 7852–7872.

15 M. Moreno-Manas, R. Pleixats, *Acc. Chem. Res.* **2003**, *36*, 638.

16 H. Bonnemann, *Fuel Cells* **2004**, *4*, 289–296.

17 S. Ozkar, R. G. Finke, *J. Am. Chem. Soc.* **2002**, *124*, 5796.

18 Y. Chauvin, B. Gilbert, I. Guibard, *J. Chem. Soc., Chem. Commun.* **1990**, 1715–1716.

19 P. A. Z. Suarez, J. E. L. Dullius, S. Einloft, R. F. DeSouza, J. Dupont, *Polyhedron* **1996**, *15*, 1217–1219.

20 Y. Chauvin, L. Mussmann, H. Olivier, *Angew. Chem. Int. Ed. Engl.* **1996**, *34*, 2698–2700.

21 Y. Chauvin, *Actual Chim.* **1996**, 44–46.

22 P. Bonhote, A. P. Dias, N. Papageorgiou, K. Kalyanasundaram, M. Gratzel, *Inorg. Chem.* **1996**, *35*, 1168–1178.

23 J. Dupont, G. S. Fonseca, A. P. Umpierre, P. F. P. Fichtner, S. R. Teixeira, *J. Am. Chem. Soc.* **2002**, *124*, 4228–4229.

24 R. R. Deshmukh, R. Rajagopal, K. V. Srinivasan, *Chem. Commun.* **2001**, 1544–1545.

25 P. Wasserscheid, T. Welton, *Ionic Liquids in Synthesis*, Wiley-VCH, Weinheim, **2003**.

26 Z. F. Fei, T. J. Geldbach, D. B. Zhao, P. J. Dyson, *Chem.-Eur. J.* **2006**, *12*, 2123–2130.

27 S. Tsuzuki, H. Tokuda, K. Hayamizu, M. Watanabe, *J. Phys. Chem. B* **2005**, *109*, 16474–16481.

28 C. Hardacre, J. D. Holbrey, S. E. J. McMath, D. T. Bowron, A. K. Soper, *J. Chem. Phys.* **2003**, *118*, 273–278.

29 J. Dupont, P. A. Z. Suarez, *Phys. Chem. Chem. Phys.* **2006**, *8*, 2441–2452.

30 S. Saha, P. K. Mandal, A. Samanta, *Phys. Chem. Chem. Phys.* **2004**, *6*, 3106–3110.

31 H. Zhao, *Phys. Chem. Liq.* **2003**, *41*, 545–557.

32 K. A. Fletcher, S. Pandey, *J. Phys. Chem. B* **2003**, *107*, 13532–13539.

33 L. Crowhurst, P. R. Mawdsley, J. M. Perez-Arlandis, P. A. Salter, T. Welton, *Phys. Chem. Chem. Phys.* **2003**, *5*, 2790–2794.

34 J. G. Huddleston, G. A. Broker, H. D. Willauer, R. D. Rogers, in *Ionic Liquids*, Vol. *818*, **2002**, pp. 270–288.

35 K. A. Fletcher, S. Pandey, *Appl. Spectrosc.* **2002**, *56*, 1498–1503.

36 S. V. Dzyuba, R. A. Bartsch, *Tetrahedron Lett.* **2002**, *43*, 4657–4659.

37 S. N. Baker, G. A. Baker, F. V. Bright, *Green Chem.* **2002**, *4*, 165–169.

38 K. A. Fletcher, I. A. Storey, A. E. Hendricks, S. Pandey, S. Pandey, *Green Chem.* **2001**, *3*, 210–215.

39 A. J. Carmichael, K. R. Seddon, *J. Phys. Org. Chem.* **2000**, *13*, 591–595.

40 J. Dupont, *J. Braz. Chem. Soc.* **2004**, *15*, 341–350.

41 J. B. Harper, R. M. Lynden-Bell, *Mol. Phys.* **2004**, *102*, 85–94.

42 S. Christie, R. H. Dubois, R. D. Rogers, P. S. White, M. J. Zaworotko, *J. Inclusion Phenom.* **1991**, *11*, 103–114.

43 F. C. Gozzo, L. S. Santos, R. Augusti, C. S. Consorti, J. Dupont, M. N. Eberlin, *Chem.-Eur. J.* **2004**, *10*, 6187–6193.

44 C. S. Consorti, P. A. Z. Suarez, R. F. de Souza, R. A. Burrow, D. H. Farrar, A. J. Lough, W. Loh, L. H. M. da Silva, J. Dupont, *J. Phys. Chem. B* **2005**, *109*, 4341–4349.

45 J. Lopes, A. A. H. Padua, *J. Phys. Chem. B* **2006**, *110*, 3330–3335.

46 I. Billard, G. Moutiers, A. Labet, A. El Azzi, C. Gaillard, C. Mariet, K. Lutzenkirchen, *Inorg. Chem.* **2003**, *42*, 1726–1733.

47 E. Solutskin, B. M. Ocko, L. Taman, I. Kuzmenko, T. Gog, M. Deutsch, *J. Am. Chem. Soc.* **2005**, *127*, 7796–7804.

48 S. Dorbritz, W. Ruth, U. Kragl, *Adv. Synth. Catal.* **2005**, *347*, 1273–1279.

49 P. J. Dyson, J. S. McIndoe, D. B. Zhao, *Chem. Commun.* **2003**, 508–509.

50 D. Nohara, T. Ohkoshi, T. Sakai, *Rapid Commun. Mass Spectrom.* **1998**, *12*, 1933–1935.

51 A. Mele, G. Romano, M. Giannone, E. Ragg, G. Fronza, G. Raos, V. Marcon, *Angew. Chem. Int. Ed.* **2006**, *45*, 1123–1126.

52 J. D. Martin, *ACS Symp. Ser.* **2002**, *818*, 413–427.

53 S. Biggin, J. E. Enderby, *J. Phys. C-Solid State Phys.* **1982**, *15*, L305–L309.

54 M. Rovere, M. P. Tosi, *Rep. Prog. Phys.* **1986**, *49*, 1001–1081.

55 H. Cang, J. Li, M. D. Fayer, *J. Chem. Phys.* **2003**, *119*, 13017–13023.

56 J. Dupont, C. S. Consorti, F. Rominger, in preparation.

57 S. M. Dibrov, J. K. Kochi, *Acta Crystallogr., Sect. C: Cryst. Struct. Commun.* **2006**, *62*, O19–O21.

58 M. Antonietti, D. B. Kuang, B. Smarsly, Z. Yong, *Angew. Chem.Int. Ed.* **2004**, *43*, 4988–4992.

59 J. M. Cao, B. Q. Fang, J. Wang, M. B. Zheng, S. G. Deng, X. J. Ma, *Prog. Chem.* **2005**, *17*, 1028–1033.

60 Y. Zhou, *Curr. Nanosci.* **2005**, *1*, 35–42.

61 S. M. Chen, Y. D. Liu, G. Z. Wu, *Nanotechnology* **2005**, *16*, 2360–2364.

62 S. Y. Gao, H. J. Zhang, X. M. Wang, W. P. Mai, C. Y. Peng, L. H. Ge, *Nanotechnology* **2005**, *16*, 1234–1237.

63 K. S. Kim, D. Demberelnyamba, H. Lee, *Langmuir* **2004**, *20*, 556–560.

64 D. B. Zhao, Z. F. Fei, T. J. Geldbach, R. Scopelliti, P. J. Dyson, *J. Am. Chem. Soc.* **2004**, *126*, 15876–15882.

65 H. Itoh, K. Naka, Y. Chujo, *J. Am. Chem. Soc.* **2004**, *126*, 3026–3027.

66 B. S. Lee, Y. S. Chi, J. K. Lee, I. S. Choi, C. E. Song, S. K. Namgoong, S. G. Lee, *J. Am. Chem. Soc.* **2004**, *126*, 480–481.

67 G. T. Wei, Z. S. Yang, C. Y. Lee, H. Y. Yang, C. R. C. Wang, *J. Am. Chem. Soc.* **2004**, *126*, 5036–5037.

68 Y. Z. Wu, X. L. Zhao, F. Tian, X. P. Hao, J. X. Yang, *J. Rare Earths* **2006**, *24*, 115–118.

69 R. Tatumi, H. Fujihara, *Chem. Commun.* **2005**, 83–85.

70 S. M. Zhang, C. L. Zhang, Z. S. Wu, Z. J. Zhang, H. X. Dang, W. M. Liu, Q. J. Xue, *Acta Chim. Sin.* **2004**, *62*, 1443–1446.

71 S. M. Zhang, C. L. Zhang, J. W. Zhang, Z. J. Zhang, H. X. Dang, Z. S. Wu, W. M. Liu, *Acta Phys.-Chim. Sin.* **2004**, *20*, 554–556.

72 B. S. Lee, S. Lee, *Bull. Korean Chem. Soc.* **2004**, *25*, 1531–1537.

73 K. S. Kim, S. Choi, J. H. Cha, S. H. Yeon, H. Lee, *J. Mater. Chem.* **2006**, *16*, 1315–1317.

74 M. A. Firestone, M. L. Dietz, S. Seifert, S. Trasobares, D. J. Miller, N. J. Zaluzec, *Small* **2005**, *1*, 754–760.

75 R. Marcilla, M. L. Curri, P. D. Cozzoli, M. T. Martinez, I. Loinaz, H. Grande, J. A. Pomposo, D. Mecerreyes, *Small* **2006**, *2*, 507–512.

76 Y. Wang, H. Yang, *J. Am. Chem. Soc.* **2005**, *127*, 5316–5317.

77 D. B. Zhao, Z. F. Fei, W. H. Ang, P. J. Dyson, *Small* **2006**, *2*, 879–883.

78 V. Mevellec, B. Leger, M. Mauduit, A. Roucoux, *Chem. Commun.* **2005**, 2838–2839.

79 C. C. Cassol, A. P. Umpierre, G. Machado, S. I. Wolke, J. Dupont, *J. Am. Chem. Soc.* **2005**, *127*, 3298–3299.

80 E. T. Silveira, A. P. Umpierre, L. M. Rossi, G. Machado, J. Morais, G. V. Soares, I. L. R. Baumvol, S. R. Teixeira, P. F. P. Fichtner, J. Dupont, *Chem.-Eur. J.* **2004**, *10*, 3734–3740.

81 M. A. Watzky, R. G. Finke, *J. Am. Chem. Soc.* **1997**, *119*, 10382.

82 J. A. Widegren, J. D. Aiken, S. Ozkar, R. G. Finke, *Chem. Mater.* **2001**, *13*, 312.

83 M. A. Watzky, R. G. Finke, *Chem. Mater.* **1997**, *9*, 3083.

84 G. S. Fonseca, G. Machado, S. R. Teixeira, G. H. Fecher, J. Morais, M. C. M. Alves, J. Dupont, *J. Colloid Interface Sci.* **2006**, *301*, 193–204.

85 L. S. Ott, M. L. Cline, M. Deetlefs, K. R. Seddon, R. G. Finke, *J. Am. Chem. Soc.* **2005**, *127*, 5758–5759.

86 L. M. Rossi, G. Machado, P. F. P. Fichtner, S. R. Teixeira, J. Dupont, *Catal. Lett.* **2004**, *92*, 149–155.

87 G. S. Fonseca, A. P. Umpierre, P. F. P. Fichtner, S. R. Teixeira, J. Dupont, *Chem.-Eur. J.* **2003**, *9*, 3263–3269.

88 A. P. Umpierre, G. Machado, G. H. Fecher, J. Morais, J. Dupont, *Adv. Synth. Catal.* **2005**, *347*, 1404–1412.

89 C. W. Scheeren, G. Machado, S. R. Teixeira, J. Morais, J. B. Domingos, J. Dupont, *J. Phys. Chem. B* **2006**, *110*, 13011–13020.

90 B. J. Hornstein, R. G. Finke, *Chem. Mater.* **2004**, *16*, 3972.

91 B. J. Hornstein, R. G. Finke, *Chem. Mater.* 2004, *16*, 139.

92 C. Besson, E. E. Finney, R. G. Finke, *J. Am. Chem. Soc.* 2005, *127*, 8179–8184.

93 C. Besson, E. E. Finney, R. G. Finke, *Chem. Mater.* 2005, *17*, 4925.

94 E. J. W. Verwey, J. T. G. Overbeek, *Theory of the Stability of Lyophobic Colloids*, Dover, New York, 1999.

95 H. Modrow, S. Bucher, J. Hormes, R. Brinkmann, H. Bonnemann, *J. Phys. Chem. B* 2003, *107*, 3684–3689.

96 S. Bucher, J. Hormes, H. Modrow, R. Brinkmann, N. Waldofner, H. Bonnemann, L. Beuermann, S. Krischok, W. Maus-Friedrichs, V. Kempter, *Surf. Sci.* 2002, *497*, 321–332.

97 H. Bonnemann, W. Brijoux, R. Brinkmann, E. Dinjus, T. Joussen, B. Korall, *Angew. Chem. Int. Ed. Engl.* 1991, *30*, 1312–1314.

98 L. S. Ott, R. G. Finke, *Coord. Chem. Rev.* 2007, *251*, 1075–1100.

99 J. P. Hansen, I. R. McDonald, *Theory of Simple Liquids*, 2nd. edn., Academic Press, London, 1986.

100 M. Bostrom, D. R. M. Williams, B. W. Ninham, *Phys. Rev. Lett.* 2001, *8716*.

101 B. W. Ninham, *Adv. Colloid Interface Sci.* 1999, *83*, 1–17.

102 M. A. Gelesky, A. P. Umpierre, G. Machado, R. R. B. Correia, W. C. Magno, J. Morais, G. Ebeling, J. Dupont, *J. Am. Chem. Soc.* 2005, *127*, 4588–4589.

103 G. S. Fonseca, J. B. Domingos, F. Nome, J. Dupont, *J. Mol. Catal. A: Chem.* 2006, *248*, 10–16.

104 K. R. Seddon, A. Stark, M. J. Torres, *Pure Appl. Chem.* 2000, *72*, 2275–2287.

105 J. Dupont, J. Spencer, *Angew. Chem. Int. Ed.* 2004, *43*, 5296–5297.

106 L. S. Ott, R. G. Finke, *Inorg. Chem.* 2006, *45*, 8382–8393.

107 B. A. Korgel, D. Fitzmaurice, *Phys. Rev. B* 1999, *59*, 14191–14201.

108 J. C. Liu, B. X. Han, H. L. Zhang, G. Z. Li, X. G. Zhang, J. Wang, B. Z. Dong, *Chem.-Eur. J.* 2002, *8*, 1356–1360.

109 A. I. Frenkel, C. W. Hills, R. G. Nuzzo, *J. Phys. Chem. B* 2001, *105*, 12689–12703.

110 B. G. Trewyn, C. M. Whitman, V. S. Y. Lin, *Nano Lett.* 2004, *4*, 2139–2143.

111 Y. Zhou, J. H. Schattka, M. Antonietti, *Nano Lett.* 2004, *4*, 477–481.

112 C. W. Scheeren, G. Machado, J. Dupont, P. F. P. Fichtner, S. R. Texeira, *Inorg. Chem.* 2003, *42*, 4738–4742.

113 J. D. Scholten, G. Ebeling, B. Ferrera, J. Dupont, Manuscript in preparation 2006.

114 D. Bacciu, R. Kingsley, K. Cavell, I. A. Fallis, L.-L. Ooi, *Angew. Chem. Int. Ed.* 2005, *44*, 5282–5284.

115 X. D. Mu, J. Q. Meng, Z. C. Li, Y. Kou, *J. Am. Chem. Soc.* 2005, *127*, 9694–9695.

116 X. D. Mu, D. G. Evans, Y. A. Kou, *Catal. Lett.* 2004, *97*, 151–154.

117 S. D. Miao, Z. M. Liu, B. X. Han, J. Huang, Z. Y. Sun, J. L. Zhang, T. Jiang, *Angew. Chem. Int. Edit.* 2006, *45*, 266–269.

118 J. Huang, T. Jiang, B. X. Han, W. Z. Wu, Z. M. Liu, Z. L. Xie, J. L. Zhang, *Catal. Lett.* 2005, *103*, 59–62.

119 J. Huang, T. Jiang, B. X. Han, H. X. Gao, Y. H. Chang, G. Y. Zhao, W. Z. Wu, *Chem. Commun.* 2003, 1654–1655.

120 J. H. Davis, *Chem. Lett.* 2004, *33*, 1072–1077.

121 D. B. Zhao, Z. F. Fei, R. Scopelliti, P. J. Dyson, *Inorg. Chem.* 2004, *43*, 2197–2205.

122 Z. M. Liu, Z. Y. Sun, B. X. Han, J. L. Zhang, J. Huang, J. M. Du, S. D. Miao, *J. Nanosci. Nanotechnol.* 2006, *6*, 175–179.

123 G. S. Fonseca, J. B. Domingos, F. Nome, J. Dupont, *J. Mol. Catal. A: Chem.* 2006, *248*, 10–16.

124 G. S. Fonseca, A. P. Umpierre, P. F. P. Fichtner, S. R. Teixeira, J. Dupont, *Chem.-Eur. J.* 2003, *9*, 3263–3269.

125 G. S. Fonseca, J. D. Scholten, J. Dupont, *Synlett* 2004, 1525–1528.

126 A. J. Bruss, M. A. Gelesky, G. Machado, J. Dupont, *J. Mol. Catal. A: Chem.* 2006, *252*, 212–218.

127 J. Sirieix, M. Ossberger, B. Betzemeier, P. Knochel, *Synlett* 2000, 1613–1615.

128 R. Crabtree, *Acc. Chem. Res.* 1979, *12*, 331–338.

129 J. Huang, T. Jiang, H. Gao, B. Han, Z. Liu, W. Wu, Y. Chang, G. Zhao, *Angew. Chem. Int. Ed.* 2004, *43*, 1397–1399.

130 J. Horiuti, M. Polanyi, *J. Mol. Catal. A: Chem.* 2003, *199*, 185–197.

131 G. S. Fonseca, E. T. Silveira, M. A. Gelesky, J. Dupont, *Adv. Synth. Catal.* **2005**, *347*, 847–853.

132 X. Z. Xue, T. H. Lu, C. P. Liu, W. L. Xu, Y. Su, Y. Z. Lv, W. Xing, *Electrochim. Acta* **2005**, *50*, 3470–3478.

133 V. Calo, A. Nacci, A. Monopoli, A. Detomaso, P. Iliade, *Organometallics* **2003**, *22*, 4193–4197.

134 V. Calo, A. Nacci, A. Monopoli, A. Fornaro, L. Sabbatini, N. Cioffi, N. Ditaranto, *Organometallics* **2004**, *23*, 5154–5158.

135 V. Calo, A. Nacci, A. Monopoli, *J. Organomet. Chem.* **2005**, *690*, 5458–5466.

7
Carbon and Silicon Carbide Nanotubes Containing Catalysts

Cuong Pham-Huu, Ovidiu Ersen, and Marc-Jacques Ledoux

7.1
Introduction

Catalysis has been closely associated to the development of our technological modern society for more than 100 years and continues to be so today [1]. The two first worldwide examples of catalysis intervention were first, the Fischer–Tropsch synthesis, allowing the transformation of coal into valuable liquid fuels, which provided an energetic power supply to the German army during World War II [2], where the nine plants in operation have a combined capacity of about 660×10^3 t per year; the second concerns the improvement of the octane number, still during World War II, to improve the performance of the allied aircraft during the battle of Britain [3]. The British lost 915 planes compared to the 1733 lost by the Germans. This was because of their superior maneuverability with 50% faster bursts of acceleration from their 100-octane fuel compared to the German 87-octane fuel. Since then an exponential increase in the participation of catalytic processes has been observed in the development of the industrialised countries. A catalyst allows the transformation of reactants into desired products with high selectivity reducing at the same time the cost linked with waste disposal. Many types of materials can be used as catalysts. These include metals, compounds, i.e. metal oxides, sulfides, carbides, nitrides, etc., organometallic complexes, and enzymes. Most of the industrially employed catalysts are dispersed on high surface area supports (alumina, silica or activated carbon) in order to increase the number of active sites.

The application and development of an industrial catalyst starts with research and control of materials at the molecular level followed by laboratory evaluation and a microplant reactor which pave the way to the real industrial catalyst. All of these processes cover several domains of research, from molecular chemistry to transport phenomena and solid state studies. Today the understanding of the macroscopic phenomena involved in catalysis seems to be relatively well under control for most of the different catalytic processes whereas the comprehension of what happens at a nanoscopic scale remains to be improved, in order to design a new generation of catalysts with better activity and selectivity. The direct way to

Nanoparticles and Catalysis. Edited by Didier Astruc
Copyright © 2008 WILEY-VCH Verlag GmbH & Co. KGaA, Weinheim
ISBN: 978-3-527-31572-7

get access to this knowledge is to reduce as much as possible the size of the active site ensemble in order to provide a detailed understanding of the active phase at a chemical level. Decreasing the size of the active phase also significantly increases the effective contact surface of the reactants with it, which contributes to lowering the cost of the catalyst per unit weight of active phase. All this contributes to the inclusion of catalysis in the new and increasingly developed nanotechnology science. However, during the catalytic reaction the starting size of the catalytic site is modified by aggregation which makes size control a real technological and scientific challenge.

One of the most common ways to avoid aggregation is to deposit the active phase inside nanocages or nanochannels, i.e. zeolites or mesoporous silica. However, the diffusion and the accessibility of the reactants to these constrained catalytic sites can be strongly affected by the tortuous structure of the support and consequently there can be a significant effect on the selectivity of the reaction, rendering such a system unable to operate under severe conditions [4, 5]. In addition, it is thought that the tailoring of nanostructured catalysts could lead to new electronic and catalytic properties for the improvement of catalytic activity and selectivity in view of the envisaged legislation on waste reduction. High activity would also reduce the catalyst weight per reactor volume unit and consequently reduce the waste generated during the catalyst manufacturing while the high selectivity would reduce the generation of undesired by-products, costly in terms of atoms lost and in terms of their negative environmental effect.

Recently, the new development in the field of one-dimensional (1D) and conductive materials, i.e. carbon and silicon carbide, sheds a new light on the use of these materials in catalysis in the place of traditional macroscopic supports [6–12]. Metals supported on these carbon nanostructures exhibit unusual catalytic activity and selectivity patterns when compared to those encountered with traditional catalyst supports such as alumina, silica or activated carbon. The extremely high external surface area displayed by these nanoscale materials significantly reduces the mass transfer limitations, and the peculiar interaction between the deposited active phase and the surface of the support has been advanced to explain their catalytic behaviour. Many of their fundamental and remarkable physical properties are now well-known and present efforts are largely devoted to producing carbon nanotubes with high selectivity and in a large volume in order to decrease their price and allow their exploitation in a wide range of applications.

This chapter aims to present some general background on the development of the catalytic synthesis of 1D carbon and carbide materials, and on their application as catalyst supports for nanoparticles system or as catalysts [10, 11, 13]. Nanoparticles are an ensemble of atoms with an average size of a few nanometers and hold a large number of metastable atoms with respect to the bulk structure. It is thought that these metastable atoms could play an important role during catalysis while their stabilization, i.e. through electronic or chemical interactions with the support surface, allows one to explore more new catalytic systems with peculiar activity and selectivity than ever before. In view of the large body of scientific and industrial reports in the literature on this exciting field, the references cited in this chapter are not meant to be exhaustive, but are merely representative of the subject to date. The most representative reviews published in this field are included in the references which allows the reader to get more insight into their specific field of interest. The first part of the chapter will be devoted to the synthesis of the different nanomaterials along with their characterization. It is worth noting that for potential applications of carbon nanotubes it is of interest to develop synthesis methods allowing the production of large quantities of high purity carbon nanotubes. The second part will deal with their catalytic applications. For the sake of clarity the catalytic activity of these nanostructured catalysts will be compared with that obtained on conventional catalysts and the advantages will be highlighted.

7.2
Carbon and SiC Nanotubes/Nanofibers

7.2.1
Carbon Nanotubes and Nanofibers

Recently, 1D carbon forms, i.e. carbon nanotubes (CNTs) and carbon nanofibers (CNFs), have attracted large scientific interest since the discovery of multi-walled carbon nanotubes in arc discharge materials [14]. Carbon nanotubes, single-walled (SWNTs) and multi-walled (MWNTs), consist of rolled graphite basal planes with low surface energy and an open channel in the middle (Fig. 7.1A). Typically, the length of a nanotube varies from 1 to several micrometers, and the diameter from ca. 1 nm (for single-walled carbon nanotubes) to about 100 nm (for multi-walled carbon nanotubes, depending on the number of walls). The inner diameter of the carbon nanotubes can vary from a few to several dozen nanometers, depending on the synthesis conditions. For the multi-walled carbon nanotubes the distance between two neighboring layers of graphene is close to 0.34 nm which is similar to the interplanar distance of the d_{0002} plane of graphite. In fact this distance is slightly higher than 0.34 nm due to the existence of constraints induced by the curvature of the graphene planes [12]. The electronic density of the carbon nanotube is also modified compared to that of the planar graphite. It is expected that the rolling up of the graphene sheet to form the tube induces a rehybridization of the carbon orbital with a concomitant modification of the electronic density (π-density).

Carbon nanofibers are similar in shape and length to carbon nanotubes but consist of graphene planes oriented at an angle with respect to the fiber axis with no tubule in the middle, i.e. a pile of "chinese hats", which by projection shows a fishbone arrangement (Fig. 7.1B) [11, 12]. The angle orientation of the graphene planes with respect to the fiber axis can also be finely tuned by modifying the synthesis conditions [15]. The graphene planes inside the carbon nanofibers can also adopt a different arrangement, i.e. platelet with the graphene planes oriented perpendicularly to the fiber axis or a fishbone arrangement with a hollow channel periodically close along the fiber axis, depending on the nature of the metal catalyst

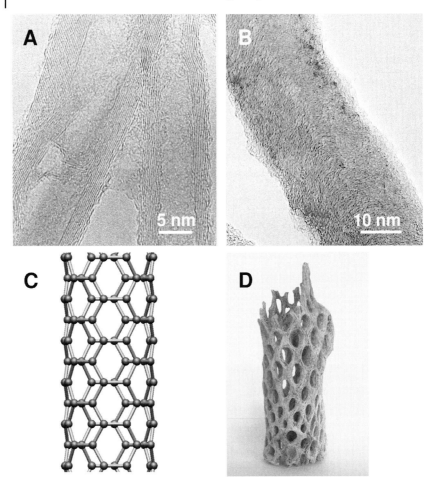

Fig. 7.1 TEM micrographs showing the microstructural arrangement of the graphene planes in (A) carbon nanotubes: parallel to the tube axis and in (B) carbon nanofibers: fishbone arrangement along the fiber axis. The thin amorphous carbon layer visualized on the topmost layers of the materials is attributed to the low-temperature decomposition of the starting carbon source during the cooling process. Such a layer can be efficiently removed by air oxidation. (C) Schematic representation of a rolled graphene sheet in the carbon nanotube. (D) Optical image of a dry cactus showing the same organization as observed in the hexagonal arrangement in the graphene carbon sheet. (Courtesy of Dr. J. Boksquet).

and the reactants mixture [16–20]. The exposed external surface of the carbon nanofibers is constituted of prismatic planes with high reactivity.

These nanomaterials are also characterized by their extremely high aspect ratio. The aspect ratio is defined as the length of the major axis divided by the width or diameter of the minor axis. According to this definition, spheres have an aspect ratio of 1, while carbon nanotubes or nanofibers have an aspect ratio ranging from a few tenths to several thousands. The high aspect ratio of these materials confers to them a significant external surface area compared to classical materials. These 1D carbon nanostructures have been extensively studied due to their unique chemical and mechanical properties which render them attractive for numerous potential applications [6–12, 21–23]. Among these the fields of composites (high mechanical resistance due to the exclusive presence of covalently bonded carbon atoms) and catalysis seem to be the most promising according to the work published to date in the open literature [11–13].

Recently, carbon nanotubes have been extensively studied as catalyst supports in several fields of catalysis from gas-phase to liquid-phase processes. Carbon nanotubes hold several advantages when they are used as a catalyst support: (i) Their small dimension (nanometers in diameter) which significantly increases the external surface area of the catalyst, especially for liquid-phase reactions where diffusion becomes a limiting step. The diffusion pathlength in the liquid phase is several orders of magnitude lower than in the gas phase, which significantly enhances the mass transfer limitation. (ii) Their high thermal conductivity provides a homogeneous and rapid heat transfer of the reaction's heat to the catalyst support which avoids the formation of a surface hot spot which could modify to a great extent the overall selectivity of the reaction. (iii) Their structure, which exhibits a high specific surface area, generally $>100\,m^2\,g^{-1}$, along with a large pore volume without micropores. The existence of an electronic interaction (π-interaction) between the carbon nanomaterial surface and the deposited species could also affect significantly the final properties of the supported species.

The first carbon nanotubes structure was discovered in 1991 by Iijima in the carbon soot produced by a carbon arc machine [14]. The first attempt to produce carbon nanotubes with high yield was initiated by Ebbesen and Ayajan by modifying the deposition parameters of the arc discharge technique [24]. Generally, to produce carbon nanotubes, an arc is maintained between two cylindrical electrodes under a He atmosphere. The tubes are formed as a soft sooty fibrous core on the cathode. However, it is worth noting that the deposition temperature in the arc discharge is extremely high, ca. 3700 °C, and the precise control of the different parameters is not straightforward. Another drawback of the arc discharge method is the low selectivity toward carbon nanotubes and the large amount of carbon nanoparticles or soot which are formed needing a purification which is time and yield consuming [25–27]. Therefore it is of interest to find new synthesis methods which can overcome these drawbacks. The most promising method to produce carbon nanotubes with high selectivity and yield consists in the catalytic assisted synthesis with which many research groups are involved [28, 29]. The catalytic process consists in contacting a gaseous carbon source, i.e. hydrocarbons, CO, organic complexes, etc., with a catalytic active phase, Ni, Co, Fe, etc, at a temperature ranging from 600 to 1200 °C. The ability of these metals to generate carbon nanostructures from gaseous carbon sources is related to their catalytic activity for the decomposition of gaseous carbon compounds, their possibility to be transformed during the course of the reaction into metastable carbide forms and their high affinity to form solid solution with carbon allowing its high diffusion through the metal particle to form ordered carbon nanotubes (or nanofibers) [30, 31]. The catalytic route allows the formation of large amounts of carbon nanotubes with respect to the starting weight of the catalyst, along with a high selectivity which avoid the post-synthesis purification steps to remove undesirable products such as carbon nanoparticles or soot. Pure carbon nanotubes can be recovered after a simple acidic or basic treatment of the final product to remove the catalytic metal and the support [32, 33].

Work carried out in the laboratory on the catalytic synthesis of carbon nanotubes using a mixture of C_2H_6 and H_2 and a Fe/Al_2O_3 catalyst led to the synthesis of several grams of carbon nanotubes per gram of catalyst per hour [33]. The quartz reactor tube was rapidly filled with a black fluffy solid after two hours of reaction indicating the extremely high activity of the catalyst to form carbon nanotubes (Fig. 7.2).

The selectivity towards carbon nanotubes was investigated by statistical SEM observation of the as-synthesized material. Representative SEM images are presented in Fig. 7.3A and B. The carbon nanotubes formed were extremely pure as no trace of other carbon nanoparticles was observed inside the sample. On the other hand, the carbon nanotubes were extremely homogeneous in diameter, around 40 nm, which is not the case for the carbon nanotubes formed by arc discharge. The average aspect ratio value experimentally determined from statistical SEM analysis was higher than 1000. Synthesis carried out at higher temperature, i.e. 800 °C instead of 680–750 °C, keeping other reaction conditions similar was less selective and led to the formation of carbon nanoparticles inside the sample.

High-magnification TEM images of the carbon nanotubes are presented in Fig. 7.3C and D. Close observation reveals that the tube wall has a relatively low crystal-

Fig. 7.2 Optical micrographs showing the reactor filled with a Fe/Al_2O_3 catalyst (1 g) and the same after growing carbon nanotubes under a mixture of C_2H_6/H_2 at 680 °C for 2 h. The carbon yield was about 6000 wt.% with respect to the iron catalyst weight.

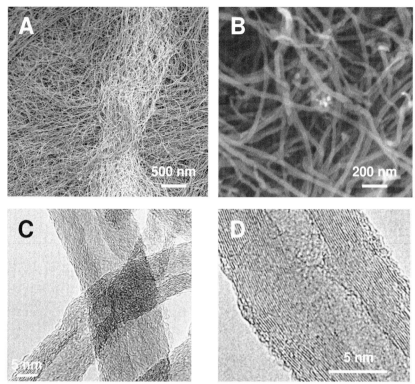

Fig. 7.3 (A, B) SEM micrographs of the carbon nanotubes synthesized via CVD of a mixture of C_2H_6/H_2 at 680 °C. Statistical SEM analysis confirms the absence of any carbon nanoparticles inside the sample which highlights the high selectivity of the synthesis method towards carbon nanotubes formation. (C, D) The corresponding TEM micrographs show the relatively homogeneous diameter of the carbon nanotubes along with defects in the graphene planes due to the low synthesis temperature.

linity and contains a high number of defects within the graphene plane. High resolution TEM micrographs also show the presence of a thin layer of amorphous carbon on the topmost surface of the tube. The amorphous carbon layer was expected to be formed during the cooling of the system [34]. This phenomenon could be avoided or significantly diminished by rapid evacuation of the reaction atmosphere at the end of the synthesis as reported by Emmenegger *et al.* [34]. It is also noteworthy that bamboo-like carbon tubes were completely absent from the sample. Ermakova *et al.* [35] have reported that the formation of these carbon impurities was strongly linked to the interaction between the active phase and the support. It is thought in our case that the interaction between the active phase and the support was relatively low to allow the avoidance of these bamboo-like carbon nanotubes.

TEM micrographs also reveal that some iron nanoparticles were located inside the carbon nanotubes. This is consistent with the proposed mechanism of reaction found in the literature. The authors think that during the carbon diffusion through the iron lattice an elevation of temperature occurs, causing oscillation of the metal lattice along with a temperature increase. As a result, the iron is transformed from the solid state into a quasi-liquid state, despite the fact that the synthesis temperature is far lower than the melting temperature of iron [35]. The quasi-liquid iron phase is slowly extracted during the CNTs growth and, remains in as an encapsulated iron nanoparticle within the carbon nanotube channel. This gradual encapsulation process occurring in the case of iron catalyst explains why the yield of carbon nanotubes is significantly lower than the yield of carbon nanofibers over nickel-based catalyst where only a very small amount of the nickel is transformed into a quasi-liquid nickel phase [36], leaving a large amount of nickel free for catalysis.

The porosity of the multi-walled carbon nanotubes was mainly mesoporous: an inner hollow channel with an average pore size ranging between 5 and 60 nm and aggregated pores ranging from 20 to 100 nm. No micropores are found in these carbon nanomaterials, contrary to what is observed in activated charcoal where generally half or one-third of the specific surface area comprised a micropore network [6–10].

7.2.2
Direct Macroscopic Shaping of Carbon Nanotubes

Processing carbon nanotubes with a controlled macroscopic shape to obtain materials with practical use is nowadays a major challenge in the field of advanced materials. The synthesis and the use of free nanotubes or fibers could be hazardous for the health of users and manipulators. The fluffy powder form can be shaped with the use of a binder but this does not solve the risk during the initial synthesis of the tubes. In addition the binder can block the accessibility of the channel, if this channel is to be used for different applications. The macroscopic shaping allows direct use of these 1D carbon materials in a conventional catalytic reactor which could further be optimized as an integrated process. The direct use of nanoscopic materials, even with exceptional performance, in conventional reactors is indeed not simple and generally impossible due to the problems linked with handling and pressure drop. Carbon nanotubes with controlled macroscopic shapes have been efficiently synthesized in the laboratory by modifying the shape of the reactor (constraint synthesis) or by modifying the nature of the carbon source and the catalyst (FeCp pyrolysis) or by patterning in a periodic structure the growth catalyst before contacting the gaseous carbon source. It is expected that the macroscopic shaping of such nanostructured materials will open up a real opportunity for their use as catalyst supports in competition with the traditional catalyst carriers such as alumina, silica or activated carbon.

7.2.2.1 Self-supported Carbon Nanotubes through Constraint Synthesis
In the case of constraint synthesis the catalyst consisted of 20 wt.%Fe/Al$_2$O$_3$, deposited inside a tubular reactor, where the two ends were closed with a permeable carbon felt allowing the gaseous reactants to pass through the reactor channel

Fig. 7.4 (A) Optical micrograph showing the improvement on the apparent density of the carbon nanotubes synthesized without and under constraint. The total weight of each sample was 7 g. (B) Examples of the macroscopic size of the self-supported carbon nanotubes obtained by varying the diameter of the reactor.

whilst at the same time confining the growth space of the carbon nanotubes [37, 38]. The carbon nanotubes synthesized under these conditions displayed a significantly higher density compared to those obtained without constraint, as shown by the optical image in Fig. 7.4A. The synthesis under constraint allowed the apparent density to significantly increase up to a value typically close to 200 kg m^{-3} (compared to 20 kg m^{-3} in the absence of confinement). SEM and TEM analysis confirm the similarity in the morphology and microstructure of the carbon nanotubes synthesized under constraint and those obtained in an open reactor. In the former case the carbon nanotubes were more dense due to the constraint of the reactor, in good agreement with the high density observed. The shape can be varied by changing the size of the reactor. Some examples are presented in Fig. 7.4B where cylinders of self-supported carbon nanotubes with different diameters are generated.

The TEM analysis of the MWNTs showed the complete absence of any amorphous carbon in the sample, confirming the high selectivity of the synthesis. Samples were prepared after dispersion of MWNTs in ethanol, then deposited on a copper grid and dried. The diameter of MWNTs typically ranged between 30 and 50 nm, with length reaching several micrometers. A relatively high number of structural defects was observed along the tube walls. These defects were attributed to the low synthesis temperature of the CVD method which could lead to the insertion of pentagons (positive curvature) or heptagons (negative curvature) inside the hexagon structure of the graphene layers. The statistical TEM observations indicated that most of the nanotubes were closed tips, even after NaOH and HNO$_3$ treatments. This was confirmed by the relatively stable value of the specific surface area before and after purification (typically between 150 and 200 m^{2}g^{-1}).

Fig. 7.5 Optical micrographs illustrating the isotropic shrinkage of the cylinder of self-supported multi-walled carbon nanotubes after wetting with water and slow drying in air at room temperature.

The great advantage of the present synthesis compared to those previousely reported in the literature [39–45] lies in the simplicity of the process which allows easy scaling-up with minimum cost investment and avoids adding foreign agents during the shaping. The density of the as-synthesized self-supported carbon nanotubes can be modified by wetting them with water or an organic solvent, followed by slow evaporation of the solvent. Surface tension effects during the liquid evaporation shrink the cylinder into a denser cylinder with smaller dimensions, as illustrated in Fig. 7.5. Similar results have also been reported by Smalley and coworkers [44] and by Zhang *et al.* [45] during their processing of macroscopic carbon nanotube objects. It is noteworthy that the piece of self-supported carbon nanotubes is strong enough to be easily manipulated after synthesis and cut to obtain the desired shape for downstream applications.

7.2.2.2 Parallel and Patterned Carbon Nanotubes by Pyrolysis of Organic Compounds

The synthesis of the patterned macroscopic shape consists in the growth of carbon nanotubes in a parallel direction from a flat substrate surface. This was achieved using a mixture of toluene containing ferrocene compounds or directly with ethylene and steam etching [46–49]. The carbon nanotubes were formed as a thick layer covering the silica reactor wall. The sample could easily be recovered by scratching the dark deposit with a razor blade.

In Fig. 7.6 are presented the typical SEM images with different magnifications of the carbon nanotubes forest generated via the pyrolysis of ferrocene in a toluene flow realized in the laboratory.

The synthesis was carried out at 850 °C with a flow rate of 100 ml min^{-1} of toluene and 5 vol.% ferrocene. The ferrocene was located at the entrance of the reactor in a low temperature zone, ca. 500 °C. The macroscopic pieces of carbon nanotubes were recovered from the quartz substrate located in the hottest zone of the reactor. According to the SEM observation the thickness of the aligned carbon nanotubes forest was relatively high, approaching almost 250 µm, while the diameter was relatively homogeneous at around 50 nm. The height of the carbon nanotubes

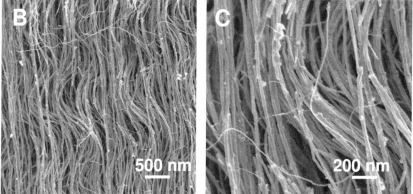

Fig. 7.6 (A) Sheet of parallel carbon nanotubes forest with a height of 250 μm. The height of the carbon nanotubes forest can be easily controlled from ca. 100 μm up to 500 μm. (B, C) High-resolution SEM images showing the details morphology of the patterned carbon nanotubes with a homogeneous diameter centered at around 50 nm. The aspect ratio value of these patterned carbon nanotubes was around 5000.

carpet can also be significantly raised by increasing the synthesis time, i.e. >1 mm for 6 h of synthesis. The as-synthesized carbon nanotubes pattern can be recovered from the reactor as a self-supporting plate or cylinder by dissolving the silica reactor in a diluted HF solution.

SEM images of the top and the bottom of the CNTs carpet are show in Fig. 7.7. The carbon nanotubes at the top of the carpet were well aligned and almost no nanoparticles were detected, which highlighted the selectivity of the synthesis method to form carbon nanotubes (Fig. 7.7A). The bottom was less pure and a thin layer of amorphous compounds was observed (Fig. 7.7B).

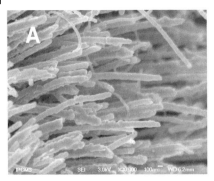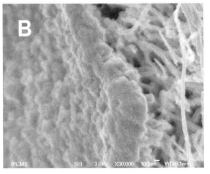

Fig. 7.7 SEM micrographs showing the morphology of the carbon nanotubes carpet generated by pyrolysis of a ferrocene and toluene mixture at 800 °C. **A**: top view, **B**: bottom view.

TEM analysis indicates that the tubes formed have open ends with a relatively small inner channel, i.e. 6 nm (Fig. 7.8A). High resolution TEM images (Fig. 7.8B and C) show that the graphene layers forming the tube walls were extremely well crystallized, which can be attributed to the relatively high synthesis temperature and the use of toluene instead of linear hydrocarbons.

In some tubes an encapsulated iron phase was also observed (Fig. 7.8C). Indeed, the iron particle was well encapsulated by several layers of graphene and thus, was not accessible for the subsequent catalytic use of the material. Some amorphous phase was also detected on the upper surface layer of the sample which could be due to the cooling period of the synthesis where the reaction temperature was still high enough to decompose the gaseous hydrocarbon but not sufficiently high to allow the segregation of ordered solid carbon. Similar observations have been reported by other groups [34, 36].

7.2.2.3 Carbon Nanotubes Growth on Periodically Patterned Structure

Silicon nanocrystals(Si-ncs), coated with iron, were introduced by spin coating into the periodical holes of a Si(100) substrate [n-type Si(100) heavily doped (Sb) with an electrical resistivity of $3 \text{ m}\Omega \cdot \text{cm}$] created by a lithographic process which is described elsewhere [50]. The holes have a diameter of 7 µm, a thickness of 300 nm and a mean hole distance of 3 µm). The clusters of Fe/Si-ncs introduced into these Si grooves were further dried at 100 °C and calcined in air at 350 °C for 2 h. The flat piece of Si(100)-based sample was fixed vertically between quartz wool in the growth reactor. The catalyst was reduced *in situ* in flowing hydrogen at 400 °C for 1 h. The carbon nanotubes deposition was conducted at 800 °C in a mixture of C_2H_6/H_2 for 5 min. The representative SEM micrographs of the final sample are shown in Fig. 7.9. The formation of discrete bundles of carbon nanotubes separated from each other on the Si-based substrate is observed. During the growth process the Si-ncs support was extracted from the Si substrate and remain stuck on the tip of the CNTs (Fig. 7.9C). A more detailed description of these hybrid nanostructures has been reported elsewhere [51, 52]. One can speculate that such

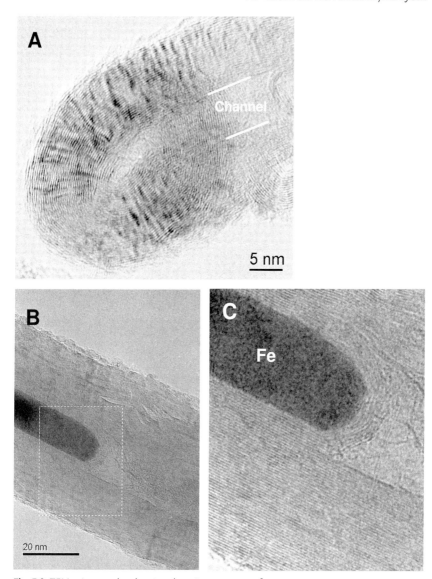

Fig. 7.8 TEM micrographs showing the microstructure of the carbon nanotubes synthesized by ferrocene pyrolysis. (A) Tube with open end, (B, C) high magnification showing the presence of straight graphene planes in the tube wall. High magnification TEM images show the presence of some encapsulated iron particles inside the nanotube channel.

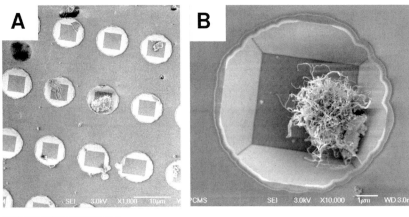

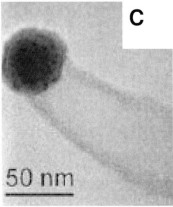

Fig. 7.9 (A, B) SEM micrographs with different magnification showing the presence of a separate bunch of carbon nanotubes located inside the Si holes. (C) TEM micrograph showing the presence of a Si nanocrystal on the tip of the formed carbon nanotube.

structures could be useful as catalysts through photon absorption and electrical transfer from the Si-ncs to the conductive carbon nanotubes.

7.2.3
Structural Ordering Assisted by Thermal Treatment

Carbon nanotubes synthesized by the catalytic route exhibit a less well-defined graphene structure compared to those formed by the arc-discharge method due to the low synthesis temperature. This can be rectified by submitting these nanotubes to a high temperature treatment in the range of 1800 to 2600 °C under an inert atmosphere. The heat-treated carbon nanotubes exhibit a more ordered graphene structure compared to their counterpart obtained after the low-temperature synthesis via catalytic route (Fig. 7.10A and B).

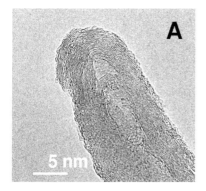
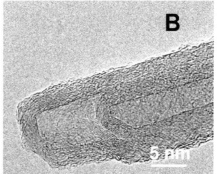

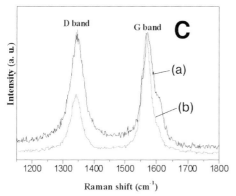

Fig. 7.10 High-resolution TEM images of multi-walled carbon nanotubes: (A) After synthesis at 680 °C and (B) after heat treatment at 2600 °C in an argon atmosphere for 2 h. The higher graphitization degree of the graphene planes is clearly visible on the heat-treated sample. (C) Raman spectra of the carbon nanotubes before (a) and after (b) heat treatment showing the significant increase in the I_G/I_D band ratio.

The TEM images clearly indicate that the graphene planes containing defects of the as-synthesized carbon nanotube were transformed after heat treatment into more ordered graphene planes, without any observable modification of the shape of the carbon nanotube. The more ordered structure of the heat-treated carbon nanotubes was also confirmed by comparing the Raman spectra of the carbon nanotubes before and after heat treatment (Fig. 7.10C). The G band at 1580 cm^{-1} is attributed to the graphite E_{2g} symmetry and corresponds to the structural intensity of the sp^2-hybridized carbon atoms of the nanotube. The second Raman peak named the D band, at 1350 cm^{-1}, shows the presence of disordered carbon atoms. The area ratio (I_G/I_D) increases after the thermal treatment, indicating a decrease in disordered carbon in the MWNTs structure. The heat treatment also allows the complete removal of oxygenated species on the topmost surface of the tube, according to the XPS analysis, as a function of the heat treatment temperatures (not shown). Apparently, post-synthesis heat treatment could be efficiently employed to remove structural defects in the tube along with a significant modification of the surface nature through the removal of oxygenated groups.

Another problem linked with the subsequent use of these carbon nanotubes is finding an effective method to open their tips. In general, single-walled carbon nanotubes and small multi-walled carbon nanotubes were formed with a closed tip which renders useless the exploration of the properties of metal nanoparticles catalyst located inside their channel. Several methods have been proposed in the literature to selectively open the tube, i.e. acidic treatments [53–56] or oxidation in air or oxygen at temperatures ranging between 600 and 700 °C [57, 58]. According to these methods, the tips of the tubes incorporating reactive five-membered carbon rings have a higher curvature and higher strain than the rest of the tubes and thus should be preferentially attacked, leading to the opening of the tubes [59].

7.2.4
SiC Nanotubes and Nanofibers

The homologue 1D ceramic material, i.e. SiC nanotubes or nanofibers, can be efficiently prepared by reacting carbon nanotubes or nanofibers with SiO vapor at relatively low temperature, 1200 to 1300 °C [60–63]. The starting solid carbon source can be either randomly bulk 1D carbon or patterned 1D carbon grown by CVD using a porous alumina template as mold. The carburization proceeded through an atomic replacement of C by Si with a concomitant release of CO as gaseous product. The CO formed was actively pumped from the reaction zone allowing the rate of carbide formation to increase.

During the reaction the starting carbon nanomaterials can be transformed into SiC nanotubes or SiC nanofibers, depending on the conditions and also on the diameter of the starting carbon source. The microstructure of the as-synthesized SiC nanofibers observed by SEM is presented in Fig. 7.11A and B. The SiC nanofiber was built by stacking several SiC nanoparticles (ca. 20 to 30 nm in diameter) along the fiber axis. The smooth surface previously observed for the carbon nanofibers was no longer observed but was replaced by a more tortuous morphology consisting of a stacking of several SiC nanoparticles along the fiber axis giving a rough surface (Fig. 7.11B). The loss of carbon atoms as CO gas along with the density change when going from C to SiC was probably responsible for the formation of the rough surface. The purity of the sample was confirmed by powder X-ray diffraction which only exhibited the diffraction lines corresponding to the β-SiC phase, crystallized in a face centered cubic structure.

The high-resolution TEM micrograph reveals the presence of a high density of stacking faults along the SiC fiber and also the existence of a thin amorphous layer covering part of the fiber surface (Fig. 7.11C). XPS analysis indicated that the thin amorphous layer was a mixture of SiO_2 and $SiOxCy$ phases [64].

The 1D SiC material generally displayed a diameter bigger than the starting carbon parent. This could be attributed to the epitaxial post-growth gas-phase

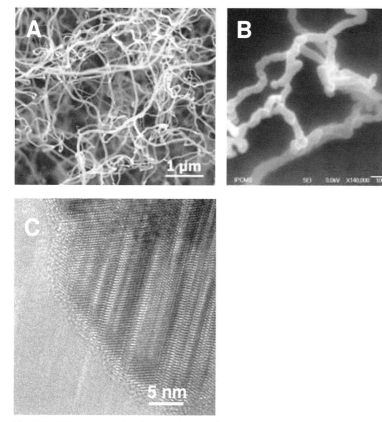

Fig. 7.11 SEM micrographs showing the complete microstructure retention when going from carbon nanofibers (A) to SiC nanofibers (B). The high-resolution TEM micrograph of the 1D SiC is presented in (C) and reveals the presence of a relatively high density of stacking faults along the growth axis and a thin amorphous layer on the surface.

mechanism on the pristine surface of the SiC nanowires according to the following chemical reaction: SiO (gas) + 3 CO (gas) → SiC (solid) + 2 CO$_2$ (gas). A similar observation has already been reported by Hu *et al.* [65] during the gas–solid reaction between the multi-walled carbon nanotubes and SiO vapor.

7.3
One-dimensional Conductive Materials for Catalysis

Catalytic sequences generally occur at an atomic scale on the defect sites, i.e. terrace sites, edge sites and kink sites, of the active phase nanoparticles. Often the true active sites are generated during the catalytic reaction by surface modification in the presence of the reactants and products. In these cases, the surface nature of the active sites can be different from the related crystalline bulk, and can make the real interpretation of the catalytic performance difficult. However, when the active phase is in strong interaction with the support, i.e. alumina or silica, the surface modification of the active phase during the catalytic reaction can be significantly modified due to the existing interaction.

In the case of an exothermic reaction, i.e. Fischer–Tropsch synthesis or catalytic partial oxidation of methane, the security of the whole system can be drastically compromised due to temperature runaway. One strategy to get rid of this problem is to prepare a catalyst where the active sites are deposited on a support with high chemical inertness and high thermal conductivity. The high chemical inertness allows one to ease the active phase surface modification during the catalytic reaction, while the thermal conductivity plays the role of a heat well to rapidly disperse the heat from a single site throughout the entire matrix of the support avoiding hot spots formation, in the case of exothermic reactions, whereas for the case of endothermic reactions, the heat conductivity of the support eases the supply of heat from the external oven to the active sites.

It is expected that these 1D materials (carbon and/or carbide) will be extensively employed in numerous fields in the near future, from energy to healthcare to environment. Some examples of the catalytic application of these 1D materials realized in the laboratory are detailed below. For the details of the numerous reactions that involve 1D materials the reader can refer to the recent reviews published in this field [5–10].

7.3.1
Carbon Nanotubes Containing Nanoparticle Catalysts

Carbon nanotubes supporting catalysts have been employed in several catalytic reaction processes and some examples are presented below to illustrate the peculiar catalytic properties of these catalysts. The catalytic performance is compared with that obtained on state-of-the-art traditional catalysts.

7.3.1.1   Pd/CNTs Catalyst Characteristics

The catalyst consisted of homogeneous palladium nanoparticles dispersed inside the carbon nanotubes. The palladium was introduced into the carbon nanotubes support by means of incipient wetness impregnation using an aqueous solution containing palladium nitrate. After impregnation, the wet solid was allowed to dry at room temperature and then at 100 °C for 2 h. The solid was further reduced in flowing hydrogen at 400 °C for 2 h. TEM analysis revealed that the palladium particles were relatively well dispersed on the carbon nanotube surface with an average particle size of around 5 nm (Fig. 7.12). Statistical TEM analysis only revealed a small fraction of palladium particles located on the outer surface of the carbon nanotubes which indicated the relatively high selectivity of filling.

The preferential location of the metal particles inside the tubes could be explained by the fact that during the impregnation process, the liquid completely filled the

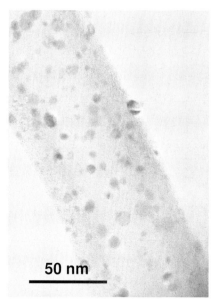

Fig. 7.12 TEM micrograph showing the location of palladium particles inside the carbon nanotube channel.

inner part of the tube by capilarity [66]. Results reported in the literature have shown that CNTs could easily be filled with low surface tension liquids, i.e. <190 mN m^{-1} [67]. Water, with a surface tension of 72 mN m^{-1}, would thus wet and rapidly fill the inner part of the tube. The filling is also strongly favored by the presence of surface oxygenated groups which render the tube surface more hydrophilic.

The ability of the low surface tension liquid to fill the carbon nanotubes also depends strongly on the tube diameter according to work reported by Ugarte *et al.* [68]: the smaller the inner channel, the lower the filling efficiency. Ma *et al.* [69] have reported filling results with platinum on carbon nanotube with two different inner diameters, i.e. <10 nm and 60–100 nm, by incipient wetness filling of an alcoholic solution of platinum salt. They observed that a complete filling was achieved with nanotubes having an inner diameter of 60–100 nm whereas no filling was observed with a tube having smaller inner diameter, i.e. <10 nm.

However, it is noteworthy that because the classical TEM image is only a 2D projection of the entire sample, one cannot attribute with certitude the exact location of the palladium particles with respect to the carbon nanotube surface. The resolution of the individual metal particles was also difficult to solve due to the projection of a relatively thick 3D sample on a planar surface.

The precise location of the palladium nanoparticles was further evaluated by a TEM 3D technique (tomogram) which is a promising technique due to its high resolution level which is not accessible by other 3D methods, such as X-ray/ XANES tomography, SIMS or microtome/AEM sectioning methods. It is noteworthy that the electron tomogram (or 3D TEM) was extensively employed by de Jong and coworkers [70, 71] to characterize the morphology and location of small metal particles inside a zeolite or mesoporous network whereas such study devoted to metal particles encapsulated inside the carbon nanotubes tubule has not yet been reported.

The tilt series was acquired in bright-field mode on a F20 TECNAI microscope (FEI Company) using an accelerated voltage of 200 kV and a high tilt sample holder. Several series of 2D-TEM images were acquired using a 2048 × 2048 pixel cooled CCD array detector, in different places of the grid containing typical open-ended nanotubes. Each series of projections was collected over a tilt range of −65° to 65°, with an image recorded every 2° between −40° and 40° and every 1° elsewhere, giving a total of 91 images. During the acquisitions, the TEM parameters such as defocus, horizontal specimen shift and specimen tilt were controlled automatically. No evidence of irradiation damage in the sample was observed during the acquisition.

3D TEM slices along the tube thickness are presented in Fig. 7.13 which allows one to accurately observe the location of the palladium particles with respect to the tube morphology. The presence of some amorphous-like phase inside the channel was attributed to the ethanol used for the preparation of the sample.

The volume reconstructed sample also allows observation of the distribution of the palladium particles in the plane section along the tube axis, as shown in Fig. 7.14. These possibilities allow one to get more insight into the location of the metal or oxide particles with respect to the tube inner diameter and could be an extremely useful technique for determining the direct relationship between the location of the active phase on or in the carbon nanotube and its catalytic activity and/or selectivity.

The digital 3D images reconstruction of the sample is presented in Fig. 7.15 which clearly shows the location of the palladium particles along the tube axis. It is noteworthy that, in the case of large inner diameter, i.e. 30 nm, almost 60% of the palladium particles were located inside the channel. The high filling was attributed to the large inner diameter which favors the penetration of the fluid containing the salt precursor inside the channel by capillarity forces.

According to the reconstructed image, the Pd particles were extremely homogeneous in size, regardless of their location with respect to the tube surface. This indicates that the interaction between the tube surface, either outer or inner, and the deposited metal mainly originated from the π interaction and the presence of the oxygenated surface groups, which render the tube surface more hydrophilic according to the results reported in the literature [72, 73]. Similar results have also been reported by Winter *et al.* [72] during their study of the filling and particle size distribution of Pt and Co on the hollow carbon nanofibers support. Literature data reports the relatively low wetting and absorption of water on glassy carbon which is heat treated at 2000 °C [74]. Gogotsi and coworkers [73] have observed that the surface of as-synthesized carbon nanotubes with high hydrophobic character was not able to attract water inside the nanotube channel because the low

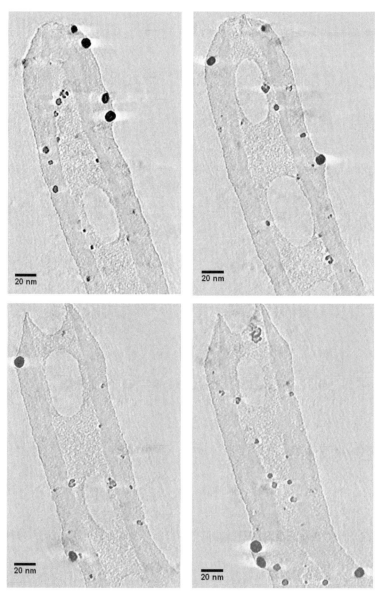

Fig. 7.13 Transverse sections along the carbon nanotube thickness from the reconstructed volume showing the location of palladium particles on the same plane as that of the carbon nanotube. The round-shaped particles located on the topmost surface of the liquid layer covering the nanotube with an average diameter of around 10 nm are the gold particles deposited before 3D TEM observation. The gold particles position allows one to perform a least-squares tracking procedure to the re-alignment of the different TEM images taken at different angular positions. These gold markers are picked or automatically located on the pre-aligned images and thus the program tests theirs respective positions and allows the changes in the values of tilt angles and tilt axis position for each image. This test is crucial and has to be run many times in order to obtain a 2D image series well aligned with corrected tilt angles for the subsequent volume reconstruction.

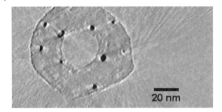

Fig. 7.14 Planar TEM view along the tube axis deduced from the volume reconstructed image of the 3D TEM.

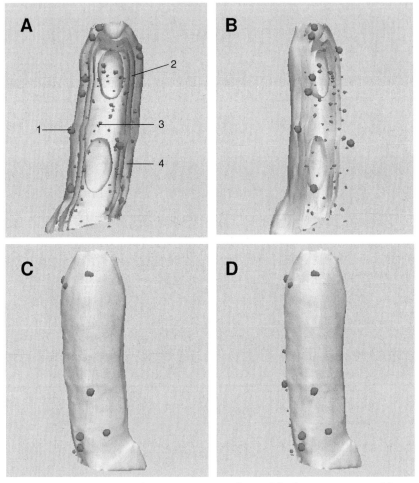

Fig. 7.15 3D TEM reconstructed images showing the exclusive location of palladium particles inside the carbon nanotube channel. The palladium nanoparticles have an average size of 1 to 2 nm. The round-shaped particles with larger diameter, i.e. 5 nm, are the gold particles used for the reconstruction of the tomogram. 1. Gold nanoparticles used as markers, 2. Carbon nanotuber walls, 3. Pd nanoparticles, 4. Residue species from ethanol.

wetting of the surface by water results in a high contact angle. The same tubes after treatment in NaOH were readily filled by water due to the increase in the surface hydrophilicity. A similar observation has also been reported by Winter *et al.* [72] in the case of cobalt active phase deposited on hollow carbon nanofibers after synthesis and without subsequent acidic treatment to create surface oxygenated groups.

7.3.1.2 Hydrogenation of the C=C Bond in the Liquid Phase

The liquid-phase selective hydrogenation of the C=C bond in α,β-unsaturated cinnamaldehyde was studied to show the benefit of the use of the CNTs support versus the traditional powder activated charcoal-based catalyst [75, 76]. The catalyst consisted of homogeneous palladium nanoparticles dispersed inside the carbon nanotubes. The characteristics of the Pd-based catalyst have already been detailed above.

The reaction was carried out at 80 °C. The catalyst was suspended in a dioxane solution kept in a continuously stirred reactor while the hydrogen supply was bubbled into the liquid. The hydrogenation performance (expressed in terms of conversion) and the selectivity of the Pd/SiC nanotubes catalyst are presented in Table 7.1 and compared with those obtained on a commercial Pd/AC (activated charcoal).

The carbon-nanotubes-based catalyst exhibits a higher hydrogenation conversion than that observed on the activated charcoal-based catalyst under similar reaction conditions. According to the TEM analysis, which did not show any significant differences in the metal particles size distribution, the higher hydrogenation observed on the CNT-based catalyst was explained by the high surface-to-volume ratio compared to that of the AC-based catalyst due to the lower mass transfer phenomenon, especially in the liquid medium. According to the inner diameter of the carbon nanotubes channel, i.e. 40 to 60 nm, which mostly lies in the mesopore range, one should expect the absence of any diffusion problems which are generally linked with activated charcoal micropores, i.e. <5 nm. As discussed above the carbon nanotubes can be efficiently filled by capillarity force when the solvent has a low surface tension, i.e. <200 mN m^{-1}. The high capillarity forces existing inside the mesopores and the open ended carbon nanotubes could also greatly favor the penetration of fluid into the channel. Work carried out on a palladium

Table 7.1 Cinnamaldehyde conversion and the C=C bond hydrogenated product yield as a function of time on stream on the Pd/CNT and Pd/AC catalysts.

Catalyst	Pd/CNT			Pd/AC		
Time on stream (h)	1	18	32	1	12	39
Conversion (%)	4	44	99	2	30	95
Yield for C=C hydrogenated bond (%)	4	34	88	1	18	50

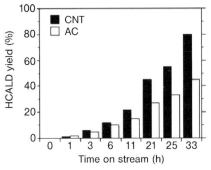

Fig. 7.16 Yield of the C=C bond hydrogenation product (HCALD) over the Pd/CNT and Pd/AC catalysts.

supported carbon nanotubes catalyst with extremely small inner diameter, i.e. <10 nm, and thus, with all the palladium particles located on the outer surface of the tube, for the liquid-phase hydrogenation of cinnamaldehyde exhibits a slightly lower hydrogenation activity. Such a difference could be attributed to the strong interaction between the graphite surface and the deposited metal active phase inside the channel which in turn would modify the hydrogenation activity. The behavior of dissolved hydrogen inside the carbon nanotube channel could also play a role in such activity.

A large difference was observed in the product selectivity, i.e. almost exclusively C=C hydrogenation on the CNT-based catalyst whereas both saturated alcohol (corresponding to the hydrogenation of C=O and C=C bonds) and aldehyde were observed in similar amounts on the AC-based catalyst (Fig. 7.16). The absence of microporosity and the high mass transfer rate in the CNT-based catalyst were probably responsible for the significant improvement of the C=C bond hydrogenation selectivity. In the AC-based catalyst the large amount of micropores could lead to diffusion problems and successive hydrogenation giving the saturated alcohol. The presence of impurities such as sulfur, nitrogen, oxygen or chloride on the activated charcoal surface could also be responsible for the difference in selectivity. Similar results have also been reported elsewhere [9] during the hydroformylation of 1-hexene over Rh-based catalyst where the MWNT catalyst exhibits higher catalytic performance compared to its homolog supported on the high specific surface area activated charcoal. Again, the lower mass transfer process due to the microporous character of the activated charcoal was advanced to explain the observed results.

7.3.1.3 Selective Hydrogenation of Nitrobenzene to Aniline in the Liquid Phase

In this reaction the catalyst consisted of 5 wt.% Pd/MWNTs with 60% of the palladium particles being located inside the carbon nanotube channel (detailed characterization was presented in Section 7.3.1.1). Details concerning the catalyst preparation can be found in Section 7.3.1.1. The catalytic activity obtained in a batch mode at room temperature is presented in Table 7.2. The results obtained

Table 7.2 Liquid-phase hydrogenation of nitrobenzene over Pd/CNT and Pd/AC catalysts at room temperature and under 20 bar total pressure of hydrogen. The initial nitrobenzene concentration in dioxane was 2 mol l^{-1}.

Catalyst	Pd/MWNT				Pd/AC			
Time (min)	100	270	350	450	100	270	350	450
Conversion (%)	32	72	85	93	22	61	70	75

Table 7.3 Oxidative dehydrogenation of 9,10-dihydroanthracene ito anthracene in the liquid phase over carbon-based catalysts.

Nature of the carbon catalyst	Reaction time (h)	Anthracene (Yield (%))	9,10-Anthracenediquinone (Yield (%))
CNTs as synthesized	120	90	Traces
CNTs treated at 2600 °C	24	99	Traces
Exfoliated carbon	170	50	Traces

are compared with those of the commercial Pd/activated charcoal catalyst operating under the same reaction conditions. The MWNT-based catalyst exhibits a higher conversion than the commercial catalyst, especially when the initial concentration of nitrobenzene inside the reactor is increased. The high hydrogenation activity observed on the MWNT-based catalyst could be attributed to the small, nanoscopic, size of the catalyst, and the complete absence of micropores which provided an efficient contact surface between the reactants and the active phase, particularly in a liquid medium where the mass transfer phenomenon becomes very significant. In the case of the activated charcoal-based catalyst the large part of the active phase located inside the micropores was probably not accessible, resulting in a lower activity despite the large difference observed in the specific surface area between the two catalysts, i.e. 900 m^2 g^{-1} for the activated charcoal and 10 m^2 g^{-1} for the carbon nanotubes catalyst.

7.3.1.4 Selective Oxidative Dehydrogenation (ODH) of Dihydroanthracene to Anthracene

In this experiment the metal-free carbon nanotubes acted directly as a support and active phase for the liquid-phase ODH of dihydroanthracene to anthracene at low temperature. The results are compared with those obtained on other catalysts. Oxidative dehydrogenation of dihydroanthracene in the liquid-phase was carried out over different CNT-based materials. The results are presented in Table 7.3. The test was also performed on an exfoliated graphite material for comparison. The results clearly showed that the reaction was chemically selective whatever catalyst was used: only traces of the diquinone were formed. The catalytic dehydrogenation activity was relatively low on the as-synthesised carbon nanotubes. The same catalyst after heat treatment at high temperature exhibits a huge dehydrogenation activity increase, i.e. 99% instead of 30% for the same period of test (24 h) which indicated a positive effect of the graphitization on the dehydrogenation activity. The catalytic activity improvement when going from low graphitized carbon nanotubes to a higher graphitization level could be explained by better adsorption of the reactants on the catalyst surface and desorption of the products from the active site.

The improvement in the dehydrogenation activity could also be attributed to the extremely low oxygen concentration on the heat treated carbon-based catalyst, as shown by XPS measurements on the sample treated at different temperatures. It was thought that the presence of oxygenated groups on the carbon nanotubes surface could modify greatly the adsorption strength and nature of the reactants, which in turn could have an influence on the overall catalytic activity. A previous study has demonstrated that the presence of a large amount of oxygen-containing functional groups prevents the transport of molecules on the carbon nanotubes surface [77]. During the course of the reaction, part of the carbon nanotubes was covered by the oxygen present in the liquid-phase medium. However it was expected that such oxygen content was much lower than that present on the as-synthesised carbon nanotubes. The better catalytic activity observed on the heat-treated carbon nanotubes versus the as-synthesised material could be due to enhancement of the the adsorption abilities of the heat-treated nanotubes due to the lower adsorbed oxygen on their surface, which could increase the surface density of the reactants, and thus increase the conversion. It was shown that both O_2 and aromatic compounds (like anthracene) can easily be adsorbed on the carbon nanotubes surface with an adsorption energy of 18.5 kJ mol^{-1} for O_2 [78] and 9.4 kJ mol^{-1} for C_6H_6 [79]. The adsorption of aromatic molecules is favored by the delocalization and hybridization of π electrons between the nanotubes and the adsorbed molecules. Previous studies on the ODH of ethylbenzene to styrene over various kinds of carbon-based materials have shown that, during the reaction, part of the catalyst's surface, i.e. the defect surface, plays the role of oxygen adsorption site and thus generates strong basic surface oxygen groups such as quinoidic groups [80] which participate in the ODH reaction. The following mechanism was advanced to explain the ODH sequence on the carbon-based catalysts: the hydrocarbon was first adsorbed on the available surface next to the basic oxygen groups and underwent dehydrogenation with concomitant formation of surface OH groups [81, 82], the basic quinoidic groups were generated by reacting the surface OH groups with the dissolved oxygen with a release of water into the reaction medium (Fig. 7.17). The increase in the catalytic ODH activity could also be linked to the more ordered structure of the carbon catalyst after heat treatment. It has been observed in previous studies that disordered carbon was not active for the ODH reaction, and only well-organized graphite planes with adsorbed oxygen groups were active.

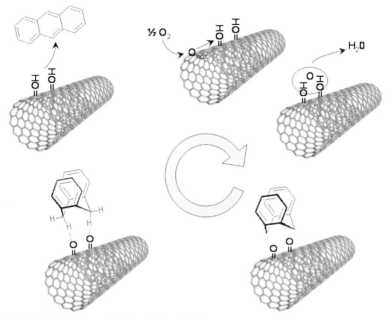

Fig. 7.17 Tentative mechanism of the oxidative dehydrogenation of dihydroanthracene to anthracene over heat-treated carbon nanotubes.

On the as-grown CNTs the carbon surface is partly decorated with numerous oxygen groups such as carboxylic groups (–COOH) and hydroxy groups (–OH). These oxygen groups represent high reactive centers for adsorption of water molecules. In turn, the adsorbed water molecules become secondary adsorption centers for other water molecules by means of hydrogen bonds [83]. In the presence of adsorbed water the adsorption ability of the CNT surface for other molecules could be significantly affected due to the steric hindrance. Adsorbed water molecules lower the access to the adsorption sites and weaken the interactions between the carbonyl groups and organic molecules (donor–acceptor complex formation) which again lowers the surface activity of the carbon nanotubes. The increase in the dehydrogenation activity when going from the as-grown CNTs to the graphitized CNTs was attributed to the reduction in the number of oxygen groups on the CNT surface, in agreement with the XPS results not reported here.

The catalytic activity on untreated or treated MWNTs could be due to a curvature effect, even if it is very weak because of the relatively large diameter of the tubes (30–50 nm). This is confirmed by the lower catalytic activity observed on exfoliated carbon, which has not the same morphologic properties (absence of curvature). Similar results have been obtained by Nakamichi *et al.* [84]. They observed a conversion reaching 93% over activated charcoal in the same reaction conditions.

Nevertheless, the catalytic tests were carried out at 120 °C and the activated charcoal used had a very high specific surface area (1500 m^2 g^{-1}), compared to the MWNTs (130 m^2 g^{-1}), mainly constituted by micropores. However, the micropores can absorb the reactants and the products, and thus lower the conversion. The use of MWNTs as catalysts avoids this drawback, because of the lower specific surface area and absorption capability. It is also possible that the adsorption strength of organic molecules becomes weaker on the graphitized surface because of the reduction in the number of defects with high adsorption energy. This would favor the desorption of the product, leading to a higher turn-over per active site. The high amount of defects on the as-grown CNTs could more strongly retain the product molecules which would inhibit the turn-over frequency of the active site. The peculiar interaction between the adsorbed molecules and the defects on the tube wall could also modify to a great extent their migration or desorption. A similar observation has been also reported by Ulbricht *et al.* [85]. Such interaction should be reduced on the graphitized CNTs surface.

7.3.2
SiC Nanotubes Containing Nanoparticles Catalysts

7.3.2.1 Selective Oxidation of H$_2$S to Elemental Sulfur in a Trickle-bed

The catalytic selective oxidation of H$_2$S to elemental sulfur has been extensively studied in order to reduce the release of this toxic and corrosive compound into the atmosphere [86–88]. The reaction should be carried out at relatively low temperature, 40–60 °C, in order to increase the selectivity towards solid sulfur instead of SO$_2$, and relatively high space velocity to also avoid SO$_2$ formation by a too long time residence of the sulfur produced by the reaction. This catalytic process is useful in the ultimate removal of trace amounts of H$_2$S from low-temperature acid gas from a stationary source. The use of SiC as a support for the NiS$_2$ active phase has been a great improvement in this process compared to conventional alumina or alumina-doped supports [89, 90]. The catalytic performance of the NiS$_2$/SiC nanotubes catalyst is compared with that of the NiS$_2$/SiC in a grains form and also to the NiS$_2$/CNTs, in order to highlight the advantage of using 1D material as catalyst support instead of a traditional macroscopic shapes support.

The catalytic results obtained at 60 °C and with a weight hourly space velocity (WHSV) of 0.02 h^{-1} are presented in Fig. 7.18. At high WHSV the conventional catalyst supported on SiC grains was rapidly deactivated within a few hours. The deactivation was attributed to the fact that such a high space velocity decreases or even inhibits the transformation of H$_2$S into elemental sulfur according to the kinetic laws. In addition, at the low reaction temperature tested here the solid sulfur formed from the reaction was steadily deposited on the catalyst surface, which in turn inhibited the accessibility of the reactants to the active sites. The high catalytic performance of the NiS$_2$/SiC-nanotubes could be attributed to the high external surface area of the support which provides high accessibility of the reactants to the active sites and to the presence of a tubule inside the material which artificially increased the reactants partial pressure, i.e. nanocondensation

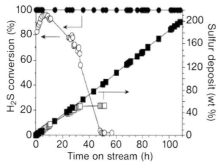

Fig. 7.18 (A) H₂S conversion and the corresponding solid sulfur deposit on the surface of the NiS₂/SiC-grains (open symbols) and NiS₂/SiC-nanotubes (filled symbols) catalysts as a function of time on stream at 60 °C. The sulfur selectivity at 60 °C was the total on the two catalysts. (B) SEM micrographs of the SiC nanotubes supported NiS₂ catalyst after more than 100 h of reaction with a solid sulfur loading of 200 wt.%.

or nanocapillarity, during their travel through the tube. On the CNTs-based catalyst the amount of solid sulfur deposited on the catalyst surface after about 100 h was almost 200 wt.% whereas no deactivation by active site encapsulation was observed, indicating that the solid sulfur was progressively removed from the active site during the course of the reaction [91]. The improvement in the solid sulfur storage capacity of the CNTs-based catalyst was attributed to the increase in the free volume of the nanotube material as compared to the traditional grain size material.

Experiments carried out on the same SiC nanofibers show similar catalytic results. The SEM observation of the sample after the desulfurization test shows the presence of inhomogeneous large solid sulfur particles wrapping the SiC structure where a large part of the catalyst surface was still accessible (Fig. 7.19). The remaining free surface of the catalyst allows one to explain the high desulfurization activity despite the relatively high solid sulfur deposition on the catalyst surface.

7.4
Conclusion

In summary, the catalytic route provides an interesting way to prepare carbon nanotubes with high yield and selectivity at relatively low synthesis temperature. The easy scale-up of the catalytic method allows one to ensure the mass production of these 1D materials with reasonable cost for large scale applications. The development of different methods to produce carbon nanotubes in a controlled macroscopic shape and size, i.e. constraint synthesis, patterned and aligned carbon nanotubes forest by pyrolysis of organic compounds, avoids the formation of fines

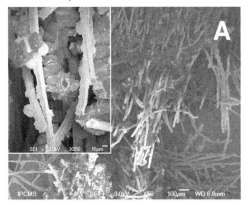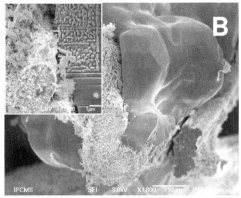

Fig. 7.19 SEM micrographs of the NiS$_2$/SiC nanofibers composite after the desulfurization test. The solid sulfur formed during the test was present in the form of large particles on the surface of the catalyst.

and favors easy handling and practical use in conventional catalytic reactors. The as-synthesized carbon nanotubes, in bulk or supported forms, can be effectively transformed into SiC which display higher mechanical and oxidative resistance for subsequent use in an oxidative medium.

Metal nanoparticles can be easily introduced into the carbon nanotubes tubule or onto the surface of carbon nanotubes and SiC nanotubes/nanofibers by a simple wetness impregnation followed by classical thermal treatments. It is thought that such new nanoscopic encapsulated materials will display unusual properties as compared to those of their unfilled counterparts and could open new opportunities in the catalysis field.

In hydrogenation or dehydrogenation reactions, especially in the liquid phase, the carbon nanostructured material exhibits an interesting catalytic behavior compared to that encountered with a commercial activated charcoal-based catalyst, despite the large difference in terms of the support specific surface area. In some new developments carbon nanotubes, free of metal-based active phase, can be effectively employed. The high selectivity obtained over the CNT-based catalysts was attributed to the complete absence of micropores and residual acidic sites in the support matrix. The great advantage of the direct use of carbon nanotubes without additional metal active phase is the reduction in cost of the catalyst along with the lack of environmental concerns on the disposal and recovery of waste linked with the leaching of the metal. The possibility to synthesize these 1D materials with a macroscopic shape also significantly improves their use in conventional reactors by facilitating the handling and recovery.

The low-temperature selective oxidation of H$_2$S coming from the Claus plant unit using SiC nanotubes/nanofibers is probably not the most appropriate process for a potential industrial application for the next years to come. However, the results obtained allow one to get more insight about the confinement effect operating inside the channel of these 1D materials which could be useful for other applications in catalysis.

7.5
Outlook

The different catalytic applications presented above confirm the potential of carbon nanomaterials to be used as catalyst supports, either in the gas phase or the liquid phase, in place of the traditional macroscopic supports. We believe that our understanding concerning the role of the carbon nanofiber surface on the catalytic performance represents the most simple but reasonable starting point, although future work is undoubtedly required to quantify and allow a better understanding of the advantages that these carbon nanomaterials can bring to the catalysis field. The most promising catalytic field for these 1D carbon and carbide materials is expected to be where a lot of catalytic processes still need to be improved in order to meet the increasingly strict environmental legislation and also to reduce as much as possible the generation of waste. It is thought that supplementary work is needed to rationally control the active phase deposition in order to increase the final catalytic activity and selectivity. Considerable basic research is also needed to allow the synthesis of these 1D materials with controlled macroscopic shape for downstream applications. It is worth noting that although the macroscopic shape is similar to that used today in conventional catalytic processes, the scale of the constituted material has dramatically changed from micrometers to nanometers, with a large increase in the external surface area.

It is also worth noting that, apart from the direct use of these nanomaterials as catalyst supports, other useful applications, especially for self-supported carbon nanotubes, may also be expected in the synthesis of several one-dimensional materials (metal, oxide or ceramic) which could find some new applications not only in catalysis but also in several areas of the emerging nanotechnology field. It is thought that the intensive development of an optimized manufacturing process allowing the production of carbon nanotubes on a large scale with high purity and reproducibility and low cost will enable these nanomaterials to become as popular as silicon is today in the industrialized countries. However, one should keep in mind that the influence of such nanomaterials on human health should be determined with accuracy in order to allow them to be fully employed in the nanotechnology field.

Acknowledgments

The work presented here has issued from fruitful collaborations with a large number of people both inside and outside the laboratory and also through large financial support from several industrial companies such as: CnpQ (Brazil), Lurgi GmbH (Germany), Ademe (France) and Sicat (France). The authors would like to address their acknowledgement to Dr. Ricardo Vieira (InpQ, Brazil), Drs. Izabela

Janowska, Jean-Mario Nhut, Gauthier Winé, Thierry Dintzer (LMSPC), Dr. Patrick Nguyen (Sicat, France), Dr. Laurie Pesant (IFP, France), Dr. Jean-Philippe Tessonnier (Inorganic Chemical Department, Berlin, MPG), Dr. Claude Estournès (PNF², Toulouse, France), Dr. Raymond Ziessel (LCM, Strasbourg, France). The high-resolution TEM experiments were performed at the Institut de Physique et Chimie des Matériaux de Strasbourg (IPCMS, UMR 7504 du CNRS, Strasbourg, France) and the Fritz-Haber Institut (Inorganic Chemical Department, MPG, Berlin, Germany).

References

1 R. L. Burwell, Jr., *ACS Symp. Ser.* **1983**, *222*, 3–12.

2 R. B. Anderson, *The Fischer-Tropsch Synthesis*, Academic Press, New York, **1984**.

3 Catalysis Looks to the Future, NRC, National Academic Press, **1992**.

4 G. C. Bond, *Heterogeneous Catalysis: Principles and Applications*, Oxford Science Publications, Clarendon Press, Oxford, **1987**.

5 R. J. Farrauto, C. H. Bartholomew, *Fundamentals of Industrial Catalytic Processes*, Blackie Academic & Professional, London, **1997**.

6 (a) M. S. Dresselhaus, G. Dresselhaus, P. C. Eklund (Eds.), *Science of Fullerenes and Carbon Nanotubes*, Academic Press, London, **1996**; (b) T. W. Ebbesen (Ed.), *Carbon Nanotubes: Preparation and Properties*, CRC Press, Boca Raton, **1997**; (c) P. J. F. Harris (Ed.), *Carbon Nanotubes and Related Structures, New Materials for the Twenty-First Century*, Cambridge University Press, Cambridge, **2000**; (d) P. M. Ajayan *Chem. Rev.* **1999**, *99*, 1797.

7 R. Andrews, D. Jacques, D. Qian, T. Rantell, *Acc. Chem. Res.* **2002**, *35*, 1008.

8 Y. P. Sun, K. Fu, Y. Lin, W. Huang, *Acc. Chem. Res.* **2002**, *35*, 1096.

9 Ph. Serp, M. Corrias, Ph. Kalck, *Appl. Catal. A* **2003**, *253*, 337.

10 J. M. Nhut, R. Vieira, L. Pesant, J. P. Tessonnier, N. Keller, G. Ehret, C. Pham-Huu, M. J. Ledoux, *Catal. Today* **2002**, *76*, 11.

11 J. M. Nhut, L. Pesant, J. P. Tessonnier, G. Winé, J. Guille, C. Pham-Huu, M. J. Ledoux, *Appl. Catal. A: General* **2003**, *254*, 345.

12 K. P. de Jong, J. W. Geus, *Catal. Rev.-Sci. Eng.* **2000**, *42(4)*, 481.

13 M. J. Ledoux, C. Pham-Huu, *Catal. Today*, **2005**, *102–103*, 2–14.

14 S. Iijima, *Nature (London)* **1991**, *354*, 56.

15 C. Park, N. M. Rodriguez, M. S. Kim, R. T. K. Baker, *J. Catal.* **1997**, *169*, 212.

16 I. Martin-Gullon, J. Vera, J. A. Conesa, J. L. Gonzalez, C. Merino, *Carbon* **2006**, *44*, 1572.

17 G. B. Zheng, K. Kouda, H. Sano, Y. Uchiyama, Y. F. Shi, H. J. Quen, *Carbon* **2004**, *42*, 635.

18 N. M. Rodriguez, A. Chambers, R. T. K. Baker, *Langmuir* **1995**, *11*, 3862.

19 S. Helveg, C. Lopez-Cartes, J. Sehested, P. L. Hausen, B. S. Clausen, J. R. Rostrup-Nielsen, *Nature* **2004**, *427*, 426.

20 S. H. Yoon, S. Lim, S. H. Hong, W. Qiao, D. D. Whitehurst, I. Mochida, *Carbon* **2005**, *43*, 1828.

21 H. J. Dai, *Acc. Chem. Res.* **2002**, *35*, 1045.

22 K. T. Lau, D. Hui, *Composite: Part B* **2002**, *33*, 263.

23 E. T. Thostenson, Z. Ren, T. W. Chou, *Comput. Sci. Tech.* **2001**, *61*, 1899.

24 T. W. Ebbesen, P. M. Ayajan, *Nature* **1992**, *358*, 220.

25 J. M. Bonard, L. Forro, D. Ugarte, W. A. de Herr, A Châtelain, *Eur. Chem. Chronicle* **1998**, *1*, 9.

26 T. W. Ebbesen, P. M. Ayajan, H. Hiura, K. Tanigaki, *Nature*, **1993**, *367*, 519.

27 A. Cao, X. F. Zhang, C. L. Xu, J. Liang, D. H. Wu, B. Q. Wei, *J. Mater. Res.* **2001**, *16*, 3107.

28 C. Pham-Huu, N. Keller, V. V. Roddatis, G. Mestl, R. Schlögl, M. J. Ledoux, *Phys. Chem. Chem. Phys.* **2002**, *4*, 514.

29 M. J. Toebes, F. F. Prinsloo, J. H. Bitter, A. J. van Dillen, K. P. de Jong, *J. Catal.* **2003**, *214*, 78.

30 V. I. Zaikovskii, V. V. Chesnokov, R. A. Buganov, *Kinet. Catal.* **1999**, *40*, 612.

31 C. Pham-Huu, R. Vieira, B. Louis, A. Carvalho, J. Amadou, Th. Dintzer, M. J. Ledoux, *J. Catal.* **2006**, *240*, 194–202.

32 A. G. Rinzler, J. Liu, H. Dai, P. Nikolaev, C. B. Huffman, F. J. Rodriguez-Macias, P. J. Boul, A. H. Lu, D. Heymann, D. T. Colbert, R. S. Lee, J. E. Fischer, A. M. Rao, P. C. Eklund, R. E. Smalley, *Appl. Phys. A*, **1998**, *67*, 29.

33 G. Gulino, R. Vieira, J. Amadou, P. Nguyen, M. J. Ledoux, S. Galvagno, G. Centi, C. Pham-Huu, *Appl. Catal. A: General* **2005**, *27*, 89.

34 C. Emmenegger, J. M. Bonard, P. Mauron, P. Sudan, A. Lepora, B. Grobety, A. Zuttler, L. Schapbach, *Carbon* **2003**, *41*, 539.

35 M. A. Ermakova, D. Y. Ermakov, A. L. Chuvilin, G. G. Kuvshinov, *J. Catal.* **2001**, *201*, 183.

36 C. Park, M. A. Keane, *J. Catal.* **2004**, *221*, 386.

37 M. J. Ledoux, C. Pham-Huu, G. Ulrich, R. Vieira, P. Nguyen, J. Amadou, J. P. Tessonnier, D. Bégin, R. Ziessel, French Pat. Appl. No. 05-01107, assigned to the CNRS and the Université Louis Pasteur of Strasbourg, **2005**.

38 J. Amadou, D. Bégin, P. Nguyen, J. P. Tessonnier, Th. Dintzer, E. Vanhaecke, M. J. Ledoux, C. Pham-Huu, *Carbon* **2006**, *44*, 2587.

39 H. W. Zhu, C. L. Xu, D. H. Wu, B. Q. Wei, R. Vajtai, P. M. Ajayan, *Science* **2002**, *29*, 884.

40 K. Jiang, Q. Li, S. Fan, *Nature* **2002**, *419*, 801.

41 B. Vigolo, A. Pénicaud, C. Coulon, C. Sander, R. Pailler, C. Journet, P. Bernier, Ph. Poulin, *Science* **2000**, *290*, 1331.

42 Y. L. Li, I. A. Kinloch, A. H. Windle, *Science* **2004**, *304*, 276.

43 K. Hata, D. N. Futaba, K. Mizuno, T. Namai, M. Yumura, S. Iijima, *Science* **2004**, *306*, 1362.

44 L. M. Ericson, H. Fan, H. Peng, V. A. Davis, *Science* **2004**, *305*, 1447.

45 P. G. Whitten, G. M. Spinks, G. G. Wallace, *Carbon* **2005**, *43*, 1891.

46 C. Singh, M. S. P. Shaffer, A. H. Windle, *Carbon* **2003**, *41*, 359.

47 X. Li, A. Cao, Y. J. Jung, R. Vajtai, P. M. Ayajan, *Nano Lett.* **2005**, *5*(*10*), 1997.

48 M. Pinault, V. Pichot, H. Khodja, P. Launois, C. Reynaud, M. Mayne-L'Hermite, *Nano Lett.* **2005**, *5*(*12*), 2394.

49 L. Zhu, Y. Xiu, D. W. Hess, C. P. Wong, *Nano Lett.* **2005**, *5*(*12*), 2641.

50 I. Kleps, A. Angelescu, M. Miu, *Mater. Sci. Eng. C* **2002**, *19*, 219.

51 V. Svrcek, C. Pham-Huu, J. Amadou, D. Bégin, M. J. Ledoux, F. Le Normand, O. Ersen, S. Joulie, *J. Appl. Phys.* **2006**, *99*, 064306.

52 V. Svrcek, F. Le Normand, S. Joulie, O. Ersen, C. Pham-Huu, M. J. Ledoux, *Appl. Phys. Lett.* **2006**, *88*, 033112.

53 D. Y. Kim, C. M. Yang, Y. S. Park, K. K. Kim, S. Y. Jeong, J. H. Han, Y. H. Lee, *Chem. Phys. Lett.* **2005**, *413*, 135.

54 J. Sloan, J. Hammer, M. Zwiefka-Sibley, M. L. H. Green, *Chem. Commun.* **1998**, 347.

55 S. C. Tsang, Y. K. Chen, P. J. F. Harris, M. L. H. Green, *Nature* **1994**, 372.

56 J. Liu, A. A. Rinzler, H. Dai, J. H. Hafner, R. K. Bradley, P. J. Boul, A. Lu, T. Iverson, K. Shelimov, C. B. Huffman, F. R. Macias, Y. S. Shon, T. R. Lee, D. T. Colbert, R. E. Smalley, *Science*, **1998**, *280*, 125.

57 J. M. Moon, K. H. An, Y. H. Lee, *J. Phys. Chem. B* **2001**, *105*, 5677.

58 C. Li, D. Wang, T. Liang, X. Wang, J. Wu, X. Hu, J. Liang, *Powder Technol.* **2004**, *142*, 175.

59 D. T. Colbert, *Science* **1994**, *266*, 1218.

60 M. J. Ledoux, S. Hantzer, C. Pham-Huu, J. Guille, M. P. Desaneaux, *J. Catal.* **1988**, *114*, 176.

61 N. Keller, C. Pham-Huu, G. Ehret, V. Keller, M. J. Ledoux, *Carbon* **2003**, *41*, 2131.

62 M. J. Ledoux, C. Pham-Huu, *CaTTech* **2002**, *4*, 226.

63 M. J. Ledoux, R. Vieira, C. Pham-Huu, N. Keller, *J. Catal.* **2003**, *216*, 333.

64 N. Keller, C. Pham-Huu, M. J. Ledoux, C. Estournès, G. Ehret, *Appl. Catal. A: General* **1999**, *187*, 255.

65 L. Hu, Y. X. Li, X. X. Ding, C. Tang, S. R. Qi, *Chem. Phys. Lett.* **2004**, *397*, 271.

66 H. Ye, N. Naguib, Y. Gogotsi, A. G. Yazicioglu, C. M. Megaridis, *Nanotechnology* **2004**, *15*, 232.

67 T. W. Ebbesen, *Acc. Chem. Res.* **1998**, *31*, 558.

68 D. Ugarte, A. Châtelain, W. A. de Herr, *Science 274*, **1897** (1996).

69 H. Ma, L. Wang, L. Chen, C. Dong, W. Yu, T. Huang, Y. Qian, *Catal. Commun.* **2007**, *8*, 452.

70 A. H. Janssen, C. M. Yang, Y. Wang, F. Schuth, A. J. Koster, K. P. de Jong, *J. Phys. Chem. B* **2003**, *107*, 10552.

71 C. J. Gommes, K. P. de Jong, J. P. Girard, S. Blacher, *Langmuir* **2005**, *21*, 12378.

72 F. Winter, G. Leendert Bezemer, C. van der Spek, J. D. Meeldijk, A. J. Van Dillen, J. W. Geus, K. P. De Jong, *Carbon* **2005**, *43*, 327.

73 D. Mattia, H. H. Bau, Y. Gogotsi, *Langmuir* **2006**, *22*, 1789.

74 H. O. Pierson, *Handbook of Carbon, Graphite*, Diamond and Fullerenes, Noyes, Park Ridge, NJ, **1993**.

75 C. Pham-Huu, N. Keller, L. J. Charbonnière, R. Ziessel, M. J. Ledoux, *Chem. Commun.* **2000**, *19(11)*, 1871.

76 C. Pham-Huu, N. Keller, G. Ehret, L. J. Charbonnière, R. Ziessel, M. J. Ledoux, *J. Mol. Catal. A: Chem.* **2001**, *170*, 155.

77 A. Kuzuetsiva, D. B. Mawhinney, V. Naumenko, J. T. Yates, J. Liu, R. E. Smalley, *Chem. Phys. Lett.* **2000**, *391*, 292.

78 H. Ulbricht, G. Moos, T. Hertel, *Phys. Rev. B* **2002**, *66*, 075404.

79 J. Zhao, J. P. Lu, *Appl. Phys. Lett.* **2003**, *82*, 3746.

80 H. Ago, T. Kuglet, F. Cacialli, W. R. Salaneck, M. S. P. Shaffer, A. H. Windle, R. H. Friend, *J. Phys. Chem. B* **1999**, *103*, 8116.

81 G. Mestl, N. I. Maksimova, N. Keller, V. V. Roddatis, R. Schlögl, *Angew. Chem. Int. Ed.* **2001**, *113(11)*, 2122.

82 G. Mestl, N. I. Maksimova, N. Keller, V. V. Roddatis, R. Schlögl, *Angew. Chem. Int. Ed.* **2002**, *40(11)*, 2066.

83 M. M. DubinIn, In *Chemistry and Physics of Carbon*, P. L. Walker Jr. (Ed.), Marcel Dekker, New York, **1996**.

84 N. Nakamichi, K. Hirotoshi, H. Masahiko, *J. Org. Chem.* **2003**, *68*, 8272.

85 H. Ulbricht, G. Moos, T. Hertel, *Phys. Rev. Lett.* **2003**, *90*, 095501.

86 J. Wieckowska, *Catal. Today* **1995**, *24*, 405.

87 Sulphur **1997**, *250*, 45.

88 A. Pieplu, O. Saur, J. C. Lavalley, *Catal. Rev. – Sci. Eng.* **1998**, *40*, 409.

89 N. Keller, C. Pham-Huu, C. Estournès, M. J. Ledoux, *Catal. Lett.* **1999**, *61*, 151.

90 M. J. Ledoux, C. Pham-Huu, N. Keller, S. Savin-Poncet, J. B. Nougayrède, W. Boll, R. Morgenroth, *Sulphur* **2000**, *249*, 41.

91 J. M. Nhut, P. Nguyen, C. Pham-Huu, N. Keller, M. J. Ledoux, *Catal. Today* **2004**, *91–92*, 91.

8
Size-selective Synthesis of Nanostructured Metal and Metal Oxide Colloids and Their Use as Catalysts

Manfred T. Reetz

8.1
Introduction

8.1.1
General Comments on Catalysis using Transition Metal Nanoparticles

During the last 2–3 decades a vast amount of knowledge regarding the preparation and characterization of nanostructured transition metal colloids in the zerovalent form has accumulated [1, 2]. The emphasis has been largely on the development of methods for the control of size and, more recently, even of shape [3]. In the majority of cases application in catalysis was not pursued systematically, i.e., generally only a simple model reaction such as the hydrogenation of cyclohexene was studied.

Application of transition metal nanoparticles can occur in two fundamentally different forms: (i) The use of the colloidal "solutions" of preformed nanoparticles; or (ii) the use of immobilized forms of the colloidal nanoparticles on solid supports. As an alternative to the former procedure, nanoparticles can be formed from a molecular precursor (salt or complex) *in situ* during the actual catalytic reaction (e.g., Heck or Suzuki coupling). For a thorough discussion of these aspects, the reader is referred to Chapter 10. In the case of immobilization on a solid carrier prior to catalysis, the two-step process has a positive feature in that one can rely on the availability of well-defined species prepared by a size-selective synthesis. Such syntheses are now available which opens up new perspectives, especially in view of the known fact that the size of metal nanoparticles has a strong influence on their catalytic profiles. If one focuses on applications, it should be pointed out that the potential advantage of size-selective methods for metal colloid preparation still needs to demonstrated in a general way and compared to the traditional methods which have been used reliably in industrial heterogeneous catalysis for decades [4]. The much older methods include, *inter alia*, various forms of impregnation such as the incipient wetness method and/or the process used in the preparation of, for example, eggshell catalysts which require several steps following initial impregnation (drying, calcining, activating). Thus, the question

Nanoparticles and Catalysis. Edited by Didier Astruc
Copyright © 2008 WILEY-VCH Verlag GmbH & Co. KGaA, Weinheim
ISBN: 978-3-527-31572-7

arises as to the true practical value of size- and shape-selective methods of metal nanoparticle preparation beyond their academic interest in basic research. There is reason to be optimistic.

In contrast to nanoscaled transition metal colloids in the zerovalent form, less is known regarding the analogous *metal oxides*, i.e., colloidal MO_n (M=metal; n=1, 2, 3, etc.). Of course, such species in solid bulk form or as immobilizates on surfaces or in solid carriers have been known for a long time in heterogeneous catalysis [4], in semiconductor technology and other areas [3, 5]. More recently special methods have been developed for their preparation in constrained environments [6], in microemulsions [7], or as one- or two-dimensional nanostructures [5d, 8].

This chapter focuses mainly on the author's own contributions in the area of *aqueous nanoscaled transition metal oxides*. Specifically, the preparation in high concentration and application of aqueous colloids comprising metal oxide and mixed metal oxide nanoparticles will be described. The chapter begins with a short overview of the author's earlier work on methods for size- and shape-selective preparation of transition metal (zerovalent) colloids, because this led to the development of simple and practical ways to prepare the corresponding aqueous transition metal oxide colloids.

8.2
Size- and Shape-selective Preparation of Metal Nanoparticles in the Zerovalent Form

It is well known that transition metal colloids are accessible in a variety of ways, the most general one involving the reduction of a transition metal salt by such reducing agents as H_2, formaldehyde, ethanol, hydrazine, borohydrides or aluminum hydrides [1, 2]. In order to prevent undesired agglomeration, these processes are generally performed in the presence of stabilizers such as polar polymers (e.g., poly(vinylpyrrolidone)), surfactants such as tetraalkylammonium salts, or special ligands (e.g., phenanthroline) which exert stabilization by way of electrostatic and/or steric effects. If one scrutinizes all of this work, it becomes clear that the size of the nanoparticles (most often Pd or Pt in model studies) generated by the various methods appears to depend, *inter alia*, upon the nature of the reducing agents. Since it is of considerable theoretical interest to understand the source of size-selectivity, one might expect the vast amount of data to provide some clues regarding particle growth. However, direct comparisons between the systems are difficult, because not only the nature of the reductant varies, but also the conditions (e.g. solvent, concentration, temperature, etc.). This makes a sound theoretical analysis of comparative particle growth in the respective systems essentially impossible.

Detailed studies of size-selective syntheses of transition metal colloids in a systematic manner are not as common as one would wish [1, 2]. In a series of classical papers Turkevich postulated that the relative rates of two crucial processes, namely nucleation and growth, determine the size of metal particles [9]. Accordingly, fast nucleation relative to growth results in small particle size. In its most basic form this postulate is still accepted today [1, 2, 9d–f]. The problem is then reduced to the question of how to identify and how to control the factors which govern the kinetic processes. Nucleation and growth are themselves complicated processes which have been illuminated mechanistically only in a few select cases, as in Henglein's classical studies concerning silver and palladium clusters using pulse radiolysis [10]. Accordingly, charged dimers Ag_2^+ (or Pd_2^+) are the first intermediates to be formed. However, in bulk preparations of transition metal colloids using other methods it is generally not clear at what particular size the nucleus begins to constitute a new phase. Moreover, following formation of a stable nucleus, various mechanisms for further growth are possible. For example, an "atom by atom" buildup has been discussed, as in photographic emulsions [11]. Such surface catalytic growth also appears to be the primary principle in other processes as well [12]. This also pertains to the formation of the interesting class of polyanion-stabilized metal colloids, which has been termed by Finke as "living metal polymerization" [13]. Other than the atom by atom mechanism, growth can also be imagined to occur by collision of unstable assemblies of metal atoms (agglomerative growth). The operation of both mechanisms in a given metal colloid preparation is also conceivable.

With regards to purely synthetic aspects, several new approaches to size-selectivity have been described during the last 20 years. For example, seminal work by the groups of Nagy, Fendler, Boutonnet and others on the preparation of colloidal metals in constrained environments [1] has been elaborated and extended with the emergence of efficient processes, as in the reduction of transition metal salts in micelles [14], surfactant emulsions [15], liquid crystals [16], or polymerized vesicles [17]. In other important work thiolate-encapsulated Au nanoparticles were prepared size-selectively in the range 2–6 nm by manipulating the thiol-to-gold ratio during the reduction process [18]. In yet a different approach the nature of alcohols used as reductants has been systematically varied [19, 20]. Upon going from methanol to ethanol to isopropanol, the size of the Pd or Pt nanoparticles increases, although the actually observed range is rather small [19]. It turned out that the alcohol concentration plays a more dominant role. Oddly enough, the opposite trend, with respect to the nature of the alcohol, was observed upon switching from palladium or platinum to rhodium [20]. One might expect that nanoparticle size should correlate with the reducing power of the system, strong reducing agents leading to fast nucleation relative to growth and therefore to smaller particles. However, this is not evident in these or other systems due to a variety of possible overriding effects [1, 2, 21]. The hydrogenation of [Ru(cod)(cot)] (cod=1,5-cyclooctadiene; cot=1,3-cyclooctatriene) in mixtures of THF/methanol affords novel types of spongy Ru nanoparticles in the size range 12 to >500 nm, depending upon the solvent composition [22]. Ligand stabilized giant metal clusters, the size of which were originally postulated to be characterized by the so-called magic numbers, constitute a special class of nanoparticles [1, 2, 23]. The discussion concerning the actual factors which govern the size of these monodisperse particles is still controversial [1, 2, 23, 24].

In the 1990s we described an electrochemical method for the preparation of $R_4N^+X^-$-stabilized transition metal colloids [25]. It had been known since Grätzel's work that such ammonium salts stabilize Pt colloids [26a], an observation that Boutonnet [26d, e], Toshima [26c], Esumi [26f] and others [26] made use of in later work. Moreover, electrochemical processes had been employed in the preparation of insoluble metal powders, usually in aqueous acidic media [27]. More recently, controlled electrochemical deposition of nanostructured Pd and Cu particles has been reported by Hempelmann [28a, b]. This work is related to that of Penner concerning the electrochemical deposition of silver nanocrystallites on an atomically smooth graphite basal plane [28c], and to the report of Searson and Chien concerning the electrochemical deposition of Ni and Co in polycarbonate templates of nanometer-sized pores created by nuclear track etching [28d]. Other forms of electrochemical metal deposition have been reported by Wiley [28e] and Bartlett [28f]. In addition to these important innovations, we developed other approaches. Initially a sacrificial anode as the metal source in a simple electrolysis cell was employed, $R_4N^+X^-$ serving as the electrolyte and as the stabilizer in an organic solvent (Scheme 8.1) [25]. This means that the bulk metal at the anode is oxidized to metal cations which then migrate to the cathode where reduction occurs with formation of ad-atoms. These form clusters which are trapped by the surfactant, a process which results in stabilized colloids rather than insoluble metal powders as in traditional electrochemical metal deposition (Fig. 8.1).

We observed an interesting degree of size-selectivity upon varying the current density, specifically in the model system comprising palladium as the transition metal [25]. Other parameters such as the duration of electrolysis and distance between electrodes were not considered nor were they specifically controlled. Later we published a systematic study of the effect of these and other factors, including solvent polarity and temperature [29]. It was demonstrated that all of these parameters must be considered in order to obtain synthetically useful results in a simple and reproducible manner [29, 30]. Thus, the size of the Pd nanoparticles in the range 1.2–5 nm is readily adjustable. Characterization of the Pd colloids was performed using transmission electron microscopy (TEM), small angle X-ray scattering (SAXS) and X-ray powder diffractometry (XRD) evaluated by Debye-function analysis (DFA) [29]. Possible mechanisms of particle growth were also addressed. Experiments directed toward the size-selective electrochemical fabrication of

Scheme 8.1 Electrochemical fabrication of nanosized transition metal colloids (the stabilizer is usually a tetraalkylammonium salt) [25].

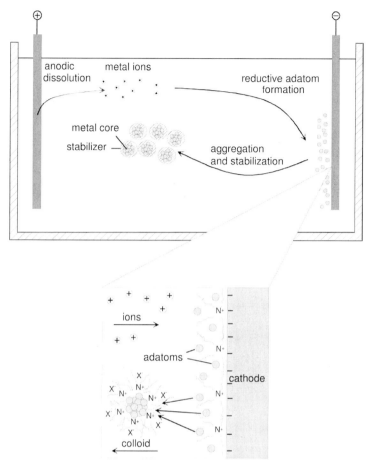

Fig. 8.1 Schematic representation of the electrochemical formation of $R_4N^+X^-$-stabilized transition metal colloids. The exact position of the counter ions X^- was not stipulated [25b], and may be at the surface of the metal core [2a, 13].

(n-C_6H_{13})$_4$N$^+$Br$^-$-stabilized nickel colloids showed similar effects [29]. Finally, a new strategy for preparing bimetallic colloids such as Pt/Pd nanoparticles electrochemically was presented, based on the use of a preformed colloid (e.g., (n-C_8H_{17})$_4$N$^+$Br$^-$-stabilized Pt particles) and a sacrificial anode (e.g., Pd sheet) in a type of grafting process [29].

In a combined TEM and STM study the ammonium surfactant and the Pd core of the colloid was visualized (Fig. 8.2) [31]. The value of this report lies in the fact that the previously often postulated existence of a stabilizing layer around a metal cluster was proven experimentally for the first time.

It turned out that, as expected, the thickness of the stabilizer shell depends upon the length of the alkyl group in the tetraalkylammonium ion R_4N^+. The position

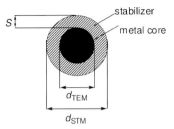

Fig. 8.2 Schematic diagram of the geometric parameters of $R_4N^+Br^-$-stabilized Pd colloids [31].

of the bromide (Br^-) could not be located by this method, nor was this question addressed (see Fig. 8.1). It has been pointed out that these anions occur most likely near the surface of the Pd core due to electrostatic reasons [2a, 13] (see also Fig. 8.1). This may well be the case, but it is the bulky ammonium cations which are particularly instrumental in steric stabilization and which make up most of the thickness S of the stabilizing mantle (Fig. 8.2). For example, it was observed that "small" ammonium bromides (or chlorides) such as tetraethylammonium salts fail to prevent agglomeration and undesired precipitation of Pd metal [32]. A question that still needs to be addressed is the possibility that the $R_4N^+Br^-$-stabilized metal nanoparticle could pick up an electron (or more) at the cathode. Such controlled reduction would reduce the number of bromide ions on the surface.

The mechanism of electrochemically induced particle growth is of significant mechanistic interest [29]. Although detailed mechanistic studies regarding colloid formation were not carried out, all of the present observations are consistent with a continuous growth mechanism involving $R_4N^+X^-$-stabilized particles. However, some formation of new nucleation points during electrolysis, and therefore of new colloids, may well occur parallel to growth [29]. In order to gain more information on the structural nature of the Pd colloids and possibly on their mode of formation, XRD was applied. Important details of the analysis can be found in the original study [29]. Moreover, by using water-soluble ammonium salts such as betaines, aqueous solutions of Pd colloids were obtained [25]. In such a medium supramolecular aggregates were observed by static and dynamic X-ray scattering [33].

The electrochemical method is not restricted to palladium. For example, palladium can be replaced by nickel as the sacrificial anode [29, 30, 32]. In this particular case $(n-C_6H_{13})_4N^+Br^-$-stabilized Ni colloids were obtained having a TEM diameter (inner metal core) of 1.7 nm. Yasuhara and Sakomoto have also applied the electrochemical method to the preparation of Ni colloids, specifically by using $(n-C_4H_9)_4N^+BF_4^-$ as the stabilizing surfactant [34a]. Exploratory experiments employing other sacrificial anodes such as Co [35], Fe [32], Cu [32], Ag [32], Ti [36], and Au [32] were likewise successful, although we did not (yet) focus on size-selectivity. In some cases the colloids were not as stable as those derived from palladium or nickel. Ma and Wang have replaced ammonium salts as stabilizers by a polar solvent (PVP) in the electrochemical formation of Ag colloids [34b].

The control of *shape-selectivity* is more difficult. A detailed study of the use of Au sheets as sacrificial anodes and special ammonium salts showed that shape-selectivity in the form of rodlike clusters is possible [37]. The electrochemical method has also been applied successfully in the preparation of gold nanorods [28g] and gold nanocubes [28h].

In another electrochemical approach we used inert electrodes and a transition metal salt such as $PtCl_2$ as the source of metal [38]. Reduction of Pt^{2+} at the cathode has to be compensated by some oxidative process at the anode. Therefore, we substituted tetraalkylammonium *halides* by the analogous *acetates* $R_4N^+CH_3CO_2^-$, hoping that they would fulfill three purposes, namely to function as the electrolyte, as the stabilizer and as the reductant (Kolbe-like):

$$\text{Cathode:} \quad Pt^{2+} + 2e^- \quad \rightarrow \quad Pt$$

$$\text{Anode:} \quad 2CH_3CO_2^- \quad \rightarrow \quad 2CH_3CO_2^\bullet + 2e^-$$

Indeed, when using an excess of $(n\text{-}C_4H_9)_4N^+OAc^-$, this worked well. Surfactant-stabilized Pt colloids in THF having a size of 2.5 nm or 5 nm were obtained, the specific size depending upon the current density used. Similarly, $RhCl_3$, $RuCl_3$, $OsCl_3$, $Pd(OAc)_2$, $Mo_2(OAc)_4$ were employed with formation of the corresponding nanosized colloids, although size-selectivity was not studied in these exploratory experiments. Moreover, bimetallic colloids of the type Pt/Sn, Pd/Cu and Pd/Pt were also obtained [38].

The two versions of the Mülheim electrochemical process provide colloidal solutions (e.g., in THF) of a variety of transition metal or bimetallic nanoparticles, and constitute a simple, clean and reliable alternative to chemical processes such as reduction by borohydrides in which the excess reducing agent and/or the oxidized form thereof have to be removed from the product (in fact, boron originating from boron hydrides is sometimes incorporated in the nanoparticles) [26]. But are these methods of any use in catalysis? One possibility is immobilization on solid carriers, delivering materials having islands of metal clusters of a predefined size. Moreover, they allow for the design of heterogeneous catalysts with well-defined compositional and structural features on a macroscopic and microscopic level.

Generally, it is desirable to achieve metal dispersion preferentially on the outer surface of the support, because this minimizes diffusion problems [4]. This is the reason why shell catalysts, especially those of the eggshell type, are traditionally preferred. Accordingly, mm-sized pellets are treated in special impregnation procedures using metal salts, followed by reduction, which finally leads to a metal-containing outer region several hundred μm thick. However, it is likely that within this *macroscopically* visible region the metal particles are not exclusively fixed on the respective outer surface in a *microscopic* sense, but are also distributed deep in the pores of the solid support where they are not readily reached by substrates. In order to prevent this, preformed and size-selectively prepared metal clusters on solid supports were immobilized [39]. By considering pore diameter and cluster size, it is possible to design catalysts in which metal particles of a specified size are found solely on the outer microscopic surface of the support. At the same time, eggshell morphology macroscopically is likewise achieved. Accordingly, upon stirring solutions of these preformed colloids at room temperature in the presence of solid supports such as SiO_2 or Al_2O_3 in the form of powders or pellets, immobilization occurs without any change in cluster size [39]. This means that metal crystallites of a defined size and elemental composition can be specifically deposited on such supports. When using pellets, shell catalysts [39] are easily produced without the need for the special preparation techniques necessary in conventional impregnation processes, such as pretreatment of the solid support, use of viscous solvents or additives, or special drying procedures. For example, impregnation of Al_2O_3 pellets (Johnson Matthey 11838) with a 0.5 M THF solution of 3 nm sized (by TEM) $(n\text{-}C_8H_{17})_4N^+Br^-$-stabilized Pd clusters leads to rapid carrier fixation. Eggshell catalysts are obtained within 1–4 s, as shown by the photographs of pellet cross sections (Fig. 8.3a, b). The stabilizer can then be washed off.

Since, on a microscopic scale, the metal was found to be only on the surface, as demonstrated by TEM analyses of ultramicrotometrically cut samples, such catalysts have been dubbed *"cortex catalysts"* to distinguish them from traditional eggshell catalysts. In the particular case of 3.5 nm-sized $(n\text{-}C_4H_9)_4N^+Br^-$-stabilized Pd colloids immobilized on Al_2O_3, analysis showed that the nanoparticles had penetrated the Al_2O_3 matrix to the extent of less than 10 nm. Preliminary application in the hydrogenation of cyclooctene catalyzed by a Pd cortex catalyst showed an increase in activity by a factor of three relative to the traditional eggshell catalyst [39]. More work is necessary in this field. Another way to immobilize the metal colloids is the encapsulation in sol–gel materials followed by removal of the stabilizer [40]. In this case Pd nanoparticles of defined size are trapped in spherical "caves" of the silicate, allowing chemoselective olefin hydrogenation [40].

Colloidal solutions of $R_4N^+Br^-$-stabilized Pd and Ni nanoparticles have also been used directly as catalysts in Heck couplings (e.g., **1** + **2** → **3**) [41], Suzuki reactions (e.g., **4** + **5** → **6**) [41] and [3+2]cycloaddition reactions (e.g., **7** + **8** → **9**) [42]. Immobilized forms can also be utilized in these reactions [41]. However, although these observations are of mechanistic interest [43], they are not of practical significance

| *a* | *b* | *c* | *d* |

Fig. 8.3 Cross sections of Al_2O_3 pellets (diameter 3.2 mm) after penetration with a 0.5 M THF solution of (n-$C_8H_{17})_4N^+Br^-$-stabilized Pd clusters for (a) 1 s, (b) 4 s, (c) 10 s. (d) After penetration with an aqueous 0.5 M solution of betaine-stabilized Pd clusters for 5 h [39].

since it is possible to use "homeopathic" amounts of a Pd salt in ligandless systems [44–46].

CO$_2$Bu + [aryl iodide] →(Pd-colloid, NaHCO$_3$/Bu$_4$NBr, DMF/30 °C/14 h)→ [product] CO$_2$Bu

1 **2** **3** (99%)

[phenylboronic acid with B(OH)$_2$] **4** + X—[aryl]—R **5** →(Pd-colloid)→ [biphenyl]—R **6** (80–100%)

[bicyclic ketone] **7** + CO$_2$CH$_3$ **8** →(Ni-colloid)→ [cyclopentane product] CO$_2$CH$_3$ **9** (35%)

An intriguing application of these Pd nanoparticles in basic research concerns the question of the solubility of H$_2$ in such materials relative to bulk palladium [47]. Hydrogen concentration–pressure isotherms of surfactant-stabilized palladium clusters and polymer-embedded palladium clusters with diameters of 2, 3 and 5 nm were measured with the gas sorption method at room temperature. The results show that, compared to bulk palladium, the hydrogen solubility in the α-phase of the clusters is enhanced fivefold to tenfold, and the miscibility gap is narrowed. Both results can be explained by assuming that hydrogen occupies the subsurface sites of the palladium clusters. The Pd–H isotherms of all clusters show the existence of hysteresis, even though the formation of misfit dislocations is unfavorable in small clusters. Compared to surfactant-stabilized clusters, the polymer-embedded clusters show slow absorption and desorption kinetics. Moreover, evidence for a cubic-to-icosahedral transition of quasi-free Pd–H clusters by the hydrogen content was reported [47c].

Another application concerns the fabrication of metallic and bimetallic nanostructures on surfaces by electron beam induced metallization of the R$_4$N$^+$Br$^-$-stabilized Pd and Pd/Pt colloids [48]. In a simple three-step process, namely dip-coating, electron beam writing of patterns (e.g. lines) and rinsing, lines having a thickness of only 30 nm were achieved. This technique is of potential interest in lithography and catalysis. Finally, tetraalkylammonium salt-stabilized transition metal nanoparticles have a strong tendency to self-assemble on carbon surfaces with formation of hexagonal close packed structures [49].

As mentioned earlier, tetraalkylammonium acetates $R_4N^+OAc^-$ were used as stabilizers, as reductants at the anode and as electrolytes in the electrochemical preparation of metal colloids starting from the corresponding salts. This research led to an interesting serendipitous discovery. It was observed that *prior* to electrolysis, THF solutions of $PdCl_2$ (or $Pd(OAc)_2$) in the presence of the surfactant (n-$C_4H_9)_4N^+OAc^-$ turn black if left to stand for a day [50a]. Analysis of the mixture showed that partial reduction had occurred with formation of 2 nm-sized colloidal Pd particles. It was then shown that the reductant is acetate ($CH_3CO_2^-$) in a type of chemical "Kolbe process" with CO_2 being the side-product. Of course, it had been known for a long time that heating solid $Pd(OAc)_2$ results in the formation of a black product (Pd bulk) and a gas (probably including CO_2). In our case the process occurs under controlled conditions in solution in the presence of a surfactant which traps and stabilizes the Pd nanoparticles as they grow. Therefore, we optimized the process using $Pd(NO_3)_2$ in the presence of excess (n-$C_8H_{17})_4N^+(CH_3CO_2^-)$ in THF at 66 °C (4h). Under these mild conditions a quantitative yield of Pd colloids having a size of 2–2.5 nm was achieved [50b]. The material can be isolated in solid form by evaporation of the THF, and redispersed in a variety of organic solvents if so desired. The reaction was generalized in that essentially any carboxylate RCO_2^- can be employed in this novel chemical process [50b]:

$$Pd(NO_3)_2 + (n\text{-}C_8H_{17})_4\,N^+(RCO_2^-) \xrightarrow[\text{reflux}]{\text{THF}} Pd\text{-colloid}$$

This led to the consideration of size-control, because the reducing capability of carboxylates RCO_2^- can be expected to depend on the nature of the R-group. Since electron transfer from RCO_2^- to the metal is irreversible due to rapid decarboxylation ($RCO_2^- \rightarrow RCO_2\cdot \rightarrow R\cdot + CO_2$), only the peak potentials $E_{p(Ox)}$ of the oxidation of RCO_2^- are experimentally accessible. These were measured and shown to correlate with the size of the Pd nanoparticles obtained in the chemical preparation (Table 8.1) [50b]. For example, dichloroacetate with its two electron-withdrawing chlorine atoms has a high $E_{p(Ox)}$ value and is thus a poor chemical reductant,

Table 8.1 Particle size of Pd colloids, pKa values, and peak potentials of the carboxylates studied [50b].

Reductant (n-Oct)$_4$N$^+$-salt of	$E_{p(Ox)}$ [V]	pK$_a$ of the acid	d [nm] after 50 min	d [nm] after 4 h
Dichloracetate	1.44	1.5	5.4	6.8
Lactate	1.16	3.1	2.8	_[a]
Glycolate	1.17	3.8	2.9	5.0
Acetate	1.07	4.8	2.0	2.5
Pivalate	1.08	5.0	2.2	2.5

a Not determined.

leading to relatively large Pd particles (5.4–6.8 nm). In contrast, acetate and pivalate are much better reductants and indeed lead to smaller Pd nanoparticles (2.5 nm) [50b].

A clear correlation exists between particle size and peak potentials of the carboxylates used in the colloid synthesis (Fig. 8.4). This is the first study in which the long-discussed theoretical relationship between "reducing power" and metal particle size [1, 2] was unambiguously demonstrated [50]. As noted earlier, it is difficult to compare reductants such as H_2, formaldehyde, ethanol and borohydrides because the experimental conditions vary greatly, and the "reducing power" of the structurally vastly different reductants can hardly be quantified. Our results clearly corroborate the classical theory of particle growth in colloids [9, 13]: Fast initiation by a powerful reductant results in fast nucleation relative to further growth, and this in turn enforces the formation of small particles. Shape-selectivity is more difficult to control [3], but some progress has been made using tetraalkyl-ammonium α-hydroxy-carboxylates as stabilizers, leading to trigonal morphology of colloidal Ni particles [50c]. Finally, heating certain transition metal acetates (e.g., $Pd(OAc)_2$) in polar solvents such as propylene carbonate (PC) affords PC-stabilized colloids even in the absence of ammonium salts [50d].

Subsequently, this important insight led to another serendipitous discovery. We had previously shown that the electrochemical preparation of colloidal metal nanoparticles, usually carried out in organic solvents such as THF, can also be performed in *water*, provided H_2O-soluble ammonium salts of the betaine type are

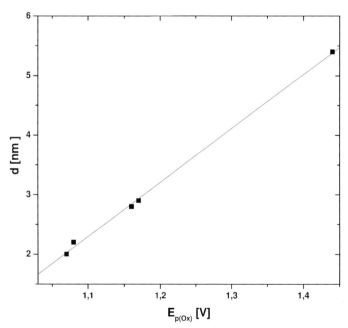

Fig. 8.4 Particle size of Pd colloids as a function of peak potentials $E_{p(Ox)}$ of the carboxylates [50b].

used [25]. Therefore, in an attempt to expand the method using carboxylates RCO_2^- as the chemical reductants, H_2O-soluble betaines of the type $R(CH_3)_2N^+(CH_2)_nCO_2^-$ were tested in an aqueous medium. The goal was to develop an environmentally benign system, useful for industrial applications. We hoped that gentle warming would lead to the reduction of the metal salt by the carboxy-function of the betaine, and that excess betaine would stabilize the metal particles in the zerovalent form. However, to our surprise reduction was not observed, although the transition metal salt (e.g., $PtCl_2$ or $PtCl_4$) was consumed [51]. Analysis of the brown (not black) colloidal solution showed that PtO_2 nanoparticles, stabilized by the betaine, had formed. Since the betaine is basic, an unexpected base-induced hydrolysis/ condensation had occurred which is faster than the alternative reduction. We therefore embarked on a study directed toward the controlled preparation of aqueous colloidal metal oxides (Section 8.3) [52].

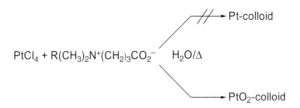

8.3
Preparation and Application of Aqueous Colloids of Metal Oxide and Multimetal Oxide Nanoparticles

As delineated in the Introduction, numerous research groups have focused on the preparation and application of transition metal nanoparticles in the zerovalent form during the last two decades but, parallel to this development, research in the area of nanosized metal *oxides* has also increased. The areas of application of metal oxide nanoparticles range from catalysis [4] to semiconductors (e.g., ZnO, ZnS, CdSe) [5]. In the area of heterogeneous catalysis, many different preparative methods have been described [4]. Other methods are based on the hydrolysis of transition metal salts in microemulsions [3–8]. These and other approaches have been reviewed and will not be specifically treated here. Rather, the main focus is on the author's own research.

As described at the end of the last section, we came upon the study of metal oxide nanoparticles in aqueous solution accidentally. Our only earlier experience with this class of nanoparticles, although in organic solvents such as THF, was the investigation of the controlled O_2-mediated oxidation of tetraalkylammonium bromide stabilized cobalt colloids prepared size-selectively by the electrochemical method (Fig. 8.5) [53].

Considerably easier is the preparation of aqueous colloidal forms of transition metal oxides, which was optimized with the emergence of a straightforward method. The first example pertains to a water-soluble colloidal form of Adams catalyst (PtO_2) [52].

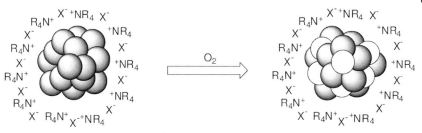

Fig. 8.5 Idealized representation of the oxidation of (n-$C_6H_{17})_4N^+X^-$-stabilized cobalt clusters with O_2 (dark spheres, metal atoms; light spheres, oxygen atoms) [53].

$$PtCl_4 \xrightarrow[\text{stabilizer}]{H_2O/\text{base}} PtO_2\text{-colloid}$$

Two types of water-soluble surfactants were used, carbobetaines **10a,b** and sulfobetaines **11**.

$$C_{12}H_{25}(CH_3)_2N^+(CH_2)_nCO_2^- \quad C_{12}H_{25}(CH_3)_2N^+(CH_2)_3SO_3^-$$

$$\textbf{10 a } n = 3 \qquad\qquad\qquad \textbf{11}$$
$$\textbf{b } n = 1$$

Upon stirring an aqueous solution of $PtCl_4$ and carbobetaine **10a** (molar ratio 1:4) in the presence of NaOH at 50 °C, complete consumption of the yellow platinum salt was observed with formation of a deep red–brown colloidal solution of PtO_2/**10a**. The reaction was monitored by UV/Vis spectroscopy (disappearance of the $PtCl_4$ absorption at 250 nm and appearance of a plasmon absorption in the range 200–800 nm) and by TEM [52]. High resolution TEM analysis of the final colloid showed the presence of 1.8 ± 0.3 nm-sized particles. The existence of nanoparticles in solution was also demonstrated by SAXS [52]. The use of stabilizer **10b** also gave 1.8 nm-sized PtO_2 particles. The condensation process can also be performed at reflux temperature for 2.5 h with Li_2CO_3 as the base, resulting in concentrated 1.9 ± 0.4 nm-sized PtO_2/**10b** colloids. High resolution TEM analysis of the samples revealed the existence of lattice planes, demonstrating the nanocrystalline character of the particles, as in the case of PtO_2/**10b**.

An analogous process is possible using the sulfobetaine **11** and Li_2CO_3 as the base, which provides 1.7 ± 0.3 nm-sized PtO_2/**11** colloids. Other $Pt_{(IV)}$ salts such as H_2PtCl_6 were also employed for the preparation of PtO_2 colloids [51, 52]. The addition of excess acetonitrile to the colloidal solution results in the essentially complete precipitation of the PtO_2 colloid as a brown solid, and this can be used to store the material. The precipitation of the PtO_2/**10b** colloid afforded a solid with a platinum content of 15.2%, which is completely redispersible in water at high concentration (0.1–0.5 M). Such high concentration is crucial for many applications. Alternative approaches for the workup of the crude colloidal solution are a dialysis process or ion exchange which removes any chloride [51, 52]. Platinum contents of more than 30% can be attained by a combination of these workup procedures if so desired.

Further characterization showed that the colloids are indeed composed of PtO_2 particles. For example, X-ray photoelectron spectroscopy (XPS) of the prepared colloids and of commercial Adams catalyst clearly showed the presence of $Pt_{(IV)}$ species, as demonstrated by the similarities in the Pt 4f regions [52]. The binding energy E_B amounts to 74.5 eV, both in colloidal PtO_2 and in the original Adams catalyst. Analysis of colloidal PtO_2 and commercial Adams catalyst by extended X-ray absorption fine structure (EXAFS) spectroscopy [52] also gives evidence for a very similar composition and structure of the particles. For example, measurements of the PtO_2 colloid at the Pt L_{III} edge revealed oxygen backscatterer at 2.0, 3.6 and 3.8 Å and a platinum backscatterer at 3.1 Å, with a coordination number of 6 for platinum. These values are in full agreement with those obtained for commercial Adams catalyst in the present study and those reported in the literature for pure α-PtO_2. Of course, the presence of hydroxy moieties on the surface of the oxidic nanoparticles is likely.

The aqueous colloidal Adams catalyst can be immobilized conveniently on solid carriers such as Al_2O_3 by adding the calculated amount of solid material to the aqueous colloidal solution and stirring. However, prior reduction to the zerovalent form followed by immobilization is actually more efficient [54]. After washing with methanol several times to remove the surfactant, the heterogeneous catalysts were examined by TEM which showed that essentially no undesired agglomeration had occurred (1.9 nm particles).

In order to test for catalytic activity, the reductive amination of benzaldehyde **13** by n-propylamine **12** was chosen as a model reaction [52].

Using the new catalyst PtO_2/Al_2O_3, commercially available Pt/Al_2O_3 (5%) and Adams catalyst, reductive amination was carried out at a substrate to Pt ratio of 1000:1 in methanol as the solvent at room temperature and atmospheric pressure. Catalyst activities were determined by measuring the uptake of H_2 per g Pt per minute. The results show that immobilized colloidal PtO_2 (1060 mL $H_2 g^{-1} min^{-1}$) is about 4–5 times more active than the commercial catalysts (253 and 196 mL $H_2 g^{-1} min^{-1}$, respectively) [52]. Selectivity in favor of the desired mono-benzylated product is >99%. The PtO_2 colloid is also effective in the hydrogenation of carbonyl compounds or olefins, either in solution or in immobilized form [51, 54].

In the same original study, preliminary experiments directed toward the preparation of water-soluble colloidal bimetallic oxides were shown to be successful [52]; this was later optimized and generalized [54, 55]. Upon stirring the aqueous solution of $PtCl_4$ and $RuCl_3$ (1:1 molar ratio) in the presence of sulfobetaine **11** and Na_2CO_3 at 80 °C for 18 h, a black colloidal solution of $PtRuO_x$ was obtained [52].

$$PtCl_4 + RuCl_3 \xrightarrow[\text{stabilizer}]{H_2O/base} PtRuO_x\text{-colloid}$$

Dialysis was used to purify the colloid, which was then characterized by TEM, demonstrating the presence of 1.5 ± 0.4 nm-sized particles [52]. Elemental analysis revealed 6.4% Pt and 4.0% Ru by weight, corresponding to a Pt:Ru molar ratio of about 1:1. Energy dispersive X-ray (EDX) spot analysis proved the presence of both metals in individual nanoparticles. According to an XPS analysis, platinum occurs as $Pt_{(II)}$ and $Pt_{(IV)}$ in similar amounts, in addition to $Ru_{(IV)}$. This means that $Ru_{(III)}$ reduces about 0.5 equivalents of $Pt_{(IV)}$ to $Pt_{(II)}$ with formation of $Ru_{(IV)}$, as expected from the redox behavior of the two metals $(Pt^{4+} + 2Ru^{3+} \rightarrow Pt^{2+} + 2Ru^{4+})$. The PtRuO$x$/**11** colloid is stable for months in an aqueous medium. Upon employing different ratios of the two metal precursors, the metal composition (Pt:Ru) of the colloids can be adjusted in the range from 1:4 to 4:1 on an optional basis [52]. Moreover, if so desired the colloidal Pt/Ru oxides can be converted into the corresponding zerovalent metal colloids by treating the aqueous colloidal solutions or the immobilized forms with H_2 at room temperature. No significant changes in particle size or Pt/Ru composition occurred under these conditions [54]. These Pt and Pt/Ru colloids were used successfully as electrocatalysts in the oxidation of methanol, as studied mechanistically using differential electrochemical mass spectrometry [55].

In a follow-up study the scope and limitation of the method was defined [54, 56]. We were particularly interested in fabricating colloidal mixed-metal oxides containing more than two metals. Of special interest was the work of Smotkin and Mallouk, who had shown that quaternary alloys composed of Pt, Ru, Os and Ir in the zerovalent form, prepared by other means, are highly active electrocatalysts for potential use in fuel cells [57]. Therefore, we tested our method in the preparation of a mixed metal oxide containing these four metals [56]. The hydrolysis of an aqueous mixture of H_2PtCl_6, $RuCl_3$, $OsCl_3$ and H_2IrCl_6 under basic conditions in the presence of the stabilizing surfactant **11** afforded aqueous colloidal solutions of the mixed metal oxide PtRuOsIrOx displaying a particle size of 1.3–1.6 nm [56].

$$H_2PtCl_6 + RuCl_3 + OsCl_3 + H_2IrCl_6 \xrightarrow[\underset{\mathbf{11}}{Li_2CO_3}]{H_2O} PtRuOsIO_x\text{-colloids (1.3 - 1.6 nm)}$$

It was possible to control the ratio of the four metals by choosing the appropriate relative amounts of starting metal salts. The initial experiments focused on the use of 1:1:1:1 amounts of the four salts, and indeed, a colloid was obtained displaying a Pt/Ru/Os/Ir ratio of 27/20/23/30 as evidenced by elemental analysis [56]. TEM analysis showed the average size to be 1.3 ± 0.4 nm. In order to prepare the mixed metal oxide analog of the Smotkin/Mallouk material, the relative amounts of the starting metal salts were adjusted to 44/41/10/5. Gratifyingly, a similar material resulted having a composition of Pt(40)/Ru(41)/Os(16)/Ir(3) and a size of 1.6 ± 0.3 nm, as shown by TEM analysis. Working at a four-fold higher concentration resulted in a similar material Pt(44)/Ru(41)/Os(11)/Ir(4) ($d_{TEM} = 1.2 \pm 0.3$),

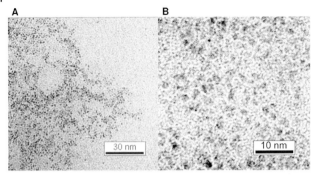

Fig. 8.6 High-resolution TEM images of the 11-stabilized colloid Pt(27)/Ru(20)/Os(23)/Ir(30)O$_x$ at normal (A) and high (B) magnification [56].

which demonstrates that absolutely identical conditions are not necessary for adequate reproducibility [56].

We observed that the high resolution TEM images of the samples are all rather similar. Typical TEM images of the Pt(27)/Ru(20)/Os(23)/Ir(30)-sample are shown in Fig. 8.6. At high magnification lattice planes can be seen (Fig. 8.6b), which demonstrates the nanocrystalline nature of the material. Application of EDX proved the presence of all four metals, although a quantitative analysis was not possible due to partial overlap of peaks (e.g., the L_{21}-transition of osmium overlaps with the peaks due to copper present in the TEM grid) [56].

In order to shed light on the electronic nature of this novel mixed metal oxide, e.g., the oxidation state of the metals therein, XPS was applied [56]. The Pt binding energies found (spectrum shown in the original paper) indicate the presence of Pt(II) and Pt(IV) species. This was expected on the basis of our previous experience with Pt/Ru mixed metal oxides in which platinum occurs as Pt(II) and Pt(IV). Indeed, the redox properties of the two metals predict oxidation of ruthenium by platinum according to the following stoichiometry:

$$Pt^{4+} + 2Ru^{3+} \rightarrow Pt^{2+} + 2Ru^{4+}$$

This means that if the salts are chosen in a 1:1 proportion, some of the platinum will remain in the +4 oxidation state. A peak with binding energy of 71.9 eV was assigned to hydroxy-platinum species of the type Pt(OH)$_2$ by comparison with reference data (72.3 eV, recalibrated). The peak at ca. 74.2 eV was more difficult to assign because bulk PtO$_2$ and Pt(OH)$_4$ have similar Pt binding energies (74.7 and 74.3 eV, respectively). It may well be that in the mixed metal oxide both Pt(IV) oxide and hydroxide species are present, but the observation that reduction during data acquisition leads to Pt(II) hydroxide species favors Pt(IV) hydroxide [56].

For the other metals, the experimental binding energies agree with those of bulk oxides (IrO$_2$, 61.7–62.0 eV, OsO$_2$, 51.7–52.0 eV, RuO$_2$, 462.7 eV and 463.6 eV), but the presence of hydroxide cannot be ruled out due to the lack of reference data.

We also applied XRD in order to gain further information regarding the structure, size and size distribution. Specifically, X-ray wide-angle scattering (WAXS) in combination with DFA was employed. The reader is referred to the study for details [56]. It turns out that the Pt(27)/Ru(20)/Os(23)/Ir(30) material following heat treatment is best characterized as consisting of a fraction of fcc particles (60% by mass), and hcp particles (40%). The average size is approximately 2 nm.

In order to test the electrocatalytic properties of the tetrametal oxide colloid PtRuOsIrO$_x$ having a metal ratio of 42/44/9/5, which is close to the Smotkin/Mallouk alloy, it was necessary to support it on high surface conducting carbon (e.g., Vulcan XC 72). In the present case the colloid was first reduced with H$_2$ and the immobilizate (15% metal content) was then incorporated in a conventional membrane-electrode-assembly (MEA) which is part of the PEFC. This material as well as the previously prepared Pt/Ru-Vulcan and the usual commercial E-TEK Pt/Ru-Vulcan (which is normally employed in experiments of this kind) were first tested in H$_2$ oxidation in the absence of CO [56]. The current voltage curves are shown in Fig. 8.7. It is evident that the new Pt/Ru/Os/Ir catalyst and the commercial Pt/Ru catalyst show similar performance. However, if the electrocatalytic experiments are performed in the presence of varying amounts of CO, the catalysts behave differently. Upon applying 50 ppm CO (or more) the performance of the commercial catalyst is reduced drastically. In sharp contrast, the catalytic profile of the Pt/Ru/Os/Ir catalyst is more promising (Fig. 8.7). It shows a significantly improved tolerance toward CO. At 50 ppm CO contamination and at a voltage of 0.4 V, electrical performance is about 50% higher with the Pt/Ru/Os/Ir catalyst than with the commercial E-TEK Pt/Ru-Vulcan system. Even at 1039 ppm CO the fuel cell still displays appreciable performance, which is important in potential

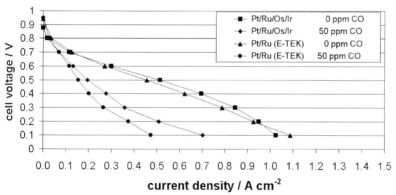

Fig. 8.7 Current–potential curves of PEFC single cells under ambient pressure operation with H$_2$/air and H$_2$, 50 ppm CO/air; $T=80\,°C$; active area: 17.4 cm^2. Anode: Pt(40)/Ru(41)/Os(16)/Ir(3) catalyst supported on Vulcan XC 72 (0.20 mg cm^{-2}) and commercial PtRu-ETEK supported on Vulcan XC 72 (0.20 mg cm^{-2}). Cathode: commercial E-TEK electrode (0.5 mg cm^{-2} Pt) [56].

applications. Thus, the results of Smotkin and Mallouk [57] concerning the Pt/Ru/Os/Ir combination using different synthetic methods are substantiated. Of course, this particular combination may well not be industrially viable, but it shows that a combinatorial search [57] combined with our method [51, 52, 54, 56] may pave the way to practical applications.

In further work other transition metals and their combinations as bimetallic and trimetallic aqueous colloidal metal oxides were tested [51, 54]. Although optimization was not strived for at that stage, the general protocol used previously for PtO_2, $PtRuO_x$ and $PtRuOsIrO_x$ colloids was successful in a number of cases, i.e., aqueous colloidal solutions containing 1–2 nm-sized nanoparticles were obtained. In addition to betaines, other stabilizers such as poly(vinylpyrrolidone), polyethylene glycol (PEG), poly(asparagin acid Na salt), bovine serum albumin (BSA), and poly(oxyethylene-20-cetyl) (NIO) were also employed successfully [54]. It was possible to prepare such interesting oxides as FeO_x, SnO, WO_x, CoO_x, RhO_x, RuO_x, OsO_x, IrO_x, $PtMoO_x$, $PtOsO_x$, $PtIrO_x$, $IrMoO_x$, $PtRuWO_x$, $PtRuMoO_x$, $PtRuOsO_x$, $PtRuSnO_x$ and $PtRuIrO_x$ as colloids in water [51, 54]. In some cases the transition metal nitrates were used as precursors which means that the metal oxide colloids are halogen-free [54]. It needs to be pointed out that these oxides constitute the result of exploratory experiments, and full characterization beyond TEM analyses have not yet been carried out. Sometimes the TEM pictures of the colloidal solutions reveal loose aggregation of the 1–2 nm-sized nanoparticles in the form of suprastructures, in other cases slow precipitation begins after a certain period of time, which limits the method (unless further optimization is performed). Moreover, in some cases (e.g., Cu), immediate precipitation is observed. This simple preparative method affords a multitude of fascinating materials, which remain to be studied with regard to electronic and catalytic properties. For example, as expected the color of the colloidal solutions covers a large range, depending upon the nature of the metal (e.g., PtO_2: yellow–orange; PdO: red; $IrOx$: blue). Unfortunately, it has not been possible to vary the size of the nanoparticles in the range1–5 nm. Rather, in almost all cases the average size amounts to 1–2.5 nm [51, 54].

The chemistry described here also opens the door for a new approach to the production of the analogous sulfides, selenides and tellurides [3, 5]: Initial experiments showed that it is possible to transform aqueous colloidal PtO_2 into colloidal forms of PtS and PtS_2 by treatment with Na_2S [58]. Mixed metal oxides/sulfides are intermediates, but so far no attempts have been made to isolate and characterize such unusual aqueous colloids.

In another important development it was demonstrated that the colloidal transition metal oxides can be produced and immobilized on appropriate surfaces without the presence of such stabilizers as the betaines **10/11** or other surfactants or polymers [54, 59]. This strategy, dubbed "instant method", has a number of advantages, because it is a one-step *in situ* procedure. Due to the absence of a stabilizer, it is also cheaper. Moreover, the usual necessity to remove the stabilizer following immobilization no longer pertains. The instant method was first demonstrated using H_2PtCl_6 and various forms of carbon black as the carrier [59]. The usual basic hydrolysis in the presence of carbon black leads to 1–2 nm-sized PtO_x nanoparticles which are immediately trapped by the solid carrier before undesired bulk PtO_2 can precipitate. Unusually high metal loading is easily achieved (up to 60%), which is necessary for application as catalysts in fuel cell technology. The solids can then be reduced chemically or electrochemically to the corresponding Pt/C materials:

$$H_2PtCl_6 + C \xrightarrow{Li_2CO_3/H_2O} PtO_x/C \xrightarrow{Reduction} Pt/C$$
$$(1\text{-}2 \text{ nm})$$

A surprising discovery was made, namely that the nature of the type of carbon black has a marked influence on the efficiency of electrocatalysis as measured by the activity of oxygen reduction [59]. For this purpose rotating disc electrode measurements were performed. Following reduction the current was measured at a potential at which the oxygen reduction occurs in a kinetically controlled process (at 0.9 V vs. NHE). It turned out that the carbon black used in ink jet technology, Printex XE2, leads to dramatically more active electrocatalysts than Vulcan XC72 (ETEK) which is traditionally employed in fuel cells. The increase in electrocatalytic activity upon using Printex XE2 was measured to be 34% (Fig. 8.8) [59]. High resolution TEM analysis of the carbon black revealed the reason for this novel effect. We discovered that it is the roughness of Printex XE2 which inhibits uncontrolled particle growth and agglomeration under operating conditions. The surface of this particular carbon black resembles a landscape composed of round "bath-tubs", the size of which corresponds approximately to the size of the Pt-nanoparticles. These are trapped at these sites, making undesired agglomeration

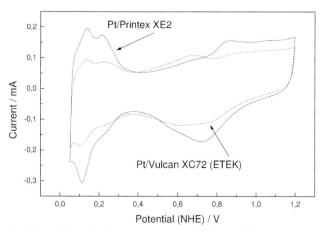

Fig. 8.8 Solid line: Cyclic voltammogram of Pt/Printex XE2 catalyst (63% metal loading, 2.128 μg Pt) prepared by the instant method. Dotted line: Commercial Pt/Vulcan XC72 catalyst from ETEK (58% metal loading, 2.32 μg Pt). Scan rate 100 mV s^{-1}, electrolyte is Ar-saturated 0.5 M perchloric acid [59].

(which leads to lower electrocatalytic activity) under operating conditions less likely. In contrast, Vulcan XC72 has a fairly smooth surface, allowing detrimental agglomeration to occur much more readily. Later even better results were achieved [60]. We suggest that researchers in this field should use carbon blacks such as Printex XE2 which have the optimal surface morphology. This may also apply to chemical catalysts composed of carbon black and transition metal clusters.

Finally, we addressed the question whether conditions can be found which allow stable aqueous colloidal solutions of transition metal oxides to be produced *without* any stabilizer being present. Such a protocol would not only constitute an alternative to the instant method, it could also enable certain applications not possible with the *in situ* procedure. This seems to contradict all of the above lines of thought, yet we discovered one case in which it actually works, namely IrO_x [61]. Upon dissolving an iridium salt such as $IrCl_3 \cdot H_2O$ in water and adding a base such as KOH, a blue–violet solution was obtained without any sign of undesired precipitation of bulk IrO_x, even at high concentrations. TEM analysis shows the presence of 2 nm-sized particles [61]. Importantly, the solution is stable for weeks. Thus, this approach is probably simpler and cheaper than our previous method [51, 54, 56] using stabilizers or procedures reported by other groups [62]. We assume that in this special case electrostatic factors contribute to unusual stabilization, but currently do not understand why iridium appears to be a unique case. The same treatment using other metals such as Pt or Pd leads to immediate formation of insoluble bulk metal oxide.

$$IrCl_3 \cdot x\ H_2O \xrightarrow[H_2O]{KOH} IrO_x\text{-colloid}$$

$$(1\text{-}2\ nm)$$

This form of IrO*x* colloid [61] has obvious advantages over analogs which incorporate organic stabilizers [13, 51, 54, 62]. Many different applications appear likely, including deposition on metal surfaces for corrosion protection. Thus far catalytic applications in colloidal or immobilized form have not been explored.

8.4
Conclusion

The Mülheim electrochemical method of producing $R_4N^+X^-$-stabilized transition metal colloids is a viable alternative to the traditional chemical process. The preparation of aqueous colloidal solutions of nanosized transition metal oxides or multimetal oxides (1–3 nm) is possible by an unusually simple procedure, namely hydrolysis of the corresponding metal salts under basic conditions in the presence of water-soluble stabilizers. High concentrations (0.1–0.5 M) are usually possible, which is crucial for industrial applications. In many cases an *in situ* method for immobilization on a solid carrier such as carbon black is possible in the absence of a stabilizer. In the rare case of IrO_x, the colloidal solution (2 nm) is stable for weeks in the absence of a stabilizer. The metal oxides as aqueous colloids or in immobilized form on solid supports can be reduced to the zerovalent form if so desired. Applications in chemical catalysis, electrocatalysis and in corrosion protection of metal surfaces are emerging.

References

1 Early reviews of metal nanoparticle preparation and catalysis: (a) G. Schmid, *Chem. Rev.* **1992**, *92*, 1709–1727; (b) L. N. Lewis, *Chem. Rev.* **1993**, *93*, 2693–2730; (c) J. S. Bradley, in *Clusters and Colloids: From Theory to Applications* (Ed.: G. Schmid), VCH, Weinheim, **1994**, Ch. 6, pp. 459–544; (d) L. N. Lewis, in *Catalysis by Di- and Polynuclear Metal-Cluster Complexes* (Eds.: R. D. Adams, F. A. Cotton), Wiley-VCH, New York, **1998**, pp. 373–394; (e) N. Toshima, in *Fine Particles Science and Technology. From Micro to Nanoparticles* (NATO ASI Series) (Ed.: E. Pelizzetti), Kluwer, Dordrecht, **1996**, pp. 371–383; (f) N. Toshima, T. Yonezawa, *New J. Chem.* **1998**, *22*, 1179–1201.

2 Recent reviews: (a) D. Astruc, F. Lu, J. R. Aranzaes, *Angew. Chem.* **2005**, *117*, 8062–8083; *Angew. Chem. Int. Ed.* **2005**, *44*, 7852–7872; (b) *Nanoparticles* (Ed.: G. Schmid), Wiley-VCH, Weinheim, **2004**; (c) *Metal Nanoparticles: Synthesis, Characterization and Applications* (Eds.: D. L. Feldheim, C. A. Foss, Jr.), Marcel Dekker, New York, **2001**; (d) *Nanoparticles and Nanostructured Films. Preparation, Characterization and Applications* (Ed.: J. H. Fendler), Wiley-VCH, Weinheim, **1998**; (e) G. Schmid, in *Nanoscale Materials in Chemistry* (Ed.: K. J. Klabunde), Wiley-Interscience, New York, **2001**, pp. 15–59; (f) A. Roucoux, J. Schulz, H. Patin, *Chem. Rev.* **2002**, *102*, 3757–3778; (g) M. Moreno-Manas, R. Pleixats, *Acc. Chem. Res.* **2003**, *36*, 638–643; (h) R. M. Crooks, M. Zhao, L. Sun, V. Chechik, L. K. Yeung, *Acc. Chem. Res.* **2001**, *34*, 181–190, (i) B. L. Cushing, V. L. Kolesnichenko, C. J. O'Connor, *Chem. Rev.* **2004**, *104*, 3893–3946.

3 (a) Review of chemistry and properties of nanocrystalline metals, metal oxides, sulfides and selenides with emphasis on shape-selectivity (see also [2i]): C. Burda, X. Chen, R. Narayanan, M. A. El-Sayed, *Chem. Rev.* **2005**, *105*, 1025–1102; see also (b) S. H. Im, Y. T. Lee, B. Wiley, Y. Xia, *Angew. Chem.* **2005**, *117*, 2192–2195; *Angew. Chem. Int. Ed.* **2005**, *44*, 2154–2157; (c) N. Zettsu, J. M. McLellan, B. Wiley, Y. Yin, Z.-Y. Li, Y. Xia, *Angew. Chem.* **2006**, *118*, 1310–1314; *Angew. Chem. Int. Ed.* **2006**, *45*, 1288–1292; (d) D. A. Zweifel, A. Wei, *Chem. Mater.* **2005**, *17*, 4256–4261.

4 (a) J. L. G. Fierro, *Metal Oxides: Chemistry and Applications*, CRC Press/Taylor & Francis Group, Boca Raton, **2006**; (b) G. A. Somorjai, *Introduction to Surface Chemistry and Catalysis*, Wiley, New York, **1994**; (c) J. M. Thomas, W. J. Thomas, *Principles and Practice of Heterogeneous Catalysis*, VCH, Weinheim, **1997**; (d) *Handbook of Heterogeneous Catalysis* (Eds.: G. Ertl, H. Knözinger, J. Weitkamp), VCH, Weinheim, **1997**; (e) G. A. Somorjai, R. M. Rioux, *Catal. Today* **2005**, *100*, 201–215; (f) R. J. Puddephatt in *Metal Clusters in Chemistry, Vol. 2* (Eds.: P. Braunstein, L. A. Oro, P. R. Raithby), Wiley-VCH, Weinheim, **1999**, pp. 605–615; (g) C. R. Henry, *Appl. Surf. Sci.* **2000**, *164*, 252–259; (h) T. P. St.Clair, D. W. Goodman, *Top. Catal.* **2000**, *13*, 5–19; (i) M. Bowker, R. A. Bennett, A. Dickinson, D. James, R. D. Smith, P. Stone, *Stud. Surf. Sci. Catal.* **2001**, *133*, 3–17; (j) M. Kralik, A. Biffis, *J. Mol. Catal. A* **2001**, *177*, 113–138; (k) J. M. Thomas, R. Raja, *Chem. Rec.* **2001**, *1*, 448–466; (l) C. Mohr, P. Claus, *Sci. Prog. (Northwood, U.K.)* **2001**, *84*, 311–334; (m) J. M. Thomas, B. F. G. Johnson, R. Raja, G. Sankar, P. A. Midgley, *Acc. Chem. Res.* **2003**, *36*, 20–30; (n) K. A. Zemski, D. R. Justes, A. W. Castleman, Jr., *J. Phys. Chem. B* **2002**, *106*, 6136–6148; (o) C. Feldmann, H.-O. Jungk, *Angew. Chem.* **2001**, *113*,

372–374; *Angew. Chem. Int. Ed.* **2001**, *40*, 359–362; (p) J. Kiwi, M. Grätzel, *Angew. Chem.* **1979**, *91*, 659–660; *Angew. Chem., Int. Ed. Engl.* **18**, 624–626; (q) P. A. Christensen, A. Harriman, G. Porter, P. Neta, *J. Chem. Soc., Dalton Trans. 2* **1984**, *80*, 1451–1464; (r) Y. Yamashita, K. Yoshida, *Chem. Mater.* **1999**, *11*, 61–66; (s) D. M. Kaschak, S. A. Johnson, D. E. Hooks, H.-N. Kim, M. D. Ward, T. E. Mallouk, *J. Am. Chem. Soc.* **1998**, *120*, 10887–10894.

5 (a) H. Weller, *Angew. Chem.* **1993**, *105*, 43–55; *Angew. Chem., Int. Ed. Engl.* **1993**, *32*, 41–53; (b) M. P. Pileni, in *Nanoscale Materials in Chemistry* (Ed.: K. Klabunde), Wiley-Interscience, New York, **2001**; (c) A. L. Rogach, D. V. Talapin, H. Weller, in *Colloids and Colloid Assemblies* (Ed.: F. Caruso), Wiley-VCH, Weinheim, **2004**, pp. 52–95; (d) Organization of nanoparticles into complex nanostructures and superstructures: N. A. Kotov, *Nanoparticles Assemblies and Superstructures*, CRC Press/Taylor and Francis Group, Boca Raton, **2006**; (e) M. A. White, J. A. Johnson, J. T. Koberstein, N. J. Turro, *J. Am. Chem. Soc.* **2006**, *128*, 11356–11357; (f) Z. P. Xu, Q. H. Zeng, D. Q. Lu, A. B. Yu, *Chem. Eng. Sci.* **2006**, *61*, 1027–1040; (g) D. L. Huber, *Small* **2005**, *1*, 482–501; (h) Y. M. Huh, Y. W. Jun, H. T. Song, S. Kim, J. S. Choi, J. H. Lee, S. Yoon, K. S. Kim, J. S. Shin, J. S. Suh, J. Cheon, *J. Am. Chem. Soc.* **2005**, *127*, 12387–12391.

6 See for example: R. Tsukamoto, K. Iwahori, M. Muraoka, I. Yamashita, *Bull. Chem. Soc. Jpn.* **2005**, *78*, 2075–2081.

7 Examples: (a) D. de Caro, J. S. Bradley, *Langmuir* **1997**, *13*, 3067–3069; (b) T. O. Ely, C. Amiens, B. Chaudret, E. Snoeck, M. Verelst, M. Respaud, J.-M. Broto, *Chem. Mater.* **1999**, *11*, 526–529; (c) N. Moumen, M. P. Pileni, *Chem. Mater.* **1996**, *8*, 1128–1134; (d) J. B. Nagy, A. Claerbout, *Surfactants Solution*, **1991**, *11*, 363–382; (e) A. Claerbout, J. B. Nagy, *Stud. Surf. Sci. Catal.* **1991**, *63*, 705–716.

8 Examples: (a) D. Kisailus, B. Schwenzer, J. Gomm, J. C. Weaver, D. E. Morse, *J. Am. Chem. Soc.* **2006**, *128*, 10276–10280; (b) X. Wang, Y. Li, *Pure Appl. Chem.*

2006, *78*, 45–64; (c) L. Vayssieres, *Int. J. Nanotechnol.* **2004**, *1*, 1–41.

9 (a) J. Turkevich, P. C. Stevenson, J. Hillier, *Discuss. Faraday Soc.* **1951** (11), 55–75; (b) J. Turkevich, *Gold Bull.* **1985**, *18*, 86–91; (c) J. Turkevich, G. Kim, *Science* **1970**, *169*, 873–879; (d) A. I. Kirkland, P. P. Edwards, D. A. Jefferson, D. G. Duff, *Annu. Rep. Prog. Chem., Sect C: Phys. Chem.* **1991**, *87*, 247–304; (e) D. G. Duff, A. Baiker, P. P. Edwards, *Langmuir* **1993**, *9*, 2301–2309.

10 (a) A. Henglein, *Top. Curr. Chem.* **1988**, *143*, 113–180; (b) M. Michaelis, A. Henglein, *J. Phys. Chem.* **1992**, *96*, 4719–4724; (c) A. Henglein, D. Meisel, *Langmuir* **1998**, *14*, 7392–7396; (d) A. Henglein, M. Giersig, *J. Phys. Chem. B* **1999**, *103*, 9533–9539; (e) see also Papirer's study of the thermolysis of dicobalt octacarbonyl in a constrained environment in which the kinetics of nucleation and growth were monitored: E. Papirer, P. Horny, H. Balard, R. Anthore, C. Petipas, A. Martinet, *J. Colloid Interface Sci.* **1983**, *94*, 220–228.

11 (a) R. B. Pontius, R. G. Willis, *Photogr. Sci. Eng.* **1973**, *17*, 326–333; (b) G. Bredig, *Angew. Chem.* **1898**, *11*, 951–954.

12 (a) D. G. Duff, A. Baiker, in *Preparation of Catalysts VI: Scientific Bases for the Preparation of Heterogeneous Catalysts* (Eds.: G. Poncelet, J. Martens, B. Delmon, P. A. Jacobs, P. Grange), *Stud. Surf. Sci. Catal.*, Vol. 91, Elsevier, Amsterdam, **1995**, pp. 505–512; (b) K. R. Brown, M. J. Natan, *Langmuir* **1998**, *14*, 726–728.

13 (a) M. A. Watzky, R. G. Finke, *J. Am. Chem. Soc.* **1997**, *119*, 10382–10400; (b) R. G. Finke, in *Metal Nanoparticles: Synthesis, Characterization and Applications* (Eds.: D. L. Feldheim, C. A. Foss, Jr.), Marcel Dekker, New York, **2001**, pp. 17–54.

14 (a) M. P. Pileni, *Ber. Bunsen-Ges.* **1997**, *101*, 1578–1587; (b) M. Antonietti, F. Gröhn, J. Hartmann, L. Bronstein, *Angew. Chem.* **1997**, *109*, 2170–2173; *Angew. Chem., Int. Ed. Engl.* **1997**, *36*, 2080–2083; (c) A. Taleb, C. Petit, M. P. Pileni, *Chem. Mater.* **1997**, *9*, 950–959; (d) K. Kurihara, J. H. Fendler, I. Ravet, J. B. Nagy, *J. Mol. Catal.* **1986**, *34*, 325–335; (e) S. Puvvada, S. Baral, G. M. Chow, S. B. Qadri, B. R.

Ratna, *J. Am. Chem. Soc.* **1994**, *116*, 2135–2136; (f) A. B. R. Mayer, J. E. Mark, R. E. Morris, *Polym. J.* **1998**, *30*, 197–205.

15 T. Hanaoka, T. Hatsuta, T. Tago, M. Kishida, K. Wakabayashi, *Appl. Catal., A* **2000**, *190*, 291–296.

16 J. H. Ding, D. L. Gin, *Chem. Mater.* **2000**, *12*, 22–24.

17 M. A. Markowitz, G.-M. Chow, A. Singh, *Langmuir* **1994**, *10*, 4095–4102.

18 (a) M. M. Maye, W. Zheng, F. L. Leibowitz, N. K. Ly, C.-J. Zhong, *Langmuir* **2000**, *16*, 490–497; (b) M. Brust, M. Walker, D. Bethell, D. J. Schiffrin, R. Whyman, *J. Chem. Soc., Chem. Commun.* **1994**, 801–802; (c) J. Fink, C. J. Kiely, D. Bethell, D. J. Schiffrin, *Chem. Mater.* **1998**, *10*, 922–926; (d) R. H. Terrill, T. A. Postlethwaite, C. Chen, C.-D. Poon, A. Terzis, A. Chen, J. E. Hutchison, M. R. Clark, G. Wignall, J. D. Londono, R. Superfine, M. Falvo, C. S. Johnson, Jr., E. T. Samulski, R. W. Murray, *J. Am. Chem. Soc.* **1995**, *117*, 12537–12548; (e) M. J. Hostetler, C.-J. Zhong, B. K. H. Yen, J. Anderegg, S. M. Gross, N. D. Evans, M. Porter, R. W. Murray, *J. Am. Chem. Soc.* **1998**, *120*, 9396–9397.

19 (a) T. Teranishi, M. Miyake, *Chem. Mater.* **1998**, *10*, 594–600; (b) T. Teranishi, I. Kiyokawa, M. Miyake, *Adv. Mater.* **1998**, *10*, 596–599; (c) T. Teranishi, M. Hosoe, T. Tanaka, M. Miyake, *J. Phys. Chem. B* **1999**, *103*, 3818–3827.

20 (a) G. W. Busser, J. G. van Ommen, J. A. Lercher, in *Advanced Catalysts and Nanostructured Materials: Modern Synthetic Methods* (Ed.: W. R. Moser), Academic Press, San Diego, **1996**, pp. 213–230; (b) G. W. Busser, J. G. van Ommen, J. A. Lercher, *J. Phys. Chem. B* **1999**, *103*, 1651–1659; (c) H. Hirai, Y. Nakao, N. Toshima, *J. Macromol. Sci. Chem.* **1979**, *13*, 727–750.

21 (a) R. G. DiScipio, *Anal. Biochem.* **1996**, *236*, 168–170; (b) K. Esumi, M. Shiratori, H. Ishizuka, T. Tano, K. Torigoe, K. Meguro, *Langmuir* **1991**, *7*, 457–459.

22 (a) O. Vidoni, K. Philippot, C. Amiens, B. Chaudret, O. Balmes, J.-O. Malm, J.-O. Bovin, F. Senocq, M.-J. Casanove, *Angew.*

Chem. **1999**, *111*, 3950–3952; *Angew. Chem. Int. Ed.* **1999**, *38*, 3736–3738; (b) A. Duteil, R. Quéau, B. Chaudret, R. Mazel, C. Roucau, J. S. Bradley, *Chem. Mater.* **1993**, *5*, 341–347.

23 (a) G. Schmid, *Inorg. Synth.* **1990**, *27*, 214–218; (b) M. N. Vargaftik, V. P. Zagorodnikov, I. P. Stolarov, I. I. Moiseev, D. I. Kochubey, V. A. Likholobov, A. L. Chuvilin, K. I. Zamaraev, *J. Mol. Catal.* **1989**, *53*, 315–348.

24 (a) D. H. Rapoport, W. Vogel, H. Cölfen, R. Schlögl, *J. Phys. Chem. B* **1997**, *101*, 4175–4183; (b) D. van der Putten, R. Zanoni, C. Coluzza, G. Schmid, *J. Chem. Soc., Dalton Trans.* **1996**, 1721–1725.

25 (a) M. T. Reetz, W. Helbig, *J. Am. Chem. Soc.* **1994**, *116*, 7401–7402; (b) M. T. Reetz, W. Helbig, S. A. Quaiser, in *Active Metals: Preparation, Characterization, Applications* (Ed.: A. Fürstner), VCH, Weinheim, **1996**, pp. 279–297.

26 Surfactant stabilizers of the type $R_4N^+X^-$ have been used particularly often; see for example: (a) J. Kiwi, M. Grätzel, *J. Am. Chem. Soc.* **1979**, *101*, 7214–7217; (b) Y. Sasson, A. Zoran, J. Blum, *J. Mol. Catal.* **1981**, *11*, 293–300; (c) N. Toshima, T. Takahashi, H. Hirai, *Chem. Lett.* **1985**, 1245–1248; (d) M. Boutonnet, J. Kizling, P. Stenius, G. Maire, *Colloids Surf.* **1982**, *5*, 209–229; (e) M. Boutonnet, J. Kizling, R. Touroude, G. Maire, P. Stenius, *Appl. Catal.* **1986**, *20*, 163–177; (f) K. Meguro, M. Torizuka, K. Esumi, *Bull. Chem. Soc. Jpn.* **1988**, *61*, 341–345; (g) J. Wiesner, A. Wokaun, H. Hoffmann, *Prog. Coll. Polym. Sci.* **1988**, *76*, 271–277; (h) N. Satoh, K. Kimura, *Bull. Chem. Soc. Jpn.* **1989**, *62*, 1758–1763; (i) H. Bönnemann, W. Brijoux, R. Brinkmann, E. Dinjus, T. Joussen, B. Korall, *Angew. Chem.* **1991**, *103*, 1344–1346; *Angew. Chem., Int. Ed. Engl.* **1991**, *30*, 1312–1314.

27 (a) N. Ibl, *Chem. Ing. Tech.* **1964**, *36*, 601–612; (b) R. Walker, *Chem. Ind. (London)* **1980**, 260–264.

28 (a) H. Natter, T. Krajewski, R. Hempelmann, *Ber. Bunsen-Ges.* **1996**, *100*, 55–64; (b) H. Natter, R. Hempelmann, *J. Phys. Chem.* **1996**, *100*, 19525–19532; (c) J. V. Zoval, R. M. Stiger, P. R. Biernacki, R. M. Penner, *J. Phys. Chem.* **1996**, *100*, 837–

844; (d) T. M. Whitney, J. S. Jiang, P. C. Searson, C. L. Chien, *Science* **1993**, *261*, 1316–1319; (e) L. Xu, W. L. Zhou, C. Frommen, R. H. Baughman, A. A. Zakhidov, L. Malkinski, J.-Q. Wang, J. B. Wiley, *Chem. Commun.* **2000**, 997–998; (f) P. N. Bartlett, P. R. Birkin, M. A. Ghanem, *Chem. Commun.* **2000**, 1671–1672; (g) Y.-Y. Yu, S.-S. Chang, C.-L. Lee, C. R. C. Wang, *J. Phys. Chem. B* **1997**, *101*, 6661–6664; (h) C.-J. Huang, P.-H. Chiu, Y.-H. Wang, W. R. Chen, T. H. Meen, *J. Electrochem. Soc.* **2006**, *153* (8), D129–D133.

29 M. T. Reetz, M. Winter, R. Breinbauer, T. Thurn-Albrecht, W. Vogel, *Chem.-Eur. J.* **2001**, *7*, 1084–1094.

30 M. A. Winter, Dissertation, Ruhr-Universität Bochum (Germany), **1998**.

31 M. T. Reetz, W. Helbig, S. A. Quaiser, U. Stimming, N. Breuer, R. Vogel, *Science* **1995**, *267*, 367–369.

32 (a) W. Helbig, Dissertation, Ruhr-Universität Bochum (Germany), **1994**; (b) S. A. Quaiser, Dissertation, Ruhr-Universität Bochum (Germany), **1995**.

33 T. Thurn-Albrecht, G. Meier, P. Müller-Buschbaum, A. Patkowski, W. Steffen, G. Grübel, D. L. Abernathy, O. Diat, M. Winter, M. G. Koch, M. T. Reetz, *Phys. Rev. E* **1999**, *59*, 642–649.

34 (a) A. Yasuhara, A. Kasano, T. Sakamoto, *Organometallics* **1998**, *17*, 4754–4756; (b) B. Yin, H. Ma, S. Wang, S. Chen, *J. Phys. Chem. B* **2003**, *107*, 8898–8904.

35 (a) J. A. Becker, R. Schäfer, R. Festag, W. Ruland, J. H. Wendorff, J. Pebler, S. A. Quaiser, W. Helbig, M. T. Reetz, *J. Chem. Phys.* **1995**, *103*, 2520–2527; (b) J. A. Becker, R. Schäfer, R. Festag, J. H. Wendorff, F. Hensel, J. Pebler, S. A. Quaiser, W. Helbig, M. T. Reetz, *Surf. Rev. Lett.* **1996**, *3*, 1121–1126.

36 M. T. Reetz, S. A. Quaiser, C. Merk, *Chem. Ber.* **1996**, *129*, 741–743.

37 (a) Y.-Y. Yu, S.-S. Chang, C.-L. Lee, C. R. C. Wang, *J. Phys. Chem. B* **1997**, *101*, 6661–6664; see also (b) M. B. Mohamed, Z. L. Wang, M. A. El-Sayed, *J. Phys. Chem. A* **1999**, *103*, 10255–10259.

38 M. T. Reetz, S. A. Quaiser, *Angew. Chem.* **1995**, *107*, 2461–2463; *Angew. Chem., Int. Ed. Engl.* **1995**, *34*, 2240–2241.

39 M. T. Reetz, S. A. Quaiser, R. Breinbauer, B. Tesche, *Angew. Chem.* **1995**, *107*, 2956–2958; *Angew. Chem., Int. Ed. Engl.* **1995**, *34*, 2728–2730.

40 M. T. Reetz, M. Dugal, *Catal. Lett.* **1999**, *58*, 207–212.

41 (a) M. T. Reetz, R. Breinbauer, K. Wanninger, *Tetrahedron Lett.* **1996**, *37*, 4499–4502; (b) M. Beller, H. Fischer, K. Kühlein, C.-P. Reisinger, W. A. Herrmann, *J. Organomet. Chem.* **1996**, *520*, 257–259.

42 M. T. Reetz, R. Breinbauer, P. Wedemann, P. Binger, *Tetrahedron* **1998**, *54*, 1233–1240.

43 M. T. Reetz, E. Westermann, *Angew. Chem.* **2000**, *112*, 170–173; *Angew. Chem. Int. Ed.* **2000**, *39*, 165–168.

44 M. T. Reetz, E. Westermann, R. Lohmer, G. Lohmer, *Tetrahedron Lett.* **1998**, *39*, 8449–8452.

45 Review: M. T. Reetz, J. G. de Vries, *Chem. Commun.* **2004**, 1559–1563.

46 I. P. Beletskaya, A. V. Cheprakov, *Chem. Rev.* **2000**, *100*, 3009–3066.

47 (a) A. Pundt, C. Sachs, M. Winter, M. T. Reetz, D. Fritsch, R. Kirchheim, *J. Alloys Compd.* **1999**, *293–295*, 480–483; (b) C. Sachs, A. Pundt, R. Kirchheim, M. Winter, M. T. Reetz, D. Fritsch, *Phys. Rev. B* **2001**, *64*, 075408/1–075408/10; (c) A. Pundt, M. Dornheim, M. Guerdane, H. Teichler, H. Ehrenberg, M. T. Reetz, N. M. Jisrawi, *Eur. Phys. J. D.* **2002**, *19*, 333–337; (d) M. Suleiman, N. M. Jisrawi, O. Dankert, M. T. Reetz, C. Bähtz, R. Kirchheim, A. Pundt, *J. Alloys Compd.* **2003**, *356–357*, 644–648; (e) A. Pundt, M. Suleiman, C. Bähtz, M. T. Reetz, R. Kirchheim, N. M. Jisrawi, *Mater. Sci. Eng., B* **2004**, *108*, 19–23.

48 (a) M. T. Reetz, M. Winter, G. Dumpich, J. Lohau, S. Friedrichowski, *J. Am. Chem. Soc.* **1997**, *119*, 4539–4540; (b) G. Dumpich, J. Lohau, E. F. Wassermann, M. Winter, M. T. Reetz, *Mater. Sci. Forum* **1998**, *287–288*, 413–415; (c) J. Lohau, S. Friedrichowski, G. Dumpich, E. F. Wassermann, M. Winter, M. T. Reetz, *J. Vac. Sci. Technol., B* **1998**, *16*, 77–79.

49 M. T. Reetz, M. Winter, B. Tesche, *Chem. Commun.* **1997**, 147–148.

50 (a) M. Maase, Dissertation, Ruhr-Universität Bochum (Germany), **1999**;

(b) M. T. Reetz, M. Maase, *Adv. Mater.* **1999**, *11*, 773–777; (c) J. S. Bradley, B. Tesche, W. Busser, M. Maase, M. T. Reetz, *J. Am. Chem. Soc.* **2000**, *122*, 4631–4636; (d) M. T. Reetz, G. Lohmer, *Chem. Commun.* **1996**, 1921–1922.

51 (a) M. Koch, Dissertation, Ruhr-Universität Bochum (Germany), **1999**; (b) M. T. Reetz, M. G. Koch, WO 2000/29332, May 25, 2000.

52 M. T. Reetz, M. G. Koch, *J. Am. Chem. Soc.* **1999**, *121*, 7933–7934.

53 M. T. Reetz, S. A. Quaiser, M. Winter, J. A. Becker, R. Schäfer, U. Stimming, A. Marmann, R. Vogel, T. Konno, *Angew. Chem.* **1996**, *108*, 2228–2230; *Angew. Chem., Int. Ed. Engl.* **1996**, *35*, 2092–2094.

54 (a) M. Lopez, Dissertation, Ruhr-Universität Bochum (Germany), **2002**; (b) M. T. Reetz, M. Lopez, WO 2003/078056 A1, September 25, 2003.

55 H. Wang, C. Wingender, H. Baltruschat, M. Lopez, M. T. Reetz, *J. Electroanal. Chem.* **2001**, *509*, 163–169.

56 M. T. Reetz, M. Lopez, W. Grünert, W. Vogel, F. Mahlendorf, *J. Phys. Chem. B* **2003**, *107*, 7414–7419.

57 (a) E. Reddington, A. Sapienza, B. Gurau, R. Viswanathan, S. Sarangapani, E. S. Smotkin, T. E. Mallouk, *Science* **1998**, *280*, 1735–1737; (b) B. Gurau, R. Viswanathan, R. Liu, T. J. Lafrenz, K. L. Ley, E. S. Smotkin, E. Reddington, A. Sapienza, B. C. Chan, T. E. Mallouk, S. Sarangapani, *J. Phys. Chem. B* **1998**, *102*, 9997–10003.

58 M. Krein, Dissertation, Ruhr-Universität Bochum (Germany), **2002**.

59 M. T. Reetz, H. Schulenburg, M. Lopez, B. Spliethoff, B. Tesche, *Chimia* **2004**, *58*, 896–899.

60 M. T. Reetz, H. Schulenburg, M. Lopez, B. Spliethoff, B. Tesche, in preparation.

61 (a) M. T. Reetz, H. Schulenburg, WO 2005/095671, October 13, **2005**; (b) M. T. Reetz, H. Schulenburg, in preparation.

62 (a) A. Harriman, J. M. Thomas, G. R. Millward, *New J. Chem.* **1987**, *11*, 757–762; (b) M. Hara, C. C. Waraksa, J. T. Lean, B. A. Lewis, T. E. Mallouk, *J. Phys. Chem. A* **2000**, *104*, 5275–5280; (c) M. Hara, T. E. Mallouk, *Chem. Commun.* **2000**, 1903–1904.

9
Multimetallic Nanoparticles Prepared by Redox Processes Applied in Catalysis

Florence Epron, Catherine Especel, Gwendoline Lafaye, and Patrice Marécot

9.1
Introduction

Since the introduction of Pt–Re catalysts for naphtha reforming, bimetallic and multimetallic catalysts have been attracting much attention with regard to improving the selectivity and stability of metallic active sites. However, the catalytic performances of the supported nanoparticles strongly depend on the preparation method. Whereas classical co-impregnation and successive impregnation techniques lead to an unpredictable deposition of the two metals, different methods have been developed to favor the metal–metal interactions, such as surface organometallic chemistry, chemical vapor deposition, deposition via the colloidal route, redox reactions in the aqueous phase etc. This latter technique allows one to obtain bimetallic particles in smooth temperature conditions.

This chapter deals with the preparation of multimetallic nanoparticles based on the results of the chemistry of redox processes applied in catalysis and catalyst preparation [1–4]. First, the general aspects of this technique will be briefly described; the second part will be devoted to the practical aspects of this preparation method and then some applications of the resulting catalysts in the synthesis of organic chemicals, in environmental catalysis and in catalysis for energy will be presented.

9.2
General Aspects

Whatever the technique used, the first step always consists of the preparation of a monometallic catalyst, which is generally activated by calcination and reduction. Then, the monometallic catalyst is introduced into the preparation reactor, reduced and the liquid phase is added to obtain a suspension of the catalyst. The metal constituting the monometallic catalyst is denoted as "parent" metal. The last step of the preparation procedure consists of the introduction of the modifier in solution in the form of a precursor salt at ambient temperature. The different

Nanoparticles and Catalysis. Edited by Didier Astruc
Copyright © 2008 WILEY-VCH Verlag GmbH & Co. KGaA, Weinheim
ISBN: 978-3-527-31572-7

techniques differ according to this last step. Indeed, four preparation methods can be chosen as a function of the reversible Nernst potential of the different species involved in the preparation procedure, namely (i) direct redox reaction (process also called "cementation") between the "parent" metal and the oxidized form of the modifier, (ii) redox reaction between an adsorbed reducing agent and the oxidized form of the modifier, (iii) catalytic reduction and (iv) underpotential deposition. According to these techniques, trimetallic catalysts can also be prepared by replacing the parent monometallic catalyst with a bimetallic one. This can be extended to the preparation of multimetallic catalysts.

9.2.1
Preparation of Bimetallic Catalysts by Direct Redox Reaction

In this case, metal–metal interactions result from surface redox reactions between the prereduced parent metal M and the oxidized form of the modifier $M'^{Z'+}$.

The half reaction corresponding to the oxidation of the parent metal is the following:

$$M \rightleftharpoons M^{Z+} + ze^-$$

According to the Nernst equation, the equilibrium potential ($E_{M^{Z+}/M}$) of the reversible oxidation of M is expressed as the following:

$$E_{M^{Z+}/M} = E^\circ_{M^{Z+}/M} + \frac{RT}{zF}\ln\frac{a_{M^{Z+}}}{a_M} \tag{9.1}$$

where $a_{M^{Z+}}$ and a_M are the activities of oxidized (M^{Z+}) and reduced (M) species, respectively, $E^\circ_{M^{Z+}/M}$ is the standard reduction potential, R the gas constant, F the Faraday constant and T the temperature.

In the same way, the equilibrium potential of the reversible reduction of the modifier in oxidized form ($M'^{Z'+}$) is given by:

$$E_{M'^{z'+}/M'} = E^\circ_{M'^{z'+}/M'} + \frac{RT}{z'F}\ln\frac{a_{M'^{z'+}}}{a_{M'}} \tag{9.2}$$

If the value of the change in Gibbs free energy $\Delta(\Delta G)$ given by:

$$\Delta(\Delta G) = -(\Delta G'_{M'^{z'+}/M'} - \Delta G_{M^{z+}/M}) = -F(z'E_{M'^{z'+}/M'} - zE_{M^{z+}/M}) \tag{9.3}$$

is negative M'^{Z+} will be reduced by the metal M as follows:

$$z'M + zM'^{z'+} \rightarrow zM' + z'M^{z+}$$

The system will tend to an equilibrium that corresponds to the equality of the two reversible potentials $E_{M'^{z'+}/M'}$ and $E_{M^{z+}/M}$. In this way, surface atoms M of the metallic particles are replaced by atoms M' of the modifier. This method is used to deposit metals with high standard reduction potential, such as noble metals, onto nanoparticles of metals with a lower standard reduction potential. Many examples of the use of direct redox reactions in the preparation of bimetallic catalysts could be quoted, such as the modification of (i) copper catalysts by ruthenium [5, 6], by platinum [7, 8], by palladium [7, 8] or by gold [7], (ii) platinum catalysts by gold [9–12] or palladiumcatalysts by platinum [13]. As an example, platinum ($E^\circ_{Pt^{2+}/Pt} = 1.188\,V$ vs. SHE) can be deposited onto copper particles considering copper in the metallic state ($E^\circ_{Cu^{2+}/Cu} = 0.34\,V$ vs. SHE) according to the reaction:

$$Cu(s) + Pt^{2+}(aq) \rightarrow Cu^{2+}(aq) + Pt(s)$$

Of course, if the precursor salt is introduced in the form of a complex, the stability constant of the complex must be taken into account. For instance, the standard reduction potential of platinum in chloroplatinic acid solution ($E^\circ_{PtCl_6^{2-}/Pt} = 0.744\,V$ vs. SHE) is lower than that of the Pt^{2+}/Pt couple and corresponds to the half reaction:

$$PtCl_6^{2-}(aq) + 4e^- \rightleftarrows Pt(s) + 6Cl^-(aq)$$

9.2.2
Redox Reactions of Adsorbed Species in the Preparation of Bimetallic Catalysts

This method is called in the literature "recharge" or "refilling" method [10–17] or "adsorption of metallic ions via ionization of adsorbed hydrogen" [18–25].

In this method, the ions of the modifier are reduced by a reagent (most commonly hydrogen), which preadsorbs selectively on the metal. This requires the parent metal to chemisorb hydrogen. Then, the adsorbed reducing agent, for instance hydrogen, can be considered as a source of electrons according to the following reaction:

$$H_{ads} \rightarrow H^+ + e^-$$

Then the reduction of the modifier $M'^{z'+}$ can occur:

$$M'^{z'+} + z'H_{ads} \rightarrow M' + z'H^+$$

This process takes place in several steps [2]. The result can be either bulk or adsorbed deposit if the $M'^{z'+}$ ion is a noble metal ion.

In this method, the prereduced monometallic catalyst suspended in aqueous solution is treated with hydrogen until the surface of the catalyst is completely saturated by adsorbed hydrogen. Then, the reactor is flushed with inert gas to remove the hydrogen dissolved in the electrolyte. After outgassing, a deoxygenated solution containing the precursor salt of the modifier is added.

Many examples are reported in the literature where platinum, palladium or ruthenium are used as parent metals, such as Pt–Cu [15, 18], Pt–Au [12, 17, 19], Pt–Pd [26], Pt–Re [27, 28], Pt–Ru [22], Pt–Bi [20] , Pt–Pb [21], Pd–Pt [13], Pd–Cu [16, 23], Pd–Ge [24, 25].

9.2.3
Catalytic Reduction in the Preparation of Bimetallic Catalysts

As for the "refilling" method described in the previous section, the chemisorption properties of the parent metal are used to activate hydrogen, but the deposition of the modifier is performed under flowing hydrogen. According to the thermodynamics, molecular hydrogen in solution can reduce any metallic salt that has a redox potential greater than that of the H^+/H_2 couple. However, at room temperature, the charge transfer reaction can be the rate determining step according to the Volmer–Butler equation. Then, the catalytic properties of the metal are used to increase the rate of reduction of the additive ion by means of a reducing agent, such as hydrogen, in solution. Thus the metal plays the role of the catalyst in the redox reaction. In such conditions, the additive should be deposited on the catalytic site that is active for the reduction reaction. However, transfer of electrons from the parent metal to the metallic modifier could also favor the formation of three-dimensional agglomerates. In the same way, if the parent catalyst consists of a metal supported on a conductor, the additive will deposit competitively on the parent metal and on the support.

To date, catalytic reduction has been mainly used to modify platinum by rhenium [27, 29–32] or by copper [33–35], palladium by copper [35–37], by silver [35], by gold [35] or by tin [38], ruthenium by copper [5] or by lead [39], or rhodium by copper [40] or by germanium [41, 42]. Most recently, trimetallic catalysts have also been prepared by catalytic reduction, such as Pt–Ir–Sn [43, 44], Pt–Ir–Ge or Pt–Re–Ge [45] or Pd–Sn–Au [46].

9.2.4
Underpotential Deposition

Underpotential deposition consists of the formation of metal submonolayers at potentials more positive than the reversible Nernst potential, before bulk deposition can occur [2, 47–51]. This technique is also called "adatom deposition" [52] or "adsorption of metal atoms" [53].

When M' is deposited in a submonolayer (M'_{ML}) on the metal M, according to the reaction,

$$M'^{z'^+} + z'e^- \rightleftharpoons M'$$

the coverage θ of M by M' and then the activity of M' in the submonolayer ($a_{M'_{ML}}$) is less than one. When there is no specific interaction between the surface and the adatoms ($\theta \leq 0.2$), the Langmuir adsorption isotherm can be used according to Ref. [54]. Then the activity of M' is given by:

$$a_{M'_{ML}} = \frac{\theta}{1-\theta} \tag{9.4}$$

For the deposition of a submonolayer of metal, the equilibrium potential can be given by:

$$E_{M'^{z'+}/M'_{ML}} = E^{\circ}_{M'^{z'+}/M'_{ML}} + \frac{RT}{z'F} \ln \frac{a_{M'^{z'+}}}{a_{M'_{ML}}} \tag{9.5}$$

where $a_{M'^{z'+}}$ is the activity of the depositing metal ion and $E^{\circ}_{M'^{z'+}/M'_{ML}}$ is the standard submonolayer potential.

The equilibrium potential of the submonolayer is always more positive than the Nernst potential of a bulk deposition. As a result, an underpotential deposition (UPD) of adatoms of M' on M can occur. Underpotential deposition leads to the formation of submonolayers before the three-dimensional bulk deposition occurs and the coverage varies with the potential and time of deposition. UPD is characterized by the existence of adatoms. The simplest methods for investigating UPD are electrochemical.

Examples of the use of the UPD technique are the modification of platinum catalysts by copper [53–60], by arsenic [61], by gold [62], by iron [63], by lead or tin [64, 65], or by rhodium [66] and the modification of palladium by copper [23], by germanium [24] or by iron [67–69]. The majority of these examples concerns the modification of electrodes, which is not the subject of this chapter.

9.3
Practical Aspects

9.3.1
Stability of Supported Catalysts in the Aqueous Phase

It is mentioned in Section 9.2 that the preparation of bimetallic nanoparticles by surface redox reactions occurs under a controlled atmosphere between a "parent" monometallic catalyst and the cation of the second metal dissolved in aqueous solution. However, the immersion of the parent catalyst in the solution may induce a modification of the metal particle size before the introduction of the ion of the second metal. In an aqueous medium, water molecules could penetrate between the metal–support interface and induce the migration of metal crystallites. Moreover, water and gas, for instance hydrogen, can adsorb and dissociate on metals, leading to a modification of their electronic properties. Consequently, it is important to study the stability of supported monometallic catalysts in an aqueous environment under a gas atmosphere at room temperature. The influence of different parameters such as the initial oxidation state of the metal, the pH of the solution, the presence of chloride anions in the solution and the nature of the gas atmosphere has been reported in the literature. The stability of supported metal catalysts treated in aqueous media under gas atmosphere at room temperature depends on the initial oxidation state of the metal, i.e. preoxidized or *in situ* reduced. Indeed, immersion in the liquid phase of a prereduced platinum supported catalyst then exposed to air induces a more or less important sintering of the metal, depending on various parameters such as the nature of the support, the pH of the solution, etc. [70]. An *in situ* reduction before catalyst immersion in an aqueous solution prevents or restricts the metal accessibility loss.

The sintering of the parent metal may result from the immersion in the aqueous medium only or from both the immersion in the solution and the nature of the bubbling gas. Table 9.1 shows for example that the metallic dispersion loss of a Pt/silica catalyst is more marked in the presence of hydrogen, compared to nitrogen, whatever the initial pH of the solution.

This phenomenon results from the metal–H binding energy that can be far higher than that of the metal–metal [71, 72]. This metal–H interaction weakens the metal–support interaction and so the metallic particles become unstable and migrate onto the support, leading to platinum nanoparticle growth. This sintering is even stronger in the presence of Cl$^-$ ions (acidic solution) or OH$^-$ ions (basic solution), which can interact with metal particles, inducing also a decrease in the density of electrons in the metal [13, 70, 72]. Another explanation was proposed for the sintering of copper in Cu/SiO$_2$ catalysts, which is based on a dissolution/redeposition process [73].

These effects are all the more pronounced as the metal–support interactions of the parent catalyst are weak. In other words, when the metal–support interactions are prevailing, the metallic particles remain stable. Examples are reported in the literature for platinum [70, 72], palladium [38] and rhodium [41], showing that silica leads to a more severe sintering than alumina. Indeed, it appears that the

Table 9.1 Comparison of the treatment effects on the dispersion loss of different catalysts under hydrogen and nitrogen in aqueous media during 24 h [70].

Catalysts	Platinum dispersion loss (%): $100(D_0 - D)/D_0$		
	pH 0.7	pH 6	pH 10
1.0 wt.% Pt/SiO$_2$ (D_0=41%)			
under N$_2$	60	6	10
under H$_2$	72	38	62
1.0 wt.% Pt/Al$_2$O$_3$ (D_0=45%)			
under H$_2$	0	0	10

platinum–support interactions decrease following the sequence: $Al_2O_3 > C > SiO_2$ [74]. Furthermore, aggregates of parent metal particles were observed on catalysts supported on silica and treated in aqueous media under hydrogen atmosphere and dried at 120 °C, contrary to those on an alumina support [41, 72]. The sintering of the parent metal can occur via crystallite migration onto the support.

In the course of the preparation of bimetallic catalysts by redox reactions, the modifier itself can induce a modification of the metal particle size. Then, sintering of Pt can occur during redox reactions under a hydrogen atmosphere in the presence of Au, which has a higher standard redox potential than that of Pt [10]. This sintering is explained by the oxidation of platinum atoms of low coordination number and a further reduction by hydrogen of those oxidized forms on platinum atoms of high coordination number. On the other hand, the deposition of germanium restricts the sintering of rhodium particles which is observed when a parent rhodium/silica catalyst is immersed in an aqueous solution under a hydrogen atmosphere. An explanation of this phenomenon could be the consumption of hydrogen activated on the rhodium surface by the reduction of the germanium salt [41]. The same phenomenon was observed for palladium/alumina modified by tin [38].

In conclusion, the treatment of supported catalysts in an aqueous medium seems therefore unfavorable to the presence of well dispersed nanoparticles. Nevertheless, an *in situ* reduction of the parent catalyst allows one to prevent or restrict metal sintering, depending on the nature of the support. However, in the absence of any contact of the catalyst with air, many parameters can also influence the metal particle size. Then, for example it is better to avoid especially very acidic (pH 2) or very basic (pH 10) solutions [13, 72].

9.3.2
Study of the Deposition Reaction

9.3.2.1 On Bulk Catalysts

In the case of powder bulk catalysts, Cu Raney was modified by direct redox reaction between reduced copper and the salt of a noble metal M (Ru, Pt and Au) [5, 7]. Typically the Cu–M bimetallic catalysts were obtained by mixing a freshly prepared Cu Raney with an aqueous solution of the noble metal salt. When the amount of M is in excess compared to the number of copper surface atoms, it appears that ruthenium deposition is restricted to approximately 1/3 of the copper surface atoms. For platinum and gold, a deposit larger than a monolayer is obtained, indicating that subsurface copper atoms are involved in direct redox reaction. This result is explained by the lower potential difference between copper and ruthenium compared to those of copper and platinum or gold [7]. However, the reactions involved in the direct redox reaction may not be as simple as indicated in Section 9.2. A typical time distribution of ion concentrations in solution during the preparation of Cu–Pt is shown in Fig. 9.1. It can be observed that platinum ions disappear very rapidly from the solution while at the same time copper

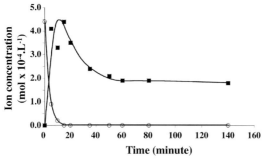

Fig. 9.1 Time distribution of (•) Cu ion and (○) Pt ion concentrations in solution during the preparation of Cu–Pt by direct redox reaction, with an initial atomic ratio $Pt^{4+}/Cu_S = 57$ (2 g of Raney Cu, 100 mL of solution). Reprinted from C. Montassier, J. C. Ménézo, J. Naja, J. Barbier, J. M. Dominguez, P. Sarrazin, B. Didillon, *J. Mol. Catal.*, **1994**, *91*, 107–117, Copyright (1994), with permission from Elsevier.

ions are formed and their concentration reaches a maximum. Later, the soluble copper fraction decreases to a plateau. The Cu^{n+} species concentrations at this plateau are not representative of the redox reaction but are determined by the solubilities of the different copper ions in the solution, depending on the anion accompanying the metal precursor [7]. Likewise, Pt Adams was modified by gold added by direct and refilling redox reactions [10].

Recently, novel catalysts were obtained by impregnation of Cu–Pt or Cu–Pd nanospheres on active carbon by the incipient-wetness technique. The bimetallic species were obtained by direct redox reaction between the surface of bulk copper nanospheres and Pt or Pd salts [8].

9.3.2.2 On Supported Catalysts
The supported catalysts can be modified either by noble metals or base metals.

For noble modifiers, two techniques were mainly used: the direct redox and the refilling method with preadsorbed hydrogen [10, 12, 13, 75]. The different supports presented in the literature are silica [10, 75], alumina [13] and active carbon [12]. It should be pointed out that the preparation of Pt–Au/SiO$_2$ bimetallic catalysts using hydrogen preadsorbed on platinum is a complex reaction involving simultaneous reduction of $AuCl_4^-$ by hydrogen and direct oxidation of metallic platinum. According to the difference in initial rates (10^{19} versus 10^{17} atoms of Au reduced per s and per m^2 of Pt) pointed out in Ref. [10], it can be assumed that during the preparation of Pt–Au bimetallic catalysts, for low gold coverages (when the quantity of hydrogen is sufficient to reduce all the introduced gold ions), gold is reduced by preadsorbed hydrogen. On the other hand, for higher gold concentrations, reduction by preadsorbed hydrogen and then by direct oxidation of platinum can occur [10].

When the modifier is a base metal, the most commonly used redox reaction is catalytic reduction. Nevertheless, the "refilling" method with preadsorbed hydrogen has been utilized in some cases. For example, platinum over alumina reforming catalyst was modified by rhenium via this technique [27]. However, the rhenium deposit is limited by the amount of hydrogen adsorbed on the surface platinum. Then, many successive "refilling" steps must be performed in order to deposit the amount of Re needed for the usual Pt–Re reforming catalysts. On the other hand, the catalytic reduction allows one to deposit large amounts of rhenium because the reaction is carried out under hydrogen flow [27, 30]. This method was also used for the preparation of different bimetallic catalysts, such as Rh–Ge [41, 42], Pt–Ru [76], Pd–Cu [35, 36], Pt–Cu [34, 35], Pd–Sn [38], Ru–Pb [39], Pt–Sn [43, 44], or trimetallic catalysts: Pt–Ir–Sn [43, 44], Pt–Ir–Ge [45], Pt–Re–Ge [45], Pd–Sn–Au [46], Pt–Ru–Sn [76] and Pt–Ru–Mo [76].

In the course of the catalytic reduction, deposition of the second metal (or the third) can occur on both the parent metal and the support, depending on the nature of the support and the operating conditions. Effectively, in the case of bimetallic Rh–Ge [41] and Pt–Sn [38] catalysts, it was observed that the amount of Ge or Sn deposited is higher on alumina than on silica supported catalysts (Fig. 9.2). Furthermore, the same experiments carried out under nitrogen flow do not lead to additive deposition on silica supported catalysts, while some deposit occurs on alumina supported catalysts. This result was explained by the adsorption of the additive anionic species in an acidic medium on the alumina support compared to a lack of adsorption on the silica support, which is in accord with their isoelectric points. Moreover, the greater additive deposition under hydrogen flow compared to that observed under nitrogen supported the idea that hydrogen activated on the parent metal particles participates in the deposition reaction.

In order to decrease the contribution of the support, different parameters can be optimized:

- The deposition on the metal by catalytic reduction being faster than that on the support, the reaction time can be limited to some minutes [27, 30, 43].

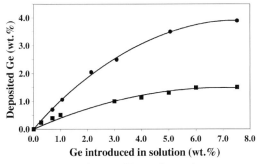

Fig. 9.2 Wt.% Ge deposited by catalytic reduction over (•) Rh/ silica and (•) Rh/alumina as a function of the wt.% Ge introduced in solution. Reprinted from G. Lafaye, C. Micheaud-Especel, C. Montassier, P. Marecot, *Appl. Catal. A,* **2002**, *230*, 19–30, Copyright (2002), with permission from Elsevier.

- A competitive adsorbate can be added to the reaction mixture. For example the addition of citric acid in the acid medium allows one to decrease tin deposit on an alumina support in the course of the preparations of Pt–Sn and Pt–Ir–Sn reforming catalysts [43, 44].
- The nature of the precursor and especially the oxidation state of the additive. For example the catalytic reduction is faster with Re^{4+} than with Re^{7+}, which promotes Re deposition on Pt for a short reaction time [27].
- The nature of the reducer [77].

Other parameters can also strongly influence the kinetics and the extent of a modifier deposition by catalytic reduction, such as the partial pressure of hydrogen, the reaction temperature, the concentration of the additive salt and the dispersion of the parent catalyst [27].

9.3.3
Characterization of the Metal–Metal Interaction

9.3.3.1 Model Reactions
Different model reactions were used in order to study the interaction between the modifier and the parent metal. It was observed that an inert additive introduced by a redox reaction generally poisons, more or less, the activity of the parent metal or strongly modifies the selectivity of the reaction, which indicates a deposition of the additive on the parent metal. For example, a decrease in activity for structure insensitive reactions, such as toluene hydrogenation [41] or cyclohexane dehydrogenation [43, 78] proves the existence of bimetallic nanoparticles. Likewise, in the case of the 2,2-dimethylpropane reaction, the modification of both the selectivity and the apparent activation energy, demonstrates an interaction between Pd and Au introduced by direct redox reaction. Conversely, no modification was observed on the catalysts prepared by incipient wetness co-impregnation [75].

Another useful probe for indicating the extent of Pt–Re alloying on reforming catalysts is cyclopentane hydrogenolysis, the alloy being intrinsically more active than either pure metal [79, 80]. In Fig. 9.3, the synergy in the activity shows a great interaction between the two metals for Pt–Re catalysts prepared by catalytic reduction [30].

9.3.3.2 Adsorption of Probe Molecules
The adsorption of probe molecules followed by different techniques allows one to prove the metal–metal interaction for catalysts prepared by redox reactions. For example, by chemisorption measurements, a decrease in the total amount of adsorbed H_2 was observed with an increasing germanium content introduced by catalytic reduction on parent rhodium catalysts [81]. As TEM characterization showed a comparable mean particle size for all the catalysts, such evolution suggests that Ge covers the Rh surface [41]. These results were consistent with those

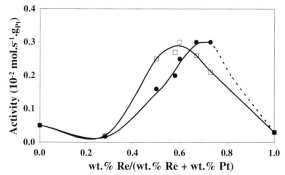

Fig. 9.3 Cyclopentane hydrogenolysis (10^{-2} mol converted per second and per gram of Pt) on Pt–Re catalysts prepared by catalytic reduction in HNO_3 medium: (□) catalysts activated by reduction (hydrogen, 8 h, 500 °C); (•) catalysts activated by calcination (air, 4 h, 450 °C) and then reduction (hydrogen, 8 h, 500 °C). Reprinted from C. L. Pieck, P. Marecot, J. Barbier, *Appl. Catal. A*, **1996**, *143*, 283–298, Copyright (1996), with permission from Elsevier.

obtained by FTIR spectroscopic studies of adsorbed CO showing a decrease in chemisorbed CO [81]. Moreover, modifications and band shifts of FTIR spectra of adsorbed CO have indicated a close contact between the two metals in Pt–Cu, Pt–Ag, Pd–Cu, Pd–Ag and Rh–Ge catalysts prepared by catalytic reduction [35, 81].

Temperature programmed reduction (TPR) under hydrogen is a technique mainly used to characterize metallic phases. In the case of M_1–M_2 bimetallic catalysts, a reduction peak usually appears between those of the pure metals M_1 and M_2, this peak being all the more significant when the interaction M_1–M_2 is high. Bimetallics prepared by redox reactions show this characteristic (Fig. 9.4) [34, 35, 42, 43]. Nevertheless, when a competitive adsorption of the additive occurs on the support, a hydrogen consumption corresponding to the reduction of the isolated species appears at higher temperature [34, 35, 42, 43]. The presence of these species on the support also allows one to explain the variations in the Brönsted and Lewis acidity of reforming Pt/alumina or Pt–Ir/alumina modified by Ge [44]. The model reaction to determine the Brönsted acidity was the 3,3-dimethylbut-1-ene isomerization while FTIR adsorption of pyridine was used to determine the number of Lewis acid sites.

9.3.3.3 Physical Techniques
Some physical techniques have also been used to characterize metallic nanoparticles prepared by surface redox reactions, for example:
- X-ray photon spectroscopy (XPS) analysis was performed on stored Cu–Au catalysts prepared by direct redox reaction with no subsequent reduction by hydrogen. After contacting the reduced copper catalyst with the Au salt, Au is in the metallic state [7].

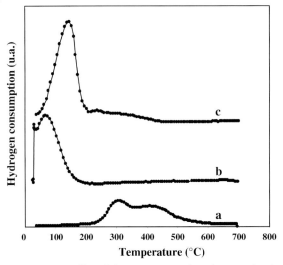

Fig. 9.4 TPR profiles of alumina supported catalysts, oxidized 12 h under flowing oxygen at 400 °C: (a) 1 wt.% Cu; (b) 3 wt.% Pt and (c) 3 wt.% Pt–1 wt.% Cu prepared by catalytic reduction. Reprinted from F. Epron, F. Gauthard, J. Barbier, Appl. Catal. A, **2002**, 253–261, Copyright (2002), with permission from Elsevier.

- X-ray diffraction (XRD), which allows one to determine the composition of crystallized compounds, indicated that bimetallic aggregates are formed for Pt–Au prepared by direct redox or the refilling method [11] and for Pt–Sn prepared by catalytic reduction [38].
- Extended X-ray absorption fine structure (EXAFS) measurements for Pd–Pt systems prepared either by direct redox or refilling method indicated the presence of platinum in the vicinity of palladium for both preparations [13].
- Transmission electron microscopy (TEM) coupled with energy dispersive X-ray (EDX) microanalysis was carried out on Pd or Pt modified by Cu, Ag and Au [35], Pd–Sn [38], Pd–Cu [36], Rh–Ge [41], Pt–Ru–Sn and Pt–Ru–Mo (Fig. 9.5) [76], Pd–Sn–Au and Pt–Sn–Au [46] to collect more information on metallic particles in terms of particle size and composition. The catalysts prepared by redox reactions showed that the parent metal is always associated to the additive, but can be found isolated for supports which are able to adsorb the modifier.
- Mössbauer spectroscopy showed that on silica supported catalysts, only Pd–Sn solid solutions or alloys are present, whereas, on alumina supported catalysts, tin is present at

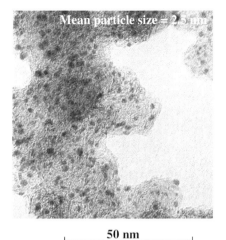

Fig. 9.5 TEM image of a 20 wt.% (Pt + Ru)–1.8 wt.% Mo/C
catalyst prepared by catalytic reduction.

the interface between the support and palladium, or alone
on the support, due to its high affinity for alumina [38]. On
trimetallic Pd–Sn–Au, a trimetallic phase was also evidenced
on a silica support [46].

All these techniques are valuable in assessing the degree of alloying. By combining some of these techniques, it was possible to go further in the determination
of the location of the deposit (M_2) on parent metal particles (M_1). Effectively, in
the course of redox reactions, the deposit exhibited a selectivity depending on the
relative redox potential of M_1^{n+}/M_1 and M_2^{n+}/M_2. Then, when the difference in
potential is low such for the Pt–Au, Pd–Pt or Cu–Ru couples, the additive was
preferentially deposited on low coordination sites (corners, edges, etc.) [6, 7, 10,
13], but when the difference in potential is high, for example with Cu–Pt, Cu–Au,
all Cu sites were involved in the redox reaction [7].

9.3.4
Influence of the Gaseous Environment on the Nanoparticle Stability

The catalytic properties are strongly influenced by the activation mode of the catalysts. Then, the influence of different gas treatments at the end of the additive
deposition by redox reactions was examined. These treatments may lead to a sintering or a segregation of the metals.

After preparation by catalytic reduction and after drying overnight at 70 °C under
nitrogen flow, Pt–Cu/Al$_2$O$_3$ catalysts were tested for the nitrate reduction as is or
after air exposure at ambient temperature [34]. The highest activity was obtained
in the latter case. This allows one to increase Pt accessibility in bimetallic catalysts,
whereas the coverage of Pt particles by Cu in the just-prepared catalysts is probably

too great. Conversely, a poor activity is obtained after oxidation at 400 °C, the segregation between the two metals being too great with a migration of copper towards the alumina support. Likewise, calcination at 450 °C, followed by reduction at 500 °C greatly diminishes the Pt–Re metal–metal interaction [30, 78]. Thus, the maximum in activity observed for cyclopentane hydrogenolysis (Fig. 9.3) is shifted to higher rhenium contents after calcination [30].

More often the samples are directly reduced under hydrogen at various temperatures after additive deposition and drying. It appears that direct reduction with a high heating-rate (~10 °C min^{-1}) under hydrogen leads to a sintering whereas a low heating-rate (~2 °C min^{-1}) prevents or restricts it [70]. Otherwise, it was observed that the hydrogenating properties of bimetallic Rh–Ge/Al$_2$O$_3$ catalysts prepared by catalytic reduction depend greatly on the reduction temperature and go through an optimum as a function of the reduction temperature which depends on the germanium content [42]. This phenomenon can be explained by both a change in the degree of reduction of germanium in contact with rhodium and a Ge enrichment of bimetallic particles resulting from migration at high temperature of germanium species deposited on the alumina support. In the case of Pt–Cu samples prepared by catalytic reduction, a treatment under hydrogen at high temperature (400 °C) induced a surface enrichment of Pt, copper atoms diffusing inwards [34].

The effect of an oxidation step after a first reduction under hydrogen at 300–500 °C was also examined. It appeared that the exposure to air of Rh–Ge/Al$_2$O$_3$ catalysts prepared by catalytic reduction results in an irreversible modification of the metallic phase, leading to a segregation of the two metals [42, 81]. Likewise, hydrogenation of isoprene and FTIR analysis have shown that air exposure of Pd–Fe catalysts prepared using preadsorbed hydrogen modifies the nature of the interaction between Fe and the Pd sites of low coordination, leading to species that are inactive for the reaction [67]. The detrimental effect of air exposure was also observed for Pt–Au catalysts, resulting in an irreversible rearrangement of the metallic surface created by the redox reaction. This rearrangement effect was evidenced by X-ray diffraction and by high resolution electron microscopy studies [17].

9.4
Applications in the Synthesis of Organic Chemicals

In the field of organic chemistry, the main applications using bimetallic catalysts prepared by surface redox reactions were for selective hydrogenation and hydrogenolysis reactions.

9.4.1
Selective Hydrogenation

The selective hydrogenation of organic molecules with different unsaturated functional groups is an important challenge in synthesis processes for the production of valuable chemical compounds.

9.4.1.1 Competition Between C=C and C=O Bonds

The carvone molecule (terpenic monocyclic ketone) is an interesting case, since it has three different functional groups which can be hydrogenated: one C=O group and two C=C groups (endocyclic and exocyclic ones), which display different chemical reactivities for hydrogenation (Fig. 9.6). For the same amount of carvone converted, bimetallic Pd–Cu/SiO$_2$ catalysts prepared by the refilling method favored the selective formation of carvotanacetone (unsaturated ketone), whereas the same systems prepared by co-impregnation led to carvomenthone (saturated ketone) [16]. This different selectivity behavior was a result of the Pd sites where the Cu was deposited and then an effect of the method of preparation.

By FTIR of adsorbed CO, a preferential deposit of Cu on Pd sites of lower coordination was observed on the bimetallic catalysts prepared by the redox reaction. On the contrary, in the co-impregnated systems, a Cu distribution occurred on the dense Pd(111) planes. It was reported that the formation of carvotanacetone is produced mainly on the Pd sites of high coordination where the desorption of the molecule is favored. The adsorption is stronger on low coordination sites, leading to a higher residence time of the molecule and then to further hydrogenation [82].

The performances of bimetallic Rh–Ge/SiO$_2$ and Rh–Ge/Al$_2$O$_3$ catalysts prepared by catalytic reduction were measured for another α,β-unsaturated aldehyde, namely citral (Fig. 9.7). The addition of Ge to Rh by the surface redox reaction promoted the hydrogenation of citral to the unsaturated alcohols (geraniol/nerol), while the saturated aldehyde (citronellal) was the main product on the monometallic Rh catalyst [41].

Moreover, these catalysts obtained by catalytic reduction were more active and selective than bimetallic Rh–Ge systems prepared by successive impregnations. The citral conversion and the selectivity to unsaturated alcohols increased with the Ge content, this effect being most obvious on alumina supported catalysts (Fig. 9.8). The selectivity enhancement on alumina in comparison with silica was explained by Ge deposited on the alumina support in the close vicinity of Rh nanoparticles during the catalytic redox reaction. These Ge species would be partially reduced during the reduction step of the catalyst preparation and would promote the activation of the carbonyl group for hydrogenation. Then, the catalytic properties of the bimetallic Rh–Ge/Al$_2$O$_3$ catalysts prepared by the redox reaction strongly depended on their reduction temperature, which must be

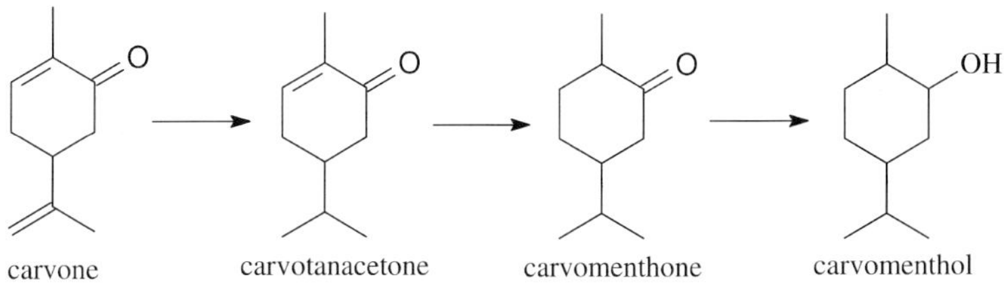

Fig. 9.6 Simplified reaction scheme for carvone hydrogenation.

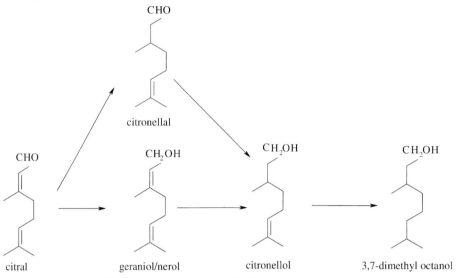

Fig. 9.7 Simplified reaction scheme for citral hydrogenation.

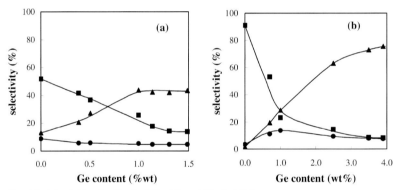

Fig. 9.8 Selectivities for Rh–Ge/SiO$_2$ (a) and Rh–Ge/Al$_2$O$_3$ (b) catalysts prepared by surface redox reaction at 50% citral conversion ($T_{reaction}$=70 °C, P_{H2}=76 bar, isopropanol as solvent): (•) citronellal; (▲) unsaturated alcohols; (•) citronellol. Reprinted from G. Lafaye, C. Micheaud-Especel, C. Montassier, P. Marecot, Appl. Catal. A, **2002**, *230*,19–30, Copyright (2002), with permission from Elsevier.

optimised according to the Ge content and the nature of solvent used for the reaction [42].

The performances of bimetallic Pd–Ge/carbon catalysts prepared by the refilling technique were also studied for the liquid phase hydrogenation of C=C bonds conjugated with carbonyl or other functional groups (hydroxy, carboxy, nitrile etc.)

[25]. The presence of the $Ge^{\delta+}$ centres resulted mainly in changes in the adsorption of substrates and/or products of the studied reaction.

9.4.1.2 Competition Between C=C Bonds

During isoprene hydrogenation, iron added by ionization of preadsorbed hydrogen promoted the activity of Pd/Al_2O_3 catalysts while the opposite effect was observed when iron was introduced by co-impregnation or successive impregnations [67, 68]. This different behavior was the result of an electron transfer from Fe to Pd, and of a deposition of Fe adatoms on the surface of Pd nanoparticles by the redox reaction while the co-impregnated catalysts would lead to alloy formation with a strong Pd surface segregation.

Other bimetallic systems containing Pd were studied for vegetable oil hydrogenation in order to produce lubricants. The main unsaturated fatty acids in vegetable oils are linolenic acid (C18:3), linoleic acid (C18:2) and oleic acid (C18:1). A partial hydrogenation of rapeseed oil to the C18:1 acid (keeping the cis-isomer form) is needed to obtain oils with a low sensitivity to oxygen. Moreover, the trans-C18:1 and the totally saturated (stearic acid, C18:0) compounds must be avoided since they are still solid at ambient temperature. Pd/SiO_2 catalysts modified by addition of Cu and Pb by catalytic reduction were investigated for the selective hydrogenation of sunflower oil at low temperature [83, 84]. The addition of Cu inhibited the hydrogenating activity of Pd, in agreement with the low hydrogenating properties of Cu at low temperature, slightly decreasing the isomerization toward trans-oleic isomers. The addition of Pb led to a significant drop in the isomerization to trans-oleic derivatives. The advantage of the preparation by the redox technique was maintenance of the high catalytic activity for C=C hydrogenation compared to the successive impregnation method.

9.4.2
Selective Hydrogenolysis

Studies on the competitive cleavages of C—C and C—O bonds in polyols such as glycerol, glucitol (sorbitol) or erythritol in aqueous phase on metals have revealed the particular properties of copper, especially those of Raney copper. Indeed, this metal is not hydrogenolysing but is able to dehydrogenate C—O bonds into carbonyls which are irreversibly adsorbed under reaction conditions. The nucleophilic action of surface hydroxyl groups on these adsorbed carbonylated derivatives leads to degradation products by cleavage of C—C and C—O bonds. Raney copper modification with a more noble metal (M=Ru, Pt, Au, Pd) than Cu by direct redox reaction allowed one to obtain bimetallic Cu–M systems favoring the selective cyclodehydration of polyols into more valuable furanic derivatives (Fig. 9.9) [5, 85].

This behavior depended on the surface atomic M/Cu ratio and resulted from an electron transfer from Cu to the second metal. Bimetallic Cu–Pt and Cu–Au nanoparticles prepared in the same way but on a charcoal support

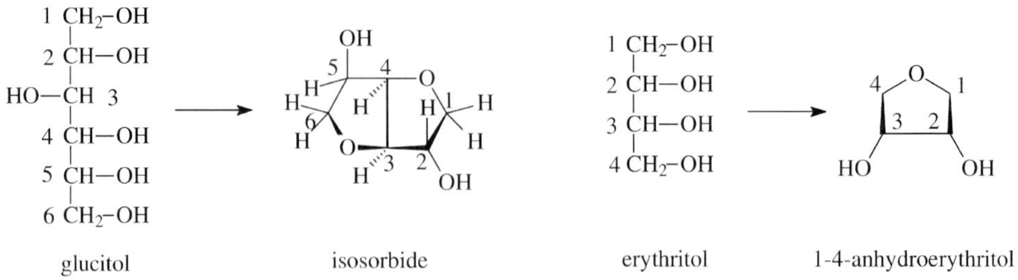

Fig. 9.9 Examples of polyol conversion to furanic derivatives.

show catalytic properties very close to those of the previous Raney copper based systems [86].

9.5
Applications in Environmental Catalysis

The preparation of bimetallic catalysts by surface redox reactions is well adapted for applications in environmental catalysis and, more particularly, in nitrate removal for water treatment. The catalytic nitrate reduction in the presence of a reducing agent such as hydrogen or formic acid leads to intermediate nitrite and to either nitrogen or an undesirable product, ammonia, as final product. Catalysts active for nitrate reduction are composed of (i) a noble metal, such as palladium or platinum and (ii) a transition metal such as copper, tin, indium or silver. For this application, the interaction between the noble metal and the promoter is of major importance. Indeed, Epron *et al.* [33] compared the activity of bimetallic Pt–Cu/Al$_2$O$_3$ catalysts with the same nominal composition but prepared according to two techniques that differed in the method of introduction of copper onto the parent Pt/Al$_2$O$_3$ catalyst. These techniques were (i) catalytic reduction, favoring the deposition of copper onto platinum and (ii) impregnation by cationic exchange with the support favoring the deposition of copper onto alumina. Figure 9.10 shows the disappearance of nitrate as a function of time under hydrogen in the presence of the two types of catalyst. This figure shows clearly that the most active catalyst is that prepared by catalytic reduction. It was inferred that, in the bimetallic catalyst, the role of copper is to reduce nitrate to nitrite according to a redox process. In this step, the interaction between copper and platinum is of major importance to maintain copper in the metallic state by way of hydrogen adsorbed on platinum. This mechanism requires that the promoter should be easily oxidized by nitrate in the reaction conditions and then reduced by hydrogen chemisorbed on the noble metal.

Then, the preparation by catalytic reduction of multimetallic catalysts for application in water denitration was extended to other bimetallic nanoparticles such as Pt–Ag [35], Pd–Cu [33, 35–37], Pd–Sn [38], Cu–Pt [8], Cu–Pd [8] or to trimetallic nanoparticles such as Pd–Sn–Au [46], deposited on various supports.

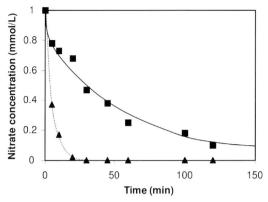

Fig. 9.10 Nitrate concentration versus time in the presence of PtCu/Al$_2$O$_3$ catalysts prepared by catalytic reduction (▲) or by impregnation by cationic exchange (■) (0.75 wt.% Pt, 0.25 wt.% Cu; reaction conditions: nitrate source Mg(NO$_3$)$_2$, T=10 °C, P_{H2}=1 bar, $m_{cat.}$=8 g L^{-1}). Reprinted from F. Epron, F. Gauthard, C. Pinéda, J. Barbier, *J. Catal.*, **2001**, *198*, 309–318, Copyright (2001), with permission from Elsevier.

9.6
Applications in Catalysis for Energy

9.6.1
Naphtha Reforming

The reforming of low-octane feedstocks is classically carried out over bifunctional catalysts made up of metallic nanoparticles deposited on an acidic phase. The reactions involved in this process are hydrogenation, dehydrogenation, isomerization, cyclization and hydrocracking. In the first type of naphtha reforming catalysts, the metal function was provided by platinum and the acidic one by the support itself, such as chlorinated alumina (Pt/Al$_2$O$_3$-Cl). The disadvantage of this catalyst was its rapid deactivation by coking. The stability was improved in the late 1960s with the second generation of catalysts consisting of bimetallic systems such as Pt–Re/Al$_2$O$_3$-Cl. However, this type of catalyst must be sulfided in order to avoid the formation of undesired gaseous products.

Numerous studies were devoted to the use of bimetallic catalysts, promoted either by an active metal such as rhenium or iridium, or an inactive one such as tin or germanium. These different catalysts were generally prepared by co-impregnation or successive impregnations. Moreover, most of the catalysts prepared by redox reactions were only evaluated in model reactions. For example, Corro *et al.* [87] prepared Pt–Sn/Al$_2$O$_3$-Cl catalysts by catalytic reduction or co-impregnation and compared their resistance to coking under cyclopentane feed. Thus, catalysts prepared by the surface redox reaction were less sensitive than the others to deactivation. This result was explained in terms of a more effective interaction between tin and platinum when tin is introduced by catalytic reduction onto a parent platinum-based catalyst. The deactivation by coke and sulfur deposition was also studied on Pt–Au/Al$_2$O$_3$ catalysts prepared by catalytic reduction [88].

Some test reactions were performed under industrial-type conditions and can be considered as representative of the behavior which would be observed in real reforming of feedstocks. For example, Pieck *et al.* [78] studied the catalytic performances of Pt–Re/Al$_2$O$_3$-Cl prepared by different methods, namely co-impregnation, successive impregnation and catalytic reduction. The catalysts were activated by calcination and then reduction. They were tested in n-heptane reforming, used as a model reaction of paraffin dehydrocyclisation. The dehydrocyclisation is the reaction to promote since it leads to aromatics for the petrochemical industry and also to a large amount of hydrogen. On the contrary, the production of gaseous products (C$_1$ to C$_4$) must be avoided. The results obtained with the different sulfided catalysts, reported in Table 9.2 demonstrate that the highest conversion and yield in toluene are obtained with the catalyst prepared by catalytic reduction, for which the interactions between Pt and Re are the highest. This result was explained by the increase in the dehydrogenation activity of Pt–Re entities by sulfidation. As sulfur adsorbs preferentially on rhenium, the higher the interaction between Pt and Re, favored by the preparation method, the higher the effect of sulfur.

Recently trimetallic catalysts have been prepared by redox reactions for application in naphtha reforming. Fürcht *et al.* modified Pt–Sn/γ-Al$_2$O$_3$ by addition of Te, Bi [89] or Au, Ir and Pd [90] by the refilling method. Their catalysts were evaluated in n-octane reforming under industrial conditions. As reported in Refs. [44, 45], it could be advantageous to replace the sulfidation step by the addition of an inactive element such as germanium or tin on a bimetallic Pt–Re or Pt–Ir catalyst. Indeed, the presence of tin or germanium instead of sulfur leads to more stable catalysts with a higher yield in n-heptane dehydrocyclisation under industrial-type conditions. For example, the conversion of n-heptane and the yield in toluene are reported in Fig. 9.11 as a function of the tin content in trimetallic Pt–Ir–Sn/Al$_2$O$_3$-Cl catalysts, and compared with those obtained with a presulfided bimetallic cata-

Table 9.2 n-Heptane (nC$_7$) conversion (at 30 min), yields in toluene and gaseous products (C$_1$ to C$_4$) obtained with sulfided Pt–Re/Al$_2$O$_3$-Cl (0.6 wt.% Pt, 0.5 wt.% Re) prepared by co-impregnation (CI), successive impregnations (SI) or catalytic reduction (CR). (Reaction conditions: T=500 °C, $\dfrac{H_2}{nC_7} \approx 6$ (molar ratio), P=10 bar, WHSV=2.9 h^{-1} [75].

Preparation method	Conversion (%)	Toluene (%)	C$_1$ to C$_4$ (%)
CR	96.5	27.5	32
SI	79.0	17.0	30
CI	85.0	21.5	25

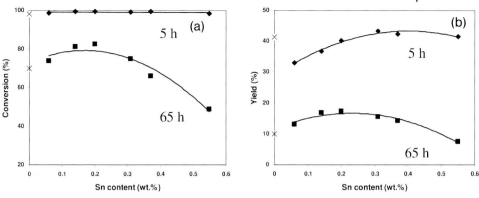

Figure 9.11 (a) n-Heptane conversion and (b) yield in toluene obtained with trimetallic Pt–Ir–Sn/Al$_2$O$_3$-Cl prepared by catalytic reduction, after 5 h (♦) and 65 h (•) or presulfided bimetallic Pt–Ir/Al$_2$O$_3$ catalyst after 5 h (×) or 65 h (✳) of reaction (0.3 wt.% Pt, 0.3 wt.% Ir; reaction conditions: $T = 500\,°C$, $\dfrac{H_2}{nC_7} = 3$ (molar ratio), $P = 5\,bar$, WHSV $= 2\,h^{-1}$).

lyst, without tin. These results show that the conversion is close to 100% for all the catalysts after 5 h but decreases to around 50% in the worst case after 65 h of reaction. This deactivation, mainly due to the presence of coke, is less important for trimetallic catalysts with a low tin content (less than 0.3 wt.%) than for the presulfided bimetallic catalyst. In the same way, a strong decrease in the toluene yield occurs between 5 h and 65 h of reaction. However, the highest toluene yields are obtained with trimetallic catalysts with less than 0.4 wt.% tin.

9.6.2
Fuel Cells

One of the drawbacks to the development of proton exchange membrane fuel cells using reformed fuels is the presence of CO in the reformate, which drastically decreases the performance by poisoning the anodic catalyst. Tolerance to a high CO concentration would certainly help to simplify the reformer and especially the hydrogen purification system. This simplification would lead to reduction in the cost of the system which is the main challenge to enter the market. Accordingly, the preparation of bi- and tri-metallic catalysts by surface redox reactions was investigated by adding modifiers (such as Ru, Sn and Mo) on a Pt/C-based system commonly used both as an anode or a cathode catalyst but having a very poor CO tolerance [76]. Among the bimetallic samples, the best results were obtained with the Pt–Sn/C systems prepared by catalytic reduction. Moreover, the modification of Pt–Ru nanoparticles by Sn or Mo showed better CO tolerance than the corresponding bimetallic catalysts, with a significant enhancement of the stability in the case of Mo.

9.7
Conclusion

The preparation of supported multimetallic nanoparticles by redox reactions allows one to favor the interaction between the different metals, in agreement with the theoretical predictions. Nevertheless, the presence of the support must be taken into account since the modifier could have a strong affinity for it. In this chapter, it is demonstrated that the operating conditions, such as the pH of the solution, the nature of the reducing agent or of the precursor salt etc., can be controlled in order to minimize the deposition of the modifier onto the support. Furthermore, it is pointed out that, once the multimetallic nanoparticles are formed, they could be sensitive to external parameters, as for example gaseous treatments, exposure to air and so on, leading to different catalytic performances. Finally, the different applications presented in this chapter demonstrate that for the same metal loading, metallic nanoparticles prepared by redox reactions present more interesting properties than their counterparts prepared by classical techniques. Illustrations are given in several areas such as catalysis for organic chemical synthesis, for environment and for energy.

References

1 J. Margitfalvi, S. Szabo, F. Nagy, *Stud. Surf. Sci. Catal.*, **1986**, *27*, 373–409.

2 S. Szabo, *Int. Rev. Phys. Chem.*, **1991**, *10*, 207–248.

3 J. Barbier, Catalytica Studies Division, *Advances in Catalysts Preparation*, study number 4191 CP, **1992**.

4 J. Barbier, in *Handbook of Heterogeneous Catalysis*, Vol.1, G. Ertl, H. Knözinger, J. Wertkamp (Eds.) Wiley-VCH, Weinheim, **1997**, pp. 257–264.

5 C. Montassier, J. C. Menezo, J. Moukolo, J. Naja, L. C. Hoang, J. Barbier, J. P. Boitiaux, *J. Mol. Catal.*, **1991**, *70*, 65–84.

6 J. Barbier, J.C. Menezo, C. Montassier, J. Naja, G. Del Angel, J. M. Dominguez, *Catal. Lett.*, **1992**, *14*, 37–43.

7 C. Montassier, J. C. Menezo, J. Naja, J. Barbier, J. M. Dominguez, P. Sarrazin, B. Didillon, *J. Mol. Catal.*, **1994**, *91*, 107–117.

8 N. Barrabes, J. Just, A. Dafinov, F. Medina, J. L. G. Fierro, J. E. Sueiras, P. Salagre, Y. Cesteros, *Appl. Catal. B*, **2006**, *62*, 77–85.

9 I. Bakos, S. Szabo, *J. Electroanal. Chem.*, **1993**, *344*, 303–311.

10 J. Barbier, P. Marecot, G. Del Angel, P. Bosch, J. P. Boitiaux, B. Didillon, J. M. Dominguez, I. Schifter, G. Espinosa, *Appl. Catal. A*, **1994**, *116*, 179–186.

11 V. Bertin, P. Bosch, G. Del Angel, R. Gomez, J. Barbier, P. Marecot, *J. Chim. Phys.*, **1995**, *92*, 120–33.

12 P. Del Angel, J. M. Dominguez, G. Del Angel, J. A. Montoya, E. Lamy-Pitara, S. Labruquere, J. Barbier, *Langmuir*, **2000**, *16*, 7210–7217.

13 C. Micheaud, M. Guerin, P. Marecot, C. Geron, J. Barbier, *J. Chim. Phys.*, **1996**, *93*, 1394–1411.

14 J. C. Menezo, M. F. Denanot, S. Peyrovi, J. Barbier, *Appl. Catal.*, **1985**, *15*, 353–356.

15 J. M. Dumas, S. Rmili, J. Barbier, *J. Chim. Phys.*, **1998**, *95*, 1650–1665.

16 R. Melendrez, G. Del Angel, V. Bertin, M. A. Valenzuela, J. Barbier, *J. Mol. Catal. A*, **2000**, *157*, 143–149.

17 G. Espinosa, G. Del Angel, J. Barbier, P. Bosch, V. Lara, D. Acosta, *J. Mol. Catal. A*, **2000**, *164*, 253–262.

18 S. Szabo, F. Nagy, *J. Electroanal. Chem.*, **1977**, *84*, 93–98.

19 S. Szabo, F. Nagy, *J. Electroanal. Chem.*, **1977**, *85*, 339–343.

20 S. Szabo, F. Nagy, *J. Electroanal. Chem.*, **1978**, *87*, 261–265.

21 S. Szabo, F. Nagy, *J. Electroanal. Chem.*, **1984**, *160*, 299–303.

22 S. Szabo, F. Nagy, *J. Electroanal. Chem.*, **1987**, *230*, 233–240.

23 S. Szabo, I. Bakos, F. Nagy, *J. Electroanal. Chem*, **1989**, *271*, 269–277; S. Szabo, I. Bakos, F. Nagy, *J. Electroanal. Chem*, **1989**, *263*, 137–146.

24 I. Bakos, S. Szabo, F. Nagy, T. Mallat, Z. Bodnar, *Electroanal. Chem.*, **1991**, *309*, 293–301.

25 Z. Bodnar, T. Mallat, I. Bakos, S. Szabo, Z. Zsoldos, Z. Schay, *Appl. Catal. A*, **1993**, *102*, 105–123.

26 S. Szabo, F. Nagy, *Isr. J. Chem.*, **1979**, *18*, 162–165.

27 C. L. Pieck, P. Marecot, J. Barbier, *Appl. Catal. A*, **1996**, *134*, 319–329.

28 S. Szabo, I. Bakos, *React. Kinet. Catal. Lett.*, **1998**, *65*, 259–263.

29 C. L. Pieck, P. Marecot, J. Barbier, *Appl. Catal. A*, **1996**, *141*, 229–244.

30 C. L. Pieck, P. Marecot, J. Barbier, *Appl. Catal. A*, **1996**, *143*, 283–298.

31 C. L. Pieck, P. Marecot, J. Barbier, *Appl. Catal. A*, **1996**, *145*, 323–334.

32 J. Barbier, P. Marecot, C. L. Pieck, *Stud. Surf. Sci. Catal.* **1997**, *111*, 327–334.

33 F. Epron, F. Gauthard, C. Pineda, J. Barbier, *J. Catal.*, **2001**, *198*, 309–318.

34 F. Epron, F. Gauthard, J. Barbier, *Appl. Catal. A*, **2002**, *237*, 253-261.

35 F. Gauthard, F. Epron, J. Barbier, *J. Catal.*, **2003**, *220*, 182–191.

36 J. Sa, S. Gross, H. Vinek, *Appl. Catal. A*, **2005**, *294*, 226–234.

37 J. Sa, H. Vinek, *Appl. Catal. B*, **2005**, *57*, 247–256.

38 A. Garron, K. Lazar, F. Epron, *Appl. Catal. B*, **2005**, *57*, 57–69.

39 J. C. Menezo, L. C. Hoang, C. Montassier, J. Barbier, *React. Kinet. Catal. Lett.*, **1992**, *46*, 1–6.

40 J. M. Dumas, C. Geron, H. Hadrane, P. Marecot, J. Barbier, *J. Mol. Catal.*, **1992**, *77*, 87–98.

41 G. Lafaye, C. Micheaud-Especel, C. Montassier, P. Marecot, *Appl. Catal. A*, **2002**, *230*, 19–30.

42 G. Lafaye, T. Ekou, C. Micheaud-Especel, C. Montassier, P. Marecot, *Appl. Catal. A*, **2004**, *257*, 107–117.

43 C. Carnevillier, F. Epron, P. Marecot, *Appl. Catal. A*, **2004**, *275*, 25–33.

44 F. Epron, C. Carnevillier, P. Marecot, *Appl. Catal. A*, **2005**, *40*, 157–169.

45 M. Boutzeloit, V. M. Benitez, C. Especel, F. Epron, C. R. Vera, C. L. Pieck, P. Marecot, *Catal. Commun.*, **2006**, *7*, 627–632.

46 A. Garron, K. Lazar, F. Epron, *Appl. Catal. B*, **2006**, *65*, 240–248.

47 D. M. Kolb, M. Pazasnyski, H. Gerischer, *J. Electroanal. Chem.*, **1974**, *54*, 25–38.

48 H. Gerischer, D. M. Kolb, M. Pazasnyski, *Surf. Sci.*, **1974**, *43*, 662–666.

49 D.M. Kolb, in *Advances in Electrochemistry and Electrochemical Engineering*, M. Gerischer, C. N. Tobias (Eds.), John Wiley and Sons, New York, **1975**.

50 E. Leiva, *Electrochim. Acta*, **1996**, *41*, 2185–2206.

51 E. Lamy-Pitara, J. Barbier, *Appl. Catal. A*, **1997**, *149*, 49–87.

52 N. Furuya, S. Motoo, *J. Electroanal. Chem.*, **1979**, *98*, 195–202.

53 M. W. Breiter, *J. Electrochem. Soc. Electrochem. Sci.*, **1967**, *114*, 1125–1129.

54 S. Swathirajan, H. Mizota, S. Bruckenstein, *J. Phys. Chem.*, **1982**, *86*, 2480–2485.

55 M. W. Breiter, *Trans. Faraday Soc.*, **1969**, *65*, 2197–2205.

56 N. Furuya, S. Motoo, *J. Electroanal. Chem.*, **1976**, *72*, 165–175

57 E. Lamy, J. Barbier, C. Lamy, *J. Chim. Phys.*, **1980**, *77*, 967–972.

58 I. Bakos, S. Szabo, *React. Kinet. Catal. Lett.*, **1990**, *41*, 53–57.

59 A. I. Danilov, E. B. Molodkina, Y. M. Polukarov, *Russ. J. Electrochem.*, **1998**, *34*, 1249–1256.

60 D. L. Lu, M. Ichihara, K. I. Tanaka, *Electrochim. Acta.*, **1998**, *43*, 2325–2330.

61 N. Furuya, S. Motoo, *J. Electroanal. Chem.*, **1977**, *78*, 243–256.

62 N. Furuya, S. Motoo, *J. Electroanal. Chem.*, **1978**, *88*, 151–160.

63 E. Lamy-Pitara, L. El Quazzani-Benhima, J. Barbier, *J. Electroanal. Chem.*, **1992**, *335*, 363–370.

64 E. Lamy-Pitara, L. El Ouazzani-Benhima, J. Barbier, M. Cahoreau, J. Caisso, *J. Electroanal. Chem.*, **1994**, *372*, 233–242.

65 I. Bakos, S. Szabo, *Electrochim. Acta*, **2001**, *46*, 2507–2513.

66 I. Bakos, S. Szabo, *J. Electroanal. Chem.*, **2003**, *547*, 103–107.

67 R. Bachir, E. Lafitte, P. Marecot, B. Didillon, J. Barbier, *J. Chim. Phys.*, **1997**, *94*, 1906–1913.

68 R. Bachir, P. Marecot, B. Didillon, J. Barbier, *Appl. Catal. A*, **1997**, *164*, 313–322.

69 R. Bachir, A. Reguig, P. Marecot, J. Barbier, *J. Soc. Algérienne Chim.*, **2002**, *12*, 233–241.

70 A. Douidah, P. Marecot, S. Labruquere, J. Barbier, *Appl. Catal. A*, **2001**, *210* (1–2), 111–120.

71 D. R. Lide (Ed.), *CRC Handbook of Chemistry and Physics*, 75th Edn., CRC Press, Boca Raton, FL, **1994/1995**.

72 A. Douidah, P. Marecot, J. Barbier, *Appl. Catal. A*, **2002**, *225* (1–2), 11–19.

73 P. Granger, J. M. Dumas, C. Montassier, J. Barbier, *J. Chim. Phys.*, **1995**, *92*, 1557–1575.

74 A. Douidah, P. Marecot, S. Szabo, J. Barbier, *Appl. Catal. A*, **2002**, *225* (1–2), 21–31.

75 M. Bonarowska, J. Pielaszek, W. Juszczyk, Z. Karpinski, *J. Catal.*, **2000**, *195* (2), 304–315.

76 S. Escribano, R. Mosdale, P. Marecot, P. Korovtchenko, G. Lafaye, J. Barbier, CO Tolerance of Multi-metallic Platinum -based Catalyst, Proceedings France -Deutschland Fuel Cell Conference 2002, Oct 7–10, p.10, Forbach, **2002**.

77 P. Marecot, E. Rohart, *Fr. Pat.* FR 2 771 310, **1999**.

78 C. L. Pieck, P. Marecot, C. A. Querini, J. M. Parera, J. Barbier, *Appl. Catal. A*, **1995**, *133* (2), 281–292.

79 S. M. Augustine, W. M. H. Sachtler, *J. Catal.*, **1987**, *106*, 417.

80 S.M. Augustine, W. M. H. Sachtler, *J. Phys. Chem.*, **1987**, *91*, 5953.

81 G. Lafaye, C. Mihut, C. Especel, P. Marecot, M. Amiridis, *Langmuir*, **2004**, *20* (24), 10612–10616.

82 G. Del Angel, R. Melendrez, V. Bertin, J. M. Dominguez, P. Marecot, J. Barbier, *Stud. Surf. Sci. Catal.*, **1993**, *78*, 171.

83 B. Nohair, C. Especel, P. Marecot, C. Montassier, L. C. Hoang, J. Barbier, *C.R. Chim.*, **2004**, *7*, 113–118.

84 B. Nohair, C. Especel, G. Lafaye, P. Marecot, L. C. Hoang, J. Barbier, *J. Mol. Catal. A*, **2005**, *229*, 117–126.

85 C. Montassier, J. C. Menezo, J. Naja, P. Granger, J. Barbier, P. Sarrazin, B. Didillon, *J. Mol. Catal.*, **1994**, *91*, 119–128.

86 C. Montassier, J. M. Dumas, P. Granger, J. Barbier, *Appl. Catal. A*, **1995**, *121*, 231–244.

87 G. Corro, P. Marecot, J. Barbier, *Stud. Surf. Sci. Catal.*, **1997**, *111*, 359–366.

88 G. Espinosa, G. Del angel, J. Barbier, P. Marecot, I. Schifter, *Stud. Surf. Sci. Catal.*, **1997**, *111*, 421–426.

89 A. Furcht, A. Tungler, S. Szabo, A. Sarkany, *Appl. Catal. A.*, **2002**, *226*, 155–161.

90 A. Furcht, A. Tungler, S. Szabo, Z. Schay, L. Vida, I. Gresits, *Appl. Catal. A.*, **2002**, *231*, 151–157.

10
The Role of Palladium Nanoparticles as Catalysts for Carbon–Carbon Coupling Reactions

Laurent Djakovitch, Klaus Köhler, and Johannes G. de Vries

10.1
Introduction

Transition metal nanoparticles have attracted a great deal of attention over the last 10 years. Their preparation, structure determination, and application are topics of current interest. Their catalytic properties have led to interesting applications. Because these are in fact ligand-free catalysts, high reactivity is a common feature of their use.

One of the most important applications concerns their use as catalyst in aromatic C–C coupling reactions, such as the Heck, Suzuki, Sonogashira, Stille and Negishi reactions (Fig. 10.1). These reactions constitute important methodology for the synthesis of complex organic molecules such as pharmaceuticals and agrochemicals. They are catalysed by palladium species in almost every available form: homogeneous complexes, stable colloids or nanoparticles, supported palladium particles, immobilised palladium complexes, palladacycles and pincers. Even extremely small amounts of palladium down to the ppb level can be sufficient to activate the aryl halides with high TOFs.

In this chapter, we discuss selected literature on palladium catalysts (or catalyst precursors) for the Heck, Suzuki and Sonogashira reactions. The review covers simple homogeneous palladium complexes, ligand-free palladium catalytic systems, stable palladium colloids and particles and supported palladium catalysts. It focuses on the role of palladium nanoparticles (as catalyst precursors or formed *in situ* during the course of the reaction) from a mechanistic point of view.

10.2
Stable Palladium Colloids and Nanoparticles

Several recent reviews deal with the subject of colloidal catalysts and the reader interested in the broader field is referred to these authoritative articles [1–4].

Nanoparticles and Catalysis. Edited by Didier Astruc
Copyright © 2008 WILEY-VCH Verlag GmbH & Co. KGaA, Weinheim
ISBN: 978-3-527-31572-7

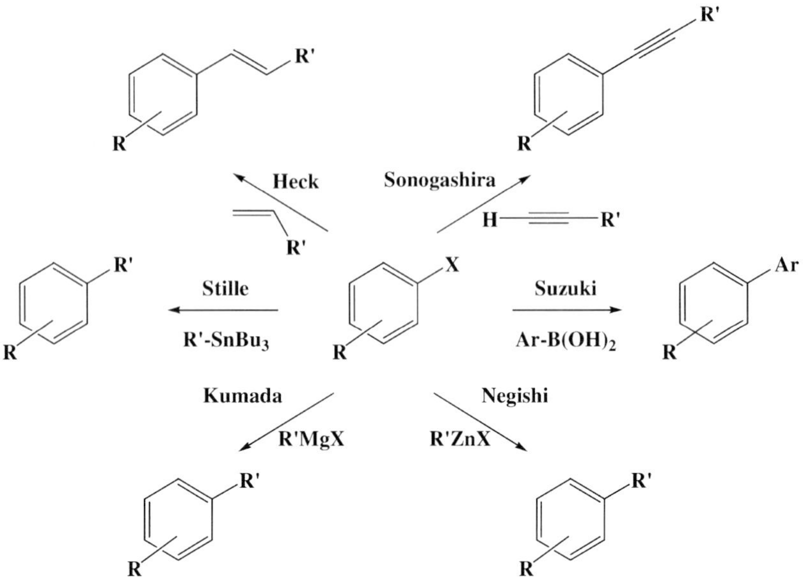

Fig. 10.1 Pd-catalysed aromatic substitution reactions (X = halide or other suitable leaving group).

The main common characteristic of colloidal particles is their small size (typically 1 to 10 nm). The size of nanoparticles in solution is dynamic and continuous redistribution in size can occur. In most cases, agglomeration leads to the formation of less active "larger" metal particles and this process may end in precipitation of larger crystals (palladium black). The per-atom catalytic efficiency of metal particles increases as the particle size decreases; however, the probability of colloid agglomeration increases as their size decreases. To prevent agglomeration (and aggregation), and to preserve the finely dispersed state of the original particles, colloids are often prepared in the presence of stabilizers that "adsorb" onto the particle surface.

Several methods have been reported for the synthesis of stable colloids: (i) chemical reduction of metal salts, (ii) thermal, photochemical or sonochemical decomposition of metal–olefin complexes, (iii) hydrogenation of coordinating olefins, (iv) vapor phase decomposition, and (v) electrochemical reduction. During the synthesis several steps have been clearly identified: (i) generation of single atoms in the zero oxidation state, (ii) nucleation to form the initial cluster of atoms, (iii) growth of the cluster to a specific size, and (iv) reaction of the atoms at the rim with the stabilisers. As stabilisers, one can discern four groups: (i) those providing electrostatic stabilisation by formation of a double electric layer (Fig. 10.2a), such as tetra-alkylammonium salts, (ii) those providing steric stabilisation by adsorption, mainly on polymers or oligomers (Fig. 10.2b), (iii) those providing electrosteric interaction using ionic surfactants, and (iv) those providing "*ligand stabilisation*" such as phosphines, thiols, amines or carbon monoxide.

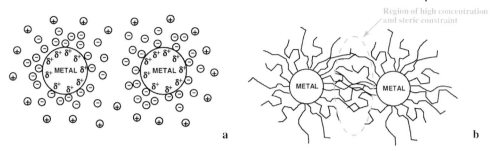

Fig. 10.2 Stabilization of metal colloids by: (a) electrostatic stabilisation, (b) steric stabilisation.

10.2.1
Palladium Colloids in the Heck Reaction

The development of Pd colloids as catalyst for C—C coupling reactions is rather recent [5]. The first example was reported by Beller et al. in 1996; they used pre-formed Pd colloids stabilised by tetra-octylammonium bromide prepared following the Bönneman procedure in the Heck arylation [6]. The colloidal system was effective for the Heck arylation of styrene or butyl acrylate by activated aryl bromides, but showed only moderate to little activity for deactivated aryl bromides and aryl chlorides. To obtain these results, the authors found that the colloidal pre-catalysts must be added slowly to the reaction mixture to avoid the formation of inactive palladium black at the beginning of the reaction.

The area has been mostly advanced by contributions from Reetz and coworkers who reported independently in 1996 the preparation of nanostructured Pd colloids prepared by electrochemical reduction [7, 8]. This method resulted in Pd colloids stabilized by propylene carbonate with a uniform size of 8–10 nm which are highly stable under the Heck reaction conditions (140–155 °C). As expected, activated aryl bromides gave the highest product yields; remarkably this colloidal system also activated chlorobenzene, giving up to 30% conversion using relatively high Pd loadings (3.5 mol%). Continuing their effort to develop efficient Pd catalysts for the Heck reaction, Reetz and coworkers described a new catalytic system combining a PdX_2 salt ($Pd(OAc)_2$, $PdCl_2(PhCN)_2$, etc.) with a tetra-arylphosphonium salt [9]. This catalytic system was even able to activate aryl chlorides, albeit with low TONs (130 for chlorobenzene). Regioselectivity in the Heck reaction between chlorobenzene and styrene was further improved by the addition of N,N-dimethylglycine. The authors suggested, without proof, that the protocol used led to the formation of solvent-stabilised Pd nanoparticles as the real catalyst. An interesting practical application of this system is the industrially relevant ethenylation of 2-bromo-6-methoxynaphthalene Scheme 10.1), a precursor of Naproxen.

Later, Reetz and coworkers pointed out the role of Pd colloids in phosphine-free Pd-catalysed Heck reactions, as typically encountered under Jeffery's conditions ($Pd(OAc)_2$, Bu_4NBr, K_2CO_3). Studying the Heck reaction of iodobenzene with ethyl

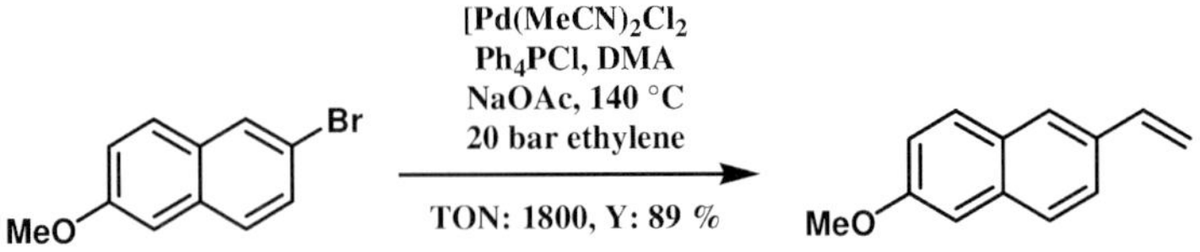

Scheme 10.1 Heck ethenylation of 2-bromo-6-methoxynaphthalene.

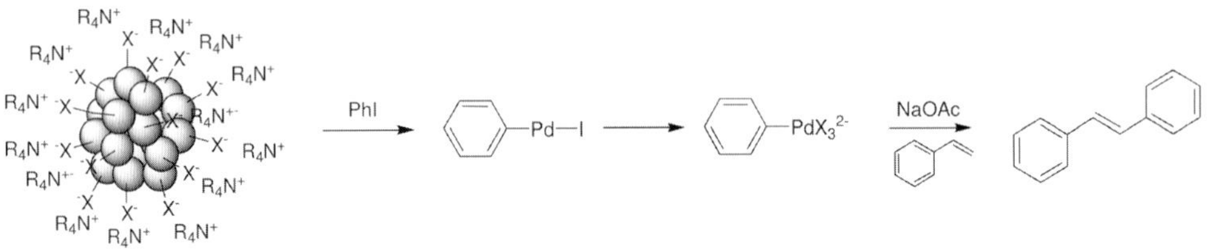

Scheme 10.2 Proposed pathway for stoichiometric Heck reaction of Pd colloids.

acrylate in *N*-methyl pyrrolidinone (NMP), the authors observed the formation of Pd nanoparticles of ca. 1.6 nm. Addition of iodobenzene to preformed Pd nanoparticles led to the formation of $[PhPdX_3]^{2-}$ suggesting an oxidative addition of iodobenzene to the colloidal palladium (Scheme 10.2) [10, 11]. It is clear that an important function of the tetra-alkylammonium salt in Jeffery's reaction is the stabilisation of the palladium clusters.

Antonietti and coworkers reported, in 1997, the preparation of nanometer-sized Pd colloids (ca. 10–100 nm) stabilised by the polystyrene-b-poly-4-vinylpyridine block copolymer micelle [12]. These hybrid materials, which are soluble under the reaction conditions in toluene, THF and cyclohexane, were applied to the Heck reaction of aryl bromides with styrene. The block copolymer-stabilised Pd colloids showed almost the same activity as $Pd(OAc)_2$ associated or not to ligands (phosphine, pyridine), but with higher stability since no palladium black was observed during these reactions. Testing different sizes of Pd colloids, the authors noticed a strong inverse correlation between activity and particle size attributed to diffusion limitation when using larger block copolymer or, alternatively, to higher reactivity of small Pd particles. No noticeable differences were observed in the size of the Pd particles after the reaction; the materials were claimed to be reusable, but no recycling experiments were reported.

Further elucidation of the reactivity of Pd colloids was reported by Bradley and Blackmond [13]. These authors prepared a series of well defined homopolymer-stabilised Pd colloids, varying the particle size, and used them in the Heck reaction between 4-bromobenzaldehyde and butyl acrylate. Interestingly, the authors found a correlation between initial rate and particle size: the initial rate increases with decreasing particle size. No correlation was reported regarding the palladium specific surface area. Whereas the precise morphology of the Pd particles could not be defined, the authors found that they were quite equidimensional, allowing a representation as regular polyhedra to apply *"surface defect site statistics"* to characterise the surface structure idealised as a cuboctahedral particle. Utilizing the knowledge that the relative abundance of high-coordination number sites (faces) and low-coordination number or "defect" sites (edges and vertices) varies with the Pd cluster size, the authors used initial rates at 10% conversion measured using reaction calorimetry as a function of the Pd particle size to analyse the reaction kinetics. The authors found a good correlation between the initial rate and the number of defect sites, concluding that the defect sites are the active centers in the Heck reaction. In view of more recent findings, we believe that the defect sites are indeed important in the Heck reaction, but more probably because these sites are more sensitive to leaching through an initial oxidative addition of the aryl halide to generate a solvated $[ArPd^{(II)}X]$ molecular species, or more likely its anionic equivalent $[ArPdX_2]^-$ or $[ArPdX_3]^{2-}$.

Prakash and Thompson used monodisperse poly(4- and poly(2-vinylpyridine) nanospheres as stabilisers to support 1–4 nm Pd nanoparticles in a study concerning the Heck, Suzuki and Stille coupling reactions [14]. The material was found to be very active. However, the reactions were limited to the coupling of the highly reactive 4-bromo-nitrobenzene with, respectively, butyl acrylate, phenylboronic acid and phenyl-trimethylstannane. No noticeable influence of the nature of the stabilisers was reported in this study. The authors argue that the material was stable under the reaction conditions, but this conclusion is only based on macroscopic TEM analyses. In view of the low Pd concentration which was required to perform effective Heck, Suzuki or Stille coupling reactions, it is not surprising that no major variation of the Pd particle size was observed.

Gin and coworkers used Pd nanoparticles stabilised by a cross-linked lyotrophic liquid crystal (LLC) in the Heck reaction of aryl bromides with styrene or butyl acrylate. The material was prepared, after preparing a LLC structured polymer (Fig. 10.3), by ion exchange followed by reduction under hydrogen to give a composite containing Pd nanoparticles (4–7 nm) that exhibited high catalytic activity (TON ≥1000). No deactivation of the catalyst was observed; however this conclusion was based on only two cycles [15].

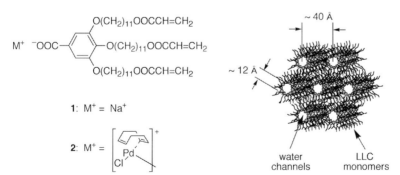

Fig. 10.3 Pd nanoparticles stabilised by a cross-linked lyotrophic liquid crystal. Reprinted from Ref. [14] with the permission from the ACS.

Crooks and coworkers reported the use of perfluorinated polyether-derivatised poly(propylene imine) dendrimer-encapsulated Pd nanoparticles (2–3 nm) (DEC) for the Heck reaction of aryl halides with acrylates (Fig. 10.4) [16, 17]. The DEC promoted the Heck arylation of butyl acrylate with aryl iodides and bromides, at a rather low temperature (90 °C), giving, after optimisation, 70% yield for the coupling of iodobenzene. No significant Pd leaching and no decomposition of the catalyst were observed during the reactions. The catalyst could be recovered although the activity decreased upon successive cycles. Interestingly, the catalyst can be used in the absence of additional base to promote the Heck reaction. Presumably the interior tertiary amines of the dendrimer act as the base. In that case the catalysts could not be recycled. In another series of experiments, Crooks and coworker reported the use of DEC Pd catalysts in supercritical CO_2 ($scCO_2$). Coupling iodobenzene with methyl acrylate resulted in the exclusive formation of methyl 2-phenylacrylate instead of the expected methyl cinnamate, which was attributed to the use of $scCO_2$ as solvent (Scheme 10.3).

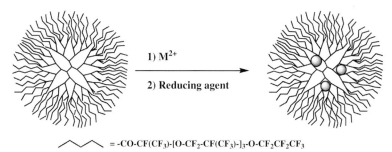

= -CO-CF(CF$_3$)-[O-CF$_2$-CF(CF$_3$)-]$_3$-O-CF$_2$CF$_2$CF$_3$

Fig. 10.4 Dendrimer-encapsulated Pd nanoparticles.

100 % methyl-2-phenylacrylate

1-2,6 mol% Pd

Et$_3$N, scCO$_2$

0 % methyl-cinnamate

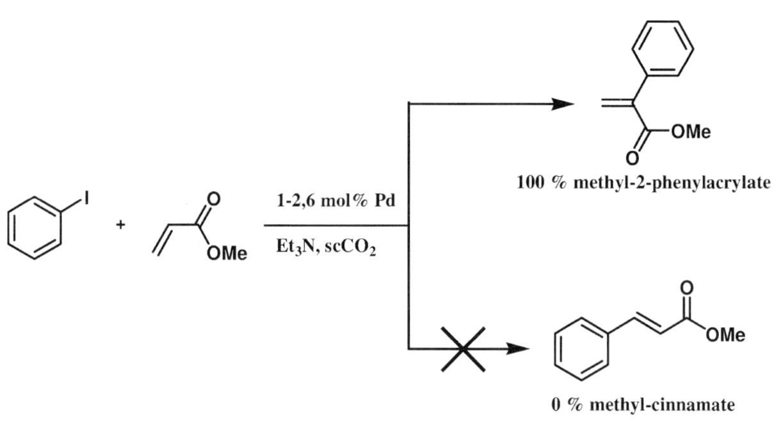

Scheme 10.3 Selectivity dependence towards the solvent for Heck arylation of methylacrylate using DEC Pd-catalyst.

Pd nanoparticles (Pd$_{60}$ and Pd$_{40}$) encapsulated in a PANAM-OH dendrimer were also used by Christensen and coworkers to promote the Heck arylation of aryl iodides with acrylic acid [18]. Using 0.025 mol/[Pd], high turnover numbers were observed (1900 < TON < 37 000); however, the catalyst's activity decreased strongly upon reuse and high Pd leaching was observed, resulting in high Pd contamination of the product. These observations were attributed to thermal degradation of the PANAM dendrimer under the reaction conditions.

More recently, Whitesell and Fox reported the use of a Pd nanoparticle (ca. Pd$_{300}$, 2 nm) cored dendrimer (Fig. 10.5) in the Heck reaction of iodobenzene with ethyl acrylate and styrene, or the Suzuki reaction of iodo- and bromobenzene with phenylboronic acid. High turnover numbers were achieved (25 000 < TON < 42 000), giving moderate to high product yield (38% to 75%) [19].

Biffis reported an elegant way to prepare stable Pd colloids (10–20 nm) by using functionalised microgel (soluble cross-linked copolymers) as stabilisers (Fig. 10.6) [20, 21]. The system is based on partially sulfonated macromolecules prepared by copolymerisation of methyl methacrylate, ethylene dimethacrylate and sulfoethyl methacrylate. The resulting material was exchanged with Pd(OAc)$_2$ followed by reduction using ethanol. High yields were achieved in the Heck reaction of iodobenzene and 4-bromoacetophenone with butyl acrylate. Activated aryl bromides and aryl chlorides did not react. The catalysts were claimed to be stable under the reaction conditions since no Pd-black was observed, but no data supported this interpretation. Similar Pd catalysts were applied to Suzuki coupling reactions, giving moderate to high yields in reactions with aryl bromides [22]. Recently, Biffis and coworkers reported the preparation of Pd nanoparticles stabilised by cross-linked functional organic resins containing tertiary amino, cyano, carboxyl and pyridyl groups, by immobilisation of the palladium precursor followed by chemical reduction or by metal vapor deposition [23]. The catalysts showed moderate to high activity for the coupling reaction of iodobenzene with methyl acrylate; the chemical yields depending on the nature of the stabiliser and the reaction temperature. The authors found evidence for Pd leaching, showing that at the beginning of the

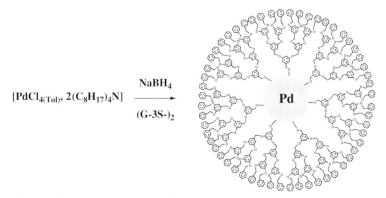

$$[PdCl_{4(Tol)}, 2(C_8H_{17})_4N] \xrightarrow[\text{(G-3S-)}_2]{\text{NaBH}_4} \text{Pd}$$

Fig. 10.5 Pd nanoparticle cored dendrimer.

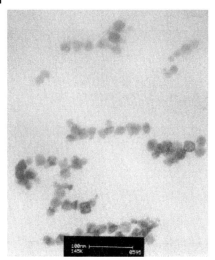

Fig. 10.6 Pd colloids stabilized by microgel. Reprinted from Ref. [19] with permission from Elsevier.

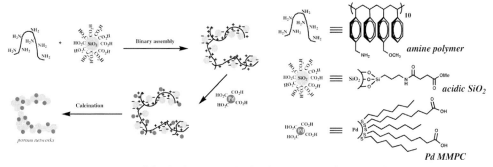

Fig. 10.7 Immobilized Pd nanoparticles by "bottom-up" self-assembled polymers.

reaction 81% of the initial palladium was found in bulk solution. After cooling, only 12–13% of the initial palladium remained in the solution, confirming the so-called dissolution–redeposition equilibrium reported by several authors.

Moreno-Mañas and Pleixats reported the use of soluble Pd nanoparticles stabilised by 1,5-bis(4,4′-bis(perfluorooctyl)phenyl)-1,4-pentadien-3-one in the fluorous phase [24]. This catalytic system was used in the Heck arylation of iodobenzene with ethyl acrylate, showing high recyclability without apparent loss of activity over five consecutive runs by simple extraction of the catalytic material with fluorous solvents.

Rotello and coworkers in an interesting study reported the use of immobilised Pd nanoparticles obtained through polymer mediated "bottom-up" self-assembly (Fig. 10.7) [25]. Pd clusters stabilised by a mixed monolayer (i.e. Pd MMPC) were added to a hybrid material, obtained from a binary assembly between an amine polymer and acid silica colloid, to give, after calcination, the pre-catalyst. This material promoted the Heck reaction between aryl bromides and styrene; activated aryl bromides giving the highest product yields.

Shi and coworkers reported the use of a thin Pd colloid layer on the pore channel surface of a mesoporous silica SBA-15, obtained through impregnation with a solution of Pd(OAc)$_2$ in THF followed by "in situ" reduction [26, 27]. The material promoted the Heck reaction of aryl iodides and activated aryl bromides with styrene or methyl acrylate, showing an excellent recyclability over at least five runs. While slight Pd leaching was detected through ICP analysis (0.4 ppm), investigations carried out with the "hot filtration" method revealed that these species were not as active as the heterogeneous material. The authors concluded that the Heck arylation runs through heterogeneous pathways. However, more probably, the leached Pd forms rapidly inactive black Pd particles in the absence of stabilisers, as often reported for homogenous systems.

Other silica-stabilised Pd nanoclusters (2–5 nm) were prepared by Park and coworkers from a mixture of Pd(PPh$_3$)$_4$ in tetraethyleneglycol and tetramethoxysilane through hydrolysis [28]. The resulting material exhibited good activity in the Heck, Suzuki, and Sonogashira coupling reactions.

Marr and coworkers described the preparation of Pd nanoparticles-stabilised aerogels using an ionic liquid route [29]. The material was prepared by inclusion of pre-formed Pd colloids (2 nm) stabilised by imidazolium salts in a silica monolith. The material was only tested in the reaction of iodobenzene with butyl acrylate, showing good activity (TON = 2092). No aggregation of Pd particles was observed by TEM; however Pd leaching, probably accounting for the activity, was detected by ICP-AES.

Trzeciak and coworkers reported an interesting study on Pd nanoparticles (19.8 nm) stabilised by polyvinylpyrrolidone (PVP) as highly active catalysts for Heck reaction of bromobenzene with butyl acrylate in [Bu$_4$N]Br media [30]. TEM studies showed that the Pd particle size decreased significantly after reaction with bromobenzene (to 17.3 nm). XPS and UV–vis spectra evidenced the formation of [Bu$_4$N]$_2$[Pd(Ph)Br$_3$] and [Bu$_4$N]$_2$[PdBr$_4$] Pd$^{(II)}$-complexes resulting from Pd dissolution. This supports an oxidative addition step of aryl halides onto the Pd particles, liberating active species in the bulk.

Zhang and Wang reported the use of Pd nanoparticles (5 nm) stabilised by poly(ethylene glycol) (PEG) as an efficient and recyclable catalyst for the Heck reaction of aryl iodide and bromides with ethyl acrylate or styrene [31]. Leaching of Pd species in solution was evidenced by ICP-AES analyses. The authors concluded from a filtration experiment made on a cold reaction mixture that the activity was due to heterogeneous systems. However, the method used allows the dissolved Pd species to redeposit on the support. That is, more probably, the activity observed with these catalysts was due to (temporarily) leached Pd species.

Calò and coworkers reported the use of Pd nanoparticles (3.3 nm) supported on chitosan as very efficient heterogeneous catalysts for the Heck reaction of aryl bromides and activated aryl chlorides with butyl acrylate in tetrabutylammonium

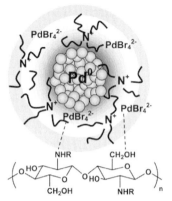

Fig. 10.8 Chitosan-supported Pd nanoparticles.
Reprinted from Ref. [31] with permission from the ACS.

ionic liquids [32]. XPS spectra evidenced the presence of two Pd oxidation states: $Pd^{(0)}$ and $Pd^{(II)}$ attributed to $[PdBr_4]^{2-}$ which was formed during the preparation of the material in the ammonium salt used to stabilise the supported Pd nanoparticles on chitosan (Fig. 10.8). The catalytic material could be recycled, however, with deactivation and apparent Pd leaching.

Dupont and coworkers described a study of the behavior of Pd nanoparticles (1.7 nm) stabilised in imidazolium ionic liquids as catalyst in the Heck reaction between butyl acrylate and aryl iodides [33]. The reaction was performed in an ionic liquid/organic two-phase solvent. TEM analyses performed before and after the reaction indicated substantial changes in Pd particle size with an average of 6.1 nm. In addition, ICP-AES analyses gave evidence of significant Pd leaching which varied during the course of the reaction (i.e. depending on the conversion). However, as often observed, the organic layer was inactive. These results are coherent with the hypothesis that the Pd nanoparticles act as reservoirs of active species following the so-called "dissolution–redeposition equilibrium" often reported for heterogeneous Pd catalysts. Rothenberg et al. devised an ingenious method to prove exactly this point. They built a two-chamber reactor separated by a membrane, which was impermeable for the nanoclusters (Fig. 10.9). One chamber contained the preformed nanoclusters and the other chamber contained all the other reactants, including the insoluble base. The build-up of product in the second chamber proves that low molecular weight palladium species travel from the first to the second chamber, supporting the notion that reaction occurs on monomeric or dimeric leached species [34].

McQuade and coworkers reported the use of Pd nanoparticles encapsulated in cross-linked reverse micelles in the Heck reaction between aryl iodides and methyl acrylate (Fig. 10.10) [35]. The catalyst showed high activity, comparable to that of Herrmann's catalyst; however activity remained limited to aryl iodides. Interestingly, low Pd loadings exhibited higher activities, a result consistent with catalysis by molecular species.

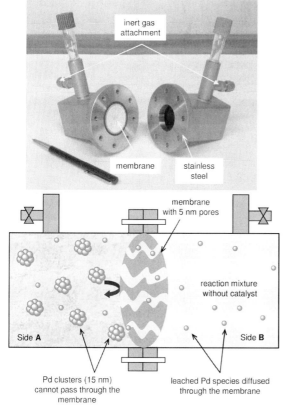

Fig. 10.9 Heck Reaction catalysed by palladium nanoparticles in a two-chamber reactor showing that palladium becomes detached from the particles during the catalytic cycle. Reprinted from Ref. [33] with permission from Wiley.

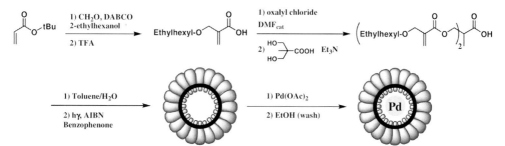

Fig. 10.10 Synthesis of Pd nanoparticles encapsulated in a cross-linked reverse micelle.

Kobayashi and coworkers reported the use of subnanometer Pd clusters stabilized within micelles for the Heck reactions of aryl iodides with ethyl acrylate [36]. The method used, called the "polymer intercalated method", is based on microencapsulation allowing the immobilisation of Pd(PPh$_3$)$_4$ as spherical micelles. After cross-linking, phosphine-free small Pd clusters (0.7 nm) stabilised through electronic interaction with the phenyl moieties present in the copolymer were obtained. Structure proof was deduced from EXAFS experiments. These catalytic systems exhibited high reactivity. No modification of the Pd particle size was observed by TEM after the reaction. The nature of the copolymer used to stabilise the Pd clusters was shown to have an important impact on the catalytic performances, mainly due to the rate of Pd leaching. The best systems, that exhibited low Pd leaching (<5 ppm), could be reused. Hot filtration tests confirmed that the clear filtrate did not act as a catalytic system, which was supported by the so-called "three-phase-test" or "Davies' test" using grafted aryl iodide as the aryl halide source. However, the validity of this test is in doubt, since according to the Davies' procedure, soluble aryl iodide must be used together with immobilised aryl iodide in order to induce the leaching process (*vide supra*). From these results, the authors concluded that the Heck reaction occurred within the polymer micelles.

Recently, Beletskaya and coworkers reported the use of Pd colloids stabilised by block-copolymer micelles formed by polystyrene-co-poly(ethylene oxide) and cetylpyridinium chloride as surfactant [37]. The material exhibited moderate to high activities and high recyclability for the Heck reaction between aryl iodides and acrylates.

10.2.2
Palladium Colloids in Other C—C Coupling Reactions

Compared to the large body of work concerning the Heck reaction, few reports deal with the use of Pd-colloids in the Suzuki coupling, the main contributions often coming from research groups involved in the general area of C—C couplings.

Reetz and coworkers reported the use of Pd and Pd/Ni clusters stabilized by tetrabutylammonium salts for the Suzuki coupling of aryl bromides and chlorides with phenylboronic acid [38]. The bimetallic Pd/Ni clusters were found generally to be more reactive, reacting also with activated aryl chlorides.

Several studies were reported by El-Sayed and coworkers, analysing mainly the influence of stabilisers in the Suzuki reaction in aqueous media [39–43]. The initial report on the Suzuki coupling of aryl iodides with phenylboronic acid using Pd nanoparticles stabilised by PVP showed that the rate of the reaction was dependent on the Pd catalyst concentration. The authors concluded that the Suzuki reaction occurred at the metallic surface. Catalyst deactivation and Pd black precipitation were observed, suggesting that Pd aggregation occurred during the reaction. Subsequent contributions compared various stabilisers: PVP, PANAM-OH dendrimers of different generations (G2-OH to G4-OH). The stability of the encapsulated Pd nanoparticles was measured by the tendency to form Pd black during the reaction, as the Suzuki reaction can be regarded as a good "acid test". G3-OH and PVP were found to be good stabilisers, also giving efficient catalysts. The G2-OH dendrimer was not suitable for stabilisation, although the catalyst was active in the reaction. This was attributed to the size of the Pd particles obtained after preparation (3.6 nm versus 1.4 nm for G3/G4-OH), which is larger than the dendrimer size itself (2.9 nm). G4-OH was found to be an effective stabiliser, but now the catalysts showed poor activity due to the strong encapsulation of the Pd particles. Further studies showed that the particle size increases during the first cycle (38% growth with PVP, 54% growth with G4-OH), and decreases when PVP is used as stabiliser in the second cycle, whereas it still increases with G4-OH. To explain the results obtained with PVP it was suggested that the larger particles aggregate and precipitate out of solution during the second cycle. The results obtained with G4-OH are consistent with strong encapsulation of the Pd nanoparticles by G4-OH slowing down the growth of the particles and resulting in comparatively high Pd leaching in solution. The phenomenon was found to be dependent on the reaction conditions: phenylboronic acid, or G4-OH, used in excess prevents the particle growth, both acting probably as capping agents. The authors propose a mechanism based on an Eley–Rideal process where the deprotonated phenylboronic acid adsorbs onto Pd and reacts with the aryl halide in solution. This is in contrast to the mechanism of the Suzuki reaction with a soluble palladium catalyst where the oxidative addition of the aryl halide precedes the transmetallation step. Pt nanoparticles were also studied, and they generally led to similar observations as the Pd nanoparticles while being less active.

Rothenberg and coworkers studied the application of mixed-metal nanoclusters in the Suzuki reaction. They made small tetra-octylammonium formate stabilised clusters (1.6–2.1 nm) based on Cu, Ru, Pd and Pt and binary and ternary combinations thereof. Of the single metals, Pd was the most active, but surprisingly, Cu and Ru also showed good activity. The mixed Cu–Pd nanocluster was as active as the Pd nanoclusters [44].

Coggan and coworkers also reported a process for producing Pd/Cu nanoparticles stabilised by tetra-octylammonium bromides and used them as catalysts in the Suzuki reaction [45, 46]. Various Pd/Cu ratios were studied, a 1:1 ratio giving the most active catalysts.

Hu and coworkers reported the use of Pd nanoparticles (1–5 nm) stabilised by poly(N,N-dihexylcarbodiimide) (PDHC) in the Suzuki coupling reaction of aryl iodides and bromides [47]. The catalytic material, that was prepared *in situ* by NaBH$_4$ reduction of H$_2$PdCl$_4$ in a mixture of PDHC, toluene and water, was highly active (TON > 1600) and relatively robust against leaching. The material could be recycled, however with decreased activity. It was also tested using microwave heating.

Kobayashi and coworkers also successfully used their subnanometer Pd clusters stabilized within micelles as catalyst in the Suzuki reaction (including recycling). It was found that the Pd leaching was mainly dependent on the reaction conditions. Addition of external phosphine ligand resulted in enhanced activity and decreased Pd leaching [48, 49].

Calò and coworkers reported the use of Pd nanoparticles in ionic liquids (quaternary ammonium salts) as catalysts for the Suzuki coupling reaction of aryl bromides and chlorides with phenylboronic acid. The catalytic system was found to be very active, and could be reused three times before deactivation occurred. The catalytic system was also efficient in the Stille coupling reaction [50].

Lately, Bradley and coworkers reported the use of Pd particles (7.4 nm) captured in cross-linked polystyrene–PEG resin as efficient, stable against leaching and recyclable catalysts in the Suzuki coupling of aryl bromides with phenylboronic acids [51].

A few articles report the Sonogashira coupling reaction using Pd colloids. Wang and coworkers reported recently an efficient procedure using Pd nanoparticles stabilised by PVP for the Sonogashira reaction of aryl iodides and bromides with phenylacetylene [52]. Almost quantitative conversions were obtained after 6 h with this catalyst in ethanol in the presence of K_2CO_3 as the base. The catalyst could be recycled by simple decantation and was reused eight times without significant loss of activity.

Rothenberg and coworkers noted the effect of the counter ion of the palladium salt used to make TBAB stabilised palladium nanoparticles, which they used as catalyst in the Sonogashira reaction of phenylacetylene with 4-cyanobromobenzene [53]. They found that the activity decreased in the order of $NO_3^- > Cl^- > OAc^-$. Since the binding strength of these anions to palladium runs in the reverse order they suggested that the rate of "leaching" of palladium from the cluster is determined by its stability. They proposed a mechanism similar to that proposed by de Vries for the Heck reaction in which palladium atoms at the rim of the cluster are subject to oxidative addition with ArBr. Assisted by the anion, a soluble $[ArPd(L)BrX]^-$ species is detached from the cluster. This species then completes the catalytic cycle of the Sonogashira reaction in solution. Alternatively the reaction is catalysed by palladium atoms that leach spontaneously from the cluster.

From the above it is clear that great progress has been made in the application of palladium colloids as catalysts for Heck and cross-coupling reactions. However, for reactions with deactivated aryl bromides and aryl chlorides, these catalysts seem to have insufficient reactivity and catalysts based on bulky trialkylphosphines seem necessary. There is a high degree of overlap between the use of colloids and ligand-free Pd catalysts (see Section 10.3.1), since most likely both types of catalytic system lead to the formation of soluble Pd species in bulk solution. In both cases, the stabilisation of active Pd species in solution is challenging: smart methods have been developed to achieve high TON by preventing the aggregation of dissolved Pd species to palladium black.

10.3
Ligand-free Palladium Catalysts

10.3.1
The Ligand-free Heck Reaction

In Heck's early publications it was already documented that the Heck reaction on aryl iodides can be carried out using $Pd(OAc)_2$ as catalyst without any added

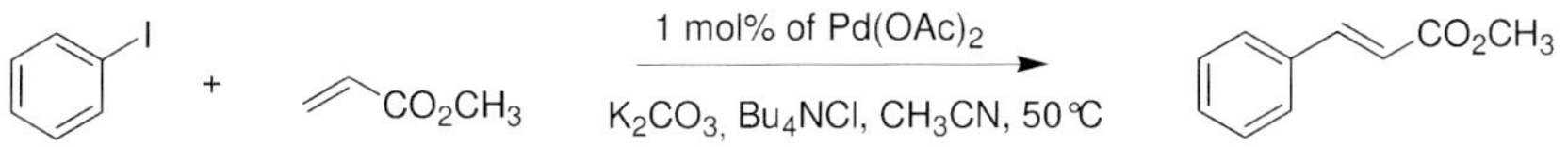

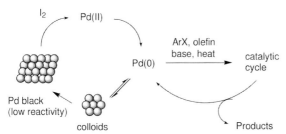

Scheme 10.4 Heck reaction on aryl iodides using Jeffery's conditions.

Scheme 10.5 Recycle of palladium in ligand-free Heck
reaction by reoxidation of precipitated Pd black.

ligands. In explaining the benefits of this catalyst over Pd/C he suggested that the
in situ reduction of Pd(OAc)$_2$ leads to "finely divided metal" [54]. Later work by
Jeffery showed the beneficial effect of the addition of tetra-alkylammonium salts
and water in these reactions, particularly when mineral bases such as K$_2$CO$_3$ are
used (Scheme 10.4) [55, 56]. Although the initial reason for the addition of the
tetra-alkylammonium salt was to act as a PTC, it soon became clear that this was
certainly not the only reason for the accelerating effect.

Using TEM, Reetz and Westermann showed that Pd nanoparticles with an
average size of 1.6 nm were present under the Jeffery conditions [11]. Thus, an
important role of the tetra-alkylammoniun salt is the stabilisation of the nanoparticles. Interestingly, in these ligand-free reactions the palladium precipitates at the
end of the reaction. Presumably, during the reaction there is a balance between
the Ostwald ripening of the colloids, which leads to their increase in size, and the
reaction of aryl iodide with the palladium atoms at the rim, which reduces the size
of the clusters. Once the starting materials are depleted, the Ostwald ripening leads
to deposition of palladium black. Using ICP-MS de Vries and coworkers measured
that indeed more than 99% of the palladium is precipitated. The precipitated catalyst can be reused, but has much lower activity than the original ligand-free palladium. Reoxidation of the catalyst which was precipitated on a solid carrier is easily
affected by addition of I$_2$ or Br$_2$, which completely restores the original activity
(Scheme 10.5) [57]. In this way they were able to perform eight recycles with the
same catalyst.

Quite a few groups have performed additional research trying to answer the
question as to whether the Heck reaction takes place on the surface of the clusters
or if it occurs entirely or partially in solution. De Vries and coworkers applied
electrospray MS (negative mode) on the Heck reaction of aryl iodides with butyl
acrylate. They uncovered a number of monomeric anionic species which led them
to propose the catalytic cycle depicted in Scheme 10.6 [57, 58].

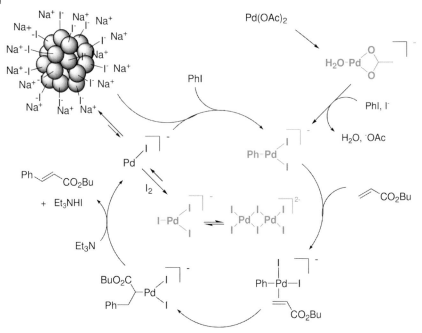

Scheme 10.6 Catalytic cycle of ligand-free Heck reaction between aryl iodides and acrylate ester (Species in grey detected by ES-MS or EXAFS).

In the initial stages of the reaction they noted the presence of $[Pd(H_2O)OAc]^-$, which may explain the importance of added water in the Jeffery chemistry. At a later stage two dominant species are $[PdI_3]^-$ and $[PhPdI_2]^-$. Presumably, the latter species is the resting state of the catalyst in this reaction, suggesting that olefin complex formation or olefin insertion is the rate-determining step. Evans et al. analysed the same reaction using EXAFS and found that the dominant species was the dimeric $[Pd_2I_6]^-$ [59]. The origin of this species is not clear, presumably iodide is oxidised to iodine, which suggests the presence of oxygen. Since the oxidative addition is fast, the presumption is that most palladium is indeed in the oxidised form and little of it is in the clusters, depending on the S/C ratio. In this mechanistic proposal the other function of Jeffery's tetra-alkylammonium salt becomes clear: the tetra-alkylammonium group serves as a cation for the anionic palladium species, securing maximum solubility in organic solvents and possibly inhibiting dimer formation. Amatore and Jutand have already established the importance of anionic palladium species in Heck reactions catalysed by phosphine ligated palladium [60]. Working with pre-formed palladium clusters, Dupont et al. [33] and Rothenberg et al. [34] proved, independently, that these colloids act as reservoir of active palladium species. It seems likely that the species which becomes detached from the cluster is $[ArPdX_2]^-$ [58].

In most applications of the ligand-free Heck reaction under Jeffery conditions 1 mol% of palladium catalyst or more is used. If the same conditions are applied to the Heck reaction on aryl bromides very poor conversions are obtained. The reason for this is clearly visible: precipitation of palladium. Indeed, precipitation of palladium is a well-known phenomenon in all Pd-catalysed reactions, even with palladium catalysts containing phosphorus ligands, at substrate/catalyst ratios of 100 or less. The discrepancy is due to the oxidative addition of aryl bromides on palladium being much slower than that of aryl iodides, allowing the colloids to grow to a size where precipitation occurs. To make this Heck reaction feasible, the growth of the clusters must be suppressed somehow. The challenge is thus to prevent the formation of palladium black by stabilising the colloids or by slowing down their growth (see Section 10.2). Although these solutions certainly work and allow the Heck reactions to be performed on aryl bromides without the use of phosphorus ligands, they have lost the attraction of the simple ligand-free catalyst system that was so easy to separate and recover at the end of the reaction.

De Vries and coworkers found another solution based on kinetic considerations of the reactions outlined in Scheme 10.5: The formation and growth of the nanoparticles must be higher order in the palladium concentration, whereas the catalytic cycle is first order or even half order (if the catalyst resting species is dimeric) in palladium. Thus, increasing the substrate catalyst ratio will slow down the former process more than the latter [61]. Indeed this non-obvious solution is the crux to what has been dubbed "homeopathic" palladium by Beletskaya [62] and by de Vries [61]. The essence is that the ligand-free palladium catalyst becomes more active at lower doses. The concept becomes immediately clear upon observation of Fig. 10.11. At 1 mol% of palladium the catalyst almost immediately precipitates as palladium black, leading to low conversions even after prolonged periods.

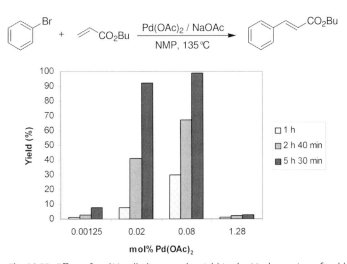

Fig. 10.11 Effect of mol% palladium on the yield in the Heck reaction of aryl bromides [61].

At values between 0.1 and 0.01 mol% the catalyst remains in solution as the size of the clusters does not increase since the rate of the oxidative addition is sufficiently high at these temperatures. Lower amounts of palladium are of course possible, but now the reaction becomes impractically slow. The catalytic cycle for this reaction is presumably similar to that for aryl iodides as depicted in Scheme 10.6, with one notable difference: Since oxidative addition of the bromoarene is the rate-determining step, most palladium will be locked up in the colloids. Indeed, examination with ES-MS revealed only small amounts of $[PdBr_3]^-$; no other species were observable either in the positive or in the negative mode.

One tell-tale sign of the involvement of colloids in catalytic reactions is the dependence of the turnover frequency on the S/C ratio. This is already apparent from Fig. 10.11. A calculation of the turnover frequencies of these reactions reveals the following values after 2 h 40 min: 309 (0.08 mol%), 787 (0.02 mol%), 900 (0.00125 mol%) h^{-1}. This increase can be easily explained because at higher S/C ratios the size of the nanoparticles is reduced and the ratio between available palladium at the surface vs. total palladium increases [63].

Earlier, Reetz and coworkers had seen a similar effect when testing the effect of the additive *N,N*-dimethylglycine on the Heck reaction between bromobenzene and styrene [10]. This additive supposedly stabilises the colloids, leading to good yields when between 0.0009 and 1.5 mol% Pd was used. They noted that only at the very low catalyst concentrations is there no need for the additive and rates are the same as without.

Two other groups have independently reported similar homeopathic ligand-free palladium catalysed reactions of aryl bromides. Schmidt followed the same analysis and increased the amount of aryl bromide, relative to the olefin by a factor of 6 [64, 65]. He also noted a beneficial effect of the addition of formate. He explained this effect by assuming that this keeps all palladium active in the Pd(0) state. However, this is at odds with the finding of de Vries that very little Pd(II) is present in bromoarene Heck reactions. Maybe formate is a superior stabiliser of the colloids or increases the solubility of the anionic palladium species. Schmidt also established that in the Heck reaction of bromobenzene with styrene catalysed by ligand-free $PdCl_2$ the rate of the reaction is diffusion limited [66]. Thus higher rates were observed upon increasing the stirrer speed. Yao and coworkers have reported similar results; they find K_3PO_4 is a superior base for these ligand-free Heck reactions on aryl bromides [67].

Recently, Leadbeater reported a ligand-free Heck reaction of bromarenes in water in the presence of TBAB using microwaves as the heating source. Here the temperature is around 170 °C, which allowed him to lower the mol% $Pd(OAc)_2$ to 0.0019 obtaining yields of styrenes and cinnamates between 30 and 83% [68].

Ligand-free palladium has been used as catalyst for many different arylating agents. Aryldiazonium salts are sufficiently reactive to prevent formation of palladium black even at 1–2 mol% $Pd(OAc)_2$ [69]. Aromatic carboxylic anhydrides can also be used as arylating agents using 0.1 mol% of ligand-free $PdCl_2$ in the presence of some extra halide such as NaCl or NaBr [70]. The cation did not have a strong influence in this case. Some mechanistic information was obtained by de

Vries and coworkers, using TEM, EXAFS, EDX and ES-MS. Colloids were clearly visible with TEM, and electrospray-MS showed the presence of anionic palladium species such as $[ArPdCl_2]^-$ [58]. The authors also noted the presence of small amounts of butyl chloroacrylate, which suggests that Pd(II) is reduced to Pd(0) by a Wacker type mechanism.

There are a number of cases reported where 1 mol% or more of ligand-free palladium catalyst could be used in the Heck reaction of aryl bromides. The medium and the base play an important role in these cases. Bumagin reported the use of 1 mol% of ligand free $Pd(OAc)_2$ in the Heck reaction of acrylic acid with a number of substituted bromarenes in water, using Na_2CO_3 as base [71]. Buchwald reported the Heck reaction of bromoarenes using 1–4 mol% of $Pd(OAc)_2$ with Et_4NCl in DMA under relatively mild conditions (85–100 °C) [72]. In his chemistry the use of the hindered bases dicyclohexylamine or methyldicyclohexylamine was essential. Presumably both the hindered base as well as the aqueous solvent plays a role in stabilising the colloids.

Several reports exist on the use of ionic liquids as solvent for ligand-free Heck reactions. Zhang found that 2-n-butyl-1,1,3,3-tetramethylguanidinium acetate was an excellent solvent for a ligand-free Heck reaction with 0.16 mol% of $PdCl_2$. Remarkably, this reaction needs no extra base, presumably since the solvent can act as a base [73]. Earlier studies found slower reactions using imidazolium type ionic liquids as solvent when compared to conventional media [33, 74]. Presumably, this is due to formation of palladium carbine-type complexes [75].

As stated earlier, it is very important to keep the size of the clusters down to prevent formation of palladium black. Apart from the kinetic approach described above a mechanical approach is also possible [76]. Two groups have described an ultrasound promoted ligand-free Heck reaction of aryl iodides [77, 78]. In one case, the yield of product was found to be 8-fold higher than under comparable thermal conditions.

Not surprisingly, most industrial applications of the Heck reaction use ligand-free palladium [78, 79]. Reetz and de Vries have reviewed the use of ligand-free palladium in the Heck reaction [80].

10.3.2
Ligand-free Palladium as Catalyst in Other C—C Bond Forming Reactions

Much research has been carried out on the ligand-free palladium-catalysed Suzuki reaction. Already in 1989, Beletskaya reported a ligand-free Suzuki reaction in water, between iodobenzoates and phenylboronic acid using $Pd(OAc)_2$ as catalyst [81]. Later, Novak took up the quest to develop a highly active catalyst for the Suzuki reaction and since he noted that this reaction suffers from phosphine inhibition he decided to test three ligand-free palladium catalyst precursors: $Pd(OAc)_2$, $[(\eta^3–C_3H_5)Pd PdCl]_2$, and $Pd_2(dba)_3.C_6H_6$ [82]. All three catalysts performed well in the Suzuki reaction between 4-nitro-iodobenzene and phenylboronic acid. In the reaction with 4-nitro-bromobenzene the first and last catalyst were clearly superior with yields of 96–98% (Scheme 10.7). Novak suggests that

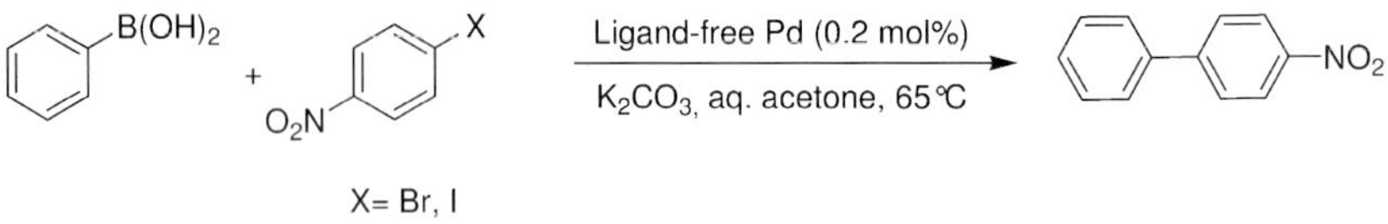

X= Br, I

Scheme 10.7 Ligand-free Suzuki reaction.

these reactions are indeed catalysed by palladium clusters without providing supporting evidence.

Badone extended this chemistry by performing the reaction in water and adding 2 eq. of TBAB. Using instead 10 mol% of this additive led to a much lower yield. Dupont used $PdCl_2(Et_2S)_2$ and $Pd(OAc)_2$ as catalysts (0.02–0.5 mol%) in DMF and K_3PO_4 as base and found that the first catalyst can be used for the Suzuki reaction of bromarenes and activated chloroarenes. The second catalyst needed 20 mol% of Bu_4NBr as additive to give the same performance [83]. Gong and coworkers examined the use of ligand-free palladium acetate and palladium chloride (0.5–2 mol%) as catalysts in the Suzuki cross-coupling of arylboronic acid with aryl and vinyl bromide in ethanol at room temperature and with exposure to air. Good yields of biaryls were obtained using TBAB as additive [84].

Several variants have been published. Welton used ionic liquids such as $[bnim][BF_4]$ as solvent [85]. The product is extracted with ether and the salt by-products are removed with an aqueous wash. After this treatment the catalyst solution could be reused three times without loss in activity. Dyson developed a novel nitrile-containing ionic liquid 4-pyridinium butyronitrile chloride $[PyC_3CN]^+Cl^-$, which upon reaction with $PdCl_2$ formed $[PyC_3CN]^+PdCl_4^{2-}$ [86]. This catalyst and other similar ones were used in the Suzuki reaction (1.2 mol%) using $[PyC_3CN]^+[Tf_2N]^-$ as solvent. Leadbeater developed a microwave version which allowed for the rapid preparation of biaryls in good yield, which is important for use in combinatorial chemistry [87]. Later he found out that under these conditions no palladium was necessary if 3.8 equivalents of the base Na_2CO_3 were used [88]. In addition, the reaction only worked with phenylboronic acid and not with other boronic acids. The reaction mixture was screened for the presence of fortuitous transition metals and none were found in amounts over 0.5–1 ppm. These results were put in doubt when de Vries and coworkers published the application of their homeopathic palladium protocol to the Suzuki reaction, which showed that good yield of biaryls could be obtained from bromarenes at 90 °C with catalyst loadings as low as 0.005 mol% [89, 90]. It should be kept in mind that the microwave reactions were performed at 150 °C. At this temperature sub-ppm amounts are capable of catalysing the Suzuki reaction. Re-examination by Leadbeater finally pinpointed the cause: The sodium carbonate used as base in his "uncatalysed" reaction contained 50 ppb of palladium contamination, which was sufficient to catalyse the Suzuki reaction at the high temperatures attained under the microwave conditions [91]. Indeed, upon repeating these experiments using K_2CO_3 as base, which contains much lower levels of palladium, good activities were found at 100–250 ppb palladium. Thus, turnover numbers in excess of 1 million were reached. At the ppb levels, the microwave-promoted reactions gave higher yields than the thermal reactions (150 °C), but at 2.5 ppm palladium the results were the same. The fact that other arylboronic acids are inactive could be traced to impurities in these compounds.

Bedford developed the Suzuki reaction on aryl chlorides using $Pd(OAc)_2$ and Bu_4NBr in water and obtained low to reasonable yields of biaryls (33–65%) with a TON up to 3200 on activated chlorides [92]. Similarly, Bhattacharya and Sengupta developed a protocol for the Heck and Suzuki reaction in water using 1 mol% of $Pd(OAc)_2$ and cetyltrimethylammonium bromide (50 mol%) as surfactant and found this gave better results than with Bu_4NBr. They performed UV–vis and TEM studies and were able to show the presence of palladium nanoparticles with an average size of 5 nm [93]. Other variants that have been published used $PdCl_2$ (0.3 mol%)/K_2CO_3/pyridine (reflux) [94]; $Pd(OAc)_2$ (2 mol%)/DABCO/KOH/PEG (80 °C) [95]; Pd(OAc) (0.5 mol%), Na_2CO_3, H_2O/acetone, 35 °C [96] and $Pd(OAc)_2$ (0.1 mol%)/KOH/H_2O (base added last) 20 → 75 °C) [97].

Instead of aryl halides, arenediazonium salts are also excellent arylating agents in the Suzuki coupling, although more hindered arylboronic acids did not react. The reaction is catalysed by several sources of ligand-free palladium such as $Pd(OAc)_2$, $Pd_2(dba)_3$ and Pd/C at room temperature in dioxane without any added base [98]. Use of potassium aryl trifluoroborate salts also allowed the introduction of more sterically hindered aryl groups [99].

Li and coworkers reported the use of $Pd(OAc)_2$/DABCO/air/CH_3CN for a copper-free Sonogashira reaction using only 0.01 mol% [100].

Ligand-free palladium has found some use in other C—C bond forming reactions, like oxidative homocoupling of alkynes [100], Negishi and Kumada reactions [90] or the methoxycarbonylation of iodobenzene [30, 101].

10.4
Palladacycles, Pincers and Other Palladium Complexes as Precursors of Palladium Nanoclusters

The area of palladium-catalysed coupling reactions got a new impulse when in 1995 Herrmann and Beller published their work on the discovery of a palladacycle as a catalyst for the Heck reaction (Fig. 10.12) [102, 103]. It had been known earlier that tri-*ortho*-tolylphosphine is a very good ligand for the palladium-catalysed Heck

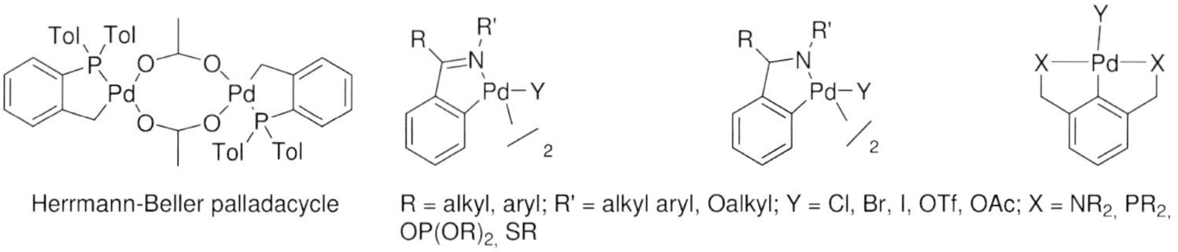

Herrmann-Beller palladacycle

R = alkyl, aryl; R' = alkyl aryl, Oalkyl; Y = Cl, Br, I, OTf, OAc; X = NR_2, PR_2, $OP(OR)_2$, SR

Fig. 10.12 Palladacycles and pincers used as pre-catalysts in the Heck and Suzuki reactions.

reaction, but Herrmann and Beller showed that treatment of Pd(OAc)$_2$ with this ligand for 3 min at 50 °C gave excellent yields of their palladacycle. The catalyst turned out to be highly active at higher temperatures, which not only allowed its use at low mol% but also enabled the Heck reaction of activated chloroarenes. This exciting new development brought a stream of new researchers into this field as many phosphorus, nitrogen, oxygen or sulfur ligands containing aromatic compounds are able to form palladacycles or pincer complexes [104–107]. Many of these compounds turned out to be excellent catalysts for the C—C bond forming reactions discussed in this chapter, resulting in over 400 publications.

This discovery also started an interesting debate about the mechanism of the Heck reaction, catalysed by these palladacycles. If one assumes that the structure of the catalyst remains intact, a Pd(II)/Pd(IV) type mechanism becomes almost inevitable [102, 103, 108, 109]. Although this proposal was initially supported by many, doubts were voiced when Hartwig showed that palladacycles are easily transformed into Pd(0) species, albeit under slightly different conditions from those during the Heck reaction [110]. Other mechanistic proposals included mono-ligated palladium in a Pd(0)/Pd(II) cycle [111], anionic Pd(0) species [112] and involvement of a dimeric [ArPd(L)Br]$_2$ species [113]. Although none of these mechanisms has been sufficiently proven nor disproven, the following experiment sheds an entirely different light on the matter.

When Nowotny et al. immobilised the imine-based palladacycle developed by Milstein et al. [114] and applied it in the Heck reaction between bromobenzene and styrene, the recycling experiment did not deliver the expected results. The catalyst was duly filtered off and reused, but the solid lost its activity after the second use [115]. On the other hand, when bromobenzene and styrene were added to the filtrate the reaction proceeded at the same rate (Scheme 10.8).

This clearly shows that the catalyst becomes detached from its ligand, presumably by reduction, yet it remains highly active. The authors also noted that whereas

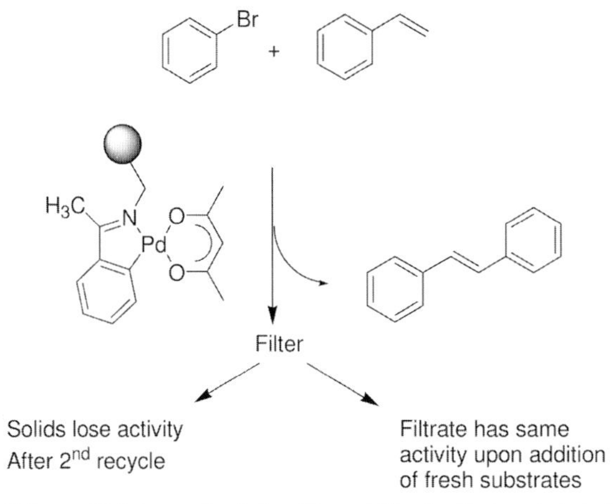

Scheme 10.8 Loss of activity of immobilised palladacycle is indicative of ligand-free palladium.

the Heck reaction using the immobilised palladacycle showed a clear induction period, this was not the case for the filtrate containing the soluble catalyst. Based on these data, the authors concluded that the palladium in solution probably was in the form of palladium nanoparticles.

Beletskaya and coworkers synthesised a range of nitrogen-containing palladacycles and measured their activity in the Heck reaction of iodoarenes and bromoarenes. They noted an induction period with all catalysts, which can be shortened through the addition of NaBH$_4$. Based upon this observation and the observation of sigmoidal kinetic curves they proposed that the palladacycles are all precursors of a low ligated Pd(0) species. Also in this study an increase in TOF upon increasing the S/C ratio was found [116]. The differences in rate between the palladacycles were explained by the different rates of Pd(0) release from the palladacycle. Another attempt to create a recyclable catalyst was similarly revealing. Thus, Gladysz in an attempt to recycle an imine-based palladacycle carrying fluorous ponytails also found activity transferred to the non-fluorous phase and was able to prove the presence of Pd colloids using TEM [117, 118].

Dupont and coworkers performed extensive research on the use of a chloropalladated propargyl amine as catalyst in the Heck reaction (Scheme 10.9) [119, 120]. This catalyst is exceptionally active and allows the Heck reaction to be performed on iodobenzene at 30 °C. The catalyst is inhibited by addition of a 300-fold excess of Hg. Hammett correlation of the reaction on *para*-substituted bromobenzenes gave a $\rho = 2.7$, which is in accord with an oxidative addition on Pd(0). TEM showed the presence of nanoparticles in a solution of the catalyst which was pre-treated with n-butyl acrylate, NaOAc and TBAB. After addition of iodobenzene nanoparticles could no longer be detected showing that the oxidative addition of iodobenzene to the colloids is indeed a very fast reaction. Reaction with haloarenes and acrylate esters attached to a solid support gave the Heck products in excellent yields, proving that the catalyst is not heterogeneous in nature.

De Vries and coworkers showed that the Heck reaction between bromobenzene and butyl acrylate catalysed by Pd(OAc)$_2$ had a similar kinetic profile as the Heck reaction catalysed by the Herrmann–Beller catalyst [61]. Analysis of the reaction mixture before and during the reaction using ES-MS showed the almost

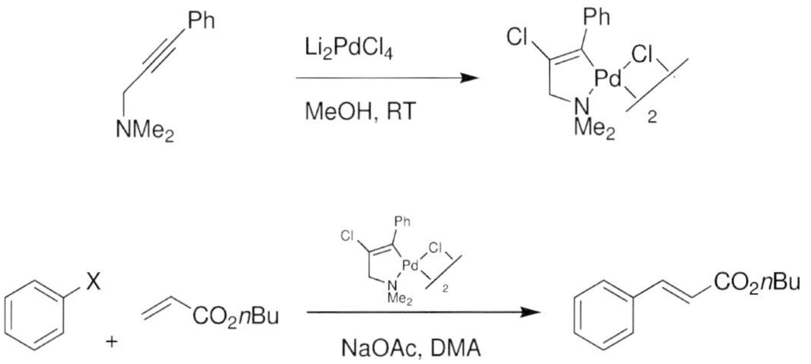

Scheme 10.9 Chloropalladated propargyl amine as catalyst precursor in the Heck reaction.

immediate complete disappearance of the palladacycle once the reaction was started. During the reaction only small amounts of PdBr$_3^-$ were detected in both reactions. In contrast with the iodobenzene reaction, no oxidative addition product was found. This seems to support the idea that most of the palladium is in the form of nanoclusters in the reaction with bromoarenes.

Several studies have appeared that show beyond doubt that PCP and SCS pincer complexes also decompose during the Heck reaction and lead to the formation of colloidal palladium [121–124]. Evidence was based on immobilisation studies and on the application of the extensive Hg poisoning protocol developed by Finke, which proved the presence of palladium colloids [125].

Navarro, Urriolabeitia and coworkers showed that palladacycles based on imi-nophosphoranes decompose to nanoparticles, as observed by TEM by simple reflux in DMF for 24 h [126]. These solutions of nanoparticles catalysed the Heck reaction without an induction period, whereas the palladacycles themselves showed induction periods of around 10 min. A mercury poison test was positive.

Palladacycles have also been used extensively as catalysts in other C—C bond forming reactions such as the Suzuki, Sonogashira and Stille couplings. Particularly active in this field are the groups of Bedford [107] and Nájera (Fig. 10.13) [127, 128]. Usually these catalysts perform better in the presence of Bu$_4$NBr. Surprisingly, much less mechanistic work has been performed in this area.

Liu studied the activity of cyclopalladated imines in the Suzuki reaction and found them to be extremely active (Fig. 10.13) [129]. He was able to show that reaction of his palladacycle with phenylboronic acid in the presence of K$_2$CO$_3$ and ethanol led to the formation of palladium nanoparticles with a diameter range of 50–60 nm (TEM). Upon addition of *p*-bromoanisole to this solution *p*-methoxybi-phenyl was formed quantitatively. Reaction of the same complex with ethanol produced acetaldehyde and also led to the formation of nanoparticles, however, in this case palladium black was soon formed. It thus would appear that aromatic boronic acids are capable of stabilising the palladium nanoparticles.

Corma and coworkers attached Nájera's palladacycle to polyethylene glycol and used this catalyst in the Suzuki and Sonogashira couplings [130]. They observed the formation of nanoparticles. In this case the virgin catalyst had a higher activity than the nanoparticles obtained from an earlier reaction.

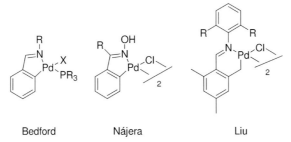

Bedford   Nájera   Liu

Fig. 10.13 Palladacycles used as catalysts in the Suzuki reaction.

Other examples include the use of palladacycles in the Stille reaction [110, 131, 132], or the Hiyama reaction between aryl halides and arylsiloxanes [133]. In most cases the authors observed the formation of nanoparticles.

Several excellent reviews have appeared recently in which published results regarding C—C bond forming reactions catalysed by several sources of palladium, including palladacycles and pincers, are critically evaluated with emphasis on mechanistic aspects [58, 105, 134–136].

It seems highly likely that all palladium-catalysed reactions that commence with an oxidative addition as the first step of the catalytic cycle proceed though a Pd(0)/Pd(II) mechanism. Thus one needs to conclude that all palladacycles and pincers are converted to some form of Pd(0) in these reactions. In many cases this was shown to be in the form of palladium nanoparticles. However, with the more reactive iodoarenes it is possible that most of the catalyst is in the form of an anionic or neutral monomeric or dimeric palladium species.

Pd(0) complexes ligated with monodentate phosphines are not very stable in solution. Even at room temperature without any light, heating, shaking or stirring, solutions of $Pd(PPh_3)_4$ will turn black and palladium nanoparticles were detected using TEM [137]. It thus seems likely that in high temperature (>120 °C) Heck reactions using palladium phosphine catalysts nanoparticles will form, in particular in reactions with aryl bromides and chlorides. Asymmetric Heck reactions do exist, testifying to the fact that the classical Heck cycle in which one or two phosphorus ligands are attached to the metal during the entire catalytic cycle is not obsolete [138, 139]. However, in all reported cases bidentate phosphine ligands were used and the reaction took place at temperatures from 0 °C up to 100 °C. It is thus very likely that a gray area exists were the reaction is catalysed both by palladium phosphine complexes and by the nanoparticles.

At this point, there are still some gaps in our understanding, particularly concerning the reaction on chloroarenes where palladacycles seem to perform better than $Pd(OAc)_2$, with or without added TBAB. This is unexpected if one assumes that both mechanisms proceed though a Pd(0)/Pd(II) mechanism in which the Pd(0) is in the form of palladium nanoparticles.

In conclusion, for C-C bond forming reactions on aryl bromides and iodides there is no need for the use of complexes, palladacycles or pincers as simple ligand-free $Pd(OAc)_2$ will perform as well. For C—C bond forming reactions on aryl chlorides some palladacycles may be the catalyst of choice. For problematic cases, the palladium complexes based on bulky electron-rich phosphines or catalysts based on carbene ligands are likely to give the best results.

10.5
Palladium Supported on Solids as Catalysts for Carbon–Carbon Coupling Reactions

The main driving forces for the development of supported Pd catalyst systems for C—C coupling reactions were recovery, recycling and reuse of the catalyst. Due to the enormous number of publications on the subject in the last few years, this chapter will focus therefore on chosen C–C coupling reactions of Heck, Suzuki and Sonogashira type only. In addition, particular emphasis is given to palladium (nanoparticles) supported on solids like metal oxides or carbons. In continuation of the previous chapters, such systems can be regarded as palladium colloids that are stabilized by the solid support. (For a recent review about progress in catalysis with Pd species supported on polymers and dendrimers see Ref. [140]).

One of the main and (in our opinion) most important differences between reported homogeneous and supported systems (including colloidal systems) is the clearly lower activity of the latter catalysts (often by orders of magnitude). This is reflected by the model systems investigated. Typical reports of heterogeneous catalysts focus on aryl iodides and activated aryl bromides as substrates. However, several new approaches allowed an efficient activation of bromobenzene and of aryl chlorides too. The following discussion will focus therefore on this limited number of particularly interesting substrates.

10.5.1
Heck Coupling by Supported (Solid) Pd Catalysts – General Motivation

Both approaches, homogeneous as well as heterogeneous catalysis, have their advantages and limitations. In fact, some of the problems in Heck catalysis by palladium complexes in solution could be solved – in principle – by heterogeneous catalysis: (i) In general, easy separation and recovery of the catalyst from the reaction mixture are given as first arguments. (ii) Recycling and reuse of the catalyst as it is or after regeneration procedures represent a second argument. In fact, the majority of papers on Heck reactions represent experimental results on this subject. (iii) Solid, supported catalysts are readily amenable to continuous processing. This argument, which is directly connected to the possibility of separation, (point (i) has been addressed and investigated in only a few papers [141, 142]) (iv) The problem of palladium black formation during the Heck reaction could be prevented by a heterogeneous system provided that the reaction occurs at the surface of the immobilized Pd species. If, however, Pd leaching into the liquid phase occurs during the reaction this approach should not be successful (see below). (v) Heterogeneous catalysts have been widely applied in industry also in liquid phase reactions (in particular for hydrogenations). The experience and understanding of these systems would form a sound foundation for their use in C–C coupling reactions. (vi) A variety of practical arguments can favor heterogeneous Heck catalysts: Heterogeneous catalysts are often easily prepared at low cost. They may even be commercially available, more stable at reaction conditions allowing e.g. higher reaction temperatures (often important for the activation of aryl chlorides) and can easily be stored over longer times. The use of expensive, sensitive ligands and inert atmosphere could be avoided.

There are also problems that can arise in heterogeneous Heck catalysis: The most important one is possible Pd leaching into the liquid phase which is more or less directly connected with all of the above mentioned arguments. In fact, nearly all papers in the last few years that investigated this subject report a more or less significant metal loss in heterogeneous Heck reactions. Points (i)–(vi) are directly dependent on this phenomenon. On the other hand, if the dissolution of palladium from the support is only temporary and palladium is re-precipitated onto the support at the end of the reaction (time dependent dissolution–redeposition processes), new possibilities and catalytic activation methods can arise. This too has been reported as a successful approach, allowing the activation of even "inert" aryl chlorides by heterogeneous catalysts [143].

In most cases, classical preparation methods well known from heterogeneous catalysis have been applied for the preparation of the Pd particles/species supported on solids. However, the palladium particle size was investigated in detail in only a few reports in the literature. A strong influence of the preparation method, Pd particle size (dispersion) and of the chemical nature of the Pd nanoparticles (e.g. metal or oxide) has been reported in several papers. A rather complete overview of the literature on heterogeneous palladium systems applied to Heck catalysis including detailed discussion on mechanistic aspects is given in reviews by Biffis et al. [20] and Jones et al. [136] for the time before 2001 and 2006, respectively.

10.5.2
Progress in Heck Reactions Catalyzed by Palladium Supported on Solids – Activation of Bromobenzene and Aryl Chlorides

As mentioned, aryl iodides are quite easy to activate because of the relatively low bond energy of the C–I bond. Activated aryl bromides have an electron-withdrawing group in the *para*-position, which weakens the C–Br bond. About 90% of the heterogeneous Heck papers use standard systems like non- or *para*-substituted aryl iodides and bromides (iodobenzene, 4-bromoacetophenone, bromobenzene; Scheme 10.10). The variety of alkenes is limited to styrene and different acrylic esters (methyl, ethyl or butyl acrylate). Non-activated aryl bromides, and in particular aryl chlorides, are of highest interest, not only for industrial application but also because mechanistic investigations using these substrates can usually be generalized. Aryl iodides can even be activated by traces of palladium, which mostly cannot be detected (during the reaction). The following survey highlights therefore chosen examples with bromobenzene and aryl chlorides as substrates.

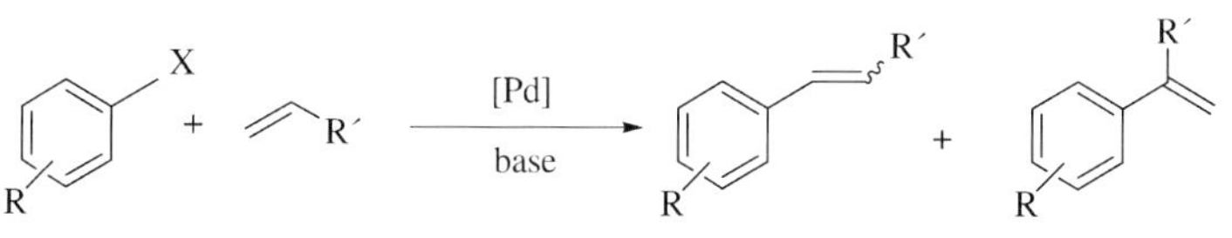

X = Cl, Br, I
R = CH$_3$CO, NO$_2$, H, CH$_3$, OCH$_3$
R′ = Aryl, COOAlkyl

Scheme 10.10 General scheme of the Heck reaction.

In the literature, a broad variety of materials has been reported as supports for palladium: Polymers as well as organic and inorganic hybrids have been utilized, for example, to stabilize Pd colloids or complexes and to avoid their agglomeration. Mesoporous silica materials have the advantage that (molecular) Pd complexes can be anchored covalently via surface Si—OH groups. In addition, they feature a very high specific surface area that allows the generation of highly dispersed palladium (very small Pd nanoparticles). Catalysis within the pores of zeolite structures also seems to be a promising approach to avoid agglomeration and deactivation of Pd species. The main benefits of common metal oxides are their availability and stability. Activated carbon is used as a catalyst support due to its high specific surface area and the resulting high Pd dispersion, and the advantageous easy filterability of this material. In addition, the noble metal can be recovered by simple burning of the support.

10.5.2.1 Heck Reactions of Non-activated Aryl Bromides (Table 10.1)

Many heterogeneous catalysts that are suitable for reactions of aryl iodides and activated bromides (like bromoacetophenone or bromonitrobenzene) fail in reactions of the non-activated bromobenzene. Apart from homogeneous catalysts an increasing number of heterogeneous catalyst systems have been developed which allow conversion of bromobenzene. Thus, this reaction system represents a suitable standard reaction for comparison of performances found with different catalysts. Activation of bromobenzene is possible using a huge variety of support materials if optimized reaction conditions are applied. In the following, different classes of carriers are discussed consecutively, although often almost no influence of the support can be observed.

Mesoporous materials and *zeolites* were mainly used to immobilize homogenous Pd complexes. Venkatesan and Singh prepared a series of Pd catalysts supported on MCM-41 [144]. In the reaction of bromobenzene and styrene they obtained TONs up to 2000 within 5 h reaction time at 150 °C. The same MCM-41 material was used by Tsai et al. to anchor a palladium bipyridyl complex. The resulting catalyst showed high TON (153 000) in the reaction of bromobenzene and butyl acrylate. However, substitution of butyl acrylate by the less reactive styrene caused a decrease of the TON by 2 orders of magnitude. In both cases very long reaction times (48 and 72 h) and high temperatures (170 °C) had to be applied [145]. Sugi and coworkers utilized a different mesoporous material, FSM-16, which was modified by pyridine and quinoline-carboimine ligands. Subsequently an immobilized Pd complex was generated. Both catalysts led to about 30% conversion of bromobenzene within 24 h at a catalyst concentration of 0.2 mol% [146].

Investigations of a variety of zeolites by Djakovitch et al. showed that the structure or the Si/Al ratio had little influence on the activity of the catalyst. All catalysts achieved complete conversions of bromobenzenes in 20 h with 0.2 mol% Pd at 140 °C [147]. However, kinetic experiments showed that monomeric Pd complexes (in particular $[Pd(NH_3)_4]^{2+}$) incorporated into the zeolite cages have reaction rates orders of magnitude higher than Pd particles. The immobilized complex converts the aryl halides within a few minutes. Recycling experiments showed that the catalyst could be easily separated and reused without loss in activity. Corma et al. also used a broad variety of different types of zeolites (e.g. Beta, KY, KX, CsX) as supports for the immobilization of PdCl$_2$. In contrast to the results above they did observe an influence of the zeolite type [148]. Other work of the same group focused on alkali-exchanged sepiolites. Again, an influence of the support was observed: Sodium sepiolites showed better catalytic results than those exchanged by potassium or cesium [149].

A new type of support was recently presented by the group of Kaneda. In their work they described the controlled synthesis of hydroxyapatite-supported Pd complexes and their application in Heck reactions of bromobenzene with styrene and butyl acrylate [150]. In both cases high TON (up to 47000) were achieved using low catalyst concentrations (2 × 10^{-3} mol%) and moderate temperatures (120 °C) within 24 h.

Metal oxides. Beside mesoporous silica materials "conventional" metal oxides have been applied for the synthesis of solid Pd catalysts for C–C coupling reactions. Alper and coworkers [151] and Singh and coworkers [144] immobilized Pd complexes on a modified silica surface modified by covalent anchoring. The resulting solid catalyst was tested in reactions of several non- and deactivated aryl bromides (bromobenzene, *p*-bromotoluene, *p*-bromoanisole) and butyl acrylate or styrene. Only TONs around 100 were achieved after 1 day reaction.

An earlier study of Wagner et al. dealt with a broad variety of oxide supports. Beside Al$_2$O$_3$ and SiO$_2$, that are typical supports for heterogeneous catalysts, MgO, TiO$_2$, ZnO and ZrO$_2$ were also investigated [157]. Pd catalysts were generated by impregnation of Pd(acac)$_2$ and subsequent reduction in a H$_2$ stream. All catalysts showed similar activity (30–50% conversion after 20 h at 140 °C; 0.1 mol% Pd) in the reaction of bromobenzene and styrene. However, later investigations demonstrated that different Pd dispersion, reduction degree and leaching tendency (varying with the support) are responsible for the differences in activity.

Table 10.1 Selected heterogeneously catalyzed Heck reactions of bromobenzene.

Entry	Catalyst	Cat. Conc. [mol%]	Alkene	Solvent	Base	T [°C]	t [h]	Yield [%]	Ref.
1	Pd/MCM-41	5 × 10^{-4}	BA	NMP	NBu$_3$	170	48	72	145
2	Pd/Beta	0.34	styrene	DMF	K$_2$CO$_3$	150	2.5	40	148
3	Pd/NaY	0.1	styrene	DMA	NaOAc	140	20	85	147
4	Pd/hydroxyapatite	2 × 10^{-3}	styrene	NMP	K$_2$CO$_3$	130	24	94	150
5	Pd/SiO$_2$	0.8	BA	NMP	Na$_2$CO$_3$	140	24	86	151
6	Pd/C	0.1	styrene	NMP	NaOAc	140	20	91	152
7	Pd/C	0.005	styrene	NMP	NaOAc	140	2	82	153
8	Pd/C	0.0025	styrene	NMP	NaOAc	140	2	83	154
9	Pd/C	2.5	EA	A336	NEt$_3$	100	3.5	100	155
10	Pd/C	3	BA	[OMIm]BF$_4$	NBu$_3$	MW[a]	0.025	80	156

a Microwave heating.

Carbon. In particular activated carbons are another interesting class of support materials for Pd nanoparticles used as catalyst for C—C coupling reactions. Detailed investigations on the optimization of a Pd/C catalyst have been performed by Heidenreich et al. Studies on the influence of the preparation method (oxidation state and dispersion of Pd, water content) resulted in highly active catalysts that achieved 90% conversion within 2 h using only 0.0025 mol% Pd (TON = 36 000; TOF = 18 000 h^{-1}) [158]. This optimized catalyst is commercially available. Particular interest has also been focused on the understanding of the reaction mechanism. Kinetic investigations showed a clear correlation between reaction rate and the content of dissolved palladium in the reaction mixture during the reaction [153, 154]. Leached palladium was found to be re-precipitated onto the support at the end of the reaction (after complete conversion of the substrates). In a recent work, Schmidt et al. compared a Pd/C catalyst to ligand free homogeneous PdCl$_2$ [64]. While the reaction catalyzed by PdCl$_2$ led to 95% yield of stilbene after 10 min, almost 4.5 h were necessary to achieve the same yield with a Pd/C catalyst. In both cases the use of an excess of bromobenzene and addition of sodium formate as a reducing agent had a strong promoting effect on catalyst activity. Perosa et al. used ionic liquid Aliquat 336, which is also known as a phase transfer catalyst, as solvent for the reaction of bromobenzene and ethyl acrylate. Applying 0.05 mol% Pd they obtained 55% stilbene yield after 1.5 h and 100% after 3.5 h [155]. Application of microwave heating led to minimized reaction times. These studies were performed by Xie et al. who achieved 80% yield after 1.5 min in the reaction of bromobenzene and butyl acrylate [156].

10.5.2.2 Reactions of Aryl Chlorides (Table 10.2)

All catalysts presented up to now were strictly limited to aryl bromides. Most of them show lower activity than pure Pd(OAc)$_2$. In the following, heterogeneous

Table 10.2 Selected heterogeneously catalyzed Heck reactions of chlorobenzene.

Entry	Catalyst	Cat. conc. [mol%]	Alkene	Solvent	Base	T [°C]	t [h]	Yield [%]	Ref.
1[a]	Pd/polymer	0.007	styrene	DMA	NBu$_3$	150	90	30	167
2	Pd/polymer	1.5	styrene	dioxane	Cs$_2$CO$_3$	80	3	89	168
3	Pd/TMS11	0.1	BA	DMA	NEt$_3$	170	32	6	159
4[a]	Pd/LDH	3	styrene	NBu$_4$Br	NBu$_3$	130	30	98	169
5[a]	Pd/LDH	3	styrene	NBu$_4$Br	NBu$_3$	MW[b]	0.5	95	169
6	Pd/SAPO-31	4	MA	DMA	NEt$_3$	120	70	67	161
7[a]	Pd/NaY	0.05	styrene	NMP	Ca(OH)$_2$	160	6	83	143
8	Pd/C	0.15	MA	NMP	NaOAc	160	16	6	86
9	Pd/C	0.7	styrene	water	NaOH/NaOOCH	100	5.5	48	165

a Addition of NBu$_4$$^+$ salts necessary.
b MW: microwaves heating.

catalysts are presented that (in contrast to Pd(OAc)$_2$) allow conversion of aryl chlorides (Scheme 10.10).

In 1997 Ying and coworkers reported that Pd-grafted mesoporous materials are active catalysts for Heck reactions. The Pd-TMS11 catalyst (prepared by gas phase reaction of volatile Pd complexes and MCM-41) was one of the very first heterogeneous catalysts that succeeded in the activation of aryl chlorides. With a catalyst amount of 0.1 mol% Pd at least 16% of chlorobenzene could be converted (TON = 64). However, 170 °C and 32 h reaction time were necessary. Only 40% selectivity to the Heck coupling products (= 6% yield) could be achieved [152, 159]. Another catalyst was prepared by Srivastava et al. by ion exchange using SAPO-31 and (NH$_3$)$_4$PdCl$_2$·H$_2$O. Up to 70% of chlorobenzene could be converted at 120 °C within 70 h (67% yield) [160].

A Pd-modified zeolite NaY was utilized as a heterogeneous catalyst for Heck reactions of 4-choroacetophenone [161, 162]. Entrapped [Pd(NH$_3$)$_4$]$^{2+}$ gave yields of about 60% of stilbene derivatives with palladium amounts of 0.1 mol% (170 °C, 20 h). Addition of NBu$_4$Br had a promoting effect on the reaction. Non- and deactivated aryl chlorides could not be converted under the same reaction conditions. However, under optimized reaction conditions by changing solvent (NMP instead of DMA), base (Ca(OH)$_2$ instead of NaOAc) and atmosphere (O$_2$ instead of argon) 85% of chlorobenzene could be converted by only 0.05 mol% Pd within 6 h at 160 °C (83% yield of stilbene). Even 4-chlorotoluene and 4-chloroanisole could be activated and selectively converted using the same reaction conditions [143].

Ley et al. used Pd(OAc)$_2$ which was stabilized by microencapsulation in polyurea in Heck reactions of various aryl bromides with butyl acrylate. Their optimized reaction conditions (supercritical CO$_2$ as solvent, 0.4 mol% Pd, 100 °C, Bu$_4$NOAc as base) also allowed the conversion of 4-chloronitrobenzene within 16 h (58% yield). This system seems to be limited to strongly activated aryl chlorides [163].

Sasson and coworkers reported the conversion of several non- and deactivated aryl chlorides using 0.7 mol% Pd/C catalyst at 100 °C within 5.5 h [164]. Although high conversion (up to 80%) was achieved, the catalyst system suffers from low selectivity; a broad variety of side products (including benzene and biphenyl derivates) is detected. Addition of reducing agents like sodium formate or zinc powder may be responsible for the formation of the side products. Independently, Choudary et al. [165], and Figueras et al. [166] reported on the use of layered double hydroxides as supports for nano-Pd catalysts (LDH-Pd(0)). Various aryl chlorides could be converted by 3 mol% Pd almost quantitatively within 1 h if microwave heating was applied. A commercial 10 wt.% Pd/C catalyst was utilized by Zhao and Arai for the reaction of chlorobenzene and methyl acrylate [86]. They observed 34% conversion and 18% selectivity to methyl cinnamate after 16 h at 160 °C using 0.15 mol% Pd. However, benzene (56% selectivity) and biphenyl (26%) were generated as undesired side products. The authors reported a correlation of the palladium content dissolved from the support and the activity. Therefore they proposed a "quasi-homogeneous" mechanism for the reaction.

10.5.3
Conclusions from the Literature Reports

10.5.3.1 Properties of the Catalyst

As shown above, several heterogeneous catalysts are capable of an efficient activation of bromobenzene and of aryl chlorides. The results obtained with palladium supported on carbon and on metal oxides revealed that the best catalyst performance was achieved if the following rules are observed: (i) Pd should be highly dispersed on the support surface. (ii) It should be present as Pd(II) (oxide or hydroxide). The classical pre-reduction in hydrogen at elevated temperature decreases the activity significantly. Low activity of several reported Pd/C catalysts can be explained in this manner. (iii) The catalysts should not be dried before use. Some water content is found to be advantageous [153].

Taking this into account, the influence of the support is of minor importance for reactions of all aryl bromides. Similar good results can be obtained with activated carbon, MgO, TiO_2, Al_2O_3, SiO_2 etc. Differences found in the literature are probably due to different Pd dispersion, palladium reduction degree or water content. For the activation of non-activated aryl chlorides a "fine-tuning" of the support and the reaction conditions may be necessary.

Pd complexes immobilized in zeolite cages turned out to be among the very best catalysts for Heck reactions. Originally, it was assumed that the isolation of molecular palladium in zeolite pores was responsible for its stabilization against agglomeration and that reaction takes place within the zeolite cages. Recent investigations have shown, however, that the cycle shown in Fig. 10.17 (later) is also valid for these highly active catalyst systems [170]. The reaction takes place outside the zeolite pores, Pd is leached into bulk solution, catalyzes the Heck reaction and diffuses back into the pores. Obviously the zeolite pore system allows a particularly efficient equilibration of all processes and cycles given in Fig. 10.17. Highly active palladium atoms or complexes are delivered continuously and agglomeration is effectively prevented.

A similar mode of operation, i.e. delivery of highly active palladium and efficient prevention of Pd agglomeration can be assumed for other heterogeneous catalyst systems that effectively activate aryl chlorides, like layered double hydroxides (LDH) or chosen polymers. Recent reports on interesting catalytic results using polymers as support unfortunately do not give clear and sufficient experimental results on mechanistic aspects or on metal leaching [163, 171].

10.5.3.2 Importance of Reaction Conditions

The complex mechanism and the various processes (Fig. 10.17) are responsible for the strong influence of the reaction conditions and parameters. Additives play an important role. Thus the performance of different catalysts can be improved by (defined) small amounts of oxidizing as well as reducing agents [64, 143, 172], by a variety of additives (in particular by alkylammonium halides) and is strongly influenced by the nature of solvent and base. Note, that these parameters also influence the Pd leaching in an analogous manner. In particular for the activation of aryl chlorides, a very careful choice of solvent, base, additives, temperature but also catalyst is necessary.

10.5.3.3 Potential for Practical Applications

The proven presence of homogeneous catalysis by leached palladium species does not necessarily mean that supported Pd catalysts are without practical use. It has been demonstrated that the control of the leaching process represents the basis for the jump in activity of catalysts for aryl chlorides [143]. In addition the understanding of the leaching process and the related parameters may represent an effective tool for the minimization of palladium loss [154]. Interesting approaches have been published for reusable catalysts and for simple reactivation procedures [57]. If, however, extensive metal leaching, catalyst restructuring or structural damage occur, significant activity losses will prevent effective reuse. Despite all success in the last years the nature of re-precipitation and metal phase restructuring is widely unknown and must be the subject of further investigations. A continuous process based on these catalysts would probably run into serious problems caused by the leaching.

Finally there are a number of very practical reasons which make supported palladium metal systems an attractive catalyst for organic synthesis in the laboratory and industrial fine chemical synthesis: The most interesting catalysts do not need additional ligands like phosphines. It is not necessary to work under an inert atmosphere. The high thermal stability of supported palladium metal catalysts allows the use of higher temperatures for Heck reactions. The resulting higher reaction rate makes it possible to activate less reactive aryl bromides and aryl chlorides in some cases.

10.5.4
Supported Palladium Catalysts in Other Coupling Reactions

Palladium species immobilized on various supports have also been applied as catalysts for Suzuki cross-coupling reactions of aryl bromides and chlorides with phenylboronic acids. Polymers, dendrimers, micro- and meso-porous materials, carbon and metal oxides have been used as carriers for Pd particles or complexes for these reactions. Polymers as supports were applied by Lee and Valiyaveettil et al. (using a particular capillary microreactor) [173] and by Bedford et al. (very efficient activation of aryl chlorides by polymer bound palladacycles) [174]. Buchmeiser et al. reported on the use of bispyrimidine-based Pd catalysts which were anchored onto a polymer support for Suzuki couplings of several aryl bromides [171]. Investigations of Corma et al. [130] and Plenio and coworkers [175] focused on the separation and reusability of Pd catalysts supported on soluble polymers. Astruc and Heuze et al. efficiently converted aryl chlorides using diphosphino Pd(II)-complexes on dendrimers [176].

Several authors successfully demonstrated the potential of micro- and mesoporous silica materials as well as zeolites as carriers. Corma et al. anchored carbapalladacycles on high surface area inorganic supports and converted aryl bromides and chlorides [177–179]. Similar approaches and catalytic results have been presented by Gürbüz et al. with imidazolidinium complexes of Pd on amorphous silica [180]. Artok et al. investigated Pd-exchanged Y zeolites as catalysts and found high activity in Suzuki coupling reactions of aryl bromides without addition of ligands [181]. Choudary et al. prepared a new type of supported catalyst by using an exchange reaction of $PdCl_4^{2-}$ in the presence of a layered double hydroxide resin [165]. Recently, Figueras and coworkers reported the preparation of a Pd/MgLa mixed oxide catalyst for the Suzuki reaction of aryl halides, including chlorides, and benzylic bromides with boronic acids in ethanol [182].

The rather classic catalyst palladium on activated carbon has been applied by Sun and Sowa et al. [183] and Heidenreich et al. [158] without additional ligands. This simple system was able to convert (mainly activated) aryl chlorides in mixtures of water and an organic solvent (DMA and NMP, respectively). Lysén et al. were able to convert aryl chlorides in pure water and without addition of any ligand [184, 185]. Suzuki reactions using Pd/C in an aqueous medium were also reported by Arcadi et al. (in the presence of surfactants) [186] and the group of Leadbeater, who applied microwave techniques [187]. Microwave (as well as ultrasound) conditions were also employed by Cravotto and Palmisano et al. [188] The substrate scope of Suzuki-type reactions in the presence of Pd/C was extended to halopyridines and haloquinolines by Tagata and Nishida [189].

Supported Pd catalysts were also successfully tested in Sonogashira reactions of aryl halides and alkynes. Most of these efforts concentrated on the development of polymer-based [130, 171, 175, 190] or dendrimer [191–193] supported catalysts. Beside these systems, a layered double hydroxide supported nano-Pd catalyst was found to be suitable for the activation of aryl chlorides [165, 194]. A variety of different supports (including zeolites, mesoporous materials, metal oxides and fluorides) were applied for Sonogashira couplings by Djakovitch and coworkers; however, the available substrate range was limited to aryl iodides and bromides in this case [195–197]. Cai and coworkers reported reactions of aryl iodides at room temperature in the presence of an MCM-41-supported Pd catalyst [198]. Application of ligand-free Pd/C in the Sonogashira reaction of aryl iodides was reported by Heidenreich et al. [158], Kotschy and coworkers [199] and Zhang [200].

The majority of the papers on Suzuki and Sonogashira reactions catalysed by supported palladium do not focus on mechanistic aspects and the corresponding contributions are thus limited in this respect. Nevertheless, it is reasonable to assume from the limited number of such investigations that the principal conclusions from detailed mechanistic studies of Heck reactions by supported catalysts are valid also for this group of coupling reactions: the active species is Pd leached into solution during the reaction and re-deposited at the end (see next section).

10.5.5

Mechanistic Aspects of Heck (and Related) Reactions by (Supported) Nanoparticles: Homogeneous or Heterogeneous Catalysis?

Very soon after the discovery of the reaction by Heck and Mizoroki, palladium black was identified as an active catalyst for the conversion of (exclusively) iodobenzene (see also Section 10.3.1) [54, 201]. Already Mizoroki had concluded from experimental results that Pd black was not the actual catalyst but only some kind of precursor or reservoir for the active (molecular) Pd species. Only some reports on heterogeneous catalysts for the Heck reaction followed over decades (see Ref. [20]). In particular, starting from the mid 1990s, the application of heterogeneous catalysts to the Heck coupling became very popular and detailed investigations on all aspects of the heterogeneous or homogeneous character of the reaction and the mechanism have been reported. Three main mechanistic approaches are imaginable and all three have been proposed in the literature:

1. The (flat) palladium surface activates a reacting molecule (the aryl halide) for the attack of the reaction partner (alkene) as illustrated in Fig. 10.14.
2. Heck catalysis occurs at defective sites on the metal surface (highly coordinatively unsaturated Pd surface atoms), see Fig. 10.15.
3. The actual active species is palladium in solution (molecularly) dissolved by leaching from the solid material. Consequently the mechanism would be a homogeneous one, where the solid catalyst serves only as a reservoir for active Pd species in solution as already proposed by Mizoroki. This general problem of metal leaching from supported catalysts in liquid–solid phases is well known and e.g. illustrated by Sheldon et al. [202].

The first proposal has never been supported by experimental results [203]. In contrast, the second mechanism has been extensively investigated and supported first by Augustine and O'Leary for reactions of aroyl chlorides [204, 205]. In a modified form, this defective site model has been transferred to typical Heck couplings by a variety of researchers investigating Heck catalysis by Pd colloids [6, 7,

Palladium

Fig. 10.14 Fictitious activation of an aryl halide on the Pd surface.

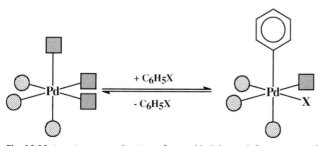

Fig. 10.15 Imaginary coordination of an aryl halide on defective sites of a Pd surface.

12, 13, 21]. These studies contributed much to the understanding of the reaction mechanism. Nevertheless it seems to be clear today that all effects observed and discussed for Heck catalysis by colloids can also be interpreted by molecular Pd species dissolved in solution. In principle, the role of Pd species of very small size (dinuclear or polynuclear Pd complexes or small clusters) cannot be excluded and is of course worth discussion and investigation.

Although there were still some articles in recent years claiming a truly heterogeneous (surface) mechanism for the Heck reaction, a careful interpretation of the literature gives clear evidence for the third mechanism (3), i.e. a quasi-homogeneous one. Probably the first clear experimental results were reported by Schmidt and Mametova [206] who performed kinetic investigations of Heck reactions with aryl iodides using different catalysts. They concluded that Pd species dissolved from the catalyst surface are responsible for catalysis. The presence of aryl halide and halide ligands is reported to be necessary for leaching and formation of active Pd species. The leaching process itself with typical supported Pd catalysts was first studied in detail by Arai et al. and reported in a series of papers [86, 207–210]. They demonstrated that the rate of reaction between iodobenzene and methyl acrylate correlates with the amount of Pd in solution and that palladium is re-deposited onto the support at high conversion of iodobenzene. The dissolution and redeposition of palladium depends on a variety of reaction parameters such as temperature, base and support. Biffis et al. and several other groups confirmed these results with different catalysts in the Heck reaction of aryl iodides and bromides [20, 64, 153, 207–212]. In principle, this pseudo-homogeneous mechanism can be concluded and has been proposed for practically all types of heterogeneous catalysts: for typical metal oxide supported palladium, for activated carbon catalysts, for different kinds of polymer supports as well as for Pd colloids (with the dissolution step, i.e. the oxidative addition of the aryl halide to a palladium surface atom, there is of course at least one heterogeneous step in the mechanistic cycle).

Some of the (probably) wrong conclusions in the literature during this period are due to the fact that measurable amounts of palladium are often observed in solution only during the reaction. Due to redeposition, the palladium concentration in solution is very small or not detectable at the end of the reaction. In addition, several experimental tests on the "heterogeneity" of the reaction were obviously not suitable for the Heck reaction [136]. And, very importantly, most mechanistic investigations were performed with aryl iodides (or activated aryl bromides) only. It has been shown in Sections 10.3 and 10.4 that extremely low (even not detectable) amounts of palladium at the ppm or sub-ppm level are able to convert these substrates easily [58].

Summarizing the literature data, there is also convincing evidence that Heck reactions carried out with palladium metal catalysts, supported or as colloids, are catalyzed by soluble Pd species leached out from the starting solid material. Palladium is dissolved from the catalyst surface during the reaction and partially or completely re-precipitated onto the support at the end of the reaction. Figure 10.16 illustrates this mechanism from an experimental point of view and correlates pal-

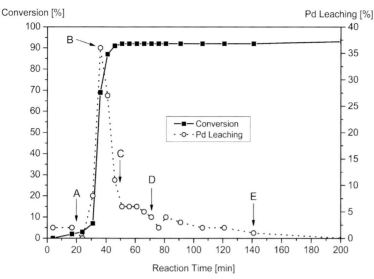

Fig. 10.16 Time-dependent correlation of conversion and Pd leaching (percentage of the total Pd amount) in the Heck reaction of bromobenzene and styrene in the presence of Pd/TiO$_2$ (Reaction conditions: 180 mmol bromobenzene, 270 mmol styrene, 216 mmol NaOAc, 0.2 mol% Pd catalyst, 180 mL NMP, 140 °C) [143]. Arrows A, B, C, etc. mark typical events during comparable experiments reported up to now (A: Pd dissolution starting at reaction temperature; B: maximum amount of Pd in solution/highest reaction rate; C: substantial redeposition of Pd onto the support with increasing conversion; D: (far-reaching) completion of Pd redeposition; E: complete redeposition of even Pd traces by increased temperature/reducing agents)

ladium concentration in solution with conversion during a typical Heck reaction. In such an experiment samples were taken continuously from the reaction mixture and Pd trace analyses were performed (AAS/ICP-OES). The schematic drawing visualizes the results of a large number of corresponding experiments, which are now available for reactions of bromobenzene, chloroacetophenone and other aryl chlorides [143, 153].

The reaction of bromobenzene and styrene (points A to C) is complete within a few minutes at 140 °C (without microwave heating!) with the best supported Pd-catalysts (e.g. palladium on carbon, on metal oxides or in zeolites; Fig. 10.16). The dissolution of palladium from the support (point A) starts (only!) at reaction temperature (140 C for bromobenzene) and depends on a variety of parameters (substrate, catalyst, solvent, temperature, base, additives, atmosphere). The maximum palladium concentration in solution correlates well with the highest reaction rate (turning point B of conversion curve). Palladium concentration in solution decreases with continued conversion (point C; decrease of aryl halide concentration) leading to (nearly) complete re-precipitation (point D) of the originally dissolved palladium (fortunately) onto the support. Increased temperature and/or excess of a reducing agent (e.g. sodium formate) added at the end of the reaction reduces the palladium concentration in solution to traces (1 ppm or sub-ppm level) [154]. This is of relevance for practical applications in the pharmaceutical industry, where palladium contamination of the organic product must be very strictly prevented [213]. The disadvantage is that the high Pd dispersion on the support can be reduced (particle agglomeration) by this drastic procedure, as demonstrated by TEM investigations of fresh and spent Pd/activated carbon catalysts. As a result the activity of the recycled/reused catalysts is reduced [86, 153]. On the other hand, work-up procedures (partial re-oxidation of Pd(0) by I_2 or Br_2) have been reported [57] that open other opportunities for reactivation of the solid catalyst. Later developments showed that a variety of catalysts can be reused several times without significant loss in activity.

One additional argument for a common mechanism with homogenous and heterogeneous (supported) palladium catalysts in carbon–carbon couplings concerns the selectivity of the reactions. In the majority of publications proposing homogenous or heterogeneous mechanisms, respectively, the same product selectivity to Heck products has been observed. One should expect rather significant differences between reactant activation on a solid surface and on a single metal complex in solution. The only difference found is the varying contribution of dehalogenation of the aryl halide (besides aryl–aryl coupling). Dehalogenation is found at increased temperature (e.g. with aryl chlorides) and under reducing conditions [154, 164, 212]. In both cases agglomeration (and precipitation) of Pd is strongly promoted (suppressing Pd leaching nearly completely). However, conversion of aryl bromide is observed (completed after a few hours). The main product found is benzene. Accordingly, it seems reasonable to assume that (only) dehalogenation of aryl halides occurs over Pd metal particles and via a truly heterogeneous surface mechanism.

10.5.5.1 Mechanistic Cycle

A possible reaction cycle for Heck catalysis with supported palladium catalysts is illustrated in Fig. 10.17. It reflects that a considerable number of processes involving metallic Pd (particles, colloids, or clusters) on the support or in solution, molecular Pd(0) and Pd(II) species supported or in solution and aryl halides complete the classical Heck cycle. Although the dissolution and redeposition of palladium has been proven by a number of authors and experiments, some of the single steps as e.g. the immediate dissolution "reaction" are still under discussion. Several experiments indicate that the oxidative addition of the aryl halide to a surface Pd(0) atom introduces the actual dissolution step for aryl bromides (and activated aryl chlorides) in the absence of additional ligands (NBu$_4$Br) [206, 214]. In these cases palladium dissolved in solution can be detected only at temperatures as high as the reaction temperature (140 °C for aryl bromides, 160 °C for aryl chlorides). No leaching is observable at lower temperatures. The sequence of oxidative addition and palladium dissolution can be different in the presence of additional ligands like bromide ions (from NBu$_4$Br): possibly palladium dissolution is not initiated by oxidative addition but by complex formation with the halide in this case [143].

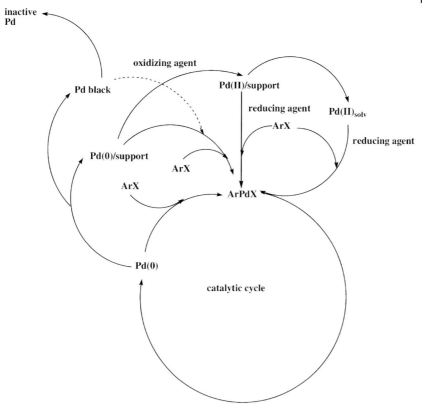

Fig. 10.17 Possible reaction pathways, processes and Pd species occurring during the Heck reaction using supported Pd catalysts.

Before oxidative addition of the aryl halide is possible, the metal (ions) must be present as, or reduced to, Pd(0). The best heterogeneous palladium catalysts contain, however, Pd(II) oxide particles on the support surface (or alternatively Pd(II)-complexes in zeolite cages) in their initial state. Possibly the oxidative addition/dissolution step is easier when a Pd(0) surface atom is bound to a Pd(II) oxide or hydroxide surface compared to metallic Pd(0). Pre-reduction of Pd(II) by educts (alkene) understandably occurs only on the outermost surface of a Pd particle. Alternatively a system containing a higher Pd(II) concentration, in comparison to Pd(0), is assumed to be more stable against Pd agglomeration and Pd black formation and thus against deactivation. The latter interpretation would also explain why working under (moderate) oxidative conditions (air or oxygen instead of inert atmosphere) increases the catalytic performance of Pd/metal oxide or zeolite catalysts.

The dissolution of palladium atoms from supported catalysts probably generates Pd species that are extremely active in the classic Heck cycle. The intrinsic nature of these species is still unknown. For aryl bromides (and to some extent for activated aryl chlorides) these systems represent (largely) true ligand-free catalysts. The reaction is fast enough to be finished before Pd agglomeration or deactivation occurs to a significant extent. For aryl chlorides additional halide ligands (preferentially Br$^-$) and counter ions (Bu$_4$N$^+$) for anionic Pd species are necessary in order to achieve a comparable situation. Conversions of chlorobenzene and chlorotoluene have been possible accordingly with addition of NBu$_4$Br only. The situation becomes more similar to homogeneous Pd complex catalysis. As mentioned above, it is more probable that Pd leaching is caused by the additional ligands and followed by oxidative addition of the aryl chloride to the dissolved Pd bromide complex.

Palladium colloids as active species can be excluded according to a series of experiments [20]. They act as a reservoir for molecular Pd species in solution. In addition, we do not see any reason why the surface of small Pd (nano)particles in solution should be active, the surface of the same Pd particles supported on a solid carrier, however, is not.

10.6
Conclusions

In the last decade, a considerable number of publications have demonstrated the potential of Pd nanoparticles and ligand-free palladium systems as catalysts for C–C coupling reactions. While the latter have been applied in many of the industrial Heck reactions, the use of stable Pd colloids is still in its infancy. Stabilisation of colloids is necessary to keep the size of the nanoparticles down to prevent the formation of less-active palladium black. It can be achieved by adding tetra-alkylammonium salts or polymers. Working at very high substrate/catalyst ratios also has the effect of keeping the size of the clusters down. This is known as "homeopathic" palladium. Other innovations came from the area of heterogeneous supported palladium catalysts. The advantages of heterogeneous catalysis (catalyst separation, recycling, reuse) have been shown for various reactions and catalyst systems. Palladium supported on carbon, on oxides, on/in zeolites, on mesoporous materials, on polymers, as well as pure Pd colloid systems were able to activate aryl iodides and chosen aryl bromides. New successful approaches and strategies led to the activation of deactivated aryl bromides, and in a few cases also of aryl chlorides, by heterogeneous catalysts. However, for problematic cases, like the coupling of poorly reactive aryl chlorides, palladacycles or pincers may work, but palladium complexes based on bulky electron-rich phosphines or carbene ligands are still the best choice.

Mechanistically, all catalytic systems, including heterogeneous catalysts, palladacycles and pincers operate through the formation of soluble Pd(0) nanoparticles in solution. In the catalytic cycle the aryl halide oxidatively adds to the palladium atoms at the rim, leading to the formation of an anionic aryl-palladium dihalide complex, or possibly its dimer. This anionic species then completes the Heck cycle.

After beta-hydride elimination the Pd(0) species will likely associate with the cluster again. While this cycle is slowly gaining acceptance in the case of Pd colloids, ligand-free catalytic systems and palladium complexes, there are still some recent articles on heterogeneous catalysts claiming a true heterogeneous (surface) mechanism for the Heck reaction. However, the palladium dissolution–redeposition processes correlate very well with the reaction rate and the selectivity generally observed. It is clear that Pd leaching is a prerequisite for high activity and selectivity of heterogeneous catalysts in Heck reactions. This leaching is often overlooked as even very small amounts (<1 ppm) of palladium are capable of catalyzing the Heck and Suzuki reactions.

References

1 G. Schmid, *Chem. Rev.* **1992**, *92*, 1709.
2 A. Roucoux, J. Schulz, H. Patin, *Chem. Rev.* **2002**, *102*, 3757.
3 M. Moreno-Mañas, R. Pleixats, *Acc. Chem. Res.* **2003**, *36*, 638.
4 R. M. Crooks, M. Zhao, L. Sun, V. Chechik, L. K. Yeung, *Acc. Chem. Res.* **2001**, *34*, 181.
5 D. Astruc, F. Lu, J. Ruiz Aranzaes, *Angew. Chem. Int. Ed.* **2005**, *44*, 7852.
6 M. Beller, H. Fischer, K. Kühlein, C. P. Reisinger, W. A. Herrmann, *J. Organomet. Chem.* **1996**, *520*, 257.
7 M. T. Reetz, G. Lohmer, *Chem. Commun.* **1996**, 1921.
8 M. T. Reetz, G. Lohmer, Studiengesellschaft Kohle GmbH **2001**, US 6224739 B1.
9 M. T. Reetz, G. Lohmer, R. Schwickardi, *Angew. Chem. Int. Ed.* **1998**, *37*, 481.
10 M. T. Reetz, E. Westermann, R. Lohmer, G. Lohmer, *Tetrahedron Lett.* **1998**, *39*, 8449.
11 M. T. Reetz, E. Westermann, *Angew. Chem. Int. Ed.* **2000**, *39*, 165.
12 S. Klingelhoefer, W. Heitz, A. Greiner, S. Oestreich, S. Foerster, M. Antonietti, *J. Am. Chem. Soc.* **1997**, *119*, 10116.
13 J. Le Bars, U. Specht, J. S. Bradley, D. G. Blackmond, *Langmuir* **1999**, *15*, 7621.
14 S. Pathak, M. T. Greci, R. C. Kwong, K. Mercado, S. G. K. Prakash, G. A. Olah, M. E. Thompson, *Chem. Mater.* **2000**, *12*, 1985.
15 J. H. Ding, D. L. Gin, *Chem. Mater.* **2000**, *12*, 22.
16 L. K. Yeung, R. M. Crooks, *Nano Lett.* **2001**, *1*, 14.
17 L. K. Yeung, C. T. Lee Jr., K. P. Johnston, R. M. Crooks, *Chem. Commun.* **2001**, 2290.
18 E. H. Rahim, F. S. Kamounah, J. Frederiksen, J. B. Christensen, *Nano Lett.* **2001**, *1*, 499.
19 K. R. Gopidas, J. K. Whitesell, M. A. Fox, *Nano Lett.* **2003**, *3*, 1757.
20 A. Biffis, M. Zecca, M. Basato, *J. Mol. Catal. A: Chem.* **2001**, *173*, 249.
21 A. Biffis, *J. Mol. Catal. A: Chem.* **2001**, *165*, 303.
22 A. Biffis, E. Sperotto, *Langmuir* **2003**, *19*, 9548.
23 A. M. Caporusso, P. Innocenti, L. A. Aronica, G. Vitulli, R. Gallina, A. Biffis, M. Zecca, B. Corain, *J. Catal.* **2005**, *234*, 1.
24 M. Moreno-Mañas, R. Pleixats, S. Villarroya, *Organometallics* **2001**, *20*, 4524.
25 T. H. Galow, U. Drechsler, J. A. Hanson, V. M. Rotello, *Chem. Commun.* **2002**, 1076.
26 L. Li, J.-L. Shi, J.-N. Yan, *Chem. Commun.* **2004**, 1990.
27 L. Li, L.-X. Zhang, J.-L. Shi, J.-N. Yan, J. Liang, *Appl. Catal. A: Gen.* **2005**, *283*, 85.
28 N. Kim, M. S. Kwon, C. M. Park, J. Park, *Tetrahedron Lett.* **2004**, *45*, 7057.
29 K. Anderson, S. Cortinas Fernandez, C. Hardacre, P. C. Marr, *Inorg. Chem. Commun.* **2004**, *7*, 73.

30 A. Gniewek, A. M. Trzeciak, J. J. Ziólkowsky, L. Kepinski, J. Wrzyszcz, W. Tylus, *J. Catal.* **2005**, *229*, 332.

31 C. Luo, Y. Zhang, Y. Wang, *J. Mol. Catal. A: Chem.* **2005**, *229*, 7.

32 V. Calo, A. Nacci, A. Monopoli, A. Fornaro, L. Sabbatini, N. Cioffi, N. Ditaranto, *Organometallics* **2004**, *23*, 5154.

33 C. C. Cassol, A. P. Umpierre, G. Machado, S. I. Wolke, J. Dupont, *J. Am. Chem. Soc.* **2005**, *127*, 3298.

34 M. B. Thatagar, J. E. Ten Elshof, G. Rothenberg, *Angew. Chem. Int. Ed.* **2006**, *45*, 2886.

35 K. E. Price, D. T. McQuade, *Chem. Commun.* **2005**, 1714.

36 K. Okamoto, R. Akiyama, H. Yoshida, T. Yoshida, S. Kobayashi, *J. Am. Chem. Soc.* **2005**, *127*, 2125.

37 I. P. Beletskaya, A. N. Kashin, A. E. Litvinov, V. S. Tyurin, P. M. Valetsky, G. van Koten, *Organometallics* **2006**, *25*, 154.

38 M. T. Reetz, R. Breinbauer, K. Wanninger, *Tetrahedron Lett.* **1996**, *37*, 4499.

39 Y. Li, X. M. Hong, D. M. Collard, M. A. El-Sayed, *Org. Lett.* **2000**, *2*, 2385.

40 Y. Li, M. A. El-Sayed, *J. Phys. Chem. B* **2001**, *105*, 8938.

41 R. Narayanan, M. A. El-Sayed, *J. Phys. Chem. B* **2004**, *108*, 8572.

42 R. Narayanan, M. A. El-Sayed, *Langmuir* **2005**, *21*, 2027.

43 R. Narayanan, M. A. El-Sayed, *J. Catal.* **2005**, *234*, 348.

44 M. B. Thathagar, J. Beckers, G. Rothenberg, *J. Am. Chem. Soc.* **2002**, *124*, 11858.

45 J. A. Coggan, N.-X. Hu, H. B. Goodbrand, T. P. Bender, Xerox Corporation **2006**, *CA 2 513 979 A1*.

46 J. A. Coggan, N.-X. Hu, H. B. Goodbrand, T. P. Bender, Pillsbury Winthrop Shaw Pittman, LLP **2006**, *US 2006/0025303 A1*.

47 Y. Liu, C. Khemtong, J. Hu, *Chem. Commun.* **2004**, 398.

48 K. Okamoto, R. Akiyama, S. Kobayashi, *Org. Lett.* **2004**, *6*, 1987.

49 R. Nishio, M. Sugiura, S. Kobayashi, *Org. Lett.* **2005**, *7*, 4831.

50 V. Calo, A. Nacci, A. Monopoli, F. Montingelli, *J. Org. Chem.* **2005**, *70*, 6040.

51 J. K. Cho, R. Najman, T. W. Dean, O. Ichihara, C. Muller, M. Bradley, *J. Am. Chem. Soc.* **2006**, *128*, 6276.

52 P. Li, L. Wang, H. Li, *Tetrahedron* **2005**, *61*, 8633.

53 M. B. Thathagar, P. J. Kooyman, R. Boerleider, E. Jansen, C. J. Elsevier, G. Rothenberg, *Adv. Synth. Catal.* **2005**, *347*, 1965.

54 J. P. Nolley, Jr., R. F. Heck, *J. Org. Chem.* **1972**, *37*, 2320.

55 T. Jeffery, in *Advances in Metal-Organic Chemistry, Vol. 5*, L. S. Liebeskind (Ed.), JAI Press, Greenwich, CT, USA, **1996**, p.153.

56 T. Jeffery, *Tetrahedron* **1996**, *52*, 10113.

57 A. H. M. de Vries, F. J. Parlevliet, L. Schmieder-van de Vondervoort, J. H. M. Mommers, H. J. W. Henderickx, M. A. M. Walet, J. G. de Vries, *Adv. Synth. Catal.* **2002**, *344*, 996.

58 J. G. De Vries, *Dalton Trans.* **2006**, 421.

59 J. Evans, L. O'Neill, V. L. Kambhampati, G. Rayner, S. Turin, A. Genge, A. J. Dent, T. Neisius, *J. Chem. Soc., Dalton Trans.* **2002**, 2207.

60 C. Amatore, A. Jutand, *Acc. Chem. Res.* **2000**, *33*, 314.

61 A. H. M. de Vries, J. M. C. A. Mulders, J. H. M. Mommers, H. J. W. Henderickx, J. G. de Vries, *Org. Lett.* **2003**, *5*, 3285.

62 I. P. Beletskaya, A. V. Cheprakov, *Chem. Rev.* **2000**, *100*, 3009.

63 A. F. Schmidt, V. V. Smirnov, *Top. Catal.* **2005**, *32*, 71.

64 A. F. Schmidt, V. V. Smirnov, *J. Mol. Catal. A: Chem.* **2003**, *203*, 75.

65 A. F. Schmidt, V. V. Smirnov, *Kinet. Catal.* **2005**, *46*, 47.

66 A. F. Schmidt, A. Al-Halaiqa, V. V. Smirnov, *J. Mol. Catal. A: Chem.* **2006**, *250*, 131.

67 Q. Yao, E. P. Kinney, Z. Yang, *J. Org. Chem.* **2003**, *68*, 7528.

68 R. K. Arvela, N. E. Leadbeater, *J. Org. Chem.* **2005**, *70*, 1786.

69 S. Sengupta, S. Bhattacharya, *J. Chem. Soc., Perkin Trans. I* **1993**, 1943.

70 M. S. Stephan, A. J. J. M. Teunissen, G. K. M. Verzijl, J. G. de Vries, *Angew. Chem. Int. Ed.* **1998**, *37*, 662.

71 N. A. Bumagin, V. V. Bykov, L. I. Sukhomlinova, I. P. Beletskaya, *J. Organomet. Chem.* **1995**, *486*, 259.

72 C. Gürtler, S. L. Buchwald, *Chem. Eur. J.* **1999**, *5*, 3107.

73 S. Li, Y. Lin, H. Xie, S. Zhang, J. Xu, *Org. Lett.* **2006**, *8*, 391.

74 A. J. Carmichael, M. J. Earle, J. D. Holbrey, P. B. McCormac, K. R. Seddon, *Org. Lett.* **1999**, *1*, 997.

75 L. Xu, W. Chen, J. Xiao, *Organometallics* **2000**, *19*, 1123.

76 M. A. Gelesky, A. P. Umpierre, G. Machado, R. R. B. Correia, W. C. Magno, J. Marais, G. Ebeling, J. Dupont, *J. Am. Chem. Soc.* **2005**, *127*, 4588.

77 R. R. Desmukh, R. Rajagopal, K. V. Srinivasan, *Chem. Commun* **2001**, 1544.

78 Z. Zhang, Z. Zha, C. Gan, C. Pan, Y. Zhou, Z. Wang, M. M. Zhou, *J. Org. Chem.* **2006**, *71*, 4339.

79 J. G. de Vries, *Can. J. Chem.* **2001**, *79*, 1086.

80 M. T. Reetz, J. G. de Vries, *Chem. Commun.* **2004**, 1559.

81 N. A. Bumagin, V. V. Bykov, I. P. Beletskaya, *Bull. Acad. Sci. USSR, Div. Chem. Sci.* **1989**, *38*, 2206.

82 T. I. Wallow, B. Novak, *J. Org. Chem.* **1994**, *59*, 5034.

83 D. Zim, A. L. Monteiro, J. Dupont, *Tetrahedron Lett.* **2000**, *41*, 8199.

84 Y. Deng, L. Gong, A. Mi, H. Liu, Y. Jiang, *Synthesis* **2003**, 337.

85 C. J. Matthews, P. J. Smith, T. Welton, *Chem. Commun.* **2000**, 1249.

86 F. Zhao, M. Arai, *React. Kinet. Catal. Lett.* **2004**, *81*, 281.

87 N. E. Leadbeater, M. Marco, *J. Org. Chem.* **2003**, *68*, 888.

88 N. E. Leadbeater, M. Marco, *Angew. Chem. Int. Ed.* **2003**, *42*, 1407.

89 J. G. de Vries, A. H. M. de Vries, *Eur. J. Org. Chem.* **2003**, 799.

90 A. Alimardanov, L. Schmieder-van de Vondervoort, A. H. M. de Vries, J. G. de Vries, *Adv. Synth. Catal.* **2004**, *346*, 1812.

91 R. K. Arvela, N. E. Leadbeater, M. S. Sangi, V. A. Williams, P. Granados, R. D. Singer, *J. Org. Chem.* **2005**, *70*, 161.

92 R. B. Bedford, M. E. Blake, C. P. Butts, D. Holder, *Chem. Commun.* **2003**, 466.

93 S. Bhattacharya, A. Srivastava, S. Sengupta, *Tetrahedron Lett.* **2005**, *46*, 3557.

94 X. Tao, Y. Zhao, D. Shen, *Synlett* **2004**, 359.

95 J.-H. Li, X.-C. Hu, Y. Liang, Y.-X. Xie, *Tetrahedron* **2006**, *62*, 31.

96 L. Liui, Y. Zhang, B. Xin, *J. Org. Chem.* **2006**, *71*, 3994.

97 D. N. Korolev, N. A. Bumagin, *Tetrahedron Lett.* **2006**, *47*, 4225.

98 S. Darses, T. Jeffery, J.-P. Genêt, J.-L. Brayer, J.-P. Demoute, *Tetrahedron Lett.* **1996**, *37*, 3857.

99 S. Darses, J.-P. Genêt, J.-L. Brayer, J.-P. Demoute, *Tetrahedron Lett.* **1997**, *38*, 4393.

100 J.-H. Li, Y. Liang, Y.-X. Xie, *J. Org. Chem.* **2005**, *70*, 4393.

101 A. M. Trzeciak, W. Wojtków, J. J. Ziółkowski, M. Zawadzki, *New J. Chem.* **2004**, *28*, 859.

102 M. Beller, H. Fischer, W. A. Herrmann, K. Öfele, C. Broßmer, *Angew. Chem. Int. Ed. Engl.* **1995**, *34*, 1848.

103 W. A. Herrmann, C. Broßmer, C.-P. Reisinger, T. H. Riermeier, K. Öfele, M. Beller, *J. Am. Chem. Soc.* **1997**, 1357.

104 J. Dupont, M. Pfeffer, J. Spencer, *Eur. J. Inorg. Chem.* **2001**, 1917.

105 I. P. Beletskaya, A. V. Cheprakov, *J. Organomet. Chem.* **2004**, *689*, 4055.

106 J. Dupont, C. S. Consorti, J. Spencer, *Chem. Rev.* **2005**, *105*, 2527.

107 R. B. Bedford, C. S. J. Cazin, D. Holder, *Coord. Chem. Rev.* **2004**, *248*, 2283.

108 B. L. Shaw, *New J. Chem.* **1998**, 77.

109 B. L. Shaw, S. D. Parera, E. A. Staley, *Chem. Commun.* **1998**, 1361.

110 J. Louie, J. F. Hartwig, *Angew. Chem. Int. Ed. Engl.* **1996**, *35*, 2359.

111 M. Beller, T. H. Riermeier, *Eur. J. Inorg. Chem.* **1998**, 29.

112 V. P. W. Böhm, W. A. Herrmann, *Chem. Eur. J.* **2001**, *7*, 4191.

113 T. Rosner, J. Le Bars, A. Pfaltz, D. G. Blackmond, *J. Am. Chem. Soc.* **2001**, *123*, 1848.

114 M. Ohff, A. Ohff, D. Milstein, *Chem. Commun.* **1999**, 357.

115 M. Nowotny, U. Hanefeld, H. van Koningsveld, T. Maschmeyer, *Chem. Commun.* **2000**, 1877.

116 I. P. Beletskaya, A. N. Kashin, N. B. Karlstedt, A. V. Mitin, A. V. Cheprakov, M. Kazankov, *J. Organomet. Chem* **2001**, *622*, 89.

117 C. Rocaboy, J. A. Gladysz, *Org. Lett.* **2002**, *4*, 1993.

118 C. Rocaboy, J. A. Gladysz, *New J. Chem.* **2003**, *27*, 39.

119 C. S. Consorti, M. L. Zanini, S. Leal, G. Ebeling, J. Dupont, *Org. Lett.* **2003**, *5*, 983.

120 C. S. Consorti, F. R. Flores, J. Dupont, *J. Am. Chem. Soc.* **2005**, *127*, 12054.

121 A. Taskinen, E. Toukoniitty, V. Nieminen, D. Y. Murzin, M. Hotokka, *Catal. Today* **2005**, *100*, 373.

122 W. J. Sommer, K. Q. Yu, J. S. Sears, Y. Y. Ji, X. L. Zheng, R. J. Davis, C. D. Sherrill, C. W. Jones, M. Weck, *Organometallics* **2005**, *24*, 4351.

123 M. R. Eberhard, *Org. Lett.* **2004**, *6*, 2125.

124 D. E. Bergbreiter, P. L. Osburn, J. D. Frels, *Adv. Synth. Catal.* **2005**, *347*, 172.

125 J. A. Widegren, R. G. Finke, *J. Mol. Catal. A: Chem.* **2003**, *198*, 317.

126 R. Bielsa, A. Larrea, R. Navarro, T. Soler, E. P. Urriolabeitia, *Eur. J. Inorg. Chem.* **2005**, 1724.

127 L. Botella, C. Nájera, *J. Organomet. Chem.* **2002**, *663*, 46.

128 L. Botella, C. Nájera, *Angew. Chem. Int. Ed.* **2002**, *41*, 179.

129 C.-L. Chen, Y.-H. Liu, S.-M. Peng, S.-T. Liu, *Organometallics* **2005**, *24*, 1075.

130 A. Corma, H. Garcia, A. Leyva, *J. Catal.* **2006**, *240*, 87.

131 D. A. Albisson, R. B. Bedford, P. N. Scully, S. E. Lawrence, *Chem. Commun.* **1998**, 2095.

132 D. Olsson, P. Nilsson, M. El asnouy, O. F. Wendt, *Dalton Trans.* **2005**, 1924.

133 E. Alacid, C. Nájera, *Adv. Synth. Catal.* **2006**, *348*, 945.

134 V. Farina, *Adv. Synth. Catal.* **2004**, *346*, 1553.

135 F. Alonso, I. P. Beletskaya, M. Yus, *Tetrahedron* **2005**, *61*, 11771.

136 N. T. S. Phan, M. Van Der Sluys, C. W. Jones, *Adv. Synth. Catal.* **2006**, *348*, 609.

137 E. Ye, H. Tan, S. Li, W. Y. Fan, *Angew. Chem. Int. Ed.* **2006**, *45*, 1120.

138 M. Shibasaki, E. M. Vogl, T. Ohshima, *Adv. Synth. Catal.* **2004**, *346*, 1533–1552.

139 M. Shibasaki, E. M. Vogl, *J. Organomet. Chem.* **1999**, *576*, 1.

140 Y. Uozumi, *Top. Curr. Chem.* **2004**, *242*, 77.

141 W. Solodenko, H. Wen, S. Leue, F. Stuhlmann, G. Sourkouni-Argirusi, G. Jas, H. Schönfeld, U. Kunz, A. Kirschning, *Eur. J. Inorg. Chem.* **2004**, 3601.

142 S. Liu, T. Fukayama, M. Sato, I. Ryu, *Org. Proc. Res. Dev.* **2004**, *8*, 877.

143 S. Pröckl, W. Kleist, M. A. Gruber, K. Köhler, *Angew. Chem. Int. Ed.* **2004**, *43*, 1881.

144 C. Venkatesan, A. P. Singh, *J. Catal.* **2004**, *227*, 148.

145 F.-Y. Tsai, C.-L. Wu, C.-Y. Mou, M.-C. Chao, H.-P. Lin, S.-T. Liu, *Tetrahedron Lett.* **2004**, *45*, 7503.

146 J. Horniakova, T. Raja, Y. Kubota, Y. Sugi, *J. Mol. Catal. A: Chem.* **2004**, *217*, 73.

147 L. Djakovitch, K. Koehler, *J. Mol. Catal. A: Chem.* **1999**, *142*, 275.

148 A. Corma, H. Garcia, A. Leyva, A. Primo, *Appl. Catal. A: Gen.* **2003**, *247*, 41.

149 A. Corma, H. Garcia, A. Leyva, A. Primo, *Appl. Catal. A: Gen.* **2004**, *257*, 77.

150 K. Mori, K. Yamaguchi, T. Hara, T. Mizugaki, K. Ebitani, K. Kaneda, *J. Am. Chem. Soc.* **2002**, *124*, 11572.

151 R. Chanthateyanonth, H. Alper, *J. Mol. Catal. A: Chem.* **2003**, *201*, 23–31.

152 C. P. Mehnert, J. Y. Ying, *Chem. Commun.* **1997**, 2215.

153 K. Köhler, R. G. Heidenreich, J. G. E. Krauter, J. Pietsch, *Chem. Eur. J.* **2002**, *8*, 622.

154 R. G. Heidenreich, J. G. E. Krauter, J. Pietsch, K. Köhler, *J. Mol. Catal. A: Chem.* **2002**, *182–183*, 499.

155 A. Perosa, P. Tundo, M. Selva, S. Zinovyev, A. Testa, *Org. Bio. Chem.* **2004**, *2*, 2249.

156 X. Xie, J. Lu, B. Chen, J. Han, X. She, X. Pan, *Tetrahedron Lett.* **2004**, *45*, 809.

157 M. Wagner, K. Köhler, L. Djakovitch, S. Weinkauf, V. Hagen, M. Muhler, *Top. Catal.* **2000**, *13*, 319.

158 R. G. Heidenreich, K. Köhler, J. G. E. Krauter, J. Pietsch, *Synlett* **2002**, 1118.

159 C. P. Mehnert, D. W. Weaver, J. Y. Ying, *J. Am. Chem. Soc.* **1998**, *120*, 12289.

160 R. Srivastava, N. Venkatathri, D. Srinivas, P. Ratnasamy, *Tetrahedron Lett.* **2003**, *44*, 3649–3651.

161 L. Djakovitch, K. Koehler, *J. Am. Chem. Soc.* **2001**, *123*, 5990.

162 L. Djakovitch, H. Heise, K. Köhler, *J. Organomet. Chem.* **1999**, *584*, 16.

163 S. V. Ley, C. Ramarao, R. S. Gordon, A. B. Holmes, A. J. Morrison, I. F. McConvery, I. M. Shirley, S. C. Smith, M. D. Smith, *Chem. Commun.* **2002**, 1134.

164 S. Mukhopadhyay, G. Rothenberg, A. Joshi, M. Baidossi, Y. Sasson, *Adv. Synth. Catal.* **2002**, *344*, 348.

165 B. M. Choudary, S. Madhi, N. S. Chowdari, M. L. Kantam, B. Sreedhar, *J. Am. Chem. Soc.* **2002**, *124*, 14127.

166 A. Cwik, Z. Hell, F. Figueras, *Adv. Synth. Catal.* **2006**, *348*, 523.

167 M. R. Buchmeiser, K. Wurst, *J. Am. Chem. Soc.* **1999**, *121*, 11101.

168 T. Seçkin, S. Köytepe, S. Demir, I. Özdemir, B. Cetinkaya, *J. Inorg. Organomet. Polym.* **2003**, *13*, 223.

169 M. Beller, A. Zapf, W. Mägerlein, *Chem. Eng. Technol.* **2001**, *24*, 575.

170 S. S. Pröckl, W. Kleist, K. Köhler, *Tetrahedron* **2005**, *61*, 9855.

171 M. R. Buchmeiser, T. Schareina, R. Kempe, K. Wirst, *J. Organomet. Chem.* **2001**, *634*, 39.

172 A. F. Schmidt, V. V. Smirnov, O. V. Starikova, A. V. Elaev, *Kinet. Catal.* **2001**, *42*, 199.

173 C. Basher, F. S. J. Hussain, H. K. Lee, S. Valiyaveettil, *Tetrahedron Lett.* **2004**, *45*, 7297.

174 R. B. Bedford, S. J. Coles, M. B. Hursthouse, V. J. M. Scordia, *Dalton Trans.* **2005**, 991.

175 A. Datta, K. Ebert, H. Plenio, *Organometallics* **2003**, *22*, 4685.

176 J. Lemo, K. Heuzé, D. Astruc, *Org. Lett.* **2005**, *7*, 2253.

177 A. Corma, D. Das, H. Garcia, A. Leyva, *J. Catal.* **2005**, *229*, 322.

178 C. Baleizao, A. Corma, H. Garcia, A. Leyva, *J. Organomet. Chem.* **2004**, *69*, 439.

179 C. Baleizao, A. Corma, H. Garcia, A. Leyva, *Chem. Commun.* **2003**, 606.

180 N. Gürbüz, I. Özdemir, T. Seçkin, B. Çetinkaya, *J. Inorg. Organomet. Polym.* **2004**, *14*, 149.

181 L. Artok, H. Bulut, *Tetrahedron Lett.* **2004**, *45*, 3881.

182 A. Cwik, Z. Hell, F. Figueras, *Org. Bio. Chem.* **2006**, *3*, 4307.

183 C. R. LeBlond, A. T. Andrews, Y. Sun, J. R. Sowa, Jr., *Org. Lett.* **2001**, *3*, 1555.

184 M. Lysén, K. Köhler, *Synlett* **2005**, 1671.

185 M. Lysén, K. Köhler, *Synthesis* **2006**, 692.

186 A. Arcadi, G. Cerichelli, M. Chiarini, M. Correa, D. Zorzan, *Eur. J. Org. Chem.* **2003**, 4080.

187 R. K. Arvela, N. E. Leadbeater, *Org. Lett.* **2005**, *7*, 2101.

188 G. Cravotto, M. Beggiato, A. Penoni, G. Palmisano, S. Tollari, J.-M. Leveque, W. Bonrath, *Tetrahedron Lett.* **2005**, *46*, 2267.

189 T. Tagata, M. Nishida, *J. Org. Chem.* **2003**, *68*, 9412.

190 A. Köllhofer, H. Plenio, *Chem. Eur. J.* **2003**, *9*, 1416.

191 K. Heuzé, D. Méry, D. Gauss, D. Astruc, *Chem. Commun.* **2003**, 2274.

192 K. Heuzé, D. Méry, D. Gauss, J.-C. Blais, D. Astruc, *Chem. Eur. J.* **2004**, *10*, 3936.

193 D. Astruc, K. Heuzé, S. Gatard, D. Méry, S. Nlate, L. Plault, *Adv. Synth. Catal.* **2005**, *347*, 329.

194 A. Cwik, Z. Hell, F. Figueras, *Tetrahedron Lett.* **2006**, *47*, 3023.

195 L. Djakovitch, P. Rollet, *Tetrahedron Lett.* **2004**, *45*, 1367.

196 L. Djakovitch, P. Rollet, *Adv. Synth. Catal.* **2004**, *346*, 1782.

197 P. Rollet, W. Kleist, V. Dufaud, L. Djakovitch, *J. Mol. Catal. A: Chem.* **2005**, *241*, 39.

198 M. Cai, Q. Xu, P. Wang, *J. Mol Catal. A: Chem.* **2006**, *250*, 199.

199 Z. Novák, A. Szabó, J. Répási, A. Kotschy, *J. Org. Chem.* **2003**, *68*, 3327.

200 G. Zhang, *Synlett* **2005**, 619.

201 K. Mori, T. Mizoroki, A. Ozaki, *Bull. Chem. Soc. Jpn.* **1973**, *46*, 1505.

202 R. A. Sheldon, M. Wallau, I. W. C. E. Arends, U. Schuchardt, *Acc. Chem. Res.* **1998**, *31*, 485.

203 K. Kaneda, M. Higuchi, T. Imanaka, *J. Mol. Catal.* **1990**, *63*, L33.

204 R. L. Augustine, S. T. O'Leary, *J. Mol. Catal.* **1992**, *72*, 229.

205 R. L. Augustine, S. T. O'Leary, *J. Mol. Catal. A: Chem.* **1995**, *95*, 277.

206 A. F. Schmidt, L. V. Mametova, *Kinet. Katal.* **1996**, *37*, 406.

207 F. Zhao, K. Murakami, M. Shirai, M. Arai, *J. Catal.* **2000**, *194*, 479.

208 F. Zhao, B. M. Bhanage, M. Shirai, M. Arai, *Chem. Eur. J.* **2000**, *6*, 843.

209 F. Zhao, M. Shirai, M. Arai, *J. Mol. Catal. A: Chem.* **2000**, *154*, 39.

210 F. Zhao, M. Shirai, Y. Ikushima, M. Arai, *J. Mol. Catal. A: Chem.* **2002**, *180*, 211.

211 K. Yu, W. Sommer, M. Weck, C. W. Jones, *J. Catal.* **2004**, *226*, 101.

212 L. Djakovitch, M. Wagner, C. G. Hartung, M. Beller, K. Koehler, *J. Mol. Catal. A: Chem.* **2004**, *219*, 121.

213 H. U. Blaser, A. Indolese, A. Schnyder, H. Steiner, M. Studer, *J. Mol. Catal. A: Chem.* **2001**, *173*, 3.

214 A. Biffis, M. Zecca, M. Basato, *Eur. J. Inorg. Chem.* **2001**, 1131.

11
Rhodium and Ruthenium Nanoparticles in Catalysis

Alain Roucoux, Audrey Nowicki, and Karine Philippot

11.1
Introduction

The last decade has evidenced an ever-increasing interest in the nanometric size chemical species area. More particularly, a great number of studies have been based on the synthesis and applications of noble transition metal colloids [1, 2]. The development of soluble metal nanoparticles and colloids as highly active nanocatalysts has been the focus of considerable effort and several research groups have contributed significantly to this central field of nanosciences and nanotechnology [3–5]. Due to their particular matter state, between homogeneous and heterogeneous, these frontier species are sometimes called "semi-heterogeneous" or "nanoheterogeneous" catalysts. From now on, soluble noble metal nanoparticles are considered as an unavoidable family of catalysts under mild conditions [6–10].

Contrary to classical heterogeneous catalysts, metal nanocatalysts – generally defined as particles between 1 and 10 nm in size – can be obtained by the bottom-up approach in two strategic ways according to the nature of the precursor, namely metal salts and organometallic compounds. Moreover, noble metal nanoparticles can be synthesized by a variety of methods according to the "organic" or "aqueous" nature of the media and the type of stabilizers used. The choice of the reaction parameters can provide some degree of control of particle size and particle composition, which is important for high activity and selectivity in catalytic applications. Finally, this field has also allowed, from recent kinetic and mechanistic studies, noble transition metal colloids to be registered as effective catalysts.

Over the past decade, nanoparticles, sometimes also called giant clusters, nanoclusters or colloids have been investigated as nanocatalysts in various catalytic applications such as hydrogenation, C–C coupling and other original reactions. This chapter reviews the recent progress in catalysis with ruthenium and rhodium nanoparticles as soluble nanocatalysts in various liquid media, organized according to the catalytic reaction type.

Nanoparticles and Catalysis. Edited by Didier Astruc
Copyright © 2008 WILEY-VCH Verlag GmbH & Co. KGaA, Weinheim
ISBN: 978-3-527-31572-7

11.2
Generalities on the Synthesis and the Stabilization Modes of Nanoparticles

The formation of zerovalent nanocatalysts can be reached following a variety of methods operating in various media (aqueous, organic, or mixture). The stabilization of nanoparticles during their synthesis is often reported in terms of electrostatic, steric or electrosteric effects in conformity with the fine interactions between the protective agent and the nanoparticle's surface. According to the media, but also related to the expertise of numerous laboratories, various stabilizers are used, such as polymers, surfactants, ligands, cyclodextrins and also dendrimers and ionic liquids [3, 4, 7, 11–12]. Polymers, surfactants, dendrimers and cyclodextrins are generally considered as steric stabilizers allowing particles entrapment, displaying then no or weak interactions with the particle's surface, while ligands can coordinate onto the surface of the particles and then have influence on the catalytic activity or selectivity.

To summarize, two efficient and facile chemical routes in nanoparticles synthesis are reported according to the nature of the precursor: (i) mild chemical reduction of transition metal salt solutions, (ii) metal atom extrusion starting from organometallic compounds able to decompose in solution under mild conditions (Fig. 11.1). From a rational point of view, each approach presents many advantages but also some drawbacks. The chemical reduction of commercial and various transition metal salts such as rhodium or ruthenium chloride species is probably the most common synthetic route to colloids and a great variety of reducing agents such as alcohols, dihydrogen, borohydrides or boranes are usually investigated in several media (organic or water) [4, 13]. This method provides some degree of control over particle size producing efficient and reproducible nanocatalysts in terms of activity and selectivity in catalytic applications (hydrogenation, C–C coupling etc.). Nevertheless in these processes, salts, water and by-products often remain in contact with the surface of the particles, thus passivating them, leading to the production of surface oxides or hydroxides, and potentially modifying their reactivity in catalysis.

By contrast, the organometallic approach developed by the team of Chaudret [12, 14] is based on ligands removal from an organometallic compound in the mildest possible conditions and with the minimum of potentially polluting reactants. Purely olefinic complexes such as $Ru(\eta^6\text{-}C_8H_{10})(\eta^4\text{-}C_8H_{12})$ or allylic complexes like $Rh(\eta^3\text{-}C_3H_5)_3$ are the most attractive precursors since, upon hydrogenation, they give rise to alkanes, which are inert towards the surface of the particles. This strategy displays several specificities in terms of control of the particles dispersity, their size, their shape, their organization and the nature of the chemical species present at their surface. The synthesis can be performed using as stabilizers classical ligands of organometallic chemistry (amines, thiols, phosphines etc.) or polymers. The amount of added ligand allows control of the growth of the particles and therefore their size. Nevertheless, this approach requires the delicate and sometimes tedious synthesis of the organometallic precursor in anaerobic conditions. Finally,

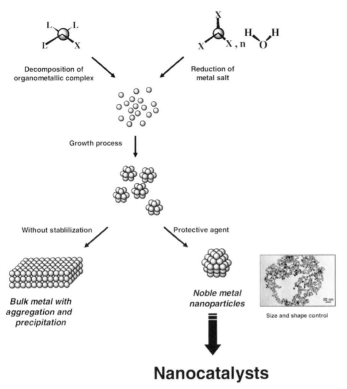

Fig. 11.1 Usual methods for synthesis of nanocatalysts.

aqueous colloidal suspensions are more difficult to obtain with this organometallic method and to our knowledge have never been described.

To conclude, whatever the strategy, nanospecies display a very rich potential for reactivity with good reproducibility in catalytic applications.

11.3
Rh and Ru Nanoparticles as Catalysts in Hydrogenation Reactions

Rh and Ru nanoparticles destined for catalytic applications are mostly used as catalysts in hydrogenation reactions, and most particularly in olefin and alkyne hydrogenation. Moreover, soluble noble metal nanoparticles are considered as a reference in monocyclic arene catalytic hydrogenation under mild conditions and several stabilized systems have also been carefully reported. Finally, interesting results in ketone hydrogenation are also described. That is why a detailed part of this chapter is devoted to hydrogenation reactions which are classified according to two main types: hydrogenation of unsaturated hydrocarbons and hydrogenation of C=O compounds. For clarity, the papers dealing with the hydrogenation of unsaturated hydrocarbon are presented according to the stabilizing mode retained for the nanoparticles.

11.3.1
Hydrogenation of Unsaturated Hydrocarbons

Soluble Rh and Ru nanoparticles are commonly used as catalysts in olefin and alkyne hydrogenation reactions. Various catalytic systems are investigated, depending on the metal precursor and the stabilizer.

11.3.1.1 Polymer Stabilized Rh and Ru Nanoparticles

Organic polymers are very often used for the protection of metal nanoparticles, providing a steric stabilizing effect. Due to this embedding effect, it is generally considered that the diffusion of the substrates through the polymer matrix can be limited. However, interesting results have been obtained.

11.3.1.1.1 Hydrogenation of Compounds with C=C Bonds Hirai and Toshima have published several papers on the synthesis of transition metal nanoparticles by alcoholic reduction of metal salts in the presence of a polymer such as polyvinyl alcohol (PVA), poly(methyl vinyl ether) (PMVE) or polyvinylpyrrolidone (PVP) in methanol (or ethanol)/water mixtures or in pure alcohols. This simple and reproducible process has been applied for the preparation of rhodium nanoparticles from rhodium (III) trichloride [15]. The particles size of metallic fcc rhodium is distributed in a narrow range, 3–7 nm and the average diameter is 4 nm. The PVP-stabilized Rh nanoparticles are more stable. These nanomaterials are efficient catalysts for olefin and diene hydrogenation in mild conditions (30 °C; $P_{H2} = 1$ bar) as shown in Table 11.1.

Table 11.1 Catalytic activity of Rh colloidal dispersions for hydrogenation of olefins and dienes. Adapted from Ref. [15].

	Catalytic activity (H_2 mole/Rh g-atom s^{-1})[a]		
Substrate	Rh-PVP-MeOH/H$_2$O	Rh-PVP EtOH	Rh-PVP-MeOH/NaOH
1-hexene	15.8	14.5	16.9
2-hexene	4.1	9.5	12.8
cyclohexene	5.5	10.3	19.2
styrene	1.9	2.5	3.2
cyclooctene	0.6	1.1	1.2
1,3-cyclooctadiene	3.7	9.8	17.5
1,5-cyclooctadiene	2.6	3.7	3.7

a Hydrogenation conditions: 30 °C, 1 bar/H_2, [Rh] = 0.01 mM, [substrate] = 25 mM, solvent = methanol (20 mL).

All the three tested colloidal dispersions exhibited roughly the same catalytic activity for hydrogenation of terminal olefins but differences were observed for internal and cyclic olefins.

Delmas et al. produced PVP-stabilized rhodium nanoparticles following the method reported by Hirai performing catalytic hydrogenation of oct-1-ene in a two liquid phase system [16].They investigated the effect of various parameters on their stability and activity under more or less severe conditions. They have shown that PVP/Rh colloids could be reused twice or more without loss of activity.

11.3.1.1.2 Hydrogenation of Compounds with C≡C Bonds 1-Hexyne could be hydrogenated using a PVP-stabilized Rh nanoparticles dispersion previously described in Section 11.3.1.1.1 [15].

Recently, Chaudhari has compared the activity of dispersed nanosized metal particles prepared by chemical or radiolytic reduction and stabilized by various polymers (PVP, PVA or poly(methyl vinyl ether)) with that of conventional sup-ported metal catalysts in the partial hydrogenation of 2-butyne-1,4-diol in butene-1,4-diol. Several transition metals such as Pd, Pt, Rh, Ru and Ni were prepared according to conventional methods and investigated [17]. Generally, the catalysts prepared by the chemical reduction method were more active than those prepared by radiolysis and, in all cases, aqueous colloids showed a higher catalytic activity (up to 40 times more) in comparison with the corresponding conventional supported catalysts. The results obtained with Rh and Ru species are reported in Table 11.2.

Catalytic studies and kinetic investigations of PVP-embedded rhodium nanopar-ticles in the hydrogenation of phenylacetylene were performed by Choukroun and Chaudret [18]. The rhodium colloids were prepared from the reaction of $[RhCl(C_2H_4)_2]_2$ in THF with two equivalents of Cp_2V as a reducing agent and in the presence of PVP. The formation of soluble Rh nanoparticles with sizes in the 2–3 nm range was confirmed by TEM. These rhodium nanoparticles were used in biphasic conditions or as heterogeneous catalysts (neat conditions) at 60 °C under a hydrogen pressure of 7 bar with a [substrate]/[catalyst] molar ratio of 3800. Total hydrogenation into ethylbenzene was observed after 6 h giving rise to a TOF (turn-over frequency is defined as [mol product][mol metal]$^{-1}$ h^{-1}) of 630 h^{-1}. The kinetics

Table 11.2 Comparison of colloidal and heterogeneous Rh and Ru catalysts in the hydrogenation of 2-butyne-1,4-diol. Adapted from Ref. [17].

Catalyst	Size (nm)	Selectivity (%)	TOF ($\times 10^{-5}$ h^{-1})	TOF$_{Mt\text{-}PVP}$/ TOF$_{Mt\text{-}CaCO3}$
Rh/PVP	5	96	4.2	—
Rh/CaCO$_3$	—	85	0.1	42
Ru/PVP	4.8	95.2	5.1	—
Ru/CaCO$_3$	—	75	0.14	36

of the hydrogenation was found to be zero-order in respect to the alkyne compound and the reduction of styrene to ethylbenzene depends on the concentration of phenylacetylene still present in solution. Additional experiments were performed in the presence of phosphine and showed that the selectivity in styrene increased and that the formation of ethylbenzene versus styrene decreased. These results suggest that the coordination of the phosphine at the surface of the particles tunes their reactivity.

11.3.1.1.3 Hydrogenation of Aromatic Compounds The previously described PVP-stabilized rhodium nanoparticles of Choukroun and Chaudret also led to interesting results in the hydrogenation of benzene in a biphasic mixture [18]. In water/benzene biphasic conditions at 30 °C and under 7 bar H_2, complete benzene hydrogenation was observed at a substrate/catalyst ratio of 2000 after 8 h giving rise to a TOF of 675 h^{-1} (related to H_2 consumed).

A similar water-soluble colloidal system has been described by James and coworkers [19]. Rhodium colloids were classically produced by reducing $RhCl_3 \cdot 3H_2O$ with ethanol in the presence of PVP and triethylamine. The monophasic hydrogenation of various substrates such as benzyl acetone, propyl-phenol and benzene derivatives was performed under mild conditions (25 °C and 1 bar H_2). The nanoparticles were poorly characterized but benzyl acetone is reduced with 50 TTO (total number of turnovers) in 43 h.

11.3.1.2 Surfactant-stabilized Rh and Ru Nanoparticles

The formation and stabilization of noble metal colloids in the aqueous phase are widely known. Platinum and palladium are most widely used in hydrogenation of C=C bonds but some results have been described with rhodium. Generally, surfactants are investigated as stabilizers for the preparation of rhodium nanoparticles for biphasic catalysis in water. In many cases, ionic surfactants, such as ammonium salts, which provide sufficiently hydrophilic character to maintain the catalytic species within the aqueous phase, are used. The obtained micelles constitute interesting nanoreactors for the synthesis of controlled size nanoparticles due to the confinement of the particles inside the micelle cores. Aqueous colloidal solutions are then obtained and can be easily used as catalysts.

11.3.1.2.1 Hydrogenation of Compounds with C=C Bonds A significant contribution has been made by Larpent and coworkers in biphasic liquid–liquid hydrogenation catalysis [20]. They studied catalytic systems based on aqueous suspensions of metallic rhodium particles stabilized by highly water-soluble trisulfonated molecules as protective agent. These colloidal rhodium suspensions catalysed octene hydrogenation in a liquid–liquid medium with TOF up to 78 h^{-1}, and it has been established that high activity and possible recycling of the catalyst can be achieved by control of the interfacial tension.

Fluorinated surfactants can also serve as micellar stabilizers for nanoparticles in water-in-supercritical CO_2 (scCO_2) microemulsions. Recently, Tsang described Ru nanoparticles as catalysts in the presence of ammonium perfluorotetradecano-

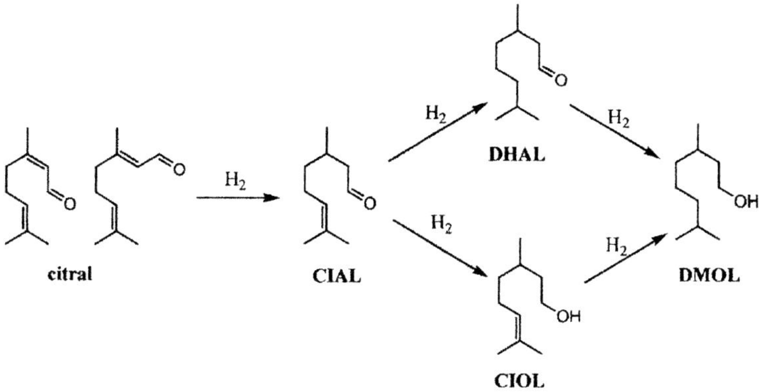

Scheme 11.1 Simplified reaction pathways of the citral hydrogenation.

Table 11.3 Influence of the Ru environment (micro-emulsion or naked) in citral hydrogenation[a].

Catalyst	Conversion (%)	CIAL (%)	DHAL (%)	DMOL (%)	CIOL (%)
naked Ru[b]	10	79.5	7.5	3.5	7.0
Ru in Me[c]	100	75.5	13.5	5.5	6.5

a   Ru (0.25 mmol), citral (0.5 mL), decane (0.1 mL), 40 °C, 10 bar H_2, 140 bar CO_2.
b   No water and fluorous surfactant.
c   Homogeneous microemulsions containing Ru nanoparticles with a diameter ca. 3–4 nm, surfactant (0.25 mmol).

ate surfactant in scCO$_2$ and reported a detailed study of the pressure effect on selective citral hydrogenation by micelle-hosted Ru nanoparticles [21]. A simplified reaction pathway of catalytic hydrogenation in supercritical carbon dioxide is presented in Scheme 11.1.

The influence of the Ru environment on the activity and product distribution was investigated. The main products were the 2,3-conjugated C=C (citronellal, CIAL), the fully saturated aldehyde (dihydrocitronellal, DHAL), the unsaturated alcohol (citronellol, CIOL), and the fully saturated alcohol (3,7-dimethyloctanol, DMOL) (Table 11.3).

The same team has also described the selective hydrogenation of cis-2-pentenenitrile with surfactant-stabilized ammonium perfluorotetradecanoate bimetallic Pd–Ru nanoparticles prepared via in situ reduction of their simple salts in reverse micelles in scCO$_2$ [22]. The optimized ratio Pd:Ru nanoparticle (1:1) shows the highest activity for the hydrogenation of functionalised alkene under mild conditions. No hydrogenation of the terminal nitrile of the molecule in amine was observed and, finally, this fluorinated micelle-hosted bimetallic catalyst gives relevant activity and selectivity in the supercritical fluid without deactivation for at least three catalytic cycles.

11.3.1.2.2 Hydrogenation of Aromatic Compounds Ammonium salts are commonly used to stabilize aqueous colloidal suspensions of nanoparticles (see Chapter 2) and probably are at the origin of significant investigation of nanoparticles in the hydrogenation of aromatic compounds and of considerable research efforts to optimise results in the last 15 years. The first study was reported in 1983–84 by Januszkiewicz and Alper who succeeded in the hydrogenation of several benzene derivatives under 1 bar H_2 and biphasic conditions starting with [RhCl(1,5-hexadiene)]$_2$ as metal source and tetraalkylammonium bromide as stabilizing agent [23, 24].

In 1993, Lemaire and coworkers reported their first results on the hydrogenation of dibenzo-18-crown-6 ether (DB18C6) to dicyclohexyl-18-crown-6 ether using $RhCl_3 \cdot H_2O$ in aqueous solution in the presence of the phase-transfer agent methyltrioctylammonium chloride and/or trioctylamine (TOA), DB18C6 being in dichloromethane solution [25]. In fact, during the course of the reaction, the formation of Rh nanoparticles was observed due to the reduction of Rh salt under dihydrogen pressure. As seen by TEM analysis, these nanoparticles display a size in the range 2–3 nm and are homogeneously dispersed on the grid. Larger nanoparticles are formed when DB18C6 is only present in the reaction medium, meaning that the phase-transfer agent not only helps the phase transfer of Rh to the organic phase but also complexes the particles thus avoiding their coalescence. In addition, these larger particles are less active in the hydrogenation of DB18C6 than the smaller ones produced in the presence of the phase transfer agent. Finally at higher pressure (50 bar), the stereoselectivity was increased up to a 95/5 ratio of the *syn/anti* isomers of the dicyclohexyl-18-crown-6 ether with a total conversion in 1 h. The same catalytic system with TOA as phase-transfer agent was further used for the stereoselective hydrogenation of disubstituted aromatic compounds [26]. Hydrogenation and hydrolysis products were obtained. The catalytic system was optimized for the 2-methylanisole reduction by a judicious choice of the amine/Rh ratio which should be high enough to stabilize very small colloidal Rh nanoparticles and low enough to avoid deactivation (Table 11.4). A high chemoselectivity and a high stereoselectivity into the (Z)-methoxy-methylcyclohexane isomer were obtained at room temperature under 1 bar or 50 bar H_2 pressure.

Moreover, enantioselective reduction of *o*-cresol derivatives (trimethylsilyl ether of *o*-cresol and 2-methylanisole) was performed with a chiral amine (dioctylcyclohexylethylamine) which combines the properties of a phase transfer agent and a chiral inductor. Enantiomeric excesses of 6% and 3%, respectively, were obtained for the two substrates under 50 bar H_2 at room temperature.

In the same way, ruthenium colloidal systems prepared by reduction of $RuCl_3$ under dihydrogen in the presence of trioctylamine allowed reduction of various substituted aromatics [27]. They studied the stereo- and chemo-selectivities of the hydrogenation of aromatics in a methanol–water system at 50 bar of H_2 and at room temperature. The results obtained are presented in Table 11.5. Reaction time to reach 100% conversion depends on the electronic and steric properties of the substituents on the aromatic ring, more electron-rich substrates giving rise to a favored reaction.

Table 11.4 Influence of the ratio TOA/Rh on conversion and selectivity in 2-methyl-anisole hydrogenation reaction after 24 h.

Molar ratio[TOA]/[Rh]	Conversion (%)	Chemoselectivity (%) $(3+4)/(2+3+4+5+6)$
0	100	89
1	69	85
1.75	43	86
2.6	95	90
3.5	97	90
4.3	75	90
5.3	53	94
7	34	87

Table 11.5 Hydrogenation of various aromatics by Ru/TOA colloids in MeOH/H$_2$O medium at room temperature under 50 bar H$_2$. Adapted from Ref. [27].

Substrates	Products	Reaction time for 100% conversion (h)	cis/trans
		9	—
		34	—
		1	—
		1	20
		1	15
		100	17

In 1997–98, James used tetrabutylammonium salts to stabilize rhodium and ruthenium nanoparticles for the hydrogenation of lignin model compounds containing the 4-propylphenol fragment in biphasic media and under various conditions (20–100 °C, 1–50 bar H_2). The best results were obtained in the hydrogenation of 2-methoxy-4-propylphenol by ruthenium nanoparticles with 300 TTO in 24 h [28–30]. Dodecyldimethylammonium propanesulfonate was also used by Albach and Jautelat who reported aqueous suspensions of Ru, Rh, Pd, Ni nanoparticles and bimetallic mixtures [31]. Benzene, cumene and isopropylbenzene were reduced in biphasic media under various conditions at 100–150 °C and 60 bar of H_2 and TTO up to 250 were obtained. More recently, Jessop and coworkers described an organometallic approach to the preparation of *in situ* rhodium nanoparticles stabilized with the surfactant tetrabutylammonium hydrogen sulphate [32]. The hydrogenation of anisole, phenol, *p*-xylene and ethylbenzoate was performed in a biphasic aqueous/supercritical ethane medium at 36 °C and 10 bar H_2. The authors report the solubility influence of the substrates on the catalytic activity. *p*-Xylene was selectively converted to *cis*-1,4-dimethylcyclohexane (53% versus 26% *trans*) and 100 TTO were obtained in 62 h for the complete hydrogenation of phenol, which is very soluble in water.

Wai and collaborators have succeeded in the hydrogenation of arenes with Rh nanoparticles in a water-in-supercritical CO_2 emulsion [33]. The catalytic system was prepared by mixing an aqueous solution of $RhCl_3$ and a mixture of surfactants containing sodium bis(2-ethylhexyl)sulfosuccinate (AOT) and a co-surfactant perfluoropolyetherphosphate ($PFPE-PO_4$). The introduction of dihydrogen in the reactor allowed the formation of Rh(0) colloids with a mean size in the range 3–5 nm which led to the reduction of naphthalene to tetraline and that of phenol to cyclohexanol as major products.

In 1999, the group of Roucoux studied a new series of easily synthesized ionic surfactants that efficiently stabilize active suspensions of rhodium colloids in the hydrogenation of arenes in a biphasic liquid–liquid medium [34–36]. The synthesis of *N,N*-dimethyl-*N*-alkyl-*N*-(2-**HydroxyEthyl)A**mmonium salts which provide an electrosterical stabilization has been obtained by one step quaternarization of *N,N*-dimethylethanolamine with the appropriate functionalized alkanes or by ion exchange. These salts **HEA-C*n*X** bear an alkyl chain containing $n = 12–18$ carbon atoms and can be prepared with various counter-anions X such as Br, Cl, I, CH_3SO_3, BF_4 (Scheme 11.2).

The surface tension measurements demonstrated that all compounds with a lipophilic alkyl chain of more than 12 carbon atoms display surfactant behavior and self-aggregate into micelles above the critical micellar concentration (cmc). The cmc values decrease from 1×10^{-2} to 2.5×10^{-4} while the hydrocarbon chain length increases and the usual linear variation of log cmc versus number of carbon atoms was observed. Moreover the variation of the counter-anion X shows the cmc values in the **HEA-C$_{16}$ series** decrease from Cl>Br>BF_4>Ms>I and finally the determination of the cmc value of the surfactant affords the possibility to optimize the efficiency of the colloidal catalyst in biphasic liquid–liquid systems (*vide infra*). The

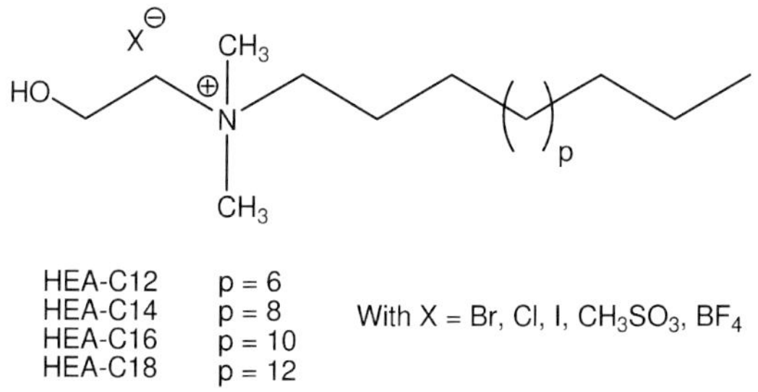

HEA-C12   p = 6
HEA-C14   p = 8   With X = Br, Cl, I, CH_3SO_3, BF_4
HEA-C16   p = 10
HEA-C18   p = 12

Scheme 11.2 Ammonium surfactants **HEA-C*n*X** derived from *N,N*-dimethylethanolamine.

catalytically active aqueous suspensions are made of metallic rhodium(0) colloids prepared at room temperature by reducing rhodium trichloride with sodium borohydride in dilute aqueous solutions of **HEA-C*n*X** salts. Nevertheless, only surfactants **HEA-C$_{16}$** and **HEA-C$_{18}$**, bearing a sufficiently lipophilic alkyl chain with 16 or 18 carbon atoms, give rise to stable monodispersed colloidal dispersions. To complete this investigation, the molecular ratio $R=$**HEA-C$_{16}$**/Rh has been optimized to prevent aggregation and to provide a good activity. Specifically, the hydrogenation of anisole in methoxycyclohexane has been studied with different ratios **HEA-C$_{16}$**/Rh up to 10. Finally, the best system was obtained for $R=2$ which gives sufficiently hydrophilic behavior to maintain catalytic species within the aqueous phase and efficiently protects the rhodium particles from aggregation.

The active suspensions are prepared at room temperature by reducing $RhCl_3 \cdot 3H_2O$ with $NaBH_4$ or under hydrogen pressure in dilute aqueous solutions of **HEAC$_{16}$X** salts. The particle sizes of the Rh-HEAC$_{16}$X systems have been determined by transmission electron cryomicroscopy observations. Samples were prepared by dropwise addition of the stabilized rhodium nanoparticles in water onto a copper sample mesh covered with carbon. The colloidal dispersion was partially removed after 1 min using cellulose before transferring to the microscope. The histograms of the size distribution were obtained on the basis of the measurement of about 300 particles. The comparative TEM studies (Fig. 11.2) show that the Rh particles in the catalytic suspension have a similar average diameter (2.1 to 2.4 nm).

A comparison of these colloidal rhodium systems Rh-**HEAC$_{16}$X** (**X**=Br, Cl, I, Ms, BF$_4$) has been made for the hydrogenation of various arene derivatives under atmospheric pressure and at room temperature [36]. In all cases, the colloidal suspension of rhodium(0) stabilized by **HEAC$_{16}$Cl** gave the best turnover frequencies compared with **HEAC$_{16}$Br, Ms** and **BF$_4$**. No hydrogenation was observed with the use of **HEAC$_{16}$I** as surfactant due to the simultaneous redox formation of iodine, a poison of nanocatalysts, during the preparation of the rhodium suspension.

Finally, the hydrogenation of various mono-, di-substituted and/or functionalized arene derivatives in optimized conditions (Rh-HEAC$_{16}$Cl in a molar ratio of

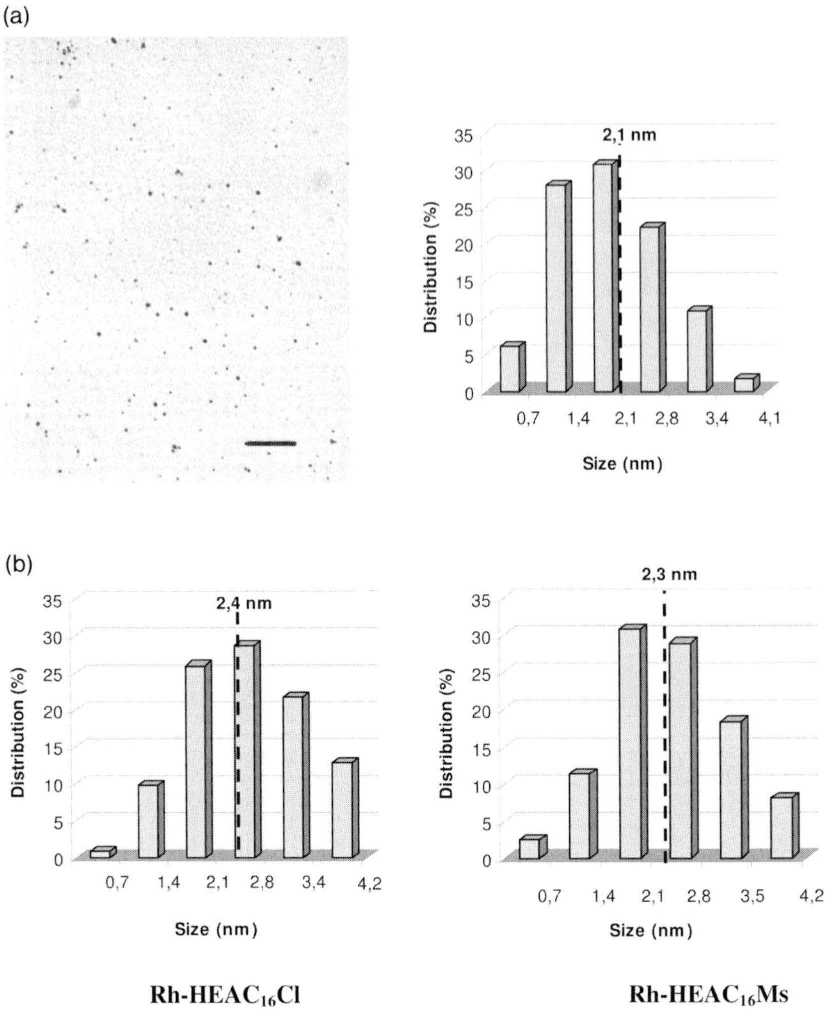

Fig. 11.2 (a) TEM micrograph (scale bar = 50 nm) and size distribution of Rh-HEAC$_{16}$Br suspension and (b) comparative size distribution histograms of Rh-HEAC$_{16}$Cl and Rh-HEAC$_{16}$Ms.

2) shows turnover frequencies up to 200 h^{-1} in pure biphasic liquid–liquid (water/substrate) media at 20 °C and 1 bar H_2 (Table 11.6).

A selective hydrogenation of di-substituted benzenes such as xylene, methylanisole, and cresol was also observed with these aqueous suspensions of rhodium(0) nanoparticles. In all cases, the *cis*-compound is the major product (near 95%). The *cis/trans* ratio decreases with the position of the substituents *o*>*m*>*p*. A comparison with a similar iridium(0) system shows that the nature of the metal does not seem to be important.

Table 11.6 Selected results for hydrogenation of various arenes under biphasic conditions[a].

Arene	Surfactant	Arene/Rh	Product (Yield %, cis/trans)[b]	Time(h)	TOF[c] (h^{-1})
benzene	HEAC$_{16}$Cl	100	cyclohexane (100)	3.6	83
benzene	HEAC$_{16}$Br	100	cyclohexane (100)	5.3	57
benzene	HEAC$_{16}$Br	200	cyclohexane (100)	6.6	91
benzene	HEAC$_{18}$Br	100	cyclohexane (100)	9.1	33
benzene	HEAC$_{16}$Ms	100	cyclohexane (100)	3.7	81
benzene	HEAC$_{16}$BF$_4$	100	cyclohexane (100)	3.7	81
ethylbenzene	HEAC$_{16}$Cl	100	ethylcyclohexane (100)	3.7	81
ethylbenzene	HEAC$_{16}$Br	100	ethylcyclohexane (100)	6.9	43
cumene	HEAC$_{16}$Cl	100	isopropylcyclohexane (100)	5.2	58
anisole	HEAC$_{16}$Cl	100	methoxycyclohexane (100)	3.6	83
anisole	HEAC$_{16}$Br	100	methoxycyclohexane (100)	5	60
anisole	HEAC$_{16}$Br	1000	methoxycyclohexane (100)	16	188
phenol	HEAC$_{16}$Cl	100	cyclohexanol (100)	5.2	58
ethyl benzoate	HEAC$_{16}$Cl	100	ethyl cyclohexanoate (100)	4.7	64
o-xylene	HEAC$_{16}$Cl	100	1,2-dimethylcyclohexane (97/3)	5.3	57
m-xylene	HEAC$_{16}$Cl	100	1,3-dimethylcyclohexane (90/10)	4.3	70
p-xylene	HEAC$_{16}$Cl	100	1,4-dimethylcyclohexane (70/30)	4.0	75
o-methylanisole	HEAC$_{16}$Cl	100	1-methoxy-2-methylcyclohexane (98/2)	5.3	57
p-methylanisole	HEAC$_{16}$Cl	100	1-methoxy-4-methylcyclohexane (94/6)	4.8	63

a Conditions : catalyst (3.8×10^{-5} mol), surfactant (7.6×10^{-5} mol), water (10 ml), substrate (3.8×10^{-3} mol), hydrogen pressure (1 bar), temperature (20 °C), stirred at 1500 min^{-1}.
b Determined by GC analysis.
c Turnover frequency defined as moles of H$_2$ per mole of rhodium per hour.

The optimized nanocatalyst could be separated by simple decantation or extraction of the product with an appropriate solvent. The durability of the catalytic system was investigated by employing it in several successive hydrogenation reactions. Similar TOFs were observed, in agreement with the absence of significant aggregation and the high water solubility of the protective agent which keeps the nanoparticles in the aqueous phase. Significant results have been obtained in the hydrogenation of anisole with 2000 TTO in 37 h.

In most cases an increase in hydrogen pressure gave rise to activation of the catalytic suspension. The efficiency of the catalytic system modified by different counter-anions of the ammonium salt was tested under high hydrogen pressure. In all cases, the hydrogenation of anisole followed by GLC analysis was usually complete after 15 min under 30 bar. The catalytic suspension Rh-HEAC$_{16}$Cl has been tested in anisole reduction under hydrogen pressure from 1 to 40 bar and has been compared with the iridium(0) system [38]. The best result was obtained with a rhodium suspension leading to a high TOF of 4000 h^{-1} under 40 bar H$_2$ (Fig. 11.3).

An interesting advantage in the use of surfactants as protective agents to stabilize nanoparticles is based on the easy modulation of their concentration which

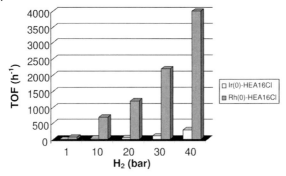

Fig. 11.3 Catalytic activities in the hydrogenation of anisole under hydrogen pressure. Comparison of aqueous suspensions of rhodium and iridium stabilized with HEA₁₆Cl.

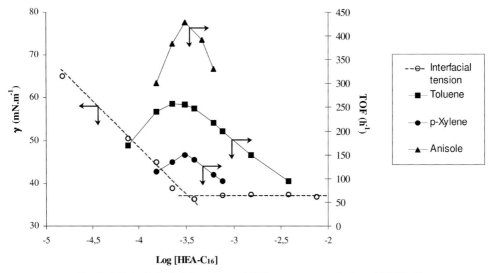

Fig. 11.4 Plot of interfacial tension and TOFs versus concentration of HEAC₁₆Cl.

fundamentally influences the turnover frequencies (TOFs). The group of Roucoux has shown that the control of the surfactant concentration and then of the interfacial tension parameters gives rise to high catalytic activities [38]. The Rh(0) stabilized nanoparticles are 2- to 5-fold more active than the standard aqueous colloidal suspension when the surfactant concentration is about the critical micellar concentration (cmc). A similar profile is obtained for various substrates such as anisole, toluene, and *p*-xylene (Fig. 11.4). In all cases, the maximum TOF is correlated with the cmc values (429, 256, 149 h⁻¹ respectively).

To summarize, the investigation of rhodium(0) colloids as nanocatalysts for the hydrogenation of a large series of aromatic compounds has shown that (i) the average particle size was $2.2 \pm 0.2\,nm$, (ii) the counter anion of the surfactant does not have a major influence on the size but significantly influences the catalytic activities, (iii) activities could be modulated on the basis of the interfacial tension and the cmc value, (iv) nanoparticle suspensions have a similar size distribution after catalysis and finally (v) the aqueous phase containing nanocatalysts could be separated from the reaction medium after an easy decantation and reused in catalysis. This aqueous-phase approach allows high activities and long catalyst lifetimes in the hydrogenation of various substrates. Moreover, this approach could also be applied to several other metals and in various catalytic applications such as C—C coupling in biphasic systems.

11.3.1.3 Ligand-stabilized Rh and Ru Nanoparticles

Today, the use of ligands as protective agents for metal nanoparticle synthesis is becoming more and more usual. The main advantage of this stabilizing mode for the nanocatalysts is the possibility to tune the surface state of the particles by the chemical influence of the ligand. Nevertheless, it is necessary to find ligands giving rise to sufficiently stable but also active nanocatalysts. The ligands usually used until now are generally simple organic ones such as thiols, amines, carboxylic acids or phosphines. More sophisticated ligands derived from homogeneous catalysis begin to be employed, examples being aminoalcohols, oxazolines, phenanthroline, or ferritines. Another advantage of using ligands as stabilizers for nanoparticles in solution is their possible asymmetric character giving access to enantioselective catalysis. Higher and maybe different enantioselective effects are expected since the coordination of the ligands onto the metal surface atoms leads to a better proximity between the metal and the chiral centers than in heterogeneous nanoparticulate systems where a chiral ligand is added after nanoparticle preparation or grafting. Despite these advantages, at present, only a few articles describe ligand-stabilized Rh and Ru nanoparticles for hydrogenation catalytic applications. In addition, almost only aromatic substrates have been used.

In 1992, Hampden-Smith and coworkers observed that aromatic hydrocarbon solutions of the hydrido $[(1,5\text{-}C_8H_{12})RhH]_4$ complex were unstable when exposed to dihydrogen, allowing the formation of 2 nm size, crystalline but agglomerated Rh nanoparticles. These nanoparticles showed catalytic properties in the hydrogenation of aromatic hydrocarbons such as benzene or toluene [39]. Similarly, Finke and Süss-Fink have proved that the true catalyst allowing benzene hydrogenation is a nanoheterogeneous $Ru(0)_n$ catalyst in the system using the supramolecular triruthenium cluster cation $[Ru_3(\mu_2\text{-}H)_3(\eta^6\text{-}C_6H_6)(\eta^6\text{-}C_6Me_6)_2(\mu_3\text{-}O)]^+$ as starting species. In addition, the data require that $Ru(0)_n$ is responsible for at least 99.97% of the observed catalysis [40].

Chaudret's team has developed the use of organometallic complexes as precursors for the synthesis of metal nanoparticles. The main advantage of this organometallic approach is the mild reaction conditions followed, the synthesis being mostly performed at room temperature and under low gas pressure (1–3 bar H_2 for example). Ideal precursors are complexes bearing olefins as ligands since they are easily decomposed under a dihydrogen atmosphere giving rise to free metal atoms and alkanes that are not able to coordinate at the nanoparticle's surface. In these conditions, nanoparticles that are well-controlled in size (narrow size distributions), morphology, composition and dispersion can be produced. The stabilization of the particles is realized through the addition of a polymer, a ligand or a ligand mixture in the reaction medium before the decomposition of the metal precursor. In the case of ruthenium nanoparticles the metal precursor is the complex $Ru(C_8H_{10})(C_8H_{12})$ (1,3,5-cyclooctatriene-1,5-cyclooctadiene ruthenium complex). Its decomposition in THF/MeOH mixtures of various compositions, in the absence of any further stabilizer, allows the formation of ruthenium nanoparticles, the size of which depends upon the reaction mixture composition [42, 43]. These nanoparticles are monocrystalline or polycrystalline (that means organized into spongy agglomerates) depending on the reaction conditions. Other alcohols such as propanol, isopropanol, pentanol [43] and heptanol [44] can also be used in a pure state or in mixtures with THF, leading to very stable colloidal solutions. These nanoparticles appear very sensitive and burn in open air, showing their potential reactivity. Catalytic tests have been carried out with Ru nanoparticles prepared in THF/MeOH 90/10 vol.% mixture. These nanoparticles display a mean size around 20 nm and are well-dispersed and very stable in solution (for at least 1 year). They were used as nanocatalysts in the hydrogenation of aromatic derivatives, namely benzene and quinoline. The catalytic experiments were carried out by addition of the substrates into the preformed colloidal solutions. In the case of benzene hydrogenation, a conversion of 82% into cyclohexane has been obtained after 14 h at 80 °C under 20 bar of dihydrogen (molar ratio [substrate]/[Ru] = 500; TON = 410; TOF = 35 h^{-1}). Similar results have been observed in the presence of added water, but no cyclohexene has been detected. When the reaction was carried out in the same conditions and in the presence of quinoline, 1,2,3,4-tetrahydroquinoline was produced with a yield of 62% (molar ratio [substrate]/[Ru] = 500; TON = 311; TOF = 16 h^{-1}). Finally, these particles are stable during catalysis since similar size and dispersion were observed by TEM study after catalysis.

More recently, chiral N-donor ligands-stabilized ruthenium nanoparticles synthesized in the same way led to interesting catalytic properties [45]. Chiral amino-alcohols and oxazolines were used as stabilizers to assure the stabilization of the particles. In these conditions, stable colloidal solutions containing ruthenium nanoparticles in a 1.6–2.5 nm size range are obtained, depending on the nature of the ligand. The catalytic activity of some of these nanoparticles has been examined in the reduction of organic prochiral unsaturated substrates. Although the asymmetric induction obtained is modest, it reveals the influence of the asymmetric ligand coordinated at the surface of the particles. The hydrogenation of *ortho-* and *para*-methylanisole performed under 40 bar H$_2$ at 50 °C for 6 h led to a more than 50% conversion of the substrate, the cyclohexane derivatives being the main products (Fig. 11.5). In both cases, the *trans*-isomer was favored up to a *trans/cis* ratio of 19/1. These results are in contrast to those traditionally obtained with hetero-

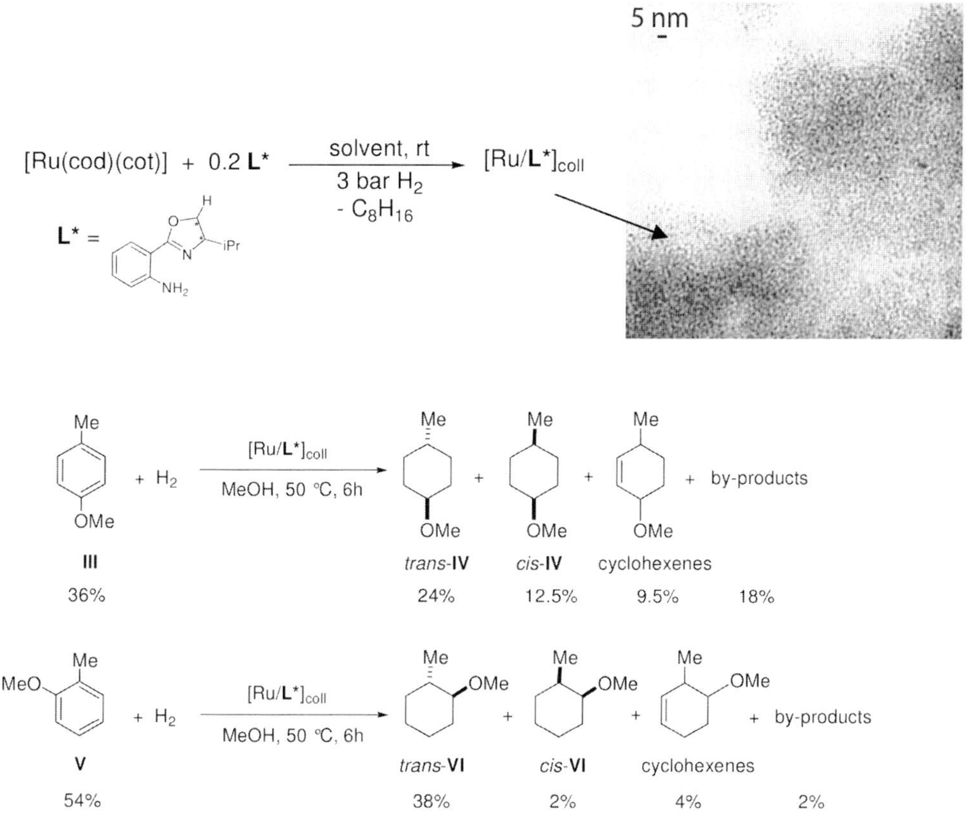

Fig. 11.5 Synthesis and TEM image of Ru/aminooxazoline nanoparticles and hydrogenation of *ortho*-and *para*-methylanisole catalyzed by a Ru/aminooxazoline colloidal system. Adapted from Ref. [44]

geneous catalytic systems which favor the *cis*-derivative. In addition, a homogeneous catalytic system prepared *in situ* from $Ru(C_8H_{10})(C_8H_{12})$ and the same ligand was not found to be active (in agreement with the well-known trend for catalytic homogeneous systems in the reduction of arenes) and the addition of free ligand to the colloidal system rendered it completely inactive.

11.3.1.4 Polyoxoanion-stabilized Rh and Ru Nanoparticles

An original approach has been developed by Finke et al. concerning the synthesis of polyoxoanion- and tetrabutylammonium-stabilized metal nanoparticles for catalytic applications. This organometallic approach allows reproducible preparation of stable nanoparticles starting from a well-defined complex in terms of composition and structure. In particular, rhodium nanoparticles were produced through hydrogen reduction of the polyoxoanion-supported Rh(I) complex $[(n-C_4H_9)_4N]_5Na_3[(1,5-COD)Rh \cdot P_2W_{15}Nb_3O_{62}]$ in acetone [45]. These Rh(0) nanoclusters are near-monodisperse with a mean size of 4 ± 0.6 nm and can be isolated as a black powder and redispersed in non-aqueous solvents such as acetonitrile.

11.3.1.4.1 Hydrogenation of Compounds with C=C Bonds Polyoxoanion-stabilized Rh nanoclusters were active in cyclohexene hydrogenation at room temperature and under dihydrogen pressure with TOF of $3650 \, h^{-1}$. In another article, Rh(0) nanocatalysts lifetimes in solution which were found to approach those of a solid-oxide-supported Rh(0) catalyst have been reported, this result being higher than any previously reported [45, 46].

11.3.1.4.2 Hydrogenation of Aromatic Compounds Finke and coworkers have also tested their polyoxoanion- and ammonium-stabilized rhodium zerovalent nanoclusters for the hydrogenation of classical benzene compounds like anisole [47, 48]. The catalytic reactions were carried out in a single phase using a propylene carbonate solution under mild conditions: 22 °C, 3.7 bar of H_2. Under these standard conditions, anisole hydrogenation with a molar ratio [Substrate]/[Rh]$=2600$ was performed in 120 h giving rise to a TTO of 1500 ± 100. The authors observed that the addition of proton donors such as $HBF_4 \cdot Et_2O$ or H_2O increases the catalytic activity and reported 2600 TTO for complete hydrogenation in 144 h at 22 °C and 3.7 bar of H_2 with a ratio $HBF_4 \cdot Et_2O/Rh$ of 10 [49]. A black precipitate of bulk Rh(0) is visible at the end of the reaction as a result of the destabilization of the nanoclusters due to the interaction of H^+ or H_2O with the basic $P_2W_{15}Nb_3O_{62}{}^{9-}$ polyoxoanion.

11.3.1.5 Ionic Liquids-stabilized Rh and Ru Nanoparticles
Efficient catalytic systems that might combine the advantages of both homogeneous (catalyst modulation) and heterogeneous catalysis (catalyst recycling) are the subject of great attention by the scientific community working on catalysis. For such purpose, ionic liquids are interesting systems since they can provide easy product separation and catalyst recycling. Ionic liquids can be used as stabilizing media for the preparation of stable nanoparticles active in catalysis, and more particularly in hydrogenation reactions. This approach is not reported in detail here since it is described in Chapter 6.

11.3.1.5.1 Hydrogenation of Compounds with C=C Bonds The group of Dupont has developed the synthesis of various metals nanoparticles (Rh, Ru, Ir, Pt and Pd) in ionic liquids media and their use in catalysis [50]. Ruthenium nanoparticles are synthesized through decomposition of the organometallic precursor Ru(COD)(COT) under molecular hydrogen (4 bar) at 75 °C in the chosen ionic liquids (1-n-butyl-3-methylimidazolium hexafluorophosphate, $BMI \cdot PF_6$, or 1-n-butyl-3-methylimidazolium tetrafluoroborate, $BMI \cdot BF_4$). Well-dispersed and stable nanometric particles ($d_m = 2$–3 nm) are then obtained and can be isolated by centrifugation and acetone washings, and further re-dispersed in the ionic liquid, acetone or used in neat conditions for respectively, liquid–liquid biphasic, homo-

Table 11.7 Hydrogenation of alkenes by Ru(0) nanoparticles under multiphase and neat conditions (75 °C and constant pressure of 4 bar, substrate/Ru = 500). Adapted from Ref. [50].

Medium	Substrate	T (h)	Conversion (%)	TON[a]	TOF (h^{-1})[b]
neat conditions	1-hexene	0.7	>99	500	714
BMI·BF$_4$	1-hexene	0.6	>99	500	833
BMI·PF$_6$	1-hexene	0.5	>99	500	1000
neat conditions	cyclohexene	0.5	>99	500	1000
BMI·BF$_4$	cyclohexene	5.0	>99	500	100
BMI·PF$_6$	cyclohexene	8.0	>99	500	62
neat conditions	2,3-dimethyl-2-butene	1.2	76	380	316

   a   Turnover number TON = mol of hydrogenated product/mol of Ru.
   b   Turnover frequency TOF = TON/h.

geneous or heterogeneous hydrogenation of alkenes and arenes under mild conditions (75 °C; 4 bar). Various olefins were used as substrates for the catalytic experiments (Table 11.7).

Similarly, Kou et al. published the synthesis of PVP-stabilized rhodium nanoparticles in BMI·PF$_6$ at room temperature [51]. These nanoparticles are produced by reduction of RhCl$_3$·3H$_2$O in the presence of PVP in a refluxing ethanol–water solution. After evaporation to dryness the residue was re-dissolved in methanol and the solution added to the ionic liquid. After methanol evaporation, ionic liquid-immobilized nanoparticles, which display a high stability and a size distribution in the range 2–5 nm, similar to those observed before their immobilization, are obtained. The catalytic performance was evaluated in the hydrogenation of olefins at 40 °C under hydrogen pressure (1 bar) in biphasic conditions giving rise to a TOF of 125 h^{-1} (measured in [mol product][mol metal]$^{-1}$h^{-1}) in the case of cyclohexene as substrate ([Substrate]/[Rh] = 250; 16 h; conversion = 100%). An advantage of this process is the easy separation of the particles from the product mixture by simple decantation or reduced pressure distillation, and their further use without loss of activity or aggregation.

11.3.1.5.2 Hydrogenation of Aromatic Compounds The reduction of RhCl$_3$·3H$_2$O in dry 1-butyl-3-methylimidazolium hexafluorophosphate (BMI·PF$_6$) ionic liquid under hydrogen pressure (4 bar) at 75 °C led to rhodium nanoparticles with a mean diameter of 2–3 nm which could be isolated by centrifugation [52]. These isolated colloids could be used as solids (heterogeneous catalysts), in acetone solution (homogeneous catalysts) or re-dispersed in BMI·PF$_6$ (biphasic system) for arene hydrogenation studies. A comparison between Ir(0) and Rh(0) nanoparticles shows that iridium colloids are much more active for the benzene hydrogenation in biphasic conditions with TOFs of 50 h^{-1} and 11 h^{-1} respectively, and 24 h^{-1} and 5 h^{-1} for *p*-xylene reduction at 75 °C and under 4 bar H$_2$.

Previously cited $BMI \cdot PF_6$ stabilized Rh(0) nanoparticles were also efficient catalysts for the complete hydrogenation of benzene (TOF$=125\,h^{-1}$) under neat conditions [50]. In a biphasic system, the authors observed a partial conversion in $BMI \cdot PF_6$ with a modest TOF of $20\,h^{-1}$ at 73% of conversion in the benzene hydrogenation.

More recently, Dupont and coworkers have studied the impact of the steric effect in the hydrogenation of monoalkylbenzenes by zerovalent nanoparticles (Ir, Rh, Ru) in the ionic liquid $BMI \cdot PF_6$ and compared the results with those obtained with the classical supported heterogeneous catalysts. They have shown a relationship between the reaction constants and the steric factors [53].

Finally, these particles generated in ionic liquids are efficient nanocatalysts for the hydrogenation of arenes, although the best performances were not obtained in biphasic liquid–liquid conditions. The main importance of this system should be seen in terms of product separation and catalyst recycling. An interesting alternative is proposed by Kou and coworkers who described the synthesis of a rhodium colloidal suspension in $BMI \cdot BF_4$ in the presence of the ionic copolymer poly[(N-vinyl-2-pyrrolidone)-co-(1-vinyl-3-butyl imidazolium chloride)] as protective agent [54]. The authors reported nanoparticles with a mean diameter around 2.9 nm and a TOF of $250\,h^{-1}$ in the hydrogenation of benzene at 75 °C and under 40 bar H_2. An impressive TTO of 20 000 is claimed after five total recycles.

11.3.1.6 Dendrimer- or Cyclodextrin-stabilized Rh and Ru Nanoparticles

Dendrimers and cyclodextrins constitute an interesting stabilizing mode in the synthesis of metal nanoparticles. They can bring selectivity in some catalytic reactions, such as olefin or arene hydrogenations, due to their specific structure.

11.3.1.6.1 Dendrimer-stabilized Rh and Ru Nanoparticles Dendrimers can act as template and protective agent for the synthesis of metal nanoparticles thanks to functionalized cavities, which can entrap and stabilize metal nanoparticles. Moreover, the dendritic branches can also act as selective gates controlling the access of small molecules to the encapsulated nanoparticles, and their terminal groups can be modulated to enhance their solubility in organic, aqueous or fluorous media. The formation of metal nanoparticles inside dendrimers can be carried out by two methods: the direct reduction of dendrimer-encapsulated metal ions or the displacement of less noble metal clusters with more noble elements [55].

Rhee et al. have described the synthesis of bimetallic Pd–Rh nanoparticles within dendrimers as nanoreactors [56]. The resulting nanoclusters efficiently promoted the partial hydrogenation of 1,3-cyclooctadiene under mild conditions with promising catalytic activity (Scheme 11.3). The dendrimer-encapsulated Pd–Rh bimetallic nanocatalysts could be reused without significant loss of catalytic activity.

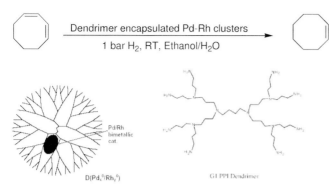

Scheme 11.3 Partial hydrogenation of 1,3-cyclooctadiene with dendrimer-encapsulated Pd–Rh nanoclusters.

11.3.1.6.2 Cyclodextrin-stabilized Rh and Ru Nanoparticles   The use of cyclodextrins offers an attractive stabilization mode for the synthesis of metal nanoparticles as these water-soluble oligosaccharides formed of 6(α), 7(β) or 8(γ) glucopyranose units could serve as protective agents for the nanoparticles and can also act as supramolecular shuttles in biphasic reactions.

Kaifer et al. were particularly interested in the catalytic properties of Pd nanoparticles derivatized with surface-attached perthiolated cyclodextrins and their use in various catalytic reactions such as Suzuki reactions [57] or the hydrogenation of alkenes [58] or allylamine [59]. The modified cyclodextrins play the role of a ligand, leading to a steric stabilization. To our knowledge, only one report describes the catalytic hydrogenation of olefins using colloidal Rh dispersions embedded by native cyclodextrins [60], generating steric stabilization via hydrophobic interactions.

Recently, the group of Roucoux has investigated the stabilization of Ru(0) colloids with classical methylated cyclodextrins, which are modulated by the cavity and the substitution degree (SD) [61]. The catalytically active aqueous suspension of metallic Ru(0) nanoparticles was prepared by chemical reduction of ruthenium chloride with sodium borohydride in dilute aqueous solutions of methylated cyclodextrins. The TEM observations show that the average particle size is about 1.5 nm with 70% of the nanoparticles between 1 and 2.5 nm (Fig. 11.6).

These nanoheterogeneous systems stabilized with various methylated cyclodextrins have shown efficient activity for the hydrogenation of olefins, and more particularly of aromatic compounds, under biphasic conditions at room temperature and atmospheric hydrogen pressure. Moreover, interesting chemoselectivities have been observed in the hydrogenation of various substituted arene derivatives (Table 11.8). The hydrogenation was easily controlled by the relevant choice of cavity and degree of methylation of the cyclodextrin.

A mechanism based on the double function of the cyclodextrins as supramolecular shuttle and protective agent was proposed (Scheme 11.4), as also described for the hydrogenation of various hydrophobic substrates using β-cyclodextrin-

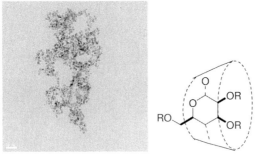

Me-Cyclodextrin (R=Me)

Fig. 11.6 TEM of Me-cyclodextrins Ru(0) nanoparticles (scale bar = 20 nm).

Table 11.8 Hydrogenation of benzene derivatives under biphasic conditions[a]. Adapted from Ref. [61].

Entry	Substrate	Cyclodextrin	Product (conv.%)[b]	Time(h)	TOF(h^{-1})[c]
1	styrene	Me-α-CD	ethylbenzene (100)	10	10
2		Me-β-CD (0.7)	ethylcyclohexane (100)	11	9
3		Me-β-CD (1.8)	ethylbenzene (100)	11	9
4		Me-γ-CD	ethylcyclohexane (100)	24	4
5	allylbenzene	Me-β-CD (0.7)	propylcyclohexane (100)	12	8
6		Me-β-CD (1.8)	propylbenzene (100)	3	34

a Conditions: catalyst (1.5×10^{-5} mol), cyclodextrin (1.5×10^{-4} mol), substrate (3.4×10^{-2} mol), hydrogen pressure (1 bar), temperature (20 °C), stirred at 1500 min^{-1}, 10 mL water.
b Determined by GC analysis.
c Turnover frequency defined as number of mol of consumed H_2 per mol of ruthenium per hour.

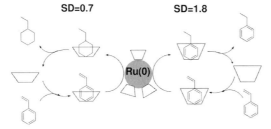

Scheme 11.4 Proposed selective mechanism of styrene hydrogenation.

modified Pd nanoparticles [62]. These nanocatalysts could be easily reused without significant loss of catalytic activity.

To conclude, cyclodextrins are attractive protective agents for stabilizing noble metal nanoparticles, leading to steric stabilization and inducing chemoselectivity in hydrogenation reactions due to the size of their cavity and their constrained environment.

11.3.2
Hydrogenation of Compounds with C=O Bonds

The reduction of aldehydes and ketones to alcohols is very easy and mainly per-formed with hydride reagents. Nevertheless, for economical and ecological reasons, catalytic hydrogenation reduction procedures based on transition metal com-pounds are more desirable. In this respect, homogeneous and heterogeneous cata-lysts have been intensively studied but only a few papers dealing with the hydrogenation of compounds bearing C=O bonds by colloidal Ru or Rh catalytic systems are available in the literature.

The selective hydrogenation of α,β-unsaturated aldehydes to the corresponding α,β-unsaturated alcohols is an important step for the synthesis of fine chemicals. Liu's team has described the hydrogenation of citronellal **1** to citronellol by using a Ru/PVP colloid prepared by the NaBH$_4$ reduction method [63]. This colloid contains small particles constituted of metallic ruthenium with a narrow size dis-tribution (TEM: 1.3–1.8 nm). This colloid exhibited a selectivity to citronellol of 95.2% with a yield of 84.2% (total conversion: 88.4%) which are good results for a monometallic catalyst (Scheme 11.5).

Liu and coworkers have also shown the incorporation influence of some metal cations such as Co^{2+} on the catalytic performances of the noble metal colloidal nanocatalysts [64]. In particular, Co^{2+} significantly increases both the activity and the selectivity of the Ru nanosized catalyst. The yield reached 97.8% and the selectivity increased to 98.8% when Co^{2+} was introduced into the catalytic system. The modification was assumed to be due to the adsorbed metal cations activating the C=O double bonds, thus accelerating the reaction rate and increasing the selectivity to unsaturated alcohols. In the same field, Liu's team reported the

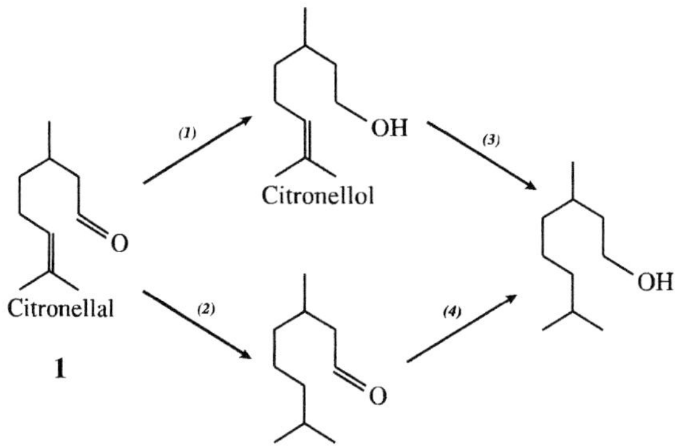

Scheme 11.5 Selective hydrogenation of citronellal to citronellol.

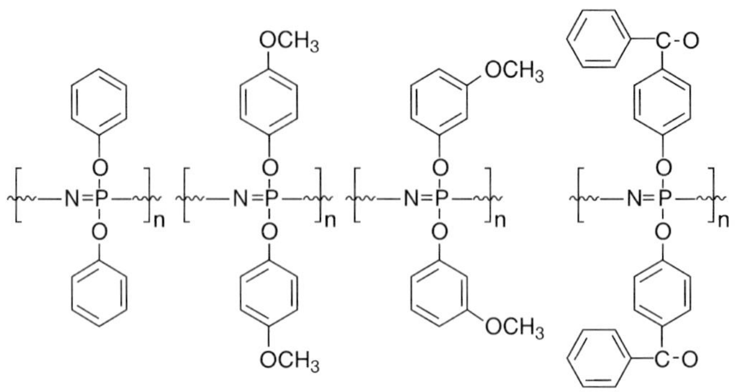

PBPP   PB*p*MPP   PB*m*MPP   PB*p*BPP

PBPP = poly [(bis(phenoxy)phosphazene]
PB*p*MPP = poly[(bis(*p*-methoxyphenoxy)phosphazene]
PB*m*MPP = poly[(bis(*m*-methoxyphenoxy)phosphazene]
PB*p*BPP = poly[(bis(*p*-benzoylphenoxy)phosphazene]

Scheme 11.6 Polyorganophospharenes: an original class of polymers containing an inorganic backbone.

hydrogenation of crotonaldehyde (2-butenal) to crotyl alcohol (2-butene-1-ol) by the PVP-stabilized ruthenium colloids [65]. The conversion of crotonaldehyde was 100% and a selectivity of 4% was obtained in absolute ethanol. The byproducts such as *n*-butanal and *n*-butanol were significant, with 79.1% and 16.9% respectively. The incorporation of metal cations such as Fe^{3+}, Co^{2+} and Ni^{2+} raised the selectivity (to 13.6%) but lowered the activity.

In 2003, Pertici et al. reported the synthesis of ruthenium nanoparticles on polyorganophosphazenes –[N=PR$_2$]$_n$– as stabilizing polymers [66]. The synthesis method consists of the decomposition of a THF solution of the organometallic $Ru(\eta^6\text{-}C_8H_{10})(\eta^4\text{-}C_8H_{12})$ complex at 45 °C under dihydrogen atmosphere in the presence of polydimethyl phosphazene (PDMP) or various polyorganophosphazenes (Scheme 11.6).

This procedure gave rise to new materials containing small ruthenium clusters which are bound to the arene groups of the polymers. Well-dispersed and very small nanoparticles (1.5±0.5 nm mean size) were observed in the polymer matrix through HRTEM analysis. These materials could be purified as fine powders containing 5 wt.% of Ru, and further dispersed in ethanol or in water, allowing their use as homogeneous catalysts. Olefins, arenes and ketones catalytic hydrogenation experiments were performed under mild conditions (25 °C; 1 bar H$_2$). A wide range of carbonyl compounds has been tested, such as cyclohexanone, ethyl acetoacetate, ethyl pyruvate and pyruvic acid as examples of aliphatic carbonyl compounds, and acetophenone as an example of a ketone bearing an aryl group. All carbonyl compounds were quantitatively reduced to the corresponding alcohols (Table 11.9).

Table 11.9 Hydrogenation of ketones with Ru/PDMP at 1 bar
H$_2$ and 25 °C. Adapted from Ref. [67].

Substrate[a]	Solvent	Time (h)	Product[b] (%)	TOF[c]
cyclohexanone	ethanol	10	cyclohexanol (100)	10
ethyl acetoacetate	ethanol	24	ethyl 3-hydroxybutyrate (100)	4.2
ethyl pyruvate	ethanol	6	ethyl lactate (100)	16.6
pyruvic acid	ethanol	8	lactic acid (100)[d]	12.5
acetophenone	ethanol	96	1-phenylethanol (100)	1
pyruvic acid	water	7	lactic acid (100)[d]	14.3
cyclohexanone[e]	ethanol	30	cyclohexanol (100)	10
ethyl acetoacetate[e]	ethanol	6	ethyl 3-hydroxybutyrate (100)	3.3
ethyl pyruvate[e]	ethanol	10	ethyl lactate (100)	16.6
pyruvic acid[e]	ethanol	8	lactic acid (100)[d]	12.5
acetophenone[e]	ethanol	96	1-phenylethanol (100)	1
pyruvic acid[e]	water	7	lactic acid (100)[d]	14.3

a   Substrate (13 mmol); catalyst Ru on PDMP (5 wt.% of Ru), 0.26 g (0.13 mg atoms Ru);
    solvent, 15 ml.
b   The composition was determined by GLC analysis.
c   Moles converted substrate per gram atoms ruthenium per h.
d   The composition was determined by ^{1}H NMR spectroscopy.
e   Catalyst recovered from runs 1–6, respectively, and reused.

The catalysts could be reused after precipitation. In addition, it has been shown by HRTEM that these catalysts are more resistant towards agglomeration since only a low degree of aggregation has been noticed after dissolution and precipitation of the catalysts. This slight agglomeration did not lead to catalyst deactivation.

The asymmetric hydrogenation of prochiral ketones is often an important step in the industrial synthesis of fine and pharmaceutical products. Several noble metal nanoparticles have been investigated for asymmetric catalysis of prochiral substrates but platinum colloids have been the most widely studied and relevant enantiomeric excesses have been reported (>95%). Nevertheless, the enantioselective hydrogenation of ethyl pyruvate catalyzed by PVP-stabilized rhodium nanocluster modified by cinchonidine and quinine was reported by Li and coworkers (Scheme 11.7) [68].

The nanocatalysts were prepared by the classical reduction in water–alcohol solution. The rhodium nanocluster is finely dispersed and can be stored in air for a long time without aggregation. The size of rhodium nanoclusters was estimated by TEM, the average diameter being 1.8 nm. The enantiomeric excess of (R)-(+) ethyl lactate and turnover frequency reach to 42.2% and 941 h^{-1}, respectively, under optimum conditions (25 °C, 50 bar H$_2$). The reported results show that cinchonidine and quinine can not only induce the enantioselectivity but can also greatly accelerate the reaction by a factor of about 50.

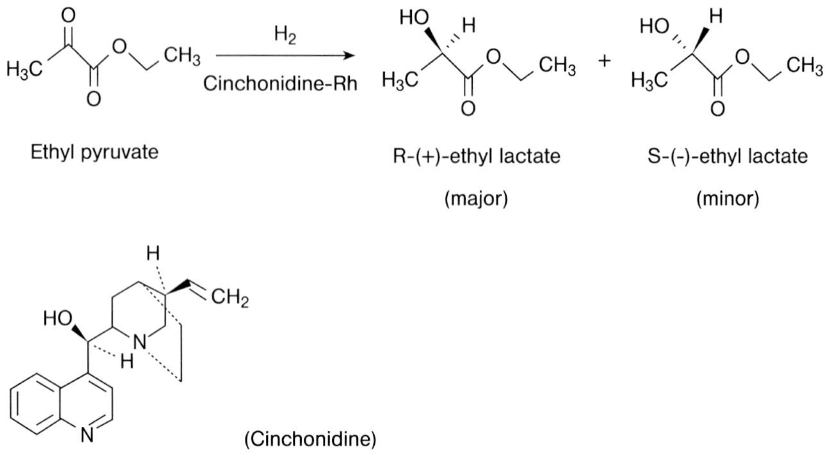

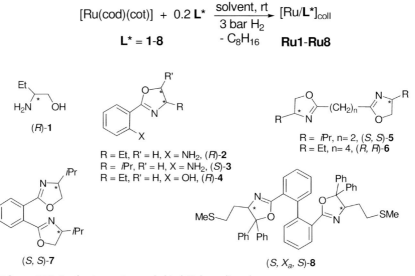

Scheme 11.7 Enantioselective hydrogenation of ethyl pyruvate.

$$[Ru(cod)(cot)] + 0.2\ \textbf{L*} \xrightarrow[\substack{3\ bar\ H_2 \\ -\ C_8H_{16}}]{solvent,\ rt} [Ru/\textbf{L*}]_{coll}$$

$$\textbf{L*} = \textbf{1-8} \qquad\qquad \textbf{Ru1-Ru8}$$

(R)-1

R = Et, R' = H, X = NH₂, (R)-2
R = iPr, R' = H, X = NH₂, (S)-3
R = Et, R' = H, X = OH, (R)-4

R = iPr, n= 2, (S, S)-5
R = Et, n= 4, (R, R)-6

(S, S)-7

(S, X_a, S)-8

Scheme 11.8 Synthesis reaction and chiral N-donor ligands
(1: β-aminoalcohol, 2–4: monooxazolines and 5–8:
bis(oxazolines) used for the preparation of Ru nanocatalysts.
Adapted from Ref. [44].

Chaudret and coworkers have studied the activity of chiral aminoalcohol- and oxazoline-stabilized Ru colloids in the hydrogen transfer reaction (Scheme 11.8) [44]. The reduction of acetophenone has been studied using isopropanol as hydrogen source and preformed nanocatalysts under basic conditions and at room temperature (Table 11.10).

Table 11.10 Hydrogen transfer of acetophenone catalyzed by aminoalcohol- and oxazoline-stabilized Ru colloids: effect of added ligand in the reaction mixture[a].

Entry	RuL*	Ru/L*$_{total}$[b]	Time (h)	Conv. (%)[c]	ee II (%)[c]
1	Ru1	1/0.2	12	77	0
2	Ru2	1/0.2	12	58	0
3	Ru3	1/0.2	12	65	10 (S)
4	Ru4	1/0.2	12	39	0
5	Ru5	1/0.2	72	14	0
6	Ru6	1/0.2	12	53	0
7	Ru7	1/0.2	48	12	0
8	Ru8	1/0.2	12	98	0
9	Ru1	1/1.2	12	31	0
10	Ru1	1/2.2	12	5	—
11	Ru3	1/1.2	12	15	10 (S)
12	Ru3	1/2.2	72	20	10 (S)
13	Ru8	1/1.2	12	79	—
14	Ru8	1/2.2	12	56	—

a   0.12 mmol acetophenone and 0.024 mmol tBuOK in 4 mL of isopropanol, and 6×10^{-3} mmol colloid. Results from duplicated experiments.
b   Ruthenium/ligand ratio taking into account the total amount of ligand present in the reaction, from the synthesis of colloid with the extra ligand added to the catalytic reaction.
c   Substrate conversion and ee determined by GC using a chiral column.

In these conditions, when no further free ligand was added to the catalytic medium, the systems were not selective (except the Ru colloidal system 3) but exhibited moderate to high activities (Table 11.10, entries 1–8). When 1 or 2 equivalents of free ligand (relative to ruthenium, Table 11.10, entries 9–14) were added, a remarkable decrease in activity was noticed without any effect on the selectivity whereas an increase in enantiomeric excess would be expected in the case of metal leaching to form a molecular catalyst. These results attest to the colloidal character of the active catalytic species.

These results have been compared to those obtained with molecular homogeneous catalytic systems stabilized with the same ligands (Table 11.11). It appears clearly that there is an influence of the asymmetric ligand coordinated to the surface of the particles. Aminoalcohol- and bis-oxazoline-stabilized nanocatalysts were less active than their corresponding molecular ones. This trend changes for amino(oxazoline) and thioether-bis(oxazoline). It is also noteworthy that Ru8 nanocatalyst is the fastest colloidal catalyst, even faster than its molecular

Table 11.11 Hydrogen transfer of acetophenone catalyzed by
Ru nanocatalysts (**Ru**) in comparison with molecular systems
(**RuM**)[a]. Adapted from Ref. [44].

Entry	L*	Catalyst	Ru/L*_total[b]	Time (h)	Conv. (%)[c]	ee II (%)[a]
1	1	Ru1	1/0.2	12	77	0
2	1	RuM1	1/1	1.5	93	12 (R)
3	2	Ru2	1/0.2	12	58	0
4	2	RuM2	1/1	12	15	43 (R)
5	3	Ru3	1/0.2	12	65	10 (S)
6	3	RuM3	1/1	12	55	45 (S)
7	4	Ru4	1/0.2	12	39	0
8	4	RuM4	1/1	72	0	0
9	5	Ru5	1/0.2	72	14	0
10	5	RuM5	1/1	20	88	25 (S)
11	6	Ru6	1/0.2	12	53	0
12	6	RuM6	1/1	12	90	11 (R)
13	7	Ru7	1/0.2	48	12	0
14	7	RuM7	1/1	20	42	29 (S)
15	8	Ru8	1/0.2	12	98	0
16	8	RuM8	1/1	12	62	0

a   0.12 mmol acetophenone and 0.024 mmol tBuOK in 4 mL of isopropanol; 3×10^{-3} mmol
    [Ru(p-cymene)Cl$_2$]$_2$ with 6×10^{-3} mmol **L***. for molecular catalysts generated *in situ*, **RuML***;
    or 6×10^{-3} mmol of colloid catalyst, **RuL***. Results from duplicated experiments.
b   Ruthenium/ligand ratio, taking into account the total amount of ligand present in the
    reaction.
c   Substrate conversion and ee determined by GC using a chiral column.

counterpart (Table 11.11, entry 15 vs. 16,). The observed activity behavior of these
systems indicates a specific coordination chemistry of the ligands at the metallic
surface. Finally, TEM analysis performed before and after catalysis revealed that
the size of the Ru nanoparticles remained unchanged, although some agglomeration was visible.

   To conclude, few hydrogenation studies of compounds with C=O bonds, in
particular with an asymmetric approach, have been carried out with rhodium and
ruthenium nanoparticles. The development of convenient Ru and Rh colloidal
suspensions for this important reaction remains a promising research area. Moreover, the investigation of modified colloids with chiral compounds is still a challenge for asymmetric hydrogenation of various prochiral ketones and the use of
two-phase liquid–liquid systems for the recycling of chiral colloids.

11.3.3
Hydrogenation of Aromatic Nitro Compounds

Aminoaromatics produced by the selective catalytic hydrogenation of the corresponding nitro precursors are relevant intermediates for agrochemicals, pharma-

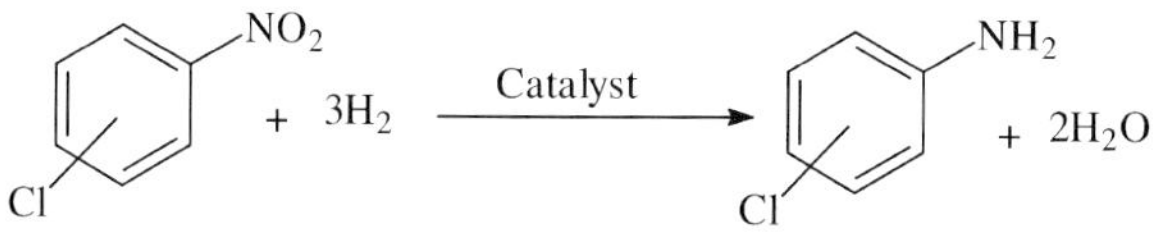

Scheme 11.9 Hydrogenation of chloronitrobenzene.

ceuticals, dyestuffs, urethanes and other industrially important products [68]. The liquid phase hydrogenation of the aromatic nitrocompounds over colloidal rhodium catalysts has been long established by kinetic methods [69]. More recently, the hydrogenation of halonitroaromatics to the corresponding haloanilines over colloidal nanocatalysts has been well investigated (Scheme 11.9).

Liu and coworkers reported the selective hydrogenation of *ortho*-chloronitrobenzene (*o*-CNB) to *ortho*-chloroaniline (*o*-CAN) without dehalogenation over finely dispersed polyvinylpyrrolidine-stabilized ruthenium colloids (PVP-Ru) [70] and ruthenium-palladium (PVP-Ru/Pd) or ruthenium-platinum (PVP-Ru/Pt) bimetallic colloids [71]. Colloidal dispersions were prepared by NaBH$_4$ reduction of metal salts or corresponding mixed-metal salts at room temperature and characterized by TEM, XPS and XRD. The particle sizes were 1.4±0.5 nm in the PVP-Ru system. The PVP-Ru/Pt and PVP-Ru/Pd nanoparticulate systems have an average diameter in the range 1.3–3.0 nm and 2.2–3.7 nm, respectively, depending on the molar ratio of Ru/Pt or Ru/Pd [72]. This team demonstrated that the PVP-Ru monometallic colloids exhibited very good catalytic performance, 100% selectivity at 100% conversion; however the reaction rate was very low (2.8 × 10^{-4} mol H$_2$/mol Ru.s). Finally, the PVP-Ru/Pt and PVP-Ru/Pd bimetallic colloids were more stable and the catalytic performance of bimetallic colloids is dependent on their composition but the activity was much higher than PVP-Ru (Table 11.12). The effect of metal cations (Co^{2+}, Fe^{3+} etc.) on the catalytic properties of various systems was also investigated.

A synergic effect of the bimetallic PVP-Pd/Ru system has also been reported by Liao who observed a remarkable increase in the selectivity for *p*-chloroaniline in the selective hydrogenation of *p*-chloronitrobenzene under atmospheric pressure and in the presence of a small amount of base [73]. To conclude, several catalytic investigations and mechanistic studies have shown the potential of colloids in this crucial industrial application.

11.4
Catalytic Formation of C–C Bonds

The use of metal nanoparticles in carbon–carbon bond formation reactions has not been widely explored yet, although these reactions are of great interest in the synthesis of polymers, agrochemicals and pharmaceutical intermediates. However, a few examples of Rh colloidal systems catalyzed hydroformylation, carbonylation

Table 11.12 Hydrogenation of o-CNB over PVP-ruthenium mono and bimetallic systems[a]. Comparison with PVP-Pt and PVP-Pd colloids. Adapted from Ref. [72].

Catalytic system	Initial rate[b] (mol H$_2$/mol Me s^{-1})	Average rate[b] (mol H$_2$/mol Me s^{-1})	Conv. (%)	Selectivity[c] (%)			
				o-CAN	AN	NB	Others
PVP-Ru[d]	0.00028	0.00028	100	>99.9	0	0	trace
PVP-Pt	1.98	0.60	100	63.9	8.3	0	27.8
PVP-Ru/Pt(4/1)	0.24	0.16	100	58.9	14.1	4.3	22.7
PVP-Ru/Pt(4/1)-CoCl$_2$	0.13	0.11	100	99.5	0.4	trace	0
PVP-Ru/Pt(1/4)	0.48	0.29	100	58.9	14.1	3.3	23.7
PVP-Ru/Pt(1/4)-CoCl$_2$	0.15	0.071	100	>99.9	0	0	trace
PVP-Pd	0.60	0.52	90.5	34.2	53.4	2.2	10.2
PVP-Ru/Pd(1/1)	0.10	0.078	88.2	29.1	41.6	0.6	28.7
PVP-Ru/Pd(1/1)-FeCl$_3$	0.18	0.12	98.2	55.5	34.9	1.7	7.9

a   Molar ratio of M : Pt = 1 : 1. Reaction conditions: 1.87×10^{-5} mol metal colloid, 1.00×10^{-3} mol o-CNB and 1.87×10^{-5} mol metal salt; 1 bar H$_2$ and 30 °C; 15.0 mL MeOH.
b   Me = Ru, Pt, Pd, or Ru + Pt, Ru + Pd.
c   AN = aniline; NB = nitrobenzene; others = mixture of o-chlorophenylhydroxylamine, o-chloronitrosobenzene, chlorobenzene, azo- and azoxy-dichlorobenzenes.
d   Reaction conditions: 40 bar H$_2$, 50 °C and 4 h.

and Pauson–Khand reactions can be found in the literature whereas ruthenium nanoparticles have been used for the Suzuki and Heck reactions.

11.4.1
Hydroformylation of Olefins

The aqueous biphasic hydroformylation of propene, namely the Ruhrchemie/ Rhone Poulenc (RCH/RP) process, has been widely used to produce n-butanal and many attempts have been proposed to improve this catalytic system, such as a thermoregulated phase transfer (TRPT) Rh(I) complex catalyst [74]. Moreover, Bönnemann et al. [75] have proved the *in situ* formation of Rh colloids when such a catalyst was applied to the aqueous biphasic hydroformylation of 1-octene.

Recently, Dupont et al. [76] have reported the hydroformylation of olefins in neat conditions using unmodified or Xantphos-modified Rh(0) nanoparticles, prepared by simple hydrogenation reduction of the chloride salts in imidazolium ionic liquids. They have shown that the particle size has a strong influence on the selectivity of the hydroformylation (Table 11.13).

When using 5.0 nm Rh(0) nanoparticles, the olefins are converted into aldehydes (Table 11.13, entries 1 and 4) and the addition of Xantphos leads to *linear/branched* selectivities up to 25 (Table 11.13, entry 2). With smaller colloids (2.7 nm), the

Table 11.13 Hydroformylation of 1-decene by Rh(0) nanoparticles with different sizes[a]. Adapted from Ref. [76].

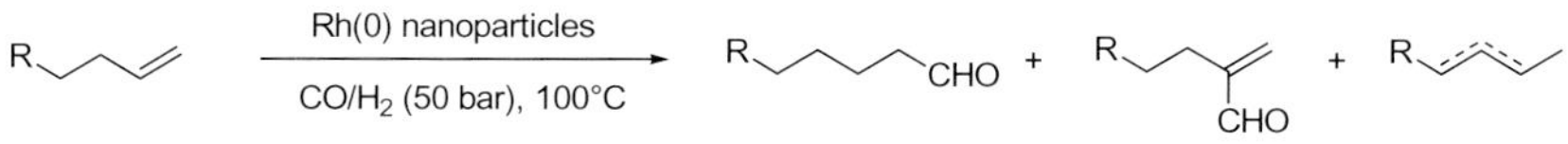

$$R\diagdown\diagup\diagup \quad \xrightarrow[\text{CO/H}_2 \text{ (50 bar), 100°C}]{\text{Rh(0) nanoparticles}} \quad R\diagdown\diagup\diagup\diagdown^{\text{CHO}} \quad + \quad R\diagdown\diagup\diagdown_{\text{CHO}} \quad + \quad R\diagdown\diagup\diagup$$

Entry	Catalyst	Olefin	t (h)	Conv.(%)[b]	l/b [c]	Isomerization (%)[d]
1	Rh(0) (5.0 nm)	1-hexene	43	100	1.8	2.5
2[e]	Rh(0) (5.0 nm)	1-hexene	34	100	25	<1
3	Rh(0) (2.7 nm)	1-decene	26	93	1.9	17
4	Rh(0) (5.0 nm)	1-decene	20.5	100	1.5	<1
5	Rh(0) (15 nm)	1-decene	49.5	72	2.0	54

a   Reaction conditions: Rh(0) (5 mg), olefin (10 mmol), CO/H$_2$ (1/1, 50 bar), T=100°C.
b   Olefin conversion.
c   Ratio linear/branched aldehydes.
d   Percentage of internal olefins.
e   Xantphos (10 mg, 0.017 mmol).

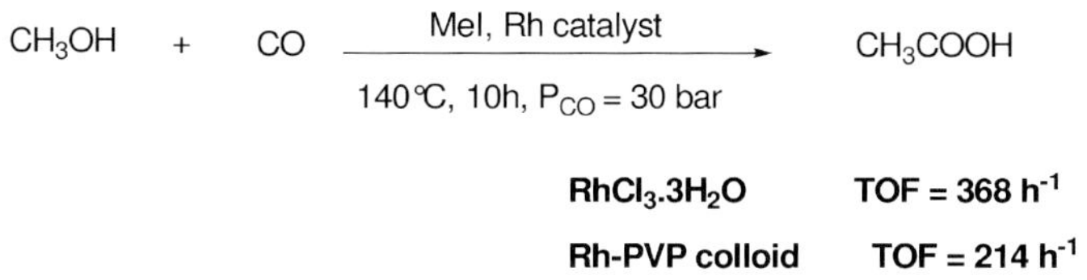

$$CH_3OH \quad + \quad CO \quad \xrightarrow[\text{140°C, 10h, } P_{CO} = 30 \text{ bar}]{\text{MeI, Rh catalyst}} \quad CH_3COOH$$

RhCl$_3$.3H$_2$O	TOF = 368 h^{-1}
Rh-PVP colloid	TOF = 214 h^{-1}

Scheme 11.10 Methanol carbonylation.

chemoselectivity decreases with around 11–17% olefin isomerization (Table 11.13, entry 3) and with larger sized ones (15 nm) only small amounts of aldehydes are produced (Table 11.13, entry 4). Moreover, TEM, XRD, IR and NMR studies have proved that the Rh(0) nanoclusters are probably transformed under the reaction conditions into soluble mononuclear Rh-carbonyl catalytically active species.

11.4.2
Methanol Carbonylation

The carbonylation of methanol in acetic acid represents an important industrial process, which has been developed by Monsanto Corporation using a homogeneous rhodium complex. Extensive investigations on the rhodium catalytic system have been carried out and Liu et al. [77] have studied the use of PVP-stabilized Rh nanoparticles for this reaction (Scheme 11.10). The stable PVP-Rh colloid presents a lower activity than Monsanto's homogeneous catalyst under the same drastic conditions (140°C, 54 bar). However, the colloidal metal catalyst could be reused several times with an increased activity (TON = 19700 cycles/atom Rh), which might presume a transformation of catalytic species. Spectroscopic studies (IR, CO adsorption, XPS) have shown that the active species were, as expected, the Rh(I) species generated by oxidative addition of methyl iodide to the rhodium colloid.

11.4.3
Coupling Reactions

Coupling reactions such as Heck, Suzuki and Sonogashira reactions have been widely described with Pd catalysts [78]. However, recently, Chang et al. [79] have reported their studies on the ruthenium-catalyzed olefination and Suzuki-type cross-coupling reactions, showing that the real active catalytic species were Ru nanoclusters when using a ruthenium complex [RuCl$_2$(p-cymene)]$_2$ as a homogeneous catalyst precursor. Thus, olefination of iodobenzene with ethyl acrylate was efficiently catalyzed by dodecylamine stabilized Ru(0) nanoparticles, which could be reused for the further runs, with almost the same activity and selectivity as those obtained with the homogeneous precursor (Fig. 11.7). These studies of the nature of catalytic species in the coupling reactions led to the development of a highly efficient, user-friendly, reusable heterogeneous ruthenium catalyst, Ru/Al$_2$O$_3$ for both the olefination and Suzuki coupling reactions.

Moreover, Rothenberg and coworkers [80] have developed copper and copper-based nanocolloids that constitute an inexpensive and eco-friendly alternative to noble metal catalysts. The cluster combinations were prepared by reduction with tetraoctylammonium formate (TOAF) in DMF of the metal chloride precursors, leading to the formation of particles of 1.6 nm to 2.1 nm, and tested in the coupling reaction of phenylboronic acid and iodobenzene (Table 11.14). These mono-, bi- or tri-nanocolloid catalysts have shown interesting catalytic activities in Suzuki cross-coupling reactions.

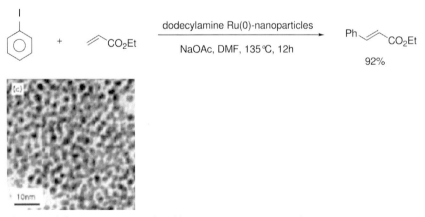

Fig. 11.7 Olefination reaction catalyzed by 2–3 nm Ru(0) nanoparticles.

Table 11.14 Cross-coupling reactions using various nanocolloid catalysts[a]. Adapted from Ref. [80].

I	B(OH)₂	Mono- or bi-metallic colloids	
		K₂CO₃, DMF, 110°C	

Entry	Catalyst (composition)	Yield (%)[b]	k_{obs} (L mol^{-1}·min^{-1})[c]
1	Cu	62	3.2×10^{-3}
2	Pd	100	5.9×10^{-2}[d]
3	Ru	40	2.0×10^{-3}
4	Pd/Ru	100	2.9×10^{-2}
5	Cu/Pd/Ru	100	3.8×10^{-2}[d]

   a   Standard reaction conditions: 0.5 mmol iodobenzene, 0.75 mmol phenylboronic acid, 1.5 mmol K₂CO₃, 0.01 mmol catalyst, 12.5 mL DMF, 110°C.
   b   GC yield after 6 h, corrected for the presence of internal standard.
   c k_{obs}: Second-order reaction rate constant.
   d   Value is the average of two repeated experiments.

The use of metal nanoparticles for catalysis in reactions generating carbon–carbon bonds is in its infancy and we could presume that this field will see a great breakthrough in the coming years.

11.4.4
Pauson–Khand Reaction

The Pauson–Khand reaction (PKR) is one of the most convergent and versatile methods for the construction of five-membered rings by condensation of alkenes, alkynes and carbon monoxide [81]. In the context of the development of new economically sustainable and environmentally responsible processes, the use of transition metal nanoparticles as heterogeneous catalysts in an atmosphere of carbon monoxide seems to be attractive and colloidal cobalt systems, supported or not, have been successfully applied for these reactions [82–86]. Recently, Chung et al. [87] have described the use of Co/Rh (Co₂Rh₂) heterobimetallic nanoparticles derived from Co₂Rh₂(CO)₁₂ supported on charcoal in the Pauson–Khand reaction. This catalytic system reacts with alkynes and α,β-unsaturated aldehydes such as acrolein, crotonaldehyde or cinammic aldehyde and leads to the formation of products resulting from [2+2+1] cycloaddition of alkyne, carbon monoxide and alkene. In these reactions, α,β-unsaturated aldehydes act as CO and alkene sources, producing 2-substituted cyclopentenones (Scheme 11.11). However, to our knowledge, all the catalytic systems based on heterobimetallic nanoparticles are supported and this will be an interesting route to explore in the near future.

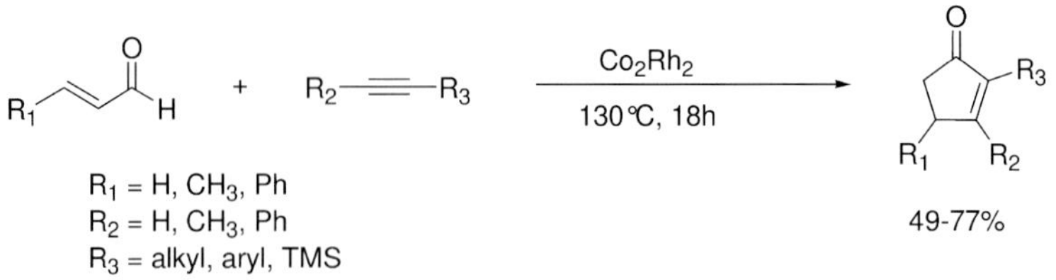

Co2Rh2
130 °C, 18h
R1 = H, CH3, Ph
R2 = H, CH3, Ph
R3 = alkyl, aryl, TMS
49-77%

Scheme 11.11 Pauson–Khand reaction.

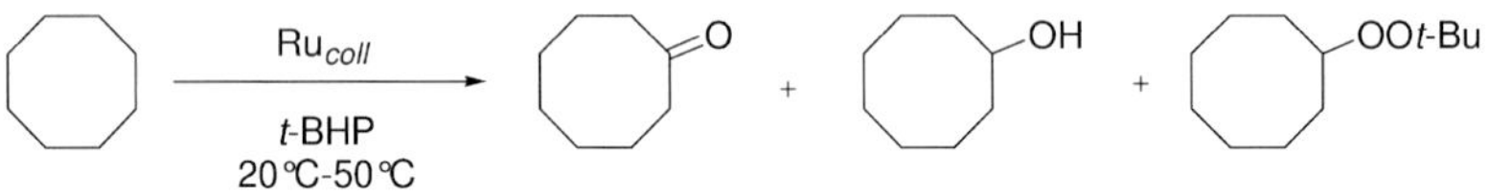

Ru coll
t-BHP
20 °C-50 °C
OH
OOt-Bu

Scheme 11.12 Cyclooctane oxidation catalyzed by Ru colloids in biphasic water–organic media.

11.5
Other Reactions

In this section, original reactions catalyzed by rhodium or ruthenium nanoparticles in various media are reported. Traditionally, these reactions were carried out in homogeneous or heterogeneous approaches.

11.5.1
Hydrocarbon Oxidation

Catalytic oxidation processes, which are of particular interest in industry, may benefit from progress in the field of nanoclusters but only few examples of oxidation reactions have been reported in the literature.

In this area, a significant result is the selective conversion of cycloalkanes to the corresponding ketones. The oxidation of cyclooctane by *tert*-butylhydroperoxide (*t*BHP) was performed with Ru colloids in biphasic water/cyclooctane media, leading to the main formation of cyclooctanol and cyclooctanone (Scheme 11.12) [88, 89]. Cyclooctanone is always the major product. Mechanistic studies with inhibition and kinetic reactions suggest ruthenium oxo species are the most probable catalysts for ketonisation. The catalyst could be recycled without any loss of activity. Finally, model extension experiments to other cycloalkanes are also investigated.

11.5.2
Dehydrocoupling of Amine–Borane Adducts

Transition metal-catalyzed approaches to boron–nitrogen compounds are potentially of significant importance. Very interestingly, Manners et al. have reported the catalytic dehydrocoupling of primary or secondary amine–borane adducts

RNH_2–BH_3 or RR NH–BH_3, which provides a mild and convenient route to cycloaminoboranes [RR N–BH_2]$_2$ and borazines [RN–BH_3]$_3$ using transition metal complexes as precatalyst [90, 91]. Finally, TEM imaging and catalyst poisoning experiments with Hg and BH_3.THF suggest that the active catalyst is colloidal [92, 93]. Analysis of an evaporated extract of a reaction mixture by TEM showed the presence of 2 nm Rh(0) colloids. In all cases, the dehydrocoupling of primary amine–borane adducts (RNH_2–BH_3 with R= Me, Ph) and ammonia–borane NH_3–BH_3 was observed in either diglyme or tetraglyme at 45 °C with the formation of the borazines (Scheme 11.13). Moderate to good yields were obtained (40–100%).

This method was extended to secondary adducts RR NH–BH_3 (R=R =Me or 1,4-C_4H_8; R= $PhCH_2$, R = Me) which afforded the dimeric species (Scheme 11.14). Reactions were performed in toluene with 0.5 mol % of organometallic pre-catalyst such as [Rh(COD)Cl]$_2$. 100% conversion was observed in 8 h at 25 °C.

Based on these results, Manners reported original tandem catalytic dehydrocoupling–hydrogenation reactions using Rh colloids and Me_2NH–BH_3 as a stoichiometric H_2 source (Scheme 11.15) [94]. The reaction was performed with cyclohexene

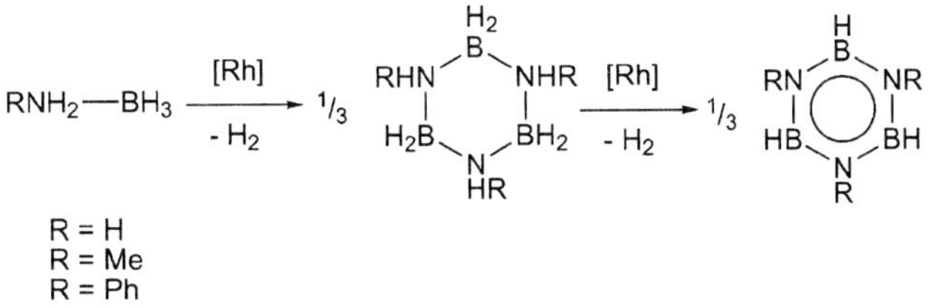

R = H
R = Me
R = Ph

Scheme 11.13 Catalytic dehydrocoupling of primary amine–borane adducts.

R = R' = Me
R = R' = (1,4-C_4H_8)
R = $PhCH_2$, R' = Me

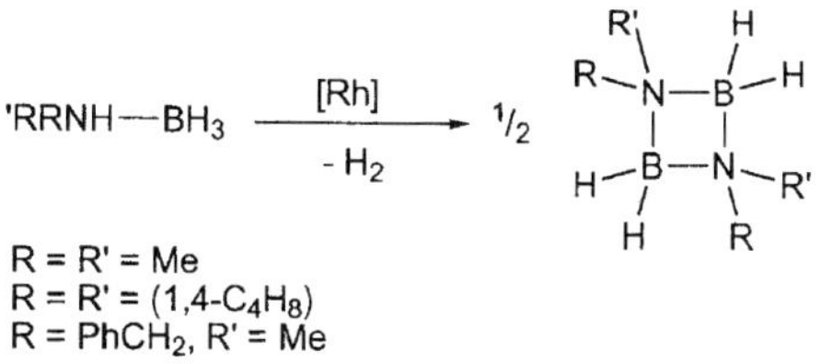

Scheme 11.14 Catalytic dehydrocoupling of secondary amine–borane adducts.

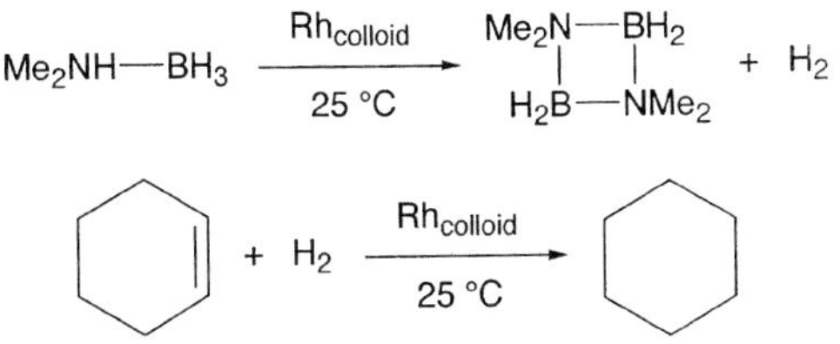

Scheme 11.15 Tandem catalytic dehydrocoupling of Me_2NH–BH_3/hydrogenation of cyclohexene.

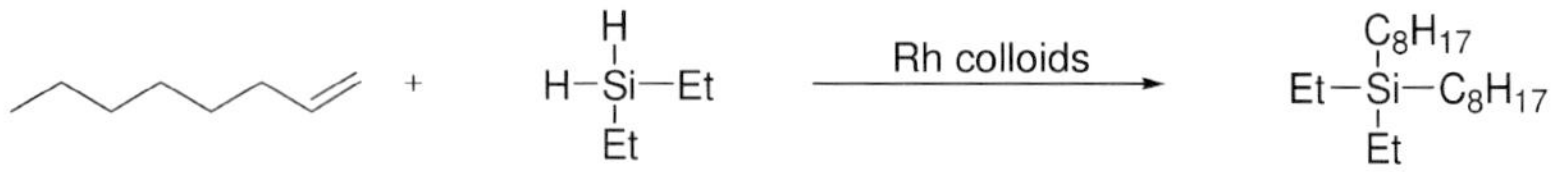

Scheem 11.16 Rh colloids catalyzed hydrosilylation reaction.

as substrate and several rhodium catalytic systems such as the precatalysts [Rh(COD)Cl]$_2$,RhCl$_3$, Rh/Al$_2$O$_3$ and Rh$_{colloid}$/[Oct$_4$N]Cl system at 25 °C. The catalytic hydrogenations of cyclohexene were all quantitative after 24 h but the hydrogenation reached only 81% (for 99% dehydrocoupling) after 24 h with stabilized Rh(0) colloids due to increased aggregation giving a lower activity.

Finally, these recent investigations show the high potential of the colloidal approach in the development of catalytic routes for the formation of new bonds between inorganic elements, although the dehydrocoupling of the analogous phosphine–borane adduct Ph$_2$PH–BH$_3$ proceeds by a homogeneous mechanism [92].

11.5.3
Hydrosilylation

The hydrosilylation reaction, leading to the formation of an alkylsilane by addition of a hydrosilane unit (Si–H) to a double bond, is widely used in the production of silicon polymers, paper release coatings and pressure-sensitive adhesives. Many homogeneous organometallic complexes based on Co, Ni, Pd, Rh or Pt have been used to catalyze this reaction, but strong evidence has proved that the really active species were metal colloids [95–97]. Moreover, it has been demonstrated that Pt colloids were the most effective catalysts for SiH addition to terminal olefins [98]. However, Lewis et al. [99] have reported Rh colloids-catalyzed hydrosilylation (Scheme 11.16). The particles were prepared by the reduction of rhodium chloride by the silane reagents. The authors have shown that Rh colloids present an interesting activity in the addition of di- and trihydrides to olefins, compared to platinum colloids.

Finally, most of the recent studies on the hydrosilylation reaction are based on Pt colloids [100] or Pt/Au or Pt/Pd bimetallic colloids supported on alumina [101].

11.6
Conclusion

This chapter gives a "nearly" exhaustive overview of the performance of Rh and Ru nanoparticles as soluble catalysts. Rh and Ru nanoparticles are mainly used in traditional hydrogenation reactions, in particular in the hydrogenation of carbon–carbon multiple bonds and aromatic compounds, but a few examples of other substrates containing carbonyl groups or other reducible functions such as the nitro group are also reported. Finally, in the main fields of homogeneous catalysis, many catalytic reactions such as hydrosilylation, oxidation, or, more recently, C–C formation (Heck and Suzuki coupling or hydroformyaltion reactions) can now be performed by Rh and Ru nanocatalysts dispersed in liquid media. In these active research areas, relevant examples, sometimes original as the tandem dehydrocoupling-hydrogenation reaction, are also described here.

This synopsis is of crucial importance since books on homogeneous catalysis do not generally deal in detail with these topics, although it is well known that some efficient molecular homogeneous complexes are simply precursors of nano-heterogeneous catalysts. Such catalysts, when correctly characterized, constitute an interesting class of catalysts taking advantage of homogeneous and/or heterogeneous ones. Thus, the use of colloids is generally compatible with various reaction media according to the organic- or water-soluble nature of the stabilizers. Moreover, the type of protective agents such as polymer, surfactant or ligands has often allowed the modulation of the reactivity of these catalytic systems. In addition, colloids can be adapted for utilization in biphasic conditions, allowing recovery of the nanocatalysts by simple decantation/filtration and their recycling. At the moment, it is clear that stabilized colloids are still not traditional or "routine" catalysts but their performances in some cases and their potential in catalytic reactions are now commonly admitted by the scientific community. Nanoparticle systems can be considered as an interesting compromise between homogeneous and heterogeneous catalytic systems, in terms of both activity and selectivity. The number of papers and reviews dealing with the use of colloids in catalysis is significant. The interest in colloidal systems is then real and not only for hydrogenation reactions and may still increase in the near future with asymmetric investigations. Interdisciplinary, the deposition of well-defined nanoparticles on various supports by different methods should advantageously replace traditional heterogeneous catalysts in terms of performance and, finally, this promising area of research should find many more industrial applications.

References

1 G. Schmid (Ed.), *Nanoparticles. From Theory to Application*, Wiley-VCH, Weinheim, **2004**.

2 G. Schmid, in *Nanoscale Materials in Chemistry*, K. J. Blabunde (Ed.), Wiley-VCH, New York, **2001**.

3 D. L. Feldheim, C. A. Foss Jr, (Eds.), Metal Nanoparticles: Synthesis, Characterization and Applications, Marcel Dekker, New York, **2002**.

4 H. Bönnemann, W. Brijoux, in *Active Metals: Preparation, Characérization, Applications*, A. Fürstner (Ed.), Wiley-VCH, Weinheim, **1996**, p. 339.

5 A. Roucoux, K. Philippot, Hydrogenation with noble metal nanoparticles, in *Handbook of Homogenous Hydrogenation*, G. de Vries (Ed.), Wiley-VCH, Weinheim, **2007**.

6 M. A. El-Sayed, *Acc. Chem. Res.*, **2001**, *34*, 257.

7 T. Yonezawa, N. Toshima, Polymer-Stabilized Metal Nanoparticles: Preparation, Characterization and Applications, in *Advanced Functional Molecules and Polymers*, Vol. 2, H. S. Nalwa (Ed.), Accelerated Development, **2001**, Ch. 3, p. 65.

8 A. Roucoux, J. Schulz, H. Patin, *Chem. Rev.*, **2002**, *102*, 27.

9 M. Moreno-Mańos, R. Pleixats, *Acc. Chem. Res.*, **2003**, *36*, 638.

10 D. Astruc, F. Lu, J. Ruiz Aranzaes, *Angew. Chem. Int. Ed.*, **2005**, *44*, 48, 7852.

11 J. A. Widegren, R. G. Finke, *J. Mol. Catal. A: Chem.*, **1999**, *145*, 1.

12 K. Philippot, B. Chaudret, *C. R. Chimie*, **2003**, *6*, 1019.

13 H. Bönnemann, W. Brijoux, R. Brinkmann, T. Joußen, B. Korall, E. Dinjus, *Angew. Chem. Int. Ed.*, **1991**, *30*, 1312.

14 B. Chaudret, in *Synthesis and Surface Reactivity of Organometallic Nanoparticles In Topics In Organometallic Chemistry*, C. Copéret, B. Chaudret (Eds.), Springer-Verlag, **2006**, *16*, 233.

15 H. Hirai, *J. Macromol. Sci. A: Chem.*, **1979**, *13(5)*, 633.

16 A. Borsla, A. M. Wilhelm, H. Delmas, *Catal. Today*, **2001**, *66*, 389.

17 M. M. Telkar, C. V. Rode, R. V. Chaudhari, S. S. Joshi, A. M. Nalawade, *Appl. Catal. A*, **2004**, *273*, 11.

18 J. L. Pellagatta, C. Blandy, V. Collière, R. Choukroun, B. Chaudret, P. Cheng, K. Philippot, *J. Mol. Catal. A: Chem.*, **2002**, *178*, 55.

19 T. Q. Hu, B. R. James, C. L. Lee, *J. Pulp. Paper. Sci.*, **1997**, *23*, 200.

20 C. Larpent, E. Bernard, F. Brisse-le-Menn, H. Patin, *J. Mol. Catal. A: Chem.*, **1997**, *116*, 277.

21 P. Meric, K. K. Yu, A. T. S. Kong, S. C. Tsang, *J. Catal.*, **2006**, *237*, 330.

22 K. K. Yu, P. Meric, S. C. Tsang, *Catal. Today*, **2006**, *114*, 428.

23 K. R. Januszkiewicz, H. Alper, *Organometallics*, **1983**, *2*, 1055.

24 K. R. Januszkiewicz, H. Alper, *Can. J. Chem.*, **1984**, *62*, 1031.

25 P. Drognat, M. Lemaire, D. Richard, P. Gallezot, *J. Mol. Catal.*, **1993**, *78*, 257.

26 K. Nasar, F. Fache, M. Lemaire, J.-C. Béziat, M. Besson, P. Gallezot, *J. Mol. Catal.*, **1994**, *87*, 107.

27 F. Fache, S. Lehuede, M. Lemaire, *Tetrahedron Lett.*, **1995**, *36*, 885.

28 T. Q. Hu, B. R. James, S. J. Rettig, C. L. Lee, *Can. J. Chem.*, **1997**, *75*, 1234.

29 T. Q. Hu, B. R. James, S. J. Rettig, C. L. Lee, *J. Pulp. Paper. Sci.*, **1997**, *23*, 153.

30 B. R. James, Y. Wang, C. S. Alexander, T. Q. Hu, *Chem. Ind.*, **1998**, *75*, 233.

31 R. W. Albach, M. Jautelat, German Pat. DE 19 807 995, Bayer AG **1999**.

32 R. J. Bonilla, P. G. Jessop, B. R. James, *Chem Commun.*, **2000**, 941.

33 M. Ohde, H. Ohde, C. M. Wai, *Chem. Commun.*, **2002**, 2388.

34 J. Schulz, A. Roucoux, H. Patin, *Chem. Commun.*, **1999**, 535.

35 J. Schulz, A. Roucoux, H. Patin, *Chem. Eur. J.*, **2000**, *6*, 618.

36 A. Roucoux, J. Schulz, H. Patin, *Adv. Synth. Catal.*, **2003**, *345*, 222.

37 V. Mévellec, A. Roucoux, E. Ramirez, K. Philipot, B. Chaudret, *Adv. Synth. Catal.*, **2004**, *346*, 72.

38 J. Schulz, S. Levigne, A. Roucoux, H. Patin, *Adv. Synth. Catal.*, **2002**, *344*, 266.

39 Z. Duan, M. J. Hampden-Smith, *Chem. Mater.* **1992**, *4*, 1146.

40 C. M. Hagen, L. Vieille-Petit, G. Laurenczy, G. Süss-Fink, R. G. Finke, *Organometallics*, **2005**, *24*, 1819.

41 O. Vidoni, K. Philippot, C. Amiens, B. Chaudret, O. Balmes, J.-O. Malm, J.-O. Bovin, F. Senocq, M.-J. Casanove, *Angew. Chem. Int. Ed.*, **1999**, *38*, 3736.

42 K. Pelzer, O. Vidoni, K. Philippot, B. Chaudret, V. Collière, *Adv. Funct. Mater.*, **2003**, *13*, 118.

43 K. Pelzer, K. Philippot, B. Chaudret, *Z. Phys. Chem.*, **2003**, *217*, 1.

44 S. Jansat, D. Picurelli, K. Pelzer, K. Philippot, M. Gomez, G. Muller, P. Lecante, B. Chaudret., *New J. Chem.*, **2006**, *30*, 115.

45 J. D. Aiken III, R. G. Finke, *Chem. Mater.*, **1999**, *11*, 1035.

46 J. D. Aiken III, R. G. Finke, *J. Am. Chem. Soc.*, **1999**, *121*, 8803.

47 J. D. Aiken III, R. G. Finke, *J. Mol. Catal. A: Chem.*, **2003**, *191*, 187.

48 R. G. Finke, in *Metal Nanoparticles: Synthesis, Characterization and Applications*, D. L. Feldheim, C. A. Foss Jr. (Eds.), Marcel Dekker, New York, **2002**, Ch. 2, pp. 17–54.

49 J. A. Widegren, R. G. Finke, *Inorg. Chem.*, **2002**, *41*, 1558.

50 E. T. Silveira, A. P. Umpierre, L. M. Rossi, G. Machado, J. Morais, G. V.

Soares, I. J. R. Baumvol, S. R. Teixeira, P. F. P. Fichtner, J. Dupont, *Chem. Eur. J.*, **2004**, *10*, 3734.

51 X.-D. Mu, D. G. Evans, Y. Kou, *Catal. Lett.*, **2004**, *97(3–4)*, 151.

52 G. S. Fonseca, A. P. Umpierre, P. F. P. Fichtner, S. R. Teixeira, J. Dupont, *Chem. Eur. J.*, **2003**, *9*, 3263.

53 G. S. Fonseca, E. T. Silveira, M. A. Gelesky, J. Dupont, *Adv. Synth. Catal.*, **2005**, *347*, 847.

54 X. D. Mu, J. Q. Meng, Z. C. Li, Y. Kou, *J. Am. Chem. Soc.*, **2005**, *127*, 9664.

55 R. M. Crooks, V. Chechnik, B. I. Lemon, L. Sun, L. K. Yeung, M. Zhao, in *Metal Nanoparticles: Synthesis, Characterization and Applications*, D. L. Feldheim, C. A. Foss Jr. (Eds.), Marcel Dekker, New York, **2002**, p. 262.

56 Y. M. Chung, H. K. Rhee, *J. Mol. Catal. A: Chem.*, **2003**, *206*, 291.

57 L. Strimbu, J. Liu, A. E. Kaifer, *Langmuir*, **2003**, *19*, 483.

58 J. Liu, J. Alvarez, W. Ong, E. Roman, A. E. Kaifer, *Langmuir*, **2001**, *17*, 6762.

59 J. Alvarez, J. Liu, E. Roman, A. E. Kaifer, *Chem. Commun.*, **2000**, 1151.

60 M. Komiyama, H. Hirai, *Bull. Chem. Soc. Jpn*, **1983**, *56*, 2833.

61 A. Nowicki, Y. Zhang, B. Léger, J. P. Rolland, H. Bricout, E. Monflier, A. Roucoux, *Chem. Commun.*, **2006**, 296.

62 S. C. Mhadgut, K. Palaniappan M. Thimmaiah, S. A. Hackney, B. Torok, J. Liu, *Chem. Commun.*, **2005**, 3207.

63 W. Yu, M. Liu, H. Liu, X. Ma, Z. Liu, *J. Colloid Interface Sci.*, **1998**, *208*, 439.

64 W. Yu, H. Liu, M. Liu, Z. Liu, *React. Funct. Polym.*, **2000**, *243*, 120.

65 W. W. Yu, H. Liu, *J. Mol. Catal.*, **2006**, *243*, 120.

66 A. Spitaleri, P. Pertici, N. Scalera, G. Vitulli, M. Hoang, T. W. Turney, M. Gleria, *Inorg. Chim. Acta*, **2003**, *352*, 61.

67 Y. Huang, J. Chen, H. Chen, R. Li, Y. Li, L. Min, X. Li, *J. Mol. Catal.*, **2001**, *170*, 143.

68 G. Booth, *Ullmanns Encyclopedia of Industrial chemistry*, Wiley-VCH, Weinheim, Germany, **2002**.

69 H. C. Yao, P. H. Emmett, *J. Am. Chem. Soc.*, **1959**, *81*, 4125.

70 M. Liu, W. Yu, H. Liu, J. Zheng, *J. Colloid Interface Sci.*, **1999**, *214*, 231.

71 M. Liu, W. Yu, H. Liu, *J. Mol. Catal.*, **1999**, *138*, 295.

72 W. W. Yu, H. Liu, *J. Mol. Catal.*, **2006**, *243*, 120.

73 Z. Yu, S. Liao, Y. Xu, B. Yang, D. Yu, *J. Chem. Soc., Chem.Commun*, **1995**, 1155.

74 Z. L. Jin, X. L. Zheng, B. Fell, *J. Mol. Catal. A: Chem.*, **1997**, *116*, 55.

75 F. Wen, H. Bönnemann, J. Jiang, D. Lu, Y. Wang, Z. Jin, *Appl. Organomet. Chem.*, **2005**, *19*, 81.

76 A. J. Bruss, M. A. Gelesky, G. Machado, J. Dupont, *J. Mol. Catal. A: Chem.*, **2006**, *252*, 212.

77 Q. Wang, H. Liu, M. Han, X. Li, D. Jiang, *J. Mol. Catal. A: Chem.*, **1997**, *118*, 145.

78 N. T. S. Phan, M. van Der Sluys, C. W. Jones, *Adv. Synth. Catal.*, **2006**, *348*, 609.

79 Y. Na, S. Park, S. B. Ha, S. Ko, S. Chang, *J. Am. Chem. Soc.*, **2004**, *126*, 250.

80 M. B. Thathagar, J. Beckers, G. Rothenberg, *J. Am. Chem. Soc.*, **2002**, *124*, 11858.

81 For recent reviews, see: (a) S. E. Gibson, A. Stevenazzi, *Angew. Chem. Int. Ed.*, **2003**, *42*, 1800; (b) K. M. Brumond, J. L. Kent, *Tetrahedron*, **2000**, *56*, 3236.

82 S. U. Son, S. S. Lee, Y. K. Chung, S. W. Kim, T. Hyeon, *Org. Lett.*, **2002**, *4*, 277.

83 S. U. Son, K. H. Park, Y. K. Chung, *Org. Lett.*, **2002**, *4*, 3983.

84 K. H. Park, S. U. Son, Y. K. Chung, *Org. Lett.*, **2002**, *4*, 4361.

85 S. U. Son, K. H. Park, Y. K. Chung, *J. Am. Chem. Soc.*, **2002**, *124*, 6838.

86 S. W. Kim, S. U. Son, S. S. Lee, T. Hyeon, *Chem. Commun.*, **2001**, 2212.

87 K. H. Park, I. G. Jung, Y. K. Chung, *Org. Lett.*, **2004**, *6*, 1183.

88 F. Launay, H. Patin, *New J. Chem.*, **1997**, *21*, 247.

89 F. Launay, A. Roucoux, H. Patin, *Tetrahedron Lett.*, **1998**, *39*, 1353.

90 C. A. Jaska, K. Temple, A. J. Lough, I. Manners, *Chem Commun.*, **2001**, 962.

91 C. A. Jaska, K. Temple, A. J. Lough, I. Manners, *J. Am. Chem. Soc.*, **2003**, *125*, 9424.

92 C. A. Jaska, I. Manners, *J. Am. Chem. Soc.*, **2004**, *126*, 9776.

93 C. A. Jaska, T. J. Clark, S. B. Clendenning, D. Grozea, A. Turak, Z-H. Lu, I. Manners, *J. Am. Chem. Soc.,* **2005**, *127*, 5116.

94 C. A. Jaska, I. Manners, *J. Am. Chem. Soc.,* **2004**, *126*, 2698.

95 L. N. Lewis, N. Lewis, *Chem. Mater.,* **1989**, *1*, 106.

96 L. N. Lewis, *J. Am. Chem. Soc.,* **1990**, *112*, 5998.

97 L. N. Lewis, R. J. Uriarte, N. Lewis, *J. Mol. Catal.,* **1991**, *66*, 105.

98 C. Eaborn, R. W. Bott, in *The Bond to Carbon*, A. G. MacDiarmid (Ed.), Marcel Dekker: New York, **1968**, Vol. 1.

99 L. N. Lewis, R. J. Uriarte, *Organometallics,* **1990**, *9*, 621.

100 J. Stein, L. N. Lewis, Y. Gao, R. A. Scott, *J. Am. Chem. Soc.,* **1999**, *121*, 3693.

101 G. Schmidt, H. West, H. Mehles, A. Lehnert, *Inorg. Chem.,* **1997**, *36*, 891.

12
Supported Gold Nanoparticles as Oxidation Catalysts

Avelino Corma and Hermenegildo Garcia

12.1
Introduction

Platinum, palladium, iridium, copper, silver and other transition noble metals are highly active, promoting many different types of organic reactions including hydrogenations, oxidations, C—C bond formation, etc. [1, 2]. In contrast to the high catalytic activity of Pt ($Z=78$) or Ir ($Z=77$), Au ($Z=79$) as a metal has been largely considered devoid of interesting catalytic activity.

This assumption of considering gold catalytically inert changed, however, in the late 90s after the seminal contribution of Haruta, who reported that gold nanoparticles are extremely active in promoting the low-temperature, aerobic oxidation of CO to CO_2 [3, 4]. Moreover, Haruta and coworkers showed that the catalytic activity of gold appears for particles in the nanometric size range, there being a direct relationship between activity and particle diameter [5, 6].

Since this breakthrough, research has been aimed at determining the reaction types that can be efficiently catalyzed by nanometric gold particles, increasing the catalytic activity of gold by stabilizing nanoparticles against agglomeration, understanding the role of the solid support in the catalytic activity and also demonstrating the similarities and distinctive properties of gold catalysis with respect to catalysis by other noble metals.

In this chapter, we will comment on the specific properties that arise when the size of the particles is decreased in the length scale of nanometers, how to form and stabilize gold nanoparticles and then on the catalytic oxidation reactions that can be effected by gold, with special emphasis on alcohol oxidation. The high activity of gold nanoparticles supported on nanoparticulated ceria will be described and justified, together with some insights into the mechanism. A specific section will be devoted to discussion of the differences in selectivity between gold and palladium catalysts in order to show the general applicability of gold nanoparticles as a catalyst for alcohol oxidation. This chapter will conclude with some prospects for the future trends in aerobic oxidation.

Nanoparticles and Catalysis. Edited by Didier Astruc
Copyright © 2008 WILEY-VCH Verlag GmbH & Co. KGaA, Weinheim
ISBN: 978-3-527-31572-7

12.2
Nanoparticles and Their Properties

Atoms are sized in picometers and typical covalent bonds are hundreds of picometers in length. Thus, particles in the nanometer size range are constituted of clusters of tens to thousands of atoms. Typical particle sizes of many common solids are on the micrometer length scale or larger. Small particles experience intense van der Waals forces that make them agglomerate, collapsing in or behaving as single large particles. In some cases this aggregation is even accompanied by an authentic *chemical* fusion of the small components into one particle in which the surface of the individual smaller units has disappeared. This phenomenon of fusion occurs particularly easily in metal particles since, in contrast to covalent bonds that are *saturable*, metallic bonds are characterized by maximum packing and a metal atom (particularly those located at the external surface) tends to be surrounded in its coordination sphere by as many other metal atoms as possible. The model of the *"full shell"* layers to determine the number of atoms in a nanometric particle is based on this maximum packing [7–9].

The properties of most of the solids that we are accustomed to dealing with, particularly metal oxides, correspond to those materials that are constituted by particles in the micrometer size. In these micrometric particles, the number of atoms in the external surface of the particle is negligible compared to the number of internal atoms. Thus, the particle exhibits the properties of the bulk material and variations in the particle size within this length scale are not reflected in measurable changes in the physical and chemical properties of the solid. The influence of the solid external surface on the behavior of the material is undetectable or nearly negligible.

12.2.1
Surface Chemistry and Nanoparticles

In the last decades, mainly due to the ability to synthesize high surface area solids, surface chemistry has experienced an extraordinary growth and has become a mature science. Materials like zeolites [10–13] or mesoporous silicas [14, 15] have total surface areas (internal plus external) much higher than those of solids whose interior is not accessible. The fact that the solids are porous and that the interior of the particles is accessible from the exterior is essential in achieving these high values of total surface area. While the external surface areas of zeolites and mesoporous silicas are similar to those values that can be achieved in amorphous silicas, it is the internal surface that makes the difference between porous and other solids. This internal surface area of micro-/meso- porous solids can be one order of magnitude larger than the external area.

In the case of mesoporous silicas, and particularly since the first report of the synthesis of MCM-41 [14, 15], that constituted a real breakthrough in material science, it is possible to prepare materials with total surface area higher than

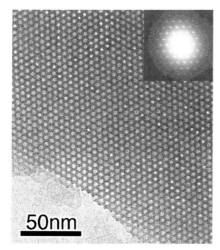

Fig. 12.1 Transmission electron microscopy of a mesoporous MCM-41. The electron diffraction pattern shown on the upper right corner reflects the structural periodicity of the material.

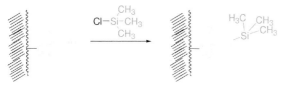

Scheme 12.1 Chemical reaction of surface silanol groups.

$1000 \, m^2 g^{-1}$. Figure 12.1 shows a typical transmission electron microscopy image showing the mesopores and the strict periodicity of the channels.

In this type of mesoporous silica, one fourth of all the silicon atoms that constitute the solid are on the surface (mainly internal). External silicon atoms are tri- or even bi- podally connected to the solid, having one or two hydroxy groups. This high ratio between external and internal silicon atoms and the fact that the surface contains hydroxy groups bonded to silicon (*silanols*) makes it possible to develop a chemistry, based on the reaction of these surface silanol groups, to modify and functionalize the solid [16–19]. Scheme 12.1 shows a typical surface reaction to protect silanol groups. Thus, surface chemistry has served, among other possibilities, to modify the hydrophilicity/hydrophobicity of the internal silica surfaces and to produce an array of hybrid organic–inorganic materials in which the organic component is covalently bonded to the inorganic framework [20–23].

The above comments serve to illustrate the possibilities offered by those solids that have a large surface area, even though in the porous solids it is mainly internal, in terms of functionalization. However, an increase in surface area also alters some intrinsic physical and chemical properties of the solids. For instance, while the stoichiometry of silica can be considered as SiO_2, it is clear that as the importance of silanol groups increases due to surface termination, the stoichiometry can change to $SiO_{2-x}(OH)_{2x}$. The presence of these hydroxy groups is reflected in the hydrophilicity of the solids, their capacity to adsorb water and their ability to form hydrogen bonds with organic molecules.

12.2.2
Properties of Nanoparticles

Analogously to the case of porous materials, when the particle size of a non-porous solid is reduced sufficiently to the point that the ratio between external to internal atoms cannot be considered zero, then the properties of the material may vary as a function of the dimensions of the particles. Considering the atomic dimensions, these effects are expected to arise in the length scale down to hundreds of nanometers, but they will be growing in importance at the size of a few nanometers. The variation in the solid properties arises from the presence of a significant proportion of atoms on the external surface, the properties and behavior of atoms on the external surface being different to those in the interior of the particle. Atoms in the interior of the particle are saturated, while those in the external surface still have the ability to bind with adsorbates due to the presence of unsaturation in their valence. Also, the preferential surface termination may require the presence of specific terminal groups (hydroxy groups in the previously mentioned case of silica) or even the absence of some atoms present in the bulk of the particle (vacancies).

One example of a non-porous metal oxide whose bulk properties vary upon reduction of particle size in the nanometric regime is ceria. The stoichiometry of bulk cerium oxide is CeO_2, but when the particle size is reduced down to a few nanometers then the particles present surface defects sites, mainly oxygen vacancies, and there is a mixture of +IV (bulk ceria) and +III (defects) oxidation states to compensate for the oxygen vacancies and maintain the electroneutrality of the solid. Thus, the stoichiometry of nanometric ceria differs from the ideal CeO_2 and becomes more like CeO_{2-x}, x being related to oxygen vacancies and the presence of Ce^{III}.

Many properties of bulk ceria change when the particle size is reduced in the nanometric regime. For instance, EPR inactive ceria becomes EPR active as a consequence of the presence of defects. Another property that arises from the nanometric regime is the change in the ceria from an insulator to a semiconductor material, the latter property being interesting in the development of photovoltaic cells [24]. The properties of nanometric ceria particles, and specifically their oxygen vacancies and the presence of Ce^{III}, will become relevant later when discussing the supports employed for the preparation of highly active supported gold.

Analagously to metal oxides, the properties of metals change when the solid is constituted by particles and the size of these particles enters the nanometric regime. Relevant to this chapter is the catalytic activity of gold nanoparticles, a property that disappears as soon as the particle size grows above tens of nanometers. Due to the characteristic of the metallic bond and the strong van der Waals interactions, we have already mentioned that noble metal nanoparticles are highly unstable and tend to collapse forming larger particles towards the micrometric size in which the proportion of highly reactive external metal atoms becomes negligible. Therefore, it is of crucial importance to devise strategies to stabilize metal nanoparticles.

The importance of gold nanoparticles in catalysis derives in part from the wealth of information and reliable experimental procedures to form these nanoparticles in different sizes and, even different shapes [7]. There are basically two different procedures to form gold nanoparticles, depending on whether they are going to be dispersed in a liquid or supported on a solid.

In spite of the fact that the formation of gold nanoparticles was known since ancient times [7], many researchers consider that the first systematic study of the formation of colloidal gold nanoparticles started with the pioneering work of Turkevich [25–27], who studied the size and shape of gold nanoparticles obtained by treating aqueous solutions of $AuCl_4^-$ with various reducing agents. Sodium citrate was a convenient reducing agent, giving narrow particle size distribution around 20 nm.

Another common experimental procedure to form colloidal gold solutions is the *two-phase* method [28–31], in which, starting from a $AuCl_4^-$ salt in aqueous solution, the reduction with hydrazine, a metal hydride, etc. is performed in the presence of an immiscible organic solvent (toluene), a phase transfer catalyst (a quaternary ammonium salt) and a ligand able to coordinate with the gold nanoparticles (phosphine). A low molar $AuCl_4^-$/ligand ratio has a positive consequence on the stability of the resulting colloid against agglomeration, but can play a negative role on the catalytic activity. Upon formation of colloidal gold in the aqueous phase, the nanoparticles will coordinate with the ligand, rendering the colloids hydrophobic and soluble in the organic solvent. This coordination will produce the phase transfer from water to the organic solvent. The colloidal suspension containing gold has to be used for further applications, since solvent removal is normally not advisable due to massive particle agglomeration. Figure 12.2 summarizes this *two-phase* methodology.

For the preparation of gold nanoparticles supported on insoluble solids, the most widely used procedure is the *precipitation–deposition* method [32–36]. Starting from an aqueous solution of $HAuCl_4$, addition of a base leads to precipitation of a mixture of $Au(OH)_3$ and related oxy/hydroxides that adsorbs into the solid and is then reduced to metallic gold by boiling the adsorbed species in methanol or any other alcohol. In this procedure, it has been established that the pH of the precipitation and the other experimental conditions (nature of the alcohol, temperature and time of the reduction, calcination procedure, etc.) can provide a certain control of the particle size of the resulting nanoparticles [3]. Figure 12.2 illustrates the steps required in the formation of supported gold nanoparticles.

To give an idea of the monodispersity that can be achieved, Fig. 12.3 shows a transmission electron microscopy image of a colloidal gold sample prepared from

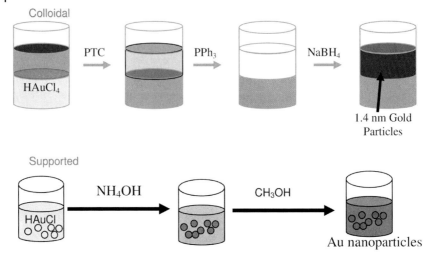

Fig. 12.2 Typical procedures for the preparation of colloidal and supported gold nanoparticles (PTC: Phase transfer catalyst).

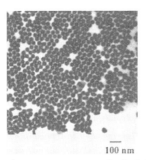

100 nm

Fig. 12.3 TEM image of collodial gold nanoparticles.

HAuCl$_4$ following the conventional *two-phase* NaBH$_4$ reduction procedure. Concerning the experimental procedures for the preparation of gold nanoparticles, the situation is that there are reliable protocols to prepare gold nanoparticles of a few nanometers and a given dispersity [37, 38].

This wealth of reproducible methods to form colloidal gold nanoparticles is, probably, in contrast to other noble metals and more specifically to palladium, In the case of palladium, even though the role of palladium nanoparticles in Suzuki and other C—C bond coupling reactions is beginning to be recognized [39–45] most of the catalytic systems still start from palladium complexes and very frequently the catalytic activity of palladium is discussed as if the initial palladium complex were the only active species, ignoring the formation of palladium nanoparticles during the course of the reaction. It would be of interest to compare the catalytic activity of these palladium complexes with that obtained starting directly from palladium nanoparticles [46, 47]. On the other hand, in contrast to the numerous reported palladium complexes including organopalladium and complexes of monoatomic palladium(0), not many organogold complexes have been reported [48–56].

Clearly, the main difference emerging between gold and palladium is that for gold the chemistry of nanoparticles is prevalent, while for palladium many catalytic reactions are based on organometallic complexes. It can be easily anticipated, however, that organometallic complexes will gain importance in the near future in the case of gold, while the role of palladium nanoparticles (even when the apparent precatalyst is an organometallic complex) will be increasingly accepted, even by those researchers who use organopalladium complexes as pre-catalysts. Thus, the area of metallic complexes and nanoparticles will converge for gold and the rest of noble metals in the near future.

Full-shell gold clusters are constructed by successively packing layers of metal atoms around a single gold atom [7]. For this reason the theoretical number of gold atoms per shell is established by $(10n^2 + 2)$, where "n" is the number of the shell being considered. Thus, there are certain gold clusters constituted by "*magic*" numbers that fit with the previous formula, like Au_{13}, Au_{55}, Au_{147}, Au_{309}, Au_{561}, Au_{1415}, etc., that are commonly encountered, particularly when working with small nanometric clusters. The number of gold atoms in a nanoparticle can be easily determined by mass spectrometry. Moreover, as mentioned earlier, these gold clusters can be obtained reproducibly fairly monodisperse, with a significant prevalence of one of these combinations.

Obviously, as the number of gold atoms grows the size of the cluster increases and the proportion of atoms on the external surface with respect to those located internally decrease. Table 12.1 lists the percentage of external atoms for those full-shell gold clusters previously mentioned. Notice that 1 and 2 nm size clusters

Table 12.1 Percentage of surface atoms as a function of the total gold atoms according to the *full-shell* theory.

Full-shell clusters (no. of shells)	Total Number of Atoms	Surface Atoms (%)
1	13	92
2	55	76
3	147	63
4	309	52
5	561	45
6	923	39
7	1415	35
8	2057	31
9	2869	28
10	3871	26

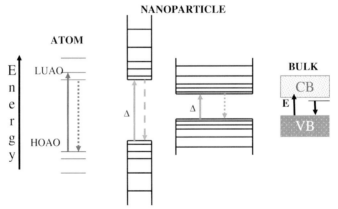

Fig. 12.4 Illustration of the transition of the electronic properties from atoms to solids that occurs in the nanometric scale. The central part represents two different situations that are better described by analogy with atoms (left) or by analogy with a solid (right). These two alternatives illustrate the rapid property changes that can be experienced by nanoparticles of slightly different sizes. LUAO and HOAO refer to the *lowest unoccupied atomic orbital and the highest occupied atomic orbital*. CB and VB refer to *conduction band and valence band*.

correspond approximately to four and six shells and that the ratio between external and internal gold atoms decreases quickly upon size increase in the first clusters, while the reduction in the proportion of surface atoms is much smoother when the cluster size increases. Also, it is more likely that the cluster will deviate from the *full-shell* prediction as the number of layers increases. As just commented, most of this gold chemistry has no equivalent in palladium and it would certainly be of interest to develop synthetic procedures for the preparation and study of the catalytic activity of analogous palladium clusters.

The reason why the properties of nanoparticles are so special and change so dramatically with relatively minor changes in size and in the number of atoms is because the remarkable transition from atomic orbitals with well defined energy states and levels to bands with occupied (valence) and unoccupied bands takes place just for clusters in this length scale. Figure 12.4 illustrates these changes from discrete orbitals with defined energies characteristic of atoms and entities with a few atoms to the electronic bands characteristic of bulk solids. In this sense, there are small nanoparticles more like the quantum mechanical description of molecules and other larger nanoparticles with electronic bands analogous to solids constituted by micrometric particles, the change occurring with minor particle size increases.

Thus, while large size variations in the micrometer regime do not affect significantly the properties of the material, there is a narrow diameter range around 2–20 nm in which dramatic property changes occurs with minor variations in the size. Beyond this reduction (below 2–3 nm) further size reduction does not much affect the properties of the nanoparticles [57].

Fig. 12.5 Visual appearance of colloidal gold nanoparticles of varying particle size. Average particle size from left to right: 4, 12, 25 and 37 nm.

These variations in the electronic structure of the nanoparticles as the size varies by a few nanometers are reflected by the appearance of colors. While bulk gold is yellow, gold nanoparticles can exhibit different colors depending on their size, varying from blueish to ruby red, and black [7, 58, 59]. Figure 12.5 shows a photograph of colloidal gold as a function of the average particle size.

The color is intense enough to justify gold nanoparticles having been utilized as red pigments in paintings and stained glass [60, 61]. The origin of this color is a band in the visible around 550 nm that is commonly called the *surface plasmon band*. This band can be simply described as arising from the confinement of electrons on a spherical bidimensional surface of nanometric dimensions [7].

12.2.3
Stabilized Gold Nanoparticles

To avoid or minimize the natural tendency to agglomerate, there are two alternative and complementary strategies, both exhibiting advantages and disadvantages with respect to each other. These two different approaches are reflected in the preparation procedure, as described earlier. The first consists in surrounding the metal cluster with a layer of ligands that insulate the nanoparticle, protecting the reactive atoms and minimizing their tendency to collapse [62–64]. Since the metal atoms in the external surface are unsaturated, they still have the possibility to form coordinative bonds and bind with some molecules, forming an aggregate constituted by a metal nanoparticle core and an external surface of molecules strongly ligated to the core. Typical ligands for gold nanoparticles are those molecules having nucleophilic atoms such as alkyl thiols, amines and phosphines or quaternary amonium cations. Polymers such as polyvinyl alcohol and polyvinyl pyrrolidones have also been frequently used to stabilize gold nanoparticles. Considering the relative size of the metal nanoparticles and the ligands, both are in the nanometric range and the nanoparticle appears to be covered by a protective layer.

While ligands have a beneficial influence on the stability of gold nanoparticles, they exert a negative influence from the catalytic activity point of view, reducing

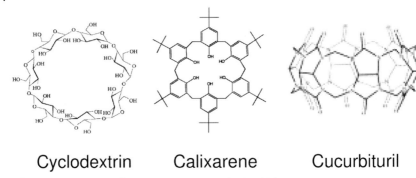

Cyclodextrin　　　Calixarene　　　Cucurbituril

Scheme 12.2 Structure of common organic capsules in which gold nanoparticles can be incorporated.

the interaction of external atoms with the substrates. For this reason, strongly bonded ligands such as thiols can be highly detrimental to the catalytic properties of the metal nanoparticles, tetralkylammonium cations are generally considered to be a good choice due to the weak, non-coordinative bond of the quaternary ammonium ligand and the nanoparticles. When using ligands to stabilize gold nanoparticles a parameter to be considered is the ligand-to-nanoparticle ratio. Generally ten or more ligand molecules per nanoparticle are needed to stabilize the colloids and avoid their agglomeration.

An interesting variant of the use of molecular ligands that still needs to be developed is the use of organic capsules as ligands (Scheme 12.2). Cyclodextrins (α-, β- and γ-), calixarenes and cucurbiturils (CBs) are hollow organic molecules that can act as hosts encapsulating guests of nanometric diameter. Meunier and coworkers reported that laser ablation of colloidal gold solutions can serve to form gold nanoparticles small enough to be incorporated inside cyclodextrins [65]. In this regard, cyclodextrin will act as a macroligand surrounding and embedding the metal nanoparticle.

However, due to the conical shape of cyclodextrins the included gold nanoparticle can slip out of the cage during the catalytic reaction and collapse into large, inactive gold particles. This problem has been solved by Corma and Garcia using pumpkin-shaped cucurbiturils as hosts to encapsulate and stabilize gold nanoparticles [66]. Cucurbiturils (CBs, from the latin word cucurbitum, meaning pumpkin) have a larger internal diameter at the center of the capsule than at the portals [67, 68]. The symmetric upper and lower portals are flanked by highly polar oxygens of carbonyl groups of glycuril units (from 5 to 8 for CB[5] to CB[8], respectively) and control the access to the interior of the empty cavity by steric and polarity effects. The barrel shape of CB is such that if gold atoms assemble inside cucurbituril to form nanoparticles (from 0.7 to 1.3 nm size, depending on the dimensions of the CB used), then the resulting embedded gold nanoparticle cannot exit the capsule and remains entrapped inside. The incarcerated gold nanoparticle can still interact with those substrates that can enter through the cucurbituril portals.

A completely different strategy to stabilize gold nanoparticles is supporting them on a high surface area solid. The origin of this stabilization is the physisorption interaction between the solid surface and the gold adsorbate that immobilizes the nanoparticle under certain conditions, making agglomeration more difficult. This approach has the advantage that it has been amply used in the development of classical supported noble metal catalysts and can be readily expanded to gold nanoparticles [63, 69]. Among the most common supports are activated carbon and metal oxides, but it can be readily assumed that carbon nanotubes, micro- and meso- porous materials as well as organic polymers and other insoluble high surface area solids will also be used in the near future.

The main advantage of supported gold is that the interaction between gold nanoparticles and the support can control and enhance to a significant extent the catalytic activities of gold. Given the importance of this point, we will deal with it in a separate section. The main disadvantage of having supported gold is the difficulty in avoiding leaching of the metal particles from the solid to the solution during the course of the catalytic reaction and the limited stability against agglomeration that can be achieved in some cases. The fact that the catalytic system becomes heterogeneous can have advantages in terms of easy separation from the reaction mixture and catalyst recovery, but also disadvantages due to inadequate reaction rate or selectivity.

Leaching of supported colloidal gold from the solid to the solution depends on the solvent used, the interaction with reaction and products and the reaction temperature. Leaching may cause depletion of gold on the solid catalyst, limiting the service life of the catalyst and its reusability. Leaching may also be responsible for the occurrence of liquid phase catalysis by the leached species that can contribute partially or totally to the observed catalytic activity of the supported gold catalyst. One important consequence of leaching is contamination of the reaction products with heavy metals. For this reason, all systems consisting of supported gold have to be carefully surveyed for the occurrence of leaching. Contamination, even by a few ppm, may cause special purification requirements to meet product specifications, particularly for those compounds that are being consumed as drugs.

The problem of leaching of supported gold catalyst has to be put into the context of the related supported palladium catalysts. It has been proposed that virtually all the solid palladium catalysts act simply by providing palladium particles that dissolve into the liquid phase, even in minute quantities, catalyzing the reaction homogeneously rather than heterogeneously [70, 71]. Typical leaching tests such as filtering the solid catalyst out from the reaction suspension at the reaction temperature at about half the value of the final conversion have been questioned, since apparently the unavoidable temperature decrease during the filtration may cause the dissolved species to reprecipitate onto the support [40, 70, 71]. This tendency to readsorb onto the solid upon minor cooling may create the false

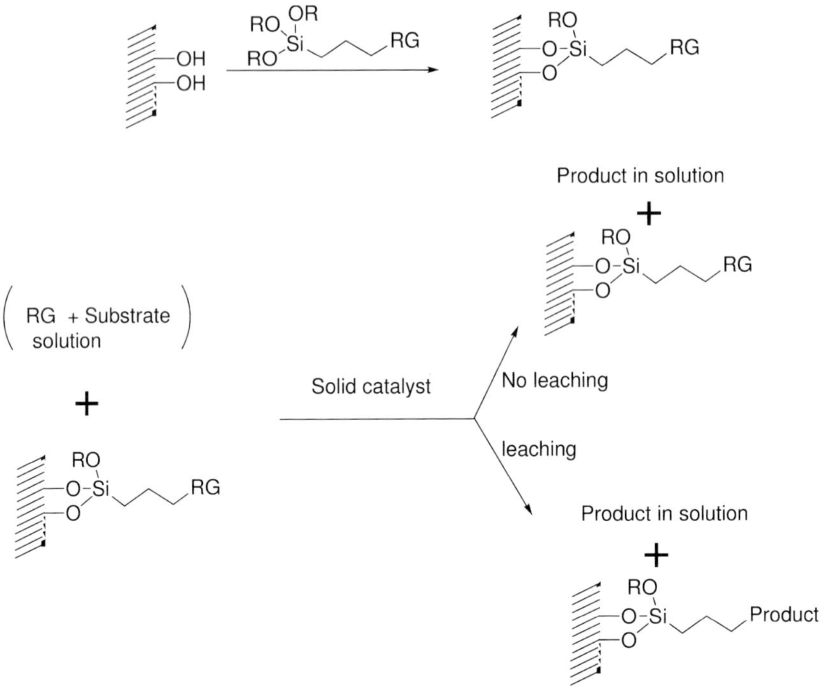

Scheme 12.3 Steps of the *three-phase test*. First it is required to attach covalently a reactive group (*RG*) to a solid surface. Then the reaction with a substrate and *RG* is carried out in solution in the presence of the solid with RG and the catalyst. If leaching occurs, in addition to observing the product in solution, the RG should be transformed into the corresponding product.

impression that no palladium was in the solution. Thus, to definitely rule out the occurrence of leaching during the reaction using palladium nanoparticles a protocol termed the "*three phase test*" has been proposed by Davies and Lipshutz [70, 71]. In this "*three-phase test*", a third component, consisting of a solid with a covalently anchored organic moiety able to act as a reagent, is added to the reaction suspension containing the supported noble metal solid (Scheme 12.3). If the reaction is truly heterogeneous no transformation of the reagent anchored to the solid should be observed. The main problem of the *three-phase* test is to detach a large percentage of the organic component covalently anchored to the solid and to obtain a firm conclusion as to the composition of the organic compounds bonded to the solid phase. Controls have to be performed to ensure that detachment from the solid and analysis of the reaction mixture is carried out satisfactorily. In any case, considering the analogies between palladium and gold, it can be anticipated that the issue of leaching should also be dealt with appropriately when using supported gold catalysts at high reaction temperatures in viscous solvents.

12.3
Influence of the Support on the Catalytic Activity of Supported Gold Nanoparticles

Interaction of highly active supported gold catalysts with probes such as C=O and XPS analyses of the external surface of gold nanoparticles point to the presence of positive gold atoms on gold nanoparticles. One of the most general ways to obtain gold nanoparticles is to start from soluble gold halides in which Au has a formal +III oxidation state and effect the reduction using appropriate chemical reducing agents such as alcohols, hydrazine, metal hydrides [7, 62]. Thus, it is widely assumed that the oxidation state of all gold atoms in the nanoparticles is 0. This is obviously an over simplification and it can be envisioned that even if most of the gold atoms in the clusters have "0" oxidation state, others, particularly at the surface, edges or corners, may have a formal positive oxidation state, as is observed in XPS and by interaction with probe molecules. Based on the high reactivity of some molecular gold species in solution, it is reasonable to assume that these positive gold atoms present in the nanoparticles play a crucial role in the catalysis.

In the case of supported gold catalysts, the density of the positive gold species must depend to a large extent on the interaction with the solid support. Thus, an ideal support should not only stabilize the gold nanoparticle size by physisorption forces, but should also provide adequate electronic density to the gold nanoparticles to stabilize positive gold atoms. Bond has proposed that these active gold atoms are mostly at the interface between the gold and the support and that the role of the support is to stabilize positively charged gold atoms at the interface [72–74].

In other cases, it has been proposed that one condition of a suitable solid support should be a metal oxide in which the metal is able to undergo redox processes in such a way that cooperate with the catalytic oxidation [33, 63, 69, 74–76]. Iron oxide, ceria and even titania are examples of supports in which the metal atoms can swing their oxidation state. In this context it would be of interest to use MnO_2 and CrO_3 as supports since these oxides have themselves been widely used as stoichiometric oxidizing reagents due to their ability to become reduced.

Oxidizability of the metal atom cannot be considered, however, a prerequisite for a solid to act as a support, since other solids, and particularly silicas, activated carbons and single wall carbon nanotubes, are suitable supports devoid of this redox property. One remarkable case is activated carbons as supports of colloidal gold, since this solid is analogous to those widely used for palladium or platinum on carbons. Notably, gold on carbon is extremely active for oxidations in aqueous media under basic pH, however, it is remarkably inert under solventless conditions. These drastic differences in activity point to the fact that the mechanism using activated carbons as support is different to those gold-catalyzed oxidations using metal oxides as supports or that there are different oxidation mechanisms depending on the reaction conditions, particularly the presence of water.

In fact, in contrast to the role ascribed to nanoparticulated ceria contributing to the overall catalytic activity of gold in solventless alcohol oxidation, "*naked*" colloidal gold particles have also been found to be active for the aerobic alcohol oxidation in water under basic conditions [77–79]. Rossi, Prati and coworkers have observed that, initially, gold nanoparticles of 3.6 nm diameter are highly active to effect the oxidation of glucose to gluconate, but the reaction stops at 21% glucose conversion and at the same time the coagulation of gold nanoparticles into large particles that precipitate leaves a clear colorless solution [80]. If the same gold nanoparticles are adsorbed onto activated carbon, TEM images show that no change in the gold particle size occurs in the adsorption, and the initial catalytic activity of npAu/C is very similar to that of "*naked*" gold colloids, indicating that all the catalytic activity of these systems derives from *naked* gold without a promotion effect derived from the interaction of gold with the support. Moreover, comparison of the time–conversion plot of two twin reactions, in one of which activated carbon is added at the time at which deactivation of colloidal gold by agglomeration occurs, shows that on addition of the activated carbon, the activity of the catalyst is extended, reaching 100% glucose conversion [80]. Taken all together these experiments have led to the conclusion that "*naked*" colloidal gold has activity for the aerobic oxidation and that the main role of activated carbon is as an inert support that avoids aggregation of gold nanoparticles. It has to be commented, however, that although this role of the support stabilizing particle size may very well also be the only one operating in the case of some common organic supports for colloidal gold such as polyvinyl alcohol, polyvinylpyrrolidone, tetrahydroxymethylphosphonium, cucurbiturils and other molecular and polymeric ligands, it is reasonable to assume that inorganic oxide supports, having a highly reactive surface, may act differently. Certainly this point deserves carefully study.

Silica was not considered a suitable support for gold nanoparticles, based on the early reports on CO oxidation and the fact that Si atoms cannot exhibit two oxidation states. This is in contrast to the fact that silica is surely the high surface area inorganic support most widely used to deposit catalytically active particles. This situation may, however, change since the recent reports in which colloidal gold has been supported on two specific silicas, the systems exhibiting high catalytic activity for oxidation. In one of the approaches, Corma and Garcia have used a task-specific ligand that is able to perform three functions [81]: (i) to coordinate and stabilize gold nanoparticles by acting as a ligand; (ii) to establish a hydrophobic interaction with cetyltrimetylammonium micelles in aqueous media; and (iii) to co-condense with tetraethyl orthosilicate during the formation of a porous silica. The major concept of this colloidal gold embedded into periodic mesoporous silica is to use an organic molecule, that acts as a ligand of colloidal gold (Scheme 12.4), which can intervene in the templating process occurring in the formation of periodic mesoporous silicas (Scheme 12.5).

As commented earlier, the synthesis of mesoporous silicas of the MCM-41 type has been a major breakthrough in material science. It is accepted that the rigid channels of this silica and their periodic arrangement originate as a replica of the flexible liquid crystal spatial arrangement created by the structure directing agent, generally cetyltrimethylammonium. When in water, the strong hydrogen bonds

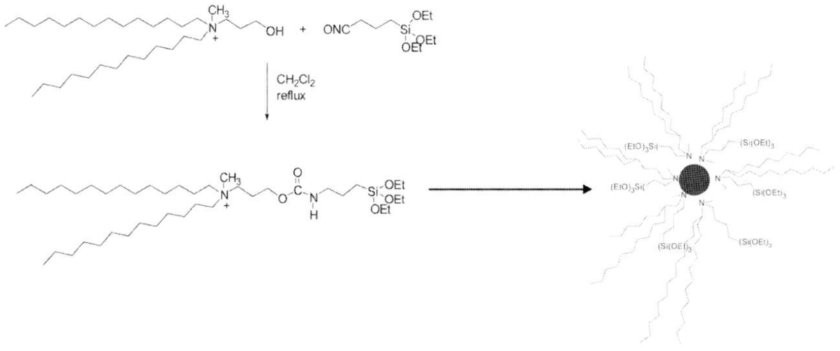

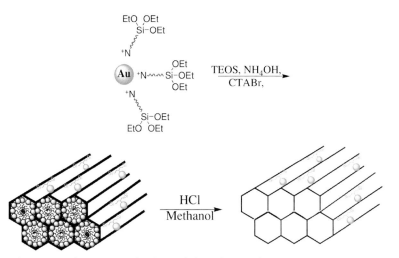

Quaternary ammonium stabilized
gold nanoparticle

Scheme 12.4 Synthesis of a task-designed ligand that serves, on the one hand, to stabilize gold nanoparticles through the quaternary ammonium moiety and on the other to co-condense with tetraethyl orthosilicate as well as to interact with the structure directing agent through the cetyl chains.

Scheme 12.5 The presence of task-specific ligands around gold nanoparticles makes it possible to template the formation of a mesoporous silica with aerobic oxidation catalytic activity.

make the apolar alkyl chains of the surfactant form positively charged micelles and eventually rods around which the negatively charged silica oligomers develop. The use of the ligand causes the colloidal gold to be arranged also at this interphase by the interaction of the surfactant while the silica is being formed.

More recently, gold nanoparticles have been obtained embedded into a highly porous amorphous silica matrix. The process starts by obtaining colloidal gold ligated to a mixture of alkylthiol and 3-mercaptopropyltriethoxysilane. These

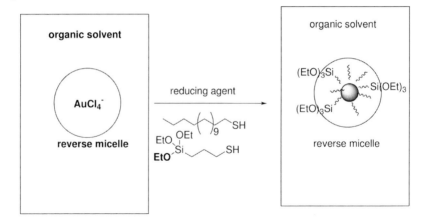

Scheme 12.6 Preparation of a porous silica containing embedded gold nanoparticles.

gold nanoparticles are added as one of the silicon sources in the formation of porous silica and the presence of the triethoxysilyl groups ensures that the co-condensation occurs around the gold nanoparticles (Scheme 12.6). Transmission electron microscopy has shown that the solid contains a grid of small gold nanoparticles surrounded by the silica matrix. Obviously, the goal that is being pursued using silica is to develop supported catalysts on a very high surface area support that can be reused without leaching.

12.4
Sustainability and Green Chemistry

There is a general concern in developing new chemistry that can eventually be sustainable and benign for the environment [82, 83]. Most of the current chemical industry is based on the use of oil and natural gas as feedstock and this situation has to change substantially in the near future. Also, sustainability has to be accompanied by the development of new processes that do not produce, or minimize up to an acceptable extent, any negative impact on the environment. For these reasons, a new scenario in chemistry has been set in which not only inefficient processes, but even the efficient ones, have to be replaced if they do not comply with certain principles that constitute the base of what has been called *Green Chemistry* [82, 83].

This situation applies not only to petrochemistry, but also to base chemicals, commodities and fine chemicals. In addition to industry, Organic Chemistry as a basic science will also be forced to change. Again, very efficient reactions will not be tolerated any longer due to their negative impact on the environment and new chemistry has to emerge. In Green Chemistry, two parameters, the E-factor [84, 85] and the Atom Efficiency (AE) [86, 87], have been established to quantify the greenness of a transformation. The E-factor indicates the kg of waste per kg of product and AE measures the percentage of the starting material that ends up in products. Both parameters are obviously interrelated. Considering a simplified overview of organic reactions, it can, in general, be stated that there are some reaction types that have a low E-factor and high AE. One example of this green general organic reaction type will be hydrogenations. Another reaction type that has a high green score is C—C couplings catalyzed by transition metals such as Suzuki, Heck, Sonogashira, Buchwald, etc.

In contrast to those two types, many other general organic reactions are very far from being ideal from the green point of view. Without trying to be exhaustive, Fig. 12.6 illustrates some of the organic reaction types to give an idea of the current situation in organic chemistry.

In this figure, it can be seen that those organic reactions using homogeneous Brönsted and Lewis acids or bases in solution are unsatisfactory from the *green chemistry* point of view. It is curious that while most of the organic reactions promoted by acids or bases are described in organic chemistry textbooks as being acid or base catalyzed, this is not totally correct since, in many cases, these reactions require over stoichiometric amounts of the homogeneous acid or base. The reasons for this are that either there is a neutralization process during the course of the reaction and the acid/base catalyst is consumed or the reagent binds strongly to the reaction products being unable to effect a turnover cycle larger than 1. A paradigmatic example of the latter is the Friedel–Crafts acylation of aromatics using AlCl$_3$, as indicated in Fig. 12.6. In any case, when using homogenous acids and bases, the reaction work-up procedure requires the neutralization of these acids and bases to isolate the reaction products. It is in this step that a considerable amount of waste products are produced. The use of heterogeneous acid and base catalysis, in which the catalysts are in a different phase than the reagents and products is one way to improve the greenness of these general reaction types [82].

The vast majority of organic oxidations at the laboratory level or in the fine chemical industry also produce a large amount of wastes. Perusal of the current

Green organic reactions:

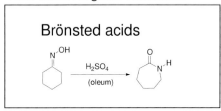

Non-Green organic reactions:

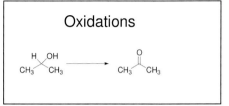

Fig. 12.6 Simplified classification of organic reactions, indicating which types require urgent attention to comply with the green chemistry principles.

organic chemistry textbooks [88] shows that all of the classical oxidation reactions, even though they give high product yields, are far from complying with the green chemistry principles, using hazardous or toxic chemicals, requiring volatile organic solvents and producing large amounts of toxic wastes. A green alternative to the conventional *stoichiometric* oxidation reactions is urgently needed. According to the green chemistry principles these novel oxidation processes must: (i) use tolerable oxidizing reagents, (ii) be undertaken in green solvents and media, (iii) be highly efficient and selective, and (iv) should not produce any noxious by-product.

12.5
Alcohol Oxidation in Organic Chemistry

Alcohol oxidation to form carbonylic compounds is one of the pivotal functional group transformations of organic chemistry [88]. Classical organic syntheses are

Table 12.2 List of some oxidizing reagents, the resulting reduced by-product and the corresponding percentage in oxygen able to be incorporated into the products.

Oxidizing reagent	Waste	Oxygen percentage
$KMnO_4$	Mn^{2+}/MnO_2	—
K_2CrO_4	Cr^{3+}	—
CH_3COOH	CH_3CO_2H	26
t-BuOOH	t-BuOH	27
ClO^-	Cl^-	30
H_2O_2	H_2O	46
O_2	H_2O	50

based on C—C bond formation by attack of organometallic reagents on carbonylic compounds to form alcohols, which subsequently can be transformed into a new organometallic reagent or into a carbonylic compound to continue the build-up of a more complex molecule. In this context, alcohol oxidation to aldehydes, ketones or carboxylic acids and derivatives are part of many synthetic routes [89, 90]. In addition, alcohol oxidation is also a process of large industrial importance, particularly in fine chemistry.

Most of the current methods for alcohol oxidation in general organic chemistry are not catalytic, but stoichiometric. By *stoichiometric* it is meant that equivalent amounts of an oxidizing reagent, such as transition metal oxides or halo-oxo acids, are needed to effect the oxidation, leading to the stoichiometric formation of wastes corrresponding to the reduced form of the oxidizing reagent. Other non-metallic oxidants that have been used as oxidizing reagents are halogens, peroxy and hydroperoxy compounds or even sulfoxides. The latter noxious reagents are transformed into equivalent amounts of sulfide in the selective Swern alcohol oxidation [91–93]. Table 12.2 shows some typical oxidizing reagents, indicating the resulting by-product and the percentage of oxygen content.

One of the principles of green chemistry is to substitute *stoichiometric* processes by *catalytic* processes. Application of this principle is particularly necessary in oxidation reactions, for which the ratio between kg of byproducts per kg of product formed is, in general, notably high. Of course, in a chemical reaction there is always the consumption of stoichiometric amounts of reagents. By transforming a *stoichiometric* into a *catalytic* process, is meant the use of unconventional, environmentally-friendly oxidizing reagents that do not readily react in a selective way with organic substrates unless suitable catalysts for the process are developed. One example of a catalytic oxidation reaction is the Meerwein–Ponndorf–Verley oxidation that is commonly carried out using Lewis acid catalysts (Scheme 2.7). Other examples are related to the used of hydrogen peroxide or organic hydroperoxides for the epoxidation of alkenes and hydroxylation of

Transition metals: MnO_4^-, MnO_2, $Cr_2O_7^{-2}$, CrO_4^{-2}, CrO_3, $Pb(OAc)_4$, Cu^+, Ag^+

Halooxo acids: IO_4^-, BrO_4^-, ClO_4^-, ClO^-

Swern reaction

Meerwein-Ponndorf-Verley disproportionation

Scheme 12.7 Some classical processes for alcohol oxidation.

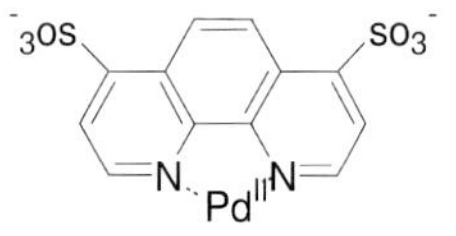

Scheme12.8 Reusable, water-soluble palladium complex that is able to promote the aerobic oxidation of alcohols [97].

aromatic compounds, the process being typically catalyzed by titanium-containing catalysts [94].

12.6
Related Precedents to the Use of Gold Catalysts for the Aerobic Oxidation of Alcohols

From the green chemistry point of view, there is no doubt that the optimum oxidizing reagent is molecular oxygen. Also to fulfill other conditions, particularly the absence of hazardous processes, it would be convenient to work at atmospheric or moderate oxygen pressure and at low temperatures. Concerning this point is has to be noted that most organic compounds form explosive mixtures with oxygen when a certain vapor composition is reached and that the use of high oxygen pressures is always a risk. Thus, mixtures below the flash or explosion point have to be used. It can be concluded that the ideal green conditions for alcohol oxidation would be the use of oxygen at atmospheric pressure in a suitable medium.

In recent years there have been important steps towards the development of these *ideal* aerobic oxidations. Thus, Mizuno and coworkers reported a new generation of ruthenium oxide that acts as a heterogeneous catalyst for the oxidation of alcohols [95, 96]. The main drawbacks of this process were, however, the need for high oxgyen pressure and the fact that the catalyst is moisture sensitive. Both disadvantages constituted a serious limitation of the process.

Sheldon and coworkers have reported a palladium complex that can effect the catalytic oxidation of alcohols by oxygen in water, the complex being reusable and of general applicability (Scheme 12.8) [97]. While the use of water in this reaction constitutes a step forward and has many advantages from the environmental point of view, it still poses problems due to the lack of water solubility of many alcohols. An additional problem arises from the need for high oxygen pressure to effect alcohol oxidation.

A milestone in the aerobic oxidation of alcohols was the report of Kaneda and coworkers describing the use of palladium supported on hydroxyapatite as a highly active and reusable solid catalyst for the atmospheric-pressure aerobic oxidation of alcohols [98]. Hydroxyapatite is a synthetic calcium phosphate phase obtained reproducibly by precipitation starting from soluble calcium nitrate and ammonium hydrogenophosphate salts. The hydroxyapatite-supported palladium was water insensitive and exhibited unprecedentedly high turnover numbers for alcohol oxidation, particularly high for 1-phenylethanol. The catalyst showed a general activity for primary, secondary, aliphatic and aromatic alcohols.

Triggered by Haruta's report on gold-catalyzed CO oxidation, there was an interest in determining whether or not the exceedingly high catalytic activity of gold nanoparticles could also serve to develop alcohol oxidation catalysts. Initial work using gold catalysts was carried out by the groups of Rossi and Hutchings working in basic aqueous conditions [77–79, 99–107]. The preferred supports for the colloidal gold particles were inorganic oxides (TiO_2, Fe_3O_4, CeO_2) and activated carbons. Metal oxides have been widely used as support for metals. In the particular case of gold, it has been reported that oxides in which the metal can have redox properties may be more suitable as the support for gold-catalyzed oxidations given that the reaction mechanisms should imply cycles involving different metal oxidation states. In this regard, Fe_3O_4 and TiO_2 have been used as supports for colloidal gold. Activated carbons, on the other hand, combine a very large surface area on which noble metal nanoparticles can be dispersed with a remarkable adsorption capacity for gases, including oxygen. Both properties, together with the ample use of activated carbons as supports for Pt, Pd and other noble metals, makes activated carbons an obvious choice as support for colloidal gold.

12.7
Gold Nanoparticles Supported on Ceria Nanoparticles

After the observation that gold nanoparticles supported on ceria nanoparticles are an extremely efficient catalyst for the room-temperature CO oxidation [6, 63, 69], a logical move was to export these results to the oxidation of alcohols. Previously we have described that ceria nanoparticles have some distinctive properties with respect to bulk micrometric ceria, the most obvious one being a deviation from the CeO_2 stoichiometry. In other words, nanoparticulated ceria contains oxygen vacancies at the surface of the crystals and the presence of Ce in the +III oxidation state. These properties of the support are very appropriate for developing a catalytic system for the aerobic alcohol oxidation. On the one hand, nanoparticulated ceria can physisorb molecular oxygen at the surface oxygen vacancies. On the other hand, non-fully saturated cerium atoms can behave as Lewis sites against alcohols.

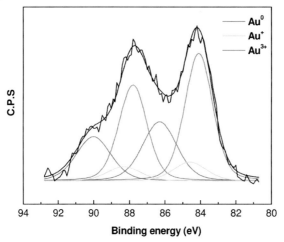

Fig. 12.7 XPS spectrum of npAu/npCeO$_2$ and a deconvolution showing the contribution of individual oxidation states (reproduced from Ref. [75]).

More importantly, nanoparticulated ceria can interact with supported gold at the interface between both solids, stabilizing positive charge density on the gold nanoparticles and positive gold atoms.

The presence of positive gold atoms at the interface with ceria can be demonstrated by XPS of npAu/npCeO$_2$ (the acronym np refers to nanoparticle) in which the broad peak corresponding to the Au $^4f_{7/2}$ core level at around 85 eV can be deconvoluted to show the contribution of three bands at 86.4, 84.7 and 84 eV which correspond to Au^{3+}, Au$^+$, Au^0 species, respectively [75]. Quantification based on the intensity of the XPS data indicates that over 10% of gold atoms have a positive oxidation state (Au$^+$). Figure 12.7 shows the XPS spectrum of npAu/npCeO$_2$ from which these values were obtained.

Moreover, IR monitoring of CO adsorbed onto npAu/npCeO$_2$ as a probe molecule reveals the presence of a band at 2155 cm^{-1} attributable to the interaction of C=O with positive gold atoms. Figure 12.8 shows one of these IR spectra recorded from npAu/npCeO$_2$.

npAu/npCeO$_2$ was found to be an extremely active catalyst for the solventless, atmospheric-pressure, aerobic oxidation of alcohols to carbonyl compounds. Table 12.3 summarizes some of the results that have been reported. Under these conditions, secondary alcohols are oxidized to ketones with essentially complete selectivity. Benzylic alcohols form the corresponding benzaldehydes. Vainillin and salicylaldehyde can be obtained in high yields using npAu/CeO$_2$ as catalyst.

Primary alcohols are more reluctant to undergo oxidation to aldehydes under solventless conditions. Typically, esters with small quantities of carboxylic acids are observed under these conditions. It has been proposed that esters derive from the oxidation of the corresponding hemiacetal, rather than from the esterification of the carboxylic acid (Scheme 12.9). This proposal is based on the observation

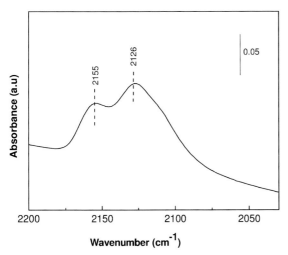

Fig. 12.8 FT-IR spectra of the CO region recorded after adsorbing CO onto npAu/npCeO$_2$ and outgassing under 10^{-6} Pa at 25 °C for 1 h. Spectrum taken from Ref. [75].

Table 12.3 Solventless aerobic alcohol oxidation catalyzed by npAu/CeO$_2$.

Substrate	Conversion (%)	Product	Selectivity (%)
3-octanol	97	3-octanone	>99
3-octanol	89	3-octanone	96
sec-phenylethanol	92	acetophenone	97
2,6-dimethylcyclohexanol	78	2,6-dimethylcyclohexanone	94
3,4-dimethoxybenzyl alcohol	73	3,4-dimethoxybenzaldehyde	83
3-phenyl-1-propanol	70	3-phenylpropyl-3-phenylpropanoate	98

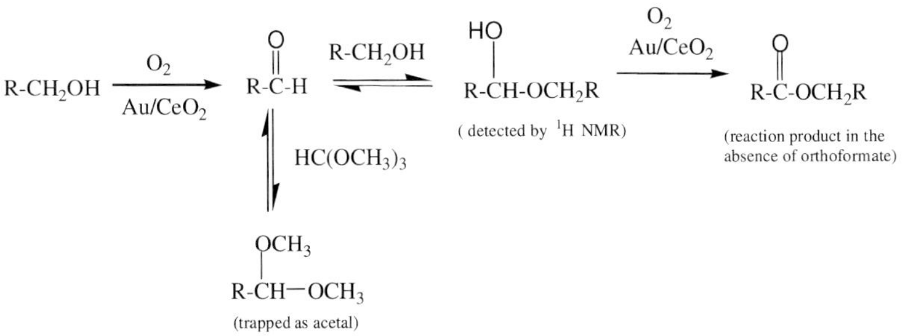

Scheme 12.9 Proposal for the formation of esters in the aerobic oxidation of primary aliphatic alcohols.

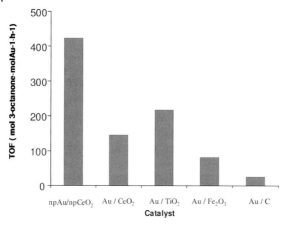

Fig. 12.9 Comparison of the turnover frequencies (moles of product per moles of gold per hour) for the solventless aerobic oxidation of 3-octanone.

that the hemiacetal is formed in significant amounts when an aliphatic aldehyde is dissolved in an excess of alcohol and also aldehydes can be intercepted as their corresponding dimethyl acetals and the formation of esters inhibited when the reaction is carried out in the presence of trimethyl orthoformate. According to Scheme 12.9, trans-acetalization of the aliphatic aldehyde, as it is being formed, by the trimethyl orthoformate will impede the formation of hemiacetal and its subsequent oxidation.

Comparison with other gold supported catalysts revealed a positive influence of nanoparticulated ceria to the success of the catalytic reaction. Thus, selecting the oxidation of 3-octanol as a reaction test, when gold was supported on micrometric size ceria, the catalytic activity decreased considerably. Also gold supported on titania, iron oxide and activated carbon exhibited lower activity than npAu/npCeO$_2$. Figure 12.9 shows a diagram comparing the activity of npAu/npCeO$_2$ with other related supported gold catalysts.

Although the high activity of npAu/npCeO$_2$ has also been confirmed with other alcohols and reaction conditions, caution has to be taken when absolute catalytic activity data is given or discussed. An illustrative case is that of gold nanoparticules supported on activated carbon whose catalytic activity is dramatically increased when the reaction is carried out in water in the presence of base. Probably in this case the reaction mechanism that operates and the species involved change in an aqueous medium. npAu/npCeO$_2$ has become recently commercially (www.upv.es/itq).

12.8
Gold vs. Palladium Catalysts for the Aerobic Oxidation of Alcohols

Palladium supported on hydroxyapatite is the benchmark heterogeneous catalyst with respect to which the catalytic activity for the aerobic oxidation of alcohols

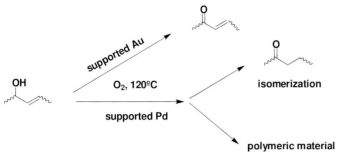

Scheme 12.10 Possible processes competing during alcohol oxidation.

should be compared [98]. Recently, it has been reported that a core/shell gold(core)/palladium(shell) catalyst supported on titania is an extremely active catalyst for the aerobic oxidation of primary aromatic and aliphatic alcohols to aldehydes [108]. As commented previously, turnover frequencies as high as 86 500 mol of product (mol of palladium)$^{-1}$h^{-1} have been reported for benzyl alcohols and it was demonstrated that the presence of a gold core is crucial to obtain this high activity [108].

However, from the organic point of view, it is of interest to develop a general catalyst that can be of use for a wide variety of alcohols containing other functional groups. The catalyst does not necessarily need to be the most active for every alcohol, but has to exhibit sufficient activity for a broad range of alcohols and has to be chemoselective for alcohol oxidation, irrespective of the presence of other oxidizable functional groups.

To study the selectivity for alcohol oxidation in the presence of other functional groups, allylic alcohols are the substrates of choice since they can undergo a variety of different reactions on each functional group and in addition they can interact with each other undergoing hydrogen rearrangement (Scheme 12.10).

It has been reported that palladium on hydroxyapatite effects selectively the oxidation of cinnamyl alcohol in trifluorotoluene [98]. This selectivity is reproducible when working in the reported solvent, conditions and substrate concentrations [98]. However, the situation is completely different under solventless conditions. Most probably, the high concentration when the oxidation is carried out in the absence of solvent makes the selectivity of the process change by favoring other processes sensitive to concentration (polymerization) and by saturating the capability of the catalytic sites to undergo reoxidation. Thus, when the noble metal to substrate ratio is of the order of 10^{-4}, most of the catalysts sites are in the reduced form and their reoxidation by oxygen is the controlling step. It has been found that, for solventless oxidations, gold catalysts exhibit a unique chemoselectivity as compared to palladium and gold/palladium catalysts, thus, rendering supported gold nanoparticles the catalyst of choice in organic chemistry due to its broad and general applicability. Table 12.4 summarizes some of the results that have been reported.

As can be seen in Table 12.4, using palladium or gold/palladium as catalysts there is a significant percentage of isomerization (giving rise to the saturated

Table 12.4 Results of the solvent-less, atmospheric-pressure aerobic oxidation of alcohols at 120 °C. PO_2 1 atm (35 mL min^{-1} flow), in the presence of heterogeneous gold or palladium catalysts. Au, Pd and AuPd catalysts refer to npAu/npCeO$_2$, Pd/hydroxyapatite and AuPd to core/shell Au(core)Pd(shell) supported on TiO$_2$ (data taken from Ref. [109]).

Substrate	Catalyst	Conversion (%) (Time) (h)	Unsat. carbonyl (%)[a]	Isomer (%)[b]	Hyd (%)[c]	Polym (%)[d]
1-octen-3-ol	Au	>99 (5)	95	—	—	—
	Pd	>99 (0.7)	49	49	2	—
	AuPd	>99 (0.3)	26	73	1	—
1-hepten-3-ol	Au	>99 (14)	93	—	—	—
	Pd	>99 (2)	24	74	2	—
	AuPd	>99 (2)	12	77	1	—
trans-carveol	Au	88 (21)	98	—	—	—
	Pd	98 (6)	40	—	—	59
3-octen-2-ol	Au	96 (21)	94	—	—	—
	Pd	98 (0.5)	19	45	—	36
2-octen-1-ol[e]	Au	51 (3)	58	—	—	37
	Pd	67 (1)	15	—	—	76
	Au	56 (1)	72	—	—	26

ketone) and polymerization. Scheme 12.10 illustrates the side reactions observed in the oxidation of allylic alcohols. In contrast, npAu/npCeO$_2$ gives rise to the formation of α,β-unsaturated ketones in complete chemoselectivity at high conversions. Minimum yields of 90% have been reported, except for 2-octen-1-ol for which the reaction has to be performed in toluene. npAu/npCeO$_2$ can be reused after the reaction, by filtration of the solid, washing with 1 M aqueous NaOH solution and drying. This chemoselectivity seems to be inherent to gold nanoparticles since npAu/TiO$_2$, although less active, behaves analogously to npAu/npCeO$_2$.

It has been proposed that the differences in the chemoselectivity of gold and palladium catalysts arises from the stability and steady-state concentration of metal hydrides (Pd-H and Au-H) during the oxidation under solventless conditions. Metal hydrides are generally accepted as reaction intermediates and in some cases detected during alcohol oxidation [75, 110]. It is accepted that the metal hydrides should be reoxidized by oxygen and their steady state concentration must be low since otherwise C=C hydrogenation and isomerization can occur. Apparently during the solventless oxidations, Pd-H does not undergo a sufficiently fast oxidation and its concentration is not negligible, causing problems with side reactions. This hypothesis was tested by performing an experiment in which 1-octen-3-ol was allowed to equilibrate in the presence of a N$_2$ atmosphere containing a low H$_2$ proportion (10%) using palladium and gold catalysts. As proposed in the case of alcohol oxidation, that from another perspective corresponds to alcohol dehydro-

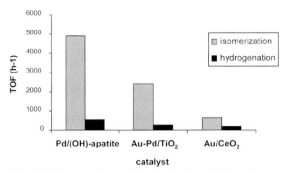

Fig. 12.10 Comparative catalytic activity of Au/CeO$_2$ and Pd/(OH)apatite to promote C=C double bond isomerization of 1-octen-3-ol in a N$_2$/H$_2$ (90/10) atmosphere.

genation, hydrogen gas should also give rise to stationary concentrations of metal hydrides, thus allowing determination of the relative reactivity of palladium and gold catalysts in the absence of oxygen. It was observed that under these special experimental conditions characterized by a controlled low concentration of hydrogen and metal hydrides, Pd/(OH)apatite and Au–Pd/TiO$_2$ promote to a significant extent C=C double bond isomerization and C=C hydrogenation. In contrast, the activity of npAu/npCeO$_2$ reacting with 1-octen-3-ol in the presence of hydrogen was significantly lower. Figure 12.10 summarizes the relative activity of palladium and gold catalysts in the presence of hydrogen. The outcome of this experiment is that gold has a lower tendency to form hydrides and/or, if formed in similar concentrations, they are less reactive than palladium hydrides.

12.9
Reaction Mechanism of Gold-catalyzed Alcohol Oxidations

The currently most widely accepted reaction mechanism for alcohol oxidation catalyzed by supported gold nanoparticles has been proposed by analogy with the assumed reaction mechanism for alcohol oxidation in palladium and platinum metals taking into consideration the kinetic data and mechanistic information obtained from gold-catalyzed CO oxidation [110]. Scheme 12.11 illustrates a reasonable mechanistic proposal.

The first step consists in the formation of a metal alcoholate by reaction of positive gold atoms with the free alcohol or the alcoholate in solution. Alternatively, the free alcohol or the alcoholate can interact with other Lewis acid sites present on the support (Scheme 12.12). Thus, the need for nanoparticle size to observe the catalytic activity will derive from the requirement of the presence of a sufficiently large population of surface Lewis acid gold atoms. Positive gold atoms are suitable species to act as Lewis acid sites. The support can also cooperate in the reaction mechanism by binding the alcoholate to other Lewis sites located on the

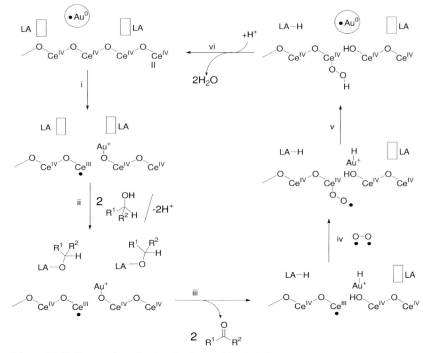

Scheme 12.11 Proposed mechanism for the aerobic alcohol oxidation.

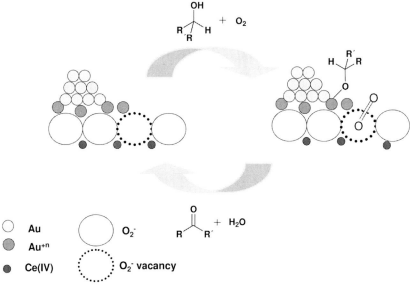

Scheme 12.12 Pictorial representation of the formation of metal-alcoholate at the surface of ceria supported gold nanoparticles.

support next to gold atoms. Alcoholates, due to their stronger basicity, interact more strongly with the positive sites than the corresponding neutral alcohols. For those reactions carried out in aqueous media and considering that water is competing with the alcohol for the Lewis sites, the presence of strong alkali conditions is usually necessary to promote the oxidation by giving rise to a sufficient concentration of alcoholate.

After formation of the metal alcoholate, hydride transfer from the O–C–H to a metal atom forming a ketone interacting with the gold and a metal hydride would occur. This is apparently the rate determining step of the alcohol oxidation process. Isotopic labelling using deuterium-labelled 1-D-benzyl alcohol has shown a moderate kinetic isotopic effect that indicates that the hydrogen at this position is being transferred in the rate determining step.

In the rate-determining step the C–H bond breaking and the metal–O bond breaking to give the carbonyl group and a metal hydride can be synchronous or can be stepwise. While in the first case, the carbon atom undergoing hydride transfer should be neutral at any instant and rather insensitive to the influence of polarity effect, if the C–H bond breaking progresses faster that the bond formation, both processes being not synchronous, a large density of positive charge will develop at the carbon atom of the carbonyl group. Therefore, in this case, the presence of electron donor substituents will increase the reaction rate by providing electron density in this transition state. Experimental data have shown that this is the case and that the initial reaction rate for para-substituted benzylic alcohols varies linearly with the Hammett constant of the substituent [111].

After its formation, the ketone is desorbed and the metal hydride is restored to the initial metal site by oxygen forming water. The actual details of this complex transformation are unknown in many cases and, certainly, this point deserves a much deeper understanding for some gold supported catalysts. The reason for this is that in some of the current mechanistic proposals, there are supports that are supposed to play a role that reasonably cannot occur with other supports. Thus, for instance, in the case of nanoparticulated ceria where, as commented earlier, oxide vacancies at the solid surface are present, molecular oxygen is initially physisorbed on a lattice oxygen vacancy, next to a Ce(III) defect. The physisorption of oxygen occupying solid lattice positions was first proposed by Bond to rationalize the mechanism of CO oxidation [73, 74]. Oxygen physisorption is converted into metal peroxide formation concurrently to the hydride transfer from a neighboring metal-hydride site. In this way, a metal hydride is converted at the support defects to a metal hydroperoxide, that subsequently decomposes to water thus recovering the initial solid defective site.

In support of this proposal, it was found by *in situ* IR spectroscopy that the formation of acetone takes place upon adsorption of isopropanol on npAu/npCeO$_2$ in the absence of oxygen. At this point, the presence of a band attributable to metal hydride is observed on the solid surface. Upon admission of oxygen, the appearance of water is recorded and the metal-hydride band disappears. Figure 12.11 shows a selection of IR spectra to illustrate the spectroscopic observations supporting the alcohol oxidation of npAu/npCeO$_2$.

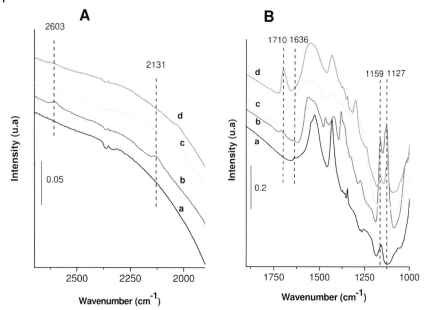

Fig. 12.11 Two regions of the FTIR spectra of the npAu/npCeO₂ catalyst of (a) activated sample; (b) after absorption of isopropanol; (c) after evacuation and consequently insertion of oxygen. In part A, the 2131 cm⁻¹ band has been attributed to metal hydride. In part B the 1710 cm⁻¹ bands have been attributed to the carbonyl group of acetone, the 1636 cm⁻¹ to the bending of water and the 1159 and 1127 cm⁻¹ bands to isopropanol.

It is obvious, that given the role of ceria in the above proposal, a clear influence of the support on the ability of gold nanoparticles to oxidize alcohols should be observed. This is the case under certain conditions, such as solventless oxidations, for which the catalytic activity depends on the support. One specific case that deserves special attention is gold supported on activated carbon (npAu/C). Here the reaction mechanism should be completely different, particularly with respect to the pathways in which the gold hydride is converted to water and the gold atom restored. A good evidence for the change in the reaction mechanism for npAu/C is that while this catalyst is rather inactive for solventless oxidations of alcohols to aldehydes, ester or ketones, npAu/C becomes among the most active catalysts for alcohol oxidations to carboxylates or ketones in aqueous media in the presence of strong bases [77, 78, 102, 104]. This remarkable change in catalytic activity as well as the influence of water and base requires additional understanding.

The role of the particle size of gold specifically in the catalytic activity for the aerobic oxidation of alcohols (as opposed to CO oxidation) has been addressed by Tsukuda and coworkers, who prepared two colloidal gold nanoparticles of 1.3 ± 0.3 and 9.5 ± 1.0 nm diameter stabilized by polyvinylpyrrolidone [112]. Particles with different sizes were obtained by rapid reduction of $AuCl_4^-$ either with $NaBH_4$ or Na_2SO_3. Data of the initial reaction rate, both uncorrected and corrected by the

Table 12.5 Results of the gas phase air oxidation of volatile alcohols catalyzed by gold nanoparticles, 1 wt.% supported on Aerosil (200 m^2 g^{-1}). The selectivity in all cases was 100%. (Data taken from Ref. [117].)

Alcohol	Temperature (°C)	Conversion (%)
1-propanol	250	27
	300	44
1-butanol	250	51
1-pentanol	250	23
	300	29
phenylcarbinol	250	50
2-propanol	100	69
2-butanol	120	34
	150	64
3-pentanol	120	85
	150	97

different specific surface area (smaller gold particles have larger surface area), show that the catalyst with the smaller gold particles has higher activity than that of the larger particle size by a factor of 384 [112]. These results show again that the particle size plays a key role in determining the catalytic activity.

Also in the case of npAu/TiO$_2$, preparation of a series of samples differing in the average gold particle size has allowed observation of a clear dependence of the initial reaction rate on the gold particle size, these results being in line with those reported by Haruta in the case of CO oxidation [69, 113–115].

Aerobic oxidation of simple volatile aliphatic alcohols, including ethanol [116], can be carried out conveniently in the gas or vapor phase. Gold nanoparticles (1%) supported on silica have been found to be a convenient catalyst [117]. Volatile primary and secondary aliphatic alcohols can be oxidized to their carbonyl compounds in the gas phase using air at atmospheric pressure in a fix-bed tubular reactor (Table 12.5) [117].

12.10
Influence of the Solvent on Aerobic Oxidation

Although gas phase oxidations can be important in the case of simple alcohols on the large scale, the most widely used oxidation processes for alcohols work in the liquid phase. The reasons for this are the low vapor pressure of alcohols due to the strong intermolecular interactions by hydrogen bonding and the fact that complex multifunctional molecules can undergo, at high temperatures, undesirable thermal reactions. There are several liquid phase oxidation procedures, the

Table 12.6 Aerobic alcohol oxidation in basic water catalyzed by npAu/CeO$_2$.

Substrate	Conversion (%)	Product	Selectivity (%)
vainillin alcohol	96	vainillin	98
2-hydroxybenzyl alcohol	>99	2-hydroxybenzaldehyde	87
3,4-dimethoxybenzyl alcohol	>99	3,4,-dimethoxybenzoic acid	>99
cinnamyl alcohol	>99	cinnamic acid	98
n-hexanol	>99	hexanoic acid	>99
Sec-phenylethanol	>99	acetophenone	51

simplest being the use of solventless conditions [75]. This experimental procedure is, however, limited to liquid or low-melting point alcohols available in multigram quantities. In other cases, solvents are needed. Considering the green chemistry principles that require the avoidance of volatile organic solvents, water appears as a solvent of choice in some cases. Gold-catalyzed aerobic oxidations in water require strong alkaline conditions, catalysis does not occur under neutral or acid conditions. The basic conditions are responsible for the fact that, in the case of primary alcohols, the reaction products are the carboxylate salts, rather than the corresponding aldehyde. Even, obtaining the carboxylic acid requires a neutralization step. Table 12.6 summarizes some of the results that have been reported for the gold catalyzed aerobic oxidation in water. Aqueous acid conditions are only able to effect oxidation of aldehydes to carboxylic acids, but they fail to promote alcohol oxidation [76]. On the other hand, the presence of bases may represent an important limitation since some of the carbonylic compounds may not be stable at these pH values and could undergo aldolic condensations, Cannizzaro disproportionation, decarboxylation and other side reactions [110].

Water is not a good solvent for alcohols that contain a high C/O atomic ratio. For this reason, most liquid-phase alcohol oxidations are still carried out using organic solvents [81, 109]. These conditions are in fact the most general since they apply to solid, water-insoluble alcohols that may not be available in multigram quantities. Supported gold nanoparticles are also efficient catalysts under these conditions, one notable issue being that no strong alkaline conditions are required in organic solvents and, therefore, aldehydes can be the final reaction products in the oxidation of primary aliphatic alcohols. Gold supported on nanocrystalline ceria has also been found to be a general and highly active catalyst under these conditions [75]. Table 12.6 summarizes some of the results that have been obtained.

Most of the early work in gold-catalyzed aerobic alcohol oxidation was carried out in water in the presence of bases. Sodium hydroxide in equimolar amounts with respect to the substrate was often used as a base. In these studies using aqueous media, the preferred substrates were glycols, glycerol and polyols that were selectively oxidized to hydroxyacids. Thus, Hutchings and coworkers found that glycerol is transformed to glyceric acid with complete selectivity at high con-

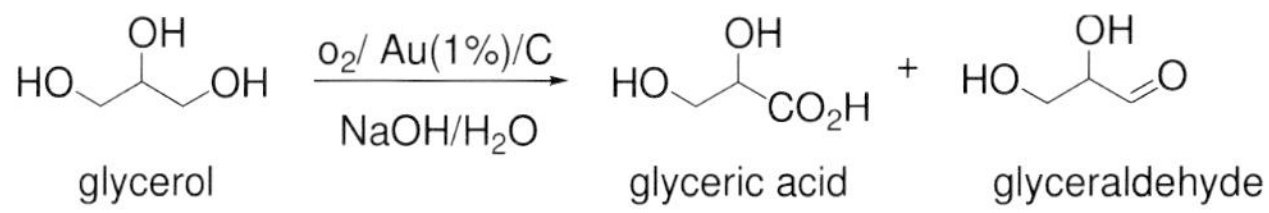

Scheme 12.13 Aerobic oxidation of glycerol in strong basic media.

versions using Au 1 wt.% supported on active carbon or graphite (Scheme 12.13) [118, 119]. Platinum as metal was found to be much less selective than gold, since the former gives rise to a significant concentration of glyceraldehyde in addition to glyceric acid.

No oxidation of glycerol was observed in the absence of NaOH and the need for a high NaOH/substrate ratio was established. Analogous results were obtained in the oxidation of phenylethane-1,2-diol and diethyleneglycol in which NaOH/substrate molar ratios between 1 and 4 were used to increase the selectivity of the oxidation towards the desired hydroxyacid [77, 102]. It is clear that it would be of great interest to understand the role of the base in gold catalysis in an aqueous medium, since it represents a clear limitation at this moment. One possibility is that the OH^- forms the alcoholate necessary to bind with positive gold atoms. However, the fact that free alcohols bind to gold particles in organic solvents in the absence of base and the lack of influence of the support (or at least the remarkable enhancement of the activity of gold supported on activated carbon) points to a more deep change in the reaction mechanism in aqueous media.

Other diols, polyols [77], amino alcohols [77] and glucose [77, 80, 120] have also been oxidized with remarkable selectivity to hydroxy- and amino- acids by oxygen in strong basic aqueous media using gold nanoparticles supported on carbon and metal oxides as catalysts (Schemes 12.14 and 12.15).

12.11
Conclusions and Future Prospects

It can be foreseen that aerobic oxidations of alcohols will replace completely other environmentally unfriendly oxidation processes in the near future. General catalysts for this purpose will be commercially available and synthetic organic chemists will use this reaction in place of the currently existing stoichiometric alcohol oxidation. In the quest for this general oxidation catalyst for aerobic alcohol oxidation, there are two alternatives, starting from noble metal complexes or gold nanoparticles. It is likely that both alternatives will develop further to the point that catalysts of a broad applicability will be independently obtained. It seems, however, that for gold metal, nanoparticles are going to be preferred, while this situation may be different for palladium. However, while reusability may not be an issue on the laboratory scale, for industrial applications it is a crucial point that will favor heterogeneous catalysis using nanoparticles. Also the uncertainty and variability in the price of precious metals can certainly influence future research.

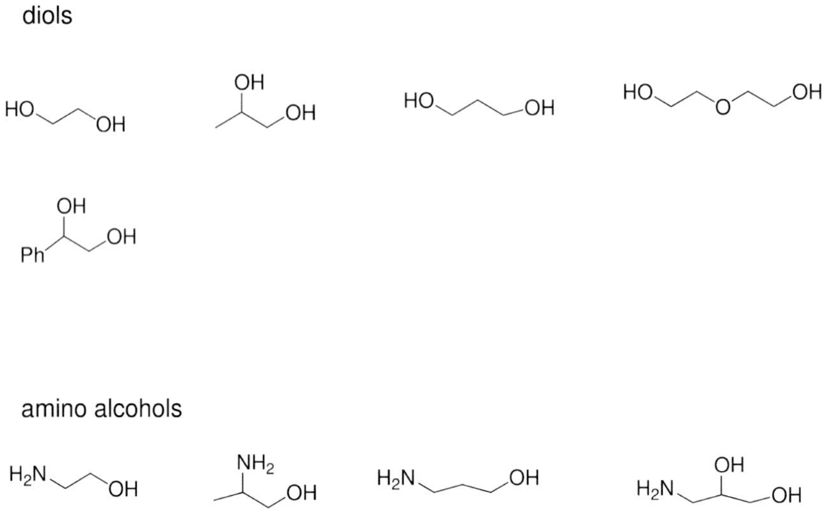

diols

amino alcohols

Scheme 12.14 Bi-/multi- functional water soluble alcohols that can be oxidized by oxygen in the presence of gold catalysts.

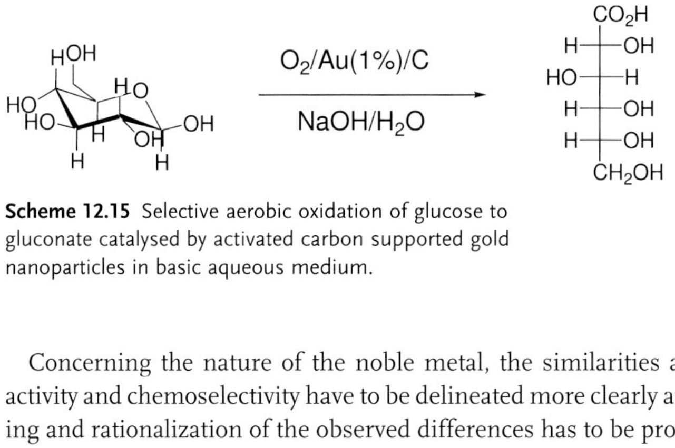

Scheme 12.15 Selective aerobic oxidation of glucose to gluconate catalysed by activated carbon supported gold nanoparticles in basic aqueous medium.

Concerning the nature of the noble metal, the similarities and differences in activity and chemoselectivity have to be delineated more clearly and an understanding and rationalization of the observed differences has to be provided. The activity of alloys and structure core/shell particles also has to be determined more precisely for a wide range of metal compositions. Given the large number of possibilities, systematic high-throughput techniques will be of great help in addressing optimization of the alloy catalytic activity.

The importance of the solid support and the way in which certain supports can influence the outcome of the catalysis also needs deeper understanding and rationalization. The interplay between media, conditions and support must also be delineated. This issue is certainly related to the operating reaction mechanism and the fact that certain pathways compete or one may prevail, depending on the conditions. The apparent contradiction about the lack of influence of the support on oxidation in aqueous media compared to solventless conditions deserves specific studies.

Aerobic oxidation is not going to be limited to alcohol oxidation. Apparently we are closer to finding suitable catalysts for alcohol oxidation, but oxidation of alkenes, aromatic hydrocarbons, alkylaromatics, imines, amines, sulfur compounds etc. will require much work in the forthcoming years. In some cases, the presence of radical initiators, even though in minor quantities, may serve to promote the oxidation catalyzed by noble metal nanoparticles, and gold in particular [56, 108]. Thus, there is no doubt that the next years will witness exciting developments in the field of metal nanoparticles for aerobic oxidation.

Acknowledgments

Financial support by the Spanish DGI (project CTQ2006-06859) is gratefully acknowledged.

References

1 W. Hermann (Ed.), *Synthetic Methods of Organometallic and Inorganic Chemistry, Volume 9: Transition Metals Part 3,* Thieme, Stuttgart, **2000**.

2 P. N. Rylander, *Organic Chemistry, Vol. 28: Organic Syntheses with Noble Metal Catalysts,* Academic Press, New York, **1973**.

3 M. Haruta, N. Yamada, T. Kobayashi, S. Iijima, *J. Catal.* **1989**, *115*, 301.

4 M. Haruta, T. Kobayashi, H. Sano, N. Yamada, *Chem. Lett.* **1987**, 405.

5 M. Haruta, *Stud. Surf. Sci. Catal.* **1997**, *110*, 123.

6 D. Cunningham, S. Tsubota, N. Kamijo, M. Haruta, *Res. Chem. Intermed.* **1993**, *19*, 1.

7 M.-C. Daniel, D. Astruc, *Chem. Rev.* **2004**, *104*, 293.

8 D. Astruc, J.-C. Blais, M.-C. Daniel, S. Gatard, S. Nlate, J. Ruiz, *Compt. Rend. Chim.* **2003**, *6*, 1117.

9 H. Haekkinen, B. Yoon, U. Landman, X. Li, H.-J. Zhai, L.-S. Wang, *J. Phys. Chem. A* **2003**, *107*, 6168.

10 R. M. Barrer, *Zeolites and Clay Minerals as Sorbents and Molecular Sieves,* Academic Press, London, **1978**.

11 H. van Bekkum, E. M. Flanigen, J. C. Jansen, *Introduction to Zeolite Science and Practice,* Elsevier, Amsterdam, **1991**.

12 D. W. Breck, *Zeolite Molecular Sieves: Structure, Chemistry and Use,* John Wiley and Sons, New York, **1974**.

13 W. M. Meier, D. H. Olson, *Atlas of Zeolite Structure Types,* Butterworths, London, **1992**.

14 J. S. Beck, J. C. Vartuli, W. J. Roth, M. E. Leonowicz, C. T. Kresge, K. D. Schmitt, C. T.-W. Chu, D. H. Olson, E. W. Sheppard, S. B. McCullen, J. B. Higgins, J. L. Schlenker, *J. Am. Chem. Soc.* **1992**, *114*, 10834.

15 C. T. Kresge, M. E. Leonowicz, W. J. Roth, J. C. Vartuli, J. S. Beck, *Nature* **1992**, *359*, 710.

16 C. Zapilko, R. Anwander, *Chem. Mater.* **2006**, *18*, 1479.

17 R. Anwander, I. Nagl, M. Widenmeyer, G. Engelhardt, O. Groeger, C. Palm, T. Roeser, *J. Phys. Chem. B* **2000**, *104*, 3532.

18 R. Anwander, C. Palm, J. Stelzer, O. Groeger, G. Engelhardt, *Stud. Surf. Sci. Catal.* **1998**, *117*, 135.

19 P. McMorn, G. J. Hutchings, *Chem. Soc. Rev.* **2004**, *33*, 108.

20 J. M. Kisler, M. L. Gee, G. W. Stevens, A. J. O'Connor, *Chem. Mater.* **2003**, *15*, 619.

21 M. Park, S. Komarneni, *Microporous Mesoporous Mater.* **1998**, *25*, 75.

22 A. Corma, M. Domine, J. A. Gaona, J. L. Jorda, M. T. Navarro, F. Rey, J. Perez-

Pariente, J. Tsuji, B. McCulloch, L. T. Nemeth, *Chem. Commun.* **1998**, 2211.

23 X. S. Zhao, G. Q. Lu, *J. Phys. Chem. B* **1998**, *102*, 1556.

24 A. Corma, P. Atienzar, H. Garcia, J.-Y. Chane-Ching, *Nature Mater.* **2004**, *3*, 394.

25 J. Turkevich, P. C. Stevenson, J. Hillier, *Discuss. Faraday Soc.* **1951**, *No. 11*, 55.

26 S. Demirci, B. V. Enustun, J. Turkevich, *J. Phys. Chem.* **1978**, *82*, 2710.

27 J. Turkevich, G. Garton, P. C. Stevenson, *J. Colloid Sci.* **1954**, 26.

28 S. Praharaj, S. K. Ghosh, S. Nath, S. Kundu, S. Panigrahi, S. Basu, T. Pal, *J. Phys. Chem. B* **2005**, *109*, 13166.

29 S. K. Ghosh, S. Nath, S. Kundu, K. Esumi, T. Pal, *J. Phys. Chem. B* **2004**, *108*, 13963.

30 J. Fink, C. J. Kiely, D. Bethell, D. J. Schiffrin, *Chem. Mater.* **1998**, *10*, 922.

31 S. Y. Kang, K. Kim, *Langmuir* **1998**, *14*, 226.

32 H. Zhu, C. Liang, W. Yan, S. H. Overbury, S. Dai, *J. Phys. Chem. B* **2006**, *110*, 10842.

33 W. Yan, S. M. Mahurin, B. Chen, S. H. Overbury, S. Dai, *J. Phys. Chem. B* **2005**, *109*, 15489.

34 S. Ivanova, C. Petit, V. Pitchon, *Appl. Catal., A: Gen.* **2004**, *267*, 191.

35 R. Zanella, S. Giorgio, C. R. Henry, C. Louis, *J. Phys. Chem. B* **2002**, *106*, 7634.

36 L. Guczi, D. Horvath, Z. Paszti, G. Peto, *Catal. Today* **2002**, *72*, 101.

37 I. Hussain, S. Graham, Z. Wang, B. Tan, D. C. Sherrington, S. P. Rannard, A. I. Cooper, M. Brust, *J. Am. Chem. Soc.* **2005**, *127*, 16398.

38 H. Hiramatsu, F. E. Osterloh, *Chem. Mater.* **2004**, *16*, 2509.

39 A. Corma, H. Garcia, A. Leyva, *J. Mol. Catal. A: Chem.* **2005**, *230*, 97.

40 A. Corma, D. Das, H. Garcia, A. Leyva, *J. Catal.* **2005**, *229*, 322.

41 A. Corma, H. Garcia, A. Leyva, *J. Catal.* **2006**, *240*, 87.

42 R. Narayanan, M. A. El-Sayed, *J. Catal.* **2005**, *234*, 348.

43 C. C. Cassol, A. P. Umpierre, G. Machado, S. I. Wolke, J. Dupont, *J. Am. Chem. Soc.* **2005**, *127*, 3298.

44 R. B. Bedford, U. G. Singh, R. I. Walton, R. T. Williams, S. A. Davis, *Chem. Mater.* **2005**, *17*, 701.

45 M. T. Reetz, J. G. de Vries, *Chem. Commun.* **2004**, 1559.

46 A. Corma, H. Garcia, A. Leyva, *Tetrahedron* **2005**, *61*, 9848.

47 P. D. Stevens, G. Li, J. Fan, M. Yen, Y. Gao, *Chem. Commun.* **2005**, 4435.

48 C. Croix, A. Balland-Longeau, H. Allouchi, M. Giorgi, A. Duchene, J. Thibonnet, *J. Organomet. Chem.* **2005**, *690*, 4835.

49 M. Ferrer, M. Mounir, L. Rodriguez, O. Rossell, S. Coco, P. Gomez-Sal, A. Martin, *J. Organomet. Chem.* **2005**, *690*, 2200.

50 M. Contel, D. Nobel, A. L. Spek, G. van Koten, *Organometallics* **2000**, *19*, 3288.

51 Y. Fuchita, H. Ieda, S. Wada, S. Kameda, M. Mikuriya, *J. Chem. Soc., Dalton Trans.* **1999**, 4431.

52 A. L. Balch, M. M. Olmstead, J. C. Vickery, *Inorg. Chem.* **1999**, *38*, 3494.

53 K.-H. Wong, K.-K. Cheung, M. C.-W. Chan, C.-M. Che, *Organometallics* **1998**, *17*, 3505.

54 M. B. Dinger, W. Henderson, *J. Organomet. Chem.* **1998**, *557*, 231.

55 T. Y. V. Baukova, L. G. Kuz'mina, N. Y. A. Oleinikova, D. A. Lemenovskii, A. L. Blumenfel'd, *J. Organomet. Chem.* **1997**, *530*, 27.

56 C. Aprile, M. Boronat, B. Ferrer, A. Corma, H. Garcia, *J. Am. Chem. Soc.* **2006**, *128*, 8388.

57 M. Valden, X. Lai, D. W. Goodman, *Science* **1998**, *281*, 1647.

58 J. Liu, Y. Lu, *J. Am. Chem. Soc.* **2004**, *126*, 12298.

59 H. Itoh, K. Naka, Y. Chujo, *J. Am. Chem. Soc.* **2004**, *126*, 3026.

60 M. Quinten, *Appl. Phys. B* **2001**, *73*, 317.

61 H. Kimura, T. Nakamichi, M. Yamada, M. Miyake, *Electrochemistry (Tokyo, Japan)* **2006**, *74*, 337.

62 D. Astruc, F. Lu, J. R. Aranzaes, *Angew. Chem., Int. Ed.* **2005**, *44*, 7852.

63 M. Haruta, *Stud. Surf. Sci. Catal.* **2003**, *145*, 31.

64 M. Brust, C. J. Kiely, *Colloid. Surf., A* **2002**, *202*, 175.

65 J.-P. Sylvestre, A. V. Kabashin, E. Sacher, M. Meunier, J. H. T. Luong, *J. Am. Chem. Soc.* **2004**, *126*, 7176.

66 A. Corma, H. Garcia, P. Montes-Navaja, A. Primo, unpublished results.

67 J. Lagona, P. Mukhopadhyay, S. Chakrabarti, L. Isaacs, *Angew. Chem., Int. Ed.* **2005**, *44*, 4844.

68 J. W. Lee, S. Samal, N. Selvapalam, H.-J. Kim, K. Kim, *Acc. Chem. Res.* **2003**, *36*, 621.

69 M. Haruta, M. Date, *Appl. Catal., A* **2001**, *222*, 427.

70 B. H. Lipshutz, S. Tasler, W. Chrisman, B. Spliethoff, B. Tesche, *J. Org. Chem.* **2003**, *68*, 1177.

71 I. W. Davies, L. Matty, D. L. Hughes, P. J. Reider, *J. Am. Chem. Soc.* **2001**, *123*, 10139.

72 F. Moreau, G. C. Bond, *Appl. Catal., A* **2006**, *302*, 110.

73 G. C. Bond, D. T. Thompson, *Gold Bull.* **2000**, *33*, 41.

74 G. C. Bond, D. T. Thompson, *Catal. Rev. – Sci. Eng.* **1999**, *41*, 319.

75 A. Abad, P. Concepcion, A. Corma, H. Garcia, *Angew. Chem., Int. Ed.* **2005**, *44*, 4066.

76 A. Corma, M. E. Domine, *Chem. Commun.* **2005**, 4042.

77 S. Biella, G. L. Castiglioni, C. Fumagalli, L. Prati, M. Rossi, *Catal. Today* **2002**, *72*, 43.

78 F. Porta, M. Rossi, *J. Mol. Catal. A* **2003**, *204–205*, 553.

79 L. Prati, M. Rossi, *J. Catal.* **1998**, *176*, 552.

80 M. Comotti, C. Della Pina, R. Matarrese, M. Rossi, *Angew. Chem., Int. Ed.* **2004**, *43*, 5812.

81 C. Aprile, A. Abad, H. Garcia, A. Corma, *J. Mater. Chem.* **2005**, *15*, 4408.

82 P. T. Anastas, M. M. Kirchhoff, *Acc. Chem. Res.* **2002**, *35*, 686.

83 M. Poliakoff, J. M. Fitzpatrick, T. R. Farren, P. T. Anastas, *Science* **2002**, *297*, 807.

84 R. A. Sheldon, *Pure Appl. Chem.* **2000**, *72*, 1233.

85 R. A. Sheldon, *J. Mol. Catal. A* **1996**, *107*, 75.

86 B. M. Trost, *Pure Appl. Chem.* **1992**, *64*, 315.

87 B. M. Trost, *Science* **1991**, *254*, 1471.

88 J. March, *Advanced Organic Chemistry: Reactions, Mechanisms and Structures*, 3rd edn., McGraw Hill, New York, **1993**.

89 W. Carruthers, *Some Modern Methods of Organic Synthesis. 3rd Ed.*, Cambridge University Press, Cambridge, **1986**.

90 J. Mathieu, R. Panico, J. Weill-Raynal, *Change and introduction of Functions In Organic Synthesis*, Mir, Moscow, **1980**.

91 L. De Luca, G. Giacomelli, A. Porcheddu, *J. Org. Chem.* **2001**, *66*, 7907.

92 J. M. Harris, Y. Liu, S. Chai, M. D. Andrews, J. C. Vederas, *J. Org. Chem.* **1998**, *63*, 2407.

93 R. E. Ireland, D. W. Norbeck, *J. Org. Chem.* **1985**, *50*, 2198.

94 A. Corma, H. Garcia, *Chem. Rev.* **2002**, *102*, 3837.

95 K. Yamaguchi, N. Mizuno, *Chem. Int. Ed.* **2002**, *41*, 4538.

96 K. Yamaguchi, N. Mizuno, *Chem. – A Eur. J.* **2003**, *9*, 4353.

97 G.-j. ten Brink, I. W. C. E. Arends, R. A. Sheldon, *Science* **2000**, *287*, 1636.

98 K. Mori, T. Hara, T. Mizugaki, K. Ebitani, K. Kaneda, *J. Am. Chem. Soc.* **2004**, *126*, 10657.

99 C. L. Bianchi, S. Biella, A. Gervasini, L. Prati, M. Rossi, *Catal. Lett.* **2003**, *85*, 91.

100 S. Biella, L. Prati, M. Rossi, *J. Catal.* **2002**, *206*, 242.

101 S. Biella, M. Rossi, *Chem. Commun.* **2003**, 378.

102 S. Biella, L. Prati, M. Rossi, *Inorg. Chim. Acta* **2003**, *349*, 253.

103 F. Porta, L. Prati, M. Rossi, G. Scari, *J. Catal.* **2002**, *211*, 464.

104 L. Prati, M. Rossi, *Stud. Surf. Sci. Catal.* **1997**, *110*, 509.

105 S. Carretin, P. McMorn, P. Johnston, K. Griffin, G. J. Hutchings, *Chem. Commun.* **2002**, *7*, 696.

106 S. Carretin, P. McMorn, P. Johnston, K. Griffin, C. Kiely, G. J. Hutchings, *Phys. Chem. Chem. Phys.* **2003**, *5*(6), 1329.

107 S. Carretin, P. McMorn, P. Johnston, K. Griffin, C. Kiely, G. A. Attard, G. J. Hutchings, *Top. Catal.* **2004**, *27*(1–4), 131.

108 D. I. Enache, J. K. Edwards, P. Landon, B. Solsona-Espriu, A. F. Carley, A. A. Herzing, M. Watanabe, C. J. Kiely, D. W.

Knight, G. J. Hutchings, *Science* **2006**, *311*, 362.

109 A. Abad, C. Almela, A. Corma, H. Garcia, *Tetrahedron* **2006**, *62*, 6666.

110 T. Mallat, A. Baiker, *Chem. Rev.* **2004**, *104*, 3037.

111 A. Abad, A. Corma, H. Garcia, unpublished results.

112 H. Tsunoyama, H. Sakurai, Y. Negishi, T. Tsukuda, *J. Am. Chem. Soc.* **2005**, *127*, 9374.

113 Y. Iizuka, H. Fujiki, N. Yamauchi, T. Chijiiwa, S. Arai, S. Tsubota, M. Haruta, *Catal. Today* **1997**, *36*, 115.

114 A. Wolf, F. Schuth, *Appl. Catal., A* **2002**, *226*, 1.

115 S. H. Overbury, V. Schwartz, D. R. Mullins, W. Yan, S. Dai, *J. Catal.* **2006**, *241*, 56.

116 S. Bilella, L. Prati, M. Rossi, *IV World Congress on Oxidation Catalysis* **2001**, *I*, 371.

117 S. Biella, M. Rossi, *Chem. Commun.* FIELD Publication Date: 2003, 378.

118 S. Carrettin, P. McMorn, P. Johnston, K. Griffin, C. J. Kiely, G. A. Attard, G. J. Hutchings, *Top. Catal.* **2004**, *27*, 131.

119 S. Carrettin, P. McMorn, P. Johnston, K. Griffin, G. J. Hutchings, *Chem. Commun.* **2002**, 696.

120 M. Comotti, C. Della Pina, E. Falletta, M. Rossi, *Adv. Synth. Catal.* **2006**, *348*, 313.

13
Gold Nanoparticles-catalyzed Oxidations in Organic Chemistry

Cristina Della Pina, Ermelinda Falletta, and Michele Rossi

13.1
Introduction

Following the rapid growth of gold nanotechnologies during the last decades [1], the application of this metal in catalysis has become an important research area [2]. However, only several years after the first report on the perspective in ethyne hydrochlorination [3] gold has been evaluated in fundamental reactions for organic synthesis such as oxidation and hydrogenation [4], and recently the use of gold catalysts for industrial application in fine chemical intermediates has been explored by academic and industrial researchers [5–8]. Compared to other metals, one of the outstanding properties of gold in catalysis is represented by the high selectivity which allows discrimination within chemical groups and geometrical positions, favoring high yields of the desired product. Thus, glycols can be converted to monocarboxylates [9], unsaturated alcohols to unsaturated aldehydes [10], unsaturated aldehydes and ketones to unsaturated alcohols [11] with selectivity approaching 100%. Considering other concomitant properties, such as biocompatibility, availability and easy recovery, gold appears as a promising "green" catalyst for sustainable processes using clean reagents, particularly O_2 and H_2, often in aqueous solution under mild conditions or in the absence of solvent. The present chapter is focused on peculiar aspects of gold catalysis applied to selective liquid phase oxidation at the carbon–oxygen bond of organic compounds carried out in our laboratories.

13.2
Catalyst Preparation

We discovered that efficient, carbon supported gold catalysts for liquid phase oxidation could be prepared starting from colloidal dispersions containing metallic gold (sol) [12a]. Thus, differently sized gold particles in the range 2–10 nm are formed by reducing chloroauric acid with $NaBH_4$ in the presence of stabilizing agents such as polyvinylalcohol (PVA), polyvinylpyrrolidone (PVP) and tetrahydroxymeth-

Nanoparticles and Catalysis. Edited by Didier Astruc
Copyright © 2008 WILEY-VCH Verlag GmbH & Co. KGaA, Weinheim
ISBN: 978-3-527-31572-7

$$HAuCl_4 \; + \quad protector \quad + \quad NaBH_4$$

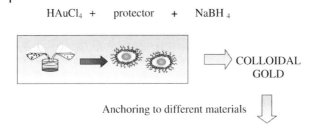

Anchoring to different materials

SUPPORTED GOLD

Fig. 13.1 Schematic representation of gold particles preparation.

ylphosphonium chloride (THMP). Au (III) concentration is a crucial factor for determining the range of particle size. Either high resolution electron transmission microscopy (HRTEM) or X-ray diffraction (XRD) techniques are used for particle size determination after immobilisation of the sol on a useful supporting material, such as a copper grid or carbon powder. TEM shows the direct image of the metal particles whereas the Scherrer equation allows the calculation of the mean diameter from the half height width of the XRD pattern [13].

According to Fig. 13.1, in our experiments colloidal gold nanoparticles were generally collected on two types of activated carbons: for catalytic tests, Au was immobilised on a coconut-derived carbon powder ($A_S = 1300 \, m^2 g^{-1}$ from Camel) at a level of 0.5–0.8% (w/w), which was chosen for the low sulfur content, whereas for XRD determination 1–2% Au (w/w) was immobilised on pyrolytic carbon powder ($A_S = 254 \, m^2 g^{-1}$ from Cabot) which was chosen for its rapid adsorption property. Figure 13.2 shows typical TEM images and size distribution of nanoparticles deposited on pyrolytic carbon powder. The sol deposition method requires particular care to ensure good reproducibility [12b].

Table 13.1 shows the best choice of protecting agents employed for preparing 2–10 nm particles in relation to the supporting material. In the case of a carbon support, differently sized gold particles were obtained using starting solutions which contained from 25 mg L^{-1} (small particles) to 500 mg L^{-1} (large particles) of gold.

13.3
Size-dependent Properties of Gold

As in the case of CO oxidation [14], the liquid phase oxidation of alcohols and aldehydes is quite sensitive to the size of the gold particles and this behavior was observed since the early tests [9]. A great contribution to this study was the discovery of the catalytic activity of unsupported gold particles. Starting from experimental evidence, it became possible to derive the simple model discussed in Section 13.3.2,

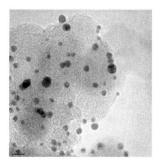 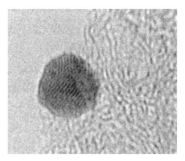

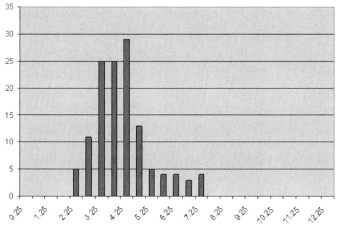

Fig. 13.2 TEM images and particle distribution of carbon supported gold particles.

Table 13.1 Protecting agents for different supporting materials.

Support	Protecting agent	d_m (nm) HRTEM
SiO$_2$	PVP	2.5–3.5
Al$_2$O$_3$	THPC	3.8–4.5
TiO$_2$	THPC	3.7–4.7
C	PVA	2.5–8.6

but a comprehensive interpretation of the data is still lacking. Historically, the first evidence of size-dependent activity related to carbon supported particles.

13.3.1
Supported Particles

In these experiments the correlation between catalytic activity and particle size was investigated using ethane-1,2-diol oxidation as a model reaction. In the presence

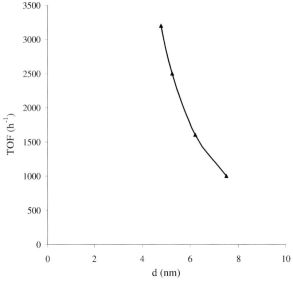

Fig. 13.3 Size-dependent catalytic activity of supported gold nanoparticles in ethane-1,2-diol (EG) oxidation. [EG]=0.5 M in water; EG/Au=3000; EG/NaOH=1; pO_2=2 bar; T=70°C; t=0.5 h.

of a 1:1 stoichiometric amount of NaOH, monocarboxylation to glycolate occurs with >98% selectivity. Owing to a shielding effect, the protecting agent must be completely removed from the supported catalyst before its use through a careful washing, in order to avoid wrong kinetic data. The size effect is outlined in Fig. 13.3, where TOF values are plotted versus the mean diameter of gold, showing a continuous decrease in TOF on increasing the gold particle size.

For the supported gold, difficulties in calculating the exposed metal area and possible metal–support interactions prevent accurate correlations, which can be more easily obtained in the absence of support, as discussed in the next section.

As a practical conclusion, gold particles over 10 nm are almost inactive in ethane-1,2-diol oxidation.

13.3.2
Unsupported Particles

One of the most problematic questions in heterogeneous catalysis is the cooperative effect of different phases present in a given catalytic system and, in particular, the so-called metal–support interaction [15]. In the case of gold catalysis, interaction of the metal with an oxidic support seems to be of fundamental importance in determining the extraordinary reactivity observed during the low temperature oxidation of CO [14].

The separate contributions of gold nanoparticles and their support to liquid phase oxidation was first clarified by studying the aerobic oxidation of glucose, because this reaction can be easily performed using either unsupported colloidal particles or supported particles under mild conditions in basic solution.

The key point for successful experiments was the possibility to avoid the use of colloid stabilizers (protecting agents), because glucose itself acts as a stabiliser. With this advantage, the high activity of unprotected particles ("naked particles" throughout the text) was discovered and extraordinary TOF values of magnitude similar to enzymatic catalysis were determined [16].

In Fig. 13.4 the catalytic activity in glucose oxidation of different metal particle dispersions (Pd, Pt, Ag, Cu, Au), having a similar size (2–5 nm), is compared. The unique behavior of gold is clearly evident showing the high TOF value of 18 043 mol of glucose (mol of gold)$^{-1}$h^{-1}, calculated with respect to the total metal amount during the first 100 s.

After about 200 s the activity dropped sharply, owing to agglomeration of gold particles, the effect can be shown by determining the mean particle size during the kinetic test, as shown in Fig. 13.5.

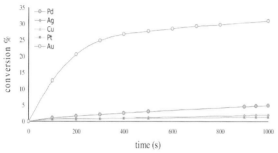

Fig. 13.4 Nanometric particles of different metals as catalysts for the aerobic oxidation of glucose. [Metal] = 10^{-4} M, [Glucose] = 0.4 M, T = 30 °C, pO_2 = 1 bar, pH = 9.5.

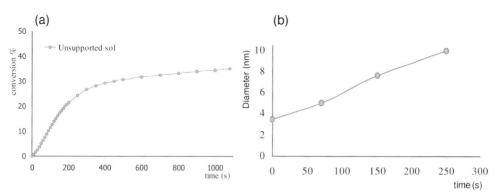

Fig. 13.5 Collapse of activity (a) and particle size (b) of colloidal gold during glucose oxidation. [Glucose] = 0.5 M, Glucose/Au = 3000, T = 30 °C, pO_2 = 1 bar, pH 9.5.

Although the instability of colloidal particles prevents their use in practical applications, the kinetic behavior during their lifetime is useful for deriving information on the intrinsic activity of gold and for catalytic comparison with supported particles. Concerning the first point, the calculation of the number of active sites can be derived on the basis of a simple model considering that (i) gold is in the form of spherical particles, as supported by TEM analysis, all having the same size corresponding to the calculated mean diameter; (ii) only external atoms are catalytically active, as normally accepted in heterogeneous catalysis; (iii) steric hindrance of the reacting molecules, diminishing the number of effective gold atoms available at the surface, is omitted.

If we perform kinetic experiments with a constant mass of metal (w) of a given density (ρ), in the form of colloidal particles having a monomodal distribution, we can correlate the radius value (r) to the activity by considering the following equations:

$$m = \text{mass of a single particle} = 4\pi r^3 / 3\rho \tag{13.1}$$

$$n = \text{number of particles} = w/m = 3w\rho/4\pi r^3 \tag{13.2}$$

$$S = \text{total external surface} = 4\pi r^2 n = 3w\rho/r \tag{13.3}$$

From the above correlations we derive that S is proportional to $1/r$ and the total number of particles n to $1/r^3$. As generally accepted, assuming that the catalyst controlled rate (v) is proportional to the exposed surface we expect from this model that $v = k/r$.

According to Bond and Thompson's diagram [17], the correlation between particle diameter, number of total atoms and dispersion is depicted in Fig. 13.6

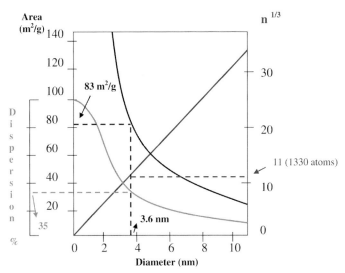

Fig. 13.6 Geometric correlation for monomodal spherical particles.

where values relative to the case discussed below of 3.6 nm particles are represented.

Applying this model to glucose oxidation, using a preformed gold sol having a mean diameter of 3.6 nm we determined in the first 70 s the TOF value of 18 043 mol of gluconate formed per mol of total gold per hour, working at 30 °C, under O_2 at $P=1$ bar and pH 9.5. As shown in Fig. 13.6, the fraction of atoms lying at the surface is 36%. It follows that the TOF value referred to the active catalytic sites, that is calculated for external gold atoms, is 50 120 h^{-1}. Obviously, this latter is a restrictive value owing to the omitted hindrance of the adsorbed molecules which lowers the availability of gold atoms.

The importance of this empirical model can be emphasized considering that no experimental methods for titrating the active sites in gold catalysts are presently known and only careful spectroscopic determinations with CO as a molecular probe allow, in some cases, the evaluation of this parameter [18].

In order to verify the kinetic model indicating the dependence of activity on particle size, we first derived an almost linear correlation between catalytic activity and gold concentration, in the range 10^{-5}–10^{-6} M, as shown in Fig. 13.7. In a second step, colloidal particles of different size (3–10 nm) were prepared by varying the chloroauric concentration from 50 to 600 mg l^{-1}, and the mean dimension of the crystallites was calculated by XRD. Using these particles we effectively observed a catalytic activity inversely proportional to the radius in the range 1.5–3.0 nm (Fig. 13.8). However, particles larger than 6 nm deviated from linearity and those larger than 10 nm were almost inactive [16].

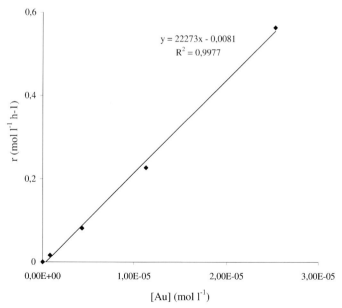

Fig. 13.7 Dependence of catalytic activity on gold particle concentration in glucose oxidation. [Glucose]=0.1 M, T=30 °C.

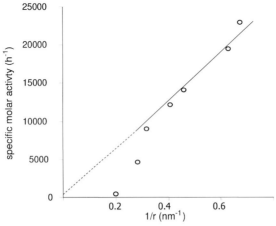

Fig. 13.8 Dependence of catalytic activity on gold particle size in glucose oxidation. [Glucose]=0.38 M, Glucose/Au=12 000, $T=30\,°C$, pH 9.5.

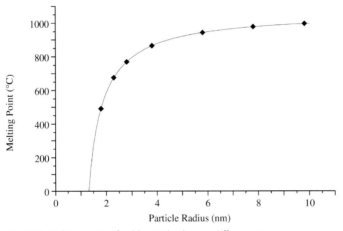

Fig. 13.9 Melting point of gold particles having different size.

The interpretation of the data of Fig. 13.8 is rather intriguing. The rapid drop in activity indicates a discontinuity in the catalytic behavior which is due not only to geometrical factors but probably also to electronic factors, influencing the electrophilic properties of the particles. We should remember that gold catalysis occurs only in basic solution where the nucleophilic properties of the reagent and the electrophilic properties of the catalyst should match the correct balance. This point will be developed in Section 13.4.2. We can observe that other properties show a sudden discontinuity in the region close to the metallic–non metallic transition of gold, as the melting point collapse indicated in Fig. 13.9.

13.4
Oxidation Mechanism

13.4.1
Metal–Support Interaction

The discovery of the catalytic activity of colloidal gold discussed in Section 13.3.2 and the ability of several porous materials to rapidly immobilize colloidal particles while conserving their dimension, discussed in Chapter 1, allow comparison between unsupported and supported particles under similar experimental conditions. Considering the conversion–time plot of glucose oxidation during the first 200 s, reported in Fig. 13.10, we can follow our experiment monitoring the kinetic test of a 10^{-4} M dispersion of Au particles having a mean diameter of 3.6 nm and evaluating the initial TOF of 18 043 h^{-1} (referred to the total gold concentration).

Then, while the naked gold particles are expressing the highest activity, 70 s, the addition of different adsorbing material, in order to generate *in situ* 0.5% w/w of supported gold is considered.

The kinetic effect of SiO_2, Al_2O_3, TiO_2 and C addition is represented by a stabilising effect of the original activity of gold, such effect increasing in the above order from silica to carbon.

The catalyst life is thus prolonged and the total conversion to gluconate can be performed with a mean TOF close to the initial one in the case of the most effective titania and carbon. Therefore, a metal–support interaction that avoids particle aggregation, favoring catalyst efficiency, is quite evident from the data collected in the long time experiments of Fig. 13.11. However, from a conventional point of view, the metal–support interaction should be considered as a synergetic effect able to increase the reaction rate. According to this interpretation, we did not observe any true rate increment upon addition of various supporting materials because 18 043 h^{-1} represents, in any case, the highest TOF due to the intrinsic activity of unsupported and supported gold.

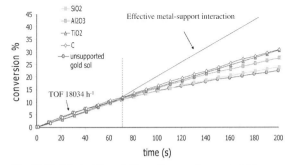

Fig. 13.10 Effect of the addition of supporting material to gold nanoparticles during glucose oxidation. [Au] = 10^{-4} M, [Glucose] = 0.4 M, T = 30 °C.

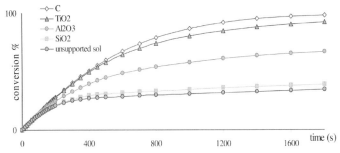

Fig. 13.11 Effect of the addition of supporting material to gold nanoparticles during glucose oxidation. [Au]=10^{-4} M, [Glucose]=0.4 M, T=30 °C.

Table 13.2 Hydrogen peroxide detection during glucose oxidation. [Glucose]=0.35 M, Glucose/Au=14 000, pH 7, T=30 °C.

t (s)	Conversion(%)	TOF (h^{-1})	[H$_2$O$_2$] found	[H$_2$O$_2$] calc.
100	0.45	2700	1.9×10^{-3}	1.7×10^{-3}
400	0.92	840	3.7×10^{-3}	3.5×10^{-3}
900	1.68	290	5.2×10^{-3}	6.4×10^{-3}

13.4.2
Kinetic Data and Molecular Mechanism

Kinetic studies on the selective liquid phase oxidation of glucose catalysed by gold were carried out using either carbon supported metal particles [19] or unsupported colloidal particles [20]. According to the first investigation, a Langmuir–Hinshelwood model was proposed where the overall reaction rate, in the presence of alkali, is limited by the surface oxidation reaction, adsorption of substrate and desorption of the product being fast processes. A negligible effect of glucose concentration on the reaction rate was observed [19].

Although not explicitly stated, water was considered to be the reduction product of dioxygen according to a dehydrogenation mechanism leading to gluconate as the oxidation product.

In the second investigation dealing with naked gold particles, the most important result was the detection of hydrogen peroxide instead of water as the reduction product of dioxygen.

In fact, during the conversion of glucose, a careful titration with permanganate allowed us to quantify H$_2$O$_2$ in a 1:1 ratio with respect to gluconate, as shown in Table 13.2 [21].

Moreover, according to kinetic experiments at low glucose concentrations (<0.1 M), a first reaction order was found, tending to an asymptote at higher concentrations (0.5 M). A first-order dependence was also found for the O_2 concentration.

An Eley–Rideal mechanism, characterised by the adsorption of glucose in its hydrated form on gold with equilibrium constant K_G, followed by reaction with oxygen coming from the liquid phase, according to Eqs. (13.4)–(13.7), can justify the result.

$$G + \sigma \leftrightarrow G\sigma \tag{13.4}$$

$$G\sigma + O_2 \rightarrow L + H_2O_2 + \sigma \tag{13.5}$$

$$r = k_s \theta_G C_{O2} = k_{cat} C_{Au} \theta_G C_{O2} \tag{13.6}$$

This mechanism gives rise to the rate equation (13.7),

$$r = k_{cat} K_G C_{Au} C_G C_{O2} / (1 + K_G C_G) \tag{13.7}$$

which justifies both the first order with respect to oxygen and the decreasing order with respect to glucose, being first order at low C_G when $K_G C_G$ is small with respect to 1 and tending to zero for large values of C_G. The detection of a relevant quantity of hydrogen peroxide beside gluconate (L), close to the calculated amount in Eq. (13.5), provides further experimental support [20].

About a previous investigation on glucose oxidation catalysed by a supported gold catalyst (0.48 wt.% Au on carbon) [18], we can observe that the rate equation suggested by the authors corresponds to a dehydrogenation–oxidation mechanism, where Au is the dehydrogenating agent and its re-oxidation is a fast step. However these authors did not investigate the dependence of the rate on C_{O2}. Considering initial rates, their rate equation can, to a good approximation, be written as Eq. (13.8), neglecting the adsorption of gluconic acid because of its minor weight, particularly in the early stages of the reaction.

$$r_0 = \frac{k_{ox} C_{cat} C_G^0}{1 + K_G C_G^0} \tag{13.8}$$

However, with the high K_G value reported by these authors (10^8) the product $K_G C_G^0$ in the denominator of Eq. (13.8) is dominant and Eq. (13.8) behaves as a zero-order equation. In our opinion, parameters k_{ox} and K_G in Eq. (13.8) are so strongly correlated that they cannot be determined separately, and the mechanism suggested by these authors is not really proven.

Owing to the scientific interest and commercial relevance of the biological glucose oxidation, presently applied in industrial plants for producing gluconates, it seemed of interest to derive accurate kinetic data for the enzymatic catalysis. Using Hyderase (a commercial glucose *oxidase* and *catalase* enzymatic preparation) under strictly similar conditions used for the gold catalysis, measurements of initial rate as a function of initial glucose concentration were interpreted by a simplified version of the already proposed reaction mechanism, having the form of a simple equation of Michaelis–Menten type. By comparing the results, the following conclusions have been reached: gold catalysis and biological catalysis are able to promote the fast and selective oxidation of glucose under mild conditions, according to the same stoichiometry, involving the formation of hydrogen peroxide as an unstable intermediate. However, the homogeneous enzymatic system and the pseudo-homogeneous colloidal system adopt different reaction mechanisms. In fact, in the case of enzymatic catalysis, the rate determining step is the oxidation of the substrate by the enzyme, which is converted into the reduced form according to a faster step and showing a zero order with respect to dioxygen, whereas the rate determining step in gold catalysis is step (13.5) where the adsorbed glucose is oxidized by dioxygen dissolved in water, according to a first-order dependence of the reaction rate on pO_2.

Despite different reaction mechanisms, similar activation energies (47.0 kJ mol^{-1} for gold catalysis and 49.6 kJ mol^{-1} for enzymatic catalysis) were found for the corresponding rate determining steps [20, 22].

Supported by kinetic studies, our knowledge on molecular aspects of glucose oxidation promoted by *Oxidase* and *Catalase* enzymes and gold nanoparticles agrees with Scheme 13.1 [23] and Scheme 13.2 [21], respectively.

Considering the molecular model for gold oxidation, experimental data concerning H_2O_2 formation, the effect of alkali in increasing the reaction rate, and the size-dependent reaction rate are the key factors. In this context, we can outline the importance of nucleophilic–electrophilic interactions between the negatively charged organic intermediate, promoted by alkali, and nanoparticles sharing electronic properties of metallic and non-metallic matter, tailored by their dimension [24]. According to the scheme, the resulting intermediate behaves as a very efficient two-electron donor towards molecular oxygen overcoming the slow step.

13.5
Gold Catalysis for Selective Oxidation

Apart from the commercial Bayer–Hoechst technology for the acetoxylation of ethene to vinylacetate, industrial processes based on gold catalysis are rather uncommon [25]. No new process employing gold is mentioned in Armor's review on new catalytic technology commercialized in the USA during the 1990s [26]. Moreover, in the rapidly expanding literature on gold chemistry, heterogeneous gold catalysis devoted to organic synthesis is scarcely represented.

Before discussing in more detail the chemical applications investigated in Milano, we outline in this paragraph some important achievements of other teams.

Beside CO oxidation, the gold promoted oxidation of propene is one of the most investigated reactions because of its industrial interest. In fact, the peculiar ability of gold to catalyse the synthesis of propene oxide by using O_2 and H_2 as reagents

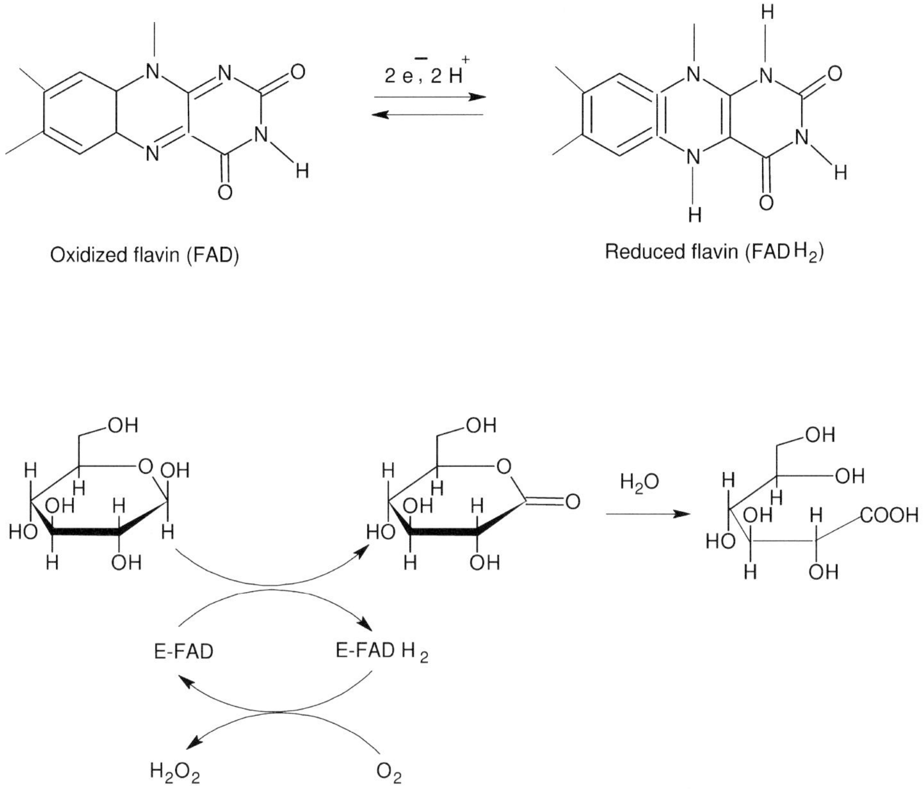
2 e⁻, 2 H⁺
Oxidized flavin (FAD)
Reduced flavin (FAD H2)
H2O
E-FAD
E-FAD H 2
H2O2
O2

Scheme 13.1 Molecular mechanism of enzymatic glucose oxidation.

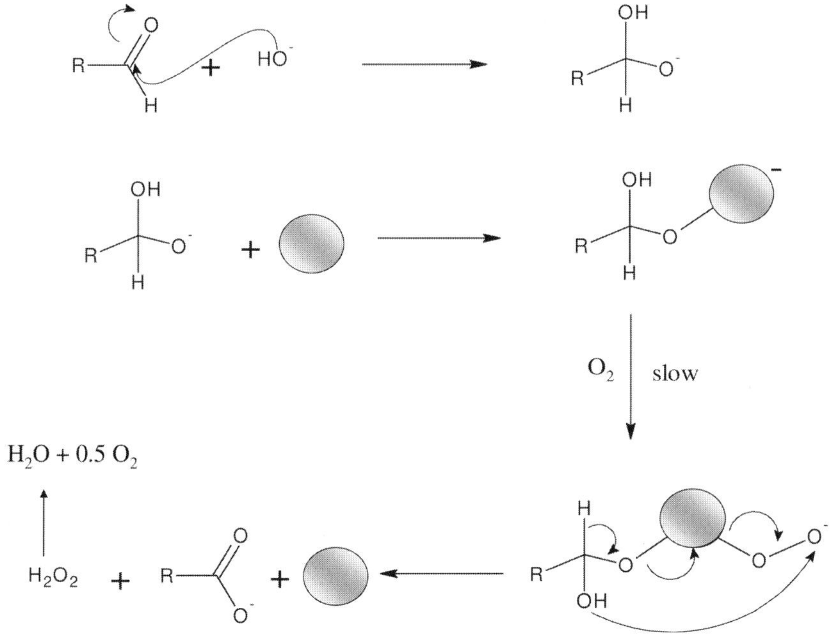
O2
slow
H2O + 0.5 O2
H2O2

Scheme 13.2 Molecular mechanism of glucose oxidation with gold nanoparticles.

has been investigated in academia [27] as well as in industrial laboratories [28–30]. Yields up to 5.6% with 80% selectivity have recently been reported [31]. A second application of great industrial interest is the oxidation of cyclohexane to the corresponding alcohol–ketone mixture (KA oil) as the intermediate for nylon 6 and 66 polymers: starting from basic research [31] industrial interest has been claimed by Solutia [32]. In particular, using gold supported on ZSM-5, Zhao et al. reported remarkable results in the solvent-free liquid phase oxidation, with TOF values superior to $3000\,h^{-1}$ [31].

An elegant study on tunable gold catalysts for the oxidation of hydrocarbons has been reported by Hutchings's group, outlining the role of catalyst and solvent in determining the selectivity: in the case of cyclohexene, Bi-modified Au/C catalysts allow the conversion mainly to cyclohexene oxide and cyclohexen-2-one in 1,2,3,5 tetramethylbenzene solution with TBHP as a radical initiator [33].

The oxidative esterification of aldehydes in the presence of alcohols to produce saturated and unsaturated aliphatic esters has been developed at the Nippon Shokubai laboratories: a great result is represented by the new synthesis of methylmethacrylate (MMA) from methacrolein, methanol and molecular oxygen, which allows yields up to $50\,mol\,MMA\,(kg\,cat)^{-1}\,h^{-1}$ using supported bimetallic gold catalysts under relatively mild conditions (80 °C and 3 MPa) [7].

13.6
Liquid Phase Oxidation of the Alcoholic Group

The first systematic study on gold catalysis for selective liquid phase oxidation has been carried out at Milano University with the aim of finding a substitute for palladium, platinum and, particularly, copper in the aerobic oxidation of the alcoholic group. In this application, metal leaching and low selectivity obstacles need to be overcome. The early experiments for testing the activity of metal gold were disappointing: whereas bulk copper quickly reacted with O_2 and ethane-1,2-diol in basic solution to produce oxoethanoate and formate derivatives [34], gold powder was totally inert towards any transformation of the glycol.

The high chemical stability of bulk gold was by-passed by discovering the exciting properties of gold nanoparticles, discussed in Section 13.3, and the logic of

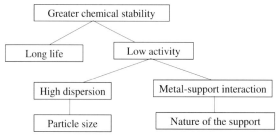

Scheme 13.3 Logical scheme for designing efficient gold catalysts.

Scheme 13.3 was then adopted by us and other different research groups also for gas-phase applications. Coprecipitation, deposition–precipitation and colloidal particles immobilisation, discussed in Section 13.2, were the preferred methods for preparing catalysts. In our case, supporting finely dispersed gold on carbon by means of metal sol immobilisation led to the discovery of an active and selective catalyst for liquid phase oxidation.

13.6.1
Oxidation of Diols

Aliphatic 1,2 diols can be converted to the corresponding monocarboxylates with O_2 under moderate pressure (1–3 bar) in the presence of the stoichiometric amount of NaOH. Compared to Pd and Pt metals, supported gold particles of 3–7 nm size represent a better catalytic system as shown in Table 13.3 [9, 35].

Developing the sol immobilisation technique described in Section 13.2 and increasing our ability in preparing small colloidal gold particles, we were able to implement the activity of the gold catalyst from a few hundred TOF units up to $3500 h^{-1}$ in the case of glycolate and $2000 h^{-1}$ in the case of lactate. Selectivity data, evaluated at 100% conversion, were exceptionally high, whereas a lower selectivity was observed in the case of Pd and Pt catalysts.

Using phenylethane-1,2-diol as a substrate, however, a new selectivity scenario appeared owing to a strong induction effect of the phenyl group. In fact, according to Scheme 13.4, two abundant by-products, namely benzoate and phenylglyoxylate, were detected beside the expected mandelate, working in the presence of NaOH.

Alkali-catalysed keto-enolic equilibrium **c** and internal Cannizzaro-type reaction **f**, represented in Scheme 13.5, should be taken into account when interpreting the experimental data.

In order to evaluate the original gold selectivity, the above reaction was carried out at the lower pH, 7, where reaction **c** and **f** as well as the overoxidation of mandelate were proven to be absent.

Under these conditions, and according to Scheme 13.6, the formation of mandelate is evidence for path **a** due to the oxidation at the terminal carbon atom,

Table 13.3 Catalytic activity and selectivity of carbon dispersed metals in the oxidation of vicinal diols.

Substrate	Catalyst	TOF (h^{-1})	Selectivity (%)
HO⌒OH	Au/C	1000–3500	98
	Pd/C	500	77
	Pt/C	475	71
OH⌒OH	Au/C	780–2000	99
	Pd/C	720	90
	Pt/C	650	89

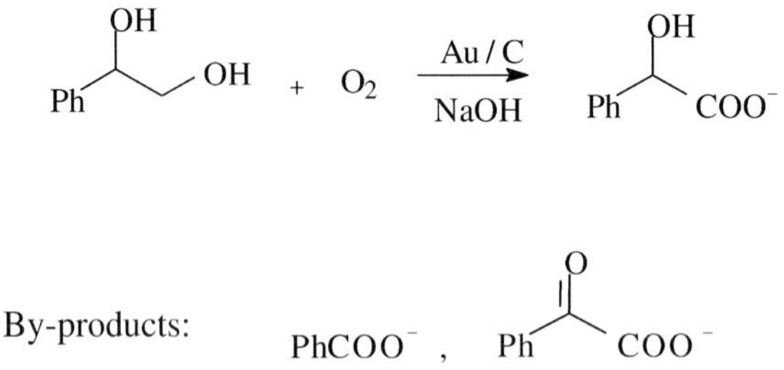

By-products: PhCOO⁻ , Ph—C(O)—COO⁻

Scheme 13.4 Reaction products detected during the oxidation of phenylethane-1,2-diol (P) with Au/C catalyst. [P]=0.4 M, P/Au=500, P/NaOH=1, T=70°C.

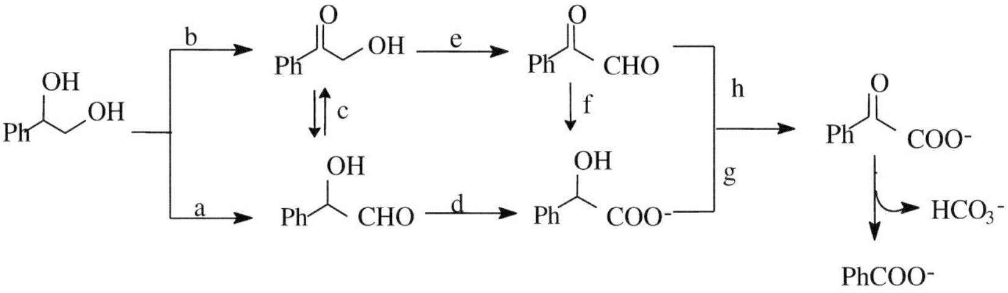

Scheme 13.5 Reaction pathway of phenylethane-1,2-diol oxidation in basic solution.

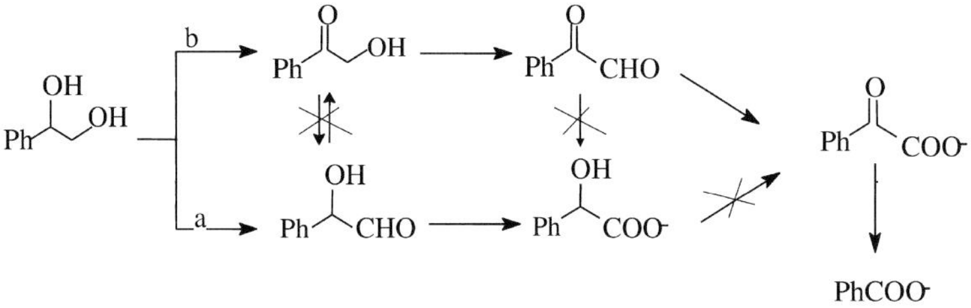

Scheme 13.6 Reaction pathway of phenylethane-1,2-diol oxidation at pH 7.

whereas phenylglyoxylate and benzoate are evidence of path **b** favoring oxidation at the internal carbon.

According to Table 13.4, the oxidation observed under these conditions concerned mainly the secondary alcoholic function, 62.5%, with respect to 37.5% of the primary one. In order to increase the selectivity towards mandelate, we had to promote reactions **c** and **f**. Table 13.5 shows how we were able to improve the selectivity from 37.5 to 83% by increasing alkali concentration and temperature [36].

Non vicinal glycols can also be selectively oxidised: the results for 1,3 propanediol and diethyleneglycol oxidation, reported in Table 13.6, indicate a lower reactivity than the vicinal diols reported in Table 13.3, whereas the selectivity to monocarboxylates was always very high.

Table 13.4 Reaction products and selectivity of phenylethane-1,2-diol oxidation at pH 7.

Conversion (%)	Mandelic (%)	Benzoic (%)	Phenylglioxylic (%)
13	4.5	5	2.5

37.5%	62.5%

Table 13.5 Optimization of phenylethane-1,2-diol (PED) oxidation for production of mandelate with Au/C catalyst.

NaOH/PED	T (K)	Conversion (%)	Selectivity (%)
1	343	52	45
2	343	100	60
2	363	100	70
4	363	100	83

Table 13.6 Oxidation of isolated diols with gold catalysts. [Substrate] = 0.4 M, substrate/Au = 100, $T = 70\,^{\circ}C$, $pO_2 = 3$ bar, pH 9.5.

Substrate	Catalyst	TOF (h^{-1})	Selectivity (%)
HO⁀⁀OH	Au/C	430	100
	Au/TiO2	490	95
HO⁀⁀O⁀OH	Au/C	240	99
	Au/TiO2	240	98

Owing to interest in the synthesis of dicarboxylate, a detailed study was done on the oxidation of diethyleneglycol in order to force the reaction towards the double oxidation, varying the amount of alkali, the nature of the catalyst and the temperature. Working under O_2 pressure at 3 bar, with a substrate: Au ratio of 1000, we found that gold on carbon produced mainly the monocarboxylate, whereas gold on titania produced up to 45% of the diacarboxylate in the presence of 2 mol of NaOH at 90 °C, as reported in Table 13.7. No other by-products were observed [37]. This represents a clear example of metal–support cooperation deserving further investigation.

Table 13.7 Influence of catalyst and experimental conditions on the oxidation of diethyleneglycol to mono- and di-carboxylate.

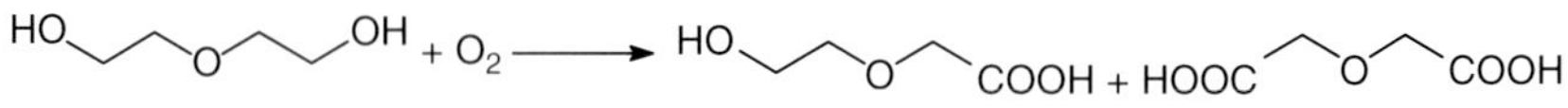

Catalyst	NaOH/Substrate	T (K)	t (h)	Conversion (%)	Monoacid (%)	Diacid (%)
1% Au/C	1	343	4	96	99	1
	2	343	4	80	97	3
		363	1	83	98	2
1% Au/TiO$_2$	1	343	4	95	98	2
		363	2	95	96	4
	2	343	3	94	70	30
		363	6	100	55	45

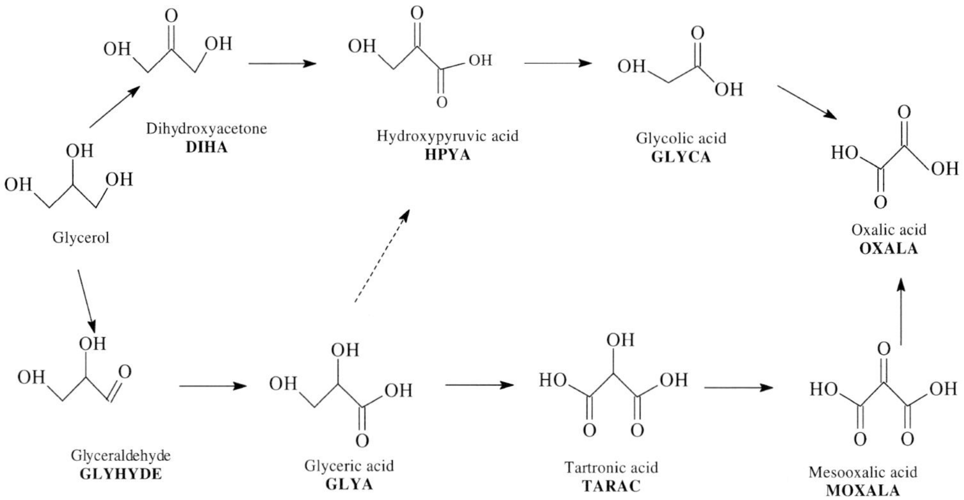

Scheme 13.7 Reaction pathway of the aerobic oxidation of glycerol.

13.6.2
Oxidation of Other Polyols

13.6.2.1 Glycerol

The large availability of glycerol as a by-product of biodiesel has stimulated, in recent years, much research in transforming this inexpensive compound into valuable chemicals. In this context, two research groups have been active in glycerol oxidation under mild conditions using gold catalysts. Despite the variety of possible reaction products, originating by the general oxidative pathway reported in Scheme 13.7, Hutchings's group has highlighted the high selectivity of gold:

using graphite as a support, in water solution at 60 °C and in the presence of NaOH, 100% selectivity to sodium glycerate could be readily achieved at 50–60% conversion [38].

A subsequent detailed investigation of glycerol oxidation has been carried out by Prati et al. in Milano. In a first study, the relationship between catalyst morphology and selectivity was explored at full conversion: it was found that larger gold particles (20 nm), supported on suitable carbons, show low TOFs but favour glycerate formation under mild conditions (30 °C, 3 bar) allowing yields up to 92% [39]. As a further development of this research, bimetallic nanoparticles were employed as supported catalyst with the aim being to improve activity and selectivity. The authors demonstrated the following points [40]:

1. Using bimetallic systems (Au–Pt) and (Au–Pd) the activity was increased significantly with respect to monometallic systems, indicating a synergistic effect between the metals. Selectivity to the desired product depends on the nature of the catalyst (particle size, alloyed phases and support) and the reaction conditions. Compared to the high selectivity to glyceric acid with pure gold, Pd addition promotes further transformation to tartronic acid, and Pt the transformation of glyceric acid to glycolic acid. Catalysts with larger metal particle size showed lower catalytic activity than catalysts with smaller particle size, while selectivity followed the opposite trend.

2. The activity and selectivity were significantly affected by the atomic ratio of the metals in $(Au_xPd_y)/C$ catalyst. In some cases, a volcano-type diagram was observed.

3. Activity and selectivity were influenced by different supports (Carbon, Graphite, TiO_2, Ti/SiO_2, SiO_2).

13.6.2.2 Sorbitol

Supported gold nanoparticles have been used in sorbitol oxidation and compared with analogous Pd and Pt catalysts [41]. The catalytic data for the monometallic catalysts are reported in Table 13.8. Using carbon as a support, the reaction took

Table 13.8 Activity and selectivity of carbon dispersed metals in glycerol oxidation.

1% M/C	Conversion (%)	Selectivity (%)		TOF (h^{-1})
		gluconate + gulonate	glucarate	
Au	33	61	8	141
Pd	14	61	11	28
Pt	37	40	4	169

place only in the presence of alkali, showing a modest TOF value. The primary alcoholic function was selectively oxidised with a marked preference for monocarboxylates (gluconate and gulonate) formation over dicarboxylate (glucarate).

Using bimetallic Au+Pd and Au+Pt catalysts, increased activity was observed which allowed full conversion to be reached, whereas selectivity to monocarboxylates at a given conversion was superior to the monometallic gold catalyst and almost independent of the nature of the second metal.

13.6.2.3 Oxidation of Other Alcohols

The gold catalysed liquid phase oxidation of the alcoholic group to carbonyl or carboxyl groups has been investigated, looking for a general method of inorganic synthesis. From the experience of different research groups, gold appears scarcely active in converting aliphatic and aromatic alcohols to aldehydes under neutral conditions, whereas in the presence of alkali the corresponding carboxylates can easily be obtained. However, important synergetic effects have recently been reported by Hutchings et al. using Au–Pd bimetallic catalysts [42] and by Corma et al. using Au-nanometric CeO_2 catalyst [43] under solvent-free conditions. A detailed analysis of this research is discussed elsewhere in this book.

13.6.2.4 Oxidation of Aminoalcohols

It is well documented that the catalytic oxidation of aminoalcohols is far from being a general method for the synthesis of aminoacids, owing to the poisoning effect of the amino group on traditional metals. We have found that gold catalysis partly overcomes the poisoning problem, as shown in Table 13.9, where nanometric particles of this metal dispersed on carbon are compared with palladium and platinum under similar conditions. It is worth noting that even in the presence of the basic amino group, alkali promotes the oxidation rate [37].

An interesting effect of the material employed as a support was detected during this investigation: gold nanoparticles of similar dimension dispersed on alumina behave better than those dispersed on carbon, as seen from the results shown in Table 13.10. Note that a similar advantage of an oxidic compound, namely titania,

Table 13.9 Catalytic oxidation of aminoalcohols with carbon dispersed metals in the absence of alkali. [Substrate] = 0.4 M, substrate/metal = 1000, pO_2 = 3 bar; T = 70 °C, t = 2 h.

Substrate	Catalyst	Conversion (%)
HO–⌒–NH₂	1% Au/C	20
	5% Pd/C	0
	5% Pt/C	0
NH₂–⌒–OH	1% Au/C	65
	5% Pd/C	0
	5% Pt/C	0

Table 13.10 Catalytic oxidation of aminoalcohols with carbon dispersed gold in the presence of alkali. [Substrate] = 0.4 M, substrate/metal = 1000, substrate/NaOH = 1, pO_2 = 3 bar, T = 70 °C, t = 2 h.

Substrate	Conversion (%)
HO—CH₂CH₂—NH₂ (HO$\diagup$NH$_2$)	23
NH₂ / OH (branched aminoalcohol)	100
HO$\diagup$NH₂	27
OH / HO$\diagup$NH₂	32

has previously been observed in relation to the double oxidation of diethyleneglycol in Section 13.6.1.

From the above experiments, we concluded that gold catalysis is quite sensitive to the nature of the substrate and further research is needed to optimising the particular synthesis we are interested in.

13.7
Oxidation of Aldehydes

As a general trend, in the oxidation of aliphatic oxygenated compounds with supported gold particles the following order of reactivity has been observed: aldehydes > primary alcohols > secondary alcohols; tertiary alcohols and carboxylic acids are almost inert under moderate conditions (up to 90 °C and 3 bar). In particular, the aerobic oxidation of aldehydes can be performed using water, organic solvents and solvent-free conditions, also in the absence of alkali. Comparing gold catalyst to the classical platinum catalyst, we found that Au showed a fairly good activity in oxidizing aldehydes in aqueous solution and, in contrast to Pt, no deactivation was observed on recycling. In the case of heavier, water insoluble compounds we observed a strong solvent effect, in particular CH_3CN showed worse behavior and CCl_4 a quite good performance. Moreover, comparing the oxidation of propanal in water and CCl_4 solution, the chlorinated solvent showed a stabilising effect, not only for gold but also for platinum, as reported in Table 13.11 [44].

The oxidation of liquid aldehydes can also be carried out in the absence of solvent. In the case of 2-methylpropanal and n-heptanal we observed that a smooth reaction occurs, with TOF values of 4000–7500 h^{-1}, using air instead of pure O_2, at low temperature (25–70 °C), thus allowing safer reaction conditions.

Table 13.11 Recycling test of propanal in water and CCl_4 solution with Au and Pt catalysts. [Substrate] = 0.23 M, substrate/metal = 1000, pO_2 = 3 bar, T = 90 °C, t = 2 h.

Run	Conversion (%)			
	Au/C		Pt/C	
	H_2O	CCl_4	H_2O	CCl_4
1	91	96	90	95
2	93	95	51	96
3	91	97	44	96
4	91	96	45	95
5	92	95	43	95

Table 13.12 Aerobic oxidation of aromatic aldehydes with gold dispersed on carbon. [Substrate] = 0.4 M, substrate/metal = 1000, pO_2 = 3 bar, t = 2 h.

Reagent	Conversion (%)	
	H_2O	CCl_4
PhCHO	25	96
p-Me-PhCHO	8	88
o-HO-PhCHO	19	0
p-HO-PhCHO	10	0

Aromatic aldehydes behave differently from the aliphatic ones and, in particular, we noted a not yet understood structure effect related to the ring substituents. In fact, comparing different substrates, a strong deactivating effect was observed for the hydroxy group in the ortho and para positions, as shown in Table 13.12.

13.8
Oxidation of Glucose

The oxidation of glucose, a cheap and renewable starting material, represents a challenging target for chemical intermediates, mainly when clean technologies can be applied.

In particular, gluconic acid and gluconates are industrial intermediates related to food chemistry cleaning agents and surfactants. The industrial production is today performed by the enzymatic method based on *aspergillus niger* mold. Using slightly different procedures, sodium gluconate or calcium gluconate are the main products. Owing to the low productivity of the fermentation process, alternative simple methods using different catalytic technologies could be of interest.

13.8.1
Oxidation to Sodium Gluconate

The catalytic oxidation of glucose to gluconate with Pt group metals has been investigated for a long time. Working under mild conditions (303–323 K, 101.3 kPa), a fast reaction resulted, allowing high conversion and good selectivity but the catalyst deactivated owing to self-poisoning and over-oxidation. In more recent studies, sophisticated bi- and tri-metallic catalysts have been proposed to overcome these effects. In particular, Bi-promoted catalysts showed the best performance [45, 46]. However, to our knowledge, no industrial application of platinum metal catalysts has been presently applied.

In our laboratories, the conversion of glucose to gluconates catalysed by gold was discovered and the outstanding features of activity, selectivity and durability were highlighted [47].

Table 13.13 summarizes the comparative tests carried out at 50 °C and atmospheric pressure for different Pd, Pt and Au catalysts indicating the superiority of gold at two different pH values.

The selectivity observed with palladium and platinum catalyst was less than 95% whereas with gold it was close to 100% at total conversion.

Detailed work on optimization followed the first results, with the aim of proposing gold as an alternative process to the biochemical route. This led to a highly efficient catalytic system. The work also took advantage of the mechanistic studies discussed in Section 13.4. Starting from TOF of a few hundred h^{-1} units, we reached the impressive value close to $60\,000\,h^{-1}$, which is similar to the behavior of enzymatic catalysis [16].

Table 13.13 Catalytic activity of different metals dispersed on carbon in glucose oxidation.

Entry	Catalyst	pH	TOF (h^{-1})
1	1% Au/C	9.5	500
2	5% Pt/C	9.5	220
3	5% Pd/C	9.5	7.5
4	5% Pd–5%Bi/C	9.5	450
5	1%Pt–4% Pd–5%Bi/C	9.5	480
6	1% Au/C	8	410
7	5% Pt/C	8	103
8	5% Pd/C	8	5
9	5% Pd–5%Bi/C	8	317
10	1%Pt–4% Pd–5%Bi/C	8	292

Table 13.14 Comparison between biological and inorganic catalysis in glucose oxidation.

Catalyst	$C_6H_{12}O_6$ (mol l^{-1})	Cat/Glucose (g kg^{-1})	pH	Stirring (rpm)	Temperature (°C)	Specific activity (h^{-1})	Gluconate productivity (kg m^{-3} h^{-1})
Hyderase	1	6	5–7	900	30	145	122
Au/C	3	5	9.5	39000	50	218	514

Owing to the outstanding properties of the gold catalysts, particular efforts were devoted to the comparison of gold catalysis with enzymatic catalysis under strictly similar conditions.

For these experiments, the following catalytic systems were identified:

1. Hyderase (from Amano Enzyme Co., U.K.) as an enzymatic preparation containing *glucose oxidase* and *catalase* as active components and flavine-adenine dinucleotide (FAD) as the rate controlling factor (1.3×10^{-6} mol g^{-1}).
2. Gold catalyst, 0.5% w/w of Au on coconut-derived carbon powder (1200 m^2 g^{-1}), prepared by sol immobilization, containing metal particles of 3.6 nm, mainly at the surface (Au/C = 1.5 from XPS data).

Kinetic data for glucose oxidation to gluconate were obtained using a glass reactor interfaced to an automatic titration device equipped with NaOH as a reagent. Magnetic stirring and a high speed turbine were alternatively used during the tests.

Considering different parameters such as pH, temperature, glucose concentration and stirring speed the optimized results reported in Table 13.14 were obtained.

According to these results, the TOF related to the molecular efficiency of the active FAD sites (6×10^5 h^{-1}) was superior to the efficiency of active external gold atoms in nanoparticles (9×10^4 h^{-1}).

However, considering the lower FAD concentration in the enzymatic extract and the three times higher glucose concentration allowed by gold, the productivity obtainable using a similar weight of catalyst was about four times greater for the gold catalyst.

13.8.2
Oxidation to Free Gluconic Acid

No direct method is yet known for the preparation of free gluconic acid, which is presently manufactured from calcium gluconate and sulfuric acid with a large amount of CaSO$_4$ as a by-product. This is due to the fact that at low pH values the enzymatic catalysis is inhibited, and also Pd, Pt, and Au catalysts are scarcely

Table 13.15 Aerobic oxidation of glucose with monometallic and bimetallic catalysts. Glucose/Au = 3000, $T = 70\,°C$, $pO_2 = 3$ bar, $t = 6.5\,h$.

Au–Me (w:w)	Conversion (%)	TOF(h^{-1})
Au	11	51
Pt	13	60
Pd	<2	<2
Rh	<2	<2
Au–Pt 1:1	64	295
Au–Pd 1:1	20	92
Au–Rh 1:1	<2	<2

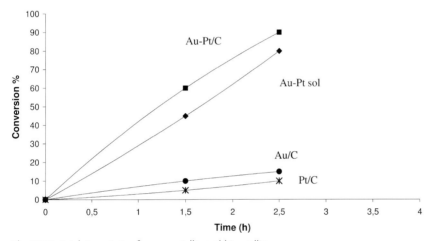

Fig. 13.12 Catalytic activity of monometallic and bimetallic nanoparticles in the synthesis of free gluconic acid.
Au : Pt = 2:1 (w:w), $pO_2 = 300\,kPa$, $T = 90\,°C$, Glucose/Au = 3000.

active. To attempt a successful direct synthesis of gluconic acid, we prepared and evaluated multicomponent catalysts and discovered a strong synergetic effect between gold and platinum (Table 13.15).

Following the initial screening of Au and Pd, Pt, Rh, the most promising Au + Pt combination was optimized, leading to a quite active catalyst for alkali-free oxidation of glucose containing gold and platinum in the ratio 2:1 (w/w).

The synergetic effect of the two metals is quite evident considering Fig. 13.12 [48].

The feasibility of the one-step synthesis of gluconic acid with promising TOF values around $1000\,h^{-1}$ is demonstrated, but critical experiments concerning the catalyst life are presently lacking.

13.9
Perspective for Gold Catalysis in Liquid Phase Oxidation

According to encouraging laboratory-scale tests, the following applications of gold catalysis for organic syntheses of industrial interest are here compared with traditional processes:

1. Sodium glycolate
   According to Scheme 13.8, the most relevant advantage of gold catalysis consists in the low cost starting material, ethylene glycol, and the clean technology that avoids the use of chlorinated reagents and carbon monoxide.

2. Sodium lactate
   No data have yet been reported for evaluating the productivity of gold catalysis with respect to the carbohydrate biological oxidation. Compared to the most important chemical alternative – the cyanohydrin hydrolysis route shown in Scheme 13.9 – the advantage of gold catalysis in using cheap reagents and avoiding the use of hydrogen cyanide is amazing.

3. Sodium gluconate
   No doubt exists that gold catalysis for glucose oxidation is one of the most appealing goals. A promising future for this new technology, mainly due to better productivity than with enzymatic catalysis, can be predicted (Scheme 13.10).

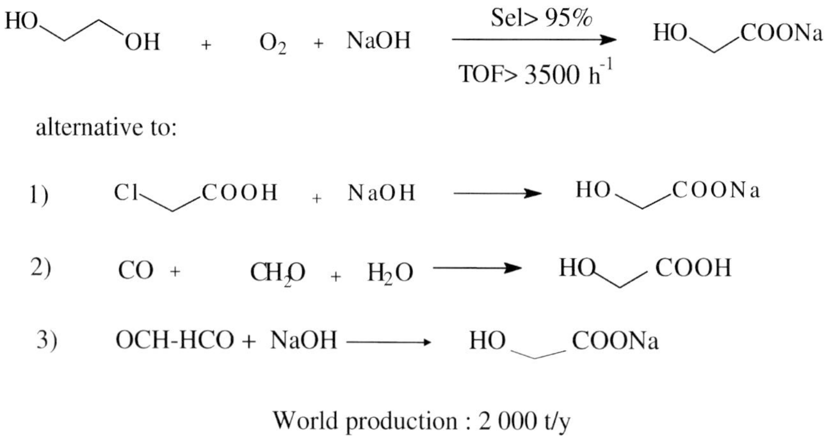

$$\text{HO}\diagdown\diagup\text{OH} \quad + \quad O_2 \quad + \quad \text{NaOH} \quad \xrightarrow[\text{TOF} > 3500 \text{ h}^{-1}]{\text{Sel} > 95\%} \quad \text{HO}\diagdown\diagup\text{COONa}$$

alternative to:

1) $\text{Cl}\diagdown\diagup\text{COOH} \quad + \quad \text{NaOH} \quad \longrightarrow \quad \text{HO}\diagdown\diagup\text{COONa}$

2) $\text{CO} + \quad \text{CH}_2\text{O} + H_2\text{O} \quad \longrightarrow \quad \text{HO}\diagdown\diagup\text{COOH}$

3) $\text{OCH-HCO} + \text{NaOH} \quad \longrightarrow \quad \text{HO}\diagdown\diagup\text{COONa}$

World production : 2 000 t/y

Scheme 13.8 Competitive gold catalysis for glycolate production.

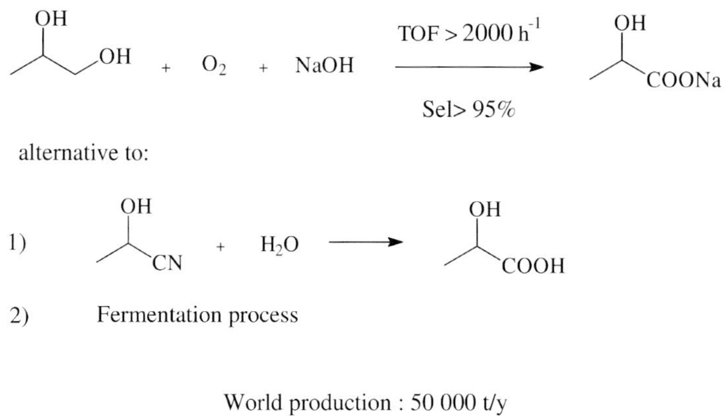

alternative to:

1)

2) Fermentation process

World production : 50 000 t/y

Scheme 13.9 Competitive gold catalysis for lactate production.

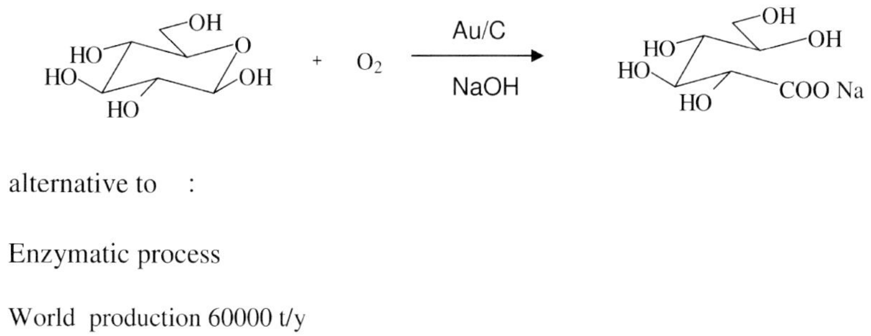

alternative to :

Enzymatic process

World production 60000 t/y

Scheme 13.10 Competitive gold catalysis for gluconate production.

13.10
Conclusions

The catalytic properties of gold nanoparticles in the selective oxidation of organic compounds is a young discovery [35], rapidly expanding in exciting applications full of promising results. Starting from liquid phase processes carried out under mild conditions, discussed in the present review, gold catalysis is fast extending to gas phase processes at higher temperature with similar successful perspectives [10, 49]. We have outlined that the outstanding properties of gold, a biocompatible non-toxic metal, can be exploited in catalysis when used in highly dispersed form. Therefore, a strong support from the imagination and ability of scientists is expected to create new, more efficient nanostructured catalysts, embodying activity, selectivity and stability, for challenging applications in organic synthesis.

Acknowledgments

This research was supported in part by the EC project "AURICAT, grant Research Training Network (HPRN-CT 2002-00174)".

We are grateful to Professor Gianmario Martra of Torino University for TEM determinations.

References

1 M. C. Daniel, D. Astruc, *Chem. Rev.* **2004**, *104*, 293.

2 M. Haruta, M. Date, *Appl. Catal. A* **2001**, *222*, 427.

3 G. Hutchings, *Catal. Today*, **2002**, *72*, 11.

4 G. Bond, D. Thompson, *Catal. Rev. Sci. Eng.* **1999**, *41*, 39.

5 Ital. Pat. IP 99A002611 to Lonza, **1999**.

6 WO pat. WO2004/099114 to Südzucker, **2004**.

7 EPA patent 1 393 800 to Nippon Shokubai Co., **2004**.

8 WO 2005/003072 to University of Milano, **2005**.

9 L. Prati, M. Rossi, *J. Catal.* **1998**, *176*, 552.

10 S. Biella, M. Rossi, *Chem. Commun.* **2003**, 378.

11 C. Milone, R. Ingoglia, M. Tropeano, G. Neri, S. Galvagno, *Chem. Commun.* **2003**, 1359.

12 (a) F. Porta, L. Prati, M. Rossi, S. Coluccia, G. Martra, *Catal. Today* **2000**, *61*, 165; (b) M. Comotti, C. Della Pina, R. Matarrese, M. Rossi, A. Siani, *Appl. Catal. A*, **2005**, *291*, 204.

13 A. L. Patterson, *Phys. Rev.* **1939**, *56*, 972.

14 G. Bond, D. Thompson, *Gold Bull.* **2000**, 41.

15 *Metal–Support Interactions in Catalysis, Sintering and Redispersion*, S. A. Stevenson, J. A. Dumesic, R. T. K. Baker, E. Ruckenstein (Eds.), Van Nostrand Reinhold, New York. **1987**.

16 M. Comotti, C. Della Pina, R. Matarrese, M. Rossi, *Angew. Chem. Int. Ed.* **2004**, *43*, 5812.

17 F. Menegazzo, M. Manzoli, A. Chiorino, F. Boccuzzi, T. Tabakova, M. Signoretto, F. Pinna, N. Pernicone, *J. Catal.* **2006**, *237*, 431.

18 Y. Önal, S. Schimpf, P. Claus, *J. Catal.* **2004**, *223*, 122.

19 P. Beltrame, M. Comotti, C. Della Pina, M. Rossi, *Appl. Catal. A* **2006**, *297*, 1.

20 M. Comotti, C. Della Pina, E. Falletta, M. Rossi, *Adv. Synth. Catal.* **2006**, *348*, 313.

21 P. Beltrame, M. Comotti, C. Della Pina, M. Rossi, *J. Catal.* **2004**, *228*, 282.

22 H. J. Hecht, H. M. Kalisz, R. D. Hendle, R. D. Schmid, D. Schomburg, *J. Mol.Biol.* **1993**, *229*, 153.

23 K. Weissermel, H. Arpe, *Industrial Organic Chemistry*, 3rd edn., VCH, Weinheim, **1997**.

24 M. Valden, X. Lai, D. W. Goodman, *Science* **1998**, *281*, 1647.

25 J. Armor, *Appl.Catal. A* **2001**, *222*, 407.

26 M. Haruta, *Proc. GOLD 2003*, Vancouver, Canada, **2003**, Sept. Oct.

27 US Pat. 2001 020 105 to Nippon Shokubai, **2001**.

28 Pat. App. WO200158887 to Bayer, **2001**.

29 Pat US 0 176 629 to Dow, **2004**.

30 A. Signa, S. Seelam, T. Tsuboda, M. Haruta, *Top. Catal.* **2004**, *29*, 95.

31 R. Zhao, D. Ji, G. Qian, L. Yan, J. Suo, *Chem Commun.* **2004**, 904.

32 US Pat. 2 004 158 103 to Solutia, **2004**.

33 M. Hughes, Yi-Jun Xu, P. Jenkins, P. McMorn, P. Landon, D. Enache, A. Carley, G. Attard, G. Hutchings, F. King, E. Stitt, *Nature* **2005**, *437*, 1132.

34 M. Lanfranchi, L. Prati, M. Rossi, A. Tiripicchio, *Chem.Commun.* **1993**, 1698.

35 L. Prati, M. Rossi, *Stud. Surf. Sci. Catal.* **1997**, *110*, 509.

36 S. Biella, L. Prati, M. Rossi, *Inorg. Chim. Acta* **2003**, *349*, 253.

37 S. Biella, G. Castiglioni, C. Fumagalli, L. Prati, M. Rossi, *Catal. Today* **2002**, *72*, 42.

38 S. Carrettin, P. McMorn, P. Johnston, K. Griffin, G. Hutchings, *Chem. Commun.* **2002**, 696.

39 F. Porta, L. Prati, *J. Catal.* **2004**, 397.

40 C. Bianchi, P. Canton, N. Dimitratos, F. Porta, L. Prati, *Catal. Today* **2005**, *102*, 203.

41 N. Dimitratos, F. Porta, L. Prati, A. Villa, *Catal. Lett.* **2005**, *99*, 181.

42 D. Enache, J. Edwards, P. Landon, B. Solsona-Espriu, A. Carley, A. Herzing, M. Watanabe, C. Kiely, D. Knight, G. Hutchings, *Science* **2006**, *311*, 362.

43 A. Abad, P. Conception, A. Corma, H. Garcia, *Ang. Chem. Int. Ed.* **2005**, *44*, 4066.

44 S. Biella, L. Prati, M. Rossi, *J. Mol. Catal. A*, **2003**, *197*, 207.

45 M. Besson, F. Lahmer, P. Gallezot, P. Fuertes, G. Fleche, *J. Catal.* **1995**, *152*, 116.

46 M. Besson, P. Gallezot, *Catal. Today* **2000**, *57*, 127.

47 S. Biella, L. Prati, M. Rossi, *J. Catal.* **2002**, *206*, 242.

48 M. Comotti, C. Della Pina, M. Rossi, *J. Mol. Catal. A* **2006**, *251*, 92.

49 M. Haruta, *Catal. Today* **1997**, *36*, 153.

14
Au NP-catalysed Propene Epoxidation by Dioxygen and Dihydrogen

Jun Kawahara and Masatake Haruta

14.1
Introduction

Propene oxide (PO, IUPAC nomenclature: 2-methyloxirane) is an important chemical feedstock, having a world annual production capacity of about 7 million tons [1]. It is processed into the major products, polyurethane polyols and propylene glycol [2]. The former are used in the manufacture of polyurethane foams and the latter for antifreeze (safer than ethylene glycol), drugs, cosmetics etc.

Currently there are two major routes, the chlorohydrin process and the organic hydroperoxide process, employed industrially for the production of PO. Both of these consist of two reaction stages, and are accompanied by the formation of by-products and co-products [3]. In 2003, Sumitomo Chemical Co., Ltd. launched a new commercial plant of $150\,000\,ty^{-1}$ capacity, in Chiba prefecture, Japan, that uses an organic hydroperoxide process combined with cumene recycling (Scheme 14.1). This process consumes dihydrogen (H_2) in addition to dioxygen (O_2) but is free from co-products other than water (H_2O). Their original mesoporous titanium silicate is used as a catalyst for the epoxidation stage [4, 5]. On the other hand, epoxidation with hydrogen peroxide (H_2O_2) has been adopted as the basis for the construction of new commercial plants by BASF/Dow and Degussa/Uhde, in Belgium and Korea, respectively, which are planned to come on stream in 2008 [1]. Degussa/Uhde uses titanosilicalite as a catalyst for the epoxidation of propene (C_3H_6, propylene) with H_2O_2 in methanol (MeOH) solvent [6, 7] (Scheme 14.1). Accordingly, the next target for PO production will shift to the direct epoxidation with O_2 alone or more feasibly with O_2 and H_2 (Scheme 14.2).

The direct epoxidation of C_3H_6 with O_2 alone is one of the most difficult reactions to achieve and is regarded as a sort of "Holy Grail" in catalysis research. Propene oxide is more reactive than propene. In contrast to ethene, C_3H_6 has allylic C–H bonds, the strength of which is the lowest in the molecule causing preferential dehydrogenation of the methyl group [8]. Furthermore, when molecular oxygen adsorbs on solid surfaces it tends to become an anionic species such as O^{2-} or O^-. Since these negatively charged oxygen species are nucleophilic in nature, they attack the carbon atom in the methyl group resulting in the production of

Nanoparticles and Catalysis. Edited by Didier Astruc
Copyright © 2008 WILEY-VCH Verlag GmbH & Co. KGaA, Weinheim
ISBN: 978-3-527-31572-7

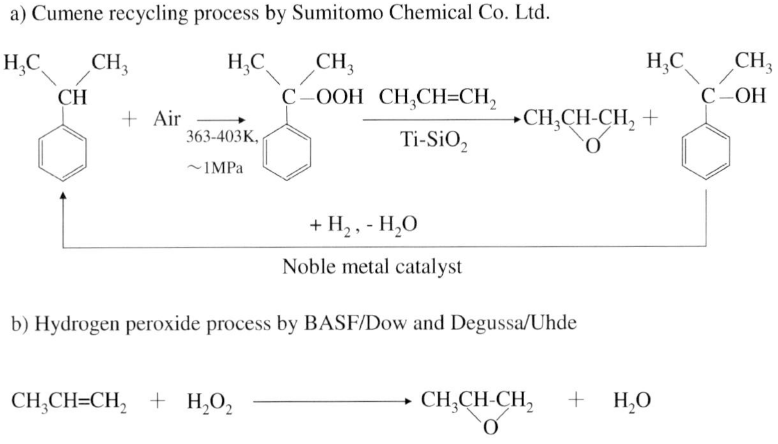

a) Cumene recycling process by Sumitomo Chemical Co. Ltd.

b) Hydrogen peroxide process by BASF/Dow and Degussa/Uhde

$$CH_3CH{=}CH_2 \; + \; H_2O_2 \longrightarrow CH_3CH\text{-}CH_2 \; + \; H_2O$$

Scheme 14.1 Recently established industrial processes for propene oxide.

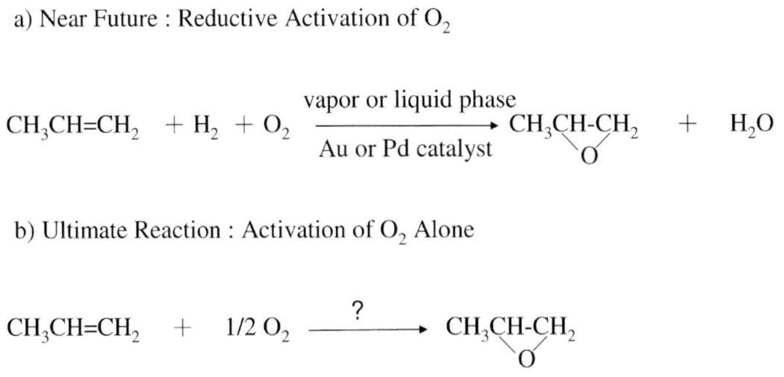

a) Near Future : Reductive Activation of O_2

$$CH_3CH{=}CH_2 \; + \; H_2 \; + \; O_2 \xrightarrow[\text{Au or Pd catalyst}]{\text{vapor or liquid phase}} CH_3CH\text{-}CH_2 \; + \; H_2O$$

b) Ultimate Reaction : Activation of O_2 Alone

$$CH_3CH{=}CH_2 \; + \; 1/2\,O_2 \xrightarrow{?} CH_3CH\text{-}CH_2$$

Scheme 14.2 Future propene oxide production processes.

acrolein or CO_2. In order to produce PO, electrophilic or neutral oxygen species are necessary to attach one oxygen atom to the electron rich C=C double bond (Scheme 14.3).

It is understandable that the majority of previous work on the vapor phase epoxidation of C_3H_6 by O_2 dealt with silver-based catalysts, because Ag/Al$_2$O$_3$ catalysts are used in commercial plants for ethene oxide production from ethene and O_2 in the gas phase. However, as far as reproducible data are concerned, PO selectivity from propene is still very low, below 50% even at a low conversion of a few %. Murata et al. reported that the use of high Si zeolitic materials as catalyst supports could give a high conversion of C_3H_6 (50–80%) but with a selectivity to PO of only around 20–30% [9].

On the other hand, reductive activation of O_2 with H_2 (*in situ* H_2O_2 production) in the liquid phase enables selective epoxidation of C_3H_6. Miyake [10] and Hoeldrich [11] reported that Pd/TS-1 and Pt-promoted Pd/TS-1 catalysts were active

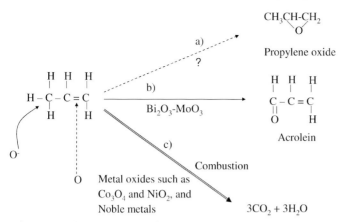

Scheme 14.3 Three major routes for the oxidation of propene with different oxygen species.

and selective in methanol and butanol solvents. A simpler and more economical process which proceeds in the absence of solvent has been developed by Hayashi. Gold nanoparticles (NPs) deposited on TiO_2 (Degussa P25, mainly anatase, partly rutile) catalyse the vapor phase reaction of C_3H_6 with O_2 and H_2 [12]. As shown in Table 14.1, selectivity to PO is as high as >99%, while C_3H_6 conversion and H_2 efficiency are not high, below 2% and below 40%, respectively. This performance motivated some chemists to replace Au with Ag and they found that catalysts such as Ag/TiO_2 [13] and Ag/TS-1 [14, 15] were also selective to PO. These silver cata-lysts are different from gold catalysts in that a characteristic feature is that oxidized silver particles with a diameter of around 8 nm are the most active. Since gold catalysts are superior to silver catalysts, this chapter focuses on the gas phase epoxidation of C_3H_6 in the presence of O_2 and H_2 over gold catalysts.

14.2
Catalyst Preparation and Catalytic Tests

Among a variety of metal oxides only TiO_2 is effective as a support for gold in the epoxidation of C_3H_6. The method used to deposit Au on TiO_2 is also crucial if high selectivity to PO is to be obtained. Figure 14.1 shows that the catalyst prepared by deposition–precipitation (DP) [16] is active at a lower temperature and produces PO with a selectivity above 90%, while that prepared by impregnation methods results in complete oxidation to form CO_2 [12]. In the DP method, at 323–363 K, the pH of $HAuCl_4$ aqueous solution is adjusted at a fixed point in the range from 6 to 10 by adding an aqueous solution of NaOH, because it can increase pH more effectively than Na_2CO_3 and lead to smaller ionic strength in the aqueous solution of $HAuCl_4$ [17]. The support material was dispersed in this solution, and was stirred for 1 h at the same temperature and pH, and then allowed to cool without stirring. The resulting solid catalyst precursor was washed, filtered, dried at 393 K

Table 14.1 Direct vapor phase epoxidation of propene over Au and Ag catalysts.

Catalyst[a]	Temp. (K), pressure (MPa)	$C_3H_6/O_2/H_2/$ Carrier, SV (h^{-1} ml g_{cat}^{-1})	C_3H_6 conversion	PO selectivity	H_2 efficiency[b]	PO STY (h^{-1} g_{PO} kg_{cat}^{-1})	Ref.
1%Au/TiO$_2$ (P25)	323, 0.1	10/10/10/70 4000	1.1%	99%<	34%	12	12
1% Au/TiO$_2$ (P25)	323, 0.1	10/10/10/70 6000	0.9%	99%	—	14	50
5% Au/TiO$_2$ (P25)	343, 0.1	33/33/33/0 1800	0.6%	83%	—	7.6	51
Pt-Au[c]/TiO$_2$/SiO$_2$	373, 0.1	10/10/10/70 3787	1.0%<[d]	90%<[d]	10%[d]	7.1<[d]	52
0.081% Au/TS-1	473, 0.1	10/10/10/70 7000	10%	76%	<30%	134 (steady state)	40
0.3% Au/Ti-SiO$_2$	423, 0.1	10/10/10/70 4000	8.5%	91%	35%	80 (steady state)	45
2%[e] Ag/TS-1	423, 0.1	5/11/17/67 4000	1.4%	94%	—	6.8	15
2.3% Ag/TiO$_2$ (P25)	323, 0.1	10/10/10/70 4000	0.4%	92%	—	3.8	13

a   Actual metal loading in wt% except for c and e;
b   defined by molar ratio of PO/H_2O;
c   atomic ratio of Pt/Au=5/95 in the starting solution;
d   estimated by the present authors;
e   metal loading in the starting solution.

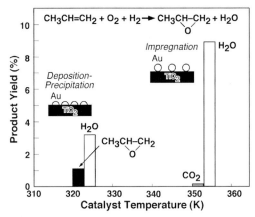

Fig. 14.1 Product yields in the reaction of propene over Au/ TiO$_2$ catalysts prepared by different methods: Catalyst, 0.5 g; Feed gas, $C_3H_6/O_2/H_2/Ar$ =10/10/10/70; space velocity, 4000 h^{-1} ml g_{cat}^{-1} [12].

overnight, and calcined at or above 573 K in air. This method is not applicable to metal oxides, the isoelectric points (isp) of which are below pH 5, for example, SiO_2 (isp = pH 2) [18].

In addition to anatase TiO_2, many kinds of Ti-containing SiO_2 (hereafter denoted as Ti-SiO_2) are effective as the support. Especially, at higher temperatures tetrahedrally coordinated Ti cations isolated from each other in the SiO_2 matrix make Au selective, while TiO_2 aggregates on SiO_2 surfaces do not. Titania modified SiO_2 supports with Ti/Si atomic ratio below 3/100 were prepared by the impregnation method [19]. Hydrothermal synthesis was used to prepare microporous TS-1 [20] and mesoporous Ti-MCM-41 and Ti-MCM-48 [17]. A three-dimensional sponge-like mesoporous titanosilicate was prepared by a modified sol–gel method [21]. A hydrophobic Ti-containing mesoporous hybrid silsesquioxane (Ti-HMM) was synthesized according to a reported procedure [22]. Figure 14.2 illustrates the pore structures of mesoporous titanosilicates used for preparing Au catalysts.

In the case of Ti-SiO_2, washing is also crucial. Complete washing is deleterious for maintaining PO selectivity, suggesting that the presence of a small amount of Na^+ and/or Cl^- are necessary for Au to be selective for the formation of PO [23] (Table 14.2). In the DP method, Au is deposited on Ti^{4+} sites but not on the surfaces of SiO_2. This selectivity of Au deposition leads to high selectivity to PO.

  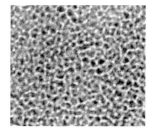

Fig. 14.2 Pore structures of mesoporous materials: (a) Ti-MCM-41 [hexagonal], (b) Ti-MCM-48 [cubic], (c) large mesoporous Ti-SiO_2 [sponge-like].

Table 14.2 Influence of preparation methods and precursor washing on the catalytic performance of Au/TS-1 [23, 24].

Prep. method	Washing	Au loading[a] (wt%)	Reaction temp. (K)	Conversion (%)		Selectivity (%)				
				C_3H_6	H_2	PO	PA	AT	AA	CO_2
DP	No	0.84	373	0.71	7.9	93	—	1.0	1.0	5
	Yes	0.17	473	0.57	21.2	3.5	73.7	—	3.5	12
GG	—	1	473	0.9	96	—	31	15	7	40

a Actual Au loading. PO: propene oxide, PA: propionaldehyde, AT: acetone, AA: acetaldehyde. Catalyst, 0.5 g; feed gas, $C_3H_6/O_2/H_2/Ar = 10/10/10/70$; space velocity, $4000\,h^{-1}\,ml\,g_{cat}^{-1}$.

In contrast, gas phase grafting (GG) of dimethyl Au(III) acetylacetonate results in Au deposition both on Ti^{4+} and on the SiO_2 surfaces, which causes poor selectivity to PO [24] (Table 14.2).

All the catalyst samples we prepared were pre-treated at 523 K, by passing 10 vol% H_2 for 30 min followed by 10 vol% O_2 diluted with Ar, prior to catalytic testing to clean up the surfaces and peripheries of the Au NPs. The catalytic activity measurements were carried out in a fixed bed reactor under atmospheric pressure. The reactant gas was composed of 10 vol% of each of the reactants, C_3H_6, O_2, H_2, diluted with an inert gas such as Ar, N_2, or He to operate outside the explosion limit. The catalyst bed temperature was varied from 303 K to 393 K for Au/TiO_2 and from 393 K to 523 K for Au/Ti-SiO_2. The reaction gases and products were analyzed by two on-line gas chromatographs equipped with a thermal conductivity detector (TCD) and a flame ionization detector (FID). A TCD detects H_2, O_2, CO_2, C_3H_6, and C_3H_8 while a FID measures C_3H_6, C_3H_8, PO and other oxygenates.

14.3
Au/TiO_2

The catalytic performance of Au/TiO_2 catalysts is markedly dependent on the methods and conditions used for their preparation, because it is mainly defined by the following three factors: selection of supports, contact structure, and size of the gold particles.

14.3.1
Effect of the Crystal Structure of TiO_2

Not only is the choice of titania as support important but also its crystal structure is critical for the production of PO (Table 14.3) [24]. Although Au deposited onto

Table 14.3 Effect of TiO_2 crystal structure on product selectivities in the reaction of propene with O_2 and H_2 over Au/TiO_2 catalysts [12, 24].

Support material	Specific surface area (m² g⁻¹)	Au loading[a] (wt%)	Reaction temp. (K)	Conversion (%)		Selectivity (%)			
				C_3H_6	H_2	PO	AT	AA	CO_2
P25	50	0.98	323	1.1	3.2	>99	—	—	—
anatase	57	1.1	313	0.26	3.0	98	2.0	—	—
rutile	39	1.0	353	0.095	16	—	3.0	—	97
amorphous	113	10.3	313	0.57	21	23	38.5	—	38.5

a   In the starting solution except for P25. PO: propene oxide, AT: acetone, AA: acetaldehyde. Catalyst, 0.5 g; feed gas, $C_3H_6/O_2/H_2/Ar$ = 10/10/10/70; space velocity, 4000 h⁻¹ ml g$_{cat}$⁻¹.

rutile TiO$_2$ support does not produce any PO at all, anatase and P25 TiO$_2$ supports make the supported Au selective for production of PO. Amorphous TiO$_2$ leads to a catalytic performance in between that for anatase and rutile, producing PO, acetone, and CO$_2$ in comparable proportions. The selectivity to PO can be explained in terms of the distance between two neighboring Ti cations in Ti–O–Ti, which is longer in anatase TiO$_2$ than the diameter of the oxygen anion (O^{2-}) and shorter in rutile TiO$_2$.

14.3.2
Contact Structure of Au Nanoparticles with the TiO$_2$ Supports

The gold NPs were well dispersed with diameters of 1–4 nm and attached to the TiO$_2$ support as hemispherical particles by use of the DP method (Fig. 14.3) [25]. The hemispherical junction can provide a longer distance for the perimeter interface between an Au NP and TiO$_2$. Detailed transmission electron micrograph (TEM) observation revealed the fine structure of the Au/TiO$_2$ catalysts, showing that Au NPs are fixed on the support with the Au{111} plane parallel to the TiO$_2${112} [26].

With this structure, PO is selectively produced, implying that the simultaneous reaction of C$_3$H$_6$ with O$_2$ and H$_2$ is the predominant reaction path. This reaction is always accompanied by H$_2$ combustion to form H$_2$O, and this results in higher yields of H$_2$O than of PO. On the other hand, use of the impregnation method for catalyst preparation gives spherical Au NPs, which are larger than 30 nm in diameter and only interact weakly with the TiO$_2$ support. With this structure PO is not

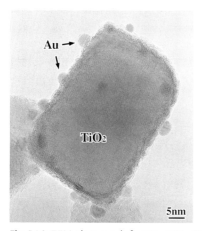

Fig. 14.3 TEM photograph for 3.3 wt%Au/TiO$_2$ catalyst prepared by deposition–precipitation followed by calcination at 673 K [25].

produced at all and CO_2 and H_2O are the only products, indicating that the reactions of C_3H_6 with O_2 and of H_2 with O_2 take place separately.

14.3.3
Size Effect of Au Particles

In most reactions, the size of the gold nanoparticles markedly affects their catalytic activity and selectivity [25, 27]. In the DP catalyst preparation method, the size of the Au particles and Au loading decrease with decrease in the concentration of $HAuCl_4$ used in the starting solution. When Au/TiO_2 is prepared with actual Au loadings below 0.1 wt%, propane (C_3H_8) is the major product, whereas above 0.1 wt% Au PO is formed selectively [12]. Similar results were previously reported by Naito and Tanimoto, i.e. that the addition of O_2 accelerated the hydrogenation of C_3H_6 in the case of a low loading Au/SiO_2 [28]. In the Au/TiO_2 catalysts which gave propane formation, the mean diameter of the Au particles was found to be below 2.0 nm, by TEM observation, while catalysts which give propene oxide have a mean diameter larger than 2.0 nm (Fig. 14.4).

When Au is deposited on TiO_2 by the DP method, Au NPs are hemispherical in shape but some are thick and others thin, depending on the Au loadings. Figure 14.5 shows the inner potential of Au NPs for different types of real Au catalysts as a function of particle thickness. The inner potential, which causes a phase shift of the electron wave (electron holography), tends to change at 4 nm and shows a sharp increase at around 2 nm. This indicates that the electronic nature of thick Au NPs as a whole changes at 2.0 nm [29]. In contrast, in the case of model catalysts prepared by vacuum deposition of Au on single crystalline rutile TiO_2, Au NPs are appreciably thinner than those in real powder catalysts [30]. The work function and energy gap deviate from those of bulk Au at 0.4 nm, i.e. 2 atom

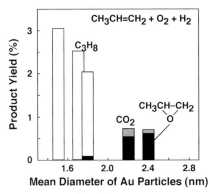

Fig. 14.4 Size effect of Au NPs on propene reaction with O_2 and H_2: Catalyst, Au/TiO_2 (P25) 0.5 g; catalyst bed temperature, 353 K; feed gas, $C_3H_6/O_2/H_2/Ar = 10/10/10/70$; space velocity, 4000 h^{-1} ml g_{cat}^{-1} [12].

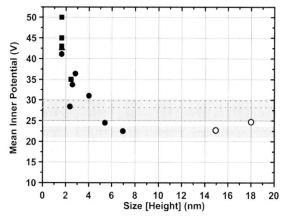

Figure 14.5 The mean inner potential of Au NPs as a function of the height. ○, spherical Au NPs on activated carbon; ▲, the line profile of the phase shift by Au NPs was linear fitted from 3D model; ■, top half of the truncated octahedron of Au on TiO$_2$; ●, spherical Au NPs on TiO$_2$ [29]. Shaded area with dotted lines shows calculated values for bulk Au. Shaded area without lines shows experimental values for bulk Au.

thickness, which corresponds to 2.0 nm in diameter. Although further investigation is needed to obtain quantitative correspondence, it can be assumed that appreciable change in electronic structure can be correlated with change in the products.

Zwijnenburg and colleagues found by X-ray photoelectron spectroscopy (XPS) and Mössbauer absorption spectroscopy (MAS) that metallic (zerovalent) Au NPs were active for the epoxidation [31]. Zhang et al. reported that Au^{3+} is stable and works as the active site in the hydrogenation of 1,3-butadiene in a diluted gas stream in the absence of O$_2$ [32]. In contrast, McFarland reported that Au particles of about 8 nm in diameter, deposited on TiO$_2$ by a micelle encapsulation method using block copolymers, favor hydrogenation in the reaction of C$_3$H$_6$ with O$_2$ and H$_2$ [33]. Gold in this catalyst was found to have no positive charge, and was zerovalent by XPS measurements. As the amount of Cl$^-$ remaining in the final Au/TiO$_2$ catalysts is different between the micelle encapsulation method and the DP method, a trace amount of Cl$^-$ may also be a key factor in determining the product selectivity.

14.3.4
Reaction Pathways for Propene Epoxidation over Au/TiO$_2$

At the interface between the gold NPs and the TiO$_2$ support, there may be a sort of equilibrium state represented by Ti^{4+} = O $\cdots$ Au^0 ⇔ Ti^{3+}–O–Au$^+$ (Scheme 14.4).

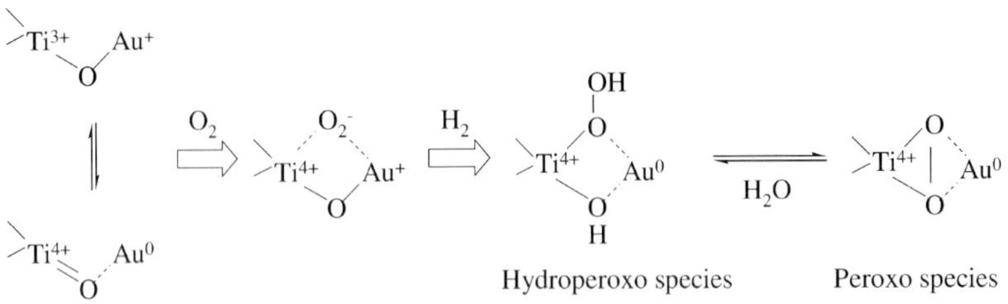

Scheme 14.4 Probable reaction pathways for propene epoxidation over Au/TiO$_2$ catalyst [12].

A smaller carbonyl peak at a higher wavenumber region in the FT-IR spectrum suggests that positively charged Au species are present [34]. Molecular oxygen is taken up by a Ti^{3+} cation oxygen deficient site and is activated, probably to an anionic oxygen species, which forms hydroperoxo- or peroxo-like species directly through reaction with H$_2$. The oxygen species adsorbed on the Ti sites at the periphery of Au NPs can react with C$_3$H$_6$ adsorbed on the surfaces of the Au NPs and of the TiO$_2$ support to produce PO.

Nijhuis et al. proposed a different reaction route, in which the role of the periphery around Au NPs is not explicitly described [35, 36]. Based on a detailed FT-IR spectroscopic investigation, they indicate that C$_3$H$_6$ reacts with the TiO$_2$ surfaces to form a bidentate propoxy species with the help of Au NPs. Over the Au surfaces, the peroxide species are formed by the reaction of O$_2$ with H$_2$, which is the rate-determining step. The bidentate propoxy species desorb with the assistance of the peroxide species to produce PO and water. This hypothesis is based on the theoretical prediction that gold is active to produce H$_2$O$_2$ from H$_2$ and O$_2$ [37] and on some recent experimental results for direct H$_2$O$_2$ synthesis [38, 39].

14.4
Au/Ti-SiO$_2$

14.4.1
Effect of Pore Structure and Pore Size of Titanium Silicate Support

In the case of titanium silicate supports, pore structure and pore size, and surface properties are crucial to the catalytic performance of the corresponding gold catalysts. Owing to its three-dimensional (3D) pore structure, Au/Ti-MCM-48 catalyst gives a higher PO yield than Au/Ti-MCM-41. The 3D pore structure has an advantage in the diffusion of reactant gases and products over the 2D pore structure of Ti-MCM-41. The yield of PO increases with the larger pore size of titanium silicates, as shown in Fig. 14.6 [40]. Mesoporous titanium silicates with large 3D pores (>7 nm) prepared using a modified sol–gel method are better than TS-1 and Ti-MCM type materials. Judging from the recent work by Delgass [41] which

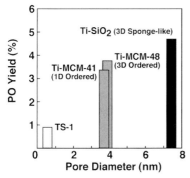

Figure 14.6 Propene oxide yield as a function of mean pore diameters of Ti-SiO₂ supports: catalyst, 0.15 g; catalyst bed temperature 423 K; feed gas, $C_3H_6/O_2/H_2/Ar$ = 10/10/10/70; space velocity, 4000 h^{-1} ml g$_{cat}^{-1}$. Actual Au loading, about 0.3 wt% [40].

succeeded in obtaining the highest PO yield, the reason why the TS-1 support appeared to be less effective might be that Au NPs could not be deposited with high dispersion because of its microporosity and hydrophobicity.

Rapid deactivation of Au/TiO₂ and Au/Ti-SiO₂ catalysts is caused by PO isomerization and cracking, followed by oligomerization over the catalyst surfaces [42]. Gold on Ti-HMM as well as Au/TS-1 has little deactivation compared with a Ti-MCM-type support owing to the hydrophobic surfaces which suppress the accumulation of PO and its derivatives [22].

14.4.2
Effect of Reactant Concentrations

The rate of PO formation is dependent on the concentration of reactant gases, especially on the concentration of H₂ [12] (Fig. 14.7). Over Au/Ti-SiO₂, at 350 K the rate increases in proportion to the H₂ concentration. The rate also increases with O₂ concentration up to 10 vol%, above which it levels off. In contrast, the PO formation rate is almost independent of C₃H₆ concentration up to 10 vol% and then tends to decrease, indicating that C₃H₆ can strongly adsorb on the catalyst surfaces. On the other hand, Delgass and coworkers have recently reported that over Au/TS-1 catalyst the reaction orders are 0.31 for O₂, 0.60 for H₂, and 0.18 for C₃H₆ at 443 K over Au/TS-1 catalyst [43].

14.4.3
Surface Treatments and Promoters

We have made a rough comparison between gas-phase propene epoxidation with O₂ and H₂ and the current industrial processes for PO production. In order to

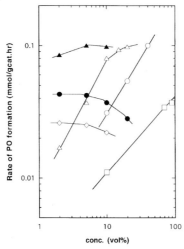

Figure 14.7 Rate constant for propene oxide formation over Au/Ti-SiO$_2$ (Ti-impregnated SiO$_2$) catalyst as a function of reactant concentrations: ○, C$_3$H$_6$/O$_2$/ H$_2$ = 5/10/10–40 (393 K); □, C$_3$H$_6$/O$_2$/H$_2$ = 5/5/10–90 (350 K); △, C$_3$H$_6$/O$_2$/H$_2$ = 5/2–20/40 (393 K); ◇, C$_3$H$_6$/ O$_2$/H$_2$ = 2–10/10/10 (350 K); ●, C$_3$H$_6$/O$_2$/H$_2$ = 2–20/10/40 (350 K); ▲, C$_3$H$_6$/O$_2$/H$_2$ = 2–10/10/40 (393 K). Each gas was diluted with Ar up to 100 vol% in total. Catalyst, 0.20 wt%Au/Ti-SiO$_2$ 0.5 g; space velocity, 4000 h^{-1} ml g$_{cat}^{-1}$ [12].

make a new direct process economically viable, the following are the preliminary targets for gold catalysts for commercial production of PO: C$_3$H$_6$ conversion of 10%, PO selectivity over 90%, H$_2$ efficiency above 50% and catalyst life longer than half a year.

Two-step titanium incorporation by grafting is effective to improve C$_3$H$_6$ conversion without changing PO selectivity and H$_2$ efficiency and suppresses catalyst deactivation owing to a decrease in the proportion of silanol sites. As mentioned in Section 14.4.1, a hydrophobic catalyst surface is an advantage. Thus, trimethylsilylation treatment, which changes surface Si–OH and Ti–OH groups to Si–O–Si–(CH$_3$)$_3$ and Ti–O–Si–(CH$_3$)$_3$ groups, raises PO yield. In the case of Ti-MCM-41, trimethylsilylation increases an optimum reaction temperature to 523 K from 423 K, while maintaining PO selectivity above 80% [44].

Although a combination of CsCl with Au catalysts depresses H$_2$ consumption, CsCl impregnation and mechanical mixing also considerably enhance the agglomeration of Au NPs [45]. Addition of Ba(NO$_3$)$_2$ or Mg(NO$_3$)$_2$ during Au deposition can improve catalytic activity. Figure 14.8 shows that a tiny amount (10–20 ppm) of trimethylamine as a gas-phase promoter appreciably improves C$_3$H$_6$ conversion, PO selectivity and H$_2$ efficiency through adsorption on Au NPs and support surfaces [46]. Interestingly, regeneration of the catalysts in the presence of trimethylamine leads to enhancement in terms of all catalytic parameters and provides a catalyst more active, selective, and stable than a fresh catalyst.

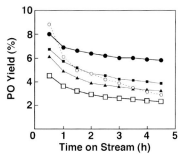

Figure 14.8 PO yields as a function of time-on-stream for Au/ Ti-SiO₂ (sponge-like mesoporous titanium silicate) catalysts with different modifications: Catalyst, 0.35 wt%Au/ Ti-SiO₂ 0.15 g; feed gas, C₃H₆/O₂/H₂/Ar = 10/10/10/70; space velocity, 4000 h⁻¹ ml g_cat⁻¹. □, Initial; ▲, trimethylsilylation; ■, trimethylsilylation + trimethylamine(15 ppm); ○, trimethylsilylation + Ba(NO₃)₂; ●, trimethylsilylation + Ba(NO₃)₂ + Trimethylamine(15 ppm)+regenerated.

In the case of TS-1, which is hydrophobic in nature, it was difficult when using the DP method to deposit the precise quantity of gold required. Recently, Delgass and coworkers reported that TS-1 treated with an NH₄NO₃ aqueous solution of ammonium nitrate (NH₄NO₃), prior to gold deposition, enables the Au to be captured more effectively and removes extra Cl⁻, and this leads to surprisingly high catalytic performance (see Table 14.1) [41].

14.4.4
Reaction Pathways for Propene Epoxidation over Promoted Au/Ti-SiO₂

A probable reaction pathway in this catalytic process is shown in Fig. 14.9. Molecular oxygen and hydrogen react to form H₂O₂ and H₂O over the surface of 2–5 nm Au NPs. The direct formation of H₂O decreases H₂ efficiency. Hydrogen peroxide formed on the Au surfaces is converted into hydroperoxo species at tetrahedrally coordinated Ti cation sites. The Ti-hydroperoxo species can subsequently react with C₃H₆ adsorbed on the SiO₂ surfaces to form PO. The existence of Ti-hydroperoxo species under reaction conditions were detected by *in situ* UV–vis spectroscopy [46] (Fig. 14.10).

Delgass and coworkers investigated Au/TS-1 by means of a kinetic study [43] and DFT calculations [47–49]. They suggested reaction pathways similar to those described above. A difference between theirs and ours comes from the sites designated for H₂O₂ generation and propene adsorption. Although in many, but not all, TS-1 supported catalysts, major Au species observed by TEM are Au NPs deposited on the outer surfaces of the TS-1, through DFT calculations it is suggested that gold clusters composed of 3–5 atoms which are confined in the microcages of TS-1 are active to form H₂O₂ from H₂ and O₂. In contrast, we assume that Au NPs are responsible for H₂O₂ formation. As for propene adsorption sites, Delgass assumes these are the Ti⁴⁺ sites whereas we assume they are the SiO₂ surfaces.

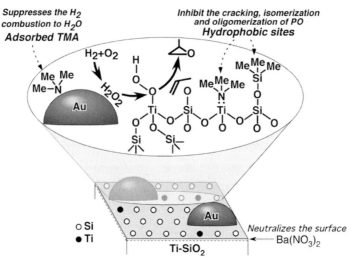

Figure 14.9 A probable reaction pathway for propene epoxidation over promoted Au/Ti-SiO₂ catalyst [46].

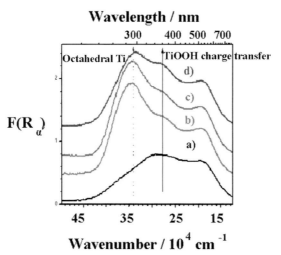

Figure 14.10 *In situ* UV–vis spectra for Au/Ti-SiO₂ catalyst before and after reaction of C_3H_6 with O_2 and H_2. (a) In a stream of Ar at 298 K before reaction; (b) in a stream of C_3H_6, O_2, H_2 and Ar during reaction at 423 K after 120 min time on stream; (c) after 270 min time on stream; (d) after reaction cooled to 298 K in a stream of Ar. Spectra are referenced to the titanium silicate support. Dotted arrow (. . .) shows the bands appearing due to the expansion of Ti coordination to octahedral during reaction. Solid arrow (—) shows the development of Ti-OOH charge transfer interactions [46].

14.5
Conclusions

The direct vapor phase epoxidation of propene using dioxygen (O_2) and dihydrogen (H_2) can be achieved using supported gold catalysts and is expected to contribute to future industrial processes for propene oxide production. At present, the following conclusions can be drawn:

1. The deposition–precipitation (DP) method of catalyst preparation is ideal for gold deposition. The gold NPs interact strongly with the support materials forming a hemispherical junction. Gold NPs are located selectively on the Ti sites of Ti-SiO_2 supports with controlled size.

2. The precise structure of the Au on TiO_2 catalysts can determine the course of the reaction, switching the process between oxidation and hydrogenation. This phenomenon is of great interest and may be caused by several factors; the dispersion of the Au NPs, the mean diameter of the Au NPs, the gold loading, the electronic properties of gold, and the amount of Cl^- remaining in the catalysts. When Au is deposited on TiO_2 by the DP method, the size of the Au NPs is crucial. Below 2 nm Au NPs promote hydrogenation of C_3H_6, whereas above 2 nm they promote epoxidation to propene oxide.

3. Titanium silicates having three-dimensionally (3D) developed pores larger than 7 nm are the most effective support material, probably owing to the ease of diffusion of reactant and product gases. On the other hand, microporous hydrophobic TS-1 can present stable gold catalysts, which also exhibit high performance when prepared by the DP method after the pretreatment of the TS-1 support with NH_4NO_3 aqueous solution.

4. In the case of Au/3D mesoporous Ti-SiO_2, trimethylsilylation of the surface hydroxy groups, treatment with $Ba(NO_3)_2$ or $Mg(NO_3)_2$ during gold deposition and addition of 10–20 ppm trimethylamine in the feed gas are effective in improving catalyst performance to a level close to industrial demands. Moreover, regeneration of the catalysts in the presence of trimethylamine improves all aspects of the catalyst performance.

5. Over Au/Ti-SiO_2 catalysts, H_2O_2 is generated on the surface of Au NPs and moves to the isolated Ti^{4+} in the silica matrix to create hydroperoxo species, which then epoxidise propene adsorbed on the silica surfaces to propene oxide. Over Au/TS-1 it is assumed that H_2O_2 is formed on small Au clusters inside the cages and reacts with propene adsorbed on the Ti^{4+} sites.

Acknowledgments

We are grateful to Dr. D.T. Thompson of the World Gold Council for his valuable comments and careful correction of English. Our thanks are also extended to Dr. M. Date of National Institute of Advanced Industrial Science of Technology and Dr. S. Ichikawa of Osaka University for drawing some figures for this chapter.

References

1 A. H. Tullo, P. L. Short, *Chem. Eng. News* 2006, *84*, 22.
2 Chemexpo.com, http://www.chemexpo.com, Chemical Profile, 9/10/2001.
3 T. A. Nijhuis, M. Makkee, J. A. Moulijn, B. M. Weckhuysen, *Ind. Eng. Chem. Res.* 2006, *45*, 3447.
4 J. Yamamotu, J. Tsuji, *Proceedings of the Catalysis Symposium on the Shift of Raw Materials and Fuels*, The Japan Petroleum Institute, Kogakuin University, Tokyo, 2nd December, 2005, pp. 7–12.
5 M. Ishino, J. Yamamotu, *Shokubai (Catalysts & Catalysis)* 2006, *48*, 511.
6 M. G. Clerici, G. Bellussi, U. Romano, *J. Catal.* 1991, *129*, 159.
7 *Elements-Degussa ScienceNewsletter*, 2004, *06*, 12–15.
8 J. R. Monnier, *Appl. Catal. A: Gen.* 2001, *221*, 73.
9 K. Murata, Y. Kiyozumi, *Chem. Commun.* 2001, 1356.
10 A. Sato, T. Miyake, *Shokubai (Catalysts & Catalysis)* 1992, *34*, 132.
11 R. Meiers, U. Dingerdissen, W. F. Hoeldrich, *J. Catal.* 1998, *176*, 376.
12 T. Hayashi, K. Tanaka, M. Haruta, *J. Catal.* 1998, *178*, 566.
13 A. L. De Oliveira, A. Wolf, F. Schuth, *Catal. Lett.* 2001, *73*, 157.
14 C. Wang, X. Guo, X. Wang, R. Wang, J. Hao, *Catal. Lett.* 2004, *96*, 79.
15 R. Wang, X. Guo, X. Wang, J. Hao, G. Li, J. Xiu, *Appl. Catal. A: Gen.* 2004, *261*, 7.
16 S. Tsubota, D. A. H. Cunningham, Y. Bandou, M. Haruta, in *Preparation of Catalysts V*, G. Poncelet, P. A. Jacobs, P. Grange, B. Delmon (Eds.), Elsevier, Amsterdam, 1991, p. 695.

17 B. S. Uphade, T. Akita, T. Nakamura, M. Haruta, *J. Catal.* 2002, *209*, 331.
18 G. C. Bond, C. Louis, D. T. Thompson, in *Catalysis by Gold*, Imperial College Press, London, 2006, Section 4.2.3, p. 79.
19 C. Qi, T. Akita, M. Okumura, M. Haruta, *Appl. Catal. A: Gen.* 2001, *218*, 81.
20 A. P. H. P. Van Der Pol, J. H. C. Van Hooff, *Appl. Catal. A: Gen.* 1992, *92*, 93.
21 A. K. Sinha, S. Seelan, M. Okumura, T. Akita, S. Tsubota, M. Haruta, *J. Phys. Chem. B*, 2005, *109*, 3956.
22 M. P. Kapoor, A. K. Sinha, S. Seelan, S. Inagaki, S. Tsubota, H. Yoshida, M. Haruta, *Chem. Commun.* 2002, 2902.
23 B. S. Uphade, S. Tsubota, T. Hayashi, M. Haruta, *Chem. Lett.* 1998, 1277.
24 M. Haruta, B. S. Uphade, S. Tsubota, A. Miyamoto, *Res. Chem. Intermed.* 1998, *24*, 329.
25 M. Haruta, S. Tsubota, T. Kobayashi, H. Kageyama, M. J. Genet, B. Delmon, *J. Catal.* 1993, *144*, 175.
26 T. Akita, K. Tanaka, S. Tsubota, M. Haruta, *J. Electron Microsc.* 2000, *49*, 657.
27 C. Bianchi, F. Porta, L. Prati, M. Rossi, *Top. Catal.* 2000, *13*, 231.
28 S. Naito, M. Tanimoto, *J. Chem. Soc., Chem. Commun.* 1988, *12*, 832.
29 K. Okazaki, S. Ichikawa, Y. Maeda, M. Haruta, M. Kohyama, *Appl. Catal A: Gen.* 2005, *291*, 45.
30 Y. Maeda, M. Okumura, S. Tsubota, M. Kohyama, M. Haruta, *Appl. Surf. Sci.* 2004, *222*, 409.
31 A. Zwijneburg, A. Goossens, W. G. Sloof, M. W. J. Craje, A. M. Van Der Kraan, L. Jos De Jongh, M. Makkee, J. A. Moulijn, *J. Phys. Chem. B* 2002, *106*, 9853.
32 X. Zhang, H. Shi, B.-Q. Xu, *Angew. Chem. Int. Ed.* 2005, *44*, 7132.

33 J. Chou, N. R. Franklin, S-H. Baeck, T. F. Jaramillo, E. W. McFarland, *Catal. Lett.* **2004**, *95*, 107.

34 F. Boccuzzi, A. Chiorino, M. Manzoli, P. Lu, T. Akita, S. Ichikawa, M. Haruta, *J. Catal.* **2001**, *202*, 256.

35 T. A. Nijhuis, T. Visser, B. M. Weckhuysen, *J. Phys. Chem. B* **2005**, *109*, 19309.

36 T. A. Nijhuis, T. Q. Gardner, B. M. Weckhuysen, *J. Catal.* **2005**, *236*, 153.

37 P. P. Olivera, E. M. Patrito, H. Sellers, *Surf. Sci.* **1994**, *313*, 25.

38 C. Sivadinarayana, T. V. Choudhary, L. L. Daemon, J. Eckert, D. W. Goodman, *J. Am. Chem. Soc.* **2004**, *126*, 38.

39 M. Okumura, Y. Kitagawa, K. Yamaguchi, T. Akita, S. Tsubota, M. Haruta, *Chem. Lett.* **2003**, *32*, 822.

40 A. K. Sinha, S. Seelan, S. Tsubota, M. Haruta, *Top. Catal.* **2004**, *29*, 95.

41 L. Cumaranatunge, W. N. Delgass, *J. Catal.* **2005**, *232*, 38.

42 G. Mul, A. Zwijnburg, B. Van Der Linden, M. Makkee, J. A. Moulijn, *J. Catal.* **2001**, *201*, 128.

43 B. Taylor, J. Lauterbach, G. E. Blau, W. N. Delgass, *J. Catal.* **2006**, *242*, 142.

44 C. Qi, T. Akita, M. Okumura, K. Kuraoka, M. Haruta, *Appl. Catal A: Gen.* **2003**, *253*, 75.

45 B. S. Uphade, O. Okumura, S. Tsubota, M. Haruta, *Appl. Catal. A: Gen.* **2000**, *190*, 43.

46 B. Chowdhury, J. J. Bravo-Suarez, M. Date, S. Tsubota, M. Haruta, *Angew. Chem. Int. Ed.* **2006**, *45*, 412.

47 D. H. Wells Jr., W. N. Delgass, K. T. Thomson, *J. Catal.* **2004**, *225*, 69.

48 A. M. Joshi, W. N. Delgass, K. T. Thomson, *J. Phys. Chem. B* **2005**, *109*, 22392.

49 D. H. Wells, Jr., W. N. Delgass, K. T. Thomson, *J. Am. Chem. Soc.* **2004**, *126*, 2956.

50 T. A. Nijhuis, T. Visser, B. M. Weckhuysen, *Angew. Chem. Int. Ed.* **2005**, *44*, 1115.

51 J. Chou, E. W. MacFarland, *Chem. Commun.* **2004**, 1648.

52 A. Zwijneburg, M. Saleh, M. Makkee, J. A. Moulijn, *Catal. Today* **2002**, *72*, 59.

15
Gold Nanoparticles: Recent Advances in CO Oxidation

Catherine Louis

15.1
Introduction

The oxidation of carbon monoxide at around room temperature is the most famous reaction known for gold catalysts. Haruta's group discovered in 1987 [1, 2] that gold is a unique catalyst for this reaction when gold metal particles are smaller than 5 nm and supported on oxides. Since then, extensive and intensive fundamental works have been published, and expanding new applications, from air purification (gas masks, gas sensors, indoor air quality control) to hydrogen purification for fuel cells (PROX, preferential selective oxidation of CO in the presence of H_2) have been developed.

Gold is indeed active in carbon monoxide oxidation at a much lower temperature ($\leq$RT) than any platinum group metal [3] (Fig. 15.1). The difference in reactivity can be due to the fact that Pt, Pd and Rh easily dissociate molecular oxygen at low temperature and bind strongly both atomic oxygen and CO. Tightly bound adsorbates have to overcome sizeable barriers to react, making the reaction rates significant only at rather high temperatures [4]. On the contrary, on gold the reactants are loosely bound, but a higher binding energy of CO on gold nanoparticles than on bulk gold may provide sufficient concentrations of CO on the surface for the reaction of oxidation to occur with negligible energy barriers. The main problem with this apparently simple reaction is that if CO is known to adsorb on low coordinated surface gold sites, the adsorption and activation of oxygen on gold particles is more puzzling (Section 15.4.3). The CO oxidation mechanism is not yet elucidated in spite of the huge number of papers published, and four main mechanisms have been suggested.

High activity in CO oxidation when gold particles are smaller than 5 nm, and supported on reducible oxides, such as iron oxide or titania, led Haruta et al. [3, 5–8] to propose the first mechanism of CO oxidation (Fig. 15.2): the reaction takes place at the interface between the gold metal particle and the oxide support, i.e., between CO adsorbed on the gold particles and O_2 activated by the oxide support. Later, based on the assumption that metal cations could be located at the metal–support interface, Bond and Thompson [9] proposed a mechanism rather similar,

Nanoparticles and Catalysis. Edited by Didier Astruc
Copyright © 2008 WILEY-VCH Verlag GmbH & Co. KGaA, Weinheim
ISBN: 978-3-527-31572-7

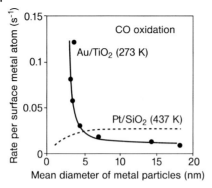

Fig. 15.1 Turnover frequency (TOF: number of reaction cycles per unit of time and surface site) for CO oxidation at 273 K over Au/TiO_2 and at 437 K over Pt/SiO_2 catalysts as a function of the diameter of the gold particles, from Ref. [14].

$$CO + Au \rightleftharpoons O\equiv C \cdots Au$$

$$O_2 + Au/TiO_2 \longrightarrow Au/TiO_2 \cdots O_2$$

$$Au/TiO_2 \cdots O_2 + [O\equiv C \cdots Au] \xrightarrow[slow]{O_2} O\equiv C \cdots Au + CO_2$$

$$\underset{O}{O\equiv C \cdots Au} \longrightarrow CO_2$$

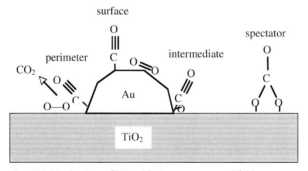

Fig. 15.2 Mechanism of CO oxidation on supported fold catalysts, as proposed by Haruta et al. [16].

but involving oxidised gold species (Au^{III}) at the gold–support interface (Fig. 15.3).

On the other hand, the study of CO oxidation performed by Goodman et al. [10] on a $Au/TiO_2(110)$ *model* catalyst, led them to propose that the whole reaction takes place on the gold particles, providing that the latter are small and bi-dimensional; they do not exhibit metallic properties any longer due to *quantum size effect*.

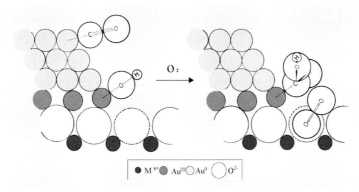

Fig. 15.3 Mechanism of co oxidation on supported fold catalysts, as proposed by Bond and Thompson [9].

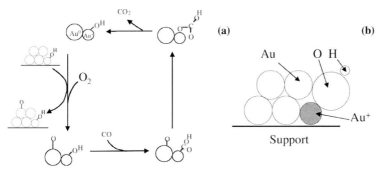

Fig. 15.4 Mechanism of CO oxidation on supported gold catalysts, as proposed by Kung et al. (a) and model ensemble of Au^0 and Au^+–OH sites (b) [11].

According to these authors, the role of the support would only be to stabilise the small gold particles. Finally Kung et al. [11, 12] explained the activity of gold on alumina, although weaker than gold on titania, by another mechanism based on the presence of Au^I species at the gold–support interface and on a reaction between O_2, which dissociatively adsorbs on Au^0, and CO, which reacts with Au^I–OH species (Fig. 15.4).

The elucidation of the reaction mechanism is complicated by the complexity of the whole system, i.e. the catalyst itself, which depends on the preparation conditions, and the conditions of reaction. This leads to contradictory experimental results, which inevitably lead to contradictory interpretations.

Several reviews have been published on CO oxidation by gold catalysts [3, 9, 13–18]. Readers can also refer to the special issue of *Applied Catalysis A* 291 (2005), and to a recent book entitled *Catalysis by Gold* [19]. The objective of the present chapter is to try to summarise the state of art of this reaction with gold catalysts, from the point of view of a researcher involved in *real* catalysis (in contrast with *model* catalysis). The objective is also to try to show that the contributions of surface science and quantum chemical calculations are very useful in providing a better understanding of the reaction. Considering the great extent of literature, it is possible that some relevant papers have been forgotten.

15.2
Preparation of Supported Gold Catalysts

It is useful first to summarise the main methods of preparation for gold catalysts supported on powdered oxides (so called *real* catalysts). Those used to prepare *model* catalysts, i.e. gold supported on single crystal oxides, are the same as for other metals, and are not repeated here.

The gold precursor most used is tetrachloroauric acid ($HAuCl_4 \cdot 3H_2O$) whose gold is in the oxidation state III. After gold deposition on a support, it is important to eliminate the chlorides because they induce gold particle sintering during the subsequent thermal treatments performed to produce metal particles [8, 20]. It is noteworthy that most of the thermal treatments are performed in air, and that oxidised forms of gold are unstable ($\Delta H_f(Au_2O_3) = +19.3 \, kJ \, mol^{-1}$), and readily lead to metallic gold.

The most simple and most conventional method of preparation is *impregnation*. It consists of wetting a support with an aqueous solution containing the chloroauric precursor, then evaporating the water. As mentioned above, the presence of chlorides in the samples leads to the formation of large gold particles (>10 nm) upon thermal treatment, and therefore to inactive catalysts. Alternatives have been recently proposed; they consist of a post-treatment with ammonia, either in the gas phase [21], or in the liquid phase [22], followed by washings with water. Chlorides can be eliminated without gold leaching, and small gold particles (3–4 nm on average) can be obtained. Bench scientists must be aware of the necessity to avoid the formation of explosive harmful *fulminating gold* when they use ammonia [23]. Washing with NaOH [24] and Na_2CO_3 are also reported [25].

Methods other than impregnation were applied by Haruta's group who succeeded in forming small gold particles, and as a consequence discovered the catalytic properties of gold. The first method used was *co-precipitation* where both the support and gold are supposed to precipitate as hydroxides [1, 26, 27]; it is, however, possible that part of the gold could be embedded in the support. The second is the famous *deposition–precipitation* method [28]. It consists of the precipitation of gold on an oxide support at a pH fixed by addition of NaOH or Na_2CO_3 to the suspension containing $HAuCl_4$ and the support. The sample is then washed with water to eliminate as much chloride and sodium ions as possible, dried under vacuum at 373 K, then calcined in air. This method works only with oxide supports, the point of zero charge (PZC) of which is higher than 5–6. Small gold particles are obtained, but a large proportion of gold is not deposited on the support, and remains in the solution. Again, an alternative procedure has been proposed [29, 30]; it consists of gradually basifying the solution through thermal decomposition of urea at 353 K ($CO(NH_2)_2 + 3 \, H_2O \rightarrow 2 \, NH_4^+ + CO_2 + 2 \, OH^-$). With this method, which includes longer time aging (4 to 20 h) all the gold is deposited on the support and, in principle, any gold loading can be reached, and small gold particles (2–3 nm on average) are obtained after thermal treatment.

Organometallic gold precursors free of chlorides can also be used. For instance, gold catalysts can be prepared by *impregnation with gold-phosphine complexes*, such as $[Au^I(PPh_3)]NO_3$ [31, 32] or $[Au_6(PPh_3)_6](BF_4)_2$ [33] that must be synthesised. Preparation must be performed in an organic medium, which is then evaporated. After calcination small gold particles are obtained with the expected gold loading. Another characteristic feature of this method is that the support must be freshly prepared metal hydroxides to provide sufficient amount of OH groups for the reaction with the phosphine complex. Dimethylgold acetylacetonate ($Me_2Au(acac)$, $(CH_3)_2Au^{III}(OCCH_3)_2CH)$) that is commercial, can be used either for *chemical vapor deposition* (CVD) [34] or for *impregnation* in organic solvent [35–37]. Although these preparations require the absence of moisture, and the size distribution of gold particles is wider than that by *deposition–precipitation*, an advantage is that any type of supports, including activated carbon and silica, can be used.

Other kinds of preparation involve *mixing with preformed gold colloids in the liquid phase*; gold is already in the metallic state and size-controlled. The gold particles can be stabilised by donor ligands, polymers, or surfactants [38–42], but the elimination of the stabilising agent through thermal treatment may lead to particle sintering. The gold particles may also be encapsulated within dendrimers [43, 44] before deposition. A constraint is that dendrimer synthesis is expensive. Naked gold particles of controlled size can also be deposited by *cluster beam deposition* [45] or by *magnetron sputtering* [46], but this requires UHV equipment.

15.3
Main Parameters Influencing the Catalytic Behavior of Gold Supported on Metal Oxides in CO Oxidation

15.3.1
Preliminary Remarks

As emphasised by several groups [9, 11], one complication with gold catalysts is that large variations in catalytic activity are found in the literature over catalysts of apparently similar composition. This is due to the fact that some parameters of preparation are difficult to control. For instance gold hydroxides and oxides are photo-sensitive and easily reducible, even at room temperature, and uncontrolled reduction may occur during preparation, drying and storage. Especially, it is risky to store dried samples in air for a long time (weeks or months). More and more groups [11, 21, 42, 47–50] are now aware of these problems, and perform the preparation in the dark, dry the samples at low temperature ($\leq$373 K under vacuum or at RT) and store them free of light and air or water. As mentioned in Section 15.2, it is very important to eliminate the chlorides before thermal treatment and possibly also sodium when NaOH or Na_2CO_3 are used for the preparations [25, 49].

Catalytic activity also clearly depends on the conditions of activation and of reaction, especially water content in the gas feed. Since the CO oxidation reaction is highly exothermic ($\Delta H = 283\,kJ\,mol^{-1}$), precise temperature control is also difficult in a stream containing more than 1 vol.% CO.

In spite of these difficulties, several important conclusions can be drawn from the literature data.

15.3.2
Gold Particle Size

There is general agreement that the *Turn Over Frequency* (TOF: number of cycles of reaction per unit of time and surface atom) or the *reaction rate* (per unit amount of Au) of CO oxidation drastically increases when the gold particle size decreases, and that the reaction of CO oxidation is structure sensitive [7, 8, 27, 31, 34, 51–60] (Fig. 15.5). This result is of course valid within the range of gold particle sizes that can be obtained on *real* catalysts, i.e., down to about 1–2 nm. The quasi-unique exception is the study reported by Goodman et al. with a *model* catalyst Au/TiO$_2$(110) for which they found that the TOF is maximum for a particle size of ~3 nm (Fig. 15.6) [10].

15.3.3
Nature of the Support

The presence of an oxide support, and therefore of a gold–support interface, is of definitive importance to get active catalysts. Unsupported gold particles are poorly

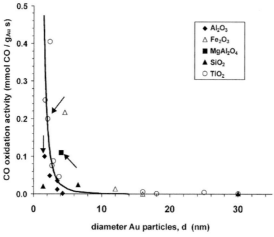

Fig. 15.5 Activities in mmol CO g$_{Au}$$^{-1}$ s^{-1}) at 273 K over different Au-based catalysts as a function of the average particle size, from Ref. [59]. The points are collected from Ref. [59] (arrows) and a number of other publications [8, 27, 31, 34, 54–57].

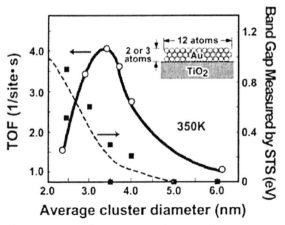

Fig. 15.6 Turnover frequency (TOF) for CO oxidation as a function of the diameter of the gold particles at 350 K over Au/TiO$_2$(110) (—); band gap of bidimensional clusters (- - -), from Ref. [10].

active or inactive, but are much larger than those present in active supported catalysts: 76 nm [61] and 10 μm [62]. The 10 μm gold crystallites become slightly active in CO oxidation when titania is deposited onto their surface. In a totally different range of size, free anionic gold clusters, Au$_2^-$ and Au$_6^-$ were found to be active in CO oxidation [63, 64].

Several groups have investigated the influence of the nature of the support on the catalyst activity. Whether the samples are prepared by *co-precipitation* [7, 26], *impregnation with phosphine gold complex* [32, 65], *deposition of gold colloids* [40], *beam deposition of gold clusters* [45], or *deposition–precipitation* [54, 66, 67], there is a general agreement that gold supported on reducible oxides is more active and stable than gold supported on non-reducible oxides (SiO$_2$, γ-Al$_2$O$_3$ and MgO); the TOF can be larger by up to one order of magnitude. It is of course possible to find a few exceptions [34, 42]. For instance, Schüth et al. [42] found that the activity of samples prepared by *deposition of gold colloids* (~3 nm) varies in an unexpected way according to the reducible or non-reducible nature of the support: TiO$_2$ > Al$_2$O$_3$ > ZrO$_2$ > ZnO.

15.3.4
Water in the Gas Feed or in the Catalyst

During catalytic reaction, the amount of water in the gas feed (typically 3–10 ppm) and in the catalyst is usually not controlled, but it has a drastic influence on the catalytic properties, even though all types of tendencies are reported. For instance, in the case of Au/TiO$_2$ catalysts the addition of water to the gas feed may have a positive effect [68, 69], no effect [70] or a negative effect [40, 62] on the activity.

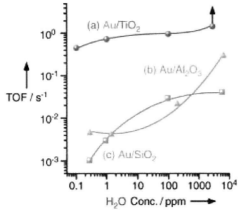

Fig. 15.7 Turnover frequency (per surface gold atom) at 273 K for CO oxidation over (a) Au/TiO$_2$; (b) Au/Al$_2$O$_3$; (c) Au/SiO$_2$ as a function of moisture concentration. Upright arrow indicates the saturation of CO conversion, from Ref. [72].

The effect of moisture in the reactant gas has been systematically studied by Haruta et al. [71, 72] on several catalysts, Au/TiO$_2$, Au/Al$_2$O$_3$ and Au/SiO$_2$ prepared by *deposition–precipitation* over a wide range of concentrations, from 80 ppb (extremely dry conditions) to 6000 ppm H$_2$O. Figure 15.7 shows that moisture can enhance the catalytic activities by more than two orders of magnitude, but this depends on the type of metal oxides. The authors anticipate that moisture plays two roles in the reaction: in the activation of oxygen and in the decomposition of carbonates, which form on basic supports. The last interpretation is supported by the Behm's group who observed by DRIFTS that water facilitates the decomposition of carbonates [73]. In Kung's mechanism (Fig. 15.4), water is effective in preventing deactivation and in regenerating deactivated Au/Al$_2$O$_3$ catalysts, and its role would be to restore the Au–OH groups, which participate in CO oxidation [12]. The promoting effect of water on Au/Al$_2$O$_3$ reported by Calla and Davis [74, 75] as well, is also consistent with an active site model that includes the coexistence of metal gold and hydroxyl groups or other species derived from H$_2$O.

Even though the effect of water addition on the activity of gold catalysts in CO oxidation is not fully understood, it is clear that the sensitivity of the catalysts to moisture may also contribute to the diversity of catalytic results reported in the literature.

15.3.5
Conditions of Activation of Gold Catalysts and State of Gold in Active Catalysts

Regarding the best conditions of activation to get active catalysts, there are enormous discrepancies between one paper and another. Often thermal activation treatments are performed in air, possibly for convenience, but also because some authors claim that gold catalysts activated under vacuum or hydrogen suffer severe deactivation during CO oxidation [76]. In contrast, others found a more active Au/TiO$_2$ catalyst after H$_2$ treatment than after air treatment at the same temperature because gold particles are smaller [53], and nowadays, pretreatments are performed under H_2 more frequently [77, 78], or there is no pre-treatment after the drying step (so-called *as-prepared* samples) [21, 79, 80].

So far, it has been assumed that gold has to be metallic to be active in the CO oxidation reaction, but one can find all kinds of statements in the literature regarding the state of gold in active catalysts (except when specified, all the catalysts mentioned below were prepared from $HAuCl_4$):

- *As-prepared* samples are more active than activated samples [21, 79, 80, 81]. In principle this means that samples containing mainly Au^{III} are more active than samples containing mainly Au^0. Several arguments against this statement can be given: (i) It is not certain that all the gold in the *as-prepared* samples is really unreduced since it is highly reducible (see Section 15.3.1), indeed some authors report that gold was partially reduced after drying [82–85]; (ii) the $CO:O_2$ gas feed may be reducing at RT [53, 77, 79, 86]; (iii) activation treatment may be accompanied by particle growth, and this makes comparison with *as-prepared* samples difficult; moreover *as-prepared* samples contain much more water than samples after thermal pretreatment, and this also may influence the activity (see Section 15.3.4); (iv) the reaction of *as-prepared* samples is non-catalytic [85].
- *As-prepared* gold in catalysts is not active [11, 50]. This implies that gold in these samples is unreduced, and that there would be no reduction under the reactant gas. Maybe the absence of reduced gold species inhibits reduction under the reactant gas as suggested in Ref. [78]. This point deserves deeper understanding.
- Gold must be fully reduced to be active. This means that after thermal treatment, only Au^0 species have been detected by techniques such as XPS [33, 40, 42, 87], XAFS [52, 78, 88] or Mössbauer [89]. However these techniques cannot detect the presence of very small amounts of unreduced gold.
- Cationic gold is necessary for activity in CO oxidation; this point is further developed in Section 15.5.3.

15.4
Properties of Gold Nanoparticles (Free or Supported)

15.4.1
Electronic Properties

As mentioned in Section 15.3.2, small gold particles give more active catalysts, but the decrease in size is accompanied by several other changes that are examined below.

Decreasing the particle size leads to an increasing proportion of low coordinated surface atoms, such as edge and corner atoms. The presence of these sites is a key factor for catalytic activity, since they are required at least for the adsorption of CO, and possibly for that of O_2 (see Section 15.4.3).

Decreasing the particle size also leads to changes in the metal structure. The fcc cuboctahedral structure of bulk gold, also found in large crystallites with specific faceted morphologies, evolves into truncated-decahedral structures for sizes between 1 and 2 nm (~250 atoms), and into specific "molecular" structures when the cluster size is smaller than 1 nm (<40 atoms) [90]. This is the same for supported gold; for instance on MgO, the gold particles between 1 and 5 nm are in two forms, fcc cuboctahedral and decahedral, while those smaller than 1 nm are fcc cuboctahedral and icosahedral [91].

The decrease in particle size also induces changes in the electronic properties [92, 93]; the particles lose metallic character, and become non-metallic or semiconductive (*quantum size effect*). Various spectroscopic techniques (XPS [94–96], HREELS [95], scanning tunneling spectroscopy (STS) [10, 98], [197]Au Mössbauer [99, 100], field emission energy [101]) indicate these changes when supported gold particles become smaller than 1–2 nm (100–300 atoms). Goodman et al. [10] showed by STS that the maximum of activity in CO oxidation obtained for Au/ $TiO_2(110)$ is due to the presence of bi-dimensional gold particles ~3 nm long and 1.0 nm high (~300 atoms) (Fig. 15.6), that exhibit a band gap of 0.2–0.6 eV, i.e. have lost their metallic character.

The changes in the electronic properties and in the structure of the gold particles can also be reflected by changes in the Au Au bond distance. When supported metal particles are smaller than 3–4 nm, there is a contraction of the metal–metal bond distance [102]. The Au Au contraction is independent of the nature of the support, and similar to that reported for free gold particles. For gold catalysts with metal dispersions near unity (~1 nm), the metallic bond length is much shorter (2.72 Å) than that of the bulk (2.88 Å).

15.4.2
Metal–Support Interactions

The necessity of having an oxide support to stabilise small gold particles, and make them active, and the importance of the nature of the support indicate that metal–support interaction is a key factor in the CO oxidation reaction.

15.4.2.1 Particle Morphology

The shape of the gold particles is an important factor for catalytic activity since it determines the relative amount of corners and edges, i.e., of low-coordinated atoms. It depends on the support through the interfacial energy; this is well known from works performed on *model* samples (Table 15.1). The higher the energy, the flatter the gold particles.

The shape of the gold particles may also depend on the activation temperature of the gold catalysts. This was observed with Au/TiO_2 prepared by *deposition–*

Table 15.1 Interfacial energy between gold particles and oxide supports.

Oxide	d_{Au} (nm)	Interfacial energy ($J\,m^{-2}$)	Ref.
Al_2O_3	1	0.27	103
MgO(100)	5	0.45	104
TiO_2(110) (anatase)	4	0.75–1.21	105

Table 15.2 Influence of the shape of gold particles on the activity of Au/TiO_2 ($T_{50\%}$: temperature corresponding to 50% conversion), from Haruta et al. [39].

Preparation	Calcination T (K)	d_{Au} (nm)	Particle shape	$T_{50\%}$ (K)
colloidal mixing	473	5	spherical	473
	873	12	hemispherical	273
deposition–precipitation	673	3.6	hemispherical	233

precipitation [106]. After reduction at 473 K under H_2, the shape of the gold particles was undefined with round parts, i.e., with both a large number of low coordinated surface atoms and a large perimeter of interface with the support, leading to high activity in CO oxidation. At a higher reduction temperature (673–773 K), the particles became facetted, i.e., contained a smaller number of low coordinated atoms. Although the particles did not drastically increase in size, the activity decreased.

Spherical gold particles are also less active than hemispherical ones because of the smaller interface perimeter (Table 15.2). This result has been correlated to the perimeter length of the gold–support interface [8], and has contributed to the establishment of the mechanism reported in Fig. 15.2.

15.4.2.2 Influence of the Oxide Support on the Electronic Properties of Gold Particles

The oxide support may also have an influence on the electronic properties of the metal gold particles, and most of the studies have been performed on *model* samples. The few examples given here compare several supports.

The reactivity of gold toward SO_2 has been used to probe metal–support interaction. Experiments of XPS and thermal desorption, and DFT calculations showed that the dissociation of SO_2 on gold particles (<5 nm) is possible when they are supported on TiO_2(110) [107], and is, in contrast, very limited on MgO(100) due to weak Au–MgO interaction [108]. The interpretation of this different reactivity is that gold interacts with oxygen vacancies of TiO_2(110) that lead to charge transfer to gold, and make it more reactive. On stoichiometric CeO_2(111) gold is inactive whereas it is active on non-stoichiometric ceria [109].

In contrast an IR study did not indicate a support effect when CO was adsorbed on gold particles (~3 nm) prepared by *vapor deposition of atoms* on FeO(111),

$Fe_3O_4(111)$ and Al_2O_3 thin film [110]; indeed the frequencies of CO vibration, and therefore CO interaction, were remarkably similar.

Moreover DFT calculations showed in the case of Au_{10} supported on reduced $TiO_2(110)$ that the electrons trapped in the O-deficient substrate are employed to furnish the bonds at the interface, but that the added charge is accumulated between Au and Ti atoms [111]. As a consequence, the Au atoms of the second layer are not affected by the presence of the support.

In the case of gold particles of the same size (~3 nm) deposited by *cluster beam deposition* on powder (hydroxylated) supports, the different binding energy of the Au $4f_{7/2}$ peak: $Au/\gamma-Al_2O_3 < Au/TiO_2 < Au/ZrO_2 <$ metallic gold, was attributed to the different extent of electron transfer from the oxides to the gold particles [45]. However, no correlation could be established with the variation of TOF for CO oxidation: $Au/TiO_2 > Au/ZrO_2 >> Au/\gamma-Al_2O_3$.

The apparent disagreements within this last work and between those obtained on *model* samples may be due to the fact that it is not possible to isolate electronic effect from morphological effect on gold particles, in *model* and *real* catalysts.

15.4.2.3 Influence of the Gold Particles on the Oxide Support Properties

Reciprocally, the presence of gold particles has an influence on the properties of the oxide supports. Gold enhances the reactivity of titania by modifying the rate of vacancy exchange between the bulk and the oxide surface, and by facilitating the migration of O vacancies from the bulk to the oxide surface [107]. The reduction under hydrogen of oxide supports such as titanium, cerium and iron oxides, occurs at a lower temperature in the presence of gold than with the pristine supports [112, 113]. This may be due to H_2 dissociation on gold particles then to H spillover onto the oxide, or changes in the electronic properties of the support. Under the $CO:O_2$ gas feed, gold also facilitates the reduction of the Fe_2O_3 support at 353 K [54]. The presence of gold retards phase transformation from anatase to rutile [114].

It is also proposed that oxygen vacancies of oxide surfaces are more abundant in the proximity of the gold particles as the direct consequence of the Schottky junction at the metal–semiconductor interface [115]. The charge polarisation induces a transfer of electrons from the support to the Fermi level of the metal, and as a consequence, for instance for titania, weakens the oxygen–titanium bond, and increases the number of oxygen vacancies in the substrate [116]. This would explain the oxygen activation on the support in close proximity to gold particles (see mechanism of Fig. 15.2), and that the presence of gold alters the band gap of titania [117] and ceria [118].

15.4.3
CO and O_2 Chemisorption

CO can chemisorb at 300 K or below on films and single crystals of gold, but only when they contain surface defect sites; CO also chemisorbs on low coordinated sites of gold particles (see for instance Ref. [15] and references therein).

In contrast, O_2 is unable to adsorb or dissociate on gold surfaces under ambient conditions [119–122]. However, free gold clusters (Au_n, $n=1$ to 20) are reactive toward molecular oxygen provided that they are anionic and contain an even number of atoms [63, 123–127]. With clusters having an odd number of electrons oxygen acquires a negative charge as O_2^-. In the case of supported gold nanoparticles, there are studies that show that they can adsorb or react with O_2 at temperature higher than RT (573 K), indicating that O_2 chemisorption is an activated process [128–130]. There is also a paper [131] reporting evidence of O_2 chemisorption at 77 K on gold nanoparticles supported on $TiO_2(110)$ and on Au(111), but only after exposure to a plasma-jet molecular oxygen beam. Recently it has been shown by XAFS that after H_2 reduction of Au/TiO_2 and Au/Al_2O_3 samples, the gold particles smaller than ~3 nm are reactive to air at 300 and 500 K, and that up to ~15% of the Au atoms can be oxidised [102, 132]. This reactivity has been correlated with the fact that the electronic properties of gold are modified when particles are so small, as exemplified by the shortening of the Au–Au lattice distance (Section 15.4.1).

In contrast, atomic oxygen easily chemisorbs on a gold surface [122, 133, 134] or particles [135, 136], and then readily reacts with CO.

15.5
Overview of the Mechanisms of Carbon Monoxide Oxidation

15.5.1
Mechanisms Involving the Oxide Support

Two mechanisms of CO oxidation reaction among those proposed in the Introduction (Section 15.1) involve reducible oxide supports: Haruta's mechanism with fully metallic gold particles (Fig. 15.2) and Bond and Thompson's mechanism involving unreduced gold species at the metal–support interface (Fig. 15.3).

15.5.1.1 Haruta's Mechanism: Metal Gold Particles
In the very first CO oxidation mechanism proposed by Haruta's group (Fig. 15.2), the reaction takes place at the interface of the gold metal particle with TiO_2, between CO adsorbed on the gold particles and O_2 activated on the support. The mechanism is based on the fact that according to the authors, (i) a maximum of activity is obtained when gold is fully reduced; (ii) reducible supports such as TiO_2 or Fe_2O_3 lead to more active catalysts than non-reducible supports such as SiO_2 or Al_2O_3 (Section 15.3.3); O_2 can be adsorbed and activated on these supports; (iii) hemispherical gold particles are more active than spherical particles (Section 15.4.2.1) because of the larger perimeter of interface between the gold and the support. Several other groups agree with this mechanism [45, 54, 137, 138, 139].

This mechanism is consistent with a *Langmuir–Hinshelwood* mechanism with a reaction between adsorbed CO and O_2 molecules as the rate-limiting step, based on kinetics studies performed on Au/TiO_2 catalysts [33, 56, 70, 73] and IR studies

[62]. In line with that, several studies of O_2 adsorption and reaction with CO on Au/TiO_2 have been performed to determine how oxygen is activated on this support and react with CO adsorbed on Au particles: $^{18}O_2$ isotope exchange by FTIR [69] and temporal analysis of products (TAP) [54, 70, 73, 140, 141]. The authors agree that the reaction mechanism on Au/TiO_2 involves adsorbed O_2, and not lattice oxygen atoms since oxygen does not exchange with the latter [70, 73]. O_2 only exchanges with CO_2 formed by the CO oxidation reaction [140, 141]; so reaction does not proceed through a *Mars and van Krevelen* mechanism. On the other hand, superoxide anion radical O_2^- identified by EPR, was proposed to be an intermediate for the reaction [54, 70, 142]. This O_2^- species would arise from the adsorption of O_2 on oxygen vacancies of the oxide surfaces, which would be more abundant in the proximity of the gold particles, as mentioned in Section 15.4.2.3. It is also proposed that activated oxygen could spillover from titania towards the gold particles [143], and that the smaller the gold particles, the faster the spillover of the reactive oxygen species from titania to gold.

According to experiments of $CO-O_2$ time-resolved titration and of isotopic exchange performed with a TAP reactor [54], gold supported on other reducible supports (iron, nickel and cobalt oxides) and calcined at 673 K (metallic state) behaves in the same way as Au/TiO_2, therefore with a mechanism consistent with Haruta's mechanism (Fig. 15.2). Iwasawa et al. also found the same type of results for calcined Au/TiO_2 and Au/Fe_2O_3 [70, 144]. However, in the case of Au/CeO_2, the support may act as an oxygen supplier with a reaction also taking place at the gold–ceria interface through a *Mars and van Krevelen* mechanism [145]. It is also possible that ceria modified by gold ions, and forming a solid solution of type $Ce_{1-x}Au_xO_{2-\delta}$, is an active catalyst [84].

It may be noted that surface formates arising from the interaction between CO and surface hydroxy groups have also been proposed as reaction intermediates [73, 146].

The carbonate species detected during CO oxidation by IR spectroscopy on the oxide supports (TiO_2, Fe_2O_3, ZnO, ZrO_2) are considered as spectators of the reaction (Fig. 15.2) [3, 6, 13, 146–148]. In the case of Au/TiO_2, when the reaction temperature is below 200 K, and the TiO_2 surface is saturated by carbonated species [6, 149], the reaction is proposed to occur only on the gold surface [3, 13, 150] whereas above 300 K Haruta's mechanism (Fig. 15.2) would apply. Between 200 and 300 K, the reaction would proceed according to the two mechanisms. This is attested by different values of the activation energies.

15.5.1.2 Bond and Thompson's Mechanism: Unreduced Gold at the Interface

In the mechanism proposed by Bond and Thompson [9] (Fig. 15.3), CO adsorbed on Au^0 reacts with a hydroxy group located on a peripheral unreduced Au^{III} species, to form the carboxylate. The latter in turn reacts with a superoxide O_2^- ion formed on the oxide support. Like Haruta, the authors emphasise that the interaction between gold and the support is particularly important, since it dictates the nature of the interface at which the reaction takes place. The hypothesis of the presence of Au^{III} species arises from the fact that for other noble metals, cationic species are located at the metal–support interface. As mentioned below in Section 15.5.3, there is evidence for the presence of cationic gold species in the presence of metallic gold, but no proof for the location at the interface.

15.5.2
Mechanisms Involving "Gold Particles Only"

A CO oxidation mechanism involving "gold particles only" is proposed especially when the support is non-reducible. For instance, Schubert et al. [54] deduced from experiments of CO–O$_2$ time resolved titration and of isotopic exchange, that when gold is supported on magnesia, silica or alumina, the O$_2$ adsorption takes place on low coordinated sites of the gold surface, but leads to a lower activity than when oxygen can be activated on a reducible support.

15.5.2.1 Goodman's Mechanism
The study on Au/TiO$_2$(110) *model* catalyst performed by Goodman et al. [10] is the first study to lead to the proposal of a reaction mechanism on "gold particles only". According to these authors when gold metal particles are small enough and bi-dimensional (Section 15.4.1), they lose their metallic properties due to the *quantum size effect* (band gap of 0.2–0.6 eV), and become an active catalyst (Fig. 15.6). The role of the support would be to stabilise the small gold particles.

   If the concept of *quantum size effect* is well established when the metal particles become smaller than 1–2 nm [92, 93] (Section 15.4.1), this does not mean that other parameters cannot influence the catalytic properties. Two studies on *model* catalysts rather lead to the conclusion that gold atoms in a low coordination state are the sites responsible for the catalytic activity in CO oxidation rather than the *quantum size effect*. One is a DFT study, which shows that a band gap is not essential for binding molecular oxygen, but that a low coordinated site is [151]. The calculations show that O$_2$ can bind to the edges of Aun clusters ($n=1$ to 7), i.e. on non-metallic clusters, and not to an Au(111) metallic surface. However when these clusters are deposited on Au(111), they become metallic, and they can also bind O$_2$. The second report is a TPD and IRAS study of adsorption/desorption of CO on bi-dimensional gold particles (1 to 5 monolayers) supported on FeO(111) thin film [152]. The authors show that the vibration frequency of CO adsorbed and the temperature of CO desorption do not depend on the particle thickness, and are the same as for larger three-dimensional particles. They conclude that the CO oxidation reaction does not arise from the *quantum size effect* as a result of particle thickness, but rather from the presence of highly uncoordinated atoms. However, it may be noted that in these two studies, co-adsorption of CO and O$_2$ has not been investigated, although it has been reported that the binding of one of the molecules may be assisted by the other [153]. It is noteworthy that in more recent papers [154, 155], Goodman et al. invoke the presence of Au$^{\delta-}$ species to explain the maximum activity for bi-layered gold particles (see Section 15.6.2).

15.5.2.2 Kung's Mechanism

According to the CO oxidation mechanism proposed by Kung et al. [11], the reaction also takes place on "gold particles only" (Fig. 15.4a), but it requires the presence of an ensemble of Au^0 and Au^I–OH sites, i.e., of not fully metallic gold particles with Au^I species located at the metal–support interface (Fig. 15.4b). In this mechanism, CO reacts with Au^I–OH and forms a hydroxycarbonyl, and oxygen is dissociatively adsorbed on the gold particles. The hydroxycarbonyl is oxidised by the adsorbed atomic oxygen into bicarbonate, which is then decarboxylated to Au^I–OH and CO_2. This mechanism is surprising since if CO is well known to adsorb on the Au^0 sites, in contrast O_2 is not known to dissociatively adsorb on gold particles (Section 15.4.3). The involvement of Au^I–OH sites was deduced from a careful study of the catalytic behavior of Au/Al_2O_3 during deactivation, regeneration with water or hydrogen, as well as studies on the influence of water in the reactant feed [12] and of the chloride content in the catalysts [20] (Section 15.3.4). This mechanism has been recently supported by a TOF-SIMS study (time-of-flight secondary ion mass spectroscopy) [66] performed on Au/TiO_2 and Au/Al_2O_3 samples prepared by *deposition–precipitation*. The spectra revealed the presence of AuO^- and AuO_2^- ion clusters after sample calcination at 623 K. According to the authors, this indicates the existence of oxidised gold in the catalysts due to either chemisorbed oxygen on gold nanoparticles or partial oxidation of gold. However the gold particles, which are small (2 to 5 nm) could have been oxidised after calcination during transfer in air to the TOF-SIMS equipment (see Section 15.4.3). One can note, however, that very recently the same group found a new IR band at 1242 cm^{-1} on Au/TiO_2 upon CO exposure that they have assigned to a hydroxycarbonyl. This species disappeared when O_2 was added [156].

15.5.3
Cationic Gold for CO Oxidation

Although some groups claim that gold must be fully reduced to be active [33, 40, 42, 52, 78, 87–89, 157], others claim that cationic gold is required for activity in CO oxidation, but that the presence of metallic gold is also necessary. This was proposed in the case of Au/Al_2O_3 [158], Au/MgO [159], Au/TiO_2 [78], Au/CeO_2 [83, 160] and Au/Fe_2O_3 [113, 161–163, 81]. For instance, a combination of techniques, HREM, STEM-XEDS, EXAFS, *in situ* XANES, XPS and Mössbauer spectroscopy, performed on *co-precipitated* Au/Fe_2O_3 after different thermal treatments led Hutchings et al. [81] to point out the important role of cationic gold in the activity of these catalysts. It is noteworthy that in these studies the location of cationic gold remains a matter of discussion [159] as does the oxidation state (I or III).

In the case of other systems prepared by adsorption of gold acetyl acetonate, it seems that only cationic gold complexes could be active in CO oxidation. In the case of Au/NaY zeolite the Au^{III} precursor is reduced to Au^I under reaction conditions at 298 K with no evidence of formation of Au^0 [164]. The catalyst shows activity although the Au^{III} complex (initial catalyst) is an order of magnitude more active than Au^I. The authors, however, report that the catalyst is less active than when it also contains Au^0. In the case of La_2O_3-supported Au^{III} complexes prepared by the same method [165], the functioning catalyst was found by XAFS to contain cationic gold and no detectable zerovalent gold, and to be highly active, and stable in CO oxidation at room temperature.

15.6
Contribution of Quantum Chemical Calculation and Surface Science to the Understanding of CO Oxidation

A number of quantum chemical calculations and experiments have been performed on metallic gold surfaces, on free gold clusters, neutral or charged, and on clusters supported on *model* surfaces of oxide supports, stoichiometric and nonstoichiometric. Two main systems have been investigated, Au/MgO(100) and Au/TiO$_2$(110), which correspond to models for non-reducible and reducible supports, respectively. These studies are especially interesting for finding out more about O$_2$ activation, and for obtaining new insights into CO oxidation.

15.6.1
Gold Supported on MgO or on Non-reducible Supports

On the basis of DFT calculations performed on gold clusters (Au$_{12}$ and Au$_{34}$) and gold surfaces with steps, Au(211) and Au(221), Liu et al. [166] proposed that CO oxidation proceeds the same way on gold supported on non-reducible supports, i.e., occurs on Au steps. It may start with a slow reaction between adsorbed CO and O$_2$ in the gas phase through the formation of an intermediate CO·O$_2$ complex, which produces CO$_2$ and an O atom on Au steps: $O_2+CO^* \rightarrow CO_2+O^*$ (slow reaction), then end with a fast reaction between the remaining oxygen atom adsorbed on gold and a second CO molecule: $O^*+CO \rightarrow CO_2$. The first reaction should be the rate-determining step because O$_2$ does not adsorb or is too weakly chemisorbed on gold particles, and therefore the reaction would not be a typical *Langmuir–Hinshelwood* mechanism. They ruled out the possibility for O$_2$ dissociation on gold particles at low temperature.

These results are consistent with other calculations published almost at the same time by Molina et al. [4, 167]. These authors focused their study on the Au–MgO interface, depicted by infinite straight boundaries between one-dimensional Au rods and the MgO support. They found that O$_2$ can bind to gold only in the presence of coadsorbed CO, but cannot dissociate. The reaction follows an *Eley–Rideal* mechanism. The most reactive site was found at the Au–MgO interface when the latter provides a small cavity, surrounded by three low-coordinated Au atoms and a Mg^{2+} cation, where CO and O$_2$ can interact and form a metastable CO·O$_2$ peroxy-like intermediate (Fig. 15.8).

These calculations point out an important feature, that the interface plays a key role for the activation of O$_2$ even though the support is non-reducible, and that the mechanism on "gold particles only" requires the assistance of the support to create

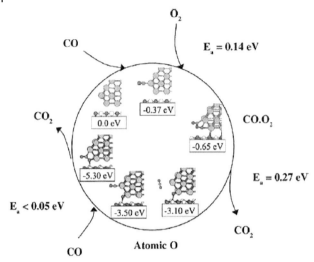

Fig. 15.8 Catalytic cycle for oxidation of CO at the type II Au/MgO interface, from Ref. [153].

favorable sites at the interface. These calculations also show that very small particles are needed to get a large number of these interface sites. However, Liu et al.
[168] note that although the CO oxidation pathway on Au/MgO described by
themselves [166] and by Molina et al. [167], possesses rather low barriers, the O_2
adsorption on gold is extremely weak (0.2 eV), so the probability of formation of
the $CO \cdot O_2$ intermediate is not high, if at all possible. Consequently, CO oxidation
in this type of system may not be favorable.

Experiments on size-selected gold clusters, Aun $(2 \leq n \leq 20)$ soft-landed on
MgO(100) thin film [169] showed that the activity in CO oxidation depends on the
number of cluster atoms, and is higher when the clusters are supported on defect-
rich MgO rather than on defect-poor MgO. *Ab initio* calculations performed with
Au_8 showed that the F centers on MgO induce a partial electron transfer to the
cluster, which promotes the binding of O_2 and CO, and activates the O—O bond to
a peroxo-like adsorbate state that can react with gas-phase CO or adsorbed CO.

Goodman et al. [170] also report a direct correlation between the increasing
concentration of F centers obtained in annealing MgO between 900 and 1300 K,
and the catalytic activity of a *real* Au/MgO catalyst with ~4 nm gold particles. The
validity of the correlation is questionable since the two magnesia supports are
different; the first is an ultrathin non-hydroxylated MgO film supported on Mo(100)
while the second is a powder, and such an oxide still containing hydroxyl groups
after thermal treatment at 900–1300 K.

Molecular beam experiments performed with ~1.5 nm gold particles supported
on an MgO(100) surface annealed in UHV at 973 K [171] showed that, after exposure to an O_2 beam then pumping, no CO_2 was produced during CO pulse titration, indicating that O_2 does not dissociate on gold before reaction with adsorbed
CO. The authors are more in favor of a mechanism at the interface, as proposed

by Molina et al. [4] (Fig. 15.8), than a mechanism involving charge transfer from the support.

15.6.2
Gold Supported on TiO$_2$ or on Reducible Supports

Experiments of thermal desorption and collision-induced desorption showed that gold particles (2–5 nm) supported on TiO$_2$(110), and populated with both O$_{ads}$ and O$_{2,ads}$ produces more CO$_2$ than the sample populated with O$_{ads}$ alone. This was interpreted as evidence that molecularly chemisorbed oxygen directly participates in the CO oxidation reaction, and that the reaction pathway does not require the dissociation of oxygen [172].

Liu et al. were the first group to publish DFT calculations on CO oxidation on the Au/TiO$_2$ rutile system, first on Au$_{16}$/TiO$_2$(110) [166] and then on a two-layer strip of gold supported on rutile TiO$_2$(110) to simulate a gold-oxide interface [168]. The CO oxidation reaction follows a *Langmuir–Hinshelwood* mechanism because of the reasonable chemisorption energy of O$_2$ at the interface. The reaction proceeds with a higher rate than on gold on non-reducible materials, and with high efficiency without O$_2$ dissociation since adsorbed O$_2$ is highly activated at the Au–TiO$_2$ interface (the O—O bond being stretched).

Another DFT study performed by Molina et al. [153, 173] came to the same type of conclusion with a Au/TiO$_2$ interface described as an infinite array of gold rod on rutile TiO$_2$(110). O$_2$ bonded to either TiO$_2$ or a Au/TiO$_2$ interface can react with CO adsorbed on the gold particle. Reaction between the two adsorbed species seems particularly facile at those sites, involving very low barriers; the reaction proceeds via CO·O$_2$ formation rather than via O$_2$ dissociation, but the CO·O$_2$ intermediate can be viewed as more fleeting than in Au/MgO (Section 15.6.1).

These authors, as well as Liu et al. [168] report charge transfer from gold particles to O$_2$ adsorbed on TiO$_2$ at the interface, and suggest that the surroundings of gold particles acts as an "O$_2$-attraction zone" from where adsorbed O$_2$ may react with CO adsorbed on gold.

In line with results that showed that free anionic gold clusters are active in CO oxidation [64, 126], several surface science studies and DFT calculations report the importance of the presence of reduced Ti defect sites at the interface between gold clusters and the TiO$_2$ surface. These defects determine the cluster shape and electronic properties of gold because charge transfer makes the electron-rich gold particles more reactive [59, 107, 174, 175].

The recent DFT study performed by Norskov et al. [111, 176] on an Au$_{10}$ cluster (according to a rigid model mimicking a flat particle) on oxygen-defective rutile TiO$_2$(110) agrees with the results of the two groups cited above [168, 173]. In addition to this, they propose that the mechanism on "gold particle only" occurs in parallel, and is common for all supports. The mechanism taking place at the metal-oxide interface enhances the activity of the system and is substrate dependent; this explains the different activity observed for different gold-based catalysts. This is also consistent with the fact that a "gold particle only" mechanism was proposed to occur on a Au/TiO$_2$ *real* catalyst when the reaction is performed at low temperature (200 K) (Section 15.5.1.1) and on a Au/TiO$_2$(110) *model* catalyst. Indeed as mentioned in Section 15.5.2.1, Goodman et al. have continued investigating the reason for the maximum of activity obtained for bilayered gold particles. They obtained the same sharp maximum of activity when a gold bilayer covered a film of TiO$_2$ [154]. They found that the key feature is the interaction of the first layer of gold with reduced Ti^{3+} atoms naturally present on the TiO$_x$ support, yielding anelectron-rich Au layer (Au$^{\delta-}$), which is crucial for O$_2$ activation; this also leads to strong CO adsorption on the top of the bilayer. In this system the support cannot be involved in the CO oxidation reaction since the gold overlayer precludes access to the Ti cation sites by the reactant.

All the calculations reported here indicate that a high concentration of low-coordinated gold atoms (i.e. small gold particle size) is required to bind the reactants, and this is the main parameter governing the reactivity of gold-based catalysts (see also Fig. 15.5).

15.6.3
Influence of Water on the Adsorption of O$_2$ on Gold Clusters

Surface science and calculation also provide new elements of understanding on the effect of water depicted in Section 15.3.4. The influence of water vapor on the adsorption of O$_2$ on anion gold clusters has been studied both experimentally and theoretically [177]. The presence of an hydroxy group on Au$_N^-$ clusters (N=2 to 5) leads to a reversed O$_2$ adsorption activity with respect to bare clusters *(Section 15.4.3)*: Non-reactive bare gold clusters, with odd-N atoms, become active when OH is bound (Au$_N$OH$^-$), whereas active bare clusters, with even-N atoms, become inactive. DFT calculation provides evidence that electron-transfer induced by the binding of an OH group enhances the reactivity toward O$_2$ for odd anionic gold clusters and suppresses it for the even ones.

Other DFT calculations have shown that co-adsorption of H$_2$O and O$_2$ on Au$_8$ clusters, free or supported on MgO(100), leads to the formation of an O$_2 \cdots$ H$_2$O complex involving partial proton sharing or proton transfer, and leading to a hydroperoxy-like complex (HO$_2$) [178]. This favors the activation of the O—O bond, i.e. the bond extension to values characteristic of a peroxo- or superoxo-like state. Consequently, the reaction with CO can occur with a small activation barrier of ~0.5 eV, either through an *Eley–Rideal* mechanism if O$_2$ is adsorbed on the top face of Au$_8$ clusters or through a *Langmuir–Hinshelwood* mechanism if O$_2$ is adsorbed on the periphery of the cluster.

It is noteworthy that STM experiments and DFT calculations performed on non-stoichiometric TiO$_2$(110) [179, 180] showed that water readily dissociates into OH groups on oxygen vacancies, which favors O$_2$ adsorption on TiO$_2$(110). In contrast, water does not dissociate on stoichiometric TiO$_2$(110), and O$_2$ cannot adsorb. Calculations performed by another group [181] showed, in addition, that the OH groups possess a long-range effect on O$_2$ adsorption on TiO$_2$(110), at least over 1 nm. The authors propose that adsorbed O$_2$ can readily diffuse along the surface Ti atoms in the presence of OH, which constitute a large reservoir of O_2 for CO oxidation on the interface of Au/TiO_2.

Another view is proposed by Lopez et al. [182]: water would affect the interaction between the gold particles and the support, modify the Au–TiO_2 interface energy and therefore the size and the shape of the gold particles and, as consequence, the number of low coordination sites.

15.6.4
How to go Further in Understanding the Mechanism(s) of CO Oxidation Thanks to *Model* Catalysts

The contributions of surface science experiments and quantum chemical calculations performed on *model* catalysts are very useful in obtaining a better understanding of the catalytic reaction, but these *model* catalysts are also very different from *real* catalysts:

- In *model* catalysts, gold is directly deposited in the oxidation state zero as gold atoms or clusters. In the case of *real* catalysts, Au^I or Au^{III} precursors are used, and a thermal treatment is needed afterwards to get gold metal particles. It is probably for this reason that unreduced gold species can remain present in these samples. However, some *model* catalysts, $Au/TiO_2(110)$ [10, 154] or Au/MgO [171] , show catalytic activity close to those of *real* catalysts.

- There is a substantial difference in size between the gold particles in *real* catalysts, which are at least 1 nm (≥ 50 atoms), and the gold clusters of the *model* catalysts, especially those "created" for calculations whose number of atoms does not exceed 20. One cannot however rule out the hypothesis that very small clusters, hardly detectable by TEM or other techniques, could be the most active entities in *real* catalysts. A consequence of these differences of size is that the gold clusters in *model* catalysts are non-metallic (*quantum size effect*) in contrast to most of the gold particles in *real* catalysts (see Section 15.4.1).

- The gold clusters in *model* catalysts can be submitted to non-negligible charge transfer from the support due to the presence of defects, such as F centers for MgO and Ti^{3+} for TiO_2. This makes them more reactive. In *real* catalysts, the fact that the oxide surface is hydroxylated may affect the nature of the surface defects, healed by OH or H_2O molecules. As a consequence this can modify the strength of the metal–support interaction, the charge transfer at the interface and the morphology of the gold particles, and ultimately the mechanism and the kinetics of CO oxidation.

For these reasons it is risky to directly transpose results obtained with *model* catalysts to *real* catalysts, even though one cannot deny that they bring new ways of thinking.

There are pending questions regarding the presence of cationic gold species in *real* catalysts (Figs. 15.3 and 15.4), on their oxidation state, optimal amount and location, at the interface or as isolated entities, and their involvement in the reaction of CO oxidation. Strong efforts are still needed to develop and make more reliable the characterisation of the gold species under working conditions. Moreover, clarification is needed in the interpretations of the spectra of IR, XANES-EXAFS and ^{197}Au Mössbauer, all this being dependent on taking more care in the preparation of the catalyst and in the catalytic measurements. Deactivation, which is an issue that has not been addressed in this chapter, also needs more investigation and a greater degree of understanding.

From the studies on *model* catalysts, several challenges can be proposed.

- We have emphasised the role of water in the CO oxidation reaction (Sections 15.3.4 and 15.6.3). *Model* catalysts are usually free of water, but a few experiments and calculations involving water are beginning to be published (Section 15.6.3). In line with this, the use of hydroxylated support surfaces could be a breakthrough in quantum chemical calculations and surface science. Several groups have recently proposed realistic models of different surfaces of γ-Al$_2$O$_3$ [183, 184], MgO [185] and TiO$_2$ anatase [186] and rutile [187], accounting for hydroxylation/dehydroxylation processes induced by temperature effects. Raybaud's group is now studying the interaction with Pd clusters [188, 189].

- Another challenge is to insert cationic gold species (Au$^{\delta+}$, Au$^+$, Au^{3+}) at the metal–support interface or on the oxide support as isolated entities, and to calculate the effect on the reactivity in CO oxidation, and work of this type is commencing. One can cite DFT calculations showing the oxidation of Au^0 to Au$^{\delta+}$ after adsorption on CeO$_2$(111) and its reactivity in the water gas shift reaction [190]. Another DFT study compares the stability of gold species in a variety of oxidation states (Au$^{0\ to\ III}$ and Au(OH)$_{1\ to\ 3}$) on ZnO(0001) [191]. Things are moving fast since a very relevant paper has recently been published by Wang and Hammer [192]: DFT calculations performed on alkaline TiO$_2$(110) having terminal OH groups show that the Au$_7$ cluster binds strongly, and becomes cationic as a result of the formation of Au–O bonds. The cationic Au$_7$ cluster is able to bind both molecular and atomic oxygen at perimeter sites next to the support and CO further away from the support. The reaction of CO oxidation is easier (solely exothermic reaction steps) than when Au_7 is supported on a pristine or reduced $TiO_2(110)$ surface. A stiel more recent study combining DFT and expeniments on *model* catalysts has shown that gold clusters bond more strongly on oxidized $TiO_2(110)$ that on reduced or hydrated $TiO_2(110)$ [193].

- It would also be very interesting to be able to perform calculations with larger gold particles that would be truly metallic, but DFT deserves much more powerful computer resources than those available nowadays. An alternative would be to use semi-empirical methods whose calculations are much faster than those of DFT, but require the use of parameters extracted from experiments and *ab initio* calculation. A paper that refers to this issue has recently been published by Goniakowski et al. [194], and deals with Pd/MgO(100). The core of the approach is a many-body potential energy surface (PES) constructed on the basis of results of *ab initio* calculations for model metal/oxide interface structures. This is an effective approach to simulate non-reactive deposition of nanoscale metal objects with several thousands of atoms on a surface of highly ionic oxide, i.e. weakly interacting systems. The drawback of this method is that it excludes the possibility of interfacial charge transfers, typical for defective interfaces, but these authors foresee future quantum treatment with a self-consistent evaluation of the interfacial charge distribution to solve this limitation.

15.7
Concluding Remarks: Attempt to Rationalise the Results on CO Oxidation

In spite of the diversity of published results, one can attempt to summarise what seems to be known about the CO oxidation reaction, but it must be kept in mind that there are always papers showing opposite results and interpretations:

1. It is the combination of gold particles and oxide support, which leads to active catalysts provided that the gold particles are smaller than 5 nm (Section 15.3.3).
2. The presence of metallic gold is a requirement for the reaction to occur; but it is possible that cationic gold species also plays a role in the reaction (Section 15.5.3).
3. The catalytic activity increases as the gold particle size decreases down to 1–2 nm, which is the size limit in *real* catalysts. The decrease in size leads to changes in the properties of the particles, (Section 15.4.1): number of low coordinated surface sites, electronic and structure

properties, and morphology, but several groups [59, 60, 151, 152] support the idea that the size-dependent activity is dominated by variation of the number of low coordination sites or of sites at the support–gold particle interface.

4. Gold supported on reducible or semiconductive oxide supports is usually more active than gold supported on non-reducible or insulating oxide supports; again some exceptions can be encountered (see Section 15.3.3).

5. There are probably two reaction pathways occurring simultaneously when the support is reducible [111, 176]: (i) a pathway on "gold particle only" according to an *Eley–Rideal* mechanism; interestingly DFT calculations suggest that the gold–support interface could be involved in providing special configurations of sites, which would favor the formation of a $CO \cdot O_2$ intermediate (Fig. 15.8) [4]; (ii) a pathway involving the oxide support on which oxygen can activate and react at the interface with CO adsorbed on gold, according to a *Langmuir–Hinshelwood*-type mechanism. On non-reducible supports, only the first pathway could be operative; this would explain the lower activity of gold on these supports.

References

1 M. Haruta, T. Kobayashi, H. Sano, N. Yamada, *Chem. Lett.* **1987**, *2*, 405.
2 M. Haruta, K. Saika, T. Kobayashi, S. Tsubota, Y. Nakahara, *Chem. Express* **1988**, *3*, 159.
3 M. Haruta, M. Daté, *Appl. Catal. A* **2001**, *222*, 427.
4 L. M. Molina, B. Hammer, *Phys. Rev. B* **2004**, *69*, 155424.
5 M. Haruta *Catal. Surveys Jpn.* **1997**, 61.
6 F. Boccuzzi, A. Chiorino, S. Tsubota, M. Haruta, *J. Phys. Chem.* **1996**, *100*, 3625.
7 M. Haruta, S. Tsubota, T. Kobayashi, H. Kageyama, M. J. Genet, B. Delmon, *J. Catal.* **1993**, *144*, 175.
8 M. Haruta, *Catal. Today* **1997**, *36*, 153.
9 C. G. Bond, D. T. Thompson, *Gold Bull.* **2000**, *33*, 41.
10 M. Valden, X. Lai, D. W. Goodman, *Science* **1998**, *281*, 1647.
11 C. K. Costello, J. H. Yang, H. Y. Law, Y. Wang, J. N. Lin, L. D. Marks, M. D.

Kung, H. H. Kung, *Appl. Catal. A* **2003**, *243*, 15.
12 C. K. Costello, M. C. Kung, H.-S. Oh, K. H. Kung, *Appl. Catal. A* **2002**, *232*, 159.
13 M. Haruta, *Cattech* **2002**, *6*, 102.
14 M. Haruta, *Chem. Record* **2003**, *3*, 75.
15 R. Meyer, C. Lemire, S. K. Shaikhutdinov, H. J. Freund, *Gold. Bull.* **2004**, *37*, 72.
16 M. Haruta, *J. New Mater. Electrochem. Syst.* **2004**, *7*, 163.
17 G. J. Hutchings, *Catal. Today* **2005**, *100*, 55.
18 M. S. Chen, D. W. Goodman *Acc. Chem. Res.* **2006**, *39*, 681.
19 G. C. Bond, C. Louis, D. Thompson *Catalysis by Gold*, Imperial College Press, London, **2006**, Vol. 6.
20 H. S. Oh, J. H. Yang, C. K. Costello, Y. M. Wang, S. R. Bare, H. H. Kung, M. C. Kung, *J. Catal.* **2002**, *210*, 375.
21 W.-C. Li, M. Comotti, F. Schüth, *J. Catal.* **2006**, *237*, 190.

22 L. Delannoy, N. El Hassan, A. Musi, N. Nguyen Le To, J.-M. Krafft, C. Louis, *J. Phys. Chem. B* **2006**, *110*, 22471.

23 J. M. Fisher, *Gold Bull.* **2003**, *36*, 155.

24 J. H. Yang, J. D. Henao, C. Costello, M. C. Kung, H. H. Kung, J. T. Miller, A. J. Kropf, J.-G. Kim, J. R. Regalbuto, M. T. Bore, H. N. Pham, A. K. Datye, J. D. Laeger, K. Kharas, *Appl. Catal. A* **2005**, *291*, 73.

25 J. M. C. Soares, M. Hall, M. Cristofolini, M. Bowker, *Catal. Lett.* **2006**, *109*, 103.

26 M. Haruta, H. Kageyama, N. Kamijo, T. Kobayashi, F. Delannay, *Stud. Surf. Sci. Catal.* **1988**, *44*, 33.

27 M. Haruta, N. Yamada, T. Kobayashi, S. Iijima, *J. Catal.* **1989**, *115*, 301.

28 S. Tsubota, D. A. H. Cunningham, Y. Bando, M. Haruta, *Stud. Surf. Sci. Catal.* **1995**, *91*, 227.

29 R. Zanella, S. Giorgio, C. R. Henry, C. Louis, *J. Phys. Chem. B* **2002**, *106*, 7634.

30 R. Zanella, L. Delannoy, C. Louis, *Appl. Catal. A* **2005**, *291*, 62.

31 Y. Yuan, K. Asakura, H. Wan, K. Tsai, Y. Iwasawa, *Catal. Lett.* **1996**, *42*, 15.

32 Y. Yuan, A. P. Kozlova, K. Asakura, H. Wan, K. Tsai, Y. Iwasawa, *J. Catal.* **1997**, *170*, 191.

33 T. V. Choudhary, C. Sivadinarayana, C. C. Chusuei, A. K. Datye, J. P. F. Jr, D. W. Goodman, *J. Catal.* **2002**, *207*, 247.

34 M. Okumura, S. Nakamura, S. Tsubota, T. Nakamura, M. Azuma, M. Haruta, *Catal. Lett.* **1998**, *51*, 53.

35 M. Okumura, M. Haruta *Chem. Lett.* **2000**, 396.

36 J. Guzman, B. C. Gates, *Nano Lett.* **2001**, *1*, 689.

37 J. Guzman, B. C. Gates, *Langmuir* **2003**, *19*, 3897.

38 F. Porta, L. Prati, M. Rossi, S. Coluccia, G. Martra, *Catal. Today* **2000**, *61*, 165.

39 S. Tsubota, T. Nakamura, K. Tanaka, M. Haruta, *Catal. Lett.* **1998**, *56*, 131.

40 J.-D. Grunwaldt, C. Kiener, C. Wogerbauer, A. Baiker, *J. Catal.* **1999**, *181*, 223.

41 J. D. Grunwaldt, M. Maciejewski, O. S. Becker, P. Fabrizioli, A. Baiker, *J. Catal.* **1999**, *186*, 458.

42 M. Comotti, W.-C. Li, B. Spliethoff, F. Schüth, *J. Am. Chem. Soc.* **2006**, *126*, 917.

43 R. W. J. Scott, O. M. Wilson, R. M. Crooks, *Chem. Mater.* **2004**, *16*, 5682.

44 H. Lang, S. Maldonado, K. J. Stevenson, B. D. Chandler, *J. Am. Chem. Soc.* **2004**, *126*, 12949.

45 S. Arii, F. Mortin, A. J. Renouprez, J. L. Rousset, *J. Am. Chem. Soc.* **2004**, *126*, 1199.

46 G. M. Veith, A. R. Lupini, S. J. Pennycook, G. W. Ownby, N. J. Dudney, *J. Catal.* **2005**, *231*, 151.

47 M. Daté, Y. Ichihashi, T. Yamashita, A. Chiorino, F. Boccuzzi, M. Haruta, *Catal. Today* **2002**, *72*, 89.

48 T. Akita, P. Lu, S. Ichikawa, K. Tanaka, M. Haruta, *Surf. Interface Anal.* **2001**, *31*, 73.

49 B. Schumacher, V. Plzak, K. Kinne, R. J. Behm, *Catal. Lett.* **2003**, *2003*, 109.

50 R. Zanella, C. Louis, *Catal. Today* **2005**, *107-108*, 768.

51 A. I. Kozlov, A. P. Kozlova, K. Asakura, Y. Matsui, T. Kogure, T. Shido, Y. Iwasawa, *J. Catal.* **2000**, *196*, 56.

52 R. Zanella, S. Giorgio, C.-H. Shin, C. R. Henry, C. Louis, *J. Catal.* **2004**, *222*, 357.

53 V. Schwartz, D. R. Mullins, W. Yan, B. Chen, S. Dai, S. H. Overbury, *J. Phys. Chem. B* **2004**, *108*, 15782.

54 M. M. Schubert, S. Hackenberg, A. C. van Veen, M. Muhler, V. Plzak, R. J. Behm, *J. Catal.* **2001**, *197*, 113.

55 M. Haruta, *Stud. Surf. Sci. Catal.* **1997**, *110*, 123.

56 S. D. Lin, M. Bollinger, M. A. Vannice, *Catal. Lett.* **1993**, *17*, 245.

57 S.-J. Lee, A. Gavriilidis, *J. Catal.* **2002**, *206*, 305.

58 R. J. H. Grisel, K.-J. Weststrate, A. Gluhoi, B. E. Nieuwenhuys, *Gold Bull.* **2002**, *35*, 39.

59 N. Lopez, T. V. W. Janssens, B. S. Clausen, Y. Xu, M. Mavrikakis, T. Bligaard, J. K. Nørskov, *J. Catal.* **2004**, *223*, 232.

60 S. H. Overbury, V. Schwartz, D. R. Mullins, W. Yan, S. Dai, *J. Catal.* **2006**, *241*, 56.

61 Y. Iizuka, T. Tode, T. Takao, K. Yatsu, T. Takeuchi, S. Tsubota, M. Haruta, *J. Catal.* **1999**, *187*, 50.

62 M. A. Bollinger, M. A. Vannice, *Appl. Catal. B* **1996**, *8*, 417.

63 L. D. Socaciu, J. Hagen, T. M. Brenhardt, L. Wöste, U. Heiz, H. Häkkinen, U. Landman, *J. Am. Chem. Soc.* **2003**, *125*, 10437.

64 W. T. Wallace, R. L. Whetten, *J. Am. Chem. Soc.* **2002**, *124*, 7499.

65 Y. Yuan, K. Asakura, H. Wan, K. Tsai, Y. Iwasawa, *Catal. Today* **1998**, *44*, 333.

66 L. Fu, N. Q. Wu, J. H. Yang, F. Qu, D. L. Johnson, M. C. Kung, H. H. Kung, V. P. David, *J. Phys. Chem. B* **2005**, *109*, 3704.

67 S. H. Overbury, L. Ortiz-Soto, H. Zhu, B. Lee, M. D. Amiridis, S. Dai, *Catal. Lett.* **2004**, *95*, 99.

68 E. D. Park, J. S. Lee, *J. Catal.* **1999**, *186*, 1.

69 F. Boccuzzi, A. Chiorino, M. Manzoli, P. Lu, T. Akita, S. Ichikawa, M. Haruta, *J. Catal.* **2001**, *202*, 256.

70 H. Liu, A. I. Kozlov, A. P. Kozlova, T. Shido, K. Asakura, Y. Iwasawa, *J. Catal.* **1999**, *185*, 252.

71 M. Daté, M. Haruta, *J. Catal.* **2001**, *201*, 221.

72 M. Daté, M. Okumura, S. Tsubota, M. Haruta, *Angew. Chem.* **2004**, *43*, 2129.

73 B. Schumacher, Y. Denkwitz, V. Plzak, M. Kinne, R. J. Behm, *J. Catal.* **2004**, *224*, 449.

74 J. T. Calla, R. J. Davis, *J. Phys. Chem. B* **2005**, *109*, 2307.

75 J. T. Calla, R. J. Davis, *Ind. Eng. Chem. Res.* **2005**, *44*, 5402.

76 D. Cunningham, S. Tsubota, N. Kamijo, M. Haruta, *Res. Chem. Intermed.* **1993**, *19*, 1.

77 W. Yan, B. Chen, S. M. Mahurin, V. Schwartz, D. R. Mullins, A. R. Lupini, S. J. Pennycook, S. Dai, S. H. Overbury, *J. Phys. Chem. B* **2005**, *109*, 10676.

78 J. H. Yang, J. D. Henao, M. C. Raphulu, Y. Wang, T. Caputo, A. J. Groszek, M. C. Kung, M. Scurrell, J. T. Miller, H. H. Kung, *J. Phys. Chem. B* **2005**, *109*, 10319.

79 F. Moreau, G. C. Bond, A. O. Taylor, *J. Catal.* **2005**, *231*, 105.

80 M. A. Debeila, R. P. K. Wells, J. A. Anderson, *J. Catal.* **2006**, *239*, 162.

81 G. J. Hutchings, M. S. Hall, A. F. Carley, P. Landon, B. E. Solsona, C. J. Kiely, A. Hercing, U. Makkee, J. A. Moulijn, A. Overweg, J. C. Fierro-Gonzalez, J. Guzman, B. C. Fates, *J. Catal.* **2006**, *242*, 71.

82 S. T. Daniells, A. R. Overweg, M. Makkee, J. A. Moulijn, *J. Catal.* **2005**, *230*, 52.

83 S. Carrettin, P. Concepcion, A. Corma, J. M. L. Nieto, V. F. Puntes, *Angew. Chem.* **2004**, *43*, 2538.

84 A. M. Venezia, G. Pantaleo, A. Longo, G. D. Carlo, M. P. Casaletto, F. L. Liotta, G. Deganello, *J. Phys. Chem. B* **2005**, *109*, 2821.

85 J. M. C. Soares, M. Bowker, *Appl. Catal. A* **2005**, *291*, 136.

86 M. C. Kung, C. K. Costello, H. H. Kung, *Catalysis* **2004**, *17*, 152.

87 R. J. H. Grisel, C. J. Weststrate, A. Goossens, M. W. J. Crajé, A. M. van der Kraan, B. E. Nieuwenhuys, *Catal. Today* **2002**, *72*, 123.

88 N. Weiher, E. Bus, R. Prins, L. Delannoy, C. Louis, D. E. Ramaker, J. T. Miller, J. A. van Bokhoven, *J. Catal.* **2006**, *240*, 100.

89 A. Goossens, M. W. J. Crajé, A. M. van der Kraan, A. Zwijnenburg, M. Makkee, J. A. Moulijn, L. J. de Jongh, *Catal. Today* **2002**, *72*, 95.

90 C. L. Cleveland, U. Landman, T. G. Schaaff, M. N. Shafigullin, P. W. Stephens, R. L. Whetten, *Phys. Rev. Lett.* **1997**, *79*, 1873.

91 D. A. H. Cunningham, W. Vogel, H. Kageyama, S. Tsubota, M. Haruta, *J. Catal.* **1998**, *177*, 1.

92 J. A. A. J. Perenboom, P. Wyder, F. Meier, *Phys. Rep.* **1981**, *78*, 173.

93 W. P. Halperin, *Rev. Mod. Phys.* **1986**, *58*, 533.

94 S. B. DiCenzo, S. D. Berry, E. H. Hartford, *Phys. Rev. B* **1988**, *38*, 8465.

95 S. T. Lee, G. Apai, M. G. Mason, R. Benbow, Z. Hurych, *Phys. Rev. B* **1997**, *23*, 505.

96 H. G. Boyen, T. Herzog, G. Kästle, F. Weigl, P. Ziemann, J. P. Spatz, M. Möller, R. Wahrenberg, M. G. Garnier, P. Oelhafen, *Phys. Rev. B* **2002**, *65*, 075412.

97 Q. Guo, K. Luo, K. A. Davis, D. W. Goodman, *Surf. Interface Anal.* **2001**, *32*, 161.

98 C. P. Vinod, G. U. Kulkarni, C. N. R. Rao, *Chem. Phys. Lett.* **1998**, *289*, 329.

99 L. Stievano, S. Santucci, L. Lozzi, S. Calogero, F. E. Wagner, *J. Non-Cryst. Solids* **1998**, *232*, 644.

100 P. M. Paulus, A. Goossens, R. C. Thiel, A. M. V. D. Kraan, G. Schmid, L. J. D. Jongh, *Phys. Rev. B* **2001**, *64*, 205418.

101 M. E. Lin, R. Reifenberger, A. Ramachandra, R. P. Andres, *Phys. Rev. B* **1992**, *46*, 15498.

102 J. T. Miller, A. J. Kropf, Y. Zha, J. R. Regalbuto, L. Delannoy, C. Louis, E. Bus, J. A. van Bokhoven, *J. Catal.* **2006**, *240*, 222.

103 D. Chatain, F. Chabert, V. Ghetta, J. Fouletier, *J. Am. Ceram. Soc.* **1993**, *76*, 1568.

104 S. Giorgio, C. Chapon, C. R. Henry, G. Nihoul, J. M. Penisson, *Philos. Mag. A* **1991**, *64*, 87.

105 S. Giorgio, C. R. Henry, B. Pauwels, G. P. Tenderloo, *Mater. Sci. Eng. A* **2001**, *297*, 197.

106 R. Zanella, C. Louis, S. Giorgio, R. Touroude, *J. Catal.* **2004**, *223*, 328.

107 J. A. Rodriguez, G. Liu, T. Jirsak, J. Hrbek, Z. Chang, J. Dvorak, A. Maiti, *J. Am. Chem. Soc.* **2002**, *124*, 5243.

108 J. A. Rodriguez, M. Pérez, T. Jirsak, J. Evans, J. Hrbek, L. Gonzalez, *Chem. Phys. Lett.* **2003**, *378*, 526.

109 J. A. Rodriguez, M. Pérez, J. Evans, G. Liu, J. Hrbek, *J. Chem. Phys.* **2005**, *122*, 241101.

110 S. Shaikhutdinov, R. Meyer, M. Naschitzki, M. Bäumer, H.-J. Freund, *Catal. Lett.* **2003**, *86*, 211.

111 I. N. Remediakis, N. Lopez, J. K. Norskov, *Appl. Catal. A* **2005**, *291*, 13.

112 L. I. Llieva, D. H. Andreeva, A. A. Andreev, *Thermochim. Acta* **1997**, *292*, 169.

113 M. Khoudiakov, M.-C. Gupta, S. Deevi, *Appl. Catal. A* **2005**, *291*, 151.

114 M. A. Debeila, M. C. Raphulu, E. Mokoena, M. Avalos, V. Petranovskii, N. J. Coville, M. S. Scurrell, *Mater. Sci. Eng. A* **2005**, *396*, 61.

115 J. C. Frost, *Nature* **1988**, *334*, 577.

116 M. Okumura, Y. Kitagawa, M. Haruta, K. Yamaguchi, *Appl. Catal. A* **2005**, *291*, 37.

117 M. A. Debeila, M. C. Raphulu, E. Mokoena, M. Avalos, V. Petranovskii, N. J. Coville, M. S. Scurrell, *Mater. Sci. Eng. A* **2005**, *396*, 70.

118 M. A. Centeno, C. Portales, I. Carrizosa, J. A. Odriozola, *Catal. Lett.* **2005**, *102*, 289.

119 W. R. MacDonald, K. E. Hayes, *J. Catal.* **1970**, *18*, 115.

120 P. C. Richardson, D. R. Rossington, *J. Catal.* **1971**, *20*, 420.

121 J. J. Pireaux, M. Chtaib, J. P. Delrue, P. A. Thiry, M. Liehr, R. Caudano, *Surf. Sci.* **1984**, *141*, 211.

122 N. Saliba, D. H. Parker, B. E. Koel, *Surf. Sci.* **1998**, *410*, 270.

123 B. E. Salisbury, W. T. Wallace, R. L. Whetten, *Chem. Phys.* **2000**, *262*, 131.

124 D. M. Cox, R. Brickman, K. Creegan, A. Kaldor, *Z. Phys. D* **1991**, *19*, 353.

125 Y. D. Kim, M. Fischer, G. Gantefor, *Chem. Phys. Lett.* **2003**, *377*, 170.

126 D. Stolcic, M. Fischer, G. Gantefor, Y. D. Kim, Q. Sun, P. Jena, *J. Am. Chem. Soc.* **2003**, *125*, 2848.

127 Q. Sun, P. Jena, Y. D. Kim, M. Fischer, G. Gantefor, *J. Chem. Phys.* **2004**, *120*, 6510.

128 H. Berndt, I. Pitsch, S. Evert, K. Stuve, M.-M. Pohl, J. Radnik, A. Martin, *Appl. Catal. A* **2003**, *244*, 169.

129 T. Fukushima, S. Galvagno, G. Parravano, *J. Catal.* **1979**, *57*, 177.

130 J. Guzman, B. C. Gates, *J. Phys. Chem. B* **2003**, *107*, 2242.

131 J. D. Stiehl, T. S. Kim, S. M. McClure, C. B. Mullins, *J. Am. Chem. Soc.* **2004**, *126*, 1606.

132 J. A. van Bokhoven, C. Louis, J. T. Miller, M. Tromp, O. V. Safonova, P. Glatzel, *Angew. Chem.* **2006**, *45*, 4651.

133 A. Outka, R. J. Madix, *Surf. Sci.* **1987**, *179*, 351.

134 D. H. Parker, B. E. Koel, *J. Vac. Sci. Technol. A* **1990**, *17*, 1717.

135 V. A. Bondzie, S. C. Parker, C. T. Campbell, *Catal. Lett.* **1999**, *63*, 143.

136 T. S. Kim, J. D. Stiehl, C. T. Reeves, R. J. Meyer, C. B. Mullins, *J. Am. Chem. Soc.* **2003**, *125*, 2018.

137 L. Guczi, D. Horvath, Z. Paszti, L. Toth, Z. E. Horvath, A. Karacs, G. Peto, *J. Phys. Chem. B* **2000**, *104*, 3183.

138 R. J. H. Grisel, B. E. Nieuwenhuys, *Catal. Today* **2001**, *64*, 69.

139 L. Delannoy, N. Weiher, N. Tsapatsaris, A. M. Beesley, L. Nchari, S. L. M. Schroeder, C. Louis, *Topies Catal.* **2007**, in press.

140 M. Olea, M. Kunitake, T. Shido, Y. Iwasawa, *Phys. Chem. Chem. Phys.* **2001**, *3*, 627.

141 M. Olea, Y. Iwasawa, *Appl. Catal. A* **2004**, *275*, 35.

142 M. Okumura, J. M. Coronado, J. Soria, M. Haruta, J. C. Conesa, *J. Catal.* **2001**, *203*, 168.

143 J.-D. Grunwaldt, A. Baiker, *J. Phys. Chem. B* **1999**, *103*, 1002.

144 H. Liu, A. I. Kozlov, A. P. Kozlova, T. Shido, Y. Iwasawa, *Phys. Chem. Chem. Phys.* **1999**, *1*, 2851.

145 A. C. Gluhoi, H. S. Vreeburg, J. W. Bakker, B. E. Nieuwenhuys, *Appl. Catal. A* **2005**, *291*, 145.

146 A. Knell, P. Barnickel, A. Bauker, A. Wokaun, *J. Catal.* **1992**, *137*, 306.

147 A. K. Tripathi, V. S. Kamble, N. M. Gupta, *J. Catal.* **1999**, *187*, 332.

148 B.-K. Chang, B. W. Jang, S. Dai, S. H. Overbury, *J. Catal.* **2005**, *236*, 392.

149 F. Boccuzzi, A. Chiorino, M. Manzoli, D. Andreeva, M. Tabakova, T. 1999. *J. Catal.* **1999**, *188*, 176.

150 M. Haruta, M. Daté, Y. Iisuka, F. Boccuzzi, *Shokubai Catalysts Catalysis* **2001**, *43*, 125.

151 G. Mills, M. S. Gordon, H. Metiu, *J. Chem. Phys.* **2003**, *118*, 4198.

152 C. Lemire, R. Meyer, S. Shaikhutdinov, H.-J. Freund, *Angew. Chem.* **2004**, *43*, 118.

153 L. M. Molina, B. Hammer, *Appl. Catal. A* **2005**, *291*, 21.

154 M. S. Chen, D. W. Goodman, *Science* **2004**, *306*, 252.

155 D. C. Meier, D. W. Goodman, *J. Am. Chem. Soc.* **2004**, *126*, 1892.

156 J. D. Henao, T. Caputo, J. H. Yang, M. C. Kung, H. H. Kung, *J. Phys. Chem. B* **2006**, *110*, 8689.

157 K. Okumura, K. Yoshino, K. Kato, M. Niwa, *J. Phys. Chem. B* **2005**, *109*, 12380.

158 H. H. Kung, M. C. Kung, C. K. Costello, *J. Catal.* **2003**, *216*, 425.

159 J. Guzman, B. C. Gates, *J. Phys. Chem. B* **2002**, *106*, 7659.

160 J. Guzman, S. Carrettin, J. C. Fierro-Gonzalez, Y. Hao, B. C. Gates, A. Corma, *Angew. Chem.* **2005**, *44*, 4778.

161 A. M. Visco, A. Donato, C. Milone, S. Galvagno, *React. Kinet. Catal. Lett.* **1997**, *61*, 219.

162 N. A. Hodge, C. J. Kiely, R. Whyman, M. R. H. Siddiqui, G. J. Hutchings, Q. A. Pankhurst, F. E. Wagner, R. R. Rajaram, S. E. Golunski, *Catal. Today* **2002**, *72*, 133.

163 R. M. Finch, N. A. Hodge, G. J. Hutchings, A. Meagher, Q. A. Pankhurst, M. R. H. Siddiqui, F. E. Wagner, R. Whyman, *Phys. Chem. Chem. Phys.* **1999**, *1*, 485.

164 J. C. Fierro-Gonzalez, B. C. Gates, *J. Phys. Chem. B* **2004**, *108*, 16999.

165 J. C. Fierro-Gonzalez, V. A. Bhirud, B. C. Gates *Chem. Comm.* **2005**, 5275.

166 Z.-P. Liu, P. Hu, A. Alavi, *J. Am. Chem. Soc.* **2002**, *124*, 14770.

167 L. M. Molina, B. Hammer, *Phys. Rev. Lett.* **2003**, *90*, 206102.

168 Z.-P. Liu, X.-Q. Gong, J. Kohanoff, C. Sanchez, P. Hu, *Phys. Rev. Lett.* **2003**, *91*, 266102.

169 A. Sanchez, S. Abbet, U. Heiz, W. D. Schneider, H. Häkkinen, R. N. Barnett, U. Landman, *J. Phys. Chem. A* **1999**, *103*, 9573.

170 Z. Yan, S. Chinta, A. A. Mohamed, J. P. J. Fackler, D. W. Goodman, *J. Am. Chem. Soc.* **2005**, *127*, 1604.

171 O. Meerson, G. Sitja, C. R. Henry, *Eur. Phys. J. D* **2005**, *34*, 119.

172 J. D. Stiehl, T. S. Kim, S. M. McClure, C. B. Mullins, *J. Am. Chem. Soc.* **2004**, *126*, 13574.

173 L. M. Molina, M. D. Rasmussen, B. Hammer, *J. Chem. Phys.* **2004**, *120*, 7673.

174 A. Vijay, G. Mills, H. Metiu, *J. Chem. Phys.* **2003**, *118*, 6536.

175 Z. Yang, R. Wu, D. W. Goodman *Phys. Rev. B* **2000**, *61*, 14066–14071.

176 I. N. Remediakis, N. Lopez, J. K. Norskov, *Angew. Chem.* **2005**, *44*, 1824.

177 W. T. Wallace, R. B. Wyrwas, R. L. Whetten, R. Mitric, V. Bonacic-koutecky, *J. Am. Chem. Soc.* **2003**, *125*, 8408.

178 A. Bongiorno, U. Landman, *Phys. Rev. Lett.* **2005**, *95*, 106102.

179 R. Schaub, P. Thostrup, N. Lopez, E. Lægsgaard, I. Stensgaard, J. K. Nørskov, F. Besenbacher, *Phys. Rev. Lett.* **2001**, *87*, 266104.

180 S. Wendt, R. Schaub, J. Matthiesen, E. K. Vestergaard, E. Wahlström, M. D. Rasmussen, P. Thostrup, L. M. Molina, E. Lægsgaard, I. Stensgaard, B. Hammer, F. Besenbacher, *Surf. Sci.* **2005**, *598*, 226.

181 L. M. Liu, B. McAllister, H. Q. Ye, P. Hu, *J. Am. Chem. Soc.* **2006**, *128*, 4017.

182 N. Lopez, J. K. Nørskov, T. V. W. Janssens, A. Carlsson, A. Puig-Molina, J.-D. Grunwaldt, *J. Catal.* **2004**, *225*, 86.

183 M. Digne, P. Sautet, P. Raybaud, P. Euzen, H. Toulhoat, *J. Catal.* **2002**, *211*, 1.

184 M. Digne, P. Sautet, P. Raybaud, H. Toulhoat, E. Artacho, *J. Phys. Chem. B* **2002**, *106*, 5155.

185 W. Langel, M. Parrinello, *J. Chem. Phys.* **1995**, *103*, 3240.

186 C. Arrouvel, M. Digne, M. Breysse, H. Toulhoat, P. Raybaud, *J. Catal.* **2004**, *222*, 152.

187 P. J. Lindan, N. M. Harrison, M. J. Gillan, *Phys. Rev. Lett.* **1998**, *80*, 762.

188 M. C. Valero, P. Raybaud, P. Sautet, *J. Phys. Chem.* **2006**, *110*, 1759.

189 M. C. Valero, P. Raybaud, P. Sautet, *Plys. Rev. B.* **2007**, *75*, 16381.

190 Z.-P. Liu, S. J. Jenkins, D. A. King, *Phys. Rev. Lett.* **2005**, *94*, 196102.

191 N. S. Phala, G. Klatt, E. V. Steen, S. A. French, A. A. Sokolb, C. R. A. Catlow, *Phys. Chem. Chem. Phys.* **2005**, *7*, 2440.

192 J. G. Wang, B. Hammer *Phys. Rev. Lett.* **2006**, *97*, 136107.

193 D. Matthey, J. G. Wang, S. Wendt, J. Matthiesen, R. Schaub, E. Laegsgaard, B. Hammer, F. Besenbacher, *Science.* **2007**, *315*, 1692.

194 J. Goniakowski, C. Mottet, C. Noguera, *Phys. Status Solidi B* **2006**, *243*, 2516.

16
NO Heterogeneous Catalysis Viewed from the Angle of Nanoparticles

Frédéric Thibault-Starzyk, Marco Daturi, Philippe Bazin, and Olivier Marie

16.1
Introduction

Burning fossil fuels (natural gas, oil etc.), burning fuel produced from sustainable sources, or even the simple strong heating of air, generates pollutants like nitrogen oxides, which need to be destroyed before they can reach the atmosphere. This chemical process, performed for example in car exhausts, is known as deNOx, and it is possible by using heterogeneous catalysis. DeNOx (and heterogeneous catalysis) was performed at the nanometer scale long before nanotechnology was invented. However, although the factors at play were the same, they were not considered from the point of view of the size of the particles, but rather considering the dispersion of the active metal phase on the support, since this was the actual parameter that was measured. Now that new techniques allow one to view or measure particles directly (with TEM for example), the size of particles is starting to be one of the main parameters considered, and it becomes possible to rationalise deNOx catalysis also in terms of nanostructures and nanoparticles.

What is the relationship between heterogeneous catalysis and nanoparticles? One clue can probably be given by turning back to the principles of heterogeneous catalysis. The simplest catalytic act can usually be decomposed into five steps, for a reaction without intermediate: the diffusion of the reactant towards the catalytic surface, the adsorption of the reactant on the active sites, its conversion into the product, the desorption of the reactant from the active site and its diffusion to the external medium, far from the surface. The key step takes place on the active site, when the reactant is transformed into the product. This involves the crossing of an energy barrier, by the formation of an activated complex, i.e. an excited state where the chemical bond(s) in the reactant is (are) broken leading to chemical bond(s) in the final species. Such a phenomenon, from the fundamental point of view, consists in an electronic transfer from molecular orbitals. The physical contribution of the catalytic surface is to polarise the electron clouds, to orient the charge density towards the crossing of the potential gap separating the reactant from the product. Depending on the electronic structure of the reactant and

Nanoparticles and Catalysis. Edited by Didier Astruc
Copyright © 2008 WILEY-VCH Verlag GmbH & Co. KGaA, Weinheim
ISBN: 978-3-527-31572-7

product, the active site can be an anion, a cation, or a metal particle, capable of increasing or diminishing the electronic charge in the molecular orbitals. The charge transfer is favored by the polarizing effect of the site, strongly linked to the charge/size ratio. This simple and schematic vision of a catalytic reaction can help our understanding of the role and peculiarities of nanoparticles in catalysis: the density of charge can be much bigger in a charged nanoparticle than in a massive crystallite, giving rise to a higher local electric field, and to a more efficient induced polarisation state, i.e. to an easier charge transfer. Of course this oversimplified model cannot explain all of heterogeneous catalysis, and other parameters can play a role, sometimes in an opposite direction (large metal or oxide particles can be needed to store one or more reactants, etc.). Nevertheless, in many cases, the active site alone can be identified as an atomic entity, or as a nanometric atomic cluster with the functions mentioned here.

Many other aspects are important for catalysts at the nanometer scale [1]. The local environment of the active site needs to be controlled to modify adsorption parameters for reactants, diffusion parameters for all products (reactants, products and poisons), and to optimise concentrations around the active site. The architecture of the active site needs to be mastered at the nanometer scale to optimise multistep reactions or to obtain good activities in broader experimental conditions. Crystalline defects in oxides or on metallic particles need to be controlled. Nanoparticles often have original catalytic properties, but the problem is to stabilise them during the reaction. Interactions between nanoparticles have to be taken into account, as well as nanoparticles–support interactions. Long-distance effects are also very important: charge transport, or transport of chemical species, diffusion of ions or electrons on the surface or in the bulk of the solid, creation of defects and dynamic restructuring of the surface, spillover, subsurface diffusion etc. The common point between all these is the order of magnitude of distances. Usually, chemistry considers matter at the level of the molecule. In heterogeneous catalysis, because of interactions between molecules and solids, matter is considered at the supramolecular level, and thus sizes and distances are nanometers. These are the meaningful distances for the influence of the direct environment of the active site on its catalytic activity.

Here, we will first present briefly the general chemistry at work in deNOx catalysis. Then we will focus on the various specific nanometric issues in the domain. NOx decomposition mainly takes place on a metal center, as do most redox reactions in heterogeneous catalysis. We will therefore focus on how nanometric size influences the role of metal particles in this domain. It is very much linked to interaction with the oxide support, which will be the subject of the following part. Zeolites are very specific oxide supports, controlling the distance between reactants and imposing chemical reactivity within the nanometer scale. Metal particles formed in zeolites have their size controlled between 0.5 and several dozens of nm, sometimes with Angstrom accuracy. We will next deal with three-way catalysts, mainly used for automotive deNOx. They involve very complex solids, where particle size is a key issue for activity. Finally, we will mention quickly new nanometric solids like carbon nanotubes, which have appeared recently in deNOx catalysis.

16.2
The Chemistry of deNOx Catalysis

We can essentially distinguish between two kinds of NOx sources due to human activities: stationary and automotive emissions. The first are ascribed to the factories producing nitrous compounds (such as fertilisers or nitric acid plants) or to industries where the chemical or heating processes lead to oxidation of the nitrogen contained in air (glass, cement, metal and power plants, for example). The other NOx sources are automotive exhaust gases. DeNOx of waste gases from stationary sources can be efficiently achieved by using the so-called SCR (selective catalytic reduction) process [2]. Nowadays, the large majority of the current industrial NOx removal is carried out by this technology. In this case ammonia injected in the waste gases is used as the reducing agent, while the industrial catalysts are based on TiO_2-supported $V_2O_5-WO_3$ and/or $V_2O_5-MoO_3$ oxides. The reaction stoichiometry in typical SCR conditions is the following:

$$4NH_3 + 4NO + O_2 \rightarrow 4N_2 + 6H_2O \tag{16.1}$$

Using isotopically labelled reactants, it has been demonstrated that on both vanadia-based catalysts and noble metals, the two nitrogen atoms of N_2 arise one from NO and the other from ammonia. V_2O_5-based catalysts also catalyse the reduction of NO_2 in the presence of oxygen:

$$4NH_3 + 2NO_2 + O_2 \rightarrow 3N_2 + 6H_2O \tag{16.2}$$

In the absence (or lack) of O_2 the reaction stoichiometry (16.1) converts into the following:

$$4NH_3 + 6NO \rightarrow 5N_2 + 6H_2O \tag{16.3}$$

In this case, four N_2 molecules are formed by the reaction between NO and NH_3 (reaction (16.1)) and the last one is formed from two molecules of NO.

The reaction between NO and ammonia can also proceed in a different way, giving rise to the unwanted product N_2O in oxygen-rich atmospheres:

$$4NH_3 + 4NO + 3O_2 \rightarrow 4N_2O + 6H_2O \tag{16.4}$$

This implies that the conversion of ammonia is obtained in other ways than reaction (16.1), i.e. is in part oxidized by oxygen instead of NO in one of the following ways:

$$2NH_3 + 3/2O_2 \rightarrow N_2 + 3H_2O \tag{16.5}$$

$$2NH_3 + 2O_2 \rightarrow N_2O + 3H_2O \tag{16.6}$$

$$2NH_3 + 5/2O_2 \rightarrow 2NO + 3H_2O \tag{16.7}$$

Alternatively, hydrocarbons can also be used as reducing agents, by applying a technology closer to that of vehicles exhaust control [3].

In the case of automotive gas exhaust emissions, the SCR process is limited by technical problems, so that it has been dealt with only recently, essentially for emission control for trucks. From the general point of view, deNOx for cars is still an unresolved problem and different processes are under investigation. The development of three-way catalysts (TWC) over the last 30 years has been a remarkable technical achievement. Current catalytic converters, based on various combinations of Pt, Pd and Rh as the active ingredients, provide a very high level of emission control for the removal of CO, NOx and unburnt hydrocarbons. Under normal working conditions for a stoichiometric gasoline engine (temperatures from about 400 to 800 °C) they have proved to be efficient and reliable.

Diesel engines and lean-burn gasoline engines produce an exhaust containing a large excess of oxygen, and are therefore a totally different case. Under these conditions the conventional TWCs are completely ineffective for NO reduction. Therefore, for lean-burn engines it was initially assumed that the platinum group metals (PGMs) could not be used for NOx reduction. Consequently, much of the early work on what is often referred to as "lean deNOx" focussed on base metal or non-metal systems [4]. Following the work on NO decomposition on Cu/ZSM-5, Held et al. [5] and Iwamoto et al. [6] first reported some success using the same material as for lean deNOx. Zeolite-based catalysts [7] can be particularly effective for the selective catalytic reduction (SCR) of NO with methane (e.g. Co or Ga/ZSM-5) or propene (e.g. Cu/ZSM-5). However, the hydrothermal resistance of these materials is usually unsatisfactory [8], although reports by Chen and Sachtler [9] and Chajar et al. [10] of water-tolerant zeolitic materials may yet change this view. While the PGMs are inefficient at moderate and high temperatures, pioneering work by Hamada et al. [11] and by Obuchi et al. [12] demonstrated that at lower temperatures (typically below 300 °C) these metals could catalyse NOx reduction by the hydrocarbons found in the exhaust (e.g. propane and propene) under lean-burn conditions. Hamada et al. [13] also reported that numerous single metal oxides and supported oxides exhibited significant activities for the same reaction, yet at higher temperatures.

Nevertheless, NOx removal by HC treatment under a continuous flow, whatever the catalyst, induces fuel over-consumption, due to the relatively large amount of hydrocarbons in the post-injection, compared to the NOx concentration. An alternative to the catalytic approach is based on NOx adsorption, in which NOx is stored in the catalyst under lean conditions and the NOx trap is regenerated for a short period under fuel-rich conditions. This technique has shown good applicability in Japan and Sweden. For most of the world, and Europe in particular, an issue remains with NOx-traps due to the high sulfur levels in fuels and to poisoning by sulfur oxides [14]. Catalyst reactivation is then needed; it is achieved by periodically desulfating the material with an *in situ* treatment at high temperatures (~700 °C). Unfortunately, the fuel amount used in this latter post-combustion operation is relatively large, inducing an over-consumption estimated to be 5%.

16.3
The Metal Center

16.3.1
Size of Nanoparticles

One of the general ideas in heterogeneous catalysis is that smaller active particles (a few atoms per cluster, and thus more around the Angstrom in size than the nanometer) lead to better contact with the reactant, and to an increased activity of the solid. Gold is now famous for something even more striking: it only has catalytic properties in very small particles, and this for electronic rather than geometric reasons. Such is not the case in deNOx, and the smaller particles are not the best active sites. Nanoparticles having a size of 10 nm or more are necessary for a good active site.

Platinum catalysts have been the subject of several studies with that perspective. Using Pt on various solids (silicas, alumina, with various porosities etc.), Praliaud et al. [15] found a very clear correlation between dispersion and catalytic activity, whatever the support and its morphology. The lower the dispersion, and thus the bigger the particle size (in the range 1–20 nm), the more active the catalyst. The selectivity (i.e. the reaction pathway itself) was not affected by particle size. It is noteworthy that the excellent correlation with metal dispersion on the solid was not observed for particle size, determined by TEM. This is probably related to the low accuracy of the measurement of size by TEM, which is a local method, whereas dispersion measurement is done by global analysis, thus providing a perfect statistical average.

The chemical nature of the oxide has a strong influence on the importance of Pt particle size. The effect is much clearer on silica than it is on γ-alumina or on zirconia [16]. It is also very dependent on the chemical reaction considered. It is true for oxidation of NO into NO_2, but much less so for its reduction, where the amount of reducing agent may have a stronger role than the initial dispersion and particle size of Pt [17].

The key to this surprising high activity of catalysts with big Pt particles probably lies in the sintering of the metal during the reaction. In a systematic study of the influence of the particle size from 3 to 22.5 nm, Martens et al. found a lower sintering of big particles [18]. The reaction was easier on small particles (the light-off temperature increased with particle size), and the maximum NOx conversion did not actually depend on the particle size. Having big enough particles can thus protect the catalyst from metal sintering.

With rhodium catalysts, a dopant can be introduced to stabilise the metal. Ce^{3+} in Rh/ZrO_2 leads to perfect correlation between the dispersion of Rh and the amount of added cerium, with a clearly marked optimum [1].

The size of the metal particles can influence the reaction itself. In the reaction between NO and methane with Rh catalyst on alumina at 450 °C, methane was oxidised to CO_2 on 5.8 nm particles, whereas it was converted into a mixture of

CO and CO_2 on smaller particles (at higher temperature, the difference disappeared) [19]. The catalyst with smaller particles deactivated quickly.

Sintering might be one reason for loss of activity during the reaction, but other deactivation processes were found, also linked to particle size. XPS has shown, during NO_2 dissociation on Pt/Al_2O_3, that the loss of activity of the catalyst was linked to the build up of Pt oxide on the solid [20]. Pt oxide is formed more easily on small metal particles, and catalysts with lower dispersion and bigger particle size are more resistant to deactivation. This was confirmed when comparing with $Pt/BaO/Al_2O_3$, which was less active, and on which Pt oxide formed more easily. Having bigger Pt particles can thus also protect the catalyst by preventing the formation of Pt oxide. This seems a general mechanism. The bigger particles do not have an intrinsic higher activity than small particles, but they are more resistant to sintering and to oxidation, and thus lead to more stable solids and to overall more active catalysts.

16.3.2
Morphology of Metal Particles

An important difference between metal particles for deNOx seems to lie more in the particle morphology than in its size. On well structured, mostly cubic Pt nanocrystals (~12 nm), the peak NOx conversion, as well as the peak conversion temperature, depends on the total amount of exposed platinum [21]. The deNOx catalytic behavior is mainly related to the shape (facet effect) and only to a lesser extent to the size (bulk effect) of the particles. The N_2/N_2O ratio is higher for the catalyst rich in Pt nanoparticles with low index facets than for the polycrystalline one. The concentration of edges, corners, kinks and surface defects of polycrystalline Pt particles only determines the catalytic activity of NOx conversion, and not the N_2/N_2O ratio. Large (24 nm) and small (2.4 nm) polycrystalline particles show the same specific activity. On Pd/MgO model catalysts with average clusters sizes of 2.8, 6.9 and 15.6 nm, prepared by epitaxial growth, the medium sized particles exhibiting mainly {111} facets are more active than big ones with mainly {100} facets [22]. This shows that the study of particle size in heterogeneous catalysis is not a simple matter. The particle morphology has to be taken into account, and whereas the shape of a particle can be determined on planar model solids, it remains very difficult on industrial catalysts.

16.4
Metals in Zeolites

Zeolites are crystalline alumino silicates with internal voids, channels or cavities, denoted, according to IUPAC, as micropores below 2 nm and mesopores between 2 and 50 nm. These pores give zeolites a place of choice for nanometric technology and chemistry. They can contain (and protect) metallic nanoparticles or nanometric molecular complexes. Zeolites can protect metals from sintering. Up to a certain

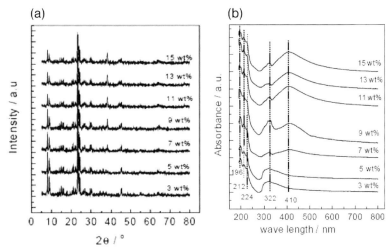

Fig. 16.1 XRD patterns (A) and UV spectra (B) of 3–15 wt.% Ag/H-ZSM-5 catalysts (from Ref. [23]).

loading, metal particles will all be located inside the pores, and their size will be limited by the pore size. Above that critical loading, or if metal particles start migrating out of the pore system, they will be able to grow bigger than the pore dimensions. Looking at deNOx catalysis on zeolites from the point of view of nanoparticles is thus particularly interesting. Interaction between the metal and the oxide can be extremely strong, and new electronic effects can happen in zeolites.

The relationship between loading and particle size can be exemplified on Ag-MFI. This solid is prepared by ionic exchange of an H-MFI zeolite by Ag^+. The pore system in MFI zeolites is rather small, and limited to 0.55 nm. At a loading higher than 7%, Ag starts being present not only as a cation inside the pores, but also as metallic clusters on the outer surface of the zeolite. These clusters are detected by XRD ($2\Theta = 38°$, Fig. 16.1A) and UV spectroscopy (band at 410 nm, Fig. 16.1B).

A correlation was observed between the amount of silver on the outer surface and the catalytic activity in NO conversion to N_2, up to 10% loading, Fig. 16.2. The optimal size for the clusters on the external surface is such that active intermediate NO_3^- species are stabilised at temperatures where CH_4 is activated on Ag^+ cations present in the pore system, but before methane combustion on metal clusters takes place.

16.4.1
Cerium for Controlling Metal Particle Size

Metals are often introduced into zeolites by ion exchange from the acidic or sodium form of the zeolite. This introduces a difficulty for interpretation since, during the process, an acidic proton or a sodium cation is removed, and the disappearance of these catalytic sites has to be taken into account. The metal can, as we have just seen with silver, be present as a cation or as a reduced metal, and some

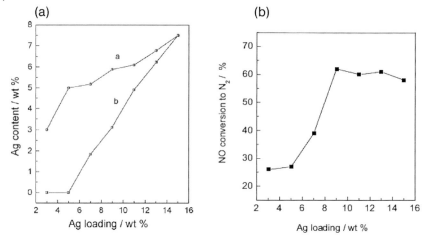

Fig. 16.2 A, Correlations of silver loading with the amount of silver in the zeolite channels (a) and deposited on the zeolite surface (b). B, Variation of NO conversion to N_2 on increasing Ag loading on the H-ZSM-5 zeolite, from Ref. [23].

sodium cation might remain. The preparation of bimetallic systems is possible, for various purposes. Cerium cations, on zeolites as well as on silica or alumina, have the great advantage of protecting catalysts from metal sintering. Cerium has been used as a promoter in the SCR of NOx by methane, improving oxidation of NO while protecting CH_4 from combustion [24]. The catalyst was prepared by first exchanging Na-MFI with cerium(III) nitrate, followed by calcinations, with further exchange by silver (I) nitrate before the final calcination. During the first calcination step, CeO_2 nanoparticles (1–3 nm) are created; they protect from sintering during the second calcination or during the catalytic reaction, and prevent the formation of big silver particles, which would be more active for the combustion of methane. This was evidenced by STEM and XPS characterisation of the samples with or without cerium.

CeO_2 coating of zeolite crystals is also a very good way to protect the solid from water, which is present in exhaust gases and usually destroys the zeolite structure at high temperature. Coating Cu-MFI with CeO_2 particles (2–5 nm) was shown to provide a better catalyst for the SCR of NOx by propene. The activity was increased by synergistic effects between copper and cerium, while at the same time the hydrophilicity of ceria attracted water out of the zeolite crystals and protected them from water damage [25].

16.4.2
Formation of Specific Metal Complexes in Nanometric Zeolite Pores

Ag^+ dimers and trimers have been observed in MFI zeolites [26]. Thermal as well as photochemical activation leads to charge transfer from the zeolite to the entrapped oligomeric ions, and UV irradiation yields efficient catalytic conversion of NO into N_2 and other nitrogen oxides, with $[Ag(NO)^+]_2$ and $[Ag(NO)^+]_3$ excimers as intermediates. The excited state is stabilised by entrapment, which provides the driving force for the deNOx reaction.

Iron nanoparticles are formed in Fe-MFI during calcination in O_2, with very short interatomic Fe—Fe distances (2.53 Å) as determined by EXAFS [27]. Although the solid is not completely reduced, this distance is close to the nearest-neighbour distance in metallic iron, and the structure of the iron cluster in the zeolite could be similar to that of the metallic center of ferrodoxin (II). A specific infrared band has been detected for NO interacting with this iron complex, which is linked to the catalytic activity (rather than the isolated iron).

16.4.3
Influence of the Zeolite Si/Al Ratio and Pore Structure

Using a zeolite support rather than a non-microporous solid like alumina can be very fruitful, and the same silver nanoclusters on alumina and H-MFI have very different activity. On zeolite, they catalyse NOx SCR by methane, but they lead to combustion of methane at temperatures as low as 300 °C [28].

The Si/Al ratio in zeolites controls acidobasic properties, and influences electronic distribution, and thus metal ions in the structure. Cu^+ in faujasite-type zeolites is present as monomers and dimers at Si/Al = 5.6, but when Si/Al reaches high values (13.9 and 390), only monomers are observed. Photocatalytic activity is therefore higher on the sample with Si/Al = 13.9, although its Cu content is much lower [29].

The crystal structure of the zeolite itself, controlling the shape of the pores, is a major specificity of these materials. With the same Si/Al ratio (25), Co-FER and Co-MFI give very different results in NOx SCR with methane [30]. MFI is the most active over the whole temperature range, Co^0 nanoclusters are formed more easily in MFI, together with a nanosized Co oxide phase. The Co^{2+}/Co^{3+} redox couple seems to be more abundant in MFI, and it is the most active species for the CH_4 SCR and for NO oxidation to NO_2. The role of the acidic OH groups in zeolite in this reaction has been shown by steady state isotopic transient kinetic analysis (SSITKA) [31]. $NO_2^{\delta+}$ species can be stabilised in the pores, on an OH group, and form with methane, confined in the neighbourhood, the intermediate complex CNOxHy, key to the actual process. The pore structure of MFI would thus ease the formation of that complex and stabilise all these by confinement in the pores.

16.4.4
Non-thermal Assisted Plasma Reaction

Nanocrystals of alkaline zeolites have interesting potential for deNOx catalysis according to Grassian [32]. The outer surface of the crystallite is the major aspect of these materials, it plays a central part in the catalysis on this material, although extra-framework aluminum seems to be the active species for deNOx. Non-thermal assisted plasma reaction on these solids is a potentially important process that does not require the presence of Pt group metals [33].

16.5
Three-way Catalysis

16.5.1
General Points

In the middle of the seventies, new environmental regulations, in California at first, and later in the rest of the United States, obliged car constructors to install catalytic devices in order to simultaneously convert CO and residual hydrocarbons to H_2O and CO_2, while NOx emissions were reduced to dinitrogen. Those heterogeneous catalysts were therefore called three-way catalysts (TWC) to highlight their triple function [34]. The first materials to be used for this application were non-noble metal catalysts (so-called base metal catalysts), largely due to concerns over the cost and availability of noble metals [35]. However, it quickly became apparent that the base metals (oxides of Ni, Cu, Co, Mn, and Cu/Cr, for example) lacked the intrinsic reactivity, durability, and poison resistance required for automotive applications [36–40]. The simultaneous oxidation/reduction of exhaust gases requires specific conditions, and the highest conversion of the pollutants is attained close to the stoichiometry. Fuel-rich (net reducing) or fuel-poor (net oxidizing) air-to-fuel (A/F) ratios severely decrease the TWCs' efficiency. CeO_2 is a very interesting additive: the Ce^{4+}/Ce^{3+} redox couple leads to oxygen buffering properties, so-called oxygen storage capacity (OSC) [41, 42]. In order to increase their performance, the conventional TWCs employed in the late 1980s mostly contained Rh and Pt as active noble metals and CeO_2 as oxygen storage component. CeO_2 has been so broadly used for automotive pollution control since the beginning of the 1980s that it is by now the most important application of rare earth oxides. CeO_2 increases CO and hydrocarbon conversion during rich periods, and NO conversion during lean periods. Pure ceria systems lack thermal resistance and low-temperature activity [43], but contacting CeO_2 with noble metals (such as Rh) can help the transfer of oxygen from the bulk to the surface [44]. With higher OSC of the catalyst, higher conversion efficiency and resistance to thermal aging are generally observed [45]. The severe operating conditions (especially high-temperature ageing) cause a loss of OSC due to sintering of ceria particles, thus reducing interaction of ceria with supported metals and limiting the Ce oxidation cycle [46].

As pointed out by Djéga-Mariadassou, NOx reduction on a three-way catalyst essentially takes place by NO decomposition assisted by CO over ceria supported reduced or oxidised rhodium species [33].

One of the characteristics of NO reduction, in either stoichiometric or lean conditions, is that the reaction temperature domain can be predicted by NO activation temperature (NO desorption temperature). A second feature summarizing all previous works, is that any given catalyst can be improved by introducing transition metal cations, via a strong metal–support interaction (for reducible supports or alumina, etc.), or by an exchange process (case of zeolites). It is generally accepted that NO dissociates over a zerovalent metal, whereas it associatively adsorbs as a ligand on a metal cation. We can however note that: (i) for the TWC process, the reaction occurs over both M^0 and M^{x+}; (ii) in the case of the deNOx reaction (lean conditions), the metal cation is the site for NO decomposition, leading to dinitrogen formation through an N_2O intermediate. The rate constants depend on the cation and on the temperature, and N_2O can desorb or transform into N_2 [47, 48].

The reducing agent itself is in fact always an oxidised hydrocarbon ($HCxOy$), such as an alcohol or an aldehyde. The oxidation of HC can follow two routes: total oxidation, or mild oxidation to $HCxOy$ compounds. NO_2 has been observed to favor this second route. The design of a three-function catalyst needs thus to address three separate catalytic cycles: oxidation of NO to NO_2, mild oxidation of HC to HC_xO_y, and decomposition–reduction of NO with total oxidation of HC_xO_y species (and recovery of the free active site). These cycles share no common adsorbed species [49] and are not kinetically coupled, but they must run in a concerted way.

16.5.2
The Metallic Phase

The active material contains a noble metal phase with palladium (Pd) or platinum (Pt) and rhodium (Rh). The main function of these metals is to simultaneously reduce the amount of the three pollutants, namely hydrocarbons (HC), carbon monoxide (CO) and nitrogen oxides (NOx) [50]. Heating the catalyst above 800 °C will cause a restructuring of the bimetallic noble metal surface and of the supporting washcoat (Fig. 16.3). The noble metals will agglomerate and sinter, lowering the overall active surface available for removal of pollutants [51–53]. The metal particles, initially smaller than 10 nm (Rh clusters are so small that they are undetected by TEM), will form larger agglomerates, ranging from 10 to 100 nm. At the same time TEM now also detects bigger crystallites of the support. Furthermore, the presence of additives in gasoline, such as calcium, silicon, magnesium, manganese, chromium, sulfur and phosphorus [54–56] will inevitably lower the catalyst efficiency by chemical poisoning or fouling of the active surface sites.

In the nineties, one-step processing was still the main approach [35]: salts of both cerium and the noble metals were impregnated simultaneously onto alumina, hoping that a significant fraction of the noble metal would end up in contact with ceria. This might have been true in the fresh catalyst, but thermal aging led to bigger particles for both the noble metal and ceria, with fewer contact points between the two. This was obvious in a 50 000-mile-aged Pt/Rh catalyst from a 1990 Crown Victoria vehicle, where the average Pt particle size increased from 1.2–1.6 nm (fresh) to 11 ± 2 nm (aged), while the ceria particle size increased from about 9 nm (fresh) to 18 nm (aged) [57]. Moreover, TPR analysis showed the change from noble metals in contact with ceria at low-temperature to pure ceria at high-temperature, in a simulation of the ageing process. Clearly, advances were needed in two

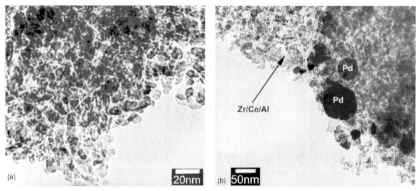

Fig. 16.3 Transmission electron micrographs of a fresh
(a) and aged (b) three-way catalyst [50].

areas: light-off performance and thermal stability of oxygen storage components. Fortunately, breakthroughs occurred in both, at the cost of increasingly complex catalytic materials and formulation strategies. Considering the light-off issue first, research on Pd as an alternative to Pt/Rh for NOx control showed that excellent NOx reduction could be achieved with Pd formulations when suitably promoted (as noted above) and at loadings higher than those for Pt/Rh [58, 59]. Simultaneously, segregating some of the Pd from ceria was recognized to yield a better low-temperature performance (including light-off) [60]. All major catalyst suppliers then used noble metals in different layers or on different support particles within a single washcoat layer. The aim was not only to avoid negative metal–metal or metal–support interactions, but also to optimize specific functions of the catalyst.

Thermal treatment under oxygen, hydrogen or chlorine might lead to smaller noble metal clusters by oxidation, splitting of oxide particles, and further reduction [50]. Both CO and HC conversion rates were improved on aged catalysts after heating between 500 and 700 °C under an oxygen-rich gas such as air. TEM suggested a decrease in the amount of very large (up to 100 nm) noble metal clusters. No efficient long-term regeneration was observed under hydrogen. A small amount of chlorine added during oxygen treatment, on the contrary, increased activity by decreasing light-off temperatures for CO, HC and NOx removal.

Surface segregation of Pd in Pd–Rh catalysts suppresses NOx reduction [61]. De Sarkar and Khanra studied the segregation difference between Pd–Rh and Pt–Rh nanoparticles, and the influence of sulfur in fuel on CO oxidation and NO. They used Monte-Carlo (MC) simulation to predict the surface composition of $Pt_{50}Rh_{50}$ and $Pd_{50}Rh_{50}$ particles (2406 atoms for 4 nm particles). They used a microkinetic model to compare the activities of both solids for reactions of $CO + O_2$, $CO + NO$ and $CO + NO + O_2$, and found that Pt and Pd segregate to the particles' surface, especially in the Pd catalyst, which is clearly better for CO oxidation, while Pt–Rh is a better catalyst for NO reduction. For both reactions, sulfur poisons the Pd–Rh catalyst more than the Pt–Rh catalyst [62].

16.5.3
The Role of Ceria

It is an ambitious task to define the role of CeO_2 in three-way catalysis since multiple effects have been attributed to this promoter [63]. Ceria was suggested to:

- promote the noble metal dispersion;
- increase the thermal stability of the Al_2O_3 support;
- promote the water gas shift (WGS) and steam reforming reactions;
- favor catalytic activity at the interfacial metal–support sites, etc.;
- promote CO removal through oxidation employing a lattice oxygen;
- store and release oxygen under lean and rich conditions, respectively.

WGS or OSC were often considered as the most significant effects. There is a high degree of synergetic interaction with noble metals (Pd, Pt, Rh) [63], which can lead to: (i) the promotion of the bulk, and the reduction of the surface, with enhanced reactivity towards CO and NO; (ii) an enhanced exchange rate between oxygen from the bulk and reactants from the gas phase, and (iii) an enhanced dispersion of the metal.

In summary, whatever the nature of the promotion effects of CeO_2 in the TWCs, these effects are only noticeable for high surface area (nanoparticles shaped solid), and low temperature reduction [45]. The increasing restrictions for automotive emissions regulations [64] have set new challenges for the development of TWCs. In fact, a major problem of the TWC converters is that significant conversions are attained only at high temperatures (>600 K), and emission of pollutants such as hydrocarbons remains high before the final operating temperature is reached. The so-called close-coupled catalysts (CCC) circumvent this problem by placing the exhaust control device very close to the engine gas collector [65]. These catalysts, being manifold mounted, experience temperatures up to 1273–1373 K, and require an extremely high thermal resistance. Accordingly, the research activity in the 1990s has been focussed mainly on the improvement of the stability of surface area in the CeO_2 promoter.

16.5.3.1 **Particle Size, Stability**
The promoting effects of CeO_2 depend strongly on the thermal stability of the oxide: as soon as significant sintering of CeO_2 particles occurs (i.e. loss of the nanoparticles typology), both OSC and metal–support interactions are inhibited, as clearly modelled by Zhdanov and Kasemo [66] (see Fig. 16.4B). These properties can be maintained by introducing other compounds into the fluorite structure. The effects of foreign cation doping, at relatively low concentrations (0.5–10 mol%), on the thermal stability of CeO_2 has been extensively investigated by Pijolat et al. [67–71]. They tried to rationalise the ability of foreign cations to stabilise CeO_2

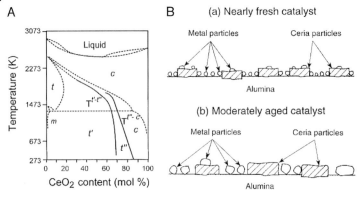

Fig. 16.4 A: Phase diagram of the CeO_2–ZrO_2 system [45]. B: Schematic views of (a) a nearly fresh catalyst surface and (b) a moderately aged catalyst with high activity, but reduced oxygen storage/release capacity.

against sintering by developing a complete set of equations based on the diffusion of cerium vacancies as the rate-limiting step in the sintering process. Among the different cations investigated (Th^{4+}, Zr^{4+}, Si^{4+}, La^{3+}, Y^{3+}, Sc^{3+}, Al^{3+}, Ca^{2+}, and Mg^{2+}), those with ionic radii smaller than that of Ce^{4+} effectively stabilised the CeO_2 against sintering.

Interestingly, insertion of a trivalent cation like lanthanum stabilises nanoparticles, but it also modifies oxygen migration and vacancy density. In this case however, OSC seems not to be affected by particle size [72].

Conversely, other authors found that for classical $Pd/CeO_2/Al_2O_3$ systems, the ceria particle size, or some related property, plays a critical role in generating the appropriate anionic vacancies for improving oxidation properties. This is probably linked with a wrong balance between capacities to generate and reoxidize those vacancies [73]. In any case, catalysts supported on ceria–zirconia/alumina showed higher activity for NO reduction, suggesting that the most active sites involve Pd interacting with highly dispersed zirconia–ceria, most likely in the form of 2D patches. The promotion of NO reduction probably involves two different effects: enhancement of the Pd reduced state and, to a lesser extent, enhancement of N–O bond breaking at the interface between Pd and Zr–Ce mixed oxide [74].

Thermal stabilisation of CeO_2 surface area was also observed in mixed oxides prepared by coprecipitation [75]. Remarkably, all doping cations larger than Ce^{4+} (e.g. La, Nd, and Y) significantly stabilise CeO_2 towards high temperature calcination. Surface M^{3+} enrichment limits the growth of ceria crystallite under oxidising conditions [76].

16.5.3.2 Role of ZrO_2 Additive: the Ceria–Zirconia Solid Solution
ZrO_2 is the most effective thermal stabiliser of CeO_2, particularly when it forms a mixed oxide with ceria [40]. By the mid-1990s, TWCs with CeO_2–ZrO_2 mixed

oxides were the most advanced technology for the development of the CCCs [40]. The phase diagram of ceria–zirconia mixed compounds presents a number of metastable phases playing a fundamental role in the material morphology and stability. The phase boundaries, as indicated in Fig. 16.4, are only approximate because distortion from the fluorite type structure is highly sensitive to the particle size in the case of the metastable tetragonal phases [45]. All possible explanations for such behavior point to a critical crystallite size, at which the tetragonal phase is favored over the monoclinic one. Consistently, by using extremely fine particles, even the cubic ZrO_2 was stabilised at room temperature [77, 78]. This kind of phase diagram immediately highlights the critical importance of the synthesis method for mixed oxides, and the relevance of the kinetic lability/inertness towards phase separation. The homogeneity of CeO_2–ZrO_2 solid solution, in fact, critically affects both the redox and textural properties. Dong et al., for example, showed that, in a solid solution, oxygen surface to bulk diffusivity, OSC, and the oxygen release rates all depend on the homogeneity of the Zr distribution. They also proposed that zirconium atoms play the additional role of carriers for oxygen transfer [79].

The amount of zirconia in the solid solution is a key parameter, and high ZrO_2 contents (>10 mol%) stabilise the surface area of CeO_2 [80, 81]. Generally speaking, the stability of the surface area increases with the ZrO_2 concentration in commercial products (up to approximately 60 mol%). At higher ZrO_2 contents, the surface area is constant. OSC nevertheless reaches a maximum for intermediate ZrO_2 contents (25–50 mol%), without any satisfactory explanation.

Remarkably high surface areas can be obtained by carrying out the synthesis in the presence of surfactants (rather than by the conventional co-precipitation) [82]. This effect, which persisted after calcination, was ascribed to the presence of an initial surface area, not to a modification of the sintering mechanism.

CeO_2–ZrO_2 can be supported on Al_2O_3 (18 to 44 wt.% mixed oxide on the support), and here again addition of zirconia can decrease the particle size from 27 to 20 nm [83]. As detected by TEM and XRD in the aged samples, the CeO_2–ZrO_2 particle size decreased as the ZrO_2 content increased from 0 to 50 mol%. This is an interesting result since the CeO_2–ZrO_2 loading was increased from 26 to 44 wt.% in these samples. The favorable effect of Al_2O_3 on CeO_2 dispersion should be less at higher loadings of the supported phase.

The efficiency of ZrO_2 for improving thermal stability of CeO_2 has long been well established, but the intrinsic role of ZrO_2 was hardly understood. Therefore, systematic studies on well-defined materials were necessary to elucidate this important point. This was the basis for a European TMR project from 1997 to 2000 (CEZIRENCAT Project, European Contract FMRX-CT-96-0060). This joint investigation, involving nine laboratories working on common samples from the same batch, was aimed at studying textural, nanostructural and chemical properties of a series of ceria–zirconia mixed oxides and the corresponding supported noble metal catalysts. It gave rise to important results:

- A complete analysis of the structural and textural properties of the solid solution, with special attention to the surface composition [84–86];

- A correlation between the zirconium content and the textural stability of doped ceria [87, 88];
- The impact of the oxidising or reducing atmosphere on the high temperature crystal growth, with a critical crystal size reached by the particles before undertaking a phase segregation process;
- The quantitative analysis of the improved redox properties of the ceria–zirconia system, particularly after ageing under reduction conditions [89, 90]. A mechanism for the redox process in ceria-based compounds was proposed [86, 91, 92]. For the first time it was also evidenced that the high temperature redox cycling can reversibly enhance/depress the redox behavior of the mixed oxides through a peculiar nanostructural modification of the outer surface layers of the samples [93].
- The contribution of the supported noble metals to the redox behavior (via spillover phenomena) was also pointed out.
- A detailed investigation on decoration and encapsulation phenomena involving metal nanoparticles was also performed.
- Catalytic performances of some selected samples were tested. The activity and selectivity of a high surface area 68/32 Ce–Zr mixed oxide TWC was studied, both fresh and after SRC-ageing (severe reducing conditions). When the experiments were carried out under cycled stoichiometric conditions (which simulated the continuous oscillation of the automotive exhaust during real operation) the activity of the samples was not directly related to their specific surface area. On the contrary, this was the case when operating in stationary stoichiometric conditions (for samples with the same compositions). Cycled operation significantly favored NO conversion, which did not take place in the steady state, while CO and C_3H_6 conversions were hardly modified when the mixed oxide was fresh, and significantly favored when it was SRC-aged. In cycled operation, both fresh and SRC-aged samples presented similar behavior towards CO, while the aged sample presented more NO conversion and less C_3H_6 conversion than the fresh sample.

For actual application in a converter, the same amounts of fresh and aged sample show similar conversions, although they present greatly different surface areas. This indicates that OSC is not only a surface property, but must be related to the bulk of the mixed oxide. Thus, the advantages of the mixed oxide compared to pure ceria in respect to its use in a TWC became clear: the mixed oxide is more resistant to ageing in the sense that it loses less surface area, even if this phenomenon does not necessarily produce a decrease in activity, particularly for NO removal. Several studies were carried out in order to understand the differences observed between fresh and SRC-aged samples, and the difference in conversions for steady state and cycled conditions. The results have shown that the behavior of the samples is related to their average oxidation degree. In particular, NO needed a certain amount of reduced sites on the sample surface to undergo dissociative adsorption and reaction. This fact could explain the different behavior in stationary versus cycled conditions. The results have also shown that selectivity to nitrogen-containing compounds is a function of how quickly oxidation of the mixed oxide takes place: the quicker the oxidation, the higher the selectivity to N_2. Thus, quicker oxidation must be favored in the catalytic converter.

When the mixed oxide is reduced, the selectivity to N_2 is increased, as well as its activity for NO reduction at very low-temperature (during start up and light-off periods). If these effects are significant for the fresh sample, they are expected to be much higher for SRC-aged samples, where the bulk of the material increases the oxidation rate [94].

In parallel to this integrated project, fundamental studies on oxygen mobility and the cerium-based compound reduction mechanism, using $^{16}O/^{18}O$ exchanges, pointed to the synergy between surface and bulk oxygen migration phenomena, indicating the role of superoxide and peroxide species [95–98].

16.5.3.3 Metal–Support Interaction

The metal–support interaction is a key parameter for understanding TWC. It influences greatly the dissociation, spillover and diffusion of adsorbed molecules, as shown by isotope substitution of oxygen by ^{18}O and of hydrogen by deuterium [99–104]. The same methodology also allowed the identification of WGS and reforming phenomena leading to hydrogen production inside the catalytic waste system [105–108]. Hydrogen sharply enhances NOx reduction properties in the catalytic converter.

A well defined epitaxial relationship between the ceria support and the metal (Rh [62, 109–113], Pt [114, 115] and Pd [116]) was detected by HREM and electron diffraction. This preferential orientation of the noble metal crystallites on ceria is probably due to metal–support interaction. For reduction temperatures up to 773 K, no significant nanostructural changes could be deduced from the HREM information. HREM also excludes any metal decoration and alloying phenomena as relevant factors for the chemical behavior of noble metals (NM) on CeO_2 reduced at 773 K or lower temperatures. In contrast, the occurrence of decoration phenomena was well established for reduction at higher temperature [117].

HREM also provided interesting information about nanostructural effects induced by reoxidation. Reoxidation temperatures ranging from 373 to 1173 K have been investigated, as well as samples pre-reduced at temperatures up to 1173 K. For catalysts reduced at $T_{red} \leq 773$ K, reoxidation up to 773 K led only to minor nanostructural changes in the catalysts. Reoxidation at 773 K does not allow any recovery from decoration or alloying phenomena for NM/CeO_2 catalysts. Much higher reoxidation temperatures are needed to achieve this objective [117].

Chemical and nanostructural properties in NM/CeO_2 catalysts are subject to the same modifications as in the SMSI effect (strong metal support interaction) in NM/TiO_2 catalysts. With increasing reduction temperature, electronic perturbations of the supported metal phase are observed first. Decoration phenomena only start in a second step, at higher reduction temperatures, especially for ceria supported materials. Likewise, the behavior towards reoxidation of the decorated samples is different. For NM/TiO_2 catalysts, 773 K is high enough a reoxidation temperature for recovery from the decorated state. Such is not the case for ceria supported materials, where electronic effects are probably the main cause for the observed deactivation [117].

Muraki et al. have underlined that fresh Rh/CeO_2 shows excellent reduction activity of NOx in a model gas test, although it deactivates significantly with thermal ageing. Rh–CeO_2 metal–support interaction is considered to play an important role in fresh Rh/CeO_2. It was found that introducing Zr into the CeO_2 crystal lattice could help significantly the stabilization of Rh–CeO_2 interaction by improving the mobility of oxygen in CeO_2 or by maintaining CeO_2 dispersion at the nanometer scale. The use of CeO_2–ZrO_2 as Rh support in actual TWC proved CeO_2–ZrO_2 to be a useful material for practical TWC applications [118].

The synthesis of nano-sized $Ce_{0.3}Zr_{0.7}Ba_{0.1}O_{2.1}$ mixed oxide is possible by modifying the surface with macromolecules. High surface area can be obtained, with novel catalytic activity and thermal stability of $Ce_{0.3}Zr_{0.7}Ba_{0.1}O_{2.1}$. TEM showed nano-sized particles ranging from 30 to 50 nm in the fresh sample calcined at 600 °C for 4 h. After treatment at 1000 °C for 4 h the powder stays in particles of 50–80 nm in diameter [119].

Doping ceria by silica has a positive effect on oxygen exchange by nanocrystalline ceria (either during reduction with hydrogen or during CO oxidation). This promotion is even clearer after redox ageing due to the formation of small ceria crystallites. The origin of this phenomenon is not any structural perturbations in the lattice of ceria induced by the dopant, as in the case of zirconia, but is in the formation of a silicate phase under reducing conditions. In fact, as evidenced by analysis of the supports [120], and by HRTEM and XRD data on Rh-containing samples [121], reduction leads to the formation of the $Ce_{9.33}(SiO_4)_6O_2$ phase which, upon reoxidation at 773 K, produces small CeO_2 crystallites and amorphous silica, as shown in Fig. 16.5. The small ceria crystallites are easily reduced, even in the bulk, as shown during TPR experiments by H_2 spillover at low temperature. Bulk reduction of ceria is in fact greatly eased by the smaller dimensions of the crystallites. By lowering the Si content, as in the presence of pure ceria, no such small crystallites are present and reduction is strongly delayed. Moreover, the structural rearrangement observed upon reoxidation of Rh/ceria–silica (Si/Ce = 0.42), which assists in the formation of CeO_2 and SiO_2 from the cerium silicate, helps avoid encapsulation of the metal. In Rh/CeO_2, decoration persists even after oxidation at higher temperatures, but here the presence of silica favors the reversal of SMSI-like decoration already after reoxidation at 773 K, where Rh is free from particles of reduced support [117, 122]. Therefore the formation of cerium silicate during the reduction step allows the development of small ceria particles upon

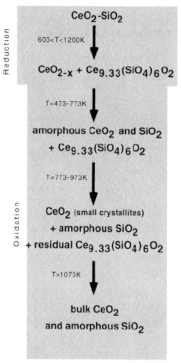

Fig. 16.5 Schematic summary of the structural transformations occurring with silica-doped ceria [120].

reoxidation. These small particles are reduced at low temperature in the presence of Rh (almost 60% of CeO_2 is reduced at a temperature lower than ~600 K). If reoxidation is carried out at 1073 K, this promotion is lost because of sintering of these small ceria crystallites.

16.6
New Nanocatalytic Materials

16.6.1
Nano-GAZ

A new nanocrystalline Ga–Al–Zn complex-oxide (designated here as nano-GAZ) has been synthesized by a hydrothermal method, with uniform nanocrystalline particles of size around 5–10 nm and very uniform 3.8 nm mesopores [123]. This new material was designed for the elimination of N-containing polycyclic aromatic compounds (NPACs) present in diesel-engine emissions, which are known to be carcinogenic and need now to be removed. For instance, the concentration of 1-nitropyrene, which is the most abundant NPAC in diesel emission extracts, is

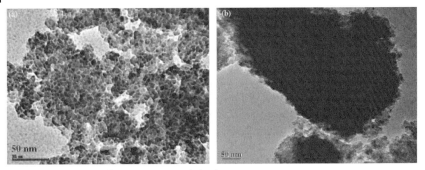

Fig. 16.6 TEM image of (a) nano-GAZ and (b) the contrasting sample prepared by precipitation (scale bar, 50 nm) (from Ref. [123]).

currently in the range of 0–10 ppm. Direct pyrolysis of NPACs generates substantial NOx emissions, as most of the nitrogen in NPACs will be oxidized to NOx, and an effective catalyst for catalytic oxidative decomposition of NPAC is desired. Nano-GAZ suppresses the formation of NOx in the decomposition of pyridine at all temperatures, whereas the simple GAZ prepared by precipitation (which does not present the same nanometric features, see Fig. 16.6) burns pyridine but generates large amounts of NOx at temperatures below 673 K (some of the NOx are removed at higher temperature). The high specific surface of nano-GAZ might not be the only factor improving activity, but the pore system probably also has a role in retaining partially oxidised products inside the core of the catalyst.

16.6.2
Nanotubes

Carbon nanotubes have also been used in deNOx. They have been claimed to be directly active, but reproduction of the results has not been possible. However, they are very good supports for nanoparticles of phosphorotungstic acids (HPW) [124], which are proven deNOx catalysts. Nanometric HPW are easily destroyed at high temperature, and are usually supported on oxides (if the isoelectric point is 7), but carbon nanotubes increase even more the NO adsorption capacity, in the form of N_2O_3. Calixarene nanotubes are also capable of trapping NO_x. Nanocavities formed by hemi-carcerands like calixarenes can trap NO^+ and form nitrates, although this has not so far been used for deNOx but rather for using the entrapped species in organic synthesis.

Interestingly, nanoparticles can be formed on deNOx catalysts during ageing. TEM revealed, on the surface of a Pd catalyst used for the equivalent of 100 000 miles, the formation of whiskers or nanotubes of $Ni_xAl_2O_{3+x}$ with Pd particles on top (Fig. 16.7) [125]. The role (if any) of these particles in the catalytic process still needs to be explained.

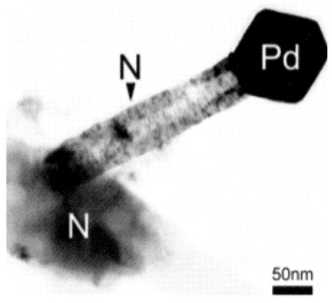

Fig. 16.7 TEM image of a Ni$_x$Al$_2$O$_{3+x}$ nanotube (N) with a Pd particle on top (from Ref. [125]).

16.7
Conclusion

This overview shows that in NOx removal catalysis (as well as in other different domains of heterogeneous catalysis) the role of nano-structured materials is fundamental. Nano-sized metal particles are most of the time necessary to ensure an efficient NO decomposition. The presence of a nano-tailored support, besides its role in keeping a high surface area for the catalyst, necessary to enhance the density of chemical interactions, also plays the role of a stabiliser for the supported metal particles, preventing their agglomeration and sintering. Only very few aspects of catalysis (the most important being perhaps OSC) can be considered without taking into account the presence of nanoparticles in the employed material. In all the new possible applications of catalysis for NOx abatement, in particular, but also in all the other domains of heterogeneous catalysis, the parameter concerning nano-shaped compounds is always present, showing that this new class of materials will impose itself on the world of chemistry for the coming years.

References

1 G. Centi, G. E. Arena, S. Perathoner, *J. Catal.* **2003**, *216*, 443.
2 G. Busca, L. Lietti, G. Ramis, F. Berti, *Appl. Catal. B* **1998**, *18*, 1.
3 J. N. Armor, *Catal. Today* **1995**, *26*, 147.
4 R. Burch, J. P. Breen, F.C. Meunier, *Appl. Catal. B* **2002**, *39*, 283.
5 W. Held, A. Koenig, T. Richter, L. Puppe, *Society of Automotive Engineers* **1990**, Paper 900496.

6 M. Iwamoto, H. Yahiro, S. Shundo, Y. Yu-u, N. Mizuno, *Shokubai (Catalyst)* **1990**, *32*, 430.

7 Y. Li, J. N. Armor, *Appl. Catal. B* **1992**, *1*, L31.

8 C. Torre-Abreu, M. F. Ribeiro, C. Henriques, F. R. Ribeiro, *Catal. Lett.* **1997**, *43*, 25.

9 H. Y. Chen, W. M. H. Sachtler, *Catal. Today* **1998**, *42*, 73.

10 Z. Chajar, P. Denton, F. Berthet de Bernard, M. Primet, H. Praliaud, *Catal. Lett.* **1998**, *55*, 217.

11 H. Hamada, Y. Kintaichi, M. Sasaki, T. Ito, *Appl. Catal.* **1991**, *75*, L1.

12 A. Obuchi, A. Ohi, M. Nakamura, A. Ogata, K. Mizuno, H. Ohuchi, *Appl. Catal.* **1993**, *2*, 71.

13 Y. Kintaichi, H. Hamada, M. Tabata, T. Yoshinari, M. Sasaki, T. Ito, *Catal. Lett.* **1990**, *6*, 239.

14 M. A. Gomez-Garcia, V. Pitchon, A. Kiennemann, *Environ. Int.* **2005**, *31*, 445.

15 P. Denton, A. Giroir-Fendler, H. Praliaud, M. Primet, *J. Catal.* **2000**, *189*, 410.

16 E. Xue, K. Seshan, J. R. H. Ross, *Appl. Catal. B* **1996**, *11*, 65.

17 J.-H. Lee, H. H. Kung, *Catal. Lett.* **1998**, *51*, 1.

18 F. Jayat, C. Lembacher, U. Schubert, J. A. Martens, *Appl. Catal. B* **1999**, *21*, 221.

19 B. Ioan, A. Miyazaki, K. I. Aika, *Appl. Catal. B* **2005**, *59*, 71.

20 L. Olsson, B. Westerberg, H. Persson, E. Fridell, M. Skoglundh, B. Andersson, *J. Phys. Chem. B* **1999**, *103*, 10433.

21 I. Balint, A. Miyazaki, K. Aika, *J. Catal.* **2002**, *207*, 66.

22 L. Piccolo, C. R. Henry, *J. Mol. Catal. A* **2001**, *167*, 181.

23 C. Shi, M. J. Cheng, Z. P. Qu, X. H. Bao, *Appl. Catal. B* **2004**, *51*, 171.

24 Z. J. Li, M. Flytzani-Stephanopoulos, *J. Catal.* **1999**, *182*, 313.

25 M. K. Neylon, M. J. Castagnola, N. B. Castagnola, C. L. Marshall, *Catal. Today* **2004**, *96*, 53.

26 S. M. Kanan, M. C. Kanan, H. H. Patterson, *Curr. Opin. Solid State Mater. Sci.* **2003**, *7*, 443.

27 R. W. Joyner, M. Stockenhuber, *Catal. Lett.* **1997**, *45*, 15.

28 X. She, M. Flytzani-Stephanopoulos, *J. Catal.* **2006**, *237*, 79.

29 M. Anpo, M. Matsuoka, K. Hanou, H. Mishima, H. Yamashita, H. H. Patterson, *Coord. Chem. Rev.* **1998**, *171*, 175.

30 C. Resini, T. Montanari, L. Nappi, G. Bagnasco, M. Turco, G. Busca, F. Bregani, M. Notaro, G. Rocchini, *J. Catal.* **2003**, *214*, 179.

31 E. M. Sadovskaya, A. P. Suknev, L. G. Pinaeva, V. B. Goncharov, B. S. Bal'zhinimaev, C. Chupin, J. Perez-Ramirez, C. Mirodatos, *J. Catal.* **2004**, *225*, 179.

32 G. H. Li, S. C. Larsen, V. H. Grassian, *Catal. Lett.* **2005**, *103*, 23.

33 G. Djéga-Mariadassou, *Catal. Today* **2004**, *90*, 27.

34 P. Degobert, *Automobile et Pollution*, Editions Technip, Paris, **1992**.

35 H. S. Gandhi, G. W. Graham, R. W. McCabe, *J. Catal.* **2003**, *216*, 433.

36 M. Shelef, K. Otto, N. C. Otto, *Adv. Catal.* **1978**, *27*, 311.

37 J. T. Kummer, *J. Prog. Energy Combust. Sci.* **1980**, *6*, 177.

38 J. T. Kummer, *J. Phys. Chem.* **1986**, *90*, 4747.

39 K. C. Taylor, *Catal. Rev. Sci. Eng.* **1993**, *35*, 457.

40 R. M. Heck, R. J. Farrauto, *Catalytic Air Pollution Control: Commercial Technology*, Van Nostrand Reinhold, New York, **1995**.

41 H. C. Yao, Y. F. Y. Yao, *J. Catal.* **1984**, *86*, 254.

42 J. C. Schlatter, P.J. Mitchell, *Ind. Eng. Chem. Prod. Res. Dev.* **1980**, *19*, 288.

43 S. J. Schmieg, D. N. Belton, *Appl. Catal. B* **1995**, *6*, 127.

44 T. Bunluesin, E. S. Putna, R. J. Gorte, *Catal. Lett.* **1996**, *41*, 1.

45 J. Kaspar, P. Fornasiero, M. Graziani, *Catal. Today* **1999**, *50*, 285.

46 A. Trovarelli, C. d. Leitenburg, M. Boaro, G. Dolcetti, *Catal. Today* **1999**, *50*, 353.

47 M. Boudart, G. Djéga-Mariadassou, *Cinétique des Réactions en Catalyse Hétérogène*, Masson, Paris, **1982**.

48 M. Boudart, G. Djéga-Mariadassou, *Kinetics of Heterogeneous Catalytic*

Reactions, Princeton University Press, Princeton, NJ, **1984**.

49 G. Djéga-Mariadassou, M. Boudart, *J. Catal.* **2003**, *216*, 89.

50 H. Birgersson, L. Eriksson, M. Boutonnet, S. G. Järås, *Appl. Catal.* **2004**, *54*, 193.

51 C. H. Bartholomew, *Appl. Catal. A* **2001**, *212*, 17.

52 P. Forzatti, L. Lietti, *Catal. Today* **1999**, *52*, 165.

53 J. R. Gonzalez-Velasco, J. A. Botas, R. Ferret, M. P. Gonzalez-Marcos, J.-L. Marc, M. A. Gutierrez-Ortiz, *Catal. Today* **2000**, *59*, 395.

54 T. N. Angelidis, S. A. Sklavounos, *Appl. Catal. A* **1995**, *133*, 121.

55 C. Battistoni, V. Cantelli, M. Debenedetti, S. Kaciulis, G. Mattogno, A. Napoli, *Appl. Surf. Sci.* **1999**, *144*, 390.

56 T. Luo, J. M. Vohs, R. J. Gorte, *J. Catal.* **2002**, *210*, 397.

57 R. K. Usmen, R. W. McCabe, G. W. Graham, W. H. Weber, C. R. Peters, H. S. Gandhi, *Society of Automotive Engineers* **1992**, Paper No. 922336.

58 J. C. Summers, W. B. Williamson, M. G. Henk, *SAE Trans.* **1988**, *97*, 158.

59 J. C. Summers, J. J. White, W. B. Williamson, *SAE Trans.* **1989**, *98*, 360.

60 J. Dettling, Z. Hu, Y. K. Lui, R. Smaling, C. Z. Wan, A. Punke, *Stud. Surf. Sci. Catal.* **1995**, *96*, 461.

61 A. D. Sarkar, B. C. Khanra, *J. Mol. Catal. A* **2005**, *229*, 25.

62 S. Bernal, F. J. Botana, J. J. Calvino, M. A. Cauqui, G. A. Cifredo, A. Jobacho, J. Pintado, J. M. Rodriguez-Izquierdo, *J. Phys. Chem.* **1993**, *97*, 4118.

63 A. Trovarelli, *Catal. Rev. Sci. Eng.* **1996**, *38*, 439.

64 Euractiv.com, 2004.

65 J. Kašpar, P. Fornasiero, N. Hickey, *Catal. Today* **2003**, *77*, 419.

66 V. P. Zhdanov, B. Kasemo, *Appl. Surf. Sci.* **1998**, *135*, 297.

67 M. Pijolat, M. Prin, M. Soustelle, O. Touret, P. Nortier, *J. Chem. Soc., Faraday Trans.* **1995**, *91*, 3941.

68 F. Gruy, M. Pijolat, *J. Am. Ceram. Soc.* **1994**, *77*, 1537.

69 M. Pijolat, M. Prin, M. Soustelle, O. Touret, *J. Phys. Phys.-Chim. Biol.* **1994**, *91*, 37.

70 M. Pijolat, M. Prin, M. Soustelle, P. Nortier, *J. Phys. Phys.-Chim. Biol.* **1994**, *91*, 51.

71 M. Pijolat, M. Prin, M. Soustelle, O. Touret, P. Nortier, *Solid State Ionics* **1993**, *63 64 65*, 781.

72 F. Deganello, A. Martorana, *J. Solid State Chem.* **2002**, *163*, 527.

73 M. Fernández-García, A. Martınez-Arias, A. Iglesias-Juez, A. B. Hungrıa, J. A. Anderson, J. C. Conesa, J. Soria, *Appl. Catal. B* **2001**, *31*, 39.

74 M. Fernández-García, A. Martınez-Arias, A. Iglesias-Juez, A. B. Hungrıa, J. A. Anderson, J. C. Conesa, J. Soria, *Appl. Catal. B* **2001**, *31*, 51.

75 J. E. Kubsch, J. S. Rieck, N. D. Spencer, in *Catalysis and Automotive Pollution Control II*, A. Crucg (Ed.), Elsevier, Amsterdam, **1991**, pp. 125.

76 P. G. Harrison, D. A. Creaser, B. A. Wolfindale, K. C. Waugh, M. A. Morris, W. C. Mackrodt, in *Catalysis and Surface Characterisation*, T. J. Dines, C. H. Rochester, J. Thomson (Eds.), The Royal Society of Chemistry, Cambridge, **1996**, pp. 76.

77 A. Chatterjee, S. K. Pradhan, A. Datta, M. De, D. Chakravorty, *J. Mater. Res.* **1994**, *9*, 263.

78 G. Stefanic, S. Popovic, S. Music, *Thermochim. Acta* **1997**, *303*, 31.

79 F. Dong, A. Suda, T. Tanabe, Y. Nagai, H. Sobukawa, H. Shinjoh, M. Sugiura, C. Descorme, D. Duprez, *Catal. Today* **2004**, *93–95*, 827.

80 A. Trovarelli, C. d. Leitenburg, G. Dolcetti, *CHEMTECH* **1997**, *27*, 32.

81 J. P. Cuif, G. Blanchard, O. Touret, A. Seigneurin, M. Marczi, E. Quémeré, SAE Paper 900496 **1997**, 970463.

82 D. Terribile, A. Trovarelli, J. Llorca, C. D. Leitenburg, G. Dolcetti, *Catal. Today* **1998**, *43*, 79.

83 M. H. Yao, R. J. Baird, F. W. Kunz, T. E. Hoost, *J. Catal.* **1997**, *166*, 67.

84 G. Colón, M. Pijolat, F. Valdivieso, H. Vidal, J. Kašpar, E. Finocchio, M. Daturi, C. Binet, J. C. Lavalley, *J. Chem. Soc., Faraday Trans.* **1998**, *94*, 3717.

85 M. Daturi, C. Binet, J. C. Lavalley, A. Galtayries, R. Sporken, *Phys. Chem. Chem. Phys.* **1999**, *1*, 5717.

86 M. Daturi, E. Finocchio, C. Binet, J. C. Lavalley, F. Fally, V. Perrichon, H. Vidal, N. Hickey, J. Kašpar, *J. Phys. Chem. B* **2000**, *104*, 9186.

87 G. Colon, M. Pijolat, F. Valdivieso, R. T. Baker, S. Bernal, *Adv. Sci. Technol.* **1999**, *16*, 605.

88 G. Colon, F. Valdivieso, M. Pijolat, R. T. Baker, J. J. Calvino, S. Bernal, *Catal. Today* **1999**, *50*, 271.

89 E. Finocchio, M. Daturi, C. Binet, J. C. Lavalley, F. Fally, V. Perrichon, H. Vidal, J. Kaspar, M. Graziani, G. Blanchard, *Stud. Surf. Sci. Catal.* **1999**, *121*, 257.

90 F. Fally, V. Perrichon, H. Vidal, J. Kaspar, G. Blanco, J. M. Pintado, S. Bernal, G. Colon, M. Daturi, J. C. Lavalley, *Catal. Today* **2000**, *59*, 373.

91 M. Daturi, E. Finocchio, C. Binet, J. C. Lavalley, F. Fally, V. Perrichon, *J. Phys. Chem. B* **1999**, *103*, 4884.

92 C. Binet, M. Daturi, *Catal. Today* **2001**, *70*, 155.

93 R. T. Baker, S. Bernal, G. Blanco, A. M. Cordon, J. M. Pintado, J. M. Rodriguez-Izquierdo, F. Fally, V. Perrichon, *Chem. Commun.* **1999**, 149.

94 J. R. González-Velasco, M. A. Gutiérrez-Ortiz, J. L. Marc, M. P. González-Marcos, G. Blanchard, *Appl. Catal. B* **2001**, *33*, 303.

95 D. Duprez, C. Descorme, T. Birchem, E. Rohart, *Top. Catal.* **2001**, *16*, 49.

96 C. Descorme, Y. Madier, D. Duprez, *J. Catal.* **2000**, *196*, 167.

97 Y. Madier, C. Descorme, A. M. Le Govic, D. Duprez, *J. Phys. Chem. B* **1999**, *103*, 10999.

98 S. Rossignol, F. Gerard, D. Duprez, *J. Mater. Chem.* **1999**, *9*, 1615.

99 S. Bedrane, C. Descorme, D. Duprez, *Catal. Today* **2002**, *75*, 401.

100 A. Holmgren, B. Andersson, D. Duprez, *Appl. Catal. B* **1999**, *22*, 215.

101 A. Holmgren, D. Duprez, B. Andersson, *J. Catal.* **1999**, *182*, 441.

102 D. Martin, D. Duprez, *J. Phys. Chem. B* **1997**, *101*, 4428.

103 D. Martin, D. Duprez, *J. Phys. Chem.* **1996**, *100*, 9429.

104 S. Kacimi, J. Barbier, R. Taha, D. Duprez, *Catal. Lett.* **1993**, *22*, 343.

105 F. Aupretre, C. Descorme, D. Duprez, *Catal. Commun.* **2002**, *3*, 263.

106 J. Barbier, D. Duprez, *Appl. Catal. B* **1994**, *4*, 105.

107 J. Barbier, D. Duprez, *Appl. Catal. B* **1993**, *3*, 61.

108 J. Barbier, D. Durpez, *Appl. Catal. A* **1992**, *85*, 89.

109 S. Bernal, J. J. Calvino, M. A. Cauqui, G. A. Cifredo, A. Jobacho, J. M. Rodriguez-Izquierdo, *Appl. Catal.* **1993**, *99*, 1.

110 S. Bernal, F. J. Botana, J. J. Calvino, C. L. Cartes, J. A. P. Omil, J. M. Rodriguez-Izquierdo, *Ultramicroscopy* **1998**, *72*, 135.

111 S. Bernal, F. J. Botana, J. J. Calvino, G. A. Cifredo, J. A. P. Omil, J. Pintado, *Catal. Today* **1995**, *23*, 219.

112 M. Pan, J. M. Cowley, R. Garcia, *Micron. Microscop. Acta* **1987**, *18*, 165.

113 M. Pan, Thesis, Arizona State University **1991**.

114 A. K. Datye, D. Kalakkad, M. H. Yao, D. J. Smith, *J. Catal.* **1995**, *155*, 148.

115 S. Bernal, J. J. Calvino, J. M. Gatica, C. Larese, C. L. Cartes, J. A. P. Omil, *J. Catal.* **1997**, *169*, 510.

116 L. Kepinski, M. Wolcyrz, *Appl. Catal. A* **1997**, *150*, 197.

117 S. Bernal, J. J. Calvino, M. A. Cauqui, J. M. Gatica, C. Larese, J. A. P. Omil, J. M. Pintado, *Catal. Today* **1999**, *50*, 175.

118 H. Muraki, G. Zhang, *Catal. Today* **2000**, *63*, 337.

119 M. Chen, P. Zhang, X. Zheng, *Catal. Today* **2004**, *93–95*, 671.

120 E. Rocchini, A. Trovarelli, J. Llorca, G. W. Graham, W. H. Weber, M. Maciejewski, A. Baiker, *J. Catal.* **2000**, *194*, 461.

121 E. Rocchini, M. Vicario, J. Llorca, C. D. Leitenburg, G. Dolcetti, A. Trovarelli, *J. Catal.* **2002**, *211*, 407.

122 S. Bernal, G. Blanco, J. J. Calvino, C. Lopez-Cartez, J. A. Pérez-Omil, J. M. Gatica, O. Stephan, C. Colliex, *Catal. Lett.* **2001**, *76*, 131.

123 S. C. Shen, K. Hidajat, L. E. Yu, S. Kawi, *Catal. Today* **2004**, *98*, 387.

124 M. A. Gomez-Garcia, V. Pitchon, A. Kiennemann, M. Corrias, P. Kalck, P. Serp, *Top. Catal.* **2004**, *30–31*, 229.

125 J. Hangas, *Catal. Lett.* **2003**, *86*, 267–272.

17
Hydrocarbon Catalytic Reactivity of Supported Nanometallic Particles

François Garin and Pierre Légaré

17.1
Catalytic Alkane Reforming on Nanometallic Particles

17.1.1
Introduction

In the book *Catalysis*, volume VI, edited by Paul H. Emmett in 1958, Ciapetta *et al.* [1] wrote "One of the most interesting and commercially important hetero-geneous catalytic processes developed during the past fifteen years is the catalytic reforming of virgin and cracked naphthas to produce high octane gasoline and pure aromatic hydrocarbons. The initial research work carried out in this field was primarily directed toward the utilization of hydrogenation–dehydrogenation catalysts such as molybdenum and chromium oxides." . . . "In the four years fol-lowing World War II, there was virtually no activity in catalytic reforming to produce motor gasoline beyond the use of several fixed bed hydroformers con-verted from the wartime production of aromatics and aviation blending stocks.". . ."In March 1949, the Universal Oil Products Company announced the Platforming Process". The platforming catalysts comprised platinum between 0.01 and 1 wt.%, deposited on alumina with combined halogen, 0.1 to 8% to provide the necessary acidity. At that time the metal was supposed to have only an activity for dehydrogenation and rehydrogenation reactions and only the acid part of the catalyst was supposed to have the power to break the carbon–carbon bond. The behavior of this catalyst was completely bifunctional with both metallic and acid functions.

The patterns of behavior in catalysis by metals were mainly focused on hydro-genation of unsaturated hydrocarbons and deuterium exchange reactions. These reactions were performed in static systems. Evaporated metal films were widely used for these reactions but the problem was to get reproducible metal surfaces, in other words to have "clean" metal catalysts [2, 3].

The first reactions performed with inert alumina supported platinum in a flow reactor were done by the group of Gault [4] in 1963. Studies directed to understand-ing the mechanism of isomerization of hexanes on platinum catalysts started with

Nanoparticles and Catalysis. Edited by Didier Astruc
Copyright © 2008 WILEY-VCH Verlag GmbH & Co. KGaA, Weinheim
ISBN: 978-3-527-31572-7

the work of Barron et al. [4]. This contribution was the first pointing out that skeletal rearrangement of alkanes can take place following a cyclic intermediate on platinum catalysts. In parallel to this study Anderson and Avery worked on oriented films which means that they tried to get well characterized crystallographic surfaces for mechanistic purposes for alkane isomerization [5]. The use of such oriented surfaces was the start of the use of single crystals and the development of the correlation between the reactivity, the selectivity and the catalyst structures.

Since 1950, studies concerning the role of the metal phase in the catalytic reactions have grown exponentially. Between 1970 and 1980, a series of very good reviews were written concerning alkanes isomerization on metal catalysts, under more or less atmospheric pressure of hydrogen plus hydrocarbon. We can cite M. Boudart [6], J.R. Anderson [7, 8], J.K.A. Clarke and J.J. Rooney [9], Z. Paàl [10], V. Ponec [11] and F.G. Gault [12]. This list is far from being exhaustive, but gives an idea of what was the pioneer work concerning the understanding of "Catalysis on Metals" at that time. The main objective of these authors was to understand the mechanism of reaction from an "organic chemist's" point of view and as Monsieur Jourdain [13] they were working with nanoparticles without speaking about nanoscience even if they were aware of the influence of particle size on the catalytic properties of supported metals with the aid of the catalytic rates.

17.1.2
Influence of the Mean Metallic Particle Sizes in Catalytic Reactions

In isomerization reactions only platinum, palladium and iridium are active metals for the skeletal rearrangement of alkanes. Hence, in the first part of this chapter we shall mainly focus on the catalytic behavior of these three metals.

It has often been observed that catalytic samples having the same qualitative composition (for instance, platinum on alumina) but differing in preparation mode, show pronounced dissimilarities in catalytic behavior. Different methods of preparation will yield catalysts differing in crystallite size, crystallite size distribution and shape of the metal crystals. Since 1969, an important work on the statistics of surface atoms and surface sites on metal crystals was undertaken by van Hardeveld and Hartog [14]. In the mean time, a considerable amount of work started to determine the specific rates, rates per unit surface area of metal, for various reactions; *and* in parallel, improvement of chemisorption techniques to determine the surface area of a metal component was achieved.

Unexpectedly, the specific rates remain constant, i.e. they do not depend on the metal dispersion [15] for most of the reactions investigated [16]. The first exceptions observed were the isomerization and hydrocracking of neopentane on platinum [17]. For these reactions the specific rates vary with dispersion by factors of 15 and 300, respectively. This result led Boudart to classify the catalytic reactions on metals as: "Facile" or "Structure insensitive" reactions, for which the specific rates do not depend upon the size of the metal particle, and "Demanding" or

"Structure sensitive" reactions for which the specific rate is highly dependent on the metal dispersion [6]. Boudart suggested that specific sites with special geometrical requirements are involved in structure sensitive reactions while all metal atoms of the surface are available for structure insensitive reactions. Such results obtained on metal supported catalysts with a polycrystalline structure for the metal crystallites corresponded to the breakthrough of Surface Science and the use of single crystals where site identification began to be possible.

Several groups around the world started working on well defined surface structure and also performed catalytic or adsorption experiments,these groups included those of G. A. Somorjai [18, 19], G. Ertl [20] and Strasbourg [21–23]. From the results obtained it was put forward that the surface structure exhibits relaxation, reconstruction, and the presence of steps and kinks on the atomic scale. Chemisorption causes adsorbate-induced restructuring of surfaces, and the substrate has a significant influence on the growth mode of the deposited material. The surface chemical bond is clusterlike, thermal activation is needed for chemical bond breaking, and rough, more open surfaces are markedly more reactive than compact surfaces with close atomic packing [24]. There are many experiments to indicate that adsorbed atoms on transition metals are held with higher binding energy at more open surface sites with lower packing density, at atomic height steps and at kinks for example [25]. At that point we have two observations on the influence of the surface structure; either we are faced with polycrystalline catalysts where the mean metallic particle sizes play a role or the crystallographic orientation of the surface may influence the reactivity. To bring together these two approaches, the use of labelled alkanes with metallic supported catalysts or model surfaces helped us to understand the mechanisms of skeletal rearrangement of alkanes.

Using carbon-13 labelled hydrocarbons it was pointed out that two Bond Shift (BS) mechanisms are taking place on inert alumina supported platinum catalysts. One involving an α,α,β triadsorbed precursor, Anderson–Avery mechanism [5], and the Garin–Gault mechanism involving a metallacyclobutane intermediate [26], Scheme 17.1.

Moreover a comparative study of the hydrogenolysis of methylcyclopentane and isomerization of hexane isomers on platinum catalysts showed the identity of the product distributions from these two reactions [27]. The results strongly suggested that a common intermediate with a five-membered ring structure was responsible for these reactions: it is the cyclic mechanism [4, 12, 21]. A careful study of the hydrogenolysis of methylcyclopentane on two catalysts of extreme dispersion showed that, in the temperature range 250°C–310°C, the product distributions were temperature insensitive on the 0.2 wt.%Pt alumina catalysts, but temperature sensitive on the 10 wt.%Pt alumina catalyst [28]. On the latter, all the observed distributions appeared as combinations of two limiting distributions, one of which included only methylpentanes and therefore corresponded to a completely selective hydrogenolysis of $-CH_2-CH_2-$ bonds; the other one contained n-hexane, but was different from the previous one.

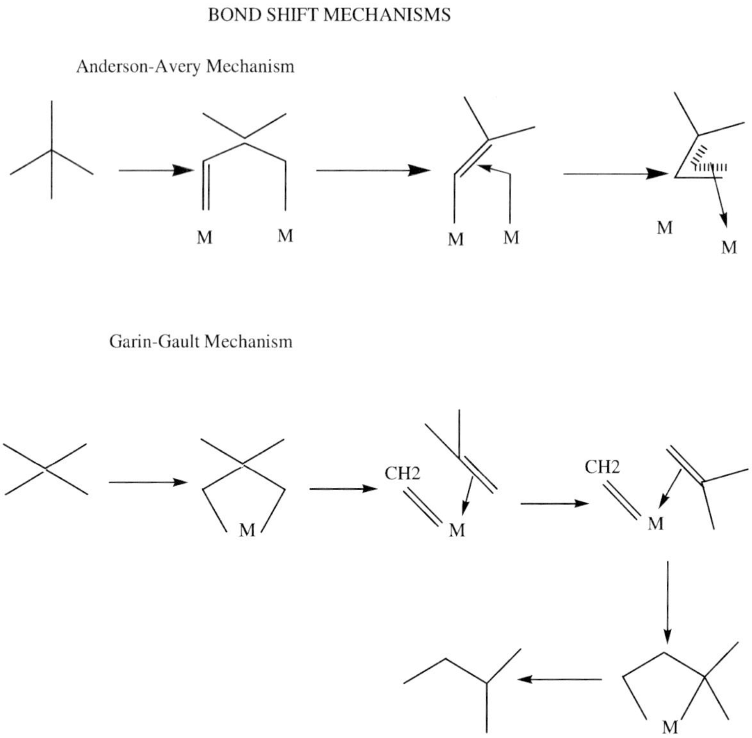

BOND SHIFT MECHANISMS

Anderson-Avery Mechanism

M   M   M   M   M

M

Garin-Gault Mechanism

CH2   CH2

M   M

M

M

Scheme 17.1

These results were interpreted by assuming that the hydrogenolysis of sub-stituted cyclopentanes can take place according to three distinct mechanisms, Scheme 17.2:

1. A non-selective mechanism (mechanism "a") occurring on highly dispersed platinum catalysts and corresponding to an equal chance of breaking any C–C bond of the ring.
2. A selective mechanism (mechanism "b") allowing only the rupture of bisecondary C–C bonds
3. A partially selective mechanism competing with mechanism "b" on catalysts of low dispersion

The most important point is to stress that the cyclic mechanism predominates on highly dispersed platinum aggregates which present a high concentration of low coordinated surface atoms, i.e. corner, step and kink atoms [22].

Another very important point is that for the same crystal sizes, around 9 nm, on Pt, Pd and Ir, the predominant mechanisms are respectively, Bond Shift (BS), Non-Selective Cyclic (NSC) and Selective Cyclic (SC) mechanisms (Table 17.1), which shows unambiguously that a strong electronic influence changes the cata-lytic selectivity.

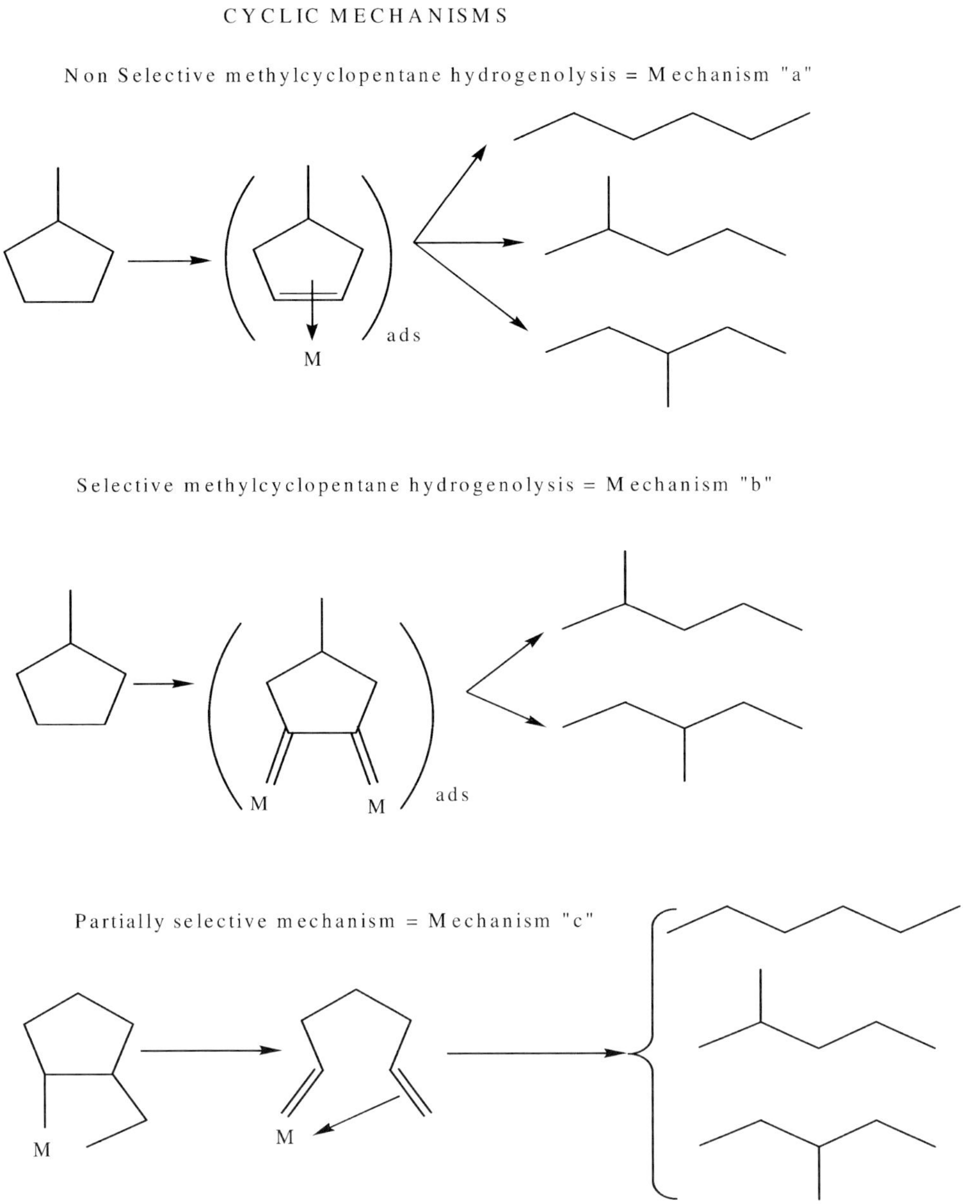

CYCLIC MECHANISMS

Non Selective methylcyclopentane hydrogenolysis = Mechanism "a"

Selective methylcyclopentane hydrogenolysis = Mechanism "b"

Partially selective mechanism = Mechanism "c"

Scheme 17.2

In Scheme 17.3 are presented the two isomerization mechanisms involved on Pt, Pd or Ir catalysts.

The relative amounts of 3-methyl pentanes 2 ^{13}C and 3 ^{13}C can be used to determine the relative contribution of the bond shift and cyclic mechanism, respectively. It is also possible, by using the ratio 3 methylpentane/n-hexane to distinguish

Table 17.1 The isomerization of 2-methylpentane 2^{13}C on Group VIII metals, following Scheme 17.3, under $P_{HC} = 5$ Torr and $P_{H2} = 760$ Torr.

Metal	Particle size (nm)	T (°C)	Ref.	Isomer Select.		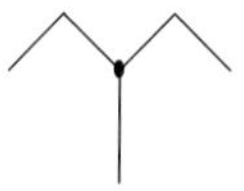 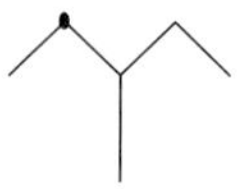 C6 products (%)			Mechanisms (%)		
						3MP	nH	MCP	BS	SC	NSC
Pd	9	270	29	30	88	12	5.4	9.7	14.6	12	0	88
Ir	2–8	160	30	31	100	0	28.1	2.0	0.5	0	100	0
Pt	9	254	31	51	16	84	34.6	12.0	4.4	84	7	9
Pt	1.5	254	31	60	83	17	20.0	34.0	6.5	17	0	83

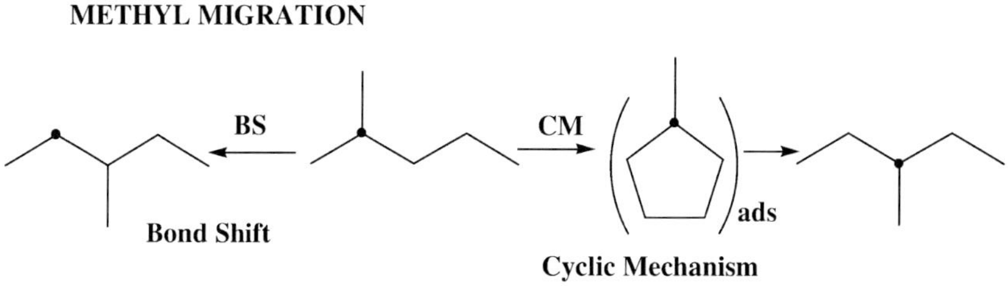

METHYL MIGRATION

BS

Bond Shift

CM

ads

Cyclic Mechanism

Scheme 17.3

between the selective (SC) and the non-selective (NSC) cyclic type isomerization, which involve the selective or the non-selective hydrogenolysis of the methylcyclopentane intermediate.

In Table 17.1 we have the results of the isomerization of 2-methylpentane 2^{13}C on Group VIII metals, following Scheme 17.3, under $P_{HC} = 5$ Torr and $P_{H2} = 760$ Torr.

If there are differences in the isomer selectivities when the metal is changed, as underlined above, there are also differences in the relative distribution of the isomers when the Pt particle sizes are changed. In Fig. 17.1 is presented the influence of the platinum particle sizes in 2-methylpentane isomerization reactions and methylcyclopentane hydrogenolysis. Notice that:

1. The percentage of the cyclic mechanism (black diamonds) is determined in the 2-methylpentane [2-^{13}C] (2-MP-2 ^{13}C) to 3-methylpentane [3-^{13}C] (3-MP-3 ^{13}C) reaction and the bond shift is determined in the 2-methylpentane [2-^{13}C] to 3-methylpentane [2-^{13}C] (3-MP-2 ^{13}C) reaction, as represented in Scheme 17.3. It is calculated following:

$$\%CM = (3\text{-MP-3 }^{13}C)/[(3\text{-MP-3 }^{13}C) + (3\text{-MP-2 }^{13}C)] \times 100$$

2. The ratio $R_{hydrog.}$ (grey squares) is equal to 3-methylpentane/n-hexane and is obtained from the methylcyclopentane (MCP) hydrogenolysis reaction.

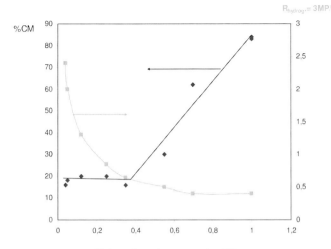

Fig. 17.1 Influence of the platinum dispersion on the relative contribution of the cyclic mechanism (CM) and on the ring opening reaction of methylcyclopentane.

3. There are no crystallites smaller than 1 nm in the low dispersion catalysts for hydrogen chemisorption on platinum particles (H/Pt) ≤ 0.5. In that case the MCP hydrogenolysis is selective, and the value of the ratio $R_{hydrog.}$ is higher than 0.5. Above (H/Pt) ≥ 0.5, the MCP hydrogenolysis is non-selective, $R_{hydrog.}$ is equal to 0.5, and crystallites smaller than 1 nm are present, as determined by electron spectroscopy and SAXS analysis.

All these observations are summarized in Fig. 17.1 [31–33].

17.1.3
Reaction Intermediates

The constancy of the percentage of cyclic mechanism in the isomerization of 2MP to 3MP when dispersion "D" is lower than 50%, or accessibility "a" = H/Pt ≤ 0.5, (mean particle size larger than 2 nm) allows one to rule out the following theory known as the mitohedrical theory. If one assumes that edge or corner atoms of normal fcc crystallites are linked with the cyclic mechanism, and face atoms with bond shift, the relative contributions of the two mechanisms should differ greatly when passing from "D" of 0.35 to 0.05. In Table 17.2 are reported, for various cubooctahedra, the total number of metal atoms N_T, the numbers of corner, edge and face atoms (N_C, N_E and N_F), the accessibility $a = N_S/N_T$ (N_S: number of surface atoms), and the "spherical diameter" d_{sph}. When the number of atoms on one edge

Table 17.2 Statistics of the superficial atoms in a regular fcc cubooctahedron.

n	N_T	N_S	N_C	N_E	N_F	$H/Pt = N_S/N_T$	d_{sph} (Å)
2	38	32	24	0	8	0.84	9
3	201	122	24	36	62	0.61	16
4	586	272	24	72	176	0.46	23
5	1 289	482	24	108	350	0.37	30
10	12 934	2 432	24	288	2 120	0.19	65
15	46 929	5 882	24	468	5 390	0.125	100
30	403 014	25 232	24	1008	24 200	0.062	204

Total number of corner atoms: N_C; edge atoms: N_E; face atoms: N_F and surface atoms N_S.
N: number of atoms on one edge of the cubooctahedron. Total number of superficial atoms:
$N_S = N_C + N_E + N_F$.

of the cubooctahedron "n" varies from 30 to 5, with a corresponding change of accessibility from 0.06 to 0.37, N_C/N_F is multiplied by 70. If one follows Anderson's argument [34], linking cyclic mechanism with corner atoms and bond shift with face atoms, the percentage of cyclic mechanism should increase from 16% to 83% in the case of regular cubooctehedra; however it remains constant. Even if the presence of defects in large crystallites maximizes the number of atoms with low coordination numbers, the cyclic mechanism should greatly increase.

The same is true if one assumes that edge atoms are responsible for the cyclic mechanism. 60% of the mechanism should be cyclic for the catalyst of high dispersion ($a=0.35$) in the case of fcc octahedral.

Therefore the sites for the bond shift and cyclic mechanisms must be topographically similar, located either both on the face or both on the edge of the metal particles. Since isomerization also takes place on very dispersed catalysts ($a>0.5$), when edge atoms predominate, we may conclude that both types of isomerization sites include edge atoms.

How can the decrease in the bond shift mechanism in the extremely dispersed catalysts be explained then? A possible explanation associates bond shift with some specific sites in fcc crystallites; these sites would include edge atoms and would disappear below a critical size. In particular, on single crystals such sites arise when incomplete layers are added over (111) and (100) faces of the fcc crystals. Of particular interest are the B_5 sites, including edge and face atoms in (110) and (311) configuration on the fcc cubooctahedron. Van Hardeveld and Hartog [14] as well as van Hardeveld and van Montfoort [35] calculated the maximum numbers N_{B5} of these B_5 sites as a function of particle size. If one assumes that the bond shift and cyclic mechanisms involve B_5 sites and isolated edge or corner atoms respectively, the percentage of bond shift should vary like the ratio $N_{B5}/N_E + N_C + N_{B5}$, reported in Table 17.3 for incomplete cubooctahedra. This ratio is equal to zero for the smallest particles, increases sharply for $n=3$ and for $n=4$ reaches a value of 0.355, not far from the limit attained in a very large crystal (0.40). This model,

Table 17.3 Statistics of surface atoms and surface sites in an incomplete fcc cubooctahedron.

n	N_T	N_S	H/Pt	$N_E + N_C$	N_{B5}	$100 \times \dfrac{N_{B5}}{N_E + N_C + N_{B5}}$	d_{sph} (Å)
1	1	1	1	1	0	0	
2	38	32	0.84	24	0	0	
3	225	138	0.61	84	24	22.2	16.8
4	688	318	0.46	174	96	35.5	24.4
5	1529	564	0.37	276	168	37.8	31.8
7	4729	1236	0.26	492	312	38.8	46.4
10	14764	2694	0.18	816	528	39.3	67.8
20	124784	11454	0.09	1896	1248	39.7	138

N_{B5}: maximum number of B_5 sites.

then, would explain the constancy of the percentage of cyclic mechanism when the metal accessibility "a" is smaller than 0.4. The fact that some residual bond shift mechanism remains even in the most dispersed catalyst would be due, in most cases, to the rather wide particle size dispersion.

However careful examination of the particle size distribution shows that the model of an incomplete cubooctahedron is not fully consistent with the observed results. The decrease in the bond shift mechanism is clearly related to the appearance of metal particles smaller than 10 Å, which is much below the critical size where the B5 sites disappear in the incomplete fcc cubooctahedron, which is around 17 Å. Theoretical calculations suggest two possible structures for these small particles.

On the one hand, pseudocrystals with a C_5 symmetry axis could be taken into account [36]. These pseudocrystals are assumed to arise from multiple twinning of fcc tetrahedral [36a, 36b] or from continuous growth on nuclei with D_{5h} or icosahedral symmetry [37, 38]. Calculations made on models consisting of icosahedra with 13 or 55 atoms showed modifications of the lattice spacing [39, 40]: radial compression and, according to the potential used, edge extension or compression. Such a change in lattice spacing could be very critical for catalytic properties.

On the other hand, polytetrahedral aggregates, suggested by Hoare and Pal [40] calculations could also be proposed to mimic the small platinum aggregates. These polytetrahedral aggregates are obtained by packing 4-atoms tetrahedra together face-to-face. Between 35 and 55 atoms (5 to 10 Å) the more stable aggregates consist of a superficial quasi-amorphous layer of highly mobile atoms surrounding a solid core with D_{5h} or icosahedral symmetry [41].

Of course such an approach, based mainly on geometrical arguments is not complete and consequently to the nanocrystallites there is an electronic influence which is linked to these small aggregates. This point will be discussed in Section 17.2.

17.1.4
Influence of the Gas Atmosphere Around the Nanoparticles

The above discussion is valid only for nanometallic aggregates which may participate under reductive atmosphere to hydrocarbon reforming catalytic reactions, in
the presence of alkanes and hydrogen only. What will be the state of these very
small aggregates under oxidative atmospheres, which is the case, for instance, in
catalytic pollution control for automotives? In gasoline engines, these catalysts are
faced with an oscillating gas composition around the stoichiometry for the classical
three-way catalysts (TWC). Now a new type of motorization has been developed,
working under lean conditions; oxidative atmospheres, for gasoline cars. The same
oxidative atmosphere is present for diesel cars. In such conditions how can we
understand the catalytic activity?

Taking as an example, the catalytic reduction of nitrogen oxides under oxidative
atmospheres, which is the challenge to solve for diesel engines, there are hundreds
of ppm of NOx in around 10% of oxygen. This means that the catalytic aggregates
are no longer reduced on the surface but at least a layer of adsorbed oxygen atoms
is present on the "metallic" surface. At that stage reconstruction of the crystallites
has to be taken into account as well as migration of crystallites and coalescence.
It is outside the scope of this chapter to develop the mechanisms of sintering, but
we have to keep this important point in mind. The wetting angle of the metallic
aggregates will be changed under an oxidative atmosphere and such an atmosphere is more liable to promote sintering than an inert or reducing atmosphere
[42]. Moreover, not only oxygen but also the reactant as NO [43] may change the
dispersion of the aggregates. It has been observed that in the presence of NO, Pt
particles sinter at temperatures higher than 200 °C; but when NO and oxygen are
admitted simultaneously to the reduced Pt catalyst, the platinum particles are not
fully oxidized up to 300 °C, contrary to what is observed when there is only oxygen
[43]. These observations point out the difficulty in making any generalisation if
the gas phase composition is not taken into account, even if the same "metallic"
catalysts are used.

Now we are going to examine the catalytic activity keeping in mind that a surface
restructuring takes place at the beginning of the catalytic reaction on a clean
surface. In fact adsorbed atoms on transition metals are held with higher binding
energy at more open surface sites with lower packing density, at atomic height
steps and at kinks and this adsorbate-induced restructuring of the clean surface
is concomitant with bond breaking which requires the displacement of the surface
metal atoms from their bare surface equilibrium positions [25].

17.1.5
Reaction Intermediates Determined from Kinetic Data

17.1.5.1 Kinetic Models
Our starting point will be the skeletal isomerization of hydrocarbons. It is not our
intention to develop the kinetic models proposed for such reactions which have

been very extensively studied since the mid-20th century [44–51]. However, a question is not yet resolved concerning the adsorption step of the reactant on a surface. Does it follow a dissociative process or an associative one?

We shall start with this observation concerning the very high negative orders versus hydrogen generally found in hydrogenolysis and isomerization of alkanes [44, 48]. One could assume that for isomerization, as for hydrogenolysis, the rate-determining step is the skeletal rearrangement of a highly dehydrogenated species, obtained by a series of consecutive dehydrogenation steps all equilibrated. According to such a process, proposed by Cimino et al. [46], the reactive species would correspond to a species which has lost a large amount of hydrogen atoms, depending on the values of the order found for hydrogen, following a dissociative mechanism as shown below:

Adsorption step followed by dehydrogenation steps on a monoatomic "free site":

$$C_nH_{2n+2} + 2 \text{ free sites} \leftrightarrow (C_nH_{2n+1})_{ads} + H_{ads}$$
$$\text{-----------------------------------}$$
$$(C_nH_{x+1})_{ads} + 1 \text{ free site} \leftrightarrow (C_nH_x)_{ads} + H_{ads}$$

and with the hydrogen adsorption equilibrium:

$$((2n+2) - x)H_{ads} \leftrightarrow \frac{1}{2}((2n+2) - x)H_2 + ((2n+2) - x) \text{ sites}$$

The global equation is:

$$C_nH_{2n+2} + 1 \text{ free site} \leftrightarrow (C_nH_x)_{ads} + \frac{1}{2}((2n+2) - x)H_2$$

where the value of the order versus hydrogen is equal to $\left[-\frac{1}{2}((2n+2) - x) \right]$.

For example, a value of –3.5 for the order versus hydrogen, found for cyclic type isomerization of pentane and hexanes [12, 26, 52] would mean that the reactive species is obtained by rupturing at least seven carbon–hydrogen bonds. Since it is difficult to imagine first that species such as $[C_5H_5]_{ads}$ or $[C_6H_7]_{ads}$ could react further to give isomer products and secondly to find a parallel with any surface organometallic complex [53], another approach has been taken.

From this dissociative mechanism a series of arguments was given in favor of another mechanism; the associative one in which the hydrocarbon reacts with an adsorbed hydrogen atom. In this mechanism, the adsorption site or "landing site" is composed of a chemisorbed hydrogen atom associated with an ensemble of "Z" potential sites. These sites are associated with the hydrogen chemisorption site. This mechanism has been introduced by Frennet et al. [51], and is shown below:

Bimolecular dehydrogenation steps are proposed, instead of the unimolecular ones previously proposed by Cimino et al. [46]. This mechanism is an associative mechanism or a reactive mechanism [51] in which an ensemble of "Z" surface atoms is involved in the catalytic reaction.

$C_nH_{2n+2} + H_{ads} + Z$ sites $= (C_nH_{2n+1})_{ads} + H_2$ — Adsorption step, with K_a the equilibrium constant

$$(C_nH_{2n+1})_{ads} + H_{ads} = (C_nH_{2n})_{ads} + H_2$$
$$(C_nH_{x+1})_{ads} + H_{ads} = (C_nH_x)_{ads} + H_2$$

(m-1) Dehydrogenation steps with K_i the equilibrium constant

$(C_nH_x)_{ads} \rightarrow iso(C_nH_x)_{ads}$ — Slow step with rate constant k

The rate is: $R = k \cdot \Theta_{C_nH_x}$

The potential surface sites can be divided into three fractions: Θ_S, Θ_H and Θ_C which represent respectively the free sites for chemisorption, those covered by hydrogen chemisorbed atoms and those covered by all hydrocarbon chemisorbed radicals; $\Theta_C \approx \Theta_{CnHx}$.

With $\Theta_S + \Theta_H + \Theta_C = 1$

To calculate Θ_{CnHx} we have:

$$K_a \prod_{i=1}^{m-1} K_i = \frac{\Theta_{C_nH_x} \cdot P_{H_2}^m}{P_{HC} \cdot \Theta_H^m \cdot \Theta_S^Z}$$

from the adsorption and dehydrogenation steps.

The rate can be written as:

$$R = k \cdot K_a \prod_{i=1}^{m-1} K_i \cdot P_{HC} \cdot \Theta_H^m \cdot \Theta_S^Z \cdot P_{H_2}^{-m}$$

For a small carbon coverage,

$$\Theta_C \sim 0, \text{ then } \Theta_S = 1 - \Theta_H \text{ and } \Theta_H = \frac{(K_{H_2} \cdot P_{H_2})^{1/2}}{1 + (K_{H_2} \cdot P_{H_2})^{1/2}}$$

The equilibrium constant K_{H_2} is related to the hydrogen adsorption isotherm. At high hydrogen coverage $\Theta_H \sim 1$ and we have:

$$\Theta_S^Z = (1 - \Theta_H)^Z = (K_{H_2} \cdot P_{H_2})^{-Z/2}$$

The rate of the reaction is

$$R = k \cdot K_a \prod_{i=1}^{m-1} K_i \cdot K_{H_2}^{-Z/2} \cdot P_{HC} \cdot P_{H_2}^{-Z/2} \cdot P_{H_2}^{-m}$$

What is important in this scheme proposed by Frennet et al. [51, 54] is the fact that the order versus hydrogen is equal to $-(Z/2+m)$ in which the term "Z" defines the size of the active site and m is the number of dehydrogenation steps, when the carbon coverage is negligible. Taking the value of −3.5 for the order versus hydrogen, $Z=7$ if $m=0$ and $Z=5$ if $m=1$. In these two cases the reactive species has lost one or two hydrogen atoms, no more. This means that the reactive species adsorbed on the surface could be σ alkyl, di σ, π adsorbed or a carbene species which have their homologues in organometallic chemistry [53].

17.1.5.2 Intermediate Species

Now, with the contribution of this associative kinetic scheme and with the help of the selectivity in the carbon–carbon bond breaking we can draw a parallel between the intermediates species involved in inorganic chemistry and those which may be proposed in heterogeneous catalysis.

For example [53, 55], Pd and Pt form numerous alkyl complexes of the type MR_2L_2, usually stabilized by phosphine ligands. Both metals, Pt^{II} and Pd^{II}, form stable π-complexes with olefins. The best known example is Zeise's salt, $K[PtCl_3(\eta^2\text{-}C_2H_4)]$. Its palladium analog is an intermediate in the oxidation of ethylene to acetaldehyde (Wacker process). Although olefin coordination to transition metals is usually strengthened by back bonding, this contribution is quite weak in complexes of Pt^{II} and Pd^{II}. The carbonyl complexes $[M(CO)Cl(\mu\text{-}Cl)]_2$ and $MCl_2(CO)_2$ were the first metal carbonyls made; (P. Schützenberger 1868–1870). Other bonds such as $M = CR_2$ and agostic $C–H\cdots M$ interactions are also proposed in heterogeneous catalysis, see Refs. [12, 21, 22, 26, 33].

At the dawn of this 21st century we have to try to fill the gap between homogeneous and heterogeneous catalysis, and the first step is to better understand the various mechanisms involved. For instance, now several questions are still not solved such as: (i) Have we in heterogeneous catalysis, as active site, an ensemble of surface atoms, deduced from the kinetic process, instead of a monoatomic site as proposed in inorganic chemistry? and (ii) Whatever the metal used, can we always find new catalytic properties when the mean metallic particle size is decreased?

17.1.6
Gold Nanoparticles

Bulk gold is generally considered to be catalytically inactive. However, recently, very small, well dispersed Au particles have been found to have high catalytic activity for various reactions. These reactions include the hydrogenation of CO and CO_2 [56–58], the reduction of NO [59–61] and the oxidation of methane [62], propene and propane [63, 64]. But the most surprising results have been achieved in low temperature CO oxidation. For example Au/TiO_2 and Au/Fe_2O_3 are capable of CO oxidation at room temperature [65–68].

The first observation from these results is that Au catalysts are able to work under reductive as well as oxidative atmosphere. The second is that the high activity of supported Au catalysts in CO oxidation depends mainly on Au particle size and the presence of suitable metal oxides. The synergy in the metal–support interaction is not well understood, but we may underline that an inert metal can be catalytically active if its size is small enough, around 2~3 nm. At this point is also underlined the very important aspect of the preparation of the catalyst, and in this field much work has to be done [69–71].

To end this first part, we can underline that two main problems have to be kept in mind. One is fundamental and addresses the question as to below which particle size the metallic properties are lost; the other is more practical and concerns the preparation and characterization of very small particles and their catalytic activity.

Now we can study the electronic structure of metal nanoparticles.

17.2
Electronic Structure of Metal Nanoparticles

In this section, we are interested in the peculiar electronic features appearing in metal nanoparticles, which could help us understand their chemical behavior. This, of course, is likely to be intimately linked with the consequences of structural features. In this respect, it could be a good idea to start our discussion from experimental data obtained with organometallic compounds where the size (i.e. the number of metallic nuclei) and the structure are well known. Hence the photoemission spectra, using both UV and soft X-ray excitation (XPS), of $Os_3(CO)_{12}$, $Ru_3(CO)_{12}$, $Ir_4(CO)_{12}$ and $Os_6(CO)_{18}$ were acquired [72, 73]. The compounds, evaporated under ultrahigh vacuum, were condensed onto metallic or poorly conducting substrates (graphite or amorphous carbon), the choice of which was revealed to be unimportant. The UPS curves ascertained that the compounds were intact as the valence electronic states, both metal d and CO-derived, could be favorably compared with gas-phase spectra. More especially, in spite of instrumental broadening, the metal d levels could be resolved. However, when switching from $Os_3(CO)_{12}$ to $Os_6(CO)_{18}$, the resolution in this region decreased. This is just the result of increasing the number of electronic states in a limited energy window, which will ultimately build the delocalised metallic valence band. Examination of the metal core levels by XPS provided another finding. When referred to the respective bulk metal peaks, these features appeared systematically shifted to high binding energies. Moreover, this shift was clearly size-dependent, increasing as the metallic cluster size decreased, amounting to 2.5 eV, as demonstrated in Fig. 17.2.

An ambiguity of these results is due to the binding of the metal atoms with the CO ligands. From the results of CO adsorption on transition metal surfaces, it is known that this may induce a high binding energy shift of the metal core levels by some 1 eV [74, 75]. However this contribution is likely to be more or less the same among the various molecules considered here. When comparing the shifts on $Os_3(CO)_{12}$ to $Os_6(CO)_{18}$, the CO/Os ratio change could hardly account for a binding energy change by 0.8 eV on the Os core levels. So, we can conclude that a high binding energy core level shift characterises small metal clusters, with a strong increase with nuclearity reduction.

Another proof in favor of this interpretation is the result of the progressive decomposition of the compounds under low energy electron bombardment [73]. It was observed that the metal core levels shift to lower binding energy as the decomposition progresses, converging finally to the bulk metal binding energy values. Moreover, this evolution was always accompanied by a loss of the original resolution of the metallic valence d states, showing that a growth process was at

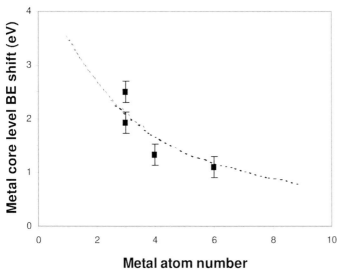

Fig. 17.2 XPS metal core level binding energy shift versus the number of metal atoms of $Ru_3(CO)_{12}$, $Os_3(CO)_{12}$, $Ir_4(CO)_{12}$ and $Os_6(CO)_{18}$. The dotted curve is a guide for the eyes.

play. It is worth stressing that the CO/metal ratio did not appear as a relevant parameter directing this evolution. Indeed, at equivalent CO/metal ratio, the binding energy shift relative to the intact compounds was more important when the substrate was crystalline graphite than amorphous carbon. Previous observations [76] have shown that, due to the low density of surface defects, diffusion of metallic species is much easier on graphite than on amorphous carbon. So the origin of the binding energy shifts is simply the following: as the decomposition progresses, the remaining species being undercoordinated become reactive, inducing a nucleation process the kinetics of which depend on the diffusion speed, which finally results in the increase in size of the metal particles. Analysis of the XPS signal evolution and *ex situ* microscopic observations showed that the final state of the process was metal particles in the size range 10–20 Å, showing that in this range, the clusters had recovered the binding energy of bulk metal. Hence the positive photoemission core level shift is restricted to the nanosize range.

Note that, although we did not discuss the fundamental origin of the phenomena (see below) this was a direct proof that in some way the electronic system of metal nanoparticles could not be identified to that of the bulk metal. Moreover, this provided a practical way to evidence nanoparticle size change during a reactive process.

The above discussed experiments could not quantitatively evaluate the contribution of the CO ligands in the observed metal core level shifts. To get rid of CO, a direct deposition of nanoparticles on amorphous carbon could appear as a solution. Hence, Ir was evaporated under vacuum and condensed on the substrate, the growth being followed by photoemission, and the results compared to those provided by $Ir_4(CO)_{12}$ [77]. Indeed, the Ir core levels were first detected at relatively high energy, converging progressively to values corresponding to bulk Ir. It was, however, shown that the shifts recorded with $Ir_4(CO)_{12}$ were greater than those obtained with the bare Ir particles by some 0.7 eV. So, this gives an upper limit of the CO contribution to the results of Fig. 17.2.

The drawback of the direct metal evaporation method used in Ref. [77] is that the photoemission results could no longer be taken as representative of a well defined nuclearity. The shape of the Ir core levels showed an asymmetry which could be interpreted as the result of a high density of electronic states at the Fermi level [77]. As shown in a comparative study of Au and Pd small deposits [78], this is a misinterpretation. This asymmetry is simply the result of the size dispersion coupled with the size-dependent electron binding energy. However, this analysis provided another interesting result. The core level linewidth of the metal atoms, in both the carbonyl and the direct metal deposition experiments, were broadened, which could be the result of a poor screening of the core hole, suggesting that the positive charge was not compensated for in the photoemission characteristic time.

Many authors have made similar observations [79–81] and various alternative interpretations were proposed [82, 83]. The main difficulty with the photoemission technique is to separate the respective contributions of the so-called "initial state effects" which characterize the ground state (i.e. resulting directly from the change in the electronic system due to size reduction) and "final state effects" (i.e. characterizing the excited state). This latter effect is generally considered to be due to the change in the extra-atomic energy which represents the contribution of the atomic environment to the screening of the positive charge left on the core ionized atom. This was evaluated by Kohiki [84] using the so-called Auger parameter which is independent of the reference level taken in the measurement. He found the extra atomic relaxation energy to decrease by 0.65 eV and 1.1 eV (with respect to the bulk) for the smallest Pd particles on α-Al_2O_3 and SiO_2, respectively. As argued by Kohiki, this applies similarly to the valence band and the core levels. This accounts for most of the positive binding energy shift reported generally for metal nanoparticles using a fixed reference scale. The supplement of positive shift considered by Kohiki as the result of an initial state effect comes from an incorrect energy reference as we show below.

There is another effect that applies specially to the photoemission of metal particles supported on a poorly conducting substrate [82]. The positive charge left by the particle surface cannot be neutralised in the photoemission characteristic time ($\sim 10^{-15}$ s) so that the sample cannot be considered as grounded (i.e. in electrostatic equilibrium with the experimental system) and the Fermi level of the analyzer is no longer a correct energy reference. This results in a positive shift of the entire photoemission spectrum to high binding energies, which is another contribution to the final state effect. When this effect is corrected for by referring the binding energy scale of the individual spectra to the valence band edge, the electron binding energies of Pd particles supported on α-alumina were found to be almost constant and somewhat lower (by ~0.3 eV) than the bulk reference for most of the growth process [85]. Moreover, for the smaller Pd particles, a further negative binding energy shift amounted to 0.3 eV. Thus from bulk Pd to Pd nanoparticles this gives a total 0.6 eV shift towards lower binding energy. As shown below, this is clearly the result of an initial state effect. To understand its physical origin, it is necessary to push the analysis further. We note first that another possible energy reference would be the valence band mean position. This is indeed a key parameter to interpret the photoemission experiments, but also the chemical behavior of metal nanoparticles, as we will see later. The valence bandwidth depends on the coordination, following approximately a square root law [86]. As a result, for metals with a more than half filled d-band, its mean position suffers a binding energy reduction when the coordination is decreased. As found in surface studies [87] the core levels, submitted to similar potential changes, follow the same trend [88]. We can check that the core levels probe the valence band mean position by considering the results presented by Kohiki [84] for Pd particles on both α-Al$_2$O$_3$ and SiO$_2$. Examination of the tables presented in Ref. [84] shows that the Pd 3d$_{5/2}$ core level shift matches almost exactly that of the valence band center, to within 0.1 eV or less, although this was not taken into account by the author. So the valence band edge and the valence band center do not behave in the same way, which requires a valence electron redistribution. This is summarised schematically in Fig. 17.3. (a) represents the photoemission spectrum of bulk metal, with the valence band and a core level peak. The Fermi level (E_f) is the binding energy (BE) scale origin. The valence band center is marked by a dotted line. The photoemission spectrum corresponding to very small particles is represented in (b). Both the core level and the reduced valence band are shifted to high binding energy by the final state effects with respect to the bulk reference. In (c), a part of the final state effect has been artificially compensated for so that the valence band center (dotted line) and the core level peaks are in the same positions as in (a). In (d), the particle Fermi

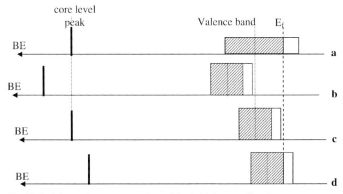

Fig. 17.3 Schematic representation of the initial state effect in small particle photoemission.

level is correctly positioned at the origin. The valence band center and the core level peak are shifted to lower binding energy.

The final conclusion of this discussion is that the core level shifts reported for small particles are the result of both initial and final state effects which could be separated only by a proper analysis.

Another parameter of importance in metal nanoparticles is the change in interatomic bond length. Extended X-Ray absorption fine structure (EXAFS) showed this on many transition metal nanoparticles with diameters less than 5 nm [89, 90]. On Pd particles with a diameter 1.4 nm, the interatomic distance reduction was about 3%. Note that this is again in line with surface observation. More generally, surfaces are submitted to tensile stress and *ab initio* calculations on 2D Pt models show a strong interatomic distance reduction by 6.6% and 9.1% for the (111) and (100) structures, respectively [91]. Of course, in nanoparticles, the structure is not purely 2D and the coordination lowering is not as strong.

This bond length reduction is not the least of the structural features specific to nanoparticles. Both experiment and theory [92] revealed unusual symmetries such as icosahedral or polyicosahedral structures. Study of free iron clusters [93] confirmed a dramatic and irregular increase in the ionisation potential for particles with less than 30 atoms. Intermediate maxima around 15 atoms indicated lower stability in this size region corresponding to incomplete atomic shells. Mass spectra of Ni and Co clusters from 50 to 800 atoms corresponded to icosahedral shell structures.

Finally, it is of primary importance to realize that, in practice, particles are always in interaction with a support. This could affect the structure, the shape, the electronic system and ultimately the chemical properties of the nanoparticles. For large particles of hundreds atoms, this interaction will affect essentially the interface atoms. This is however different for nanoparticles where all atoms are more or less under the influence of the support. With poorly interacting supports such as carbon, more or less hemispherical shapes could be expected. This is different with oxide supports. In a study of Pt particles deposited on α-alumina [94] using photoemission, we showed that some specific features in the valence band and in the O KVV Auger transition could be explained by some direct electronic sharing between Pt and oxygen in the support.

Figure 17.4 illustrates the evolution of the valence region during the Pt growth process. The peak around 10 eV resulting from metal–oxygen electronic states mixing, already present for the clean alumina surface, is enhanced under Pt deposition. Note the decrease of the O 2s peak intensity (around 23 eV) showing that the substrate is largely covered by Pt. Figure 17.5 gives a representation of the O KVV transition during the same process [94]. On a clean alumina surface, it is essentially dominated by a peak at a kinetic energy around 507 eV. After Pt deposition, the intensity of this peak decreases and a new peak arises around 518 eV.

Actually a peak around 515 eV was present on the clean surface spectrum. It originates from a crossed transition which means that the two electronic holes in the final state of the Auger process are located on bonded oxygen and aluminum atoms. The repulsive energy between them is small, resulting in a high kinetic

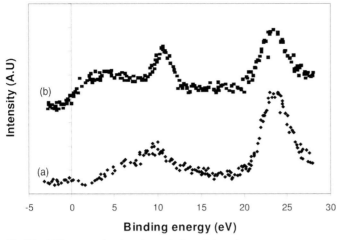

Fig. 17.4 XPS curves showing the evolution of the valence region during deposition of Pt on α-alumina. (a) Clean alumina surface, (b) Pt coverage 2×10^{15} atoms cm^{-2}.

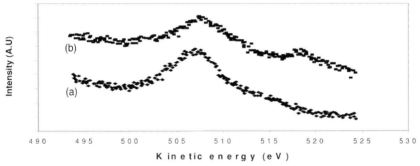

Fig. 17.5 Evolution of the X-ray induced O KVV auger transition during Pt deposition on α-alumina. (a) Clean alumina surface, (b) Pt coverage 3×10^{15} atoms cm^{-2}.

energy of the outgoing electron. The 518 eV is clearly of similar origin, involving now Pt and oxygen, thereby indicating a strong chemical interaction between them at the interface, in agreement with the valence band evolution discussed above. A similar O KVV Auger crossed transition was evidenced when depositing Mo on the same support [95] whereas it did not appear for Pd/α-alumina [84, 96]. In this respect an interesting theoretical study of a 10 Pt atoms cluster on ZrO$_2$ was recently published [97]. As could be expected, the free cluster showed the unambiguous features of a non-metallic system. However, when adsorbed on the substrate, the valence electronic states appeared delocalised with a finite density of states at the Fermi level, but this metallic character remained limited to the Pt part of the whole system. Moreover, charge accumulation between the Pt and O atoms at the interface showed a reduction in ionicity at the interface. The CO and O adsorption energy at specific sites was enhanced.

17.3
General Discussion

Now we come to the chemical consequences of the nanostructure peculiarities discussed above in relation to the catalytic properties of nanoparticles as presented in Section 17.1. It was recently suggested that a key parameter of the chemical activity of a transition metal surface was the mean valence d states relative position with respect to the molecular orbital involved in the molecule–surface interaction [98]. The higher the d-band center, the stronger the interaction with the adsorbed molecule. Moreover, it was shown that this parameter is directly related to the activation energy of the CH_3–H bond breaking on various Ni-based surfaces [99]. As could be expected, the stepped Ni(211) surfaces exhibited the lowest activation energy. We note that step sites of the fcc (211) surface are of the B5 type, as discussed in Section 17.1 in the case of Pt nanoparticles. The same trend was recently evidenced on localized defects on the Ni(111) surface [100].

This should also be of determining importance for nanoparticles of transition metals, where highly reduced coordination sites are numerous. In the previous section we demonstrated that core level photoemission could reveal the evolution of the valence band center after extraction of the initial state effect of Pd nanoparticles. Indeed, reduced atomic coordination results in local valence band reduction, inducing a low binding energy shift of the particle d states that improves the molecule–particle chemical interaction [101]. Interatomic distance reduction should somewhat counterbalance this effect but to a lesser degree. This also could be somewhat counteracted by a strong metal–support interaction. Finally, taking advantage of all these effects is a matter of subtle balance between them.

As shown in Section 17.1, the various hydrocarbon isomerization mechanisms are clearly size dependent i.e. they take place on separated surface sites. Obviously, their relative importance should be dependent on the same weighted d-band center. This is summarized schematically in Fig. 17.6 [32].

As shown in Fig. 17.6, the bond shift mechanism dominates on large Pt particles of average diameter greater than 3 nm, whereas the cyclic mechanism is the major mechanism on nanoparticles with a size below 2 nm.

Moreover, let us consider a large metal particle that has electronic levels so close that they actually form bands. The spacing between adjacent levels is approximately expressed as: $\delta \approx \varepsilon_F / N$ where ε_F is the Fermi level energy and N is the number of atoms in the particle [102]. As the spacing between the levels becomes larger than the thermal energy kT, (k: Boltzmann constant) the levels begin to behave individually and the particle may lose its metallic properties. At room temperature $kT = 2.5 \times 10^{-2}$ eV, and with ε_F of the order of 10 eV, N is calculated to be approximately 400, which corresponds to a diameter of about 2 nm. This

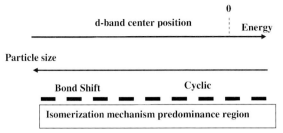

Fig. 17.6 Illustration of the isomerization mechanism dependence on the Pt particle's d-band center. The origin of the energy axis is the Fermi level.

estimation is, of course, very approximate, but a lot of data confirm that the physical and catalytic properties start to change near this size and that catalysis is one of the most sensitive means to probe the surface of such particles.

Finally, the applications of various spectroscopic and structural probes made possible the investigation of catalyst surfaces at a more microscopic level. Studies with idealized surfaces such as the faces of single crystals in an ultra-high vacuum apparatus allowed us to investigate the role of the surface in catalysis. This was completed by spectroscopic studies of supported metallic particles allowing us to characterize the size dependence of the electronic system. Specific active sites for hydrocarbon isomerization were evidenced and their appearance was linked to the lowered atomic coordination.

References

1 F. G. Ciapetta, R. M. Dobres, R. W. Baker, in *Catalysis*, Vol. VI, P. H. Emmett (Ed.), Reinhold Publishing Corporation, New York, **1958**, Ch. 6.

2 G. C. Bond, G. Webb, P. B. Wells, J. M. Winterbottom, *J. Catal.*, **1962**, *1*, 74.

3 F. G. Gault, J. J. Rooney, C. Kemball, *J. Catal.*, **1962**, *1*, 255.

4 Y. Barron, D. Cornet, G. Maire, F. G. Gault, *J. Catal.*, **1963**, *2*, 152.

5 J. R. Anderson, N. R. Avery, *J. Catal.*, **1966**, *5*, 446.

6 M. Boudart, *Adv. Catal.*, **1969**, *20*, 153.

7 J. R. Anderson, in *Structure of Metallic Catalysts, London*, Academic Press, New York, **1975**.

8 J. R. Anderson, *Adv. Catal.*, **1973**, *23*, 1.

9 J. K. A. Clarke, J. J. Rooney, *Adv. Catal.*, **1976**, *25*, 125.

10 Z. Paàl, *Adv. Catal.*, **1980**, *29*, 273.

11 V. Ponec, in *The Chemical Physics of Solid Surfaces and Heterogeneous Catalysis*, D. A. King, D. P. Woodruff (Eds.), Elsevier, Amsterdam, **1982**, Vol. 4, pp. 365–395 and *Adv. Catal.*, **1983**, *32*, 149.

12 F. G. Gault, *Adv. Catal.*, **1981**, *30*, 1.

13 in <<Le Bourgeois Gentilhomme>>, Theater play by Molière, **1670**.

14 R. van Hardeveld, F. Hartog, *Surf. Sci.*, **1969**, *15*, 189.

15 The metal dispersion is defined as the ratio between superficial and total metal atoms. It is often measured by the number of hydrogen atoms chemisorbed per metal atoms (H/Pt).

16 A. Chambellan, J-M. Dartigues, C. Corolleur, F. G. Gault, *Nouv. J. Chim.*, **1976**, *1*, 41 and references therein.

17 M. Boudart, A. W. Aldag, L. D. Ptak, J. E. Benson, *J. Catal.*, **1968**, *11*, 35.

18 B. Lang, R. W. Joyner, G. A. Somorjai, *Surf. Sci.*, **1972**, *30*, 440.

19 S. M. Davis, G. A. Somorjai, In *The chemical physics of solid surfaces and heterogeneous catalysis* (Eds: D. A. King, D. P. Woodruff), Elsevier, **1982**, Vol. 4, 217.

20 G. Ertl, *Adv. Catal.*, **2000**, *45*, 1 and references therein.

21 G. Maire, F. Garin, in *Catalysis: Science and Technology*, J. R. Anderson, M. Boudart (Eds.), Springer Verlag, Berlin, **1984**, Vol. 6, p. 162.

22 F. Garin, S. Aeiyach, P. Légaré, G. Maire, *J. Catal.*, **1982**, *77*, 323.

23 G. Maire, P. Bernhardt, P. Légaré, G. Lindauer, *Proceedings 7th International Vacuum Congress, 3rd International Conference on Solid Surfaces*, Vienna, 1977, p. 86.

24 G. A. Somorjai, *J. Phys. Chem.* **1990**, *94*, 1013.

25 R. D. Levine, G. A. Somorjai, *Surf. Sci.*, **1990**, *232*, 407.

26 F. Garin, F. G. Gault, *J. Am. Chem. Soc.*, **1975**, *97*, 4466.

27 Y. Barron, G. Maire, J. M. Muller, F. G. Gault, *J. Catal.*, **1966**, *5*, 428.

28 G. Maire, G. Plouidy, J. C. Prudhomme, F. G. Gault, *J. Catal.*, **1965**, *4*, 556.

29 M. Hajek, G. Maire, A. O'Cinneide, C. Corolleur, F. G. Gault, *J. Chim. Phys.*, **1974**, *71*, 1329.

30 F. Weisang, F. G. Gault, *J. Chem. Soc., Chem. Commun.*, **1979**, 519.

31 J. M. Dartigues, A. Chambellan, S. Corolleur, F. G. Gault, A. Renouprez, B. Moraweck, P. Bosch-Giral, G. Dalmai, *Nouv. J. Chim.*, **1979**, *3*, 591.

32 P. Légaré, F. Garin, *J. Catal.*, **2003**, *214*, 336.

33 F. G. Gault, *Gazz. Chim. Ital.*, **1979**, *109*, 255.

34 J. R. Anderson, R. J. Mc Donald, Y. Shimoyama, *J. Catal.*, **1971**, *20*, 147.

35 R. van Hardeveld, A. Van Montfoort, *Surf. Sci.*, **1966**, *4*, 396.

36 (a) S. Ino, *J. Phys. Soc.*, Japan, **1966**, *21*, 346; (b) J. G. Allpress, J. V. Sanders, *Surf. Sci.*, **1967**, *7*, 1; (c) E. Gillet, J. F. Roux, M. Gillet, *Bull. Soc. Fr. Mineral. Cristallogr.*, **1967**, *90*, 54.

37 T. Komoda, *Jpn. J. Appl. Phys.*, **1968**, *7*, 27.

38 (a) Y. Fukano, C. M. Wayman, *J. Appl. Phys.*, **1969**, *40*, 1656; (b) E. Gillet, M. Gillet, *J. Crystal Growth*, **1972**, *13/14*, 212, and *Thin Solid Films*, **1973**, *15*, 249.

39 J. J. Burton, *J. Chem Phys.*, **1970**, *52*, 345 and *Nature*, **1971**, *228*, 335.

40 M. R. Hoare, P. Pal, *Nature Phys. Sci.*, **1972**, *236*, 35 and *J. Crystal Growth*, **1972**, *17*, 77.

41 J. J. Burton, *Catal. Rev. Sci. Eng.*, **1974**, *9*, 209.

42 H. H. Lee, E. Ruckenstein, *Catal. Rev. Sci. Eng.*, **1983**, *25*, 475.

43 S. Schneider, D. Bazin, G. Meunier, R. Noirot, M. Capelle, F. Garin, G. Maire, *Catal. Lett.*, **2001**, *71*, 155.

44 K. Morikawa, W. S. Benedict, H. S. Taylor, *J. Am. Chem. Soc.*, **1936**, *58*, 1795.

45 C. Kemball, H. S. Taylor, *J. Am. Chem. Soc.*, **1948**, *70*, 345.

46 A. Cimino, M. Boudart, H. S. Taylor, *J. Phys. Chem.*, **1954**, *58*, 796.

47 D. J. C. Yates, W. F. Taylor, J. H. Sinfelt, *J. Am. Chem. Soc.*, **1964**, *86*, 2996.

48 J. H. Sinfelt, *J. Phys. Chem.*, **1964**, *68*, 344.

49 J. H. Sinfelt, D. J. C. Yates, *J. Catal.*, **1967**, *8*, 82, 348 and **1968**, *10*, 362.

50 B. S. Gudkov, L. Guczi, P. Tétényi, *J. Catal.*, **1982**, *74*, 207.

51 A. Frennet, G. Liénard, A. Crucq, L. Degols, *J. Catal.*, **1978**, *53*, 150.

52 F. Garin, F.G. Gault, G. Maire, *Nouv. J. Chim.*, **1981**, *5*, 553 and **1981**, *5*, 563.

53 F. A. Cotton, G. Wilkinson, C. A. Murillo, M. Bochmann, Advanced Inorganic Chemistry 6th Edn., **1999**.

54 A. Frennet, in *Hydrogen Effects in Catalysis*, Z. Paal, P. G. Menon (Eds.), Marcel Dekker, New York, **1988**, p. 399.

55 N. N. Greenwood, A. Earnshaw, *Chemistry of the Elements*, 2nd Edn., Butterworth Heinemann, **2001**.

56 H. Sakurai, S. Tsubota, M. Haruta, *Appl. Catal. A*, **1993**, *102*, 125.

57 H. Sakurai, M. Haruta, *Appl. Catal. A*, **1995**, *127*, 93.

58 A. Baiker, M. Kilo, M. Maciejewski, S. Menzi, A. Wokaun, in *New Frontiers in Catalysis*, L. Guczi, F. Solymosi, P.

Tétényi (Eds.), Elsevier, Amsterdam, **1992**, p. 1257.

59 A. Ueda, T. Oshima, M. Haruta, *Appl. Catal. B*, **1997**, *12*, 81.

60 T. Salama, R. Ohnishi, M. Ichikawa, *J. Chem. Soc., Faraday Trans.*, **1996**, *92*, 301.

61 M. A. P. Dekkers, M. J. Lippits, B. E. Nieuwenhuys, *Catal. Today*, **1999**, *54*, 381.

62 R. J. H. Grisel, P. J. Kooyman, B. E. Nieuwenhuys, *J. Catal.*, **2000**, *191*, 430.

63 A. C. Gluhoi, N. Bogdanchikova, B. E. Nieuwenhuys, *Catal. Today*, **2006**, *113*, 178.

64 S. Ivanova, C. Petit, V. Pitchon, *Catal. Today*, **2006**, *113*, 182.

65 D. Cunningham, S. Tsubota, N. Kamijo, M. Haruta, *Res. Chem. Intermed.*, **1993**, *19*, 1.

66 M. Haruta, S. Tsubota, T. Kobayashi, H. Kageyama, M. Genet, B. Delmon, *J. Catal.*, **1993**, *144*, 175.

67 M. Haruta, N. Yamada, T. Kobayashi, S. Iijima, *J. Catal.*, **1989**, *115*, 301.

68 S. Tanielyan, R. Augustine, *Appl. Catal. A*, **1992**, *85*, 73.

69 S. Ivanova, V. Pitchon, Y. Zimmermann, C. Petit, *Appl. Catal. A*, **2006**, *298*, 57.

70 S. Ivanova, V. Pitchon, C. Petit, H. Herschbach, A. Van Dorssalaer, E. Leize, *Appl. Catal. A*, **2006**, *298*, 203.

71 S. Ivanova, C. Petit, V. Pitchon, *Appl. Catal. A*, **2004**, *267*, 191.

72 P. Légaré, T. N. Rhodin, C. F. Brucker, *J. Vac. Sci. Technol. A*, **1983**, *1*, 1227.

73 P. Légaré, Y. Sakisaka, C. F. Brucker, T. N. Rhodin, *Surf. Sci.*, **1984**, *139*, 316.

74 R. C. Baetzold, G. Apai, E. Shustorovitch, *Phys. Rev. B*, **1982**, *26*, 4022.

75 G. F. Cabeza, P. Légaré, N. J. Castellani, *Surf. Sci.*, **2000**, *465*, 286.

76 W. F. Egelhoff Jr, G. C. Tibbetts, *Phys. Rev. B*, **1979**, *19*, 5028.

77 A. Fritsch, P. Légaré, *Surf. Sci.*, **1984**, *145*, L517.

78 A. Fritsch, P. Légaré, *Surf. Sci.*, **1985**, *162*, 742.

79 F. Parmigiani, E. Kay, P. S. Bagus, C. J. Nelin, *J. Spectrosc. Relat. Phenom.*, **1985**, *36*, 257.

80 G. K. Wertheim, S. S. DiCenzo, D. N. E. Buchanan, *Phys. Rev. B*, **1986**, *33*, 5394.

81 G. K. Wertheim, S. B. DiCenzo, *Phys. Rev. B*, **1988**, *37*, 844.

82 G. K. Wertheim, S. B. DiCenzo, D. N. E. Buchanan, P. A. Bennett, *Solid State Commun.*, **1985**, *53*, 377.

83 P. S. Bagus, C. J. Nelin, E. Kay, F. Parmigiani, *J. Spectrosc. Relat. Phenom.*, **1987**, *43*, c13.

84 S. Kohiki, *Appl. Surf. Sci.*, **1986**, *25*, 81.

85 P. Légaré, F. Finck, R. Roche, G. Maire, *Z. Phys. D*, **1989**, *12*, 19.

86 M. C. Desjonquères, D. Spanjaard, in *Concepts in Surface Physics*, Springer Verlag, Berlin, **1996**.

87 P. Légaré, G. Lindauer, L. Hilaire, G. Maire, J.-J. Ehrhardt, J. Jupille, A. Cassuto, C. Guillot, J. Lecante, *Surf. Sci.*, **1988**, *198*, 69.

88 N. J. Castellani, P. Légaré, *J. Spectrosc. Relat. Phenom.*, **1995**, *74*, 99.

89 P. A. Montano, G. K. Shenoy, E. E. Alp. W. Schulze, J. Urban, *Phys. Rev. Lett.*, *1986*, 56, **2076**.

90 R. Lambert, S. Wetjen, N. Jaeger, *Phys. Rev. B.*, **1995**, *51*, 10968.

91 P. Légaré, N. J. Castellani, G. F. Cabeza, *Surf. Sci.*, **2002**, *496*, L51.

92 T. Halicioglu, C. W. Bauschlicher Jr, *Rep. Prog. Phys.* **1988**, *51*, 883.

93 M. Pellarin, B. Baguenard, J. L. Vialle, J. Lermé, M. Broyer, J. Miller, A. Perez, *Chem. Phys. Lett.*, **1994**, *217*, 349.

94 A. Fritsch, P. Légaré, *Surf. Sci.*, **1987**, 184, L355.

95 A. Katrib, C. Petit, P. Légaré, L. Hilaire, G. Maire, *J. Phys. Chem.*, **1988**, *92*, 3527.

96 P. Légaré, B. R. Bilwes, *Surf. Sci.*, **1992**, *279*, 159.

97 A. Eichler, *Phys. Rev. B*, **2005**, *71*, 125418.

98 B. Hammer, O. H. Nielsen, J. K. Norskov, *Catal. Lett.*, **1997**, *46*, 31.

99 F. Abild-Pedersen, J. Greeley, J. K. Norskov, *Catal. Lett.* **2005**, *105*, 9.

100 M. Haroun, Thesis, Strasbourg **2007**.

101 S. J. Riley, *J. Non-Cristall. Sol.*, **1996**, *205–207*, 781.

102 M. Che, C. O. Bennett, *Adv. Catal.*, **1989**, *36*, 55

18
Surface Organometallic Chemistry on Metal:
Synthesis, Characterization and Application in Catalysis

Katrin Pelzer, Jean-Pierre Candy, Gregory Godard, and
Jean-Marie Basset

18.1
Introduction

Numerous reasons such as production cost, environmental protection or easy product recovery have led the chemical industry to search for improved or new catalysts having high chemo-, regio-, stereoselectivities, high life time, minimum ageing, and easy regeneration, altogether associated with a high activity. Ease of preparation as well as of handling is also a crucial problem. For this purpose, supported or unsupported "metallic" catalysts are very frequently used for classical reactions such as selective hydrogenation of olefins, dehydrogenation, skeletal isomerization, partial or full oxidation of alkanes, Fischer–Tropsch synthesis, H_2O_2 synthesis from oxygen and hydrogen, low temperature CO oxidation, NO_x removal or shoot residues from the automotive industry etc.

By "metallic" catalysts we mean here catalysts which are supposedly composed of "metallic particles" obtained by reducing a metallic salt, a chemical reaction which transforms them to a "zerovalent" metal. It is likely that during many cata-lytic processes (involving oxygen, or classical oxidizing agents such as H_2O_2, per-oxides, nitric oxide etc. or even hydrocarbons) the "zerovalent" metal in the particle may be transformed, most of the time reversibly, into an "oxidized" form in terms of organometallic or inorganic concepts such as bulk or surface oxide, bulk or surface carbide, bulk or surface nitride, bulk or surface sulfide, surface nitrosyl, imido, oxo, peroxo, hydride species, etc.

In order to increase catalytic performance, these "metal particles" have been modified by various kinds of additives: organic, inorganic, metallic elements. A classical example is the addition of a second metal to the "active one" frequently referred to as the "host metal". At the beginning of this strategy, the modification of a metal by a second one was purely empirical and was the result of "trial and error" experiments, "high throughput experiments", a technology already used by Bosch and Mitwasch to discover the best catalyst for ammonia synthesis [1–9].

This strategy of metal particle modification by another metal led, for example, to the synthesis of the second and third generation of reforming catalysts which

Nanoparticles and Catalysis. Edited by Didier Astruc
Copyright © 2008 WILEY-VCH Verlag GmbH & Co. KGaA, Weinheim
ISBN: 978-3-527-31572-7

included successively Pt–Re and Pt–Sn catalysts supported on alumina [10–24]. Progressively it appeared that the association of one metal to another sometimes gave spectacular results and the heterogeneous catalysis community studied in detail these bi-metallic systems and tried to correlate the observed structure with their activity and/or selectivity and/or life time. This approach led, for example, Sinfelt [25–29], Biloen and Sachtler [30–34], and Martin and Dalmon [35–41] to propose the concept of "ensemble" in catalysis by metals. Typically, hydrogenolysis of paraffins was highly dependent on the size of the ensembles but it was not clear at that time what was really the role of these "ensembles", in terms of the "molecular approach of the mechanisms".

At the same period Gault and Maire [42–51] studied the mechanism of the reactions involving hydrocarbons, and tried to understand the mechanisms of alkane transformations on the basis of the emerging mechanism of ligand rearrangement and reactivity in organometallic chemistry and homogeneous catalysis. But the attempt to make a "structure–activity" relationship was difficult. In particular, it was difficult to see how these organometallic intermediates could reasonably fit with surface structures. Often the situation was complicated by the fact that the support was not "innocent" and bi-functional catalysis was involved: for example, for the skeletal isomerization of paraffins, a dehydrogenation step was expected to occur on the metallic particle, skeletal isomerization of the olefin on the acidic support and hydrogenation of the isomerized skeletal olefin to the corresponding paraffin on the particle [52, 53]. In other instances the particle composition varyied during catalysis making the system extremely difficult to characterize. In other instances strong reducing conditions led to a partial reduction of the support, which migrated as a kind of "sub oxide" making the metallic particle partially covered by this "sub oxide" [54–60]. This phenomenon called SMSI was responsible for high chemoselectivity of the metal, for example in Fischer–Tropsch chemistry. To summarize, it was not simple to make "structure–activity" relationships in heterogeneous catalysis by metals, bimetals or supported metals.

An improvement in the understanding of the catalytic behavior of such bi-metallic systems came from the observation of a very peculiar phenomenon concerning the reactivity of organometallic and coordination compounds with oxide supported metallic particles.

Oxides supports are frequently used to support metallic particles, and stabilize low particle size with usually narrow distribution (Fig. 18.1). These oxides react at moderate temperatures (from 25 °C to several hundred degrees) with organometallic compounds of main group elements, transition metals, lanthanides or actinides. Interestingly, it was observed that the particles of "zerovalent" metal supported on various kinds of oxides react much more rapidly with the organometallics than with the oxides on which they are supported. A typical example is that of tetra-n-butyl tin which reacts with the hydroxy groups of silica at 150 °C according to the simple reaction of an "electrophilic" cleavage of a metal–carbon bond by surface silanols [61].

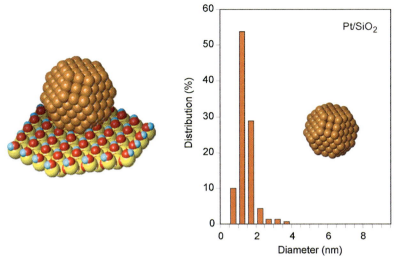

Fig. 18.1 A schematic representation of silica supported metallic particles and a real particle size distribution of platinum nanoclusters supported on silica.

$$(18.1)$$

Surprisingly, if the surface of silica supports a platinum particle covered with hydrogen, the reaction of tetra-n-butyl tin leads already at room temperature to the platinum particle reported in the following equation, on which SnR$_3$ groups are grafted and the silica support is untouched [62]. No reaction occurs on the support in such conditions.

$$(18.2)$$

These observations, as well as many others, led to the development of two fields of organometallic chemistry that we call "Surface Organometallic Chemistry on Metals" SOMC/metals and "Surface Organometallic Chemistry on Oxides"

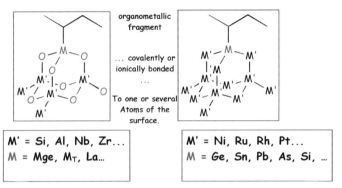

Scheme 18.1 Schematic definition of surface organometallic chemistry on oxides and on metals.

SOMC/oxides (Scheme 18.1). SOMC/metals can be defined first by the study at a very fundamental level of the synthesis, structure, stoichiometric and catalytic reactivity, of any organometallic compound with any surface of a "reduced" metal particle. In principle this field is very broad, considering the huge variety of complexes that have been made in the last century or so and the vast variety of nanoparticles that have been made or can be made using almost all the elements of the periodic table.

The field of SOMC/metals which started by means of supported particles has been recently enlarged to include unsupported particles [63], but at the moment there is no significant difference regarding the type of reactivity observed. The advantage of using "unsupported" particles, eventually called "nanoparticles", is not the small size of these "unsupported" particles nor their narrow distribution. Supported "particles" are usually smaller than the unsupported ones and they have also very narrow particle size distribution. The real advantage of using unsupported nanoparticles, is the elimination of a support effect of the type that we have previously mentioned: some bi-functional catalysis is avoided which makes catalytic interpretation simpler and allows one to grow the particles by a mechanism of crystal growth allowing the formation of certain types of surface structures, as nicely shown by Chaudret [64].

The field of SOMC/metals has proved to be very difficult from an experimental point of view due to the low percentage of metal particle on a support and the low surface area of the metal particles in comparison with the high surface area of the supports.

Various routes can be used to prepare such compounds but it appears progressively clear that surface organometallic chemistry on metals is a new way to obtain and characterize such well controlled bimetallic catalysts [65]. First, by careful control of the reaction parameters, the organometallic complex may react selectively with the "host" metal and not with the chemical functionalities present at the surface of the support, e.g. silica. Even if there is a possibility of reaction between the organometallic complex and the oxide, such reaction usually proceeds at a higher temperature (there is a difference of at least 100 °C) than that of the reaction with the metal. Secondly, surface organometallic chemistry leads to the formation of relatively well defined species (see below) allowing a better understanding of the catalytic results.

We will present here some results obtained in the laboratory on surface organometallic chemistry on metals. We will first describe the various structures achievable and then we will give some catalytic applications of these systems and emphasize the role of the structure of the supported organometallic fragments, or adsorbed metallic atoms derived from them, on the catalytic performance and on the mechanism. When possible we will try to make "structure–activity" relationships, one of the main goals of surface organometallic chemistry.

18.2
Supported Bimetallic Nanoparticles

18.2.1
Supports

18.2.1.1 Methodology and Tools
The general methodology is the grafting of a metal particle on a support. Usually this grafting occurs by impregnation of the support with a metallic salt followed by a large temperature reduction. There are a number of other strategies which do not require such severe treatment, especially when one wants to avoid the effect of counter anions such as nitrates or chlorides. In order to ensure strong grafting, the support is usually an oxide such as alumina, silica, zirconia, ceria, mesoporous silica, organic-inorganic hybrids materials, polymers etc.

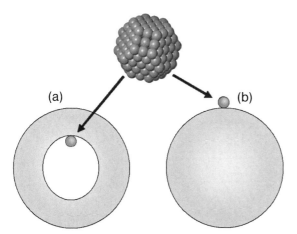

Scheme 18.2 Schematic representation of metallic nanoparticles deposited on various supports: mesoporous silica (a) and non-porous silica (b).

Metal particles can be defined as an ensemble of metal atoms of various sizes, and these atoms have different natures, depending on whether they are located in the core or at the periphery of the particles. Usually the particles correspond to a "close packing" of metal atoms with magic numbers 1, 4, 13, 55, 147, 309, 561, etc. corresponding to the filling of this close packed structure. This packing can begin with a C_5 symmetry arrangement but as the packing grows, hcp arrangement occurs.

In general, in catalysis, the active centers are located at the surface so that the properties of metal particles are related to the number of surface atoms (N_s). The particles are very often characterized by their dispersion ($D = N_s/N_t$) (N_t = total number of atoms of the metal particle). Increasing the number of surface atoms (N_s) per total number of atoms (N_t) requires particles of small sizes, typically in the nanometer range. Van Hardeveld and Hartog [66] proposed, a long time ago, that the shape of nanosized particles can be represented by cubooctahedrons. To our knowledge, even if this shape is a good model and is sometimes observed, there is no experimental proof that all particles have such a shape or preserve such structure at any temperature, especially when they are deposited on a support.

For these model structures, the surface metal atoms are located on the (111) or (100) planes. Note that a cubooctahedral particle is constituted of three kinds of atoms, located respectively on the faces (N_{face}), edges (N_{edge}) and corners (N_{corner}), and that the number of metal atoms on each edge is the same for a given cubooctahedron. For a particle with N_{edge} varying from 2 to 8, the number of atoms on the faces and corners, the number of surface atoms, the total number of atoms and the dispersion are reported in Table 18.1. From the total number of atoms (N_t) in a given metallic particle, the diameter (d) of an equivalent sphere can be determined, knowing the atomic weight (M) and the density (ρ) of the metal, by the relation: d (nm) $= 2[(3MN_t)/(4\pi N_A\rho)]^{1/3}$, where N_A is the Avogadro number. For example, the diameters of an equivalent sphere of nanoparticles of ruthenium are given in Scheme 18.3 and Table 18.1.

Table 18.1 Number of total atoms (N_t), surface atoms (N_s), faces atoms (N_{faces}), edge atoms (N_{edges}) and corner atoms (N_{corner}); dispersion ($D = M_s/M_t$) and diameter d_{Pt} of equivalent sphere, for Pt particles for which N_{edge} varies from 2 to 8.

N_{edge}	2	3	4	5	6
N_t	38	201	586	1289	2406
N_s	32	122	272	482	752
N_{faces}	8	62	176	350	584
N_{edges}	0	36	72	108	144
$N_{corners}$	24	24	24	24	24
D	0.84	0.61	0.46	0.37	0.31
d_{Pt}	1.03	1.80	2.57	3.34	4.11

$N_{edge} = 2$ $N_{edge} = 3$ $N_{edge} = 4$ $N_{edge} = 5$

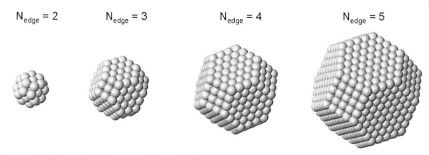

Scheme 18.3 Cubooctahedral particles with an increasing number of metal atoms on the edges [66].

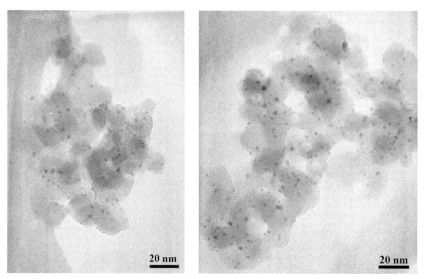

Fig. 18.2 Electron micrograph of monodispesred platinum nanoparticles of 3 nm deposited on silica (the large spheres are silica *aerosil* ® supports from Degussa).

Nano-sized metallic particles can be prepared by various routes, and mostly by reduction of a salt or an organometallic compound of the metal, under mild to strong conditions. In order to avoid coalescence of the particles (sintering), it is sometimes necessary to use organic ligands as stabilizers [63, 67–71]. The presence of ligands coordinated at the surface may prevent the particles from coalescing and allows their self-assembly onto various surfaces. Organic molecules such as amines, thiols, alcohols or alkylsilanes were used as stabilizers for these systems [64, 69, 71–74]. Another approach is to stabilize metal particles through their dispersion on an oxide support. This approach has the advantage of providing "ligand-free" metal particles (Scheme 18.2 and Fig. 18.2). Additionally, these supports provide large surface area, ca. 100 to 300 $m^2 g^{-1}$, and metal loading of about 1 wt.%

with particles of about 2 nm in diameter, as currently used, which correspond to having metal particles every 30 nm.

A great variety of oxides can be used as supports. These materials are chemically stable, but in several cases, interactions between the metallic particles and the support may occur. This interaction is very likely due to the creation of a chemical bond between the particle and the support. A model of such chemical bonding can be found in molecular cluster chemistry: when $Ru_3(CO)_{12}$ reacts with a silica surface, the silanol makes a so-called "oxidative addition" to the Ru–Ru bond of the cluster and there is formation of an $(\eta^2$-siloxy)(Ru–Ru bond) in which the surface oxygen behaves as a "3-electron ligand" in the M. L. H. Green formalism. There is no obvious reason why such reactivity would not occur when a particle of a "zerovalent" metal is adsorbed (chemisorbed) on a partially hydroxylated surface.

Thus, in the particular case of metal catalysts supported on some reducible oxides, the occurrence of so-called metal–support interaction (SMSI) effects has been reported [75]. In order to minimize the metal–support interaction, stable oxides (not reducible) such as alumina, silica or zirconia are most frequently used but as mentioned above, it is quite possible that the metal has one or several chemical bonds with the support (Scheme 18.4). For particles having a diameter of more than 2 nm the support is not believed to have a strong electronic effect on the particle because the number of metallic atoms becomes much more important than the number of chemical bonds between the particle and the support.

This is the reason why it is expected that the electronic, structural and chemical properties of metal nanoparticles adsorbed on a support may depend strongly on the size of the particle and the nature of the support. It is nevertheless likely that these effects will occur mostly on very small particles. This is an important aspect of heterogeneous catalysis [76, 77].

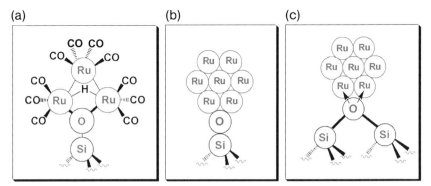

Scheme 18.4 Ruthenium carbonyl cluster chemisorbed on the silica surface (as a model of the type of bonding between a surface oxygen atom and a nanoparticle (b)) another example of bonding is represented (c) in which the surface siloxy group behaves as a 4-electron ligand to the particle.

In order to avoid such chemical interaction between the nanoparticles and the support, many chemical tools may be used: one of them consists in the passivation of surface functionalities such as hydroxy groups, defect sites or Lewis sites via the classical tools of surface organometallic chemistry on oxide [78].

18.2.1.2 Characterisation of Metallic Surfaces and Metal Nanoparticles

As already discussed, the size, the shape and the dispersion are characteristic parameters of metal particles and will determine their properties. It is possible to characterize them as follows: (i) The size of the metal particles (transmission electron microscopy, (TEM) [79–81], extended X-ray absorption fine structure (EXAFS) [82, 83]), (ii) their structures (X-ray diffraction (XRD), TEM), (iii) their chemical composition (TEM-EDX, elemental analysis), (iv) the chemical state of the surface and bulk metal atoms (X-ray photoelectron spectroscopy (XPS)), Mössbauer [84], thermo-programmed-reduction (TPR)), (v) chemisorption capacity toward probe molecules such as H_2, O_2, NO and CO (volumetric or dynamic measurements) [85–90].

In the case of metal surfaces, metallic particles can be obtained in three different states:

- Reduced and desorbed particles. In this case, the metal atoms are fully reduced under a flow of H_2 at high temperatures followed by a treatment under vacuum (or a flow of a neutral gas such as He/Ar) at high temperature. The bulk and surface atoms in the particles are represented respectively by M and M_s.
- Reduced and covered by hydrogen. For these samples, the metal atoms are fully reduced under a flow of H_2 at high temperatures then cooled under H_2. The bulk atoms are represented by M and the surface atoms, covered by chemisorbed hydrogen atoms are represented by M_s–H_x. The "stoichiometry" of the hydrogen adsorption capacity (value of x) depends on the nature of the metal, on the temperature of the adsorption and on the H_2 pressure.
- Oxidized particles. This state can be obtained by a treatment under O_2 (or air) of the reduced and desorbed particles. Depending on the nature of the metal and on the temperature of the oxidizing treatment, all the metal atoms can be oxidized, or in some cases only the surface atoms are oxidized. The oxidized metal atoms are represented by $M^{(X)}$ (bulk) or $M_s^{(X)}$-O_y (surface), where (X) is the oxidation state of the metal atom M and y is the "stoichiometry" of the adsorbed oxygen, that is the average number of oxygen atoms adsorbed by each surface metal atom.

In this study, we will focus our attention mostly on γ-Al_2O_3, SiO_2, mesoporous silica and active carbon as support materials. Their respective surface area and porosity are reported in Table 18.2.

Table 18.2 Characteristics of the supports.

Support	Al$_2$O$_3$	Al$_2$O$_3$	SiO$_2$	SiO$_2$	C
Name	γ-alumina	γ-alumina	Aerosil 200	Mesop	50S
Manufacturer	Procat.		Degussa	Grace	CECA
Surface area (m^{2}g^{-1})	198	250	214	596	1460
Porous volume (ml g^{-1})	3.3[a]		4.7[a]		0.40
Pore radius (nm)	6[a]		12[a]	3	<2

a   Intergranular porosity.

Before use, the silica and alumina supports are pretreated under flowing dry air at 500 °C for 5 h. The active carbon is pretreated following the procedure already proposed by Richard and Gallezot [91].

18.2.2
Supported Host Metal

18.2.2.1   Host Metal Deposition on the Supports

Pt and Pd supported on alumina are prepared by a dry impregnation technique using a toluene solution of platinum or palladium acetylacetonate Pt(C$_5$H$_7$O$_2$)$_2$ and Pd(C$_5$H$_7$O$_2$)$_2$, following the procedure already described [92].

Rh supported on alumina is prepared by impregnation of the support with RhCl$_3$ in aqueous solution [93].

Pt and Pd supported on silica or active carbon are prepared by a cationic exchange technique, using an aqueous solution of platinum or palladium tetramine hydroxide, Pt(NH$_3$)$_4$(OH)$_2$ and Pd(NH$_3$)$_4$(OH)$_2$, following the procedure already described [94].

Rh supported on silica is prepared by the cationic exchange technique, using an aqueous solution of chloropentamine rhodium hydroxide RhCl(NH$_3$)$_5$(OH)$_2$, following the procedure already described [93].

Ru supported on silica is prepared by grafting of triruthenium dodecacarbonyl Ru$_3$CO$_{12}$ onto the support followed by decomposition in vacuo at 373 K to form metallic ruthenium particles covered by adsorbed CO. Adsorbed CO is then removed at 673 K in vacuo. The procedure was described by Théolier et al. [95]. Ni/alumina is a commercial catalyst supplied by IFP. Ni supported on silica is prepared by the cationic exchange technique, using an aqueous solution of hexamine nickel hydroxide Ni(NH$_3$)$_6$(OH)$_2$, following the procedure already described [96]. Ir supported on silica is prepared by decomposition of Ir(Acac)$_3$ absorbed on the silica surface, following the procedure already described [97].

18.2.2.2   Characterization of the Host Monometallic Catalysts

The dispersions (number of surface metal atoms/total metal atoms) of the samples are determined by chemisorptions of hydrogen, oxygen and CO at 25 °C [98]. Prior

Table 18.3 Typical catalyst characterization by chemisorption.

Catalyst	H$_2$		O$_2$		CO		Ref.
	H/M$_s$	P$_{equ. (mbar)}$	O/M$_s$	P$_{equ. (mbar)}$	CO/M$_s$	P$_{equ. (mbar)}$	
Pt	1.8	350	1	50	1	50	98–100
Pd	1	20	1	50	1	50	101
Rh	1.2	150	—	—	—	—	102
Ru	1	20	—	—	—	—	103
Ni	1	350	—	—	—	—	104
Ir	2	350	—	—	1	100	97

to the adsorption measurements, the samples are reduced under flowing hydrogen at 380 °C or 450 °C for 3 to 5 h and then evacuated at the same temperature under vacuum (10^{-6} mbar) for 6 h. Depending on the metal, the "stoichiometries" of hydrogen, oxygen and CO adsorption (number of adsorbed hydrogen, oxygen atoms or CO molecules per surface platinum atom) and the corresponding equilibrium pressure ($P_{equ.}$) are reported in Table 18.3.

In the case of nickel, the magnetization (σ) is a good tool to evaluate particle size and oxidation state. Such magnetization was measured in an electromagnet field (H) varying from 0 to 21 kOe, at room temperature using the extraction method [105]. The magnetization at saturation σ_s, expressed in Bohr magneton per Ni (β/Ni) is obtained by extrapolation of the $\sigma = f(1/H)$ curve for $1/H \rightarrow 0$. The average metallic particle size can be estimated from the $\sigma = f(H)$ curve [106, 107] and the amount of reduced nickel atoms (Ni$^{(0)}$) per total nickel atoms (Ni) can be deduced from the magnetization at saturation, σ_s, knowing that the magnetization at saturation of one reduced nickel atom (Ni$^{(0)}$) is 0.62 β [108].

In some cases, the metallic particle size is determined by electron microscopy (JEOL 100 CX) and the dispersion is correlated to the average metallic particle size, assuming that the metallic particles have a cubo-octahedral shape [66].

With platinum and palladium catalysts, supported on silica, alumina and active carbon, both H$_2$, O$_2$ and CO probe molecules are available for dispersion measurements. For rhodium, the various values are taken from the work of Ferretti et al. [102]. For ruthenium and iridium, O$_2$ cannot be used as a probe molecule for dispersion measurements, because there is formation of bulk oxides. With nickel, only H$_2$ gives reliable results, O$_2$ and CO cannot be used as probe molecules for dispersion measurements, because there is formation of bulk oxides with O$_2$ and metal-carbonyls with CO, but the dispersion of the sample can be additionally measured by magnetic measurements.

Before any adsorption measurement, the catalysts are reduced under flowing hydrogen at 380 or 450 °C. It was demonstrated that Rh, Pt and Pd were fully reduced above 380 °C. Ru, Ir and Ni were treated under flowing hydrogen at 450 °C for 5 h. In these conditions, Ru and Ir are fully reduced and the reduction extent of nickel catalysts determined from magnetic measurements are always greater than 90%.

18.2.3
Reaction of Group XIV Organometallic Compounds with "Host" Metals

18.2.3.1 General Considerations

The reaction is usually performed at low temperature, typically 20–50 °C, in order to avoid a reaction of the tin complex with the hydroxy groups of the support, as shown by solid-state ^{13}C CP-MAS NMR. Indeed, it has been shown that tetra-n butyl tin or tributyl tin hydride reacts, above ca. 150 °C with the hydroxy groups of silica, silica-alumina or alumina with formation of a surface grafted $\equiv$M-O-Sn(n-$C_4H_9)_3$ fragment and evolution of butane or hydrogen [61, 109–111] (Scheme 18.5).

Most often, the reaction with the nanoparticles of metal is performed without solvent but some experiments were made with $Sn(n-C_4H_9)_4$ in heptane. Whatever the method and the metal, the reaction can be decomposed into several steps, quite similar to those reported for the reaction of organometallics with the surfaces of oxides [78]:

- First there is physisorption of tetrabutyl tin on the surface of the support. This is well proved, in the case of silica [61] by infrared spectroscopy which shows a strong shift to lower wavenumbers of the $v(O-H)$ band of silica from $3760\,cm^{-1}$

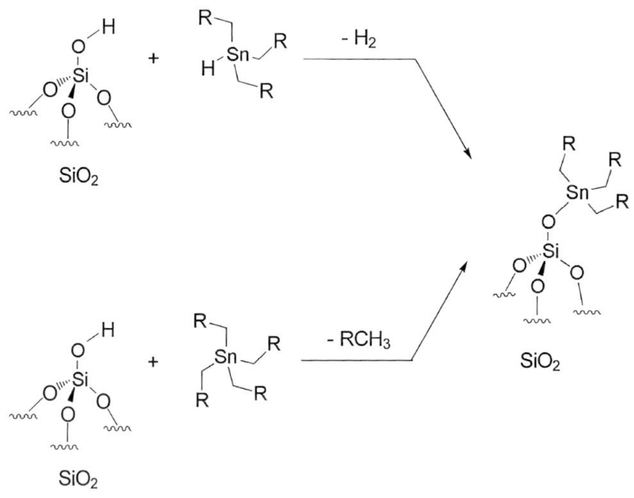

Scheme 18.5 Typical reactions between tin tetra(alkyl) or tin (tris-alkyl)(hydride) derivatives with a silica surface at temperature of 100 °C or higher leading to the same tin(tris-alkyl)(monosiloxy) derivative.

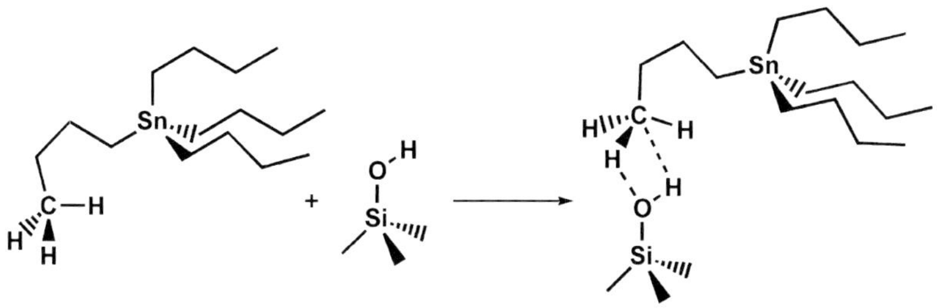

Scheme 18.6 Hydrogen bonding between the alkyl groups of Sn(n-Bu)$_4$ and the silanols.

to ca. 3700 cm^{-1}. This physisorption can be described as resulting from Van der Waals interactions between the hydroxy groups of the support and the butyl chains (Scheme 18.6).

- The next step is migration of the "physisorbed" complex from the support to the metal surface. As the Sn—C bond is weaker than the C—C and C—H bonds (Sn—C < 240 kJ mol^{-1}, C—C = 610 kJ mol^{-1}, C—H 340 kJ mol^{-1}), and since the Rh—H, Rh—C and Rh—Sn bonds are in the same range of enthalpies of formation, this Sn—C bond should be cleaved more easily. This explains why its hydrogenolysis occurs already at room temperature while the C—H and C—C bonds of alkyl ligands do not react. The mechanism of this hydrogenolysis is not yet well understood but by considering studies on the hydrogenolysis of heteroleptic tin complexes [112] one can reasonably suggest that the first step of the reaction is a pentacoordination of the tin atom followed by cleavage of the Sn—C bond with formation of a metal alkyl which is finally hydrogenated into the corresponding alkane (Scheme 18.7).

Naturally, depending on the metal and the reaction conditions, the metal-grafted tributyl tin complex can be further dehydrogenated leading then to a dibutyl fragment, a monobutyl fragment and even the totally de-alkylated form of tin, called the "naked" tin species. In this last case, the tin atoms are considered as "adatoms" chemisorbed on the metal surface. By thermal treatment or even spontaneously, depending on thermodynamic parameters, this "adatom" can migrate into the first layers of the metal particle, leading to the formation of a surface or eventually a bulk alloy if the particle is small enough.

The mechanism by which the tin–carbon bond of the grafted organometallic is cleaved by chemisorbed hydrogen is not yet fully understood but the analogy with mechanisms of molecular chemistry in solution suggests that there is a four-center "transmetallation mechanism" as in Scheme 18.8.

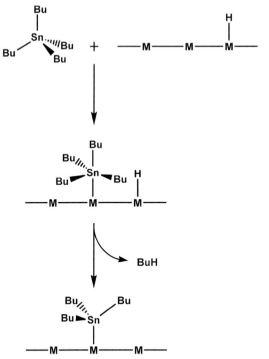

Scheme 18.7 Proposed mechanism of hydrogenolysis of tin tetra-alkyl by a Pt/silica surface covered with hydrogen [113].

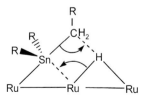

Scheme 18.8 Postulated mechanism of cleavage of the Sn—C bond of a monopodal "Ru-SnR₃" surface organometallic moiety by the ruthenium surface hydride leading to the bipodal "Ru₂(η²-SnR₂)" fragment.

18.2.3.2 Reaction of SnBu₄ under Hydrogen in Solution

18.2.3.2.1 General Considerations The reaction between the n-heptane solution of SnBu₄ and the monometallic catalysts is usually performed at room temperature in a closed glass reactor in the presence of 1 atm of hydrogen. Prior to the reaction, the catalyst is reduced at 450 °C under flowing dry hydrogen for 3 h. The reduced catalyst is introduced under hydrogen at room temperature into a Schlenk tube and the desired amount of SnBu₄ is added. The variation of the SnBu₄ concentration and the amount of butane evolved are then followed by gas chromatographic analy-

Table 18.4 Amount of tin fixed after 7 h of reaction on various "host" surfaces, at 25 °C, in n-heptane, under 1 bar of hydrogen.

Host metal	Disp.	$m_{catal.}$ (g)	M_s (mmol l^{-1})	Sn_{fsat} (mmol l^{-1})	Sn_{fsat}/M_s	Ref.
Pt/Al-685	0.41	2	7.10	5.75	0.81	114
Pt/Al-760	0.50	2	4.60	4.97	1.08	114
Pt/Al-760[a]	0.50	2	4.60	3.59	0.78	114
Pt/Al-683	0.88	2	3.00	4.20	1.40	114
Pt/Al-683[a]	0.88	2	3.00	2.10	0.70	114
Pd/Al-1	0.57	3	2.40		1.25	unpublished results
Pd/Al-2	0.25	3	1.05		0.90	unpublished results
Pt/Si-692	0.50	3	3.00	2.46	0.82	unpublished results
Pt/Si-330	0.30	3	2.10	1.49	0.71	unpublished results
Pt/Si-LACE	0.41	3	3.15	2.88	0.91	unpublished results
Pd/Si-1	0.90	1	9.75	7.31	0.75	unpublished results
Pt/CA-3	0.45	3	4.05	3.24	0.80	unpublished results
Pd/CA-2	0.48	1	6.10	4.58	0.73	unpublished results
Pd/CA-1	0.25	1	5.30	3.98	0.75	unpublished results
Rh/Si(165)	0.8	1	6.40	5.00	0.78	102
Rh/Si(120)	0.69	1	4.00	3.15	0.79	102
Rh/Si(090)	0.45	1	1.95	1.85	0.96	102
Rh/Si(045)	0.6	1	1.30	0.90	0.69	102
Rh/Al(090)	1	1	4.35	3.40	0.80	102
Ru/Si	0.6	3	3.86	3.39	0.88	unpublished results
NiD8	0.08	0.044	1.88	0.99	0.53	115
NiD18	0.18	0.101	3.92	2.29	0.58	115
NiD28	0.265	0.34	6.21	4.43	0.71	115
Ni/Al(11)	0.107	0.137	2.27	1.23	0.54	unpublished results
Ni/Si(104)	0.18	1	5.59	0.00	0.84	unpublished results
Ni/Al(Ecat)	0.22	0.2	8.04	6.36	0.79	unpublished results

a Catalysts reduced at lower temperature.

sis of, respectively, the liquid and gas phase, carried out after increasing times (t) of reaction. It has been verified that isobutane is never formed during the reaction of SnBu$_4$ with alumina supported platinum (the only product is n-butane) and that the equilibrium of isobutane and n-butane in the gas and in the liquid phase is fast.

Ferretti et al. [102] and Bentahar et al. [114] showed, by kinetics measurements, that the amount of surface organometallic fragment "SnBu$_x$" fixed on Rh and Pt catalysts reaches a plateau after about 7 h of reaction at room temperature. Increasing the amount of Sn(n-C$_4$H$_9$)$_4$ introduced per Rh or Pt atom (M), increases the amount of Sn(n-C$_4$H$_9$)$_x$ grafted fragments after 7 h, up to a saturation value (Sn$_{fsat}$) which depends on the dispersion of the sample. The same feature is observed, using silica, alumina or carbon supported Pt, Pd, Ru and Ni. The amounts of SnBu$_x$ fragments grafted after 7 h of reaction in the presence of an excess of SnBu$_4$ are reported in Table 18.4. Knowing the amount of metal surface atoms in the samples, it is possible to determine the ratio Sn$_{fsat}$/M$_s$ (Table 18.4).

The Artifact Due to Support Effects There is an interesting but "artifact" phenomenon due to a "support" effect that may lead to erroneous results regarding the "stoichiometry" of the surface reaction with supported nanoparticles. With silica and carbon as support, the maximum amount of tin fixed never exceeds the value of 0.7–0.8 Sn/surface metal, regardless of the dispersion of the metallic phase. Only alumina support leads in some particular cases to a discrepancy with this somehow "logical rule" (although a value of 0.8 would have been expected *vide infra*). In particular with well dispersed platinum (more than 80% dispersion) supported on alumina the saturation value of the fixed $SnBu_x$ fragments per surface Pt (Sn_{fsat}/Pt_s) is greater than unity [114]. With less dispersed Pt/Al_2O_3 (less than 50%), there is no appreciable difference between the amount of Sn_{fsat}/Pt using silica or alumina as support. This effcct of high temperature reduction treatment on the amount of fixed $SnBu_x$ fragment is attributed to the "activation" of the alumina support close to the metallic particles and this effect is appreciable only in the case of very small metallic particles. The same effect occurs when one uses well dispersed Pd supported on alumina (Table 18.4). This phenomenon is no longer observed if the temperature of reduction of the alumina supported "host" metal is carried out at moderate temperature (380 °C) [102]. A lower reduction temperature did not allow "activation" of the alumina support close to the metallic particle. It is suggested that high reduction temperature favors a possible transfer of a small fraction of the organotin compound from the platinum or palladium to the activated support, leading to values of Sn/Pt_s higher than unity.

This artifact effect of the support is not observed with Ni/Al_2O_3 catalysts, probably due to the large size of the particles (several nanometers).

The quantity of Sn grafted at equilibrium (Sn_{fsat}), is strictly proportional to the amount of surface metal atoms (Fig. 18.3) (the values obtained with well dispersed catalysts supported on alumina, treated at 450 °C, which lead to the artifact effect

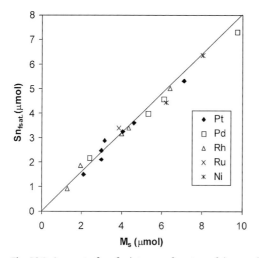

Fig. 18.3 Amount of grafted tin as a function of the number of surface atoms on various nanoparticles of different metals at 25 °C, in n-heptane, under 1 atm of hydrogen.

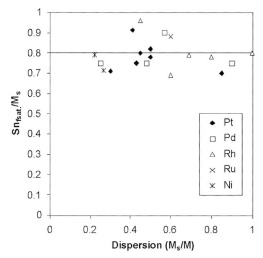

Fig. 18.4 Ratio of tin fixed/surface metal atoms as a function of metallic particles dispersion.

are removed). The slope of the curve is close to 0.8 Sn_{fsat}/M_s independent of the metal, the support and the metal loading of the catalysts: each surface metal atom seems to be able to accommodate ca. 0.8 tin alkyl fragments.

In order to determine the influence of the dispersion of the metallic particles on the ratio $Sn_{fsat.}/M_s$, these values are reported as a function of the dispersion of the samples in Fig. 18.4. It seems that the average $Sn_{fsa.}/M_s$ value is 0.8 ± 0.15, independent of the metal, support or dispersion of the samples.

18.2.3.2.2 Water-soluble Organo-tin Complexes Organometallic compound reaction with metals has always been realized in an organic solvent or without solvent. In order to generalize this method and especially to use it industrially, a method with water-soluble organometallic compounds has been developed [116]. R_3SnOH (R = Me, Et or Bu) reacts at 25 °C in water at pH 10 (KOH) with silica or alumina-deposited platinum or nickel in the presence of hydrogen. The reaction leads to methane, ethane or butane. A partially dealkylated surface species, grafted on the metallic surface, is formed. This species is completely dealkylated by heating under hydrogen at 350 °C. The resulting bimetallic catalysts present similar properties to those obtained by organic phase reaction. These results open the way to surface organometallic chemistry in water, materialized by a basic patent [117]. This domain will be developed thanks to its obvious industrial interest.

18.2.3.3 Reaction of SnBu₄ under Hydrogen in the Absence of a Solvent: Characterization of the Bimetallic Catalysts
Three types of materials can be obtained by reaction of the organometallic compound with the metal surface:
  1. A material in which a surface organometallic fragment is linked to the metallic particle via one, two, three or more "covalent" bonds.

2. A material in which the organometallic compound has lost all its organic ligands and is present on the metallic surface as "naked" adatoms.
3. A material in which the "adatoms" have been incorporated into the particle, leading to the formation of a surface alloy or bulk alloy (for very small metal particles).

A typical description of the transformation of a Pt particle supported on silica under SnBu₄ and hydrogen followed by thermal or hydrogen treatment is given in Fig. 18.5.

Naturally the above situations can coexist on the same solid, resulting in a complex situation. However, by a judicious choice of reaction conditions, it is possible to obtain, for a given metal, a single and well defined environment for tin, allowing it to be fully characterized by physico-chemical methods and above all EXAFS.

The "monografted" or "monopodal" organotin species $MSnBu_3$ was obtained by reaction of $SnBu_4$ with a Ni/SiO_2 catalyst. Its characterization by EXAFS gave one Sn–Ni bond at 0.268 nm and three Sn–C bonds at 0.217 nm in the first coordination sphere of tin [118].

Monopodalgrafted tin species; Example, with Ni:

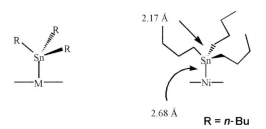

R = *n*-Bu

The "digrafted" or "bipodal" organotin species M_2SnBu_2 was prepared on rhodium particles supported on silica [119]. The EXAFS studies of this catalyst gave two Sn—Rh bonds at 0.262 nm and two Sn—C bonds at 0.212 nm.

Bipodalgrafted tin species; Example, with Rh:

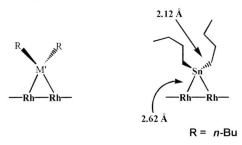

R = *n*-Bu

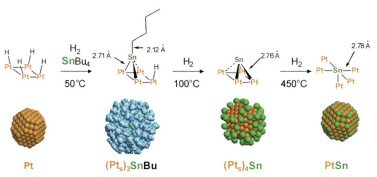

Fig. 18.5 Schematic representation of the various stages of the grafting of a surface organotin complex with a metal particle of platinum (as demonstrated by EXAFS, NMR, chemical analysis, IR. (From left to right): starting platinum particle; particle of platinum covered with the surface organometallic fragment $(Pt_s)_3(\eta^3\text{-SnBu})$; adatoms of Sn $Pt_4(\eta^4\text{-Sn})$; surface alloy [62].

The "trigrafted" or "tripodal" tin surface complex M_3SnBu was isolated on a Pt/SiO$_2$ catalyst [62]. EXAFS measurements carried out at the Sn K-edge indicated that tin was surrounded by three platinum atoms at 0.268 nm and one carbon atom at 0.211 nm (Fig. 18.6).

Tripodalgrafted tin species; Example, with Pt:

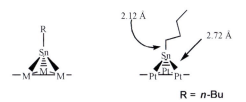

R = *n*-Bu

Naked tin "adatoms" were also obtained on a Pt/SiO$_2$ catalyst but, in order to remove all the butyl ligands the grafted organotin complex was further heated at 300 °C. In agreement with the gas evolution of 4 butane equivalents per tin, EXAFS data indicated that tin was only surrounded by four platinum atoms at the same distance of 0.276 nm (Fig. 18.7). This result clearly indicated that tin is located *on* the metal surface and not in the bulk: for example, in a bulk Pt$_3$Sn alloy, tin is surrounded by 12 platinum atoms, while in a surface alloy on a platinum bulk it is surrounded by 6 platinum atoms only. Note that such tin adatoms are always obtained when low amounts of tin are deposited on the metal. This is probably because tetrabutyl tin coordinates first on the metal atoms of the faces which are the most hydrogenolyzing. This fact will be very important in catalysis since it explains selective poisoning of metal particles (see later).

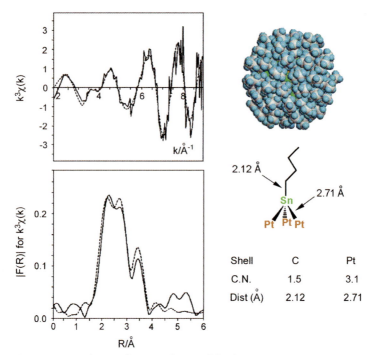

Fig. 18.6 EXAFS data (at the Sn K-edge) used for the determination of the distances between Sn and its surface neighbors and the coordination number of Sn in the surface organometallic fragment "$(Pt_s)_3(\eta^3\text{-SnBu})$".

Shell	C	Pt
C.N.	1.5	3.1
Dist (Å)	2.12	2.71

2.12 Å

2.71 Å

Sn

Pt Pt Pt

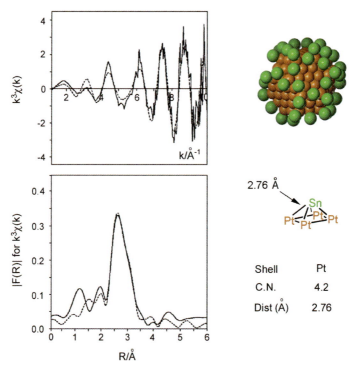

2.76 Å

Sn

Pt Pt Pt
  Pt

Shell	Pt
C.N.	4.2
Dist (Å)	2.76

Fig. 18.7 EXAFS (at the Sn K-edge) data used for the determination of the distances between Sn and its surface platinum neighbors and the coordination number of Sn in the surface adatoms "Pt_sSn".

Tin "adatoms" on "nano particles"

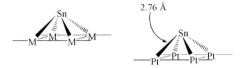

Further heating of the above sample under hydrogen at 500 °C results in an increase in the number of platinum atoms surrounding tin up to ca. 5. This can be explained by a migration of the tin atom into the platinum particle with formation of alloy (Fig. 18.8).

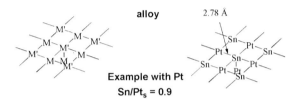

In the following, it will be demonstrated that each kind of surface species leads to a particular selectivity in catalysis:

- When the oganometallic fragments remain "coordinated" to the metal surface, chemio-, regio-, stereo-selectivity of the catalytic reaction can be modified.
- When the organometallic is transformed into "adatoms" these adatoms are expected to be present on some specific crystallographic positions of the particles and this leads to surface selective poisoning.

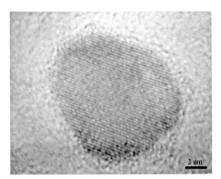

Fig. 18.8 Characterization by high resolution electron microscopy of a bimetallic nanoparticle of "PtSn" alloy obtained via the surface organometallic route.

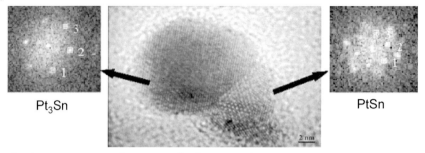

Fig. 18.9 HREM picture of a twin particle of "Pt₃Sn" and "PtSn" alloys supported on silica (with the corresponding X-ray diffraction pattern).

- When surface structures, for example surface alloys, lead to site isolation, then some reactions involving one or several surface atoms will be eventually favored.

By means of SOMC/metals, well-defined "Pt–Sn" alloys have been prepared. With silica-deposited platinum–tin particles of nanometric size, the well-defined formation of Pt₃Sn and "PtSn" alloys has been observed, as shown in the following high resolution electron micrograph (Fig. 18.9).

^{119}Sn NMR allows much better characterization of the "Pt/Sn" bimetallic nanoparticles. For example with silica-deposited platinum–tin particles of homogeneous size, (1.7 and 4.0 nm, respectively, for two different samples), it has been demonstrated that (Fig. 18.10):

- All the tin atoms are in a reduced state (isomeric displacement (δ) is always between 1.5 and 2.5 mm s^{-1}).
- One particular species with a high quadrupole splitting (Δ) and an isomeric shift of 2.3 mm s^{-1} is present in all the samples. Its relative contribution diminishes when the particles diameter is enhanced. Therefore this species could be located at the surface of the particles.

18.2.4
Reactivity of Arsenic and Mercury Organometallic Compounds with "Nanoparticles" of Nickel Covered with Hydrogen

Elimination of arsenic and mercury, present as organometallic compounds in crude-oil during extraction, is a major concern, especially for safeguarding the environment, but also for preservation of reforming catalyst activity and for plant protection from corrosion by arsenic and mercury. The usual method consists in trapping the heavy metal by a solid charge. In the IFP-RAM ® process [121, 122], this charge is constituted of alumina-supported nickel nanoparticles. *De facto*, the

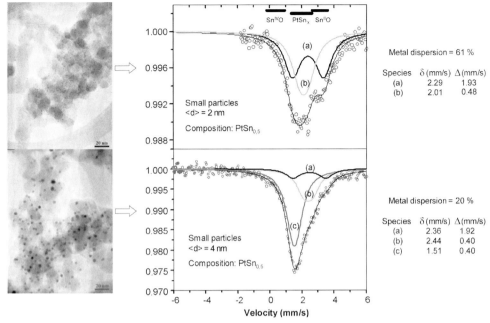

Fig. 18.10 High resolution electron micrographs and the corresponding Mössbauer spectra of small (2 nm) and larger (4 nm) nanoparticles with a PtSn$_x$ supported on silica [120] (δ is the isomer displacement and Δ is the quadrupole splitting).

removal of arsenic and mercury takes place through the rules of "surface organometallic chemistry on metals" and it is why we present these results in this particular section of this chapter. Interestingly more than 10M tons of crude oil are treated yearly by such a process licensed by Axens S.A. in countries like Thailand, Malaysia, Korea and China.

In order to understand this very peculiar "chemisorption" step which occurs in the real process of oil recovery and purification, triphenylarsine, and diphenylmercury have been used to model the real organometallic poison present in the crude oil. An industrial catalyst made of alumina-deposited nickel, containing 18 wt.% nickel as about 10 nm diameter particles, is used (this corresponds to a dispersion of 10%). In the following, we will see that the smallest "size" of the nanoparticles is no longer a requirement but the control of such a size becomes a crucial problem!

18.2.4.1 Triphenylarsine

AsPh$_3$ reaction in n-heptane solution, under hydrogen (12 bar) at a temperature between 25 and 200 °C, with Ni/Al$_2$O$_3$ only takes place on the nickel surface and is characterized by benzene (and cyclohexane, secondary product) evolvement [115]. At 80 °C, saturation of the nickel surface has been reached with As/Ni$_s$ ratio

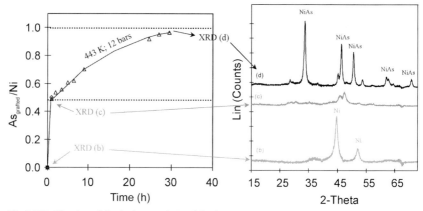

Fig. 18.11 Kinetics of the hydrogenolysis of AsPh$_3$ on Ni/SiO$_2$ (175 °C, 12 bar H$_2$) and XRD analysis of the nickel nanoparticles resulting from such reaction [123].

of 1. At 100 °C, arsenic migration from the nickel surface to the core of the particle is observed. This migration is characterized by a quick decrease in the ferromagnetism of the nickel particles which reaches a value of zero for an "As/Ni" of 0.45. At 170 °C, "NiAs" alloy formation has been highlighted by its X-ray diffraction pattern (Fig. 18.11).

At 170 °C and less than 12 bar of hydrogen, AsPh$_3$ reaction with Ni/Al$_2$O$_3$ can be broken down into two steps. The first is fast and leads to the ill-crystallized NiAs$_{0.45}$ bimetallic nanoparticles; its rate is limited by the diffusion of arsenic atom into the metallic particles. The second, much slower, leads to well-crystallized "NiAs" nanoparticles of alloy. Its rate is limited by the hydrogenolysis of the As-Ph bond. The reaction mechanism can be represented by Scheme 18.9.

18.2.4.2 Diphenylmercury

HgPh$_2$ reaction in n-heptane, under hydrogen with a Ni/Al$_2$O$_3$ surface, starts from room temperature, with the formation of biphenyl (major), benzene and cyclohexane (minor). In contrast with the arsenic case the atoms of mercury, present as "adatoms", are very mobile and can coalesce into metallic mercury. Mercury and nickel do not form alloys, but nickel is weakly soluble in mercury, which explains why, at the reaction end, the amount of nickel in the catalyst has strongly decreased, whereas mercury drops can be observed on the reactor bottom. The reaction mechanism is shown in Scheme 18.10.

In the IFP- RAM II ® process, a second "catalytic" bed constituted of sulfided nickel [124], maintained at a temperature lower than that used for the first one, allows a complete elimination of metallic mercury (see Section 18.5). Typical commercial results are excellent (see Section 18.5), less than 3 ppb of each contaminant in the process effluent as threshold limit for analysis, 1 ppb of each as expectation.

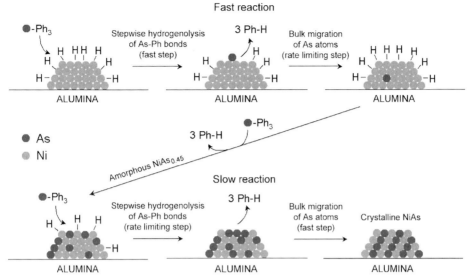

Scheme 18.9 Mechanism of the stepwise hydrogenolysis of AsPh$_3$ with Ni nanoparticles supported on SiO$_2$ (175 °C, 12 bar H$_2$) [123].

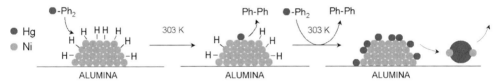

Scheme 18.10 Mechanism of HgPh$_2$ reaction with nanoparticles of Ni stabilized on SiO$_2$ (175 °C, 12 bar H$_2$) [124].

18.3
Unsupported Bimetallic Particles

18.3.1
Introduction

Nanometer-size materials have attracted remarkable academic and industrial research interest due to their fundamental properties and their potential application, ranging from fundamental studies to catalysis [125–130]. Precise control of size and chemical behavior (stability *and* reactivity) by means of the synthesis itself is a major aim due to the direct correlation of intriguing new properties with particle size, bridging the gap between molecules and bulk materials.

There is currently considerable interest in controlling the structure of a material at the microscopic level as this influences its physical and chemical properties and also its reactivity. The change in chemical behavior is due to the electronic change in the nano-sized materials. The three-dimensional size reduction of metal particles down to the range of the de Broglie wavelength changes their properties. The limited size reduces the mobility of the electrons and leads to the formation of discrete energy levels that result in quantum size effects (QSE).

Nanoparticles can be defined as isolated particles between 1 and 100 nm. They can be protected from agglomeration by protecting shells. The particles are usually stabilized by the addition of a surfactant, a polymer or a ligand to the reaction mixture to prevent undesired agglomeration and metal precipitation. Presently, much interest is also devoted to various applications of ligand protected particles for their physical properties [127, 131, 132]. For example, transition metal particles are studied as catalysts in organic and inorganic transformations [133], electrocatalysts in fuel cells [134] or materials with novel electronic [135], optical [136], and magnetic properties [137]. The catalytic and electronic properties are strongly influenced by the size of the particles, especially in the range 1–6 nm, evidencing the need for size-selective synthesis techniques. The synthesis of nanoparticles is mainly achieved by chemical reduction of transition metal salts using various reagents such as alcohols [138], borohydrides [139], or dihydrogen [140]. The drawback of these methods may be the presence of surface contaminants resulting from the reaction conditions, such as water, salts, organic residues, or even an oxide shell, which can alter the physical properties of the particles or limit access to their surface.

The preparation of unsupported metal nanoparticles was therefore achieved following an organometallic approach since this does not employ drastic reaction conditions and avoids surface contamination. In addition, the size, the shape and the surface state of the particles can be controlled using various stabilizing agents. Besides catalysis, such work can find applications in different areas such as chemical sensors [141, 142] or magnetic properties [143].

Metal clusters may also find applications in catalysis. Here the smaller the particle size, the higher the number of their surface atoms [144, 145]. This makes the particles interesting candidates for catalytic applications. Crooks used palladium clusters for hydrogenation and showed an elevated reactivity with the particles, depending on the particle size [146]. Nanoparticles are also used as the sensitive layer in selective gas detectors [147].

18.3.2
Synthesis of Unsupported Nanoparticles

There are a number of ways to synthesize metal particles with diameters between 1 and 20 nm. These include precipitation [148], organometallic preparation and deposition [149], sonochemical methods [150] and via microemulsions created using organic stabilizing agents [151]. Most routes mentioned include nucleation, growth and stabilization of the particles, which results in a range of particle sizes. For monodisperse size control, the use of organic stabilizing agents is regarded as one of the most promising routes [152, 153].

The radiation method was described by Rogninski and Schalnikoff for the first time and is based on condensation of the metal atoms after collision [154]. Reetz et al. prepared nanoparticles via electrochemical synthesis [155]. Salt reduction was developed by Bönnemann to obtain mono- and bi-metallic nanoparticles in solution [156]. Salt reduction is the most widely practised method for the synthesis of colloidal metal suspensions. Faraday synthesised gold particles by the reduction of $HAuCl_4$ [157].

Colloidal solutions of nanoparticles can also be synthesised via an organometallic approach, organometallic derivatives providing a general route for the synthesis of monodisperse nanoparticles. Using this approach, the particles are prepared via the bottom-up method [158]. The bottom-up method, together with the tools of organometallic chemistry, is used to synthesise nanoparticles [67, 159–163]. An interesting possibility to obtain ruthenium particles is the decomposition of the organometallic ruthenium precursor Ru(COD)(COT) (COD: 1,5 cyclooctadiene; COT: 1,3,5, cyclooctatriene) which is prepared according to a published procedure [164]. Yellow crystals of this complex are obtained with a yield of 70% and can be characterized by NMR spectroscopy and microanalysis.

Bradley et al. decomposed the complex Ni(COD)$_2$, at room temperature and in the presence of poly(vinyl)pyrrolidone (PVP) as stabilizer, simply using dichloromethane as solvent [173]. When they changed the ratio Ni/PVP, they observed a small variation in particle size from 2–3 nm.

Only a few studies concerning the synthesis of ruthenium particles have been published, most of them reporting the use of $RuCl_3$ as metal precursor. For example, stable colloidal solutions of monodisperse Ru particles can be obtained by reduction of $RuCl_3$ in polyols. The stabilization is then achieved by addition of poly(vinyl)pyrrolidone (PVP) (steric stabilization) or of sodium acetate (electrostatic stabilization) [165].

18.3.3
Stabilization of Unsupported Nanoparticles

In order to study the reactivity of organometallic complexes with "unsupported" monometallic nanoparticles, it is necessary to prepare these nanoparticles in a very efficient manner.

Various protective agents are necessary to stabilize nanoparticles and prevent their agglomeration. There are two main methods to achieve their stabilization: (i) stabilization by amplification of the repulsive forces by ionic additives and (ii) sterical stabilization by organic or organometallic ligands (as shown schematically in Fig. 18.12) or by polymers. For the amplification of the repulsive forces an additive is added to the solution in order to increase the electrostatic stabilization. Bönnemann et al. [140, 166] were the first to use ammonium salts (NR$_4$X) for the *in situ* decomposition of various precursors. The method used in the work of Chaudret et al. [68, 69, 167–174] to prevent the coalescence of the particles is steric stabilization which can be achieved by the adsorption of organic molecules with a long chain – frequently referred to as "organic ligands" – or "polymer ligands".

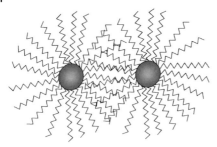

Fig. 18.12 Steric stabilization of nanoparticles via "coordinated" organic ligands.

Napper [175] was the first to study the adsorption of nanoparticles in polymers and since then, different types of polymers like vinyl polymers with polar side groups such as poly(vinyl)pyrrolidone (PVP) and poly(vinyl)alcohol (PVA) have been applied [69, 176–179].

In addition to polymers, which were the most commonly used tools, a strategy based on organic ligands coordinated at the surface of the particles has been considerably developed in the past ten years [69–71, 180–191]. The presence of organic ligands prevents the particles from coalescing and can allow their "self-assembly" onto various surfaces. The main difference between the stabilization with polymers or with organic ligands is the nature of the interaction of the stabilizing agent with the particle surface. The nature, number and arrangement of the protecting ligands may affect not only the accessibility but also the electronic properties of the particles. It follows that ligand-stabilized particles may differ considerably in their electronic properties from quasi-naked clusters or polymer stabilized ones. The interactions of ligands with the particle surface make access to the surface more difficult. There is a wide variety of ligands used for the stabilization of nanoparticles. Organic ligands can introduce true chemical metal–ligand bonds which ensure their electronic stability, while polymers are only absorbed in a less ordered way at the particle surface. The presence of ligands at the surface of the particles can then modulate their properties. In addition, when the ligand has a long alkyl chain a steric protection is also guaranteed via the possible self-assembly of the alkyl chains.

Ruthenium nanoparticles stabilized with the three different types of agents – the HDA (hexadecylamine) ligand, the PP10 (bis-1,10-diphenylphosphinedecane) ligand and PVP(polyvinylpyrrolidone) polymer have been studied by the Chaudret group [68, 69, 167–174]. The main difference between the three systems is the electronic interaction of the protecting agent with the particle surface. The PVP polymer does not bind chemically to the particle and therefore there is no electronic interaction with its surface. The HDA ligand binds through the nitrogen atom to the particle. The Ru—N bonding is known to be weak (soft acid–hard base in Pearson's HSAB principle) and therefore a weak electronic interaction between the HDA ligand and the surface is assumed. The PP10 ligand is bound to the nanoparticles through the phosphorus atom. The Ru—P bonding is known to be strong (soft acid–soft base) and the electronic interaction should be relatively strong.

18.3.4
Preparation and Characterization of Ru Nanoparticles Stabilized by Hexadecylamine

Nanoparticles can be protected by the HDA (hexadecylamine) ligand to prevent reactions between each other and in order to stabilize them in solution [69]. This stabilization mode presents several advantages such as their behavior as molecular systems, control of the size, size distribution and adjustment of the surface state (Fig. 18.13). Solution ^{13}C NMR studies evidenced a fast exchange on the NMR time scale between amine ligands "free" and "coordinated to ruthenium". In this section, the preparation of HDA-protected ruthenium nanoparticles, the study of the size of the nanoparticles and their characterization will be described.

The Ru/HDA particles are prepared by decomposition of the ruthenium precursor Ru(COD)(COT) with 3 bar of dihydrogen at 293 K in the presence of 0.2 eq HDA in THF. The particles are monocrystalline, as evidenced by HREM analysis (Fig. 18.14) and a careful observation of some electron micrographs reveals the presence of spherical "nanoparticles" which are in the process of coalescing. The size could be determined as about 1.9 nm.

The possible organisation of the amines in THF could influence the growth of the ruthenium particles in the solution and their tendency to present different forms such as spherical particles and nanorods. In the case of 0.1 eq HDA/Ru the particles are spherical, while with 0.2 eq HDA/Ru the beginning of the formation of elongated particles can be observed, continuing with time; with 0.5 eq HDA/Ru, nanorods were present.

TEM, HREM and WAXS provide useful information on the structure, the morphology and dispersion of the nanoparticles. Such studies may help in designing nanoparticles of defined shape and organisation.

The quantity of the stabilizing agent clearly has an influence on the growth of the metal particles, but the manner of its attachment to the metal atoms as well

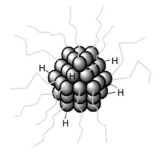

$= CH_3\text{-}(CH_2)_{15}\text{-}NH_2$

Fig. 18.13 Representation of a Nanoparticle of Ru Covered with Hexadecylamine (HDA).

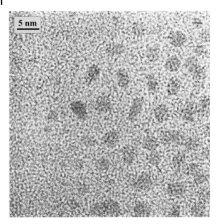

Fig. 18.14 High resolution TEM of Ru/HDA nanoparticles.

as its behavior is not yet understood. IR and NMR spectroscopy could shed some light on the coordination of the ligand at the surface of the particles since this could be a factor related to the shape of the particles. Infrared studies monitored the existence of the ligand at the surface of the particles: the corresponding bands are present in the experimental spectrum of the stabilized colloid.

By solution ^{1}H NMR studies the presence of the ligand at the surface of the particles could be confirmed. The ligand must be attached by the amine group to the metal surface since in the spectrum of the purified colloid, neither the protons in the α-, β- and γ-positions nor the —NH$_2$-group protons can be detected by NMR due to chemical shift anisotropy resulting from the slow tumbling of the particles in solution. These peaks only become visible in the presence of added ligand in the NMR tube. The observation of a peak at 4.8 ppm suggests the adsorption of H$_2$ onto the ruthenium surface. In addition, the methyl signal increases when the ligand concentration is higher. It can be suggested that this phenomenon is due to an exchange of ligands between those attached at the surface and free ligands in solution. Only the free ligand tumbling in the solution can be observed, as well as the end of the long alkyl chain when a tumbling of the chain is possible.

Solution ^{13}C NMR showed that long chain amines are present at the ruthenium surface and that they are involved in a fluxional process that means a fast exchange on the NMR time scale between free and coordinated ligand molecules.

Bradley et al. have previously reported the dynamic exchange of ^{13}CO at the surface of palladium particles [192] and Schmid et al. the phosphine exchange at the surface of gold and rhodium particles [142, 193]. The originality of the present system lies in the observation of signals for both coordinated and free ligands and for their rapid exchange. This result explains the different behavior observed for nanoparticles accommodating these ligands, for example towards the coalescence in solution into "worm-like" structures, as observed by TEM studies. The forma-tion of "worms" could result from the coordination mode of the amine at the

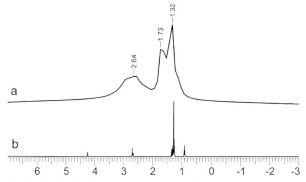

Fig. 18.15 (a) ^{1}H solid state MAS NMR (600 MHz, 5 kHz rotation) spectrum of Ru/HDA nanoparticles; (b) ^{1}H spectrum generated by "ACDLabs HNMR Predictor" for the free HDA ligand [69].

surface depending of its concentration in the reaction mixture and the possible coalescence between close particles.

Since the solubility of the nanoparticles in different solvents is very poor but their characterization is indispensable, solid state NMR studies were performed as a tool for the characterization of the raw nanoparticles, before further treatment of the particle surface. ^{1}H and ^{13}C solid state spectra of Ru/HDA particles were recorded. Additionally, adsorbed hydrogen, assumed to be present due to the reaction condition (excess H_2 during reaction), was observed directly by solid state ^{1}H MAS NMR of the solid particles. The ^{1}H spectrum, depicted in Fig. 18.15, shows only the signals of the protecting HDA ligand and no signal for chemisorbed hydrogen. Since the chemical shift of adsorbed hydrogen could be in the same region as the ^{1}H spins of the protecting ligand and the amount of hydrogen in the protecting ligand was quite high, the surface hydrogen's signal is possibly hidden behind the broad ligand signals. However the ^{1}H signals are very broad because of the intrinsically large homonuclear dipolar coupling of ^{1}H spins. A highly resolved ^{1}H spectrum can only be obtained for rotation frequencies of the order of about 15–20 kHz and larger. Technically, this rotation frequency could not be reached because of the high density of the nanoparticles. The ^{1}H spectra of dense materials such as nanoparticles, that cannot be spun fast, are therefore relatively useless. However, the chemical shift of the signals is in good agreement with those calculated for the HDA molecule. Since the H atoms of the ligand are not directly linked to the particle surface, their electronic properties, and consequently their chemical shift, should not change on coordination to the particles. This has been demonstrated by comparison of the signals of the free ligand and the ligand attached to the nanoparticles.

The ^{13}C Spectrum of the Ru/HDA particles exhibits, as the ^{1}H spectrum, signals of the protecting ligand (Fig. 18.16). Here also, the chemical shifts correspond to those of the calculated spectrum. The carbon atoms of the ligand are not directly linked to the particle surface. Therefore, the ^{13}C chemical shift of the coordinated

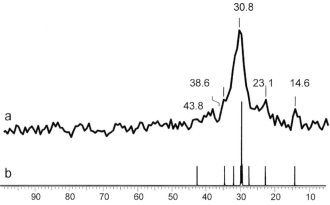

Fig. 18.16 (a) ^{13}C solid state MAS NMR spectrum (150 MHz, 5 kHz rotation) of Ru/HDA nanoparticles; (b) ^{13}C spectrum generated by "ACDLabs CNMR Predictor" for the free HAD [69].

HDA ligand should not be different from the chemical shift of the free HDA ligand, which is the case here.

A strong change in the ^{15}N chemical shift would demonstrate the direct binding of the HDA ligand by the nitrogen to the surface of the particles. Attempts to obtain a ^{15}N spectrum failed, the reason being the weak NMR sensitivity of nitrogen paired with the small mass of available sample and the small percentage of HDA ligand contained in the particles. The poor sensitivity of ^{15}N and its small natural abundance leads to an absolute sensitivity of only 3.85×10^{-6} compared to ^{1}H. For Ru/HDA the same ^{13}C signals are found as in spectra of the HDA protecting agents, since no electronic interaction is present between the carbon spins of those systems and the particles.

Whereas thiols as stabilizers are oxidatively added to ruthenium and excess thiol leads to the reductive elimination of disulfides which are released into the solution and do not exchange with the ligands present at the surface of the particles, the amine ligands exchange rapidly at the surface of the particles with free amines. This may be why the colloids adopt different structures and superstructures, even though the alkyl chains are the same and the functional groups similar. The alkyl chains of the sulfur ligands will encircle the particle and interpenetrate the alkyl chains of other ligands, either free or located at the surface of other particles, leading to the formation of superstructures. They may be removed from the colloid surface. This is in agreement with recent observations demonstrating that the removal of ligands from the surface of gold particles forces the particles to organise into superstructures [194].

In contrast, the dynamics of the amine at the surface of the particle may allow the self-organisation of these amines, which are known to produce, in water, micellar arrangements and even hexagonal phases which are used as templates in the synthesis of mesoporous materials [195, 196]. The presence of some organisation of the amines in THF could allow the growth of ruthenium particles in the channels created and then explain the "vermicular" aspects of the particles stabilized with 1-aminohexadecane. Alternatively, the dynamics of the amines could also provoke changes in the coordination sites at the surface of the particles, and therefore the preferred coordination of the amine along the growth axis of the hcp structure, perpendicular to the basal plane. In any case, the dynamics of the amine will favor the coalescence of particles that were initially spherical shaped and, consequently, the formation of worm-like particles.

Information on the mobility of the ligand can be gained by recording solid state NMR spectra at different rotation frequencies and without rotation. By rotation of the sample along the magic angle, many interactions are averaged by a "simulated" motion of the molecules and the signals narrow. By stopping the rotation of the sample, NMR signals of rigid species become very broad while those of mobile species stay relatively narrow. The line widths of the deuterated surface species at different rotating frequencies and their comparison with the values of mobile and rigid model systems show that these surface species are well attached to the surface. However, they are very mobile on the particles' surface. Since the signal for hydrides could be hidden by the hydrocarbon ligands, the experiment was carried out in ^{2}H NMR using a sample prepared as described above but treated with D_2 in solution immediately after synthesis. In this case, a peak is clearly visible at 1.59 ppm ($w_{1/2}$=120 Hz). It differs from the peak of free D_2 recorded in a blank experiment and which appears at 4.6 ppm ($w_{1/2}$=126 Hz) under similar conditions. In order to be able to attribute the peak at 1.59 ppm, this signal was compared to those corresponding to the free ligand and to that of D_2 in the gas phase at three different rotation rates, namely 4, 2 and 0 kHz. It is clear that the signal at 1.59 ppm is not due to free D_2 but to a species displaying a large mobility on the ruthenium surface i.e. surface deuteride groups. Alternatively, it could result from a fast exchange between these groups and free D_2. After observation of the behavior of amine ligands at the surface of ruthenium nanoparticles by infrared measurement, solution and solid state NMR studies, the indirect observation of hydrides at the surface was made by gas-phase NMR. The catalytic reaction of $H_2 + D_2 \rightarrow 2$ HD could be quantitatively followed by ^{1}H gas-phase NMR.

To summarize, the coupling of traditional methods for the characterisation of nanomaterials (TEM and WAXS) with simple methods of organometallic chemistry (NMR) provides useful information on the chemical reactivity of the surface of the nanoparticles and sheds light on elementary surface reaction steps such as substitution, oxidative addition and reductive elimination.

Very few NMR studies have focused on such processes as dynamic ligand exchange at the surface of metal nanoparticles, and the dynamics of coordinated ligands with the shape of the particles.

18.3.5
Preparation and Characterization of Ru Nanoparticles Stabilized by Grafted Organometallic Fragments

Nanoparticles are obtained by hydrogenation of the olefin ligand of Ru(COD)(COT), which implies decomposition of the complex. This decomposition takes

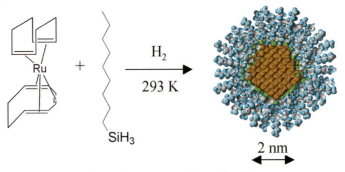

Scheme 18.11 Synthesis of Ru nanoparticles stabilized by n-octyl silyl ligands.

place under an H_2 atmosphere, which avoids possible surface oxidation. Ruthenium atoms must thus organize themselves in order to form "homodisperse" colloidal particles; these particles must be "stabilized" via one ligand. This ligand can be an alcohol, an amine or even a thiol, as proposed by the Chaudret group [68, 69, 167–174]. The use of an organometallic compound in such a process has been achieved by Bonnemann with aluminum alkyls [197, 198] and by us with an alkylsilane [63].

In the octylsilane case, the decomposition of Ru(COD)(COT) is carried out in a non-coordinating solvent such as pentane. The octylsilane is introduced into the solution containing the organometallic precursor before its decomposition under 3 bars of H_2 at room temperature (Scheme 18.11):

Four experiments have been performed with a ratio of octylsilane/ruthenium of 0.2; 0.5; 1.0 and 2.0. After 8 h under these conditions, "colloidal solutions" are obtained. The resulting nanoparticles proved to be extremely reactive toward oxygen (they are spontaneously pyrogenic in air). In each case, infrared spectroscopy reveals the presence of alkyl groups, but the $2150\,cm^{-1}$ band attributed to the Si–H bond has completely disappeared, suggesting that the reaction occurs by reaction of the Si–H with the surface hydride with H_2 reductive elimination.

Liquid and solid ^{13}C NMR study of these nanoparticles reveals that the signal of the α-carbon atom (6.2 ppm) disappears for octylsilane stabilized nanoparticles. Note that the β-carbon (27.3 ppm) is also subjected to the influence of its proximity to ruthenium. In addition, the methyl group carbon signal at 14.1 ppm is also notably enlarged, especially for Si/Ru = 0.2. This suggests that the octyl chain "back-bites" the ruthenium surface via its terminal "CH₃". Other resonances present quite the same chemical shifts and the same broadness as in free octylsilane (Fig. 18.17).

Chemical analysis of the particles confirms the presence of octylsilane fragments grafted on the particles and allows calculation of the stoichiometries of the grafted organometallic fragments: $Ru_s[Si(C_8)_{0.2}]_{0.5}$; $Ru_s[Si(C_8)_{0.3}]_{10}$; $Ru_s[Si(C_8)_{0.7}]_{1.1}$ and $Ru_s[Si(C_8)_{0.7}]_{1.9}$ for respectively Si/Ru ratios of 0.2; 0.5; 1.0 and 2.0.

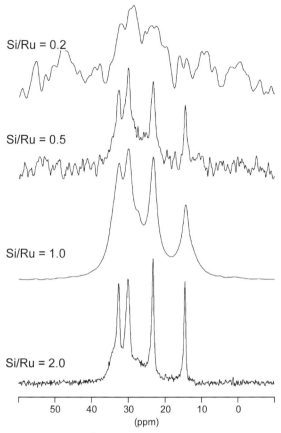

Fig. 18.17 CP-MAS ^{13}C NMR (300 MHz) spectra of Ru/Si nanoparticles stabilized with 2.0; 1.0; 0.5 and 0.2 eq. n-octylsilane.

Using electronic microscopy, well-crystallized particles have been observed, with a quite narrow size distribution, centered at around 2.3 nm when the Si/Ru ratio introduced is 1/1 (see Fig. 18.18 and 18.19). Curiously, the average size of the resulting particle increases when the Si/Ru ratio introduced is increased; it is around 3.2 nm for the Si/Ru ratio of 2.0.

18.3.6
Preparation and Characterization of Pt Nanoparticles Stabilized by Grafted Organosilyl Fragments

Platinum is considered to be a very good catalyst for enantioselective hydrogenation of prochiral ketones [199, 200] and it is considered to be less hydrogenolyzing than, for example, ruthenium [201]. Therefore, the stabilization of platinum

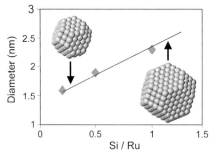

Fig. 18.18 Dependence of the average ruthenium particle size on the number of grafted organosilyl compounds.

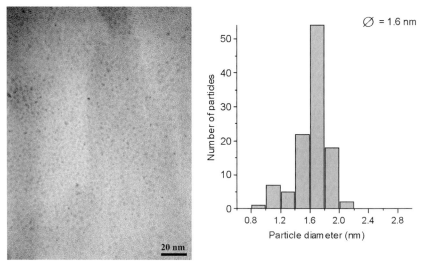

Fig. 18.19 TEM image and size histogram of Ru particles stabilized with 0.2 eq. octylsilane.

nanoparticles by organosilanes has been investigated. The organosilanes prevent the particles from agglomeration by grafting onto the metal surface. For the preparation of Pt nanoparticles, an organometallic precursor is reduced under hydrogen in the presence of organosilane ligands: there is a tendency for selective grafting of the Si on the metal surface. This stabilization of platinum particles introduces modifications in their physical and chemical properties and permits effective control of the size and dispersion of the particles. The organometallic complex bis(di-benzylideneacetone)platine[0] ($Pt(dba)_2$ or $Pt_2(dba)_3$) can be prepared following the published procedure of Moseley et al. [202]. The reaction of potassium tetrachloroplatinate and dibenzylideneacetone (dba) in the presence of sodium acetate in ethanol gives the desired complex $Pt(dba)_2$ or $Pt_2(dba)_3$ after washing and drying. The yield of the synthesis given in the literature is about 80% given but this can be improved to 85%. The color of the solid is purple; it is stable in oxygen and is easily characterized by NMR (liquid and solid), IR and microanalysis. The liquid state ^{1}H NMR spectrum of the obtained product monitors a good coherence with that described in the literature. No chemical shift is observed for the product which with the ligand in proximity to the metal core. But, in agreement with the literature, the knight shift is reduced and the spectra of the complex and the unbound ligand are similar, due to an exchange between the two in solution. The half height width is much larger for the signal of the complex, indicating a fast exchange on the NMR time scale:

$$Pt(dba)_2 + dba \rightleftharpoons Pt(dba)_3$$

Additionally, the larger half height width indicates the binding of the dba to the metal atom and shows therefore the presence of the platinum. A coherence is observed for the characteristic peaks of the dba and the half height width. This is an indication of the complexation and the exchange between the two possible forms of the platinum complex. The width of the peaks in solid state ^{13}C NMR indicates the interaction with the metal core. $Pt(dba)_2$ can be a precursor of unsupported nanoparticles via the reaction shown in Fig. 18.20. The decomposition of the complex $Pt(dba)_2$ at 293 K under 3 bar of hydrogen in dried and degassed toluene in the presence of octylsilane H_3Si-$(CH_2)_7$-CH_3 leads to platinum nanoparticles stabilized by these ligands. The synthesis can be performed with varying equivalents of the organosilane ligand (0.2, 0.5, and 1.0) per platinum. The washed and dried nanoparticles have been characterized by TEM, elemental analysis and IR analysis. The obtained colloidal solution of platinum nanoparticles exhibits various transparencies, depending on the ratio of octylsilane/metal. The higher the amount of octylsilane in solution, the darker the color; this darkening indicates the presence of larger particles. Increase in the ligand concentration also increases the stability of the colloidal solution and the precipitation of the particles becomes more difficult. In fact, the particles stabilized with 0.5 eq. of octylsilane precipitate

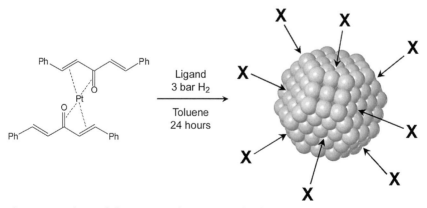

Fig. 18.20 Synthesis of platinum nanoclusters covered with octylsilyl ligands (X) from $Pt(dba)_2$.

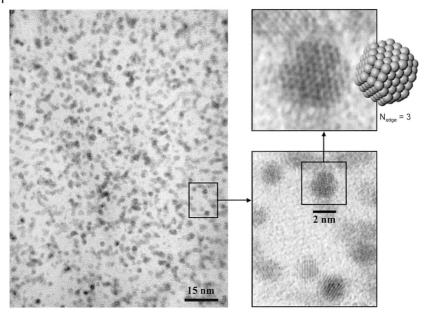

Fig. 18.21 Characterization of nanoparticles of platinum of 2 nm covered with $\equiv$Si(n-C$_8$H$_{17}$) fragments and containing ca. 201 platinum atoms and fitting with a calculated nanoparticle with a cubo-octahedral shape.

easily in cold pentane and form, after three washing and drying cycles, a black powder which was always stored under argon. An excess of ligand renders the powder sticky. No free ligand could be found in the washing solutions indicating its complete grafting but peaks originating from the hydrogenated products of released "dba" could be observed.

Interestingly the size of the nanoclusters thus obtained has a remarkably narrow distribution around 2 nm and the particles are remarkably crystalline (Fig. 18.21). The shape of the particles fits quite well with a cubooctahedron particle of 201 atoms (see the modelling data).

18.4
Some Applications of Supported Nanoparticles Modified by Organometallics

18.4.1
Introduction

It is not the remit of this chapter to cover all the domains of the catalytic properties of nanoparticles modified by organometallics. In particular we will not cover the field of Bönneman for which the metallic particles have been modified by organoaluminum compounds, a topic which is already covered in this book. Recent discoveries using size-selective nano-type Fischer–Tropsch synthesis to convert CO and H_2 into hydrocarbons are outside the scope of this chapter [203, 204]. Here there is a strong dependence of activity, selectivity and lifetime on Co particle size. Other discoveries have been made by Degussa who patented a process for H_2O_2 synthesis from molecular oxygen and molecular hydrogen with nano-size Pd particles of 6 Å [205]. We will summarize here only the results that have been obtained in the laboratory in the area of supported or unsupported nanoparticles modified by organometallic compounds: mostly we shall consider the catalytic properties of Group VIII metals covered with tin, germanium or lead.

As previously seen, after reaction (under hydrogen) of a $M'R_4$ organometallic compound (Group XIV) with the surface of the host metal ("M_s"), various species are formed, of general formula $M_s[M'R_x]_y$. Depending on x, y and treatment temperature, these species can be divided into three main groups:

- Group a, for $x \geq 1$, and $0.5 \leq y \leq 1$ and a low reaction temperature.
  In this case organometallic fragments remain on the metallic surface and are bonded to the surface via one, two or three covalent bonds, as demonstrated by EXAFS and other analytical tools (see Section 18.2.3.3). Interestingly these surface organometallic fragments play the same role as real ligands in classical molecular chemistry with possible control of the chemio-, regio- or stereo-selectivity.
- Group b, for $x=0$, small y (rather small coverage of the surface by the added metal) and low reaction temperature. In this case the grafted organometallic is completely dealkylated (species 4). These species are usually prepared at moderate temperature (e.g. around 100 °C for tin). The atoms are suspected to be localized on particular surface crystallographic sites, very likely those which are the most hydrogenolyzing with respect to the surface organometallic complex which has been grafted. These are called "adatoms" and they play the role of selective poisons.
- Group c, for $x=0$, i.e. for completely dealkylated samples (species 5), but prepared at high temperature (typically above 300 °C).
  The atoms coming from the organometallic are localized on the first metallic layer and tend to isolate the host metal atoms from each other. These catalysts will exhibit what we call a "site isolation effect" by analogy with "single site" in classical catalysis (e.g. in polymerization) [206–208] or with organometallic chemistry on oxides [209–213].

Each group allows particular catalytic processes for which some application examples realized recently are reported below.

18.4.2
Group a: Evidence for a Selective Effect in Catalysis of the Grafted "Organometallic Ligand"

18.4.2.1 **Competitive Hydrogenation of Hex-2-en-1-ol and Hex-5-en-1-ol Unsaturated Alcohols**
Toshiba and Takahashi [214] prepared nanoparticles of colloidal Pt embedded in a micelle (Fig. 18.22). With such a system the initial rate of hydrogenation of 2-undecenoic acid is almost 6 times lower than the initial rate of hydrogenation of 10-undecenoic acid. The explanation is based on the polar effect of the amphiphilic ligand whose polar head stays outside the particle. The polar head facilitates the coordination of the double bond by preventing the carboxylic function from approaching the surface of Pt but considering such a "micellar" system, one can see easily their fragility.

Hydrophobic or hydrophilic properties of nanoparticles covered by organometallic fragments can also be monitored by choosing the right ligand present on the grafted organometallic. With rhodium surfaces covered by tris n-butyl fragments the surface of the particle becomes quite hydrophobic, leading to a "kind" of "micelle" but whose core is metallic with *a kind of sigma bonded ligand*!

With such a modified rhodium particle it is possible to govern the selectivity in the double bond hydrogenation of two different unsaturated alcohols: hex-2-en-1-ol and hex-5-en-1-ol (which only differ by the double bond position, terminal or internal). The selectivity for the competitive hydrogenation of the internal versus terminal double bond was strongly modified by the presence of $-SnBu_x$ fragments at the periphery of the Rh/SiO_2 catalyst. Indeed this ratio goes from 3 with classical nanoparticles of Rh/SiO_2 to 13 with $[Rh_s]_3(\eta^3\text{-}SnBu)/SiO_2$ (Fig. 18.23) In other words hydrogenation of the internal double bond is strongly inhibited by the presence of the hydrophobic organometallic fragment present on the nanoparticles.

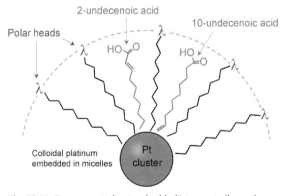

Fig. 18.22 Pt nanoparticles "embedded" in a micelle, and selective for the hydrogenation of 10-undecenoic acid, from Ref. [214].

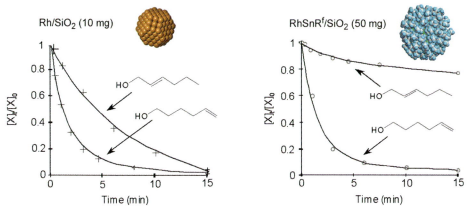

Fig. 18.23 Conversion versus time in the hydrogenation of two different olefinic alcohols (hex-2-en-1-ol and hex-5-en-1-ol) with two different catalysts Rh/SiO$_2$, RhSnR^f/SiO$_2$ in the liquid phase (autoclave, water–methanol, 20 bar of hydrogen, temperature: 25 °C) [215].

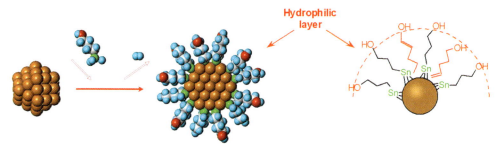

Fig. 18.24 Explanation of the selectivity for the selective hydrogenation of hex-5-en-1-ol with [Rh$_s$]$_3$(η^3-SnBu)/SiO$_2$.

With the catalyst nanoparticles covered with a hydrophobic layer, the polar alcoholic function of the substrates remains in the aqueous phase, consequently far away from the active surface rhodium atoms, and it is easier for the terminal double bond to coordinate to the surface and, consequently, this terminal olefin is hydrogenated fastest (Fig. 18.24).

18.4.2.2 Hydrogenation of α,β-Unsaturated Aldehydes

Citral is an α,β-unsaturated aldehyde containing three kinds of unsaturation: a non-conjugated trisubstituted double bond, a conjugated trisubstituted double bond and a conjugated aldehydic function, which can be either in the *Z* or *E* position with respect to the alkyl chain. Hydrogenation of this molecule can thus lead to numerous products. For example the first hydrogenation step could lead to 3,7-dimethyl-2,6-octenol (geraniol or nerol, depending on the stereochemistry of the double bond), 3,7-dimethyl-6-octenal (citronellal) and 3,7-dimethyl-2-octenal (Fig. 18.25). Further hydrogenation can then occur leading finally to 3,7-dimethyloctanol. When the hydrogenation of citral is performed with supported nanoparticles of rhodium metal, for example Rh/SiO_2 under classical conditions (liquid phase, rhodium dispersion 80% (particles in the range 1–2 nm), citral/$Rh_S = 200$, $P_{(H_2)} = 80$ bar, $T = 340$ K), the catalytic activity is very high but most of the above products are obtained and the reaction is totally non-selective, even if the major product was citronellal. The situation is totally different when the catalyst is an Rh/SiO_2 sample modified by reaction with $SnBu_4$ ($Sn/Rh_S = 0.95$). Indeed, if the catalytic activity has only slightly decreased, the selectivities are totally modified and the major product is now 3,7-dimethyl-2,6-octenol, with a selectivity of ca. 96% (Fig. 18.25).

A detailed physico-chemical study of the catalyst was performed. It showed that tin was present as two species:

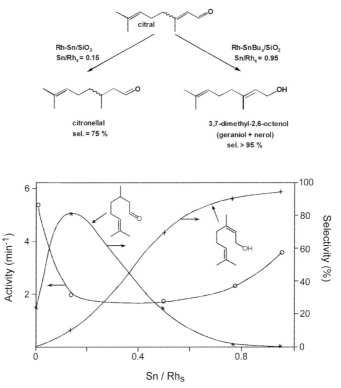

Fig. 18.25 Various pathways for the hydrogenation of citral (top); application to the hydrogenation of citral with Rh covered with various amounts of tin:(left) for Sn/Rh: 0.3, only adatoms are present; (right) for Sn/Rh: 0.8, "Rh$_3$ [≡Sn(n-Bu$_3$)]" surface organometallic fragment is present.

- adatoms (very small amount), corresponding to tin complexes having reacted with the most hydrogenalizing sites and poisoning them (see below)
- "digrafted" or "bipodal" dibutyl-tin species (high percentage).

Compared to the same system but with only the tin "adatom" species (see below), both the catalytic activity and the selectivity to (geraniol + nerol) have increased. Simple molecular modelling studies show that after grafting the SnBu fragments onto the metal particle there is a non-negligible steric hindrance due to the butyl ligands. This steric hindrance could prevent the coordination of citral via its internal double bond and could allow only a coordination of the carbonyl group, rendering the reaction highly selective (Scheme 18.12). This effect of the butyl ligands was also evidenced by changing the organometallic complex grafted onto the metal [216]. With GeBu$_4$, which leads also to the formation of a digrafted surface species, the same effect was observed while with PbBu$_4$, all butyl groups were lost and the catalytic activity became insignificant.

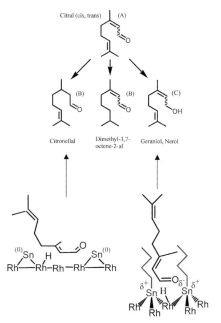

Scheme 18.12 Explanation of the selectivity in citral hydrogenation: (Left): small coverage of Rh nanoparticles by adatoms favors the η^4 coordination of the conjugated double bonds of citral. (Right): high coverage of Rh nanoparticles by $\equiv$Sn (n-butyl) favors the η^2 coordination of the carbonyl.

18.4.3
Group b: the Role of "Adatoms" in Selectivity

Several adatom effects were observed, with various reactions. In most cases, this effect can be depicted as a poisoning of undesirable sites, resulting simultaneously in a decrease in the global catalytic activity and in a significant increase in the selectivities for the desired products. We will describe here two examples, the isomerization of 3-carene into 2-carene and the dehydrogenation of butan-2-ol into methyl ethyl ketone.

18.4.3.1 Isomerization of 3-Carene into 2-Carene [217]

(+)-3-Carene (3,7,7-trimethyl-[4.1.0]-bicyclohept-3-ene) is a monoterpene present in natural compounds such as oils of turpentine. Unfortunately, its industrial applications are limited (it is only used as a solvent in coatings). In contrast, (+)-2-carene is potentially more interesting for fine chemicals, because the double bond is conjugated with the strained C—C bonds of the cyclopropyl moiety. It should be interesting to transform (+)-3-carene into (+)-2-carene (Scheme 18.13), even if, thermodynamically, the equilibrium between the two isomers corresponds to nearly the same amount of the two carenes (60% of 3-carene and 40% of 2-carene at 120 °C). Metals (Raney nickel or nickel on silica, palladium on carbon etc.) can easily catalyze this reaction but the selectivity is low, due to the hydrogenation of the two isomers into carane. Addition of very small amounts of tetrabutyl tin can completely transform the performances of these catalysts by poisoning the hydrogenation sites. For example, when a Ni^0/SiO_2 catalyst is used, the best result corresponds to a 2-carene yield of 30% but at least 30% of the carenes have been transformed into by-products. Addition of 0.04 mole of tetrabutyl tin per mole of surface nickel results in an increase in the yield of 2-carene to 37% and a decrease of the amount of by-products to less than 10%. As above, in this case tin is present as adatoms on the most hydrogenating sites (very likely those situated on the faces rather than on corners and edges).

18.4.3.2 Dehydrogenation of Butan-2-ol to Methyl Ethyl Ketone [218]

A quite similar effect was observed during the dehydrogenation of butan-2-ol to methyl ethyl ketone (MEK) on Raney nickel (Fig. 18.26). The Raney nickel is a very efficient catalyst for this reaction and leads to methyl ethyl ketone with a selectivity of ca. 90%. This result is good but for industrial applications higher selectivities are required. This can be achieved by poisoning some sites by reaction with tetrabutyl tin (the best results are obtained with a Sn/Nis ratio of 0.02). Indeed, the reaction occurs first with the sites responsible for the side reactions which are then

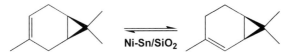

Scheme 18.13 Isomerization of 3-carene to 2-carene with nickel covered by $NisSn_{0.04}$.

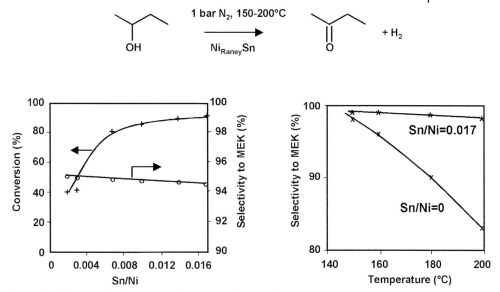

Fig. 18.26 Selective dehydrogenation of butane-2-ol to methyl ethyl ketone with $NiSn_{0.017}$.

selectively poisoned by the resulting tin "adatoms". The consequence is a slight decrease in the catalytic activity and an increase in the selectivity to MEK which can reach 99%. This catalyst, developed by IFP, has been used commercially in Japan for several years.

18.4.3.3 Selective Hydrogenation of Acetophenone to Phenyl Ethanol [219]

Acetophenone can be selectively hydrogenated to several products, depending on the ability of the catalyst to hydrogenate the aromatic ring or the carbonyl moiety or both. Pure rhodium is not selective and leads to a mixture of several compounds even if, at low conversion, there is a preference for the hydrogenation of the carbonyl rather than the aromatic ring. Interestingly, when the rhodium nanoparticles are covered with tin adatoms carefully prepared via the surface organometallic route, the system become quite selective for the hydrogenation of the carbonyl, the aromatic ring being untouched (Fig. 18.27).

The explanation given in Scheme 18.14 indicates that the aromatic ring is able to coordinate the aromatic as a η^6 ligand whereas when the surface of Rh is covered by tin adatoms, the only function which can be adapted on the surface is the carbonyl via its lone pair.

18.4.4
Group c: "Site Isolation" Phenomenon

We will present here two examples of such effect: the dehydrogenation of isobutane to isobutene and the hydrogenolysis of acids or esters to aldehydes and

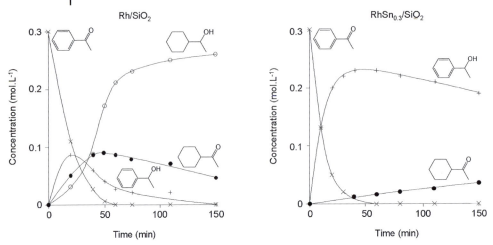

Fig. 18.27 Non-selective (left) and selective (right) hydrogenation of acetophenone on respectively Rh/silica (left) and RhSn$_{0.3}$/silica (right).

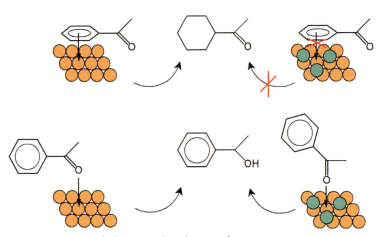

Scheme 18.14 Role of adatoms in the selectivity of acetophenone hydrogenation (on Rh and RhSn$_{0.3}$).

alcohols. In most cases the effect of tin, present as a surface alloy, will be to dilute the active sites, reducing then the yield of competitive reactions.

18.4.4.1 Dehydrogenation of Isobutane to Isobutene [220]

The dehydrogenation of isobutane to isobutene proceeds at high temperature (ca. 550 °C) and low hydrogen pressure (1 bar). In these conditions, the catalyst is very active (turnover frequency 5 s^{-1}) and moderately selective (93%) for commercial applications. The side-products are due to the hydrogenolysis properties of the metal, leading to methane and eventually coke and to skeletal isomerization to n-butane. This catalyst was then modified by addition of high amounts of tin (Sn/Pt_S between 0.5 and 1) by reaction with tetrabutyl tin. To ensure complete hydrogenolysis the solid was heated at 300 °C and then reduced under hydrogen at 500 °C. EXAFS analysis showed that, after such a treatment, tin was fully present on the surface as a surface alloy on the metal particles [62]. EXAFS analysis as well as high resolution electron microscopy indicates that the particle size distribution of platinum is slightly increased by the addition of tin and that the tin is situated, as a surface alloy, at the surface of the particle (Fig. 18.28).

This modified catalyst can be considered as totally selective for isobutene. For example, when the Sn/Pt_s ratio=0.85, the selectivity to isobutene reaches 99.5% (and the lifetime is considerably increased). This increase in selectivity for isobutene can be simply explained by the "site isolation" effect [25–41]. It is now generally admitted that the coke formation, isomerization and hydrogenolysis reactions occur mostly on large surfaces of platinum, since more than one platinum atom is involved in the reaction mechanism (Scheme 18.15) [220, 221]. Typically, this mechanism involves, after C–H bond activation, a γ-H abstraction, followed by the formation of a metallacycle and the cleavage of the C–C bonds.

The presence of tin atoms regularly distributed on the platinum surface isolates the platinum atoms by increasing the distance between two adjacent platinum atoms, as do the copper atoms on a nickel surface [222] or the tin atoms on a rhodium, platinum or nickel surface [33, 118, 223, 224]. The presence of tin thus avoids the hydrogenolysis reaction, leading to a more selective catalyst. Indeed, the formation of isobutene from isobutane involves only one platinum atom, the reaction passing through a simple mechanism of β-H elimination after the first step of C–H bond activation (Scheme 18.16).

Another consequence of the suppression of the hydrogenolysis reactions is the increase in the turnover number measured at the pseudo-stationary part of the curve giving the conversion as a function of time. Since hydrogenolysis reactions are suppressed, the coke formation, responsible for the deactivation of the catalyst is strongly reduced and the result is that the tin-modified catalysts are more active than the unmodified ones (Fig. 18.29).

In conclusion, the results strongly suggest that the concept of "single site" which is a strong step forward in "catalysis on oxides" may become as important in catalysis on metals. Scheme 18.17 illustrates these simple ideas: the control of a surface structure associated with "molecular concepts" on mechanisms explains the existing results and open a way toward the rational design of future nanoparticles in metallic catalysts.

18.4.4.2 Selective Hydrogenolysis of Esters and Acids to Aldehydes and Alcohols

Hydrogenolysis of esters to aldehydes or alcohols is difficult to achieve either by homogeneous or by heterogeneous catalysis. Indeed high temperatures and high pressures are required to achieve the reaction, leading to a non-selective hydrogenolysis with formation of acids, alcohols, CO_2, CO and hydrocarbons.

Pt size

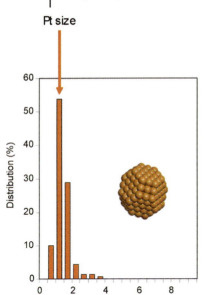

Butane evolved per Sn grafted

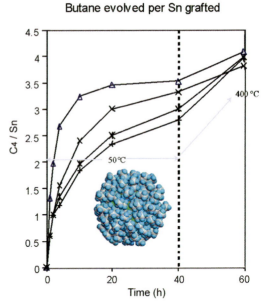

Pt/Sn size

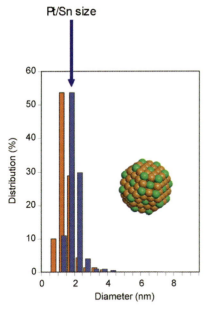

Fig. 18.28 Particle size distribution of Pt/silica (left; red) and Pt/Sn on silica (right; blue). The curves in the middle indicate the evolution of n-butane during the hydrogenolysis of Sn(n-Bu)$_4$ as a function of time of preparation and, after 40 h, as a function of temperature (from 50 °C to 300 °C).

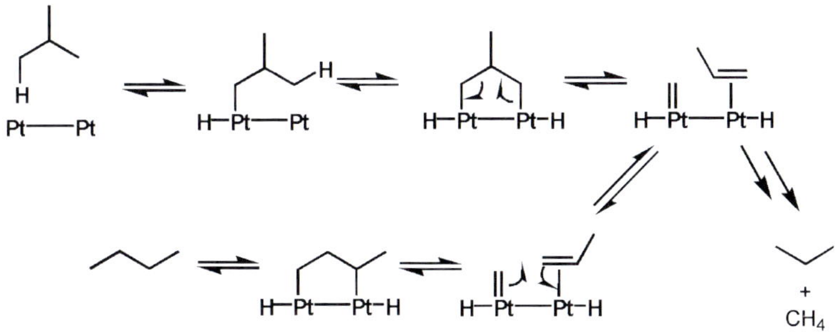

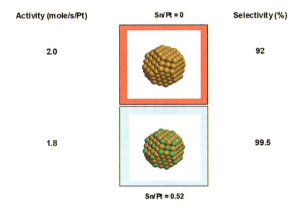

Scheme 18.15 Mechanism of isobutane hydrogenolysis and isomerisation on an "ensemble" of at least two Pt atoms.

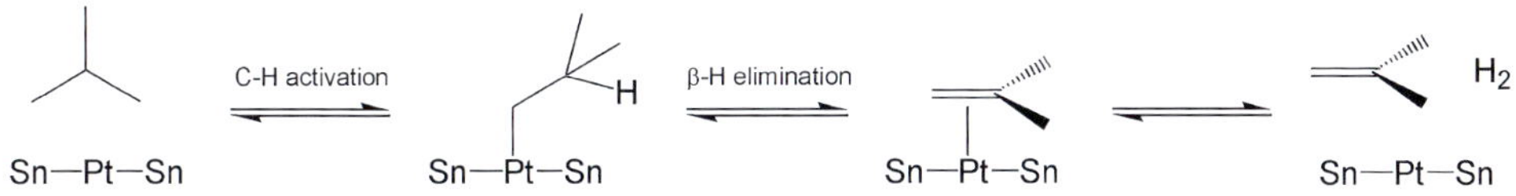

Scheme 18.16 Mechanism of isobutane dehydrogenation on "isolated" Pt atoms (surface hydrides are omitted for clarity).

Activity (mole/s/Pt)	Sn/Pt = 0	Selectivity (%)
2.0		92
1.8		99.5
	Sn/Pt = 0.52	

Isobutane dehydrogenation

Fig. 18.29 Activity and selectivity in the reaction of isobutane dehydrogenation to isobutene with nanoparticles of Pt/silica (top) and with Pt/Sn bimetallic nanoparticles / silica obtained via the organometallic route (below).

Bimetallic M–Sn alloys (M = Rh, Ru, Ni) supported on silica and prepared by reaction of the metal M with tetrabutyl tin display catalytic properties quite different from those of the monometallic catalysts. Indeed, they are very selective for the hydrogenolysis of ethyl acetate into ethanol [223, 225–228]. For example while the selectivity to ethanol is 12% with Ru/SiO_2, it increases to 90% for a $Ru-Sn/SiO_2$

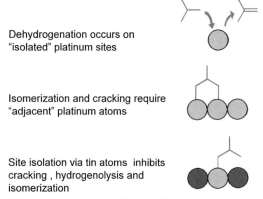

Dehydrogenation occurs on "isolated" platinum sites

Isomerization and cracking require "adjacent" platinum atoms

Site isolation via tin atoms inhibits cracking , hydrogenolysis and isomerization

Scheme 18.17 Summary of the single site concept applied to selective dehydrogenation of isobutene.

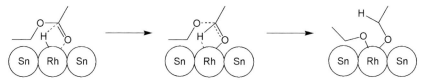

Scheme 18.18 Mechanism of esters hydrogenolysis on "isolated" Rh atoms.

catalyst with a Sn/Ru ratio of 2.5 [227]. In addition, the reaction proceeds at lower temperatures than with the classical catalysts (550 K instead of temperatures higher than 700 K).

The "proposed" reaction mechanism which may happen on these alloys is depicted in Scheme 18.18. The first step is the coordination of the ester to the alloy via its oxygen atoms (it is likely that tin favors this coordination via its electrophilicity). In a second step there is a four-center mechanism where the hydridic hydrogen makes a nucleophilic attack at the electrophilic carbon atom of the carbonyl with formation of two alkoxides which are further hydrogenated into the corresponding alcohol. The alkoxide may undergo Pt–O carbon cleavage by surface hydrides or eventually beta-H elimination leading to the aldehyde. Such β-H elimination is favored when the catalyst does not easily chemisorb hydrogen (e.g. at high tin content).

A quite similar reaction is observed with acids instead of esters. Indeed, the organic acids can undergo a hydrogenolysis reaction leading, as above to aldehydes and alcohols (Scheme 18.19).

Typically, in the case of the hydrogenolysis reaction of acetic acid, high selectivities towards ethanol (≥87%) are achieved with the Ru–Sn alloy, compared to that of the pure metal where the major product is methane. It is also possible to increase the selectivity for the corresponding aldehyde, for example by decreasing the hydrogen to acetic acid ratio in the reactants, as can be suspected if we assume

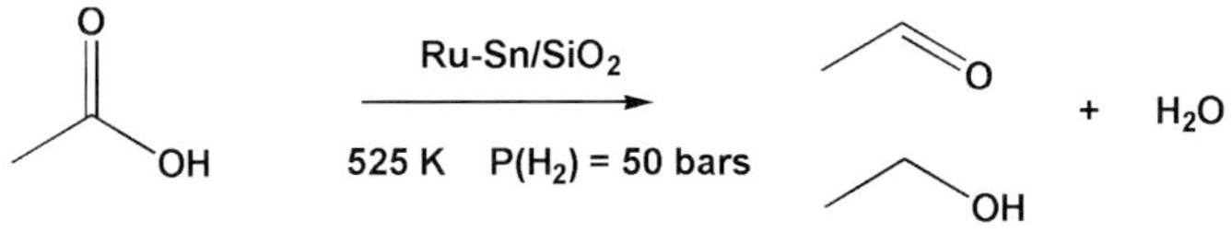

Scheme 18.19 Hydrogenolysis of carboxylic acids on Ru–Sn nanoparticles.

Table 18.5 Impact of As on catalyst cycle length (gasoline hydrogenation).

As in the Gasoline (ppb)	Cycle length (month)
10	12
50	5–7
100	2–4

that this product is a possible product of the reaction (β-H elimination from the alkoxide). These results have been patented by Rhône Poulenc who claims selectivity higher than 85% with a conversion higher than 95% for the hydrogenolysis of nonanoïc and trifluoracetic acids [229].

18.5
Application of Surface Organometallic Chemistry on Metals to the Removal of "Heavy" Metal from Contaminated Feeds

18.5.1
Application in "Heavy" Metal Elimination from Crude Oil

18.5.1.1 Introduction
During the past 20 years, refiners and petrochemical producers have experienced a serious increase in catalyst poisoning by mercury and arsenic. This phenomenon may be partially explained by the diversification of the feedstock supply resulting from the need to optimize the profitability of refining and petrochemical operations. The utilization of a more diverse feedstock supply containing metal impurities has led to operating problems such as corrosion of aluminum alloys in steam cracker cold boxes. Moreover, catalytic processes such as selective hydrogenation, working downstream of the steam cracker separation train, suffer from reduced cycle lengths and lifetimes because of poisoning (Table 18.5).

In addition, the impact of mercury, arsenic and lead on the environment and human health is a crucial issue and efforts have been focused on means to remove them.

IFP has been a pioneer in the development of such technologies, appropriate for the processing of liquid hydrocarbon feedstock boiling in the naphtha range and heavier. The RAM II and RAM III® processes (Removal of Arsenic and Mercury) now commercialized by Axens, are ideally suited for simultaneous removal of arsenic, mercury and lead (and sulfur) if any, from contaminated feedstocks upstream of aromatics units or steam cracking units. Seven RAM licenses have been awarded and are presently in operation in the Far East.

18.5.1.2 Natural Gas Condensates as Steam-cracker Feedstocks

The presence of mercury in natural gas has been detected in numerous fields for many years. Mercury found in natural gas is generally in metallic form and its concentration varies from 1 to 75 µg Nm^{-3}. Natural gas associated condensates are very different. First the mercury found in condensates is present in various chemical states: elementary, ionic and organometallic; second, arsenic is often simultaneously present; third, the concentration ranges of mercury and arsenic are generally and, respectively, in the ranges of 10 to 3000 ppb and 10 to 150 ppb. The distribution of mercury and arsenic for an Asian condensate is shown in Fig. 18.30. For this particulate condensate, the large majority of mercury is found in the C$_3$ and C$_4$ fractions, whereas, arsenic was almost exclusively found in the residue.

Contaminant Repartition in the Feedstock   As the steam-cracker is producing a wide range of hydrocarbon cuts from C$_2$ to fuel, one should have in mind that any contaminant entering the steam-cracker unit can come out in any effluent, including water. This is particularly the case for mercury (see Fig. 18.30). Mercury entering the olefin separation train is dispersed over all the effluents but it concentrates preferentially in the C$_3$ and C$_4$ cuts. It should be also pointed out that only 25% of the mercury entering the process is found in the hydrocarbon stream. This means that the rest can be retained by process internals or can go into the process water (!). The behavior of arsenic, which is the second major contaminant after mercury, is very different; on leaving the steam-cracker, it is found only in gasoline and fuel. It has never been found in the C$_3$ and C$_4$ cuts.

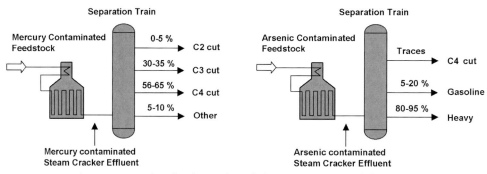

Fig. 18.30 Hg and As distribution through the majors cuts in an olefin unit.

18.5.1.3 Mercury Removal Methods

As mercury and arsenic tend to distribute throughout all the major cuts in an olefin unit (Fig. 18.30) the best or preferred location for contaminants removal for olefin plants is the feedstock before it enters the plant. This avoids future problems associated with plant contaminants such as maintenance/health concerns, the possibility of mercury passing through to selected unprotected product streams, the clean-up of regeneration effluent and plant waste streams, as well as the conventional catalyst/aluminum equipment [121, 122].

The number of streams feeding the plant, the number of contaminated streams, the mercury concentration and the type of mercury removal system considered all have a potential impact on the cost of the removal system. For overall economic assessment, one should consider the "in-plant" impact and the cost of removal systems for regeneration/waste effluent systems as well as down-stream client impacts. One should also consider other contaminants which are present in the natural gas condensate i.e. mainly arsenic but also sometimes phosphorus, lead or silicon. For these contaminants, there is no removal process which is industrially applied on the steam-cracker effluents.

18.5.1.4 Typical RAM II (Removal of Arsenic and Mercury) Process Description

The RAM II process ® (Fig. 18.31) developed by IFP is ideally suited for simultaneous removal of arsenic, mercury and lead from contaminated feedstocks, upstream of aromatics complexes or steam-cracking units. Typical results are excellent, less than 3 ppb of each contaminant in the process effluent as threshold limit for analysis, 1 ppb of each as expectation. Mercury removal from a HC cut is not as simple as the same operation from a gas or LPG feed. The very nature of the mercury compounds to be found in such a cut requires a two-step process in which the organo-mercuric species are decomposed into hydrocarbons and

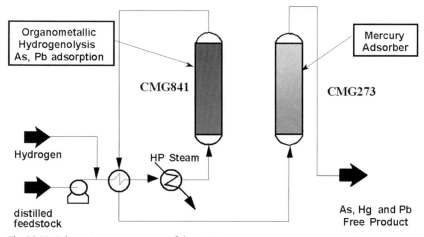

Fig. 18.31 Schematic representation of the IFP RAM II ® Process [121, 122, 230].

metallic mercury in a first step, the latter being subsequently absorbed on a trapping mass in a second step.

In the first step, hydrogen is required to hydrogenolyze the organo-mercuric compounds to be found in a HC cut in order to reach the severe specification, as low as 1 ppb of mercury in the process effluent. The feed comes directly from the distillation unit and it is considered that the feed entering the RAM II unit has been properly fractionated to be on specification. Due to the risk of the presence of free water or dissolved oxygen, the unit cannot accept in any case non-fractionated feed or feed coming from storage. At the discharge from the feed pumps, hydrogen rich gas is added under flow ratio control into the feed. Then the feed is heated in a feed/effluent exchanger; the reactor inlet temperature is achieved in the feed steam heater. Then, the mixture enters the hydrogenolysis reactor where mercury, arsenic and lead organometallic compounds are transformed into the corresponding metals; arsenic and lead are adsorbed on the catalyst. The effluent is then cooled, first in the feed/effluent exchanger and further in the cooler. This hydrogenolysis reactor effluent is then depressurized and flashed in the degassing drum.

If needed, gas effluent could be sent to a mercury absorber purge gas reactor where mercury is adsorbed on the trapping mass before release of the gas at battery limits. Liquid effluent is sent to the mercury adsorber reactor where free mercury is adsorbed and the resulting treated product is routed to the downstream unit.

18.5.1.5 **Chemical Reactions**

Arsenic The arsenic is hydrogenolyzed and adsorbed in the first step of the process, on the CMG841 catalyst (nickel on alumina). In order to predict the performances of the unit, a kinetic model has been developed at IFP, based on the following reaction, with arsenic:

$$AsR_3 + M-H \rightarrow M-AsR_2 + RH$$

This is a first order reaction with respect to As, which allows one to use the following mathematical equation:

$$C_{as} = \exp(-km/F)C_{o,as}$$

Where C_{as}: arsenic concentration in the liquid per m (g) of catalyst ($\mu g\,l^{-1}$), $C_{o,as}$: arsenic concentration of the feed ($\mu g\,l^{-1}$), k: rate constant of "adsorption" (h^{-1}), F: molar flux ($mol\,h^{-1}$).

Then, for a given catalyst slice (1/100), the arsenic concentration variation is related to the trapped As per unit of time. This allows the determination of As concentration in the product and the As profile across the bed as a function of running time (Fig. 18.32).

Mercury This model can be applied for mercury as it is hydrogenolyzed catalytically in the first step of the RAM II process into metallic mercury as follows:

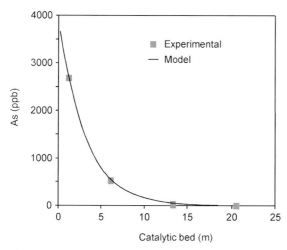

Fig. 18.32 Arsenic gradient on a catalytic bed (length) [122].

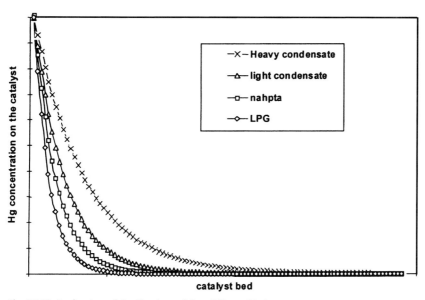

Fig. 18.33 Application of the kinetic model to different kinds of feed [122].

$$R-Hg-R + 2M-H \rightarrow Hg° + 2RH$$

In the second step of the process, the metallic mercury is adsorbed on the CMG273 trapping mass (metal sulfide supported on alumina). The mercury adsorption can be also modeled, with the same kinetic model (Fig. 18.33).

As shown in the previous figures, this predictive kinetic model has been applied in many laboratory and industrial cases with success.

18.5.1.6 **Conclusion**

The IFP RAM II and III® processes of demercurisation and dearsenification are a successful application of catalysis to the purification of a natural feedstock which utilizes nanoparticles of nickel. Interestingly the control of the size of these metal particles is crucial because the penetration of the arsenic inside the nanoparticles is a determining factor for a highly active catalyst. It must be pointed out that it is a real "catalytic" reaction which occurs at the surface of the particle, but the catalyst is slowly poisoned by arsenic as it enters inside the particle and, at the end of the process, the catalyst is dead but can be regenerated by the classical tools of mineral chemistry.

18.5.2
Application in "Heavy" Metal Ion Removal from Aqueous Effluents

SOMC/metals is based on a simple observation: when an organometallic complex reacts with a reduced supported metallic particle it generally has the tendency to react with the metallic surface to progressively form grafted organometallic species, adatoms and finally alloys. Until now, this domain has been explored in an organic medium or even in the absence of a solvent but, more recently, the SOMC/metals concept has been extended to aqueous media with hydrosoluble organometallic compounds [116] or with ionic compounds [231–233]. With hydrosoluble organo-metallic compounds, the controlled preparation of bimetallic particles has been carried out in the aqueous phase and patented by IFP as an industrial partner [117]. With ionic compounds, the strategic problem of demetallation of aqueous effluents is crucial. The principle of preparation of Pt–X bimetallic particles by reacting platinum covered with hydrogen with metallic salts was already proposed by Barbier and Lamy-Pitara [234, 235] in 1986. But for the application of this process in the field of environmental and health protection, nickel was chosen as a "host" to trap heavy metals, instead of more expensive metals such as platinum or palladium. Nickel shows an important reactivity and is able to exist in a neutral or basic medium, eventually without any support (Raney nickel).

18.5.2.1 **Cd²⁺ Removal**

The removal of solvated cadmium ions (Cd^{2+}) from water has been studied because ionic cadmium is a toxic element (itaï-itaï disease in Japan) and one of the main metallic pollutants of industrial aqueous effluents (electrodeposition, Ni–Cd batteries etc.). When alumina-deposited nickel, previously reduced under a hydrogen flow, is suspended in a Cd^{2+} solution, a quick and strong decrease in pH is observed, followed by a slow pH increase with time. The pH decrease has been interpreted by the reaction:

$$Cd^{2+} + 2Ni_s - H + 2H_2O \rightarrow (Ni_s)_2 Cd + 2H_3O^+ \tag{18.1}$$

The slow pH increase has been attributed to nickel dissolution due to the acidification of the medium following:

$$Ni^{(0)} + 2H_3O^+ \rightarrow Ni^{2+} + H_2 + 2H_2O \tag{18.2}$$

To avoid the second reaction, it is necessary to neutralize hydronium (H_3O^+) ions by controlled soda addition as soon as they appear. The general equation becomes:

$$Cd^{2+} + 2Ni_s-H + 2OH^- \rightarrow (Ni_s)_2Cd + 2H_2O \tag{18.3}$$

It is then necessary to precisely control the solution pH with an automatic device [231–233].

The hypothesis formulated in Eqs. (18.1–18.3) (see also Fig. 18.34) has been verified by measuring the soda consumption necessary to maintain the solution pH at 7, during Cd^{2+} addition to an aqueous suspension of previously reduced alumina-deposited nickel. Indeed, in these conditions, a quick consumption of 2 OH^- per added Cd^{2+} is observed, which is in agreement with the stoichiometry expected. Under an atmosphere of hydrogen, the grafting reaction of cadmium on nickel stops when nickel is fully covered by a monolayer of cadmium atoms (ca. 1 Cd/Ni_s) (Fig. 18.35). Indeed, cadmium adatoms progressively poison the nickel surface and hydrogen can then no longer adsorb on the nickel surface.

It is worth noticing that, according to the Nernst equation, that is related to the equilibrium between cadmium oxide (Cd^{2+}) and reduced cadmium, the amount of cadmium oxide (Cd^{2+}) remaining in solution at the potential of our system set by the couple (H_2/H^+), should be "theoretically" higher than 10000 ppm (Fig. 18.36). From our results, it is obvious that the cadmium concentration in solution is decreased to ppm or even ppb level. Cadmium can then be deposited at a potential more positive than that proposed by the Nernst equation. This phenomenon has already been reported in the past with other metals and has been called "underpotential deposition" [236]. In fact, when cadmium is reduced, it does not build cadmium–cadmium bonds, but cadmium–nickel bonds. Therefore, the Nernst equation with standard redox potentials (Cd/Cd^{2+}) and (H_2/H^+) does not well represent underpotential cadmium deposition. According to Kolb [237], the potential shift observed in the underpotential deposition of a metallic ion (M^{a+}) onto a metal

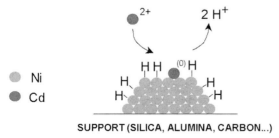

SUPPORT (SILICA, ALUMINA, CARBON...)

Fig. 18.34 Mechanism of removal of Cd^{2+} in solution by a nanoparticle of nickel covered with hydrogen.

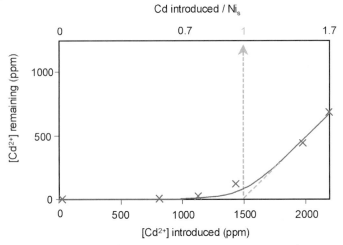

Fig. 18.35 Remaining cadmium concentration in water *versus* initial concentration.

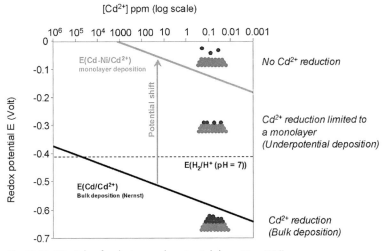

Fig. 18.36 Principle of cadmium underpotential deposition [236].

surface ($M'^{(0)}$) is related to the ionic contribution to the chemical M—M' bond due to the difference in work functions between M and M'. The formation of the cadmium–nickel bond is the driving force that allows cadmium deposition (Scheme 18.20).

However, if the cadmium–nickel interaction is beneficial for cadmium deposition, the cadmium–cadmium interaction has a harmful influence. It was proposed that when a metal ion M^{a+} is deposited as a submonolayer, the activity of the reductant $M^{(0)}$ should be less than 1 and would vary with the surface coverage [238].

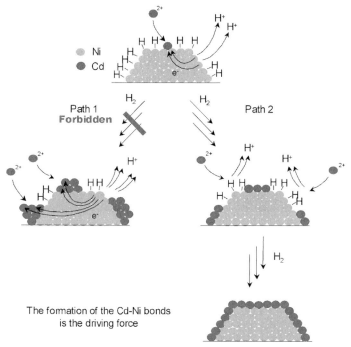

Scheme 18.20 Mechanism of cadmium deposition on a nickel nanoparticle.

If the potential of cadmium deposition is more positive at the beginning of the reaction, it would progressively decrease with the increase in the surface coverage, until it matches the regular Nernst potential for cadmium deposition for a full coverage. Therefore, the remaining concentration of cadmium in solution is lower if the cadmium coverage on the nickel surface is low. Typically, from solutions having initially 1000 ppm of cadmium, it is possible to decrease their concentrations at the ppm level for nickel coverage (Cd/Ni$_s$) close to 0.7 and as low as ppb level for coverage less than ca. 0.1 (Fig. 18.35).

18.5.2.2 Ni^{2+} and Co^{2+} Removal

Cadmium does not adsorb hydrogen. After full coverage of nickel nanoparticles by zerovalent cadmium atoms, the bimetallic Cd–Ni particle surface no longer activates hydrogen and reaction (18.1) is no longer possible. However, if the deposited atom (M′) is also able to activate the hydrogen molecule, reaction (18.1) must be continued by the formation of several M′ layers over nickel. This is indeed what has been observed when M′ = Ni or Co (Scheme 18.21).

For M′ = Ni and Co, about 80 supplementary metallic atom layers have been deposited on nickel particles and this is a nice way to prepare nanoparticles with controlled size. Obviously, the deposited atoms are completely reduced, as shown by magnetism measurements on the resulting particles and by their X-ray diffraction pattern (Scheme 18.22).

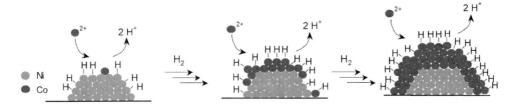

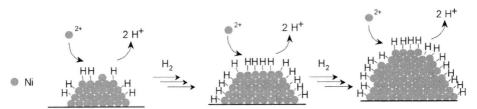

Scheme 18.21 Mechanism of nickel and cobalt deposition on a supported nickel nanoparticle.

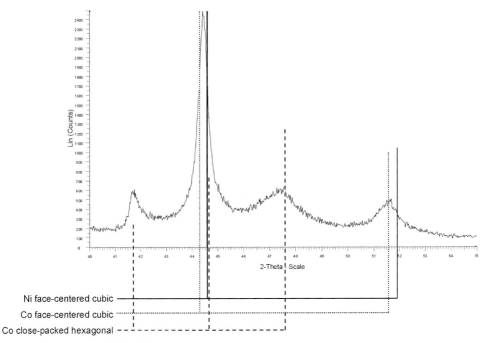

Scheme 18.22 XRD data for cobalt deposited on nickel nanoparticles.

At pH 7 under a hydrogen atmosphere, the remaining concentrations in water are about 0.5 ppm for nickel and a few ppm for cobalt, whatever the initial concentrations and the coverage rate of the nickel particles. Lower remaining concentrations can be achieved by decreasing the solution pH. Since the grafting reaction is autocatalytic for both metals, this depollution process could become highly interesting for removal of large amounts of metal such as used in nickel electroless deposition baths.

18.5.2.3 Cr^{VI} Removal

$Cr_2O_7^{2-}$ dichromates are usually employed in surface treatment processes. Cr^{VI} is highly toxic and its elimination from aqueous effluents is crucial. By applying the previous removal method based on nickel particles and hydrogen, the reaction should be described by the following equation:

$$Cr_2O_7^{2-} + 6Ni_s - H \rightarrow (Ni_s)_6 Cr_2O_3 + 2H_2O + 2OH^- \qquad (18.4)$$

As soon as a solution of sodium dichromate was added to a suspension of alumina deposited nickel, a fast and strong pH increase was observed as expected (Eq. (18.4)). The titration of hydroxide produced by regulating the pH at 7 has shown that one OH^- ion has been released per Cr atom grafted. Moreover, the amount of hydrogen consumed during chromium removal is about 3 hydrogen per atom of Cr grafted. This last measurement has been carried out by feeding the reactor with hydrogen to keep the hydrogen pressure constant. All these results are in a very good agreement with the stoichiometry of the proposed equation. Unlike the other metals (Cd, Ni and Co), Cr^{VI} is not reduced to Cr^0 but to Cr^{III}. The potential for the reduction of Cr^{III} to Cr^0 is so low that the potential of our system provided by (H_2/H^+) cannot perform such a reduction.

Since Cr^{VI} is a strong oxidant, the remaining chromium concentration in solution is very low, below the ICP-AES detection limit (ppb level). Since chromium(III) deposited as chromium oxide clusters on the nickel surface does not adsorb hydrogen, the removal is over at about 3 Cr atoms grafted per Ni_s.

18.6
Conclusion

These examples have shown that supported Group VIII metals nanoparticles modified by reaction with organometallic compounds are very useful in catalysis. These nanoparticles are catalysts in numerous applications both in fine chemicals and in petrochemicals. Depending on the reaction conditions, the Group VIII metal and the ratio of Group IV to Group VIII metals, various structures can be obtained: for the case of tin, surface organotin complexes linked to the surface and having one, two or three alkyl chains, naked tin adatoms or surface alloy. Each case leads to specific properties: the presence of alkyl ligands on tin induces steric constraints which apparently prevent the coordination of citral by its internal double bond and allow only the hydrogenation of the C=O double bond. Adatoms have numerous applications as they will selectively poison highly active sites responsible for the side-reactions (see for example the isomerization of 3-carene into 2-carene, the hydrogenation of unsaturated aldehydes and the dehydrogenation of 2-butanol to methyl ethyl ketone). In this case the amount of adatom is often very low; typically the Sn to surface metal atom ratio is lower than 0.1. After treatment at high temperature, these adatoms enter the metal, leading to the formation of surface alloys which also exhibit properties quite different from those of the starting metal. Indeed, due to the site isolation effect of the Group VIII metal atoms by tin, reactions occurring on more than one metal atom are now avoided. Two examples of such effects have been given: the dehydrogenation of isobutane to isobutene and the hydrogenation of esters and acids to aldehydes and alcohols, where the hydrogenolysis processes are suppressed.

These results show that some correlations can now reasonably be made between the structure of the nanoparticles and their catalytic properties. This should lead, in the near future, to a more rationalized design of catalysts and to the application of metals modified by surface organometallic chemistry to a larger number of reactions.

References

1 C. Bosch, A. Mittasch, H. Wolf, **1912**, *CA pat.* 139 023.

2 C. Bosch, A. Mittasch, H. Wolf, G. Stern, **1913**, *US pat.* 1 068 969.

3 C. Bosch, A. Mittasch, H. Wolf, G. Stern, **1913**, *US pat.* 1 068 967.

4 C. Bosch, A. Mittasch, H. Wolf, G. Stern, **1913**, *US pat.* 1 068 966.

5 C. Bosch, A. Mittasch, H. Wolf, G. Stern, **1915**, *US pat.* 1 148 570.

6 C. Bosch, A. Mittasch, **1915**, *US pat.* 1 158 167.

7 C. Bosch, A. Mittasch, **1917**, *US pat.* 1 244 580.

8 C. Bosch, **1921**, *US pat.* 1 386 760.

9 C. Bosch, *Chem. Fabrik* **1933**, 127.

10 R. Burch, A. J. Mitchell, *Appl. Catal.* **1983**, *6*, 121.

11 G. L. Szabo, *Stud. Surf. Sci. Catal.* **1982**, *11*, 349.

12 R. Burch, *J. Catal.* **1981**, *71*, 348.

13 R. Burch, L. C. Garla, *J. Catal.* **1981**, *71*, 360.

14 A. C. Muller, P. A. Engelhard, J. E. Weisang, *J. Catal.* **1979**, *56*, 65.

15 T. Miura, T. Nomura, K. Kubota, **1976**, *JP pat.* 51 087 489.

16 R. Bacaud, F. Figueras, *C. R. Acad. Sci., Ser. C* **1975**, *281*, 479.

17 J. E. Weisang, P. Engelhard, **1975**, *DE pat.* 2 455 375.

18 P. Duhaut, G. Martino, J. Miquel, **1975**, *DE pat.* 2 426 597.

19 R. E. Rausch, **1972**, *DE pat.* 2 104 428.

20 F. C. Wilhelm, **1972**, *DE pat.* 2 164 295.

21 J. E. Weisang, P. Engelhard, **1972**, *DE pat.* 2 139 069.

22 F. M. Dautzenberg, H. W. Kouwenhoven, **1971**, *DE pat.* 2 117 433.

23 J. C. Hayes, **1971**, *DE pat.* 2 044 630.

24 R. E. Rausch, **1970**, *DE pat.* 2 012 168.

25 J. H. Sinfelt, *Surf. Sci.* **2002**, *500*, 923.

26 J. H. Sinfelt, G. D. Meitzner, *Acc. Chem. Res.* **1993**, *26*, 1.

27 J. H. Sinfelt, *Catal. Lett.* **1991**, *9*, 159.

28 J. H. Sinfelt, J. A. Cusumano, **1975**, *US pat.* 3 901 827.

29 G. Meitzner, G. H. Via, F. W. Lytle, J. H. Sinfelt, *J. Phys. Chem.* **1992**, *96*, 4960.

30 X. Zhang, P. Biloen, *J. Catal.* **1986**, *98*, 468.

31 W. M. H. Sachtler, P. Biloen, *Am. Chem. Soc., Div. Petrol. Chem.* **1983**, *28*, 482.

32 P. Biloen, J. N. Helle, H. Verbeek, F. M. Dautzenberg, W. M. H. Sachtler, *J. Catal.* **1981**, *63*, 112.

33 M. Ichikawa, A. J. Lang, D. F. Shriver, W. M. H. Sachtler, *J. Am. Chem. Soc.* **1985**, *107*, 7216.

34 F. M. Dautzenberg, J. N. Helle, P. Biloen, W. M. H. Sachtler, *J. Catal.* **1980**, *63*, 119.

35 C. Mirodatos, J. A. Dalmon, G. A. Martin, *J. Catal.* **1987**, *105*, 405.

36 G. A. Martin, J. A. Dalmon, C. Mirodatos, *8th International Congress on Catalysis, Berlin (Germany)*, **1985**, *IV*, 371.

37 G. A. Martin, J. A. Dalmon, *J. Catal.* **1982**, *75*, 233.

38 J. A. Dalmon, G. A. Martin, *Stud. Surf. Sci. Catal.* **1981**, *7*, 402.

39 J. A. Dalmon, G. A. Martin, *J. Catal.* **1980**, *66*, 214.

40 J. A. Dalmon, J. P. Candy, G. A. Martin, *6th International Congress on Catalysis, London (U.K.)*, **1977**, *2*, 903.

41 M. F. Guilleux, J. A. Dalmon, G. A. Martin, *J. Catal.* **1980**, *62*, 235.

42 F. Garin, P. Girard, F. Weisang, G. Maire, *J. Catal.* **1981**, *70*, 215.

43 F. Garin, G. Maire, F. G. Gault, *Nouv. J. Chim.* **1981**, *5*, 553.

44 F. Garin, G. Maire, F. G. Gault, *Nouv. J. Chim.* **1981**, *5*, 563.

45 F. G. Gault, *Adv. Catal.* **1981**, *30*, 28.

46 M. Hajek, S. Corolleur, C. Corolleur, G. Maire, A. O'Cinneide, F. G. Gault, *J. Chim. Phys. Phys.-Chim. Biol.* **1974**, *71*, 1329.

47 C. Corolleur, F. G. Gault, D. Juttard, G. Maire, J. M. Muller, *J. Catal.* **1972**, *27*, 466.

48 Y. Barron, D. Cornet, G. Maire, F. G. Gault, *J. Catal.* **1963**, *2*, 152.

49 Y. Barron, G. Maire, J. M. Muller, F. G. Gault, *J. Catal.* **1966**, *5*, 428.

50 G. Maire, F. G. Gault, *Bull. Soc. Chim. France* **1967**, 894.

51 G. Maire, G. Plouidy, J. C. Prudhomme, F. G. Gault, *J. Catal.* **1965**, *4*, 556.

52 D. E. Sparks, R. Srinivasan, B. H. Davis, *J. Mol. Catal.* **1994**, *88*, 359.

53 D. E. Sparks, R. Srinivasan, B. H. Davis, *J. Mol. Catal.* **1994**, *88*, 325.

54 C. C. A. Riley, P. Jonsen, P. Meehan, J. C. Frost, K. J. Packer, *Catal. Today* **1991**, *9*, 121.

55 J. B. F. Anderson, R. Burch, *Appl. Catal.* **1986**, *25*, 173.

56 G. J. Den Otter, F. M. Dautzenberg, *J. Catal.* **1978**, *53*, 116.

57 S. J. Tauster, S. C. Fung, R. T. K. Baker, J. A. Horsley, *Science* **1981**, *211*, 1121.

58 J. A. Cairns, J. E. E. Baglin, G. J. Clark, J. F. Ziegler, *J. Catal.* **1983**, 301.

59 B. Coq, A. Bittar, R. Dutartre, F. Figueras, *Appl. Catal.* **1990**, *60*, 33.

60 S. Bernal, R. T. Baker, A. Burrows, J. J. Calvino, C. J. Kiely, C. Lopez-Cartes, J. A. Perez-Omil, J. M. Rodriguez-Izquierdo, *Surf. Interface Anal.* **2000**, *29*, 411.

61 C. Nédez, A. Théolier, F. Lefebvre, A. Choplin, J. M. Basset, J. F. Joly, *J. Am. Chem. Soc.* **1993**, *115*, 722.

62 F. Humblot, D. Didillon, F. Lepeltier, J. P. Candy, J. Corker, O. Clause, F. Bayard, J. M. Basset, *J. Am. Chem. Soc.* **1998**, *120*, 137.

63 K. Pelzer, B. Laleu, F. Lefebvre, K. Philippot, B. Chaudret, J. P. Candy, J. M. Basset, *Chem. Mater.* **2004**, *16*, 4937.

64 B. Chaudret, *Top. Organomet. Chem.* **2005**, *16*, 233.

65 J. P. Candy, B. Didillon, E. L. Smith, T. B. Shay, J. M. Basset, *J. Mol. Catal.* **1994**, *86*, 179.

66 R. Van Hardeveld, F. Hartog, *Surf. Sci.* **1969**, *15*, 189.

67 O. Vidoni, K. Philippot, C. Amiens, B. Chaudret, O. Balmes, J.-O. Malm, J.-O. Bovin, F. Senocq, M.-J. Casanove, *Angew. Chem., Int. Ed.* **1999**, *24*, 3736.

68 A. Duteil, R. Queau, B. Chaudret, R. Mazel, C. Roucau, J. S. Bradley, *Chem. Mater.* **1993**, *5*, 341.

69 C. Pan, K. Pelzer, K. Philippot, B. Chaudret, F. Dassenoy, P. M. Lecante, J. Casanove, *J. Am. Chem. Soc.* **2001**, *123*, 7584.

70 K. Pelzer, O. Vidoni, K. Philippot, B. Chaudret, V. Collière, *Adv. Funct. Mater.* **2003**, *13*, 118.

71 A. Rodriguez, C. Amiens, B. Chaudret, M.-J. Casanove, P. Lecante, J. S. Bradley, *Chem. Mater.* **1996**, *8*, 1978.

72 A. Badia, W. Gao, S. Singh, L. Demers, L. Cuccia, L. Reven, *Langmuir* **1996**, *12*, 1262.

73 M. J. Hosteler, R. G. Nuzzo, G. S. Girolami, *J. Am. Chem. Soc.* **1994**, *116*, 11608.

74 M. J. Hostetler, J. E. Wingate, C.-J. Zhong, J. E. Harris, R. W. Vachet, M. R. Clark, J. D. Londono, S. J. Green, J. J. Stokes, G. D. Wignall, G. L. Glish, M. D. Porter, N. D. Evans, R. W. Murray, *Langmuir* **1998**, *14*, 17.

75 S. A. Stevenson, J. A. Dumesic, R. T. K. Baker, E. Ruckenstein, *Metal-Support Interactions in Catalysis, Sintering and Redispersion*, Van Nostrand Reinhold, New York, **1987**.

76 T. Risse, H. J. Freund, *Top. Organomet. Chem.* **2005**, *16*, 117.

77 A. K. Santra, D. W. Goodman, *J. Phys.: Condens. Matter* **2003**, *15*, 31.

78 C. Copéret, M. Chabanas, R. Petrov Saint Arroman, J. M. Basset, *Angew. Chem., Int. Ed.* **2003**, *42*, 156.

79 P. L. Gai, E. D. Boyes, *Electron Microscopy in Heterogeneous Catalysis*, Institute of Physics Publishing, Bristol, **2003**.

80 M. J. Yacaman, J. A. Ascencio, S. Tchuacanero, M. Marin, *Top. Catal.* **2002**, *18*, 167.

81 B. Zhou, S. Hermans, G. A. Somorjai, *Nanotechnology in Catalysis*, Springer, Berlin, **2004**.

82 D. C. Koningsberger, R. Prins, *X-Ray Absorption: Principles, Applications, Techniques of EXAFS, SEXAFS, and XANES*, Wiley, New York, **1988**.

83 D. Bazin, H. Dexpert, J. Lynch, in *Series on Synchrotron Radiation Techniques and Applications*, Vol. 2, **1996**, p. 113.

84 D. Briggs, J. T. Grant, *Surface Analysis by Auger and X-ray Photoelectron Spectroscopy*, IMPublication, **2003**.

85 G. R. Wilson, W. K. Hall, *J. Catal.* **1972**, *24*, 306.

86 B. J. Kip, F. B. M. Duivenvoorden, D. C. Koningsberger, R. Prins, *J. Catal.* **1987**, *105*, 26.

87 H. Ehwald, U. Leibnitz, *Catal. Lett.* **1996**, *38*, 149.

88 G. C. Bond, L. Hui, *J. Catal.* **1994**, *147*, 346.

89 A. Bertucco, C. Bennett, *Appl. Catal.* **1987**, *35*, 329.

90 A. Frennet, P. B. Wells, *Appl. Catal.* **1985**, *18*, 243.

91 D. Richard, P. Gallezot, *Stud. Surf. Sci. Catal.* **1986**, *31*, 71.

92 E. Merlen, P. Beccat, J. C. Bertolini, P. Delichere, N. Zanier, B. Didillon, *J. Catal.* **1996**, *159*, 178.

93 J. P. Candy, A. El Mansour, O. A. Ferretti, G. Mabilon, J. P. Bournonville, J. M. Basset, G. Martino, *J. Catal.* **1988**, *112*, 201.

94 A. H. Benesis, R. M. Curtis, H. P. Studer, *J. Catal.* **1968**, *10*, 328.

95 A. Théolier, A. Choplin, L. Hilaire, L. d'Ornelas, J.-M. Basset, R. Hugo, R. Psaro, C. Sourisseau, *Polyhedron* **1983**, *2*, 119.

96 M. Agnelli, J. P. Candy, J. M. Basset, J. P. Bournonville, O. A. Ferretti, *J. Catal.* **1990**, *121*, 236.

97 F. Locatelli, J. P. Candy, B. Didillon, G. Niccolai, D. Uzio, J.-M. Basset, *J. Catal.* **2000**, *193*, 154.

98 J. P. Candy, P. Fouilloux, A. J. Renouprez, *J. Chem.Soc., Faraday Trans.I* **1980**, *76*, 616.

99 D. E. Damiani, A. J. Rouco, *J. Catal.* **1986**, *100*, 512.

100 P. B. Wells, *Appl. Catal.* **1985**, *18*, 259.

101 V. Ragaini, R. Giannantonio, P. Magni, L. Lucarelli, G. Leofanti, *J. Catal.* **1994**, *146*, 116.

102 O. A. Ferretti, C. Lucas, J. P. Candy, J. M. Basset, B. Didillon, F. Le Peltier, *J. Mol. Catal.* **1995**, *103*, 125.

103 K. Lu, B. J. Tatarchuk, *J. Catal.* **1987**, *106*, 166.

104 P. De Montgolfier, G. A. Martin, J. A. Dalmon, *J. Phys. Chem. Solids* **1973**, *34*, 801.

105 P. W. Selwood, *Chemisorption and Magnetization*, Academic Press, New York **1976**.

106 P. De Montgolfier, B. Moraweck, G. A. Martin, A. J. Renouprez, G. Dalmai-Imelik, *Fine Paricles., Int. Conf., Pap., 2nd* **1974**, 43.

107 G. A. Martin, B. Moraweck, A. J. Renouprez, G. Dalmai-Imelik, B. Imelik,

J. Chim. Phys. Physico-Chimie Biol. **1972**, *69*, 532.

108 *Hand Book of Chemistry and Physics*, 77th Edn., Lide, D. R. (Ed.), CRC Press, Boca Raton, FL, **1996**, pp. 12–115.

109 C. Nédez, F. Lefebvre, A. Choplin, J. M. Basset, E. Benazzi, *J. Am. Chem. Soc.* **1994**, *116*, 3039.

110 C. Nédez, F. Lefebvre, A. Choplin, G. P. Niccolai, J. M. Basset, E. Benazzi, *J. Am. Chem. Soc.* **1994**, *116*, 8638.

111 C. Nédez, F. Lefebvre, A. Choplin, J. M. Basset, *New J. Chem.* **1994**, *18*, 1215.

112 M. Taoufik, M. A. Cordonnier, C. C. Santini, J. M. Basset, J. P. Candy, *New J. Chem.* **2004**, *28*, 1531.

113 F. Lefebvre, J. P. Candy, J. M. Basset, *Synthesis with Supported Metal Particles by Use of Surface Organometallic Chemistry: Characterization and Some Applications in Catalysis*, Vol. 2, Wiley-VCH, Weinheim, **1999**, Ch. 2.7.

114 F. Z. Bentahar, F. Bayard, J. P. Candy, J. M. Basset, in *Fundamental and Applied Aspects of Chemically Modified Surfaces*, J. P. Blitz, C. B. Little (Eds.), The Royal Society of Chemistry, Cambridge, **1999**, p. 235.

115 V. Maurice, Y. A. Ryndin, G. Bergeret, L. Savary, J. P. Candy, J. M. Basset, *J. Catal.* **2001**, *204*, 192.

116 F. Z. Bentahar, J. P. Candy, J. M. Basset, F. Le Peltier, B. Didillon, *Catal. Today* **2001**, *66*, 303.

117 J. M. Basset, J. P. Candy, B. Didillon, F. Lepeltier, F. Clause, F. Bentahar, **2001**, *US pat.* 6 281 160.

118 P. Lesage, O. Clause, P. Moral, B. Didillon, J. P. Candy, J. M. Basset, *J. Catal.* **1995**, *155*, 238.

119 B. Didillon, C. Houtman, T. Shay, J. P. Candy, J. M. Basset, *J. Am. Chem. Soc.* **1993**, *115*, 9380.

120 J. P. Candy, E. Roisin, J. M. Basset, D. Uzio, S. Morin, L. Fischer, J. Olivier-Fourcade, J. C. Jumas, *Hyperfine Interactions*, **2005**, *165*, 55.

121 Y. A. Ryndin, J. P. Candy, B. Didillon, L. Savary, J. M. Basset, *C. R. Acad. Sci. Paris Ser. IIc* **2000**, *3*, 423.

122 B. Didillon, L. Savary, Y. A. Ryndin, J. P. Candy, J. M. Basset, *C. R. Acad. Sci. Paris Ser. IIc* **2000**, *3*, 413.

123 Y. A. Ryndin, J. P. Candy, B. Didillon, L. Savary, J. M. Basset, *J. Catal.* **2001**, *198*, 103.

124 J. P. Candy, Y. A. Ryndin, G. Bergeret, L. Savary, D. Uzio, J. M. Basset, IV *International Conference on Catalysis and Adsorption in Fuel Processing and Environmental Protection, Kudova (Poland)*, **2002**, p. 101.

125 J. H. Fendler, *Chem. Mater.* **1996**, 1616.

126 Z. L. Wang, *Adv. Mater.* **1998**, *10*, 13.

127 V. L. Colvin, M. P. Schlamp, A. P. Alivisato, *Nature* **1994**, 354.

128 C. N. R. Rao, G. U. Kulkarni, P. J. Thomas, P. P. Edwards, *Chem. Soc. Rev.* **2000**, *29*, 27.

129 G. Scmidt, *Chem. Rev.* **1992**, *92*, 1709.

130 G. Schmid, *Cluster and Colloids, From Theory to Applicaton*, VCH, Weinheim, **1994**.

131 G. Schön, U. Simon, *Colloid Polym. Sci.* **1995**, *273*, 101.

132 A. P. Alivisatos, *Science* **1996**, *271*, 933.

133 L. N. Lewis, *Chem. Rev.* **1993**, *93*, 2693.

134 M. M. Maye, Y. Lou, C.-J. Zhong, *Langmuir* **2000**, *16*, 7520.

135 M. D. Musick, C. D. Keating, M. H. Keefe, N. J. Natan, *Chem. Mater.* **1997**, *9*, 1499.

136 L. M. Liz-Marzan, P. Mulvaney, *New J. Chem.* **1998**, 1285.

137 K. A. Easom, K. J. Klabunde, C. M. Sorensen, G. C. Hadjipanayis, *Polyhedron* **1994**, *13*, 1197.

138 N. Toshima, Y. Yonesawa, *New J. Chem.* **1998**, 1179.

139 W. Yu, M. Liu, H. Liu, X. L. Z. Ma, *J. Collloid Interface Sci.* **1998**, *208*, 439.

140 H. Bönnemann, G. Braun, W. Brijoux, A. Schulze Tilling, K. Seevogel, K. Siepen, *J. Organomet. Chem.* **1996**, *520*, 143.

141 C. Nayral, E. Viala, P. Fau, F. Senocq, J.-C. Jumas, A. Maisonnat, B. Chaudret, *Chem. Eur. J.* **2000**, *6*, 4082.

142 G. Schmid, U. Giebel, W. Huster, A. Schwenk, *Inorg. Chim. Acta* **1984**, *85*, 97.

143 T. Ould Ely, C. Amiens, B. Chaudret, E. Snoeck, M. Verelst, M. Respaud, J.-M. Broto, *Chem. Mater.* **1999**, *11*, 526.

144 S. L. Braunstein, L. A. Oro, P. R. Raithby, *Metal Clusters in Chemistry*, Wiley-VCH, Weinheim, 2000.

145 B. Monika, Dissertation, UGH Essen, 2000.

146 M. Zaho, M. Crooks, *Angew. Chem.* 1999, *95*, 706.

147 L. Erades, Thesis, University Paul Sabatier, Toulouse, 2003.

148 P. Claus, A. Brückener, C. Mohr, H. Hofmeister, *J. Am. Chem. Soc.* 2000, *122*, 11430.

149 U. A. Paulus, U. Endruschat, G. J. Feldmeyer, T. J. Schmidt, H. Bonnemann, R. J. Behm, *J. Catal.* 2000, *195*, 383.

150 R. A. Salkar, P. Jeevanandam, S. T. Aruna, Y. Koltypin, A. Gedanken, *J. Mater. Chem.* 1999, *9*, 1333.

151 A. Martino, S. A. Yamanaka, J. S. Kawola, D. Loy, *Chem. Mater.* 1997, *9*, 423.

152 T. Li, J. Moon, A. A. Morrone, J. J. Mecholsky, D. R. Talham, J. H. Adair, *Langmuir* 1999, *15*, 4328.

153 K. V. Sarathy, G. U. Kulkarni, C. N. R. Rao, *Chem. Comm.* 1997, 537.

154 S. Roginsky, A. Schalnikoff, *Kolloid Z.* 1927, *43*, 67.

155 M. T. Reetz, S. A. Quaiser, *Angew. Chem. Int. Ed.* 1995, *34*, 2241.

156 H. Bönnemann, W. Brijoux, R. Brinkmann, E. Dinjus, T. Jouben, B. Korall, *Angew. Chem. Int. Ed.* 1991, *10*, 1312.

157 M. Faraday, *Trans. R. Soc.* 1857, *147*, 145.

158 S. Sun, C. B. Murray, D. Weller, L. Folks, A. Moser, *Science* 2000, *287*, 1989.

159 M. Respaud, J.-M. Broto, H. Rakoto, A. R. Fert, L. Thomas, B. Barbara, M. Verelst, E. Snoeck, P. Lecante, A. Mosset, J. Osuna, T. Ould Ely, C. Amiens, B. Chaudret, *Phys. Rev. B* 1998, *57*, 2925.

160 G. Schmid, V. Maihack, F. Lantermann, S. Peschel, *J. Chem. Soc., Dalton Trans.* 1996, *5*, 589.

161 G. Schmid, M. Bäumle, M. Geerkens, I. Heim, C. Osemann, T. Sawitowski, *Chem. Soc. Rev.* 1999, 179.

162 J. Fink, C. J. Kiely, D. Bethell, D. J. Schiffrin, *Chem. Mater.* 1998, *10*, 922.

163 S. Chen, R. W. Murray, *Langmuir* 1999, *15*, 682.

164 P. Pertuci, G. Vituli, *Inorg. Synth.* 1983, *22*, 178.

165 S. Gao, J. Zhang, Y.-F. Zhu, C.-M. Che, *New J. Chem.* 2000, *24*, 739.

166 N. Toshima, Y. Shiraishi, T. Teranishi, M. Miyake, T. Tominaga, H. Watanabe, W. Brijoux, H. Bonnemann, G. Schmid, *Appl. Organomet. Chem.* 2001, *15*, 178.

167 P. Uznanski, C. Amiens, B. Chaudret, E. Bryszewska, *Polish J. Chem.* 2006, *80*, 1845.

168 D. Wostek-Wojciechowska, J. K. Jeszka, C. Amiens, B. Chaudret, P. Lecante, *J. Colloid Interface Sci.* 2005, *287*, 107.

169 K. Pelzer, K. Philippot, B. Chaudret, *Z. Phys. Chem.* 2003, *217*, 1539.

170 M.-J. Casanove, P. Lecante, M.-C. Fromen, M. Respaud, D. Zitoun, C. Pan, K. Philippot, C. Amiens, B. Chaudret, F. Dassenoy, *Mater. Res. Soc. Symp. Proc.* 2002, *704*, 349.

171 F. Dassenoy, M. J. Casanove, P. Lecante, C. Pan, K. Philippot, C. Amiens, B. Chaudret, *Phys. Rev. B* 2001, *63*, 235407/1.

172 F. Dassenoy, M. J. Casanove, P. Lecante, M. Verelst, E. Snoeck, A. Mosset, T. O. Ely, C. Amiens, B. Chaudret, *J. Chem. Phys.* 2000, *112*, 8137.

173 T. O. Ely, C. Pan, C. Amiens, B. Chaudret, F. Dassenoy, P. Lecante, M. J. Casanove, A. Mosset, M. Respaud, J. M. Broto, *J. Phys. Chem. B* 2000, *104*, 695.

174 F. Dassenoy, K. Philippot, T. Ould Ely, C. Amiens, P. Lecante, E. Snoeck, A. Mosset, M.-J. Casanove, B. Chaudret, *New J. Chem.* 1998, *22*, 703.

175 D. H. Napper, *J. Colloid Interface Science* 1977, *58*, 390.

176 C. Lange, D. De Caro, A. Gamez, S. Storck, J. S. Bradley, W. F. Maier, *Langmuir* 1999, *15*, 5333.

177 J. S. Bradley, J. M. Millar, E. W. Hill, C. Klein, B. Chaudret, A. Duteuil, *Stud. Surf. Sci. Catal.* 1993, *75*, 969.

178 J. S. Bradley, E. W. Hill, S. Behal, C. Klein, A. Duteil, B. Chaudret, *Chem. Mater.* 1992, *4*, 1234.

179 C. Pan, F. Dassenoy, M.-J. Casanove, K. Philippot, C. Amiens, P. Lecante, A. Mosset, B. Chaudret, *J. Phys. Chem. B* **1999**, *103*, 10098.

180 G. Schmid, *Nanoparticles: From Theory to Application*, Wiley-VCH: Weinheim, Germany, **2004**.

181 G. Viau, R. Brayner, L. Poul, N. Chakroun, E. Lacaze, F. Fièvet-Vincent, F. Fièvet, *Chem. Mater.* **2003**, *15*, 486.

182 N. Zheng, J. Fan, G. D. Stucky, *J. Am. Chem. Soc.* **2006**, *128*, 6550.

183 T. Sugimoto, *Funtai Kogaku Kaishi* **2005**, *42*, 478.

184 K. Pelzer, J. P. Candy, G. Bergeret, J. M. Basset, *Eur. Phys. J.* **2007**, *43*, 197.

185 M.-C. Daniel, D. Astruc, *Chem. Rev.* **2004**, *104*, 293.

186 N. Chakroune, G. Viau, S. Ammar, L. Poul, D. Veautier, M. M. Chehimi, C. Mangeney, F. Villain, F. Fiévet, *Langmuir* **2005**, *21*, 6788.

187 K. Naka, M. Yaguchi, Y. Chuio, *Chem. Mater.* **1999**, *11*, 849.

188 W. Shenton, D. Pum, U. B. Sleytr, S. Mann, *Nature* **1997**, *389*, 585.

189 T. Cassagneau, T. E. Mallouk, J. H. Fendler, *J. Am. Chem. Soc.* **1998**, *120*, 78.

190 T. Yonezawa, T. Tominaga, D. Richard, *J. Chem. Soc., Dalton Trans.* **1996**, 783.

191 T. Sato, D. Brown, B. F. G. Johnson, *J. Chem. Soc. Chem. Commun.* **1997**, 1007.

192 J. S. Bradley, J. M. Millard, E. W. Hill, *J. Am. Chem. Soc.* **1991**, *113*, 4016.

193 G. Schmid, *Struct. Bond.* **1985**, *62*, 51.

194 G. Schmid, W. Meyer-Zaika, R. Pugin, T. Sawitowski, J.-P. Majoral, A.-M. Caminade, C.-O. Turrin, *Chem. Eur. J.* **2000**, *6*, 1693.

195 N. Ulagaappan, N. Battaram, V. N. Rraju, C. N. R. Rao, *J. Chem. Soc. Chem. Commun.* **1996**, 2243.

196 P. T. Tanev, T. J. Pinnavaia, *Science* **1996**, *271*, 1267.

197 L. Beuermann, W. Maus-Friedrichs, S. Krischok, V. Kempter, S. Bucher, H. Modrow, J. Hormes, N. Waldofner, H. Bonnemann, *Appl. Organomet. Chem.* **2003**, *17*, 268.

198 H. Bonnemann, W. Brijoux, R. Brinkmann, U. Endruschat, W. Hofstadt, K. Angermund, in *Rev. Roum. Chim.* **2000**, *44*, 1003.

199 H.-U. Blaser, B. Pugin, F. Spindler, *J. Mol. Catal. A* **2005**, *231*, 1.

200 J. T. Wehrli, A. Baiker, *J. Catal.* **1990**, *61*, 207.

201 M. Agnelli, P. Louessard, A. El Mansour, J. P. Candy, J. P. Bournonville, J. M. Basset, *Catal. Today* **1989**, *6*, 63.

202 K. Moseley, P. M. Maitlis, *Chem. Comm.* **1971**, 1065.

203 G. L. Bezemer, J. H. Bitter, H. P. C. E. Kuipers, H. Oosterbeek, J. E. Holewijn, X. Xu, F. Kapteijn, A. J. van Dillen, K. P. de Jong, *J. Am. Chem. Soc.* **2006**, *128*, 3956.

204 G. L. Bezemer, P. B. Radstake, V. Koot, A. J. van Dillen, J. W. Geus, K. P. de Jong, *J. Catal.* **2006**, *237*, 291.

205 B. Bertsch-Frank, I. Hemme, L. Von Hippel, S. Katusic, J. Rollmann, **2000**, *DE pat.* 19 912 733.

206 M. Jezequel, V. Dufaud, M. J. Ruiz-Garcia, F. Carrillo-Hermosilla, U. Neugebauer, G. P. Niccolai, F. Lefebvre, F. Bayard, J. Corker, S. Fiddy, J. Evans, J.-P. Broyer, J. Malinge, J.-M. Basset, *J. Am. Chem. Soc.* **2001**, *123*, 3520.

207 N. Millot, S. Soignier, C. C. Santini, A. Baudouin, J.-M. Basset, *J. Am. Chem. Soc.* **2006**, *128*, 9361.

208 G. Tosin, C. C. Santini, M. Taoufik, A. De Mallmann, J.-M. Basset, *Organometallics* **2006**, *25*, 3324.

209 M. Taoufik, E. Le Roux, J. Thivolle-Cazat, C. Coperet, J.-M. Basset, B. Maunders, G. J. Sunley, *Top. Catal.* **2006**, *40*, 65.

210 J. Joubert, F. Delbecq, P. Sautet, E. Le Roux, M. Taoufik, C. Thieuleux, F. Blanc, C. Coperet, J. Thivolle-Cazat, J.-M. Basset, *J. Am. Chem. Soc.* **2006**, *128*, 9157.

211 B. Rhers, A. Salameh, A. Baudouin, E. A. Quadrelli, M. Taoufik, C. Coperet, F. Lefebvre, J.-M. Basset, X. Solans-Monfort, O. Eisenstein, W. W. Lukens, L. P. H. Lopez, A. Sinha, R. R. Schrock, *Organometallics* **2006**, *25*, 3554.

212 J.-P. Candy, C. Coperet, J.-M. Basset, *Top. Organomet. Chem.* **2005**, *16*, 151.

213 F. Blanc, M. Chabanas, C. Coperet, B. Fenet, E. Herdweck, *J. Organomet. Chem.* **2005**, *690*, 5014.

214 N. Toshima, T. Takahashi, *Bull. Chem. Soc. Jpn.* **1992**, *65*, 400.

215 C. Chupin, J. P. Candy, J. M. Basset, *Catal. Today* **2003**, *79–80*, 15.

216 B. Didillon, J. P. Candy, F. Le Peltier, O. A. Ferretti, J. M. Basset, *Stud. Surf. Sci. Catal.* **1993**, *78*, 147.

217 P. Lesage, J. P. Candy, C. Hirigoyen, F. Humblot, M. Leconte, J. M. Basset, *J. Mol. Catal. A: Chem.* **1996**, *112*, 303.

218 R. Snappe, J. P. Bournonville, J. Miquel, G. Martino, **1983**, *US pat.* 4 380 673.

219 F. Humblot, M. A. Cordonnier, C. Santini, B. Didillon, J. P. Candy, J. M. Basset, *Stud. Surf. Sci. Catal.* **1997**, *108*, 289.

220 F. Humblot, J. P. Candy, F. Le Peltier, B. Didillon, J. M. Basset, *J. Catal.* **1998**, *179*, 459.

221 L. K. Loc, N. A. Gaidai, B. S. Gudkov, S. L. Kiperman, S. B. Kogan, *Kinet. Catal.* **1987**, *27*, 1184.

222 G. A. Martin, J. A. Dalmon, *J. Catal.* **1980**, *66*, 214.

223 A. El Mansour, J. P. Candy, J. P. Bournonville, O. A. Ferretti, J. M. Basset, *Angew. Chem. Int. Ed. Engl.* **1989**, *28*, 347.

224 O. A. Ferretti, J. P. Bournonville, G. Mabilon, G. Martino, J. P. Candy, J. M. Basset, *J. Mol. Catal.* **1991**, *67*, 283.

225 J. M. Basset, J. P. Candy, P. Louessard, O. A. Ferretti, J. P. Bournonville, *Wiss. Zeitschr. THLM* **1990**, *32*, 657.

226 J. P. Bournonville, J. P. Candy, G. Mabilon, **1986**, *EP pat.* 172 091.

227 P. Louessard, J. P. Candy, G. Mabilon, J. P. Bournonville, **1987**, *FR pat.* 87/03 525.

228 C. Travers, J. P. Bournonville, G. Martino, *8th International Congress on Catalysis, Berlin (Germany)*, **1985**, Vol. 4, p. 891.

229 R. M. Ferrero, R. Jacquot, **1991**, *FR pat.* 13 147 911 024.

230 B. Didillon, C. Petit-Clair, L. Savary, **2002**, *US pat.* 2 002 139 720.

231 G. Godard, J. P. Candy, J. M. Basset, **2001**, *FR pat.* 0 103 428.

232 J. P. Candy, G. Godard, J. M. Basset, *Chem. Eng. Trans.* **2003**, *3*, 787.

233 J.-M. Basset, J.-P. Candy, G. Godard, **2002**, *WO pat.* 2 002 072 483.

234 E. Lamy-Pitara, L. E. Ouazzani-Benhima, J. Barbier, M. Cahoreau, J. Caisso, *J. Electroanal. Chem.* **1994**, *372*, 233.

235 M. A. Quiroz, I. Gonzalez, H. Vargas, Y. Meas, E. Lamy-Pitara, J. Barbier, *Electrochim. Acta* **1986**, *31*, 503.

236 I. Bakos, S. Szabo, F. Nagy, T. Mallat, Z. Bodnar, *J. Electroanal. Chem. Interface Electrochem.* **1991**, *309*, 293.

237 I. Bakos, S. Szabo, *J. Electroanal. Chem.* **2003**, *547*, 103.

238 G. Kokkinidis, *J. Electroanal. Chem. Interface Electrochem.* **1986**, *201*, 217.

Index

Nanoparticles and Catalysis. Edited by Didier Astruc
Copyright © 2008 WILEY-VCH Verlag GmbH & Co. KGaA, Weinheim
ISBN: 978-3-527-31572-7